Homotopy Theory of Higher Categories

The study of higher categories is attracting growing interest for its many applications in topology, algebraic geometry, mathematical physics, logic and category theory. In this highly readable book, Carlos Simpson develops a full set of homotopical algebra techniques and proposes a working theory of higher categories.

Starting with a cohesive overview of the many different approaches currently used by researchers, the author proceeds with a detailed exposition of one of the most widely used techniques: the construction of a cartesian Quillen model structure for higher categories. The fully iterative construction applies to enrichment over any cartesian model category, and yields model categories for weakly associative n-categories and Segal n-categories. A corollary is the construction of higher functor categories, which fit together to form the $(n + 1)$-category of n-categories. The approach uses Tamsamani's definition based on Segal's ideas, iterated as in Pellissier's thesis using modern techniques due to Barwick, Bergner, Lurie and others.

CARLOS SIMPSON is Directeur de Recherches in the CNRS in Nice, France.

NEW MATHEMATICAL MONOGRAPHS

All the titles listed below can be obtained from good booksellers or from Cambridge University Press. For a complete series listing visit www.cambridge.org/mathematics

1 M. Cabanes and M. Enguehard *Representation Theory of Finite Reductive Groups*
2 J. B. Garnett and D. E. Marshall *Harmonic Measure*
3 P. Cohn *Free Ideal Rings and Localization in General Rings*
4 E. Bombieri and W. Gubler *Heights in Diophantine Geometry*
5 Y. J. Ionin and M. S. Shrikhande *Combinatorics of Symmetric Designs*
6 S. Berhanu, P. D. Cordaro and J. Hounie *An Introduction to Involutive Structures*
7 A. Shlapentokh *Hilbert's Tenth Problem*
8 G. Michler *Theory of Finite Simple Groups I*
9 A. Baker and G. Wüstholz *Logarithmic Forms and Diophantine Geometry*
10 P. Kronheimer and T. Mrowka *Monopoles and Three-Manifolds*
11 B. Bekka, P. de la Harpe and A. Valette *Kazhdan's Property (T)*
12 J. Neisendorfer *Algebraic Methods in Unstable Homotopy Theory*
13 M. Grandis *Directed Algebraic Topology*
14 G. Michler *Theory of Finite Simple Groups II*
15 R. Schertz *Complex Multiplication*
16 S. Bloch *Lectures on Algebraic Cycles (2nd Edition)*
17 B. Conrad, O. Gabber and G. Prasad *Pseudo-reductive Groups*
18 T. Downarowicz *Entropy in Dynamical Systems*
19 C. Simpson *Homotopy Theory of Higher Categories*

Homotopy Theory of Higher Categories

CARLOS SIMPSON
Université de Nice – Sophia Antipolis

CAMBRIDGE
UNIVERSITY PRESS

CAMBRIDGE
UNIVERSITY PRESS

University Printing House, Cambridge CB2 8BS, United Kingdom

One Liberty Plaza, 20th Floor, New York, NY 10006, USA

477 Williamstown Road, Port Melbourne, VIC 3207, Australia

314-321, 3rd Floor, Plot 3, Splendor Forum, Jasola District Centre, New Delhi - 110025, India

79 Anson Road, #06-04/06, Singapore 079906

Cambridge University Press is part of the University of Cambridge.

It furthers the University's mission by disseminating knowledge in the pursuit of education, learning and research at the highest international levels of excellence.

www.cambridge.org
Information on this title: www.cambridge.org/9780521516952

First published 2012

A catalogue record for this publication is available from the British Library

Library of Congress Cataloging in Publication data
Simpson, Carlos, 1962–
Homotopy theory of higher categories / Carlos T. Simpson.
p. cm. – (New mathematical monographs)
ISBN 978-0-521-51695-2 (Hardback)
1. Homotopy theory. 2. Categories (Mathematics) I. Title.
QA612.7.S56 2011
512′.62–dc23

2011026520

ISBN 978-0-521-51695-2 Hardback

Higher categories are approached by iterating Segal's method, following Tamsamani's definition of n-nerve and Pellissier's thesis. If M is a tractable left proper cartesian model category, we construct a tractable left proper cartesian model structure on the category of M-precategories. The procedure can then be iterated, leading to model categories of (∞, n)-categories.

Contents

Preface

The theory of n-categories is currently under active consideration by a number of different research groups around the world. The history of the subject goes back a long way, on separate but interrelated tracks in algebraic topology, algebraic geometry and category theory. For a long time, the crucial definition of *weakly associative higher category* remained elusive, but now we have a plethora of different possibilities available. One of the next major problems in the subject will be to achieve a global comparison between these different approaches. Some work is starting to come out in this direction, but in the current state of the theory the various different approaches remain distinct. After the comparison is achieved, they will be seen as representing different facets of the theory, so it is important to continue working in all of these different directions.

The purpose of the present book is to concentrate on one of the methods of defining and working with higher categories, very closely based on the work of Graeme Segal in algebraic topology many years earlier. The notion of "Segal category," which is a kind of category weakly enriched over simplicial sets, was considered by Schwänzl, Vogt and Dwyer, Kan and Smith. The application of this method to n-categories was introduced by Zouhair Tamsamani. It was then put into a strictly iterative form, with a general model category as input, by Regis Pellissier following a suggestion of André Hirschowitz. Our treatment will integrate important ideas contributed by Julie Bergner, Clark Barwick, Jacob Lurie and others.

The guiding principle is to use the category of simplices Δ as the basis for all the higher coherency conditions which come in when we allow weak associativity. The objects of Δ are nonempty finite ordinals

$$\begin{array}{rcl} [0] & = & \{v_0\} \\ [1] & = & \{v_0, v_1\} \\ [2] & = & \{v_0, v_1, v_2\} \\ & \ldots & \end{array}$$

whereas the morphisms are nondecreasing maps between them. In keeping with the standard notation, the zeroth object [0] of Δ is an ordinal containing one element, nonetheless we usually refer to this as something of length zero. Kan had already introduced this category into algebraic topology, when considering *simplicial sets* which are functors $\Delta^o \to \mathrm{SET}$. These model the homotopy types of CW-complexes.

One of the big problems in algebraic topology in the 1960s was to define notions of *delooping machines*. Segal's way was to consider *simplicial spaces*, or functors $\mathcal{A} : \Delta^o \to \mathrm{TOP}$, such that the first space is just a point $\mathcal{A}_0 = \mathcal{A}([0]) = *$. In Δ there are three nonconstant maps

$$f_{01}, f_{12}, f_{02} : [1] \to [2],$$

where f_{ij} denotes the map sending v_0 to v_i and v_1 to v_j. In a simplicial space $\mathcal{A}$ which is a contravariant functor on Δ, we get three maps

$$f_{01}^*, f_{12}^*, f_{02}^* : \mathcal{A}_2 \to \mathcal{A}_1.$$

Organize the first two as a map into a product, giving a diagram of the form

$$\begin{array}{ccc} \mathcal{A}_2 & \xrightarrow{\sigma_2} & \mathcal{A}_1 \times \mathcal{A}_1 \\ \Big\downarrow{\scriptstyle f_{02}^*} & & \\ \mathcal{A}_1 & & \end{array}.$$

If we require the *second Segal map* $\sigma_2 := \left(f_{01}^*, f_{12}^*\right)$ to be an isomorphism between $\mathcal{A}_2$ and $\mathcal{A}_1 \times \mathcal{A}_1$, then f_{02}^* gives a multiplication law on the space $\mathcal{A}_1$. The basic idea of Segal's delooping machine is that if we only require σ_2 to be a weak homotopy equivalence of spaces, which is the *Segal condition* at $m = 2$, then f_{02}^* gives what should be considered as a "multiplication defined up to homotopy." One salient aspect of this approach is that no map $\mathcal{A}_1 \times \mathcal{A}_1 \to \mathcal{A}_1$ is specified, and indeed if the spaces involved have bad properties there might exist no section of σ_2 at all.

Starting from a topological space with basepoint $x \in X$, the "loop space" ΩX is the space of loops in X based at x. Loops may be composed by traveling first along one, then along the other. The parametrization needs to be adjusted, so this multiplication is associative only up to homotopy. The term "delooping machine" refers to any of several kinds of further mathematical structure on the loop space ΩX, enhancing the basic composition of loops up to homotopy, which should allow one to reconstruct the space X up to homotopy. In Segal's machine the loop space appears as $\Omega X = \mathcal{A}_1$, as will be discussed further in Section 15.3. In tandem, the notion of *operad* introduced by Peter May [203] underlies the best-known and most studied family of delooping machines. There were also other techniques such as PROPs, which are starting to receive renewed interest. The various kinds of delooping machines are sources for the various different approaches to higher categories, after a multiplying effect whereby each delooping technique leads to several different definitions of higher categories. Our technical work in later parts of the book will concentrate on the particular direction of iterating Segal's approach while maintaining a discrete set of objects, but we will survey some of the many other approaches in later chapters of Part I.

The relationship between categories and simplicial objects was noticed early on with the *nerve construction* [130]. Given a category $\mathcal{C}$ its *nerve* is the simplicial set $N\mathcal{C} : \Delta^o \to \text{SET}$ such that $(N\mathcal{C})_m$ is the set of composable sequences of m arrows

$$x_0 \stackrel{g_1}{\to} x_1 \stackrel{g_2}{\to} \cdots x_{m-1} \stackrel{g_m}{\to} x_m$$

in $\mathcal{C}$. The operations of functoriality for maps $[m] \to [p]$ are obtained using the composition law and the identities of $\mathcal{C}$. The zeroth piece is just the set $(N\mathcal{C})_0 = \text{Ob}(\mathcal{C})$ of objects of $\mathcal{C}$, and the nerve satisfies a relative version of the Segal condition at each m:

$$\sigma_m : (N\mathcal{C})_m \xrightarrow{\cong} (N\mathcal{C})_1 \times_{(N\mathcal{C})_0} (N\mathcal{C})_1 \times_{(N\mathcal{C})_0} \cdots \times_{(N\mathcal{C})_0} (N\mathcal{C})_1.$$

Conversely, any simplicial set $\Delta^o \to \text{SET}$ satisfying these conditions comes from a unique category and these constructions are inverses. In other words, categories may be considered, via the nerve construction which is fully faithful, as simplicial sets satisfying the "Segal conditions." Segal [226] refers to Grothendieck [130] for this characterization.

In comparing this with Segal's situation, recall that he required $\mathcal{A}_0 = *$, which is like looking at a category with a single object.

An obvious way of putting all of these things together is to consider simplicial spaces $\mathcal{A} : \Delta^o \to \text{TOP}$ such that $\mathcal{A}_0$ is a discrete set–to be thought of as the "set of objects"–but considered as a space, and such that the Segal maps

$$\sigma_m : \mathcal{A}_m \to \mathcal{A}_1 \times_{\mathcal{A}_0} \mathcal{A}_1 \times_{\mathcal{A}_0} \cdots \times_{\mathcal{A}_0} \mathcal{A}_1$$

are weak homotopy equivalences for all $m \geq 2$. Functors of this kind are *Segal categories*. We use the same terminology when TOP is replaced by the category of simplicial sets $\mathcal{K} := \text{SET}^{\Delta^o} = \text{FUNC}(\Delta^o, \text{SET})$ (for us the term "space" will often be used interchangeably for topological spaces or simplicial sets). The possibility of making this generalization was clearly evident at the time of Segal's papers [225, 228], but was made explicit only later by Schwänzl and Vogt [224] and Dwyer *et al.* [104].

Segal categories provide a good way of considering categories enriched over spaces. However, a more elementary approach is available, by looking at categories *strictly* enriched over spaces, i.e. simplicial categories. A simplicial category could be viewed as a Segal category where the Segal maps σ_m are isomorphisms. More classically it can be considered as a category enriched in TOP or $\mathcal{K}$, using the definitions of enriched category theory. In a simplicial category, the multiplication operation is well defined and strictly associative.

Dwyer *et al.* [104] showed that we don't lose any generality at this level by requiring strict associativity: every Segal category is equivalent to a simplicial category. Unfortunately, we cannot just iterate the construction by continuing to look at categories strictly enriched over the category of simplicial categories and so forth. Such an iteration leads to higher categories with strict associativity and strict units in the middle levels. The second part of Chapter 2 will provide a detailed look at the basically well-known arguments, based on the interchange rule, which show that such full strictness is too much. Another way of seeing why a strict iteration isn't good enough[1] is to note that the Bergner model structure [39] on strict simplicial categories, doesn't satisfy one of the key properties we will need, that of being "cartesian" (see Section 7.7): the direct or cartesian products of cofibrant objects are not necessarily themselves cofibrant, a problem that will hinder both iteration and the construction of an enriched category using internal <u>HOM</u>. This suggests the need for a different construction which preserves the cartesian condition, and Segal's method works.

The iteration then says: a Segal $(n+1)$-category is a functor from Δ^o to the category of Segal n-categories, whose zeroth element is a discrete set, and such that the Segal maps are equivalences. The notion of equivalence needs to be defined in the inductive process, such as was done by Tamsamani [250]. This iterative point of view towards higher categories is the topic of our book. We emphasize an algebraic approach within the world of homotopy theory, using Quillen's homotopical algebra [215], but also paying particular attention to the process of creating a higher category from generators and relations. For me this goes back to Massey's book [201] which was one of my first references both

[1] Paoli [210] has shown that n-groupoids can be *semistrictified* in any single degree; however, one cannot get strictness in many different degrees as will be seen in Chapter 2.

for algebraic topology, and for the notion of a group presented by generators and relations.

One of the main inspirations for the recent interest in higher categories came from Grothendieck's manuscript "Pursuing stacks" [132]. He set out a wide vision of the possible developments and applications of the theory of n-categories going up to $n = \omega$. Many of his remarks continue to provide important research directions, and many others remain untouched.

The other main source of interest stems from the *Baez–Dolan conjectures*. These extend, to higher categories in all dimensions, the relationships explored by many researchers between various categorical structures and phenomena of knot invariants and quantum field theory. Hopkins and Lurie have recently proven a major part of these conjectures. These motivations incite us to search for a good understanding of the algebra of higher categories, and I hope that the present book can contribute in a small way.

The mathematical discussion of the contents of the main part of the book will be continued in more detail in Chapter 6 at the end of Part I. The intervening chapters of Part I serve to introduce the problem by giving some motivation for why higher categories are needed, by explaining why strict n-categories aren't enough, and by considering some of the many other approaches that are currently being developed.

In Part II we collect our main tools from the theory of categories, including locally presentable categories and Quillen's theory of model categories. A small number of these items, such as the discussion of enriched categories, could be useful for reading Part I. The last chapter of Part II concerns "direct left Bousfield localization," which is a special case of left Bousfield localization in which the model structure can be described more explicitly. This Chapter 9, together with the general discussion of cell complexes and the small object argument in Chapter 8, are intended to provide some "black boxes" which can then be used in the rest of the book without having to go into details of cardinality arguments and the like. It is hoped that this will make a good part of the book accessible to readers wishing to avoid too many technicalities of the theory of model categories, although some familiarity is obviously necessary since our main goal is to construct a cartesian model structure.

Part III starts the main work of looking at weakly $\mathcal{M}$-enriched (pre)categories. This part is entitled "Generators and relations" because the process of starting with an $\mathcal{M}$-precategory and passing to the associated weakly $\mathcal{M}$-enriched category by enforcing the Segal condition, should be viewed as a higher or weakly enriched analogue of the classical process of describing an algebraic object by generators and relations. We develop several aspects of this point of view, including a detailed discussion of the example of categories weakly enriched over the model category $\mathcal{K}$ of simplicial sets. We see in this case how to follow along the calculus of generators and relations, taking as an example the calculation of the loop space of S^2.

Part IV contains the construction of the cartesian model category, after the two steps treating specific elements of our categorical situation: products, including the proof of the cartesian condition, and intervals.

Part V discusses various directions going towards basic techniques in the theory of higher categories, using the formalism developed in Parts II–IV. Chapter 20 sets up the basic iteration of the model category construction finished in Part IV, leading to higher categories. Section 20.4 treats the relationship between higher groupoids and spaces.

The next few chapters of Part V contain discussions of the inversion of morphisms, limits and colimits, and adjunctions, based to a great extent on my preprint [235].

For the case of $(\infty, 1)$-categories, these topics are treated in Lurie's recent book [190] about the analogue of Grothendieck's theory of topoi, using quasi-categories.

Again following [235], we construct limits and colimits in the $(n+1)$-category $nCAT$ of n-categories, or more generally in the $\mathcal{P}$-enriched category of objects of $\mathcal{P}$, for any cartesian model category $\mathcal{P}$.

The last chapter is devoted to looking at the *Breen–Baez–Dolan stabilization hypothesis*, following my preprint [237]. This is one of the first parts of the famous *Baez–Dolan conjectures*. These conjectures have strongly motivated the development of higher category theory. Hopkins and Lurie have recently proven important pieces of the main conjectures. The stabilization conjecture is a preliminary statement about the behavior of the notion of k-connected n-category, understandable with the basic techniques we have developed here. We hope that this will serve as an introduction to an exciting current area of research.

Acknowledgements

I would first like to thank Zouhair Tamsamani, whose original work on this question led to all the rest. His techniques for gaining access to a theory of n-categories using Segal's delooping machine, set out the basic contours of the theory, and continue to inform and guide our understanding. I would like to thank André Hirschowitz for much encouragement and many interesting conversations in the course of our work on descent for n-stacks, n-stacks of complexes and higher Brill–Noether theory. And to thank André's thesis student Regis Pellissier who took the argument to a next stage of abstraction, braving multiple difficulties, not the least of which were the cloudy reasoning and several important errors in one of my preprints. We are following quite closely the main idea of Pellissier's thesis, which is to iterate a construction whereby a good model category $\mathcal{M}$ serves as input, and we try to get out a model category of $\mathcal{M}$-enriched precategories. Clark Barwick then added a further crucial insight, which was that the argument could be broken down into pieces, starting with a fairly classical left Bousfield localization. Here again, Barwick's idea serves as groundwork for our approach. Jacob Lurie continued with many contributions on different levels, most of which are beyond our immediate grasp; but still including some quite understandable innovations such as introducing the category Δ_X of finite ordered sets decorated with elements of the set X "of objects." This leads to a significant lightening of the hypotheses needed of $\mathcal{M}$. His approach to cardinality questions in the small object argument is groundbreaking, and we give here an alternate treatment which is certainly less streamlined but might help the reader to situate what is going on. These items are of course subordinated to the use of Smith's recognition principle and Dugger's notion of combinatorial model category, on which our constructions are based. Julie Bergner has gained important information about a whole range of model structures starting with her consolidation of the Dwyer–Kan structure on the category of simplicial categories. Her characterization of fibrant objects in the model structures for Segal categories, carries over easily to our case and provides the basis for important parts of the statements of our main results.

Bertrand Toën was largely responsible for teaching me about model categories. From discussions with him, I acquired the philosophy that they are a good down-to-earth yet powerful approach to higher categorical questions, and this approach is suffused throughout this work. I would like to thank Joseph Tapia and Constantin Teleman for their encouragement in this direction too, and Georges Maltsiniotis and Alain Bruguières who were able to explain the higher categorical meaning of the Eckmann–Hilton argument in an understandable way. It is a great pleasure to thank Clemens Berger, Ronnie Brown, Eugenia Cheng, Denis-Charles Cisinski, Delphine Dupont, Philip Hirschhorn, Dmitry Kaledin, Joachim Kock, Tom Leinster, Peter May, Simona Paoli and Frédéric Paugam, for many interesting and informative conversations about various aspects of this subject; and to thank my current doctoral students Samer Allouch, Hugo Bacard, Brahim Benzeghli and Chadi Taher, for continuing discussions in directions extending the present work, motivating the completion of this project. I would also like to thank my co-workers on related topics, things which if they are not directly present here, have still contributed a lot to the motivation for the study of higher categories.

I would especially like to thank Diana Gillooly of Cambridge University Press, and Burt Totaro, for setting this project in motion. Many thanks go out to the Reader for an almost overwhelming number of comments, suggestions and corrections.

Paul Taylor's diagram package is used for the commutative diagrams and even for the arrows in the text.

For the title, we have chosen something that is almost the same as the title of a special year in Barcelona a while back; we apologize for this overlap.

This research is partially supported by the Agence Nationale de la Recherche, grant ANR-09-BLAN-0151-02 (HODAG). I would like to thank the Institut de Mathématiques de Jussieu for their hospitality during the completion of this work.

To close these acknowledgements, it would be hard to over-emphasize the contribution of the CNRS in providing me with excellent working conditions, both at the Université Paul Sabatier in Toulouse, and then at the Université de Nice, a contribution surpassed only by that of Nicole, Chloé and Léo …

PART I

Higher categories

1
History and motivation

The most basic motivation for introducing higher categories is the observation that $\mathrm{CAT}_{\mathbb{U}}$, the category of $\mathbb{U}$-small categories, naturally has a structure of 2-category: the objects are categories, the morphisms are functors, and the 2-morphisms are natural transformations between functors. If we denote this 2-category by CAT^{2cat} then its truncation $\tau_{\leq 1}\mathrm{CAT}^{2cat}$ to a 1-category would have, as morphisms, the equivalence classes of functors up to natural equivalence. While it is often necessary to consider two naturally equivalent functors as being "the same," identifying them formally leads to a loss of information.

Topologists are confronted with a similar situation when looking at the category of spaces. In homotopy theory one thinks of two homotopic maps between spaces as being "the same"; however, the *homotopy category* ho(TOP) obtained after dividing by this equivalence relation doesn't retain enough information. This loss of information is illustrated by the question of diagrams. Suppose Ψ is a small category. A *diagram of spaces* is a functor $T : \Psi \to \mathrm{TOP}$, that is a space $T(x)$ for each object $x \in \Psi$ and a map $T(a) : T(x) \to T(y)$ for each arrow $a \in \Psi(x, y)$, satisfying strict compatibility with identities and compositions. The category of diagrams $\mathrm{FUNC}(\Psi, \mathrm{TOP})$ has a natural subclass of morphisms: a morphism $f : S \to T$ of diagrams is a *levelwise weak equivalence* if each $f(x) : S(x) \to T(x)$ is a weak equivalence. Letting $\mathcal{W} = \mathcal{W}_{\mathrm{FUNC}(\Psi,\mathrm{TOP})}$ denote this subclass, the homotopy category of diagrams $\mathrm{ho}(\mathrm{FUNC}(\Psi, \mathrm{TOP}))$ is defined to be the Gabriel–Zisman localization [117] $\mathcal{W}^{-1}\mathrm{FUNC}(\Psi, \mathrm{TOP})$, obtained by formally inverting the arrows of $\mathcal{W}$. There is a natural functor

$$\mathrm{ho}(\mathrm{FUNC}(\Psi, \mathrm{TOP})) \to \mathrm{FUNC}(\Psi, \mathrm{ho}(\mathrm{TOP})),$$

which is *not* in general an equivalence of categories: ho(TOP) doesn't retain enough information to recover $\mathrm{ho}(\mathrm{FUNC}(\Psi, \mathrm{TOP}))$. Consider for example the groupoid Ψ having one object whose automorphism group is $\mathbb{Z}/2\mathbb{Z}$.

The rotation action of $\mathbb{Z}/2\mathbb{Z}$ on the circle S^1 on the one hand, and the trivial action on the other, provide two functors $\Psi \to$ TOP. However, the rotation by 180° is homotopic to the identity, so both functors are the same when considered as diagrams $\Psi \to$ ho(TOP). They are different in ho(FUNC(Ψ, TOP)), indeed the homotopy quotient of the action is an invariant of the diagram up to the class $\mathcal{W}$ of levelwise weak equivalences; and the homotopy quotient of the rotation action is the double covering $S^1 \xrightarrow{2:1} S^1$, whereas the homotopy quotient of the trivial action is $S^1 \times B(\mathbb{Z}/2\mathbb{Z})$. This loss of information indicates that we need to consider some kind of extra structure beyond just the homotopy category.

The same phenomenon occurs in a number of different places. Starting in the 1950s and 1960s, the notion of *derived category*, an abelianized version of ho( ), became crucial to a number of areas in modern homological algebra and particularly for algebraic geometry. The notion of localization of a category seems to have been proposed in this context by Serre, and appears in Grothendieck's Tohoku paper [129]. A systematic treatment is the subject of Gabriel and Zisman's book [117].

As the example of diagrams illustrates, in many derived-categorical situations one must first make some intermediate constructions on the underlying categorical data, then pass to the derived category. A fundamental example of this kind of reasoning was Deligne's approach to the Hodge theory of simplicial schemes using the notion of "mixed Hodge complex" [90].

In the nonabelian or homotopical-algebra case, Quillen's notion of model category formulates a good collection of requirements that can be made on the intermediate categorical data. In [215], Quillen asked for a general structure that would encapsulate all of the higher homotopical data. In one way of looking at it, the answer lies in the notion of *higher category*. Quillen had already provided this answer with his definition of "simplicial model category," wherein the simplicial subcategory of cofibrant and fibrant objects provides a homotopy invariant higher categorical structure. As later became clear with the work of Dwyer and Kan, this simplicial category contains exactly the right information. The notion of Quillen model category is still one of the best ways of approaching the problem of calculation with homotopical objects, so much so that we adopt it as a basic language for dealing with notions of higher categories.

Bondal and Kapranov [51, 167] introduced the idea of *enhanced derived categories*, whereby the usual derived category, which is the Gabriel–Zisman localization of the category of complexes, is replaced by a *differential graded (dg) category* containing the required higher homotopy information. The notion of dg-category actually appears near the end of Gabriel and Zisman's book

[117] (where it is compared with the notion of 2-category), and it was one of the motivations for Kelly's theory of enriched categories [171]. The notion of dg-category, now further developed by Keller [170], Tabuada [246, 247], Stanculescu [240], Batanin [27], Toën [257], Moriya [208] and others, is one possible answer to the search for higher categorical structure in the k-linear case, which is pretty much analogous to the notion of strict simplicial category. The corresponding weak notion is that of *A_∞-category* used for example by Fukaya [115] and Kontsevich [176]. This definition is based on Stasheff's notion of A_∞-algebra [241], which is an example of the passage from delooping machinery to higher categorical theories.

In the far future one could imagine starting directly with a notion of higher category and bypassing the model-category step entirely, but for now this raises difficult questions of bootstrapping. Lurie has taken this kind of program a long way in [190, 191], using the notion of *quasicategory* as his basic higher-categorical object. But even there, the underlying model category theory remains important. The reader is invited to reflect on this interesting problem.

The original example of the 2-category of categories suggests using 2-categories, and their eventual iterative generalizations, as higher categorical structures. This point of view occured as early as Gabriel and Zisman's book, where they introduce a 2-category enhancing the structure of ho(TOP) as well as its analogue for the category of complexes, and proceed to use it to treat questions about homotopy groups.

Bénabou's monograph [34] introduced the notion of *weak 2-category*, as well as various notions of weak functor. These are also related to Grothendieck's notion of *fibered category* in that a fibered category may be viewed as some kind of weak functor from the base category to the 2-category of categories.

Starting with Bénabou's book, it has been clear that there would be two types of generalization from 2-categories to n-categories. The *strict n-categories* are defined recurrently as categories enriched over the category of strict $(n-1)$-categories. By the Eckmann–Hilton argument, these don't contain enough objects, as we shall discuss in Chapter 2. For this reason, these are not our main objects of study and we will use the terminology *strict n-category*. The relative ease of defining strict n-categories nonetheless makes them attractive for learning some of the basic outlines of the theory, the starting point of Chapter 2.

The other generalization would be to consider *weak n-categories*, also called *lax n-categories*, and which we usually call just "n-categories," in which the composition would be associative only up to a natural equivalence, and similarly for all other operations. The requirement that all equalities between sequences of operations be replaced by natural equivalences at one level higher, leads to a combinatorial explosion because the natural equivalences themselves are to be considered as operations. For this reason, the theory of weak 3-categories developed by Gordon, *et al.* [123], following the path set out by Bénabou for 2-categories [34], was already very complicated; for $n = 4$ it became next to impossible (see, however, Trimble [259]) and there haven't been further extensions to higher values of n. Recently more work has been done for $n = 3$ by Gurski [134] who gives a completely algebraic version of the definition, and this has been used in his work with Garner [119].

In fact, the problem of defining and studying the higher operations that are needed in a weakly associative category had been considered rather early on by the topologists who noticed that the notion of "H-space," that is to say a space with an operation which provides a group object in the homotopy category, was insufficient to capture the data contained in a loop space. One needs to specify, for example, a homotopy of associativity between $(x, y, z) \mapsto x(yz)$ and $(x, y, z) \mapsto (xy)z$. This "associator" should itself be subject to some kind of higher associativity laws, called *coherence relations*, involving composition of four or more elements.

One of the first discussions of the resulting higher coherence structures was Stasheff's notion of A_∞-algebra [241]. This was placed in the realm of differential graded algebra, but not long thereafter the notion of a "delooping machine" came out, including MacLane's notion of PROP, then May's operadic and Segal's simplicial delooping machines.

In Segal's case, the higher coherence relations come about by requiring not only that σ_2 be a weak equivalence, but that all of the "Segal maps"

$$\sigma_m : \mathcal{A}_m \to \mathcal{A}_1 \times \cdots \times \mathcal{A}_1$$

given by $\sigma_m = \left(f^*_{01}, f^*_{12}, \ldots, f^*_{m-1,m}\right)$ should be weak homotopy equivalences. His condition was inspired by Grothendieck's characterization of the nerve of a category. This delooping machine was iterated by Dunn [97]. In the operadic viewpoint, the coherence relations come from the contractibility of the spaces of n-ary operations.

By the late 1960s and early 1970s, the topologists had their delooping machines well in hand. A main theme of the present work is that these delooping machines can generally lead to definitions of higher categories, but that doesn't seem to have been done explicitly at the time. A related notion also appeared in the book of Boardman and Vogt [50], that of *restricted Kan complex*. These objects are now known as "quasicategories" thanks to Joyal's work [157]. At that time, in algebraic geometry, an elaborate theory of derived categories was being developed, but it relied only on 1-categories which were the $\tau_{\leq 1}$ of the relevant higher categories. This difficulty was worked around at all places, by techniques of working with explicit resolutions and complexes. Illusie [150] gave the definition of *weak equivalence of simplicial presheaves*, which, in retrospect, leads later to the idea of *higher stack* via the model categories of Jardine and Joyal. Somewhere in these works is the idea, which seems to have been communicated to Illusie by Deligne, of looking at the derived category of diagrams as a functor of the base category; this was later taken up by Grothendieck [133] and Cisinski [79] under the name "derivator."

In 1980, Dwyer and Kan came out with their theory of simplicial localization, allowing the association of a simplicial category to any pair $(\mathcal{M}, \mathcal{W})$, where $\mathcal{W} \subset \mathcal{M}$ is a collection of morphisms destined to be made invertible, and giving the higher categorical version of Gabriel–Zisman's theory. They developed an extensive theory of simplicial categories, including several different constructions of the simplicial localization [101, 102, 103], which inverts the morphisms of $\mathcal{W}$ in a homotopical sense. This construction provides the door passing from the world of categories to the world of higher categories, because even if we start with a plain 1-category, then localize it by inverting a collection of morphisms, the simplicial localization is in general a simplicial category which is not a 1-category. The simplicial localization maps to the usual or Gabriel–Zisman localization, but the latter is only the 1-truncation. So, if we want to invert a collection of morphisms in a "homotopically correct" way, we are forced to introduce some kind of higher categorical structure, at the very least the notion of simplicial category. Unfortunately, the importance of the Dwyer–Kan construction doesn't seem to have been generally noticed at the time.

During this period, the category theorists, and particularly the Australian school, were working on fully understanding the theory of strictly associative n-categories and ∞-categories. In a somewhat different direction, Loday [188] introduced the notion of cat^n-group, which was obtained by iterating the internal category construction in a different way, allowing categories of objects as well as of morphisms. Ronnie Brown worked on various aspects of the problem of relating these structures to homotopy theory: the strictly associative n-categories don't model all homotopy types (Brown and Higgins [67, 68]), whereas the cat^n-groups do (Brown and Loday [69, 70]).

A major turning point in the history of higher categories was Alexander Grothendieck's famous manuscript "Pursuing stacks," which started out as a collection of letters to different colleagues with many parts crossed out and rewritten, the whole circulated in mimeographed form. I was lucky to be able to consult a copy in the back room of the Princeton math library in the late 1980s, and later to obtain a copy from Jean Malgoire; a published version [132] edited by Georges Maltsiniotis should appear soon. Grothendieck introduces the problem of defining a notion of weakly associative n-category, and points out that many areas of mathematics could benefit from such a theory, explaining in particular how a theory of higher stacks should provide the right kind of coefficient system for higher nonabelian cohomology.

Grothendieck made important progress in investigating the topology and category theory behind this question. He introduced the notion of *n-groupoid*, an n-category in which all arrows are invertible up to equivalences at the next higher level. He conjectured the existence of a *Poincaré n-groupoid construction*

$$\Pi_n : \text{TOP} \to n\text{GPD} \subset n\text{CAT},$$

where nGPD is the collection of weakly associative n-groupoids. He postulated that this functor should provide an equivalence of homotopy theories between n-truncated spaces[1] and n-groupoids.

In his search for algebraic models for homotopy types, Grothendieck was inspired by one of the pioneering works in this direction, the notion of *Cat^n-groups* of Brown and Loday. This is what is now known as the "cubical" approach, where the set of objects can itself have a structure for example of $(n-1)$-category, so it isn't quite the same as the approach we are looking for, commonly called the "globular" case.[2]

Much of "Pursuing stacks" is devoted to the more general question of modeling homotopy types by algebraic objects such as presheaves on a fixed small category, developing a theory of "test categories," which has now blossomed into a distinct subject in its own right thanks to the further work of Maltsiniotis [198] and Cisinski [77]. One of the questions their theory aims to address is, which presheaf categories provide good models for homotopy theory. One could ask a similar question with respect to Segal's utilisation of Δ, namely whether other categories could be used instead. We don't currently have any good information about this. As a start, throughout the present book we will try to point out in discussion and counterexamples the main places where special properties of Δ are used.

In the parts of "Pursuing stacks" about n-categories, the following theme emerges: the notion of n-category with strictly associative composition is not sufficient. This is seen from the fact that strictly associative n-categories satisfying a weak groupoid condition do not serve to model homotopy n-types

[1] A space T is *n-truncated* if $\pi_i(T, t) = 0$ for all $i > n$ and all basepoints $t \in T$. The n-truncated spaces are the objects which appear in the Postnikov tower of fibrations, and one can define the truncation $T \to \tau_{\leq n}(T)$ for any space T, by adding cells of dimension $\geq n+2$ to kill off the higher homotopy groups.

[2] Paoli has recently defined a notion of *special Cat^n-group* [210], which imposes the globularity condition weakly.

as would be expected. Fundamentally due to the Godement relation (interchange rule) and the Eckmann–Hilton argument, this observation was refined over time by Brown and Higgins [67] and Berger [35]. We will discuss it in some detail in Chapter 2.

Since strict n-categories aren't enough, it leads to the question of defining a notion of weak n-category, which is the main subject of our book. Thanks to a careful reading by Georges Maltsiniotis [199], we now know that Grothendieck's manuscript in fact contained a definition of weakly associative n-groupoid, and that his definition is very similar to Batanin's definition of n-category [200]. Grothendieck enunciated the deceptively simple rule [132]:

Intuitively, it means that whenever we have two ways of associating to a finite family $(u_i)_{i\in I}$ of objects of an ∞-groupoid, $u_i \in F_{n(i)}$, subjected to a standard set of relations on the u_i's, an element of some F_n, in terms of the ∞-groupoid structure only, then we have automatically a "homotopy" between these built in in the very structure of the ∞-groupoid, provided it makes sense to ask for one ...

The structure of this as a definition was not immediately evident upon any initial reading, all the more so when one takes into account the directionality of arrows, so "Pursuing stacks" left open the problem of giving a good definition of weakly associative n-category.

Given the idea that an equivalence Π_n between homotopy n-types and n-groupoids should exist, it becomes possible to think of replacing the notion of n-groupoid by the notion of n-truncated space. This motivated Joyal [156] to define a model structure on the category of simplicial sheaves, and Jardine [153] to extend this to simplicial presheaves. These theories give an approach to the notion of ∞-stack, and were used by Thomason [253], Voevodsky [261], Morel and Voevodsky [207] and others in K-theory.

Also explicitly mentioned in "Pursuing stacks" was the limiting case $n = \omega$, involving i-morphisms of all degrees $0 \leq i < \infty$. Again, an ω-groupoid should correspond, via the inverse of a Poincaré construction Π_ω, to a full homotopy type up to weak equivalence.

We can now get back to the discussion of simplicial categories. These are categories enriched over spaces, and applying Π_ω (which is supposed to be compatible with products) to the morphism spaces, we can think of simplicial categories as being categories enriched over ω-groupoids. Such a thing is itself an ω-category $\mathcal{A}$, with the property that the morphism ω-categories $\mathcal{A}(x, y)$ are groupoids. In other words, the i-morphisms are invertible for $i \geq 2$, but not necessarily for $i = 1$. Jacob Lurie [189] introduced the terminology $(\infty, 1)$-*categories* for these things, where more generally an (∞, n)-category would be an ω-category such that the i-morphisms are invertible up to equivalence, for $i > n$. The point of this discussion – of notions which have not yet been defined – is to say that the notion of simplicial category is a perfectly good substitute for the notion of $(\infty, 1)$-category even if we don't know what an ω-category is in general.

This replacement no longer works if we want to look at n-categories with noninvertible morphisms at levels ≥ 2, or, somewhat similarly, (∞, n)-categories for $n \geq 2$. Grothendieck doesn't seem to have been aware of Dwyer and Kan's work, just prior to "Pursuing stacks," on simplicial categories;[3] however, he was well aware that the notions of n-category for small values of n had been extensively investigated earlier in Bénabou's book about 2-categories [34], and Gordon *et al.* on 3-categories [123]. The combinatorial explosion which is naturally inherent in these explicit theories was foreseeable, which was why Grothendieck asked for a different form of definition which could work in general.

As he foresaw in a vivid passage [132, First letter, p. 16], there are currently many different definitions of n-category. This started with Street's proposal [244], of a definition of weak n-category as a simplicial set satisfying a certain variant of the Kan condition where one takes into account the directions of arrows, and also using the idea of "thinness." His suggestion, in retrospect undoubtedly somewhat similar to Joyal's iteration of the notion of quasicategory, wasn't worked out at the time, but has recieved renewed interest, see Verity [260] for example.

The Segal-style approach to weak topological categories was introduced by Dwyer *et al.* [104] and Schwänzl and Vogt [224], but the fact that they immediately proved a rectification result relating Segal categories back to strict simplicial categories seems to have slowed down their further consideration of this idea. Applying Segal's idea seems to have been the topic of a letter from Breen to Grothendieck in 1975, see page 83 below.

Kapranov and Voevodsky [168] considered a notion of "Poincaré ∞-groupoid," which is a strictly associative ∞-groupoid but where the arrows are invertible only up to equivalence. It now appears likely that their constructions should best be interpreted using some kind of weak unit condition, such as introduced by Kock [172].

At around the same time in the mid-1990s, three distinct approaches to defining weak n-categories came out: Baez and Dolan's approach [9, 11] used *opetopes*, Tamsamani's approach [248, 250] used an iteration of the Segal delooping machine, and Batanin's approach [25, 26] used *globular operads*. The Baez–Dolan and Batanin approaches will be discussed in Chapter 4.

[3] Paradoxically, Grothendieck's unpublished manuscript is responsible in large part for the revival of interest in Dwyer and Kan's published papers!

The work of Baez and Dolan was motivated by a far-reaching program of conjectures on the relationship between n-categories and physics [8, 12, 13], which has led to important developments, most notably the recent proof by Hopkins and Lurie [193].

In relationship with Grothendieck's manuscript, as we pointed out above, Batanin's approach is the one which most closely resembles what Grothendieck was asking for, indeed Maltsiniotis [200] generalized the definition of n-groupoid which he found in "Pursuing stacks," to a definition of n-category which is similar to Batanin's one.

In the subsequent period, a number of other definitions have appeared, and people have begun working more seriously on the approach which had been suggested by Street. Batanin, in mentioning the letter from Baez and Dolan to Street [9], also points out that Hermida *et al.* [142] have worked on the opetopic ideas. M. Rosellen suggested in 1996 to give a version of the Segal-style definition, using the theory of operads. He didn't concretize this but Trimble gave a definition along these lines, now playing an important role in the work of Cheng [74]. Further ideas include those of Penon [212], Leinster's multicategories, and others. Tom Leinster has collected together ten different definitions in the useful compendium [183]. The somewhat mysterious paper of Kondratieff [175] could also be pointed out. In the simplicial direction, Rezk's complete Segal spaces [218] can be iterated as suggested by Barwick [219], indeed there are several different iteration strategies cf. Bergner's overview [48]. Joyal [158] proposes an iteration of the method of quasicategories leading to presheaves on a category Θ, and Berger has related these to homotopy types [36, 37].

We shall discuss the simplicial definitions in Chapter 5 and the operadic definitions in Chapter 4. One of the main tasks in the future will be to understand the relationships between all of these approaches. Our goal here is more down-to-earth: we would like to develop the tools necessary for working with Tamsamani's n-categories as well as the (∞, n)-categories obtained by iterating the notion of Segal category. We hope that similar tools can be developed for the other approaches, making an eventual comparison theory into a powerful method whereby the particular advantages of each definition could all be put in play at the same time.

Tamsamani [250] defined the Poincaré n-groupoid functor for his notion of n-category, and showed Grothendieck's conjectured equivalence with the theory of homotopy n-types. Some aspects of his theory will be reviewed in Sections 15.2 and 20.4.

The same has also been done partially for Batanin's theory: Berger [36] shows that the realization from Batanin ∞-groupoids to homotopy types is surjective, while Cisinski [78] shows that corresponding Poincaré ∞-groupoid functor going in the other direction is faithful and conservative.

It is interesting to note that the two main ingredients in Tamsamani's approach, the multisimplicial nerve construction and Segal's delooping machine, are both mentioned in "Pursuing stacks." In particular, Grothendieck reproduces a letter from himself to Breen dated July 1975, in which Grothendieck acknowledges having recieved a proposed definition of non-strict n-category from Breen, a definition which according to loc. cit. "... has certainly the merit of existing ...." It is not clear whether this proposed construction was ever worked out. Quite apparently, Breen's suggestion for using Segal's delooping machine must have gone along the lines of what we are doing here. Rather than taking up this direction, Grothendieck elaborated a general *Ansatz* whereby n-categories would have various different composition operations, and natural equivalences between any two natural compositions with the same source and target, an idea now fully developed in the context of Batanin's and related definitions.

Once one or more points of view for defining n-categories are in hand, the main problem which needs to be considered is to obtain – hopefully within the same point of view – an $(n+1)$-category nCAT parametrizing the n-categories of that point of view. This problem, already clearly posed in "Pursuing stacks," is one of our main goals in the more technical central part of the present book, for one model.

It turns out that Quillen's technique of model categories, subsequently deepened by several generations of mathematicians, is a great way of attacking this problem. It is by now well known that model categories provide an excellent environment for studying homotopy theory, as became apparent from the work of Bousfield, Dwyer and Kan on model categories of diagrams, and the generalization of these ideas by Joyal, Jardine, Thomason and Voevodsky, who used model categories to study simplicial presheaves under Illusie's condition of weak equivalence. In the Segal-style paradigm of weak enrichment, we look at functors $\Delta^o \to (n-1)\mathrm{CAT}$, so we are certainly also studying diagrams, and it is reasonable to expect the notion of model category to bring some of the same benefits as for the above-mentioned theories.

To be more precise about this motivation, from "Pursuing stacks" recall that nCAT should be an $(n+1)$-category whose objects are in one-to-one correspondence with the n-categories of a given universe. The structure of $(n+1)$-category therefore consists of specifying the morphism objects $\mathrm{HOM}_{n\mathrm{CAT}}(\mathcal{A}, \mathcal{B})$, which should themselves be n-categories parametrizing "functors" (in an appropriate sense) from $\mathcal{A}$ to $\mathcal{B}$.

In the explicit theories for $n = 2, 3, 4$ this is one of the places where a combinatorial explosion takes place: the functors from $\mathcal{A}$ to $\mathcal{B}$ have to be taken in a weak sense, that is to say we need a natural equivalence between the image of a composition and the composition of the images, together with the appropriated coherence data at all levels.

The following simple example shows that, even if we were to consider only strict n-categories, the strict morphisms are not enough. Suppose G is a group and V an abelian group and we set $\mathcal{A}$ equal to the category with one object and group of automorphisms G, and $\mathcal{B}$ equal to the strict n-category with only one i-morphism for $i < n$ and group V of n-automorphisms of the unique $(n-1)$-morphism; then for $n = 1$ the equivalence classes of strict morphisms from $\mathcal{A}$ to $\mathcal{B}$ are the elements of $H^1(G, V)$ so we would expect to get $H^n(G, V)$ in general, but for $n > 1$ there are no nontrivial strict morphisms from $\mathcal{A}$ to $\mathcal{B}$. So some kind of weak notion of functor is needed.

Here is where the notion of model category comes in: one can view this situation as being similar to the problem that usual maps between simplicial sets are generally too rigid and don't reflect the homotopical maps between spaces. Kan's fibrancy condition and Quillen's formalization of this in the notion of model category, provide the solution: we should require the target object to be fibrant and the source object to be cofibrant in an appropriate model category structure. Quillen's axioms serve to guarantee that the notions of cofibrancy and fibrancy go together in the right way. So, in the application to n-categories we would like to define a model structure and then say that the usual maps $\mathcal{A} \rightarrow \mathcal{B}$ strictly respecting the structure are the right ones, provided that $\mathcal{A}$ is cofibrant and $\mathcal{B}$ fibrant.

To obtain nCAT, a further property is needed, indeed we are not just looking to find the right maps from $\mathcal{A}$ to $\mathcal{B}$ but to get a morphism object $\mathrm{HOM}_{n\mathrm{CAT}}$ $(\mathcal{A}, \mathcal{B})$, which should itself be an n-category. It is natural to apply the idea of "internal $\underline{\mathrm{HOM}}$", that is to put

$$\mathrm{HOM}_{n\mathrm{CAT}}(\mathcal{A}, \mathcal{B}) := \underline{\mathrm{HOM}}(\mathcal{A}, \mathcal{B})$$

using an internal $\underline{\mathrm{HOM}}$ in our model category. For our purposes, it is sufficient to consider $\underline{\mathrm{HOM}}$ adjoint to the cartesian product operation, in other words a map

$$\mathcal{E} \rightarrow \underline{\mathrm{HOM}}(\mathcal{A}, \mathcal{B})$$

should be the same thing as a map $\mathcal{E} \times \mathcal{A} \rightarrow \mathcal{B}$. This obviously implies imposing further axioms on the model structure, in particular compatibility between $\times$ and cofibrancy since the cartesian product is used on the source side of the map. It turns out that the required axioms are already well known in the notion of *monoidal model category* popularized by Hovey [149], which is a model category provided with an additional operation $\otimes$, and certain axioms of compatibility with the cofibrant objects. In our case, the operation is already given as the cartesian product $\otimes = \times$ of the model category, and a model category which is monoidal for the cartesian product operation will be called *cartesian* (Section 7.7).

In the present book, we are concentrating on Tamsamani's approach to n-categories, which in the preprint of Hirschowitz and myself [145] was modified to "Segal n-categories" in the course of discussions with André Hirschowitz. In Tamsamani's theory an n-category is viewed as a category enriched over $(n-1)$-categories, using Segal's machine to deal with the enrichment in a homotopically weak way.

In Pellissier's thesis [211], following a question posed by Hirschowitz, this idea was pushed to a next level: to study weak Segal-style enrichment over a more general model category, with the aim of making the iteration formal. A small link was missing in this process at the end of his thesis, essentially because of an error in my preprint [234] which Pellissier discovered. He provided the correction when the iterative procedure is applied to the model category of simplicial sets. But, in fact, his patch applies much more generally if we just consider the operation of functoriality under a change of model categories.

This is what we will be doing here. But instead of following Pellissier's argument too closely, some aspects will be set into a more general discussion of certain kinds of left Bousfield localizations. The idea of breaking down the construction into several steps including a main step of left Bousfield localization, is due to Clark Barwick.

The Segal 1-categories are, as was originally proven by Dwyer *et al.* [104], equivalent to strict simplicial categories. Bergner [40] has shown that this equivalence takes the form of a Quillen equivalence between model categories. However, the model category of simplicial categories is not appropriate for the considerations described above: it is not cartesian, indeed the product of two cofibrant simplicial categories will not be cofibrant.[4] It is interesting to imagine several possible ways around this problem: one could try to systematically apply the cofibrant replacement operation, which would seem to lead to a theory very similar to the consideration of Gray tensor products by Leroy [186] and Crans [86]; or one could hope for a general construction replacing a model category by a cartesian one (or perhaps, given a model category with monoidal structure incompatible with cofibrations, construct a monoidal model category in some sense equivalent to it).

As Bergner [40] pointed out, the theories of simplicial categories and Segal categories are also equivalent to Charles Rezk's theory of *complete Segal*

[4] This remark also applies to the projective model structure for weakly enriched Segal-style categories, whereas the projective structure is much more practical for calculating maps.

spaces. As we shall discuss further in Chapter 5, Rezk requires that the Segal maps be weak equivalences, but rather than having $\mathcal{A}_0$ be a discrete simplicial set corresponding to the set of objects, he asks that $\mathcal{A}_0$ be a simplicial set weakly equivalent to the "interior" Segal groupoid of $\mathcal{A}$. Barwick pointed out that Rezk's theory could also be iterated. It turns out that there are several ways of doing this. One, discussed by Lurie [192], leads to presheaves on Δ^n, whereas the method in Rezk's recent preprint [219] involves presheaves on Joyal's category Θ^n. Rezk shows that the resulting model category is cartesian. So, this route also leads to a construction of nCAT and can serve as an alternative to what we are doing here. Bergner [48] extends her comparison results to the iterative case to obtain equivalences between the various different iterates of Rezk's theory and the iterates we consider here. If our current theory is perhaps simpler in its treatment of the set of objects, Rezk's theory has a better behavior with respect to homotopy limits.

As more different points of view on higher categories are up and running, the comparison problem will be posed: to find an appropriate way to compare different points of view on n-categories and (one hopes) to say that the various points of view are equivalent and in particular that the various $(n+1)$-categories nCAT are equivalent via these comparisons. Grothendieck gave a vivid description of this problem (with remarkable foresight, it would seem, cf Leinster's compendium [183]) in the first letter of "Pursuing stacks" [132]. He pointed out that it is not actually clear what type of general setup one should use for such a comparison theory. Various possibilities would include the model category formalism, or the formalism of $(\infty, 1)$-categories starting with Dwyer–Kan localization and moving through the theory of Joyal and Lurie.

Within the domain of simplicial theories, we have mentioned Bergner's comparison [40] between three different approaches to $(\infty, 1)$-categories. A further comparison of these theories with quasicategories is to be found in Joyal and Tierney [162], Dugger and Spivak [96] and Lurie's book [190]. Barwick and Kan [22] compare quasicategories with their new model of relative categories.

A recent result due to Cheng [74] gives a comparison between Trimble's definition and Batanin's definition (with some modifications on both sides due to Cheng and Leinster). Batanin's approach used operads more as a way of encoding general algebraic structures, and is the closest to Grothendieck's original philosophy. While also operadic, Trimble's definition is much closer to the philosophy we are developing in the present book, whereby one goes from topologists' delooping machinery (in his case, operads) to an iterative theory of n-categories. It is to be hoped that Cheng's result can be expanded in various directions to obtain comparisons between a wide range of theories, maybe using Trimble's definition as a bridge towards the simplicial theories. This should clearly be pursued in the near future, but it would go beyond the scope of the present work.

We now turn to the question of potential applications. Having a good theory of n-categories should open up the possibility to pursue any of the several programs such as that outlined by Grothendieck [132], the generalization to n-stacks and n-gerbs of the work of Breen [59], or the program of Baez and Dolan in topological quantum field theory [8]. Once the theory of n-stacks is off the ground this will give an algebraic approach to the "geometric n-stacks" considered in my preprint [233] generalizing Artin stacks [4], see Laumon *et al.* [180].

As the title indicates, Grothendieck's manuscript was intended to develop a foundational framework for the theory of higher stacks. In turn, higher stacks should be the natural coefficients for nonabelian cohomology, the idea being to generalize Giraud's work [121] to $n \geq 3$.

The example of diagrams of spaces translates, via the construction Π_n, to a notion of *diagram of n-groupoids*. This is a strict version of the notion of *n-prestack in groupoids*, which would be a weak functor from the base category Ψ to the $(n+1)$-category $nGPD$ of n-groupoids. Grothendieck introduced the notion of *n-stack*, which generalizes to n-categories the classical notion of stack. A full discussion of this theory would go beyond the scope of the present work: we are just trying to set up the n-categorical foundations first. The notion of n-stack, maybe with $n = \infty$, has applications in many areas as predicted by Grothendieck.

Going backwards along Π_n, it turns out that diagrams of spaces, or equivalently simplicial presheaves, serve as a very adequate replacement, as may be seen from the work of Brown [62, 63], Joyal [156], Jardine [153], Thomason [254], Voevodsky [261] and Morel and Voevodsky [207]. So, the notion of n-categories as a technical prerequisite for higher stacks has proven somewhat illusory. And, in fact, the model category theory developed for simplicial presheaves has been useful for attacking the theory of n-categories as we do here, and also for going from a theory of n-categories to a theory of n-stacks, as Hollander has done for 1-stacks [147], Crans [85] for 2-stacks and as Hirschowitz and I did for n-stacks [145].

In spite of these shortcuts, a full understanding of the matter requires viewing the theory of higher stacks as flowing in a natural way from the theory of higher categories. Barwick's discussion [17] is a good entry point. An n-stack on a site $\mathcal{X}$ should be seen as a morphism $\mathcal{X} \rightarrow n\mathrm{CAT}$. This requires a construction for the $(n+1)$-category $n\mathrm{CAT}$, together with the appropriate notion of morphism between $(n+1)$-categories. The latter is almost equivalent to knowing how to construct the $(n+2)$-category $(n+1)$CAT of $(n+1)$-categories. From this discussion the need for an iterative approach to the theory of n-categories becomes clear.

My own favorite application of n-stacks is that they lead in turn to a notion of *nonabelian cohomology*. Grothendieck says [132]:

> Thus n-stacks, relativized over a topos to "n-stacks over X", are viewed primarily as the natural "coefficients" in order to do (co)homological algebra of dimension $\leq n$ over X.

The idea of using higher categories for nonabelian cohomology goes back to Giraud [121], and had been extended to the cases $n = 2, 3$ by Breen [59] somewhat more recently. Breen's book motivated us to proceed to the case of n-categories at the beginning of Tamsamani's thesis work.

Another utilisation of the notion of n-category is to model homotopy types. For this to be useful one would like to have as simple and compact a definition as possible, but also one which lends itself to calculation. The simplicial approach developed here is direct, but it is possible that the operadic approaches which will be mentioned in Chapter 4 could be more amenable to topological computations. An iteration of the classical Segal delooping machine has been considered by Dunn [97] and Cohen *et al.* iterate the operadic machine [80].

The Poincaré n-groupoid of a space is a generalization of the Poincaré groupid $\Pi_1(X)$, a basepoint-free version of the fundamental group $\pi_1(X)$ popularized by Ronnie Brown [64]. Van Kampen's theorem allows for computations of fundamental groups, and as Brown has often pointed out, it takes a particularly nice form when written in terms of the Poincaré groupoid: it says that if a space X is written as a pushout $X = U \cup^W V$ then the Poincaré groupoid is a pushout in the 2-category of groupoids:

$$\Pi_1(X) = \Pi_1(U) \cup^{\Pi_1(W)} \Pi_1(V).$$

This says that Π_1 commutes with colimits.

Extending this theory to the case of Poincaré n-groupoids is one of the motivations for introducing colimits and indeed the whole model-categoric machinery for n-categories. We will then be able to write, in the case of a pushout of spaces $X = U \cup^W V$,

$$\Pi_n(X) = \Pi_n(U) \cup^{\Pi_n(W)} \Pi_n(V).$$

Of course the pushout diagram of spaces should satisfy some excision condition as in the original Van Kampen theorem, and this may be abstracted by refering to simplicial sets instead.

The homotopy theory and nonabelian cohomology motivations may be combined by looking for a higher-categorical theory of *shape*. For a space X we can define the *nonabelian cohomology n-category* $H(X, \mathcal{F})$ with coefficients in an n-stack $\mathcal{F}$ over X. This applies in particular to the constant stack A_X associated to an n-category A. The functor

$$A \mapsto H(X, A_X)$$

is co-represented by the universal element

$$\eta_X \in H(X, \Pi_n(X)_X),$$

giving a way of characterizing $\Pi_n(X)$ by a universal property. This essentially tautological observation paves the way for more nontrivial shape theories, consisting of an n-category $COEF$ and a functor $\text{Shape}(X) : COEF \to COEF$. A particularly useful version is when $COEF$ is the n-category of certain n-stacks over a site $\mathcal{Y}$, and $\text{Shape}(X)(\mathcal{F}) = \underline{\text{HOM}}(\Pi_n(X)_{\mathcal{Y}}, \mathcal{F})$, where $\Pi_n(X)_{\mathcal{Y}}$ denotes the constant stack on $\mathcal{Y}$ with values equal to $\Pi_n(X)$. This leads to subjects generalizing Malcev completions and rational homotopy theory (Quillen [216], Tanré [251], etc.), from Hain's work [135, 136, 137], such as the schematization of homotopy types by Toën [255] with Katzarkov and Pantev [169], Pridham [214] and Moriya [208], as well as de Rham shapes and nonabelian Hodge theory [230, 231, 232].

One of the main advantages to a theory of higher categories, is that the notions of *homotopy limit* and *homotopy colimit*, by now classical in algebraic topology, become *internal* notions in a higher category. Indeed, they become direct analogues of the notions of limit and colimit in a usual 1-category, with corresponding universal properties and so on. This has an interesting application to the "abelian" case: the structure of triangulated category is automatic once we know the $(\infty, 1)$-categorical structure. This was pointed out by Bondal and Kapranov [51] in the dg setting: their *enhanced triangulated categories* are just dg-categories satisfying some further axioms – the structure of triangles comes from the dg structure. Historically one can trace this observation back to the end of Gabriel and Zisman's book [117], although nobody seems to have noticed it until rediscovered by Bondal and Kapranov.

I first learned of the notion of "2-limit" from the paper of Deligne and Mumford [91], where it appears at the beginning with very little explanation (their paper should also be added to the list of motivations for developing the theory of higher stacks). Several authors have since considered 2-limits and 2-topoi, originating with Bourn [54] and continuing recently with Weber [264] for example.

The notions of homotopy limits and colimits internalized in an $(\infty, 1)$-category have now received an important foundational formulation with Lurie's work, [189, 190] on ∞-topoi (by which he means $(\infty, 1)$-topoi).

Power [213] has given an extensive discussion of the motivations for higher categories stemming from logic and computer science. He points out the role played by weak limits. Recently, Gaucher [120], Grandis [124] and others have used higher categorical notions to study directed and concurrent processes. It would be interesting to see how these theories interact with the notion of ∞-topos.

Recall that gerbs played an important role in descent theory and non-neutral tannakian categories as explained by Deligne [92]. Current developments where the notion of higher category is more or less essential on a foundational level include "derived algebraic geometry" and higher tannakian theory. It would go beyond our present scope to discuss these here, but the reader may search for numerous references.

Stacks and particular gerbs of higher groupoids have found many interesting applications in the mathematical physics literature, starting with explicit considerations for 1- and 2-gerbs. Unfortunately it would go beyond our scope to list all of these. However, one of the main contributions from mathematical physics has been to highlight the utility of higher categories which are not groupoids, in which there can be non-invertible morphisms. Explicit first cases come about when we consider *monoidal categories*: they may be considered as 2-categories with a single object. And then *braided monoidal categories* may be considered as 3-categories with a single object and a single 1-morphism, where the braiding isomorphism comes from the interchange rule as in the Eckmann–Hilton argument. These entered into the vast program of research on combinatorial quantum field theories and knot invariants – again the reader is left to fill in the references here.

John Baez and Jim Dolan [8, 10, 11, 12] provided a major impetus to the theory of higher categories, by formulating a series of conjectures about how the known relationships between low-dimensional field theories and n-categories for small values of n, should generalize in all dimensions. On the topological or field-theoretical side, they conjecture the existence of a k-fold monoidal n-category (or equivalently, a k-connected $(n + k)$-category) representing k-dimensional manifolds up to cobordism, where the higher morphisms should correspond to manifolds with corners. On the n-categorical side, they propose a notion of *n-category with duals* in which all morphisms should have internal adjoints. Then, their main conjecture relating these two sides is that the cobordism $(n + k)$-category should be the universal $(n + k)$-category with duals generated by a single morphism in degree k. The specification of a field theory is a functor from this $(n+k)$-category to some other one, and it suffices to specify the image of the single generating morphism. They furthermore go on to investigate possible candidates for the target categories of such functors, looking at higher Hilbert spaces and other such things. We will include some discussion (based on my preprint [237]) of one of Baez and Dolan's preliminary conjectures, the *stabilization hypothesis*, in Chapter 23.

The Baez–Dolan conjectures step outside of the realm of n-groupoids, so they really require an approach which can take into account non-invertible morphisms. In their "n-categories with duals", they generalize the fact that the notion of adjoint functor can be expressed in 2-categorical terms within the 2-category 1CAT. Mackaay [195] describes the application of internal adjoints to 4-manifold invariants. The notion of adjoint generalizes within an n-category to the notion of dual of any i-morphism for $0 < i < n$. At the top level of n-morphisms, the dual operation should either be: ignored; imposed as additional structure; or pushed to ∞ by considering directly the theory of ∞-categories. Of course, a morphism which is really invertible is automatically dualizable and its dual is the same as its inverse, so the interesting n-categories with duals have to be ones which are not n-groupoids.

As Cheng [73] has pointed out, in the last case one obtains a structure which looks algebraically like an ∞-groupoid, so the distinction between invertible and dualizable morphisms should probably be considered as an additional more analytic structure in itself. We don't yet have the tools to fully investigate the theory of ∞-categories. Further comments on these issues will be made in Section 3.8.

In a very recent development, Hopkins and Lurie have announced a proof of a major part of the Baez–Dolan conjectures, saying that the category of manifolds with appropriate corners, and cobordisms as i-morphisms (resp. equivalence classes of cobordisms at $i = n$), is the universal n-category with duals generated by a single element. This universal property allows one to define a functor from the cobordism n-category to any other n-category with duals, by simply specifying a single object. I hope that some of the techniques presented here can help in understanding this fascinating subject.

2
Strict n-categories

Classically, the first and easiest notion of higher category was that of *strict n-category*. We review here some basic definitions, since they introduce important notions for weak n-categories. Then starting from Section 2.3 we will point out why the strict theory is generally considered not to be sufficient. This second part of the chapter is the main motivation for introducing some kind of weak composition, but in a logical sense it isn't needed later on and could be skipped over.

In the current chapter only, *all n-categories are meant to be strict n-categories*. For this reason we try to put in the adjective "strict" as much as possible when $n > 1$; but, in any case, the very few times that we speak of weak n-categories, this will be explicitly stated. We mostly restrict our attention to $n \leq 3$.

In case that it isn't already clear, it should be stressed that everything we do in this section (as well as most of the next and even the subsequent one as well) is very well known and classical, so much so that I don't know what are the original references.[1]

To start with, a *strict 2-category* $\mathcal{A}$ is a collection of objects $\mathcal{A}_0$ plus, for each pair of objects $x, y \in \mathcal{A}_0$, a category $\mathcal{A}(x, y)$ together with functors

$$\mathcal{A}(x, y) \times \mathcal{A}(y, z) \to \mathcal{A}(x, z),$$

which form a strictly associative multiplication law in the obvious way; and such that a unit exists, that is an element $1_x \in \mathrm{Ob}(\mathcal{A}(x, x))$ with the property that multiplication by 1_x acts trivially on objects of $\mathcal{A}(x, y)$ or $\mathcal{A}(y, x)$ and multiplication by 1_{1_x} acts trivially on morphisms of these categories.

[1] An important recent reference, which we haven't had time to integrate into our presentation, is the work of Lafont *et al.* [179], in which they construct a model category for strict ∞-categories.

A *strict* 3-*category* $\mathcal{C}$ is the same as above but where $\mathcal{C}(x, y)$ are supposed to be strict 2-categories. There is an obvious notion of cartesian product of strict 2-categories, so the above definition applies *mutatis mutandis*.

For general n, the well-known definition is most easily presented by induction on n. We assume known the definition of strict $(n-1)$-category for $n-1$, and we assume known that the category of strict $(n-1)$-categories is closed under cartesian product. A *strict n-category* $\mathcal{C}$ is then a category enriched, according to Kelly [171], over the category of strict $(n-1)$-categories. This means that $\mathcal{C}$ is composed of a *set of objects* $\mathrm{Ob}(\mathcal{C})$ together with, for each pair $x, y \in \mathrm{Ob}(\mathcal{C})$, a *morphism-object* $\mathcal{C}(x, y)$ which is a strict $(n-1)$-category; together with a strictly associative composition law given in terms of $(n-1)$-functors

$$\mathcal{C}(x, y) \times \mathcal{C}(y, z) \to \mathcal{C}(x, z),$$

and a morphism $1_x : * \to \mathcal{C}(x, x)$ (where $*$ denotes the coinitial object cf. below) acting as the identity for the composition law. The *category of strict n-categories* denoted nSTRCAT is the category whose objects are as above and whose morphisms are the transformations strictly perserving all of the structures. Note that nSTRCAT admits a cartesian product: if $\mathcal{C}$ and $\mathcal{C}'$ are two strict n-categories then $\mathcal{C} \times \mathcal{C}'$ is the strict n-category with

$$\mathrm{Ob}(\mathcal{C} \times \mathcal{C}') := \mathrm{Ob}(\mathcal{C}) \times \mathrm{Ob}(\mathcal{C}'),$$

and for $(x, x'), \ (y, y') \in \mathrm{Ob}(\mathcal{C} \times \mathcal{C}')$,

$$(\mathcal{C} \times \mathcal{C}')((x, x'), (y, y')) := \mathcal{C}(x, y) \times \mathcal{C}'(x', y'),$$

where the cartesian product on the right is that of $(n-1)$STRCAT. Note that the coinitial object of nSTRCAT is the strict n-category $*$ with exactly one object x and with $*(x, x) = *$ being the coinitial object of $(n-1)$STRCAT.

The induction inherent in this definition may be worked out explicitly to give the definition as it is presented in the work of Kapranov and Voevodsky [168] for example. In doing this one finds that underlying a strict n-category $\mathcal{C}$ are the sets $\mathrm{Mor}^i(\mathcal{C})$ of *i-morphisms* or *i-arrows*, for $0 \leq i \leq n$. The set of 0-morphisms is by definition the set of objects $\mathrm{Mor}^0(\mathcal{C}) := \mathrm{Ob}(\mathcal{C})$, and $\mathrm{Mor}^i(\mathcal{C})$ is the disjoint union over all pairs x, y of the $\mathrm{Mor}^{i-1}(\mathcal{C}(x, y))$. These fit together in a diagram called a *reflexive globular set*:

$$\mathrm{Mor}^n(\mathcal{C}) \underset{t}{\overset{s}{\rightrightarrows}}\!\!\leftarrow \mathrm{Mor}^{n-1}(\mathcal{C}) \underset{t}{\overset{s}{\rightrightarrows}}\!\!\leftarrow \mathrm{Mor}^{n-2}(\mathcal{C}) \cdots \mathrm{Mor}^1(\mathcal{C}) \underset{t}{\overset{s}{\rightrightarrows}}\!\!\leftarrow \mathrm{Mor}^0(\mathcal{C}),$$

where the rightward maps are the *source* and *target* maps, and the leftward maps are the *identity* maps.[2] These may be defined inductively using the definition we have given of $\mathrm{Mor}^i(\mathcal{C})$. The structure of strict n-category on this underlying reflexive globular set is determined by further *composition laws* at each stage: the i-morphisms may be composed with respect to the j-morphisms for any $0 \leq j < i$, operations denoted by Kapranov and Voevodsky [168] as $*_j$. These are partially defined depending on iterations of the source and target maps. For a more detailed explanation, refer to the standard references, including Brown and Higgins [67], Street [244], op. cit. [168], Bénabou [34] and Gabriel and Zisman [117].

2.1 Godement relations: the Eckmann–Hilton argument

One of the most important of the axioms satisfied by the various compositions in a strict n-category is called the "Godement relation," or "interchange rule" by some authors. This is a commutativity property which comes from the fact that the composition law

$$\mathcal{C}(x, y) \times \mathcal{C}(y, z) \to \mathcal{C}(x, z)$$

is a morphism with its domain the cartesian product of the two morphism $(n-1)$-categories from x to y and from y to z. In a cartesian product, compositions in the two factors by definition are independent (commute). Thus, for 1-morphisms in $\mathcal{C}(x, y) \times \mathcal{C}(y, z)$ (where the composition $*_0$ for these $(n-1)$-categories is actually the composition $*_1$ for $\mathcal{C}$ and we adopt the latter notation), we have

$$(a, b) *_1 (c, d) = (a *_1 c, b *_1 d).$$

This leads to the formula

$$(a *_0 b) *_1 (c *_0 d) = (a *_1 c) *_0 (b *_1 d).$$

This seemingly innocuous formula takes on a special meaning when we start inserting identity maps: the "Eckmann–Hilton argument" is perfectly analogous to the proof in topology that homotopy groups π_i are commutative for $i \geq 2$. Suppose $x = y = z$ and let 1_x be the identity of x, which may be thought of as an object of $\mathcal{C}(x, x)$. Let e denote the 2-morphism of $\mathcal{C}$, identity of 1_x; which may be thought of as a 1-morphism of $\mathcal{C}(x, x)$. It acts as the identity for both compositions $*_0$ and $*_1$ (the reader may check that this

[2] The adjective "reflexive" refers to the inclusion of these leftward "identity" maps; a *globular set* without reflexivity would have only the s and t.

follows from the part of the axioms for an n-category saying that the morphism $1_x : * \to \mathcal{C}(x,x)$ is an identity for the composition).

If a, b are also endomorphisms of 1_x, then the above rule specializes to

$$a *_1 b = (a *_0 e) *_1 (e *_0 b) = (a *_1 e) *_0 (e *_1 b) = a *_0 b.$$

Thus in this case the compositions $*_0$ and $*_1$ are the same. A different ordering gives the formula

$$a *_1 b = (e *_0 a) *_1 (b *_0 e) = (e *_1 b) *_0 (a *_1 e) = b *_0 a.$$

Therefore, we have

$$a *_1 b = b *_1 a = a *_0 b = b *_0 a.$$

This argument says, then, that $\mathrm{Ob}(\mathcal{C}(x,x)(1_x, 1_x))$ is a commutative monoid and the two natural multiplications are the same.

The same argument extends to the whole monoid structure on the $(n-2)$-category $\mathcal{C}(x,x)(1_x, 1_x)$:

Lemma 2.1.1 *The two composition laws on the strict $(n-2)$-category $\mathcal{C}(x,x)$ $(1_x, 1_x)$ are equal, and this law is commutative. In other words, $\mathcal{C}(x,x)(1_x, 1_x)$ is an abelian monoid-object in the category $(n-2)$*STRCAT. □

There is a partial converse to the above observation. If the only object is x and the only 1-morphism is 1_x then nothing else can happen and we get the following equivalence of categories:

Lemma 2.1.2 *Suppose G is an abelian monoid-object in the category $(n-2)$STRCAT. Then there is a unique strict n-category $\mathcal{C}$ such that*

$$\mathrm{Ob}(\mathcal{C}) = \{x\} \quad \textit{and} \quad Mor^1(\mathcal{C}) = \mathrm{Ob}(\mathcal{C}(x,x)) = \{1_x\},$$

and such that $\mathcal{C}(x,x)(1_x, 1_x) = G$ as an abelian monoid-object. This construction establishes an equivalence between the categories of abelian monoid-objects in $(n-2)$STRCAT, and the strict n-categories having only one object and one 1-morphism.

Proof Define the strict $(n-1)$-category $\mathcal{U}$ with $\mathrm{Ob}(\mathcal{U}) = \{u\}$ and $\mathcal{U}(u,u) = G$ with its monoid structure as composition law. The fact that the composition law is commutative allows it to be used to define an associative and commutative multiplication

$$\mathcal{U} \times \mathcal{U} \to \mathcal{U}.$$

Now let $\mathcal{C}$ be the strict n-category with $\mathrm{Ob}(\mathcal{C}) = \{x\}$ and $\mathcal{C}(x, x) = \mathcal{U}$ with the above multiplication. It is clear that this construction is inverse to the previous one. □

It is clear from the construction (the fact that the multiplication on $\mathcal{U}$ is again commutative) that the construction can be iterated any number of times. We obtain the following "delooping" corollary:

Corollary 2.1.3 *Suppose $\mathcal{C}$ is a strict n-category with only one object and only one 1-morphism. Then there exists a strict $(n+1)$-category $\mathcal{B}$ with only one object b and with $\mathcal{B}(b, b) \cong \mathcal{C}$.*

Proof By the previous lemmas, $\mathcal{C}$ corresponds to an abelian monoid-object G in $(n-2)$STRCAT. Construct $\mathcal{U}$ as in the proof of 2.1.2, and note that $\mathcal{U}$ is an abelian monoid-object in $(n-1)$STRCAT. Now apply the result of 2.1.2 directly to $\mathcal{U}$ to obtain $\mathcal{B} \in (n+1)$STRCAT, which will have the desired property. □

The "stabilization hypothesis" conjectured by Breen [59] and Baez and Dolan [8] for weak n-categories is a similar delooping property which in the weak case holds when $\mathcal{C}$ is sufficiently highly connected. It will be discussed in Chapter 23.

2.2 Strict n-groupoids

Recall that a *groupoid* is a category where all morphisms are invertible. This definition generalizes to strict n-categories in a somewhat weakened way, as was pointed out by Kapranov and Voevodsky [168]. One does not require strict invertibility of morphisms, rather just that they are invertible up to "equivalence," thus the notion of strict n-groupoid considered here is more general than the notion employed by Brown and Higgins [67].

The definition requires the notion of equivalence, which in turn refers to the notion of $(n-1)$-groupoid. This provides an instructive first example of the inductive structure which permeates the theory of n-categories. So, even though strict n-categories are not the main subject of the rest of the book, it is worthwhile to give a detailed treatment.

We give a couple of equivalent versions of the definition of strict n-groupoid, with a theorem stating that they are equivalent. The definitions of n-groupoids, equivalences and truncations given below are equivalent to those of Kapranov and Voevodsky [168]. They also stated the present definitions and left the comparison as an exercise; this will be discussed in Theorem 2.2.11 below.

Our discussion in many ways parallels Tamsamani's treatment [250] of the groupoid condition for weak n-categories, to be discussed in the next chapter, and our treatment in this section comes in large part from discussions with Tamsamani about this.

A strict 0-category is just a set, all 0-categories are 0-groupoids, and an equivalence of 0-groupoids just means an isomorphism of sets.

Definition 2.2.1 The following definitions are given by mutual induction on $n \geq 1$:

- a strict n-category $\mathcal{A}$ is a *strict n-groupoid* if, for all $x, y \in \mathcal{A}$, $\mathcal{A}(x, y)$ is a strict $(n-1)$-groupoid, and for any 1-morphism $u : x \to y$ in $\mathcal{A}$, the two morphisms of composition with u
$$\mathcal{A}(y, z) \to \mathcal{A}(x, z), \qquad \mathcal{A}(w, x) \to \mathcal{A}(w, y)$$
are equivalences of strict $(n-1)$-groupoids;
- a morphism $f : \mathcal{A} \to \mathcal{B}$ between n-groupoids is *essentially surjective* if, for every $x \in \mathrm{Ob}(\mathcal{B})$, there exists $z \in \mathrm{Ob}(\mathcal{A})$ and an arrow $u \in \mathrm{Ob}(\mathcal{B}(x, f(z)))$ is nonempty;
- a morphism $f : \mathcal{A} \to \mathcal{B}$ between n-groupoids is *fully faithful* if, for all $x, y \in \mathrm{Ob}(\mathcal{A})$, the map $\mathcal{A}(x, y) \to \mathcal{B}(f(x), f(y))$ is an equivalence of $(n-1)$-groupoids; and
- a morphism $f : \mathcal{A} \to \mathcal{B}$ between n-groupoids is an *equivalence* if it is fully faithful and essentially surjective.

If $\mathcal{A}$ is an n-groupoid, say that two objects $x, y \in \mathrm{Ob}(\mathcal{A})$ are *inner equivalent* if there exists an arrow between them, that is to say if $\mathrm{Ob}(\mathcal{A}(x, y)) \neq \emptyset$. If specified, an arrow may be called an *inner equivalence* between the two objects.

Let $n\mathrm{S}\textsc{tr}\mathrm{G}\textsc{pd} \subset n\mathrm{S}\textsc{tr}\mathrm{C}\textsc{at}$ denote the full subcategory of strict n-groupoids.

Lemma 2.2.2 *Suppose $\mathcal{A}$ is an n-groupoid, and $x, y \in \mathrm{Ob}(\mathcal{A})$. If $\mathrm{Ob}(\mathcal{A}(x, y))$ is nonempty, then $\mathrm{Ob}(\mathcal{A}(y, x))$ is nonempty. Therefore the relation of inner equivalence is an equivalence relation.*

Proof Suppose $u \in \mathrm{Ob}(\mathcal{A}(x, y))$. Right multiplication by u induces an equivalence of $(n-1)$-groupoids from $\mathcal{A}(y, x)$ to $\mathcal{A}(y, y)$. The latter is nonempty, containing 1_y, so $\mathrm{Ob}(\mathcal{A}(y, x))$ is nonempty. The relation of inner equivalence says that $x \sim y$ if $\mathrm{Ob}(\mathcal{A}(x, y)) \neq \emptyset$. This is transitive by composition, and reflexive using the identity morphisms; symmetry comes from the first statement of the lemma. □

If $\mathcal{A}$ is an n-groupoid, define $\pi_0(\mathcal{A})$ to be the quotient of $\mathrm{Ob}(\mathcal{A})$ by the relation of inner equivalence. It is compatible with products:

$$\pi_0(\mathcal{A} \times \mathcal{B}) \cong \pi_0(\mathcal{A}) \times \pi_0(\mathcal{B}).$$

Alternate notation for the set $\pi_0(\mathcal{A})$, thought of as a 0-groupoid, is $\tau_{\leq 0}(\mathcal{A})$.

Suppose $\mathcal{A}$ is a strict n-category such that if $x, y \in \mathrm{Ob}(\mathcal{A})$ then $\mathcal{A}(x, y)$ is an $(n-1)$-groupoid. We can then construct a new 1-category denoted $\tau_{\leq 1}(\mathcal{A})$ with the same set of objects

$$\mathrm{Ob}(\tau_{\leq 1}(\mathcal{A})) := \mathrm{Ob}(\mathcal{A}),$$

whose morphisms sets are defined by

$$(\tau_{\leq 1}(\mathcal{A}))(x, y) := \pi_0(\mathcal{A}(x, y)).$$

Compatibility with products allows us to define a composition operation

$$(\tau_{\leq 1}(\mathcal{A}))(y, z) \times (\tau_{\leq 1}(\mathcal{A}))(x, y) \to (\tau_{\leq 1}(\mathcal{A}))(x, z),$$

as being the result of π_0 applied to the composition of $\mathcal{A}$,

$$\pi_0(\mathcal{A}(y, z) \times \mathcal{A}(x, y)) \to \pi_0(\mathcal{A}(x, z)).$$

This is easily verified to satisfy the axioms needed to define a category.

The definition of inner equivalence extends to i-morphisms for $1 \leq i \leq (n-1)$: $u, v \in \mathrm{Mor}^i(\mathcal{A})$ are said to be *inner equivalent* if $s(u) = s(v)$, $t(u) = t(v)$, and if there exists $g \in \mathrm{Mor}^{i+1}(\mathcal{A})$ with $s(g) = u$ and $t(g) = v$. For n-morphisms, equivalence is defined to be the same thing as equality.

The notion of inner equivalence will often be abbreviated to just "equivalence" if no confusion arises. It appears implicitly in the definition of essential surjectivity in Definition 2.2.1, which really says that any $x \in \mathcal{B}$ is inner equivalent to some $f(z)$.

Corollary 2.2.3 *The relation of inner equivalence between i-morphisms in an n-groupoid is an equivalence relation.*

Proof It suffices to consider $i = 0$ and apply Lemma 2.2.2. □

The above discussion of inner equivalence, and the definition of essential surjectivity, have been simplified because we are talking about n-groupoids in which all arrows are viewed as inner equivalences. In the context of a more general n-category, an inner equivalence between objects x, y is a morphism $u \in \mathrm{Ob}(\mathcal{A}(x, y))$ such that u is invertible up to equivalence. This requires additional inductive treatment, based on a refined inductive definition of truncation, to be discussed in Section 3.3. In the present section below, a simplified definition of truncation adapted to the case of strict n-groupoids will be used.

One can amplify on Lemma 2.2.2:

Lemma 2.2.4 *Suppose $\mathcal{A}$ is an n-groupoid, and suppose $u \in \mathrm{Mor}^1(\mathcal{A})$ with $x = s(u)$, $y = t(u) \in \mathrm{Ob}(\mathcal{A})$. Then u is both left and right invertible up to equivalence: there exist $v, w \in \mathrm{Ob}(\mathcal{A}(y, x))$ such that vu is inner equivalent to 1_x and uw is inner equivalent to 1_y.*

Proof By hypothesis, right multiplication by u induces an equivalence of $(n-1)$-groupoids

$$\mathcal{A}(y, x) \to \mathcal{A}(y, y).$$

Essential surjectivity of this functor, applied to $1_y \in \mathrm{Ob}(\mathcal{A}(y, y))$, says that there exists $v \in \mathrm{Ob}(\mathcal{A}(y, x))$ such that vu is inner equivalent to 1_y. Similarly, left multiplication by u is an equivalence, leading to w as required. □

Corollary 2.2.5 *Suppose $\mathcal{A}$ is a strict n-groupoid. Then the 1-category $\tau_{\leq 1}(\mathcal{A})$ is a 1-groupoid, and $\pi_0(\mathcal{A})$ is the set of isomorphism classes of $\tau_{\leq 1}(\mathcal{A})$. The left and right inverses of Lemma 2.2.4 may be chosen to be the same morphism.*

Proof The morphisms in $\tau_{\leq 1}(\mathcal{A})$ between x and y, are the inner-equivalence classes of morphisms of $\mathcal{A}$. Lemma 2.2.4 therefore says that any such morphism u has a left and right inverse. A standard argument in the 1-category $\tau_{\leq 1}(\mathcal{A})$ says that u has a two-sided inverse, and if $v \in \mathrm{Ob}(\mathcal{A}(x, y))$ is any representative for the two-sided inverse then uv and vu are both inner-equivalent to the appropriate identities. For $\pi_0(\mathcal{A})$ notice that two objects $x, y \in \mathrm{Ob}(\mathcal{A})$ are inner-equivalent if and only if $\mathcal{A}(x, y)$ is nonempty, which is the same as saying that $\tau_{\leq 1}(\mathcal{A})(x, y)$ is nonempty. Thus $\pi_0(\mathcal{A})$ is the set of isomorphism classes of $\tau_{\leq 1}(\mathcal{A})$. □

The truncation operation extends to intermediate truncations in all dimensions:

Theorem 2.2.6 *If $\mathcal{A}$ is a strict n-groupoid, for any $0 \leq k \leq n$ there exists a strict k-groupoid $\tau_{\leq k}\mathcal{A}$ whose i-morphisms are those of $\mathcal{A}$ for $i < k$ and whose k-morphisms are the equivalence classes of k-morphisms of $\mathcal{A}$ under the equivalence relation of inner equivalence. Thinking of this k-category as an n-category, the natural projection $\mathcal{A} \to \tau_{\leq k}\mathcal{A}$ is a morphism of n-categories. Truncation is compatible with the previous notations for $k = 0, 1$, and is compatible with the cartesian product:*

$$\tau_{\leq k}(\mathcal{A} \times \mathcal{B}) = \tau_{\leq k}(\mathcal{A}) \times \tau_{\leq k}(\mathcal{B}).$$

If $k \geq 1$ then $\mathrm{Ob}(\tau_{\leq k}(\mathcal{A})) = \mathrm{Ob}(\mathcal{A})$ and for two objects x, y we have

$$(\tau_{\leq k}(\mathcal{A}))(x, y) = \tau_{\leq (k-1)}(\mathcal{A}(x, y)). \tag{2.2.1}$$

For $0 \leq k \leq k' \leq n$ we have

$$\tau_{\leq k}(\tau_{\leq k'}(\mathcal{A})) = \tau_{\leq k}(\mathcal{A}). \tag{2.2.2}$$

Truncation takes equivalences to equivalences: if $\mathcal{A} \xrightarrow{f} \mathcal{B}$ is an equivalence between n-groupoids, then it induces an equivalence

$$\tau_{\leq k}(\mathcal{A}) \xrightarrow{\tau_{\leq k}(f)} \tau_{\leq k}(\mathcal{B})$$

between k-groupoids.

Proof The truncation has already been defined for $k = 0, 1$ above, and in both cases it is easy to see that it is compatible with cartesian product. Extend this to higher values of k by induction: for any $k \geq 2$, define $\tau_{\leq k}(\mathcal{A})$ to be the k-category with set of objects

$$\mathrm{Ob}(\tau_{\leq k}(\mathcal{A})) := \mathrm{Ob}(\mathcal{A}),$$

and with morphism $(k-1)$-categories defined by the formula (2.2.1) for any $x, y \in \mathrm{Ob}(\mathcal{A})$. Assuming compatibility with products for $(k-1)$-truncation of $(n-1)$-groupoids, we can define the composition operator by taking the $(k-1)$-truncation of the composition operator for $\mathcal{A}$. Compatibility with cartesian products for the k-truncation easily follows. This defines $\tau_{\leq k}(\mathcal{A})$ as a strict k-category.

To see that it is a k-groupoid, by induction on k and the definition using formula (2.2.1) it follows that the $\tau_{\leq k}(\mathcal{A})(x, y)$ are $(k-1)$-groupoids. The left and right multiplication functors of the truncation are the $(k-1)$-truncations of the corresponding functors for $\mathcal{A}$, so the fact that truncation preserves equivalences–the last statement of the theorem known inductively for $(k-1)$–implies that these are equivalences for the k-truncation. This shows that $\tau_{\leq k}(\mathcal{A})$ is a k-groupoid.

The projection morphism $\mathcal{A} \to \tau_{\leq k}(\mathcal{A})$ is also obtained from the inductive construction, and this identifies the i-morphisms of $\tau_{\leq k}(\mathcal{A})$ with those of $\mathcal{A}$, for $i < k$, and with the inner equivalence classes of those of $\mathcal{A}$ for $i = k$. The formula of composing two truncations (2.2.2) follows from this description of the i-morphisms of the truncation.

Suppose $f : \mathcal{A} \to \mathcal{B}$ is an equivalence between n-groupoids. The construction of truncation is easily seen to be functorial so f induces a functor $\tau_{\leq k}(f)$ between truncations. The fully faithful condition for $\tau_{\leq k}(f)$ follows from the present statement for $(k-1)$-truncation, and the fully faithful condition for f, in view of the formula (2.2.1). The essential surjectivity of $\tau_{\leq k}(f)$ follows from that of f, because the nonemptiness of the set of morphisms between two objects is the same in $\mathcal{B}$ or $\tau_{\leq k}(\mathcal{B})$. So truncation takes equivalences to equivalences. □

The n-groupoid condition generalizes the notion of 1-groupoid, which arises from the Poincaré 1-groupoid of a space. As discussed in the first chapter, the comparison with spaces via a higher Poincaré groupoid construction is one of the main motivations for going to n-categories. So, although it turns out that strict n-categories are not quite the right setting, nonetheless this motivates the definition of *homotopy groups* of an n-groupoid.

Suppose $\mathcal{A}$ is a strict n-groupoid. We have already defined $\pi_0(\mathcal{A}) := \tau_{\leq 0}(\mathcal{A})$ as the set of equivalence classes of objects under the relation of inner equivalence. For any "basepoint" $x \in \mathrm{Ob}(\mathcal{A})$, define $\pi_1(\mathcal{A}, x) := (\tau_{\leq 1}(\mathcal{A}))(x, x)$, which is a group since $\tau_{\leq 1}(\mathcal{A})$ is a groupoid by Corollary 2.2.5. For $2 \leq i \leq n$ define by induction $\pi_i(\mathcal{A}, x) := \pi_{i-1}(\mathcal{A}(x, x), 1_x)$. The interchange property allows us to show that this is an abelian group. These classical definitions were recalled by Kapranov and Voevodsky [168].

Unwinding the induction here, the homotopy group $\pi_i(\mathcal{A}, x)$ is more explicitly seen as the group of i-arrows from 1_x^{i-1} to itself, modulo inner equivalence, where 1_x^{i-1} denotes the $(i-1)$-arrow which is the identity of the identity of ... of the identity of x. By convention this starts with $1_x^0 := x$. Thus an element of $\pi_i(\mathcal{A}, x)$ is an equivalence class of $u \in \mathrm{Mor}^i(\mathcal{A})$ such that $s(u) = t(u) = 1_x^{i-1}$. The multiplication law is given by the composition $*_{i-1}$ with respect to the $(i-1)$-arrows. It is commutative, as was seen in the previous section. Furthermore, it is a group, as may be seen by noting that it is the same as the group defined inductively in the previous paragraph.

Lemma 2.2.7 *The homotopy groups are functorial for morphisms of strict n-groupoids. The natural projection to the truncation $\mathcal{A} \to \tau_{\leq k}(\mathcal{A})$ induces isomorphisms*

$$\pi_i(\mathcal{A}, x) \overset{\cong}{\to} \pi_i(\tau_{\leq k}(\mathcal{A}), x)$$

whenever $1 \leq i \leq k \leq n$, and similarly without basepoint for $i = 0$. For $i = k$ this identifies $\pi_i(\mathcal{A}, x)$ with the group of automorphisms of 1_x^{i-1} in $\tau_{\leq i}(\mathcal{A})$. For any $x \in \mathrm{Ob}(\mathcal{A})$ and any $i \geq 2$ we have

$$\pi i(\mathcal{A}, x) = \pi_{i-1}(\mathcal{A}(x, x), 1_x),$$

and for $i = 1$ the same statement without basepoint on the right.

Proof If $\mathcal{A} \overset{f}{\to} \mathcal{B}$ is a morphism of n-groupoids and $x \in \mathrm{Ob}(\mathcal{A})$ then $f\left(1_x^{i-1}\right) = 1_{f(x)}^{i-1}$, so for u an i-arrow of $\mathcal{A}$ with source and target 1_x^{i-1}, $f(u)$ is an i-arrow of $\mathcal{B}$ with source and target $1_{f(x)}^{i-1}$. This respects inner equivalences so it gives a map $\pi_i(\mathcal{A}, x) \to \pi_i(\mathcal{B}, f(x))$. By the characterization of truncation given at the start of Theorem 2.2.6, the projection $\mathcal{A} \to \tau_{\leq k}(\mathcal{A})$ induces an isomorphism from the set of inner equivalence classes of i-arrows in $\mathcal{A}$, to that of $\tau_{\leq k}(\mathcal{A})$ for any $i \leq k$, and furthermore for $i = k$ it induces an isomorphism to the set of k-arrows itself in the truncation. Thus it induces an isomorphism on homotopy groups. The i-arrows of $\mathcal{A}$ whose fully iterated sources and targets are x, are the same as the $(i-1)$-arrows of $\mathcal{A}(x, x)$, which gives the last statement. □

We can now see that several different possible definitions of the notion of equivalence are the same:

Proposition 2.2.8 *A morphism $f : \mathcal{A} \to \mathcal{B}$ of strict n-groupoids is an equivalence if and only if the following equivalent conditions are satisfied:*

(a) f induces a surjection $\pi_0(\mathcal{A}) \to \pi_0(\mathcal{B})$ and, for every pair of objects $x, y \in \mathrm{Ob}(\mathcal{A})$, an equivalence of $(n-1)$-groupoids $\mathcal{A}(x, y) \to \mathcal{B}(f(x), f(y))$;

(b) f induces an isomorphism $\pi_0\mathcal{A} \to \pi_0\mathcal{B}$, and for every object $a \in \mathrm{Ob}(\mathcal{A})$, the map induced by f is an isomorphism $\pi_i(\mathcal{A}, a) \overset{\cong}{\to} \pi_i(\mathcal{B}, f(a))$;

(c) if u, v are i-morphisms in $\mathcal{A}$ sharing the same source and target, and if r is an $(i+1)$-morphism in $\mathcal{B}$ going from $f(u)$ to $f(v)$, then there exists an $(i+1)$-morphism t in $\mathcal{A}$ going from u to v and an $i+2$-morphism in $\mathcal{B}$ going from $f(t)$ to r (this includes the limiting cases $i = -1$, where u and v are not specified, and $i = n-1, n$, where "$(n+1)$-morphisms" means equalities between n-morphisms and "$(n+2)$-morphisms" are not specified).

Proof Condition (a) is the same as in Definition 2.2.1, noting that $\pi_0(\mathcal{B})$ is the set of inner equivalence classes of objects, so the surjectivity on π_0 is the same thing as essential surjectivity in Definition 2.2.1.

Prove now the equivalence between (a) and (b). Inductively, the proposition will be assumed known for $(n-1)$-groupoids. Suppose $f : \mathcal{A} \to \mathcal{B}$ is a morphism of strict n-groupoids satisfying condition (a). From the formula of Lemma 2.2.7 and the inductive statement for $(n-1)$-groupoids, we get that f induces isomorphisms on the π_i for $i \geq 1$. On the other hand, the truncation $\tau_{\leq 1}(f)$ satisfies condition (a) for a morphism of 1-groupoids, and this is readily seen to imply that $\pi_0(f)$ is an isomorphism. Thus f satisfies condition (b).

Suppose, on the other hand, that $f : \mathcal{A} \to \mathcal{B}$ is a morphism of strict n-groupoids satisfying condition (b). Then of course $\pi_0(f)$ is surjective. Consider two objects $x, y \in \mathcal{A}$ and look at the induced morphism

$$f^{x,y} : \mathcal{A}(x, y) \to \mathcal{B}(f(x), f(y)).$$

We claim that $f^{x,y}$ satisfies condition (b) for a morphism of $(n-1)$-groupoids. For this, consider a 1-morphism from x to y, i.e. an object $r \in \mathcal{A}(x, y)$. By the n-groupoid condition for $\mathcal{A}$, multiplication by r induces an equivalence of $(n-1)$-groupoids

$$m(r) : \mathcal{A}(x, x) \to \mathcal{A}(x, y),$$

and furthermore $m(r)(1_x) = r$. The same is true in $\mathcal{B}$: multiplication by $f(r)$ induces an equivalence

$$m(f(r)) : \mathcal{B}(f(x), f(x)) \to \mathcal{B}(f(x), f(y)).$$

The fact that f is a morphism implies that these fit into a commutative square

$$\begin{array}{ccc} \mathcal{A}(x, x) & \longrightarrow & \mathcal{A}(x, y) \\ \downarrow & & \downarrow \\ \mathcal{B}(f(x), f(x)) & \to & \mathcal{B}(f(x), f(y)). \end{array}$$

Condition (b) for f implies that the left vertical morphism induces isomorphisms

$$\pi_i(\mathcal{A}(x, x), 1_x) \overset{\cong}{\to} \pi_i(\mathcal{B}(f(x), f(x)), 1_{f(x)}).$$

Therefore the right vertical morphism (i.e. $f_{x,y}$) induces isomorphisms

$$\pi_i(\mathcal{A}(x, y), r) \overset{\cong}{\to} \pi_i(\mathcal{B}(f(x), f(y)), f(r)),$$

and this for all $i \geq 1$. We have now verified these isomorphisms for any base-object r. A similar argument implies that $f^{x,y}$ induces an injection on π_0. On the other hand, the fact that f induces an isomorphism on π_0 implies that $f^{x,y}$ induces a surjection on π_0 (note that these last two statements are reduced to statements about 1-groupoids by applying $\tau_{\leq 1}$ so we don't give further details). All of these statements taken together imply that $f^{x,y}$ satisfies condition (b), and by the inductive statement of the theorem for $(n-1)$-groupoids this implies that $f^{x,y}$ is an equivalence. Thus f satisfies condition (a).

We now remark that condition (a) is equivalent to condition (c) for a morphism $f : \mathcal{A} \to \mathcal{B}$. Indeed, the part of condition (c) for $i = -1$ is, by the definition of π_0, identical to the condition that f induces a surjection $\pi_0(\mathcal{A}) \to \pi_0(\mathcal{B})$. And the remaining conditions for $i = 0, \ldots, n+1$ are identical to the conditions of (c) corresponding to $j = i - 1 = -1, \ldots, (n-1)+1$ for all the morphisms of $(n-1)$-groupoids $\mathcal{A}(x, y) \to \mathcal{B}(f(x), f(y))$. (In terms of u and v appearing in the condition in question, take x to be the source of the source of the source ..., and take y to be the target of the target of the target ....) Thus by induction on n (i.e. by the equivalence $(a) \Leftrightarrow (c)$ for $(n-1)$-groupoids), the conditions (c) for f for $i = 0, \ldots, n+1$ are equivalent to the conditions that $\mathcal{A}(x, y) \to \mathcal{B}(f(x), f(y))$ be equivalences of $(n-1)$-groupoids. Thus condition (c) for f is equivalent to condition (a) for f, which completes the proof of the proposition. □

Remark 2.2.9 The reader may verify that the operation of truncation, the definitions of homotopy groups and the notion of equivalence are the same as those of Kapranov–Voevodsky [168].

The notion of equivalence satisfies a key property known as "3 for 2": given a composable pair of morphisms $\mathcal{A} \xrightarrow{f} \mathcal{B} \xrightarrow{g} \mathcal{C}$, if any two of f, g and gf are equivalences, then so is the third one. This is a crucial property of the notion of weak equivalence in a Quillen model category, and will play an important role in all of the arguments in this book. It is therefore interesting to meet it in this first case. The following statement also includes another property, analogous to the statement that existence of separate left and right inverses implies invertibility.

Lemma 2.2.10 *If $f : \mathcal{A} \to \mathcal{B}$ and $g : \mathcal{B} \to \mathcal{C}$ are morphisms of strict n-groupoids and if any two of f, g and gf are equivalences, then so is the third. If*

$$\mathcal{A} \xrightarrow{f} \mathcal{B} \xrightarrow{g} \mathcal{C} \xrightarrow{h} \mathcal{D}$$

are morphisms of strict n-groupoids and if hg and gf are equivalences, then g is an equivalence.

Proof For the first part, using the fact that isomorphisms of sets satisfy the same 3 for 2 property, and using the characterization of equivalences in terms of homotopy groups (condition (b) of Proposition 2.2.8) we immediately get two of the three statements: that if f and g are equivalences then gf is an equivalence; and that if gf and g are equivalences then f is an equivalence. Suppose now that gf and f are equivalences; we would like to show that g is an equivalence. First of all it is clear that if $x \in \mathrm{Ob}(\mathcal{A})$ then g induces an isomorphism $\pi_i(\mathcal{B}, f(x)) \cong \pi_0(\mathcal{C}, gf(x))$ (resp. $\pi_0(\mathcal{B}) \cong \pi_0(\mathcal{C})$). Suppose now that $y \in \mathrm{Ob}(\mathcal{B})$, and choose a 1-morphism u going from y to $f(x)$ for some $x \in \mathrm{Ob}(\mathcal{A})$ (this is possible because f is surjective on π_0). By the definition of n-groupoid, composition with u induces equivalences in both places on the top row of the diagram

$$\begin{array}{ccccc}
\mathcal{B}(y,y) & \longrightarrow & \mathcal{B}(y,f(x)) & \longleftarrow & \mathcal{B}(f(x),f(x)) \\
\downarrow & & \downarrow & & \downarrow \\
\mathcal{C}(g(y),g(y)) & \rightarrow & \mathcal{C}(g(y),gf(x)) & \leftarrow & \mathcal{C}(gf(x),gf(x)).
\end{array}$$

Similarly composition with $g(u)$ induces equivalences in both places on the bottom row. The 3 for 2 property for $(n-1)$-groupoids, which can be assumed by induction, applied to the sequence

$$\mathcal{A}(x,x) \to \mathcal{B}(f(x),f(x)) \to \mathcal{C}(gf(x),gf(x)),$$

as well as the hypothesis that f is an equivalence, imply that the rightmost vertical arrow in the above diagram is an equivalence. Again applying the 3 for 2 to these $(n-1)$-groupoids yields first that the middle vertical arrow, and then the leftmost vertical arrow, are equivalences. In particular g induces isomorphisms

$$\pi_i(\mathcal{B},y) = \pi_{i-1}(\mathcal{B}(y,y),1_y) \xrightarrow{\cong} \pi_{i-1}(\mathcal{C}(g(y),g(y)),1_{g(y)}) = \pi_i(\mathcal{C},g(y)).$$

This completes the verification of condition (a) for the morphism g, completing the proof of the 3 for 2 property.

We now prove the second part of the lemma, starting with the notations $\mathcal{A},\mathcal{B},\mathcal{C},\mathcal{D},f,g,h$. First note that applying π_0 gives the same situation for maps of sets, so $\pi_0(g)$ is an isomorphism. Next, suppose $x \in \mathrm{Ob}(\mathcal{A})$. Then we obtain a sequence

$$\pi_i(\mathcal{A},x) \to \pi_i(\mathcal{B},f(x)) \to \pi_i(\mathcal{C},gf(x)) \to \pi_i(\mathcal{D},hgf(x)),$$

such that the composition of the first pair and also of the last pair are isomorphisms; thus g induces an isomorphism $\pi_i(\mathcal{B},f(x)) \cong \pi_i(\mathcal{C},gf(x))$. Now, by the same argument as for the 3 for 2 property above, using the hypothesis that f induces a surjection $\pi_0(\mathcal{A}) \to \pi_0(\mathcal{B})$, we get that for any object $y \in \mathrm{Ob}(\mathcal{B})$, g induces an isomorphism $\pi_i(\mathcal{B},y) \cong \pi_i(\mathcal{C},g(y))$. By condition (b) of Proposition 2.2.8 we have now shown that g is an equivalence. This completes the proof of the lemma. □

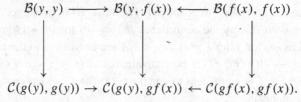

The main result of this section is an alternative characterization of the notion of n-groupoid. This characterization, involving truncation, is the one which will be generalized in the context of weak n-categories. It is conceptually simpler, but to get back to the property of Definition 2.2.1 we need to use the 3 for 2 property of Lemma 2.2.10. This is only needed for $(n-1)$-groupoids, so it would be possible to start with the condition of the following theorem as a definition and proceed in a big induction [238]. The sequencing chosen here (for which I would like to thank the Reader) renders the argument a little less complicated to follow.

Theorem 2.2.11 *A strict n-category $\mathcal{A}$ is an n-groupoid if and only if, for $x, y \in \mathrm{Ob}(\mathcal{A})$, the $(n-1)$-category of morphisms $\mathcal{A}(x, y)$ is an $(n-1)$-groupoid, and the 1-category $\tau_{\leq 1}(\mathcal{A})$, which may therefore be defined as above, is a 1-groupoid. Furthermore, the notion of n-groupoid in this definition or that of Definition 2.2.1, is equivalent to that of Kapranov–Voevodsky [168].*

Proof One direction of the equivalence is given by Corollary 2.2.5. For the other direction, suppose $\mathcal{A}$ is an n-category such that $\mathcal{A}(x, y)$ is an $(n-1)$-groupoid for any $x, y \in \mathrm{Ob}(\mathcal{A})$, and such that $\tau_{\leq 1}(\mathcal{A})$ is a 1-groupoid.

By definition $\mathcal{A}$ already satisfies the first part of Definition 2.2.1. Thus we have to show the second part, for example that for $f : x \to y$ in $\mathrm{Ob}(\mathcal{A}(x, y))$, composition with f on the right induces an equivalence

$$\mathcal{A}(y, z) \to \mathcal{A}(x, z)$$

(the other part is dual and has the same proof which we won't repeat here).

In order to prove this, we need to make a digression about the effect of composition with 2-morphisms. Suppose $f, g \in \mathrm{Ob}(\mathcal{A}(x, y))$ and suppose that u is a 2-morphism from f to g – this last supposition may be rewritten

$$u \in \mathrm{Ob}(\mathcal{A}(x, y)(f, g)).$$

Claim: Suppose z is another object; we claim that if composition with f induces an equivalence $\mathcal{A}(y, z) \to \mathcal{A}(x, z)$, then composition with g also induces an equivalence $\mathcal{A}(y, z) \to \mathcal{A}(x, z)$.

To prove the claim, suppose that h, k are two 1-morphisms from y to z. We now obtain a diagram

$$\begin{array}{ccc} \mathrm{HOM}_{\mathcal{A}(y,z)}(h, k) & \longrightarrow & \mathrm{HOM}_{\mathcal{A}(x,z)}(hf, kf) \\ \big\downarrow & & \big\downarrow \\ \mathrm{HOM}_{\mathcal{A}(x,z)}(hg, kg) & \to & \mathrm{HOM}_{\mathcal{A}(x,z)}(hf, kg), \end{array}$$

where the top arrow is given by composition $*_0$ with 1_f; the left arrow by composition $*_0$ with 1_g; the bottom arrow by composition $*_1$ with the 2-morphism $h *_0 u$; and the right morphism is given by composition with $k *_0 u$. This diagram commutes (that is the "Godement relation" or "interchange rule," cf. Section 2.1). By the $(n-1)$-groupoid property of $\mathcal{A}(x, z)$, the morphisms on the bottom and on the right in the above diagram are equivalences. The hypothesis in the claim that f is an equivalence means that the morphism along the top of the diagram is an equivalence; thus by the first part of Lemma 2.2.10 applied to the $(n-2)$-groupoids in the diagram, we get that the morphism on the left of the diagram is an equivalence. This shows that the morphism of composition with g, $\mathcal{A}(y, z) \to \mathcal{A}(x, z)$, is a fully faithful map of $(n-1)$-groupoids.

To finish the proof of the claim, we now verify the morphism of composition with g is essentially surjective, using the hypothesis that $\tau_{\leq 1}(\mathcal{A})$ is a 1-groupoid. By definition,

$$\pi_0(\mathcal{A}(y, z)) = \tau_{\leq 1}\mathcal{A}(y, z) \quad \text{and} \quad \pi_0(\mathcal{A}(x, z)) = \tau_{\leq 1}\mathcal{A}(x, z),$$

and the morphism in question here is just the morphism of composition by the image of g in $\tau_{\leq 1}(\mathcal{A})$. Invertibility of this morphism in $\tau_{\leq 1}(\mathcal{A})$ implies that the composition morphism

$$(\tau_{\leq 1}\mathcal{A})(y, z) \to (\tau_{\leq 1}\mathcal{A})(x, z)$$

is an isomorphism. The morphism

$$\pi_0(\mathcal{A}(y, z)) \to \pi_0(\mathcal{A}(x, z))$$

is the same one, so it is surjective, proving essential surjectivity of the composition functor. This completes the proof of the claim.

Return now to the proof of the condition that right composition with f induces an equivalence. The fact that $\tau_{\leq 1}(\mathcal{A})$ is a 1-groupoid implies that given f there is another morphism h from y to x such that the class of fh is equal to the class of 1_y in $\pi_0\mathcal{A}(y, y)$, and the class of hf is equal to the class of 1_x in $\pi_0\mathcal{A}(x, x)$. This means that there exist 2-morphisms u from 1_y to fh, and v from 1_x to hf. By the above claim (and the fact that the compositions with 1_x and 1_y act as the identity and in particular are equivalences), we get that composition with fh is an equivalence

$$\{fh\} \times \mathcal{A}(y, z) \to \mathcal{A}(y, z),$$

and that composition with hf is an equivalence

$$\{hf\} \times \mathcal{A}(x, z) \to \mathcal{A}(x, z).$$

Let

$$\psi_f : \mathcal{A}(y, z) \to \mathcal{A}(x, z)$$

be the morphism of composition with f, and let

$$\psi_h : \mathcal{A}(x, z) \to \mathcal{A}(y, z)$$

be the morphism of composition with h. We have seen that $\psi_h\psi_f$ and $\psi_f\psi_h$ are equivalences. By the second statement of Lemma 2.2.10 applied to $(n-1)$-groupoids, these imply that ψ_f is an equivalence.

The proof for left composition with f is the same; this completes the proof that $\mathcal{A}$ is an n-groupoid.

We point out briefly how to compare these two versions of the n-groupoid condition with Kapranov–Voevodsky's definition [168]; however, the reader is asked to refer there for their definition. The comparison between these definitions is the content of Kapranov–Voevodsky's proposition 1.6 [168], whose proof was "left to the reader" there.

Suppose $\mathcal{A}$ is an n-groupoid according to Definition 2.2.1, which we now know is the same as the condition stated in the theorem. To get to the Kapranov–Voevodsky definition, in question are the conditions $GR'_{i,k}$ and $GR''_{i,k}$ $(i < k \leq n)$ of Definition 1.1, p. 33 of *loc. cit.* [168]. By the inductive version of the present equivalence for $(n-1)$-groupoids and by the requirement that the $\mathcal{A}(x, y)$ are $(n-1)$-groupoids, we obtain the conditions $GR'_{i,k}$ and $GR''_{i,k}$ for $i \geq 1$. Thus we may now restrict our attention to the conditions $GR'_{0,k}$ and $GR''_{0,k}$. For a 1-morphism f from x to y, the conditions $GR'_{0,k}$ for all k with respect to f, are the same as the condition that for all w, the morphism of left multiplication by f

$$\mathcal{A}(w, x) \times \{f\} \to \mathcal{A}(w, y),$$

is an equivalence according to the version (c) of the notion of equivalence (Proposition 2.2.8). Thus, condition $GR'_{0,k}$ follows from the condition of Definition 2.2.1 that left multiplication is an equivalence. Similarly condition $GR''_{0,k}$ follows from the condition for right multiplication. This shows that Definition 2.2.1 implies the definition of Kapranov–Voevodsky.

On the other hand, suppose $\mathcal{A}$ is a strict n-category which is an n-groupoid according to *loc. cit.* [168]. Their condition is compatible with truncation, so $\tau_{\leq 1}(\mathcal{A})$ satisfies their condition for 1-categories, in turn is equivalent to the standard condition of being a 1-groupoid. Therefore $\tau_{\leq 1}(\mathcal{A})$ is a 1-groupoid. On the other hand, the Kapranov–Voevodsky conditions [168] for i-arrows, $1 \leq i \leq n$, include the same conditions for the $(i-1)$-arrows of $\mathcal{A}(x, y)$ for any $x, y \in \mathrm{Ob}(\mathcal{A})$. Thus by the inductive statement of the present theorem for strict $(n-1)$-categories, $\mathcal{A}(x, y)$ is a strict $(n-1)$-groupoid. Therefore, $\mathcal{A}$ satisfies the condition of the present theorem, which implies that it is an n-groupoid according to Definition 2.2.1 as we have shown above.

The n-groupoid condition, which has allowed us to consider a natural intrinsic notion of homotopy groups, is essential in the discussion that follows. Indeed, it is well known that any homotopy type can be realized by a strict

1-category but which is usually not a groupoid. McDuff [194] even shows that one can use a category with a single object, i.e. a monoid. □

2.3 The need for weak composition

In the rest of the chapter, we turn towards the problem of realization of homotopy 3-types. It is here that the phenomenon of "weak composition" shows up first, in that strict 3-groupoids are not sufficient to model all 3-truncated homotopy types.

The classical Eckmann–Hilton argument, originally used to show that the homotopy groups π_i are abelian for $i \geq 2$, applies in the context of strict n-categories to give a vanishing of certain homotopy operations. Indeed, not only the π_i but the i-th loop spaces are abelian objects, and this forces the Whitehead products to vanish. This observation, which I learned from G. Maltsiniotis and A. Bruguières, had been used by many people to argue that strict n-categories do not contain sufficient information to model homotopy n-types, as soon as $n \geq 3$. See, for example, Brown [65], with Gilbert [66] and with Higgins [67, 68]; Grothendieck's discussion of this in various places [132]; and the paper of Berger [35]. Batanin [29] discusses this phenomenon for higher operads.

In R. Brown's terminology, strict n-groupoids correspond to *crossed complexes*. While a nontrivial action of π_1 on the π_i can occur in a crossed complex, the higher Whitehead operations such as $\pi_2 \otimes \pi_2 \rightarrow \pi_3$ must vanish. The Eckmann–Hilton argument applies to strict n-categories because of the "interchange rule" or "Godement relation." This effect occurs when one takes two 2-morphisms a and b both with source and target a 1-identity 1_x. There are various ways of composing a and b in this situation, and comparison of these compositions leads to the conclusion that all of the compositions are commutative. In a weak n-category, this commutativity would only hold up to higher homotopy, which leads to the notion of "braiding"; and in fact it is exactly the braiding which leads to the Whitehead operation. However, in a strict n-groupoid, the commutativity is strict and applies to all higher arrows, so the Whitehead operation is trivial.

The same may be said in the setting of 3-categories not necessarily groupoids: there are some examples (which G. Maltsiniotis pointed out to me) in Gordon *et al.* [123] of weak 3-categories not equivalent to strict ones. This in turn is related to the difference between braided monoidal categories and symmetric monoidal categories, see for example the nice discussion in Baez and Dolan [8].

This chapter gives a variant on these observations; it is a modified version of the discussion in my preprint [238]. We will show that one cannot obtain all homotopy 3-types by any reasonable realization functor from strict 3-groupoids (i.e. groupoids in the sense of Kapranov–Voevodsky [168]) to spaces. More precisely we show that one does not obtain the 3-type of S^2. This constitutes a minor generalization of Berger's theorem [35], which concerned the standard realization functor. We define the notion of possible "reasonable realization functor" in Definition 2.4.1 to be any functor $\mathfrak{R}$ from the category of strict n-groupoids to TOP, provided with a natural transformation r from the set of objects of $\mathcal{G}$ to the points of $\mathfrak{R}(\mathcal{G})$, and natural isomorphisms $\pi_0(\mathcal{G}) \cong \pi_0(\mathfrak{R}(\mathcal{G}))$ and $\pi_i(\mathcal{G}, x) \cong \pi_i(\mathfrak{R}(\mathcal{G}), r(x))$. This axiom is fundamental to the question of whether one can realize homotopy types by strict n-groupoids, because one wants to read off the homotopy groups of the space from the strict n-groupoid. The standard realization functors satisfy other properties beyond this minimal one.

In order to apply Definition 2.4.1, the interchange argument is written in a particular way. We get a picture of strict 3-groupoids having only one object and one 1-morphism, as being equivalent to abelian monoidal objects $(\mathcal{G}, +)$ in the category of groupoids, such that $(\pi_0(\mathcal{G}), +)$ is a group. In the case in question, this group will be $\pi_2(S^2) = \mathbb{Z}$. Then comes the main part of the argument. We show that, up to inverting a few equivalences, such an object has a morphism giving a splitting of the Postnikov tower (Proposition 2.7.1). It follows that for any realization functor respecting homotopy groups, the Postnikov tower of the realization (which has two stages corresponding to π_2 and π_3) splits. This implies that the 3-type of S^2 cannot occur as a realization, Theorem 2.7.2.

2.4 Realization functors

Recall that nSTRGPD is the category of strict n-groupoids as defined in Section 2.2. Let TOP be the category of topological spaces. The following definition encodes the minimum of what one would expect for a reasonable realization functor from strict n-groupoids to spaces:

Definition 2.4.1 A *realization functor for strict n-groupoids* is a functor

$$\mathfrak{R} : n\text{STRGPD} \to \text{TOP},$$

together with the following natural transformations:

$$r : \text{Ob}(\mathcal{A}) \to \mathfrak{R}(\mathcal{A});$$

$$\zeta_i(\mathcal{A}, x) : \pi_i(\mathcal{A}, x) \to \pi_i(\mathfrak{R}(\mathcal{A}), r(x)),$$

the latter including $\zeta_0(\mathcal{A}) : \pi_0(\mathcal{A}) \to \pi_0(\Re(\mathcal{A}))$; such that the $\zeta_i(\mathcal{A}, x)$ and $\zeta_0(\mathcal{A})$ are isomorphisms for $0 \leq i \leq n$, such that ζ_0 takes the isomorphism class of x to the connected component of $r(x)$, and such that the $\pi_i(\Re(\mathcal{A}), y)$ vanish for $i > n$.

Theorem 2.4.2 *There exists a realization functor $\Re$ for strict n-groupoids.*

Kapranov and Voevodsky [168] construct such a functor. Their construction proceeds by first defining a notion of "diagrammatic set"; they define a realization functor from n-groupoids to diagrammatic sets (denoted $Nerv$), and then define the topological realization of a diagrammatic set (denoted $|\cdot|$). The composition of these two constructions gives a realization functor

$$\mathcal{G} \mapsto \Re_{KV}(\mathcal{G}) := |Nerv(\mathcal{G})|$$

from strict n-groupoids to spaces. Note that this functor $\Re_{KV}$ satisfies the axioms of 2.4.1 as a consequence of Propositions 2.7 and 3.5 of *loc. cit.* [168].

One obtains a different construction by considering strict n-groupoids as weak n-groupoids in the sense of Tamsamani [250] (multisimplicial sets) and then taking the realization of *loc. cit.* [250] which we will discuss further in Chapters 15 and 20 below. This construction is actually due to the Australian school many years beforehand – see Berger [35] – and we call it the *standard realization* $\Re_{\mathrm{std}}$. The properties of 2.4.1 can be extracted from *loc. cit.* [250] (although again they are classical results).

We don't claim here that any two realization functors must be naturally equivalent (although one suspects so). This is why we shall work, in what follows, with an arbitrary realization functor satisfying the axioms of 2.4.1.

Proposition 2.4.3 *If $C \to C'$ is a morphism of strict n-groupoids inducing isomorphisms on the π_i then $\Re(C) \to \Re(C')$ is a weak homotopy equivalence. Conversely if $f : C \to C'$ is a morphism of strict n-groupoids which induces a weak equivalence of realizations then f was an equivalence.*

Proof Apply version (b) of the equivalent conditions in Proposition 2.2.8, together with the property of Definition 2.4.1. □

2.5 *n*-groupoids with one object

Let C be a strict n-category with only one object x. Then C is an n-groupoid if and only if $C(x, x)$ is an $(n-1)$-groupoid and $\pi_0 C(x, x)$ (which has a structure of monoid) is a group. This is the equivalent version of the definition of groupoid given in Theorem 2.2.11. Iterating this remark one more time we get the following statement:

Lemma 2.5.1 *The construction of 2.1.2 establishes an equivalence of categories between the strict n-groupoids having only one object and only one 1-morphism, and the abelian monoid-objects $\mathcal{G}$ in $(n-2)$*STRGPD *such that the monoid $\pi_0(\mathcal{G})$ is a group.*

Proof Lemma 2.1.2 gives an equivalence of categories between the category of abelian monoid-objects in $(n-2)$STRCAT, and the category of strict n-categories having only one object and one 1-morphism. The groupoid condition for the n-category is equivalent to saying that $\mathcal{G}$ is a groupoid, and that $\pi_0(\mathcal{G})$ is a group. □

Corollary 2.5.2 *Suppose $\mathcal{C}$ is a strict n-category having only one object and only one 1-morphism, and let $\mathcal{B}$ be the strict $(n+1)$-category of 2.1.3 with one object b and $\mathcal{B}(b, b) = \mathcal{C}$. Then $\mathcal{B}$ is a strict $(n+1)$-groupoid if and only if $\mathcal{C}$ is a strict n-groupoid.*

Proof Keep the notation $\mathcal{U}$ of the proof of 2.1.3. If $\mathcal{C}$ is a groupoid this means that $\mathcal{G}$ satisfies the condition that $\pi_0(\mathcal{G})$ be a group, which in turn implies that $\mathcal{U}$ is a groupoid. Note that $\pi_0(\mathcal{U}) = *$ is automatically a group; so applying the observation 2.5.1 once again, we get that $\mathcal{B}$ is a groupoid. In the other direction, if $\mathcal{B}$ is a groupoid then $\mathcal{C} = \mathcal{B}(b, b)$ is a groupoid by Definition 2.2.1. □

2.6 The case of the standard realization

Before getting to our main result which concerns an arbitrary realization functor satisfying Definition 2.4.1, we take note of an easier argument which shows that the standard realization functor cannot give rise to arbitrary homotopy types.

Definition 2.6.1 A collection of realization functors for n-groupoids $\Re^n$ $(0 \le n < \infty)$ satisfying 2.4.1 is said to be *compatible with looping* if there exist transformations

$$\varphi(\mathcal{A}, x) : \Re^{n-1}(\mathcal{A}(x, x)) \to \Omega^{r(x)}\Re^n(\mathcal{A}),$$

where $\Omega^{r(x)}$ means the space of loops based at $r(x)$, such that:

- φ is a natural transformation between functors defined on pairs $(\mathcal{A}, x)$ where $\mathcal{A}$ is an n-groupoid $\mathcal{A}$ and $x \in \mathrm{Ob}(\mathcal{A})$; and

- for $i \geq 1$ the following diagram commutes:

$$\begin{array}{ccc} \pi_i(\mathcal{A}, x) = \pi_{i-1}(\mathcal{A}(x,x), 1_x) & \longrightarrow & \pi_{i-1}(\Re^{n-1}(\mathcal{A}(x,x)), r(1_x)) \\ \downarrow & & \downarrow \\ \pi_i(\Re^n(\mathcal{A}), r(x)) & \longleftarrow & \pi_{i-1}(\Omega^{r(x)}\Re^n(\mathcal{A}), \mathrm{cst}(r(x))). \end{array}$$

Here the top arrow is $\zeta_{i-1}(\mathcal{A}(x,x), 1_x)$, the left arrow is $\zeta_i(\mathcal{A}, x)$, $\mathrm{cst}(r(x))$ is the constant loop at $r(x)$, the right arrow is induced by $\varphi(\mathcal{A}, x)$, and the bottom arrow is the canonical arrow from topology. When $i = 1$, suppress the basepoints in the π_{i-1} in the diagram.

Remark The arrows on the top, the bottom and the left are isomorphisms in the above diagram, so the arrow on the right is an isomorphism and we obtain as a corollary of the definition that the $\varphi(\mathcal{A}, x)$ are actually weak equivalences.

Remark The collection of standard realizations $\Re^n_{\mathrm{std}}$ for n-groupoids, is compatible with looping. We leave this as an exercise.

Recall the statements of 2.1.3 and 2.5.2: if $\mathcal{A}$ is a strict n-category with only one object x and only one 1-morphism 1_x, then there exists a strict $(n+1)$-category $\mathcal{B}$ with one object y, and with $\mathcal{B}(y, y) = \mathcal{A}$; and $\mathcal{A}$ is a strict n-groupoid if and only if $\mathcal{B}$ is a strict $(n+1)$-groupoid.

Corollary 2.6.2 *Suppose $\{\Re^n\}$ is a collection of realization functors 2.4.1 compatible with looping 2.6.1. Then if $\mathcal{A}$ is a 1-connected strict n-groupoid (i.e. $\pi_0(\mathcal{A}) = *$ and $\pi_1(\mathcal{A}, x) = \{1\}$), the space $\Re^n(\mathcal{A})$ is weak-equivalent to a loop space.*

Proof Let $\mathcal{A}' \subset \mathcal{A}$ be the sub-n-category having one object x and one 1-morphism 1_x. For $i \geq 2$ the inclusion induces isomorphisms

$$\pi_i(\mathcal{A}', x) \cong \pi_i(\mathcal{A}, x),$$

and in view of the 1-connectedness of $\mathcal{A}$ this means (according to Proposition 2.2.8(b)) that the morphism $\mathcal{A}' \to \mathcal{A}$ is an equivalence. It follows (by Definition 2.4.1) that $\Re^n(\mathcal{A}') \to \Re^n(\mathcal{A})$ is a weak equivalence. Now $\mathcal{A}'$ satisfies the hypothesis of 2.1.3, 2.5.2 as recalled above, so there is an $(n+1)$-groupoid $\mathcal{B}$ having one object y such that $\mathcal{A}' = \mathcal{B}(y, y)$. By the definition of "compatible with looping" and the subsequent remark that the morphism $\varphi(\mathcal{B}, y)$ is a weak equivalence, we get that $\varphi(\mathcal{B}, y)$ induces a weak equivalence

$$\Re^n(\mathcal{A}') \to \Omega^{r(y)}\Re^{n+1}(\mathcal{B}).$$

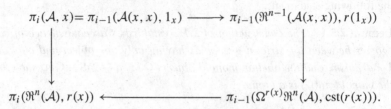

Thus $\Re^n(\mathcal{A})$ is weak-equivalent to the loop-space of $\Re^{n+1}(\mathcal{B})$. □

The following corollary is due to C. Berger [35] (although the same statement appears without proof in Grothendieck [132]). See also R. Brown and coauthors [65, 66, 67, 68].

Corollary 2.6.3 (C. Berger [35]) *There is no strict 3-groupoid $\mathcal{A}$ such that the standard realization $\Re_{\mathrm{std}}(\mathcal{A})$ is weak-equivalent to the 3-type of S^2.*

Proof We have $\pi_0(S^2) = *$ and $\pi_1(S^2) = \{1\}$, but the 3-type of S^2 is not a loop-space. By the previous corollary (and the fact that the standard realizations are compatible with looping, which we have above left as an exercise for the reader), it is impossible for $\Re_{\mathrm{std}}(\mathcal{A})$ to be the 3-type of S^2. □

2.7 Nonexistence of strict 3-groupoids of 3-type S^2

The present discussion aims to extend Berger's negative result to *any* realization functor satisfying the minimal requirements of Definition 2.4.1.

The first step is to prove the following statement (which contains the main part of the argument). It basically says that the Postnikov tower of a simply connected strict 3-groupoid $\mathcal{C}$, splits. The intermediate $\mathcal{B}$ is not really necessary for the statement but corresponds to the technique of proof.

Proposition 2.7.1 *Suppose $\mathcal{C}$ is a strict 3-groupoid with an object c such that $\pi_0(\mathcal{C}) = *$, $\pi_1(\mathcal{C}, c) = \{1\}$, $\pi_2(\mathcal{C}, c) = \mathbb{Z}$ and $\pi_3(\mathcal{C}, c) = H$ for an abelian group H. Then there exists a diagram of strict 3-groupoids*

$$\mathcal{C} \stackrel{g}{\leftarrow} \mathcal{B} \stackrel{f}{\leftarrow} \mathcal{A} \stackrel{h}{\rightarrow} \mathcal{D}$$

with objects $b \in \mathrm{Ob}(\mathcal{B})$, $a \in \mathrm{Ob}(\mathcal{A})$, $d \in \mathrm{Ob}(\mathcal{D})$ such that $f(a) = b$, $g(b) = c$, $h(a) = d$. The diagram is such that g and f are equivalences of strict 3-groupoids, and such that $\pi_0(\mathcal{D}) = *$, $\pi_1(\mathcal{D}, d) = \{1\}$, $\pi_2(\mathcal{D}, d) = \{0\}$, and such that h induces an isomorphism

$$\pi_3(h) : \pi_3(\mathcal{A}, a) = H \stackrel{\cong}{\rightarrow} \pi_3(\mathcal{D}, d).$$

Proof Start with a strict groupoid $\mathcal{C}$ and object c, satisfying the hypotheses of Proposition 2.7.1.

The first step is to construct $(\mathcal{B}, b)$. We let $\mathcal{B} \subset \mathcal{C}$ be the sub-3-category having only one object $b = c$, and only one 1-morphism $1_b = 1_c$. We set

$$\mathrm{HOM}_{\mathcal{B}(b,b)}(1_b, 1_b) := \mathrm{HOM}_{\mathcal{C}(c,c)}(1_c, 1_c),$$

with the same composition law. The map $g : \mathcal{B} \to \mathcal{C}$ is the inclusion.

Note first of all that $\mathcal{B}$ is a strict 3-groupoid. This is easily seen by refering to the Kapranov–Voevodsky definition [168] and using the equivalence of Theorem 2.2.11. We can also verify it using the equivalent condition of Theorem 2.2.11. Of course $\tau_{\leq 1}(\mathcal{B})$ is the 1-category with only one object and only one morphism, so it is a groupoid. We have to verify that $\mathcal{B}(b, b)$ is a strict 2-groupoid. For this, we again apply the condition of Theorem 2.2.11. Here we note that

$$\mathcal{B}(b, b) \subset \mathcal{C}(c, c)$$

is the full sub-2-category with only one object $1_b = 1_c$. Therefore, in view of the definition of $\tau_{\leq 1}$, we have that

$$\tau_{\leq 1}(\mathcal{B}(b, b)) \subset \tau_{\leq 1}(\mathcal{C}(c, c))$$

is a full subcategory. A full subcategory of a 1-groupoid is again a 1-groupoid, so $\tau_{\leq 1}(\mathcal{B}(b, b))$ is a 1-groupoid. Finally, $\text{HOM}_{\mathcal{B}(b,b)}(1_b, 1_b)$ is a 1-groupoid since by construction it is the same as $\text{HOM}_{\mathcal{C}(c,c)}(1_c, 1_c)$ (which is a groupoid by Definition 2.2.1 applied to the strict 2-groupoid $\mathcal{C}(c, c)$). This shows that $\mathcal{B}(b, b)$ is a strict 2-groupoid an hence that $\mathcal{B}$ is a strict 3-groupoid.

Next, note that $\pi_0(\mathcal{B}) = *$ and $\pi_1(\mathcal{B}, b) = \{1\}$. On the other hand, for $i = 2, 3$ we have

$$\pi_i(\mathcal{B}, b) = \pi_{i-2}\left(\text{HOM}_{\mathcal{B}(b,b)}(1_b, 1_b), 1_b^2\right),$$

and similarly

$$\pi_i(\mathcal{C}, c) = \pi_{i-2}\left(\text{HOM}_{\mathcal{C}(c,c)}(1_c, 1_c), 1_c^2\right),$$

so the inclusion g induces an equality $\pi_i(\mathcal{B}, b) \stackrel{=}{\rightarrow} \pi_i(\mathcal{C}, c)$. Therefore, by version (b) of equivalence in Proposition 2.2.8, g is an equivalence of strict 3-groupoids. This completes the construction and verification for $\mathcal{B}$ and g.

Before getting to the construction of $\mathcal{A}$ and f, we analyze the strict 3-groupoid $\mathcal{B}$ in terms of the discussion of 2.1.2 and 2.5.1. Let

$$\mathcal{G} := \text{HOM}_{\mathcal{B}(b,b)}(1_b, 1_b).$$

It is an abelian monoid-object in the category of 1-groupoids, with abelian operation denoted by $+ : \mathcal{G} \times \mathcal{G} \to \mathcal{G}$ and unit element denoted $0 \in \mathcal{G}$ which is the same as 1_b. The operation $+$ corresponds to both of the compositions $*_0$ and $*_1$ in $\mathcal{B}$.

The hypotheses on the homotopy groups of $\mathcal{C}$ also hold for $\mathcal{B}$ (since g was an equivalence). These translate to the statements that $(\pi_0(\mathcal{G}), +) = \mathbb{Z}$ and $\mathcal{G}(0, 0) = H$.

We now construct $\mathcal{A}$ and f via 2.1.2 and 2.5.1, by constructing a morphism $(\mathcal{G}', +) \to (\mathcal{G}, +)$ of abelian monoid-objects in the category of 1-groupoids. We do this by a type of "base-change" on the monoid of objects, i.e. we will first define a morphism $\mathrm{Ob}(\mathcal{G}') \to \mathrm{Ob}(\mathcal{G})$ and then define $\mathcal{G}'$ to be the groupoid with object set $\mathrm{Ob}(\mathcal{G}')$ but with morphisms corresponding to those of $\mathcal{G}$.

To accomplish the "base-change," start with the following construction. If S is a set, let $\mathbf{codsc}(S)$ denote the groupoid with S as set of objects, and with exactly one morphism between each pair of objects. If S has an abelian monoid structure then $\mathbf{codsc}(S)$ is an abelian monoid object in the category of groupoids.

Note that for any groupoid $\mathcal{U}$ there is a morphism of groupoids

$$\mathcal{U} \longrightarrow \mathbf{codsc}(\mathrm{Ob}(\mathcal{U})),$$

and by "base change" we mean the following operation: take a set S with a map $p : S \to \mathrm{Ob}(\mathcal{U})$ and look at

$$\mathcal{V} := \mathbf{codsc}(S) \times_{\mathbf{codsc}(\mathrm{Ob}(\mathcal{U}))} \mathcal{U}.$$

This is a groupoid with S as set of objects, and with

$$\mathcal{V}(s,t) = \mathcal{U}(p(s), p(t)).$$

A similar construction will be used later in Chapter 10 under the notation $\mathcal{V} = p^*(\mathcal{U})$. For the present purposes, note that if $\mathcal{U}$ is an abelian monoid object in the category of groupoids, if S is an abelian monoid and if p is a map of monoids then $\mathcal{V}$ is again an abelian monoid object in the category of groupoids.

Apply this as follows. Starting with $(\mathcal{G}, +)$ corresponding to $\mathcal{B}$ via 2.1.2 and 2.5.1 as above, choose objects $a, b \in \mathrm{Ob}(\mathcal{G})$ such that the image of a in $\pi_0(\mathcal{G}) \cong \mathbb{Z}$ corresponds to $1 \in \mathbb{Z}$, and such that the image of b in $\pi_0(\mathcal{G})$ corresponds to $-1 \in \mathbb{Z}$. Let N denote the abelian monoid, product of two copies of the natural numbers, with objects denoted (m, n) for nonnegative integers m, n. Define a map of abelian monoids

$$p : N \to \mathrm{Ob}(\mathcal{G})$$

by

$$p(m,n) := m \cdot a + n \cdot b := a + a + \cdots + a \, + \, b + b + \cdots + b.$$

Note that this induces the surjection $N \to \pi_0(\mathcal{G}) = \mathbb{Z}$ given by $(m, n) \mapsto m - n$.

Define $(\mathcal{G}', +)$ as the base-change

$$\mathcal{G}' := \mathbf{codsc}(N) \times_{\mathbf{codsc}(\mathrm{Ob}(\mathcal{G}))} \mathcal{G},$$

with its induced abelian monoid operation +. We have

$$\mathrm{Ob}(\mathcal{G}') = N,$$

and the second projection $p_2 : \mathcal{G}' \to \mathcal{G}$ (which induces p on object sets) is fully faithful, i.e.

$$\mathcal{G}'((m, n), (m', n')) = \mathcal{G}(p(m, n), p(m', n')).$$

Note that $\pi_0(\mathcal{G}') = \mathbb{Z}$ via the map induced by p or equivalently p_2. To prove this, say that: (i) N surjects onto $\mathbb{Z}$ so the map induced by p is surjective; and (ii) the fact that p_2 is fully faithful implies that the induced map $\pi_0(\mathcal{G}') \to \pi_0(\mathcal{G}) = \mathbb{Z}$ is injective.

We let $\mathcal{A}$ be the strict 3-groupoid corresponding to $(\mathcal{G}', +)$ via 2.1.2, and let $f : \mathcal{A} \to \mathcal{B}$ be the map corresponding to $p_2 : \mathcal{G}' \to \mathcal{G}$ again via 2.1.2. Let a be the unique object of $\mathcal{A}$ (it is mapped by f to the unique object $b \in \mathrm{Ob}(\mathcal{B})$).

The fact that $(\pi_0(\mathcal{G}'), +) = \mathbb{Z}$ is a group implies that $\mathcal{A}$ is a strict 3-groupoid (2.5.1). We have $\pi_0(\mathcal{A}) = *$ and $\pi_1(\mathcal{A}, a) = \{1\}$. Also,

$$\pi_2(\mathcal{A}, a) = (\pi_0(\mathcal{G}'), +) = \mathbb{Z}$$

and f induces an isomorphism from here to $\pi_2(\mathcal{B}, b) = (\pi_0(\mathcal{G}), +) = \mathbb{Z}$. Finally (using the notation (0, 0) for the unit object of $(N, +)$ and the notation 0 for the unit object of $\mathrm{Ob}(\mathcal{G})$),

$$\pi_3(\mathcal{A}, a) = \mathcal{G}'((0, 0), (0, 0)),$$

and similarly

$$\pi_3(\mathcal{B}, b) = \mathcal{G}(0, 0) = H;$$

the map $\pi_3(f) : \pi_3(\mathcal{A}, a) \to \pi_3(\mathcal{B}, b)$ is an isomorphism because it is the same as the map

$$\mathcal{G}'((0, 0), (0, 0)) \to \mathcal{G}(0, 0)$$

induced by $p_2 : \mathcal{G}' \to \mathcal{G}$, and p_2 is fully faithful. We have now completed the verification that f induces isomorphisms on the homotopy groups, so by version (b) of the the notion of equivalence in Proposition 2.2.8, f is an equivalence of strict 3-groupoids.

We now construct $\mathcal{D}$ and define the map h by an explicit calculation in $(\mathcal{G}', +)$. First of all, let $[H]$ denote the 1-groupoid with one object denoted 0, and with H as group of endomorphisms:

$$[H](0, 0) := H.$$

Because H is an abelian group, this has a structure of abelian monoid-object in the category of groupoids, denoted $([H], +)$. Let $\mathcal{D}$ be the strict 3-groupoid corresponding to $([H], +)$ via 2.1.2 and 2.5.1. We will construct a morphism $h : \mathcal{A} \rightarrow \mathcal{D}$ via 2.1.2 by constructing a morphism of abelian monoid objects in the category of groupoids,

$$h : (\mathcal{G}', +) \rightarrow ([H], +).$$

We will construct this morphism so that it induces the identity morphism

$$\mathcal{G}'((0, 0), (0, 0)) = H \rightarrow [H](0, 0) = H.$$

This will insure that the morphism h has the property required for 2.7.1.

The object $(1, 1) \in N$ goes to $0 \in \pi_0(\mathcal{G}') \cong \mathbb{Z}$. Thus we may choose an isomorphism $\varphi : (0, 0) \cong (1, 1)$ in $\mathcal{G}'$. For any k let $k\varphi$ denote the isomorphism $\varphi + \cdots + \varphi$ (k times) going from $(0, 0)$ to (k, k). However, H is also the automorphism group of $(0, 0)$ in $\mathcal{G}'$. The operations $+$ and composition coincide on H. Finally, for any $(m, n) \in N$ let $1_{m,n}$ denote the identity automorphism of the object (m, n). Then any arrow α in $\mathcal{G}$ may be uniquely written in the form

$$\alpha = 1_{m,n} + k\varphi + u$$

with (m, n) the source of α, the target being $(m + k, n + k)$, and where $u \in H$.

We have the following formulae for the composition $\circ$ of arrows in $\mathcal{G}'$. They all come from the basic rule

$$(\alpha \circ \beta) + (\alpha' \circ \beta') = (\alpha + \alpha') \circ (\beta + \beta'),$$

which in turn comes simply from the fact that $+$ is a morphism of groupoids $\mathcal{G}' \times \mathcal{G}' \rightarrow \mathcal{G}'$ defined on the cartesian product of two copies of $\mathcal{G}$. Note in a similar vein that $1_{0,0}$ acts as the identity for the operation $+$ on arrows, and also that

$$1_{m,n} + 1_{m',n'} = 1_{m+m',n+n'}.$$

Our first equation is

$$(1_{l,l} + k\varphi) \circ l\varphi = (k + l)\varphi.$$

To prove this note that $l\varphi + 1_{0,0} = l\varphi$ and our basic formula says

$$(1_{l,l} \circ l_{\varphi}) + (k\varphi \circ 1_{0,0}) = (1_{l,l} + k\varphi) \circ (l\varphi + 1_{0,0}),$$

but the left side is just $l\varphi + k\varphi = (k + l)\varphi$.

Now our basic formula, for a composition starting with (m, n), going first to $(m+l, n+l)$, then going to $(m+l+k, n+l+k)$, gives

$$\begin{aligned}(1_{m+l,n+l} + k\varphi + u) \circ (1_{m,n} + l\varphi + v)\\ = (1_{m,n} + 1_{l,l} + k\varphi + u) \circ (1_{m,n} + l\varphi + v)\\ = 1_{m,n} \circ 1_{m,n} + (1_{l,l} + k\varphi) \circ l\varphi + u \circ v\\ = 1_{m,n} + (k+l)\varphi + (u \circ v)\end{aligned}$$

where of course $u \circ v = u + v$.

This formula shows that the morphism h from arrows of $\mathcal{G}'$ to the group H, defined by

$$h(1_{m,n} + k\varphi + u) := u$$

is compatible with composition. This implies that it provides a morphism of groupoids $h : \mathcal{G} \to [H]$ (recall from above that $[H]$ is defined to be the groupoid with one object whose automorphism group is H). Furthermore, the morphism h is obviously compatible with the operation $+$ since

$$\begin{aligned}(1_{m,n} + k\varphi + u) + (1_{m',n'} + k'\varphi + u')\\ = (1_{m+m',n+n'} + (k+k')\varphi + (u+u'))\end{aligned}$$

and once again $u + u' = u \circ u'$ (the operation $+$ on $[H]$ being given by the commutative operation $\circ$ on H).

This completes the construction of a morphism $h : (\mathcal{G}, +) \to ([H], +)$ which induces the identity on $\text{HOM}(0, 0)$. This corresponds to a morphism of strict 3-groupoids $h : \mathcal{A} \to \mathcal{D}$ as required to complete the proof of Proposition 2.7.1. □

We can now give the nonrealization statement. A more streamlined proof of this theorem has been given by Ara [5] based on the results of Lafont *et al.* [179].

Theorem 2.7.2 *Let $\Re$ be any realization functor satisfying the properties of Definition 2.4.1. Then there does not exist a strict 3-groupoid C such that $\Re(C)$ is weak-equivalent to the 3-truncation of the homotopy type of S^2.*

Proof Suppose for the moment that we know Proposition 2.7.1; with this we will prove 2.7.2. Fix a realization functor $\Re$ for strict 3-groupoids satisfying the axioms of 2.4.1, and assume that C is a strict 3-groupoid such that $\Re(C)$ is weak homotopy-equivalent to the 3-type of S^2. We shall derive a contradiction.

Apply Proposition 2.7.1 to C. Choose an object $c \in \text{Ob}(C)$. Note that, because of the isomorphisms between homotopy sets or groups 2.4.1, we have

$\pi_0(\mathcal{C}) = *$, $\pi_1(\mathcal{C}, c) = \{1\}$, $\pi_2(\mathcal{C}, c) = \mathbb{Z}$ and $\pi_3(\mathcal{C}, c) = \mathbb{Z}$, so 2.7.1 applies with $H = \mathbb{Z}$. We obtain a sequence of strict 3-groupoids

$$\mathcal{C} \stackrel{g}{\leftarrow} \mathcal{B} \stackrel{f}{\leftarrow} \mathcal{A} \stackrel{h}{\rightarrow} \mathcal{D}.$$

This gives the diagram of spaces

$$\mathfrak{R}(\mathcal{C}) \xleftarrow{\mathfrak{R}(g)} \mathfrak{R}(\mathcal{B}) \xleftarrow{\mathfrak{R}(f)} \mathfrak{R}(\mathcal{A}) \xrightarrow{\mathfrak{R}(h)} \mathfrak{R}(\mathcal{D}).$$

The axioms of 2.4.1 for $\mathfrak{R}$ imply that $\mathfrak{R}$ transforms equivalences of strict 3-groupoids into weak homotopy equivalences of spaces. Thus $\mathfrak{R}(f)$ and $\mathfrak{R}(g)$ are weak homotopy equivalences and we get that $\mathfrak{R}(\mathcal{A})$ is weak homotopy equivalent to the 3-type of S^2.

On the other hand, again by the axioms of 2.4.1, we have that $\mathfrak{R}(\mathcal{D})$ is 2-connected, and $\pi_3(\mathfrak{R}(\mathcal{D}), r(d)) = H$ (via the isomorphism $\pi_3(\mathcal{D}, d) \cong H$ induced by h, f and g). By the Hurewicz theorem $\pi_3(\mathfrak{R}(\mathcal{D})) \cong H_3(\mathfrak{R}(\mathcal{D}), \mathbb{Z})$ (see Whitehead [265]) so there is a class $\eta \in H^3(\mathfrak{R}(\mathcal{D}), H)$, which induces an isomorphism

$$\mathbf{Hur}(\eta) : \pi_3(\mathfrak{R}(\mathcal{D}), r(d)) \stackrel{\cong}{\rightarrow} H.$$

Here

$$\mathbf{Hur} : H^3(X, H) \rightarrow \text{HOM}(\pi_3(X, x), H)$$

is the dual of the Hurewicz map for any pointed space (X, x); and the cohomology is singular cohomology (in particular it only depends on the weak homotopy type of the space).

Now look at the pullback of this class

$$\mathfrak{R}(h)^*(\eta) \in H^3(\mathfrak{R}(\mathcal{A}), H).$$

The hypothesis that $\mathfrak{R}(u)$ induces an isomorphism on π_3 implies that

$$\mathbf{Hur}(\mathfrak{R}(h)^*(\eta)) : \pi_3(\mathfrak{R}(\mathcal{A}), r(a)) \stackrel{\cong}{\rightarrow} H.$$

In particular, $\mathbf{Hur}(\mathfrak{R}(h)^*(\eta))$ is nonzero so the class $\mathfrak{R}(h)^*(\eta)$ is nonzero in $H^3(\mathfrak{R}(\mathcal{A}), H)$. This is a contradiction because $\mathfrak{R}(\mathcal{A})$ is weak homotopy-equivalent to the 3-type of S^2, and $H = \mathbb{Z}$, but $H^3(S^2, \mathbb{Z}) = \{0\}$.

This contradiction completes the proof of the theorem. □

We have seen that for any realization functor $\mathfrak{R} : n\text{STRGPD} \rightarrow \text{TOP}$ preserving homotopy groups, topological spaces with nontrivial higher Whitehead products cannot be weakly equivalent to any $\mathfrak{R}(\mathcal{A})$. As was discussed by Grothendieck in "Pursuing stacks" [132], this result motivates the search for a notion of higher category weaker than the notion of strict n-category. Following the yoga described by Lewis [187], it appears to be sufficient to weaken any single particular aspect.

Even though the rest of the book will be concerned with the notion of *weak higher category*, our first look at the case of strict n-categories serves nevertheless as a guide to the outlines of any general theory of weak n-categories. The inductive procedure which has shown up here is basic to the structure of our construction of a theory of n-categories.

3
Fundamental elements of n-categories

The observation that the theory of strict n-groupoids is not enough to give a good model for homotopy n-types, led Grothendieck to ask for a theory of n-categories with *weakly associative composition*. This will be the main subject of our book, in particular we use the terminology *n-category* to mean some kind of object in a possible theory with weak associativity, or even a composition which is only defined up to homotopy, or perhaps some other type of weakening (as will be briefly discussed in Chapter 4).

There are a certain number of basic elements expected of any theory of n-categories, and which can be explained without refering to a full definition. It will be useful to start by considering these. Our discussion follows Tamsamani's paper [250], but really sums up the general expectations for a theory of n-categories which were developed over many years starting with Bénabou and continuing through the theory of strict n-categories and Grothendieck's manuscript.

For this chapter, we will use the terminology "n-category" to mean any object in a generic theory of n-categories. We will sometimes use the idea that our generic theory should admit cartesian products and disjoint sums.

3.1 A globular theory

We saw that a strict n-category has, in particular, an underlying globular set. This basic structure should be conserved, in some form, in any weak theory.

(OB)–An n-category $\mathcal{A}$ should have an *underlying set of objects* denoted $\mathrm{Ob}(\mathcal{A})$. If $i = 0$ then the structure $\mathcal{A}$ is identified with just this set $\mathrm{Ob}(\mathcal{A})$, that is to say a 0-category is just a set.

(MOR)–If $i \geq 1$ then for any two elements $x, y \in \mathrm{Ob}(\mathcal{A})$, there should be an $(n-1)$-*category of morphisms from x to y* denoted $\underline{\mathrm{Mor}}_{\mathcal{A}}(x, y)$. From these two things, we obtain by induction a whole family of sets called the *sets of i-morphisms of* $\mathcal{A}$ for $0 \leq i \leq n$.

(PS)–With respect to cartesian products and disjoint sums, we should have $\mathrm{Ob}(\mathcal{A} \times \mathcal{B}) = \mathrm{Ob}(\mathcal{A}) \times \mathrm{Ob}(\mathcal{B})$ and $\mathrm{Ob}(\mathcal{A} \sqcup \mathcal{B}) = \mathrm{Ob}(\mathcal{A}) \sqcup \mathrm{Ob}(\mathcal{B})$.

The set of i-morphisms of $\mathcal{A}$ can be defined inductively by the following procedure. Put

$$\underline{\mathrm{Mor}}[\mathcal{A}] := \coprod_{x,y \in \mathrm{Ob}(\mathcal{A})} \underline{\mathrm{Mor}}_{\mathcal{A}}(x, y).$$

This is the $(n-1)$-*category of morphisms of* $\mathcal{A}$.

By induction we obtain the $(n-i)$-category of i-morphisms of $\mathcal{A}$, denoted by

$$\underline{\mathrm{Mor}}^i[\mathcal{A}] := \underline{\mathrm{Mor}}[\underline{\mathrm{Mor}}[\cdots[\mathcal{A}]\cdots]].$$

This is defined whenever $0 \leq i \leq n$, with $\underline{\mathrm{Mor}}^0[\mathcal{A}] := \mathcal{A}$ and $\underline{\mathrm{Mor}}^n[\mathcal{A}]$ being a set.

Define

$$\mathrm{Mor}^i[\mathcal{A}] := \mathrm{Ob}(\underline{\mathrm{Mor}}^i[\mathcal{A}]).$$

This is a set, called the *set of i-morphisms of* $\mathcal{A}$.

From the above definitions we can write

$$\underline{\mathrm{Mor}}^i[\mathcal{A}] = \coprod_{x,y \in \mathrm{Mor}^{i-1}[\mathcal{A}]} \underline{\mathrm{Mor}}_{\underline{\mathrm{Mor}}^{i-1}[\mathcal{A}]}(x, y),$$

and, by compatibility of objects with coproducts,

$$\mathrm{Mor}^i[\mathcal{A}] = \coprod_{x,y \in \mathrm{Mor}^{i-1}[\mathcal{A}]} \mathrm{Ob}(\underline{\mathrm{Mor}}_{\underline{\mathrm{Mor}}^{i-1}[\mathcal{A}]}(x, y)).$$

In particular, we have maps s_i and t_i from $\mathrm{Mor}^i[\mathcal{A}]$ to $\mathrm{Mor}^{i-1}[\mathcal{A}]$ taking an element $f \in \mathrm{Mor}^i[\mathcal{A}]$ lying in the piece of the coproduct indexed by (x, y), to $s_i(f) := x$ or $t_i(f) := y$ respectively. These maps are called *source* and *target* and, if no confusion arises, the index i may be dropped.

If $u, v \in \mathrm{Mor}^i[\mathcal{A}]$, let $\mathrm{Mor}^{i+1}_{\mathcal{A}}(u, v)$ denote the preimage of the pair (u, v) by the map (s_{i+1}, t_{i+1}). It is nonempty only if $s_i(u) = s_i(v)$ and $t_i(u) = t_i(v)$, and when using the notation $\mathrm{Mor}^{i+1}_{\mathcal{A}}(u, v)$ we generally mean to say that these conditions are supposed to hold. Similarly, we get $(n-i-1)$-categories, denoted $\underline{\mathrm{Mor}}^{i+1}_{\mathcal{A}}(u, v)$.

In this way, starting just from the principles (OB) and (MOR) together with the compatibility with sums in (PS), we obtain from an n-category a collection of sets

$$\mathrm{Mor}^0[\mathcal{A}] = \mathrm{Ob}(\mathcal{A}); \ \mathrm{Mor}^1[\mathcal{A}], \ldots, \mathrm{Mor}^n[\mathcal{A}],$$

together with pairs of maps

$$s_i, t_i : \mathrm{Mor}^i[\mathcal{A}] \to \mathrm{Mor}^{i-1}[\mathcal{A}].$$

They satisfy

$$s_i s_{i+1} = s_i t_{i+1}, \quad t_i s_{i+1} = t_i t_{i+1}.$$

These elements make our theory of n-categories into a *globular theory*.

Among other things, starting from this structure we can draw pictures in a way which is usual for the theory of n-categories. These pictures explain why the theory is called "globular." A 0-morphism is just a point, and a 1-morphism is pictured as an arrow:

A 2-morphism is pictured as

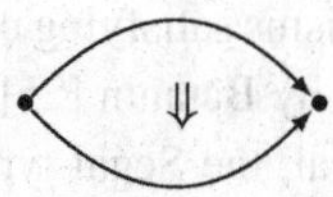

Whereas a 3-morphism should be thought of as a sort of "pillow," which might be pictured as

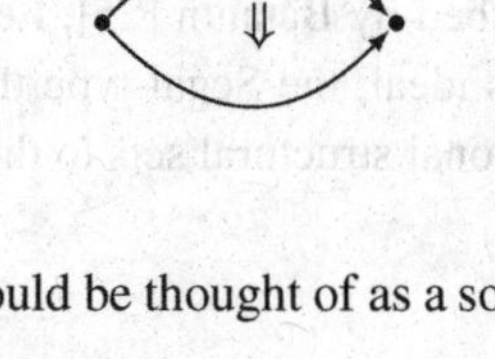

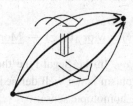

3.2 Identities

For each $x \in \mathrm{Ob}(\mathcal{A})$ there should be a natural element $1_x \in \mathrm{Mor}_{\mathcal{A}}(x, x)$, called the *identity of x*. One can envision theories in which the identity is not well defined but exists only up to homotopy, see Joyal and Kock [159], Kock [172] and Hirsh and Millès [146]. However, the theory considered here will have canonical identities.

Following the same inductive procedure as in the previous section, we get morphisms for any $0 \leq i < n$,

$$e_i : \mathrm{Mor}^i[\mathcal{A}] \to \mathrm{Mor}^{i+1}[\mathcal{A}],$$

such that $s_{i+1}e_i(u) = u$ and $t_{i+1}e_i(u) = u$. We call $e_i(u)$ the *identity* $(i+1)$-*morphism of the i-morphism u*.

Some authors introduce a *category of reflexive globules* $\mathbb{G}^n_R$, having objects M_i for $0 \leq i \leq n$, with generating morphisms $s_i, t_i : M_i \to M_{i-1}$ and $e_i : M_i \to M_{i+1}$ subject to the relations

$$s_i s_{i+1} = s_i t_{i+1}, \quad t_i s_{i+1} = t_i t_{i+1}, \quad s_{i+1} e_i = 1_{M_i}, \quad t_{i+1} e_i = 1_{M_i}.$$

A *reflexive n-globular set* is a functor $\mathbb{G}^n_R \to$ SET. Any n-category $\mathcal{A}$ induces an *underlying reflexive globular set* constructed as above. Other authors (such as Batanin) use the the subcategory $\mathbb{G}^n$ obtained by throwing out the identities e_i, and an *n-globular set* is a functor defined on this subcategory. We shall use that notation in Section 4.3.

The first and basic idea for defining a theory of n-categories is that an n-category should consist of an underlying (perhaps reflexive) globular set, plus additional structural morphisms satisfying certain properties. Whereas the Batanin-type theories described by Batanin [25], Leinster [183] and Maltsiniotis [200] are closest to this ideal, the Segal-type theories we consider in the present book will add additional structural sets to the basic globular set of $\mathcal{A}$.

3.3 Composition, equivalence and truncation

For objects $x, y, z \in \mathrm{Ob}(\mathcal{A})$ there should be some kind of morphism of $(n-1)$-categories

$$\underline{\mathrm{Mor}}_{\mathcal{A}}(x, y) \times \underline{\mathrm{Mor}}_{\mathcal{A}}(y, z) \to \underline{\mathrm{Mor}}_{\mathcal{A}}(x, z) \tag{3.3.1}$$

corresponding to *composition*. In the Segal-type theories considered in this book, the composition morphism is not well defined and may not even exist, rather existing only in some homotopic sense.

Nonetheless, in order best to motivate the following discussion, assume for the moment that we know what composition means, particularly how to define $g \circ f \in \mathrm{Mor}^1_{\mathcal{A}}(x, z)$ for $f \in \mathrm{Mor}^1_{\mathcal{A}}(x, y)$ and $g \in \mathrm{Mor}^1_{\mathcal{A}}(x, y)$.

We can then inductively define a notion of *equivalence*. Tamsamani [250] calls this *inner equivalence* to emphasize that we are speaking of arrows in our n-category $\mathcal{A}$ which are equivalences internally in $\mathcal{A}$. To be more precise, we will define what it means for $f \in \mathrm{Mor}^1_{\mathcal{A}}(x, y)$ to be an *inner equivalence between x and y*. If such an f exists, we say that *x and y are equivalent* and write $x \sim y$.

Inductively we suppose known what this means for $(n-1)$-categories, and in particular within the $(n-1)$-categories $\underline{\mathrm{Mor}}_{\mathcal{A}}(x, x)$ or $\underline{\mathrm{Mor}}_{\mathcal{A}}(y, y)$.

The definition then proceeds by saying that $f \in \mathrm{Mor}^1_{\mathcal{A}}(x, y)$ is an inner equivalence between x and y, if there exists $g \in \mathrm{Mor}^1_{\mathcal{A}}(y, x)$ such that $g \circ f$ is equivalent to 1_x in $\underline{\mathrm{Mor}}_{\mathcal{A}}(x, x)$ and $f \circ g$ is equivalent to 1_y in $\underline{\mathrm{Mor}}_{\mathcal{A}}(y, y)$.

This notion should be transitive in the sense that if f is an equivalence from x to y and g is an equivalence from y to z, then $g \circ f$ should be an equivalence from x to z. The relation "$x \sim y$" is therefore a transitive equivalence relation on the set $\mathrm{Ob}(\mathcal{A})$.

Define the *truncation* $\tau_{\leq 0}(\mathcal{A})$ to be the quotient set $\mathrm{Ob}(\mathcal{A})/\sim$.

We can go further and define the 1-categorical truncation $\tau_{\leq 1}(\mathcal{A})$, a 1-category, as follows:

$$\mathrm{Ob}(\tau_{\leq 1}(\mathcal{A})) := \mathrm{Ob}(\mathcal{A}),$$

$$\mathrm{Mor}^1_{\tau_{\leq 1}(\mathcal{A})}(x, y) := \tau_{\leq 0}(\underline{\mathrm{Mor}}_{\mathcal{A}}(x, y)).$$

In other words, the objects of $\tau_{\leq 1}(\mathcal{A})$ are the same as the objects of $\mathcal{A}$, but the morphisms of $\tau_{\leq 1}(\mathcal{A})$ are the equivalence classes of 1-morphisms of $\mathcal{A}$, under the equivalence relation on the objects of the $(n-1)$-category $\underline{\mathrm{Mor}}_{\mathcal{A}}(x, y)$.

Composition of morphisms in $\tau_{\leq 1}(\mathcal{A})$ should be defined by composing representatives of the equivalence classes. One of the main requirements for our theory of n-categories is that this composition in $\tau_{\leq 1}(\mathcal{A})$ should be well defined, independent of the choice of representatives, and indeed independent of the choice of notion of composition morphism introduced at the start of this section.

Denote also by $\sim$ the equivalence relation obtained in the same way on the objects of the $(n-i)$-categories $\underline{\mathrm{Mor}}^i[\mathcal{A}]$. Noting that it is compatible with the source and target maps, we get an equivalence relation $\sim$ on $\mathrm{Mor}^i_{\mathcal{A}}(u, v)$ for any $(i-1)$-morphisms u and v.

The above discussion presupposed the existence of some kind of composition operation, but in the Segal-style theory we consider in this book, such a composition morphism is not canonically defined. Thus, we restart the discussion without assuming existence of a composition morphism of $(n-1)$-categories. The first fundamental structure to be considered is thus:

(EQUIV)–on each set $\mathrm{Mor}^i[\mathcal{A}]$ we have an equivalence relation $\sim$ compatible with the source and target maps, giving the *set of i-morphisms up to equivalence* $\mathrm{Mor}^i[\mathcal{A}]/\sim$. For $i = n$ this equivalence relation should be trivial. The induced relation on $\mathrm{Mor}^i_{\mathcal{A}}(u, v)$ is also denoted $\sim$.

We can then consider the structure of composition, which is well-defined up to equivalence, in other words it is given by a map on quotient sets.

(COMP)–for any $0 < i \leq n$ and any three $(i-1)$-morphisms u, v, w sharing the same sources and the same targets, we have a well-defined composition map

$$\left(\mathrm{Mor}^i_{\mathcal{A}}(u, v)/\sim\right) \times \left(\mathrm{Mor}^i_{\mathcal{A}}(v, w)/\sim\right) \to \mathrm{Mor}^i_{\mathcal{A}}(u, w)/\sim$$

which is associative and has the classes of identity morphisms as left and right units.

These two structures are compatible in the sense that composition is defined after passing to the quotient by $\sim$. As a matter of simplifying notation, given $f \in \mathrm{Mor}^i_{\mathcal{A}}(u, v)$ and $g \in \mathrm{Mor}^i_{\mathcal{A}}(v, w)$ then denote by $g \circ f$ any representative in $\mathrm{Mor}^i_{\mathcal{A}}(u, w)$ for the composition of the class of g with the class of f. This is well defined up to equivalence and by construction independent, up to equivalence, of the choices of representatives f and g for their equivalence classes.

Equivalence and composition also satisfy the following further compatibility condition, expressing the notion of equivalence in terms which closely resemble the classical definition of equivalence of categories.

(EQC)–for any $0 \leq i < n$ and $u, v \in \mathrm{Mor}^i_{\mathcal{A}}$ sharing the same source and target (i.e. $s_i(u) = s_i(v)$ and $t_i(u) = t_i(v)$ in case $i > 0$), then $u \sim v$ if and only if there exist $f \in \mathrm{Mor}^{i+1}_{\mathcal{A}}(u, v)$ and $g \in \mathrm{Mor}^{i+1}_{\mathcal{A}}(v, u)$ such that $g \circ f \sim 1_u$ and $f \circ g \sim 1_v$.

With these structures, we can define the 1-categories $\tau_{\leq 1}\underline{\mathrm{Mor}}^i_{\mathcal{A}}(u, v)$, having objects the elements of $\mathrm{Mor}^i_{\mathcal{A}}(u, v)$ and as morphisms between $w, z \in \mathrm{Mor}^i_{\mathcal{A}}(u, v)$ the equivalence classes $\mathrm{Mor}^{i+1}_{\mathcal{A}}(w, z)/\sim$. The composition of (COMP) gives this a structure of 1-category, and $w \sim z$ if and only if w and z are isomorphic objects of $\tau_{\leq 1}\underline{\mathrm{Mor}}^i_{\mathcal{A}}(u, v)$. At the bottom level we obtain a 1-category denoted $\tau_{\leq 1}(\mathcal{A})$ and called the *1-truncation of* $\mathcal{A}$, whose set of objects is $\mathrm{Ob}(\mathcal{A})$ and whose set of morphisms is $\mathrm{Mor}^1[\mathcal{A}]/\sim$. These constructions are compatible with the induction in the sense that $\tau_{\leq 1}\underline{\mathrm{Mor}}^i_{\mathcal{A}}(u, v)$ is indeed the 1-truncation of the $(n - i)$-category $\underline{\mathrm{Mor}}^i_{\mathcal{A}}(u, v)$..

Suppose $x, y \in \mathrm{Mor}^{i-1}[\mathcal{A}]$ and $u, v \in \mathrm{Mor}^i_{\mathcal{A}}(x, y)$. An element $f \in \mathrm{Mor}^{i+1}_{\mathcal{A}}(u, v)$ is said to be an *internal equivalence* between u and v, if its class is an isomorphism in $\tau_{\leq 1}\underline{\mathrm{Mor}}^i_{\mathcal{A}}(x, y)$. This is equivalent to requiring the existence of $g \in \mathrm{Mor}^{i+1}_{\mathcal{A}}(v, u)$ such that $g \circ f \sim 1_u$ and $f \circ g \sim 1_v$.

An n-category $\mathcal{A}$ is an *n-groupoid* if all arrows are inner equivalences. As was discussed for strict n-categories in the previous chapter, this is equivalent to the inductive definition obtained by requiring that the $\underline{\mathrm{Mor}}^1_{\mathcal{A}}(x, y)$ be $(n - 1)$-groupoids, and $\tau_{\leq 1}(\mathcal{A})$ be a 1-groupoid. Again following the discussion of the previous chapter, if $\mathcal{A}$ is an n-groupoid, define $\pi_0(\mathcal{A})$ to be the set of isomorphism classes of $\tau_{\leq 0}(\mathcal{A})$, and for $i \geq 1$ and $x \in \mathrm{Ob}(\mathcal{A})$, let $\pi_i(\mathcal{A}, x)$ be the group of automorphisms of 1^{i-1}_x (the iterated identity) in $\tau_{\leq 1}\underline{\mathrm{Mor}}^{i-1}(\mathcal{A})\left(1^{i-2}_x, 1^{i-2}_x\right)$ (resp. in $\tau_{\leq 1}(\mathcal{A})$ for $i = 1$). The homotopy groups will be commutative for $i \geq 2$ in any theory which has at least a weak version of the interchange rule.

3.4 Enriched categories

The natural first approach to the notion of n-category is to ask for $(n - 1)$-categories of morphisms $\underline{\mathrm{Mor}}_{\mathcal{A}}(x, y)$, with composition operations (3.3.1) which are strictly associative and have the 1_x as strict left and right units. This gives a structure of *category enriched over* $(n-1)$*-categories*. In an intuitive sense the reader should think of an n-category in this way. However, if the definition is applied inductively over n, that is to say that the $(n - 1)$-categories $\underline{\mathrm{Mor}}_{\mathcal{A}}(x, y)$ are themselves enriched over $(n - 2)$-categories and so forth, one gets to the notion of strict n-category considered in the previous chapter. But, as was pointed out at the end of the previous chapter, the strict n-categories are not sufficient to capture all of the homotopical behavior we want for $n \geq 3$.

Paoli [210] has shown that homotopy n-types can be modelled by *semistrict* n-groupoids, in other words n-groupoids which are strictly enriched over weak $(n-1)$-groupoids. Bergner showed a corresponding strictification theorem for Segal categories, and the analogous strictification from A_∞-categories to dg categories has been known to the experts for some time. Lurie's technique [190] for constructing the model category structure we consider here gives additionally the strictification theorem generalizing Bergner's result. So, as we shall discuss briefly in Section 19.5, the objects of our Segal-type theory of n-categories can always be assumed equivalent to semistrict ones, that is to categories strictly enriched over the model category for $(n-1)$-precategories. This doesn't mean that we can go inductively towards strict n-categories because the strictification operation is not compatible with cartesian product, so if applied to the enriched morphism objects it destroys the strict enrichment structure. As Paoli notes [210], semistrictification at one level is as far as we can go.

3.5 The $(n+1)$-category of n-categories

One of the main goals of a theory of n-categories is to provide a structure of $(n+1)$-category on the collection of all n-categories. Of course, some discussion of universes is needed here: the collection of all n-categories in a universe $\mathbb{U}$ should form an $(n+1)$-category in a bigger containing universe $\mathbb{V} \ni \mathbb{U}$. This precision will be dropped from most of our discussions below.

Recall that the notion of 2-category was originally introduced because of the familiar observation that "the set of all categories is actually a 2-category," a 2-category to be denoted 1CAT. Its objects are the 1-categories (in the smaller universe); the 1-morphisms of 1CAT are the functors between categories; and the 2-morphisms between functors are the natural transformations.

In general, we hope and expect to obtain an $(n+1)$-category denoted nCAT, whose objects are the n-categories. The 1-morphisms of nCAT are the "true" functors between n-categories, and we obtain all of the $\mathrm{Mor}^i[n\mathrm{CAT}]$ for $0 \leq i \leq n+1$ which are higher analogues of natural transformations and so on.

The notion of internal equivalence within nCAT itself, yields the notion of *external equivalence* between n-categories: a functor $f : \mathcal{A} \to \mathcal{B}$ of n-categories, by which we mean in the most general sense an element of $\mathrm{Mor}^1_{n\mathrm{CAT}}(\mathcal{A}, \mathcal{B})$, is said to be an *external equivalence* if it is an internal equivalence considered as a 1-morphism in nCAT.

In practice, a theory of n-categories will usually involve defining some kind of mathematically structured set or collection of sets, which naturally generates a *usual* 1-*category* of n-categories, which we can denote by $n\mathrm{C{\scriptsize AT}}$. We expect then that $\mathrm{Ob}(n\mathrm{C{\scriptsize AT}}) = \mathrm{Ob}(n\mathrm{CAT})$, but that there is a natural projection $\mathrm{Mor}^1[n\mathrm{C{\scriptsize AT}}] \to \mathrm{Mor}^1[n\mathrm{CAT}]$ compatible with composition, indeed it should come from a morphism of $(n+1)$-categories $n\mathrm{C{\scriptsize AT}} \to n\mathrm{CAT}$ (which is to say, a 1-morphism in the $(n+2)$-category $(n+1)\mathrm{CAT}$!).

However, the notion of external equivalence in $n\mathrm{C{\scriptsize AT}}$ will not generally speaking have the same characterization as in nCAT: if $f \in \mathrm{Mor}^1[n\mathrm{C{\scriptsize AT}}]$ projects to an equivalence in nCAT it means that there is $g \in \mathrm{Mor}^1[n\mathrm{CAT}]$ such that fg and gf are equivalent to identities; however, the essential inverse

g will not necessarily come from a morphism in nCAT. For precisely this reason, one of the main tasks needed to get a theory of n-categories off the ground is to give a different definition of when a usual morphism $f \in \mathrm{Mor}^1_{n\mathrm{CAT}}(\mathcal{A}, \mathcal{B})$ is an external equivalence. Of course it is to be expected and–one hopes–later proven that this standalone definition of external equivalence should become equivalent to the above definition once we have nCAT in hand.

A morphism of n-categories $f : \mathcal{A} \to \mathcal{B}$ should induce a morphism of sets $\mathrm{Ob}(\mathcal{A}) \to \mathrm{Ob}(\mathcal{B})$ (usually denoted just by $x \mapsto f(x)$), and for any $x, y \in \mathrm{Ob}(\mathcal{A})$ it should induce a morphism of $(n-1)$-categories $\underline{\mathrm{Mor}}_{\mathcal{A}}(x, y) \to \underline{\mathrm{Mor}}_{\mathcal{B}}(f(x), f(y))$. Just as was the case for composition, the morphism part of f needn't necessarily be very well defined, but it should be well defined up to an appropriate kind of equivalence.

It is now possible to state, by induction on n, the second or "standalone" definition of external equivalence. A morphism $f : \mathcal{A} \to \mathcal{B}$ is said to be *fully faithful* if for every $x, y \in \mathrm{Ob}(\mathcal{A})$ the morphism $\underline{\mathrm{Mor}}_{\mathcal{A}}(x, y) \to \underline{\mathrm{Mor}}_{\mathcal{B}}(f(x), f(y))$ is an external equivalence between $(n-1)$-categories. And f is said to be *essentially surjective* if it induces a surjection $\mathrm{Ob}(\mathcal{A}) \twoheadrightarrow \mathrm{Ob}(\mathcal{B})/\sim$. Then f is said to be an *external equivalence* if it is fully faithful and essentially surjective.

We can state the required compatibility between the two notions:

(EXEQ)–a morphism $f : \mathcal{A} \to \mathcal{B}$, an element of $\mathrm{Mor}^1_{n\mathrm{CAT}}(\mathcal{A}, \mathcal{B})$, is an external equivalence (fully faithful and essentially surjective), if and only if it is an inner equivalence in nCAT (i.e. has an essential inverse g such that fg and gf are equivalent to the identities).

Note that the fully faithful condition implies (by an inductive consideration and comparison with the truncation operation) that if $f : \mathcal{A} \to \mathcal{B}$ is an equivalence in either of the two equivalent senses, and if a, b are i-morphisms of $\mathcal{A}$ with the same source and target, then the set of inner equivalence classes of $(i+1)$-morphisms from a to b in $\mathcal{A}$, is isomorphic via f to the set of inner equivalence classes of $(i+1)$-morphisms from $f(a)$ to $f(b)$ in $\mathcal{B}$.

When developing a theory of n-categories, we therefore expect to be in the following situation: having first obtained a 1-category of n-categorical structures denoted nCAT, we obtain a notion of when a morphism f in this category, or first kind of functor, is an external equivalence. On the other hand, in the full $(n+1)$-category, f should have an essential inverse g. So, one of the main steps towards construction of nCAT is to formally invert the external equivalences. This is a typical localization problem. Furthermore, nCAT should be closed under limits and colimits, so it is very natural to use Quillen's theory of model categories, and all of the localization machinery that is now known to go along with it, as our main technical tool for going towards the construction of nCAT.

To finish this section, note one of the interesting and important features of nCAT: it is, in a certain sense, *enriched over itself*. In other words, for two objects $\mathcal{A}, \mathcal{B} \in \mathrm{Ob}(n\mathrm{CAT})$, we get an n-category of morphisms $\underline{\mathrm{Mor}}_{n\mathrm{CAT}}(\mathcal{A}, \mathcal{B})$, which, since it is an n-category (and furthermore in the same universe level as $\mathcal{A}$ and $\mathcal{B}$), is itself an object of nCAT:

$$\underline{\mathrm{Mor}}_{n\mathrm{CAT}}(\mathcal{A}, \mathcal{B}) \in \mathrm{Ob}(n\mathrm{CAT}).$$

This is the motivation for using the theory of cartesian model categories with internal $\underline{\mathrm{HOM}}$ as a preliminary model for nCAT.

3.6 Poincaré n-groupoids

An n-category is said to be an *n-groupoid* if all i-morphisms are inner equivalences. More generally, we say that $\mathcal{A}$ is *$^{>}k$-groupic* if all i-morphisms are inner equivalences for $i > k$. Lurie introduces the notation *(n, k)-category* for a $^{>}k$-groupic n-category, mostly used in the limiting case $n = \infty$.

Fundamental to Grothendieck's vision in "Pursuing stacks" was the *Poincaré n-groupoid of a space*. If X is a topological space, this is to be an n-groupoid denoted by $\Pi_n(X)$, with the following properties:

$$\mathrm{Ob}(\Pi_n(X)) = X,$$

for $0 \leq i < n$

$$\mathrm{Mor}^i[\Pi_n(X)] = C^0_{\mathrm{glob}}([0,1]^i, X),$$

where the right-hand side is the subset of maps of the i-cube into X satisfying certain *globularity conditions* (explained below), and at $i = n$ we have

$$\mathrm{Mor}^n[\Pi_n(X)] = C^0_{\mathrm{glob}}([0,1]^n, X)/\sim,$$

where $\sim$ is an equivalence relation similar to the one considered above (and indeed, it is the same in the context of $\Pi_k(X)$ for $k > n$).

The globularity condition is automatic for $i = 1$, thus the 1-morphisms in $\Pi_n(X)$ are continuous paths $p : [0,1] \to X$ with source $s_1(p) := p(0)$ and target $t_1(p) := p(1)$. In the limiting case $n = 1$, the 1-morphisms in $\Pi_1(X)$ are homotopy classes of paths with homotopies fixing the source and target, and $\Pi_1(X)$ is just the classical Poincaré groupoid of the space X.

The *globularity condition* is most easily understood in the case $i = 2$: a 2-morphism in $\Pi_n(X)$ should be a homotopy between paths, that is to say it should be a map $\psi : [0, 1]^2 \to X$ such that $\psi(0, t)$ and $\psi(1, t)$ are independent of t.

For $2 \le i \le n$, the globularity condition on a map $\psi : [0, 1]^i \to X$ says that for any $0 \le k < i$ and any $z_1, \ldots, z_{k-1} \in [0, 1]$, the functions

$$(z_{k+1}, \ldots, z_i) \mapsto \psi(z_1, \ldots, z_{k-1}, 0, z_{k+1}, \ldots, z_i)$$

and

$$(z_{k+1}, \ldots, z_i) \mapsto \psi(z_1, \ldots, z_{k-1}, 1, z_{k+1}, \ldots, z_i)$$

are constant in $(z_{k+1}, \ldots, z_n)$. The *source* and *target* of ψ are defined by

$$s_i \psi : (z_1, \ldots, z_{i-1}) \mapsto \psi(z_1, \ldots, z_{i-1}, 0),$$

$$t_i \psi : (z_1, \ldots, z_{i-1}) \mapsto \psi(z_1, \ldots, z_{i-1}, 1).$$

Grothendieck's fundamental prediction was that this globular set should have a natural structure of n-groupoid denoted $\Pi_n(X)$; that there should be a *realization construction* taking an n-groupoid G to a topological space $|G|$; that these should respect homotopy groups as was considered for the strict case in Definition 2.4.1; and that these two constructions should set up an equivalence of homotopy theories between n-truncated spaces (i.e. spaces with $\pi_i(X) = 0$ for $i > n$) and n-groupoids. This would generalize the classical correspondence between 1-groupoids and their classifying spaces which are disjoint unions of Eilenberg–MacLane $K(\pi, 1)$-spaces.

3.7 Interiors

A useful notion which should exist in a theory of n-categories is the notion of $^{>}k$*-groupic interior* denoted $\mathbf{Int}^k(\mathcal{A})$. This is the largest sub-n-category of $\mathcal{A}$ which is $^{>}k$-groupic, and we should have

$$\mathrm{Mor}^i[\mathbf{Int}^k(\mathcal{A})] = \mathrm{Mor}^i[\mathcal{A}], \quad i \le k,$$

whereas for $i > k$ the i-morphisms of the interior $\mathrm{Mor}^i[\mathbf{Int}^k(\mathcal{A})]$ should consist only of those i-morphisms of $\mathcal{A}$ which are equivalences, i.e. invertible up to equivalence. Specifying exactly the structure of $\mathbf{Int}^k(\mathcal{A})$ will depend on the particular theory of n-categories, but in any case it should be a $^{>}k$-groupic n-category, i.e. an (n, k)-category.

The usual case is for $k = 1$, and sometimes the index 1 will then be forgotten. Thus, if $\mathcal{A}$ is an n-category then we get a $^{>}1$-groupic n-category $\mathbf{Int}(\mathcal{A})$.

As will be discussed below in the next section, and more extensively in Chapter 5, the notion of $^{>}1$-groupic n-category is well modeled by the notions of simplicial category, Segal category, Rezk complete Segal space or quasicategory. Simplicial categories typically arise as Dwyer–Kan localizations $L_{DK}(\ )$ of model categories, and one feature of the model category $\mathcal{PC}^n(\mathrm{SET})$ constructed in this book is that

$$\mathbf{Int}(n\mathrm{CAT}) = L_{DK}(\mathcal{PC}^n(\mathrm{SET})),$$

see Section 22.2.3. This adds to the motivation for why it is interesting and important to use model categories as a substrate for the theory of n-categories: we get a calculatory model for an important piece of nCAT namely its interior.

3.8 The case $n = \infty$

Constructing a theory of ∞-categories in general, represents a new level of difficulty and to do this in detail would go beyond the scope of the present book. We include here a few comments about this problem, largely following Cheng's observation [73]. The example considered in her paper shows that, in an algebraic sense, any ∞-category whose i-morphisms have duals at all levels looks like an ∞-groupoid.

However, it is clear that we don't want to identify such ∞-categories with duals, and ∞-groupoids. Indeed, they occupy complementary positions in the general theory as predicted by Baez and Dolan: the first being related to quantum field theory and the second to topology. Hermida *et al.* [142] have discussed these issues.

From this paradox one can conclude that the notion of "equivalence" in an ∞-category is not merely an algebraic one. At the level of finite n, the notion of equivalence of n-categories is defined by a top-down induction, so it cannot be generalized directly to $n = \infty$.

One way around this problem would be to include the notion of equivalence in the initial structure of an ∞-category: it would be a globular set with additional structure similar to the structure used for the case of n-categories; but also with the information of a subset of the set of i-morphisms which are to be designated as "equivalences." These subsets would be required to satisfy a compatibility condition similar to (EXEQ) above.

Cheng's exemple consists, roughly speaking, of a strictly associative ∞-category $\mathcal{A}$ in which any i-morphism u has a morphism going in the other direction v and $(i+1)$-morphisms $uv \to 1$ and $vu \to 1$. Taking the point of view that invertibility should be considered as an *additional structure*, we could either declare all morphisms to be invertible, in which case v would be the inverse of u since the $(i + 1)$-morphisms going to the identities would be invertible–then we get an ∞-groupoid; or we could designate only some (or potentially none other than the identities) as equivalences, yielding an "∞-category with duals" that is more like what Baez and Dolan are looking for. Both choices would be reasonable, and would lead to *different* ∞-categories sharing the same underlying algebraic structure $\mathcal{A}$.

In view of these problems, it is tempting to take a shortcut towards consideration of certain types of ∞-categories. The shortcut is motivated by Grothendieck's Poincaré n-groupoid correspondence, which he says should also extend to an equivalence between ∞-groupoids and the homotopy theory of all CW-complexes. Turning this on its head, we can use that idea to *define* the notion of ∞-groupoid as simply being a homotopy type of a space. The iterative enrichment procedure yields the notion of n-categories when started from 0-categories being sets. If instead we start with ∞-groupoids being spaces, then iterating gives a notion of (∞, n)-categories which are $^{>}n$-groupic ∞-categories, i.e. ones whose i-morphisms are supposed and declared to be weakly invertible for all $i > n$.

At the first stage, an $(\infty, 1)$-category is therefore a category enriched over spaces. In Part II we will consider in detail many of the different current approaches to this theory. At the n-th stage, in the Segal-type theory pursued here, we obtain the notion of *Segal n-category*. The possibility of doing an iterative definition in various different cases, motivates our presentation here of a general iterative construction of the theory of categories weakly enriched over a cartesian model category. The cartesian condition, to be discussed in Section 7.7, corresponds to the fact that the composition morphism goes out from a product, so the model structure should have a good compatibility property with respect to cartesian products. When the iteration starts from the model category of sets, we get the theory of n-categories; and starting from the model category of simplicial sets leads to the theory of Segal n-categories which is one approach to (∞, n)-categories. See Bergner [48] for a discussion of other approaches and some comparisons.

Looking only at (∞, n)-categories rather than all ∞-categories is compatible with the notion of nCAT, in the sense that (∞, n)CAT, the collection of all (∞, n)-categories, is expected to have a natural structure of $(\infty, n + 1)$-category. In the theory presented here, this will be achieved by constructing a cartesian model category for Segal n-categories. The cartesian property thus shows up at the output side of the iteration, and at the input side because we need to handle products in order to talk about weak composition morphisms.

As our procedure makes clear, and as came out in my work with Hirschowitz [145] and Pellissier's thesis [211], this iteration can start with any model category–using the category $\mathcal{K}$ of simplicial sets yields the theory of Segal n-categories. Thus, one of our main goals is to construct an iteration step starting with a cartesian model category $\mathcal{M}$ and yielding a cartesian model category $\mathcal{PC}(\mathcal{M})$ representing the homotopy theory of weakly $\mathcal{M}$-enriched categories.

To go towards a theory allowing non-invertible morphisms in all dimensions, imagine a theory of weak ∞-categories in which the information of which morphisms are invertible is somehow present as suggested above. Then it becomes reasonable to define the *truncation operations* $\tau_{\leq n}$ as Tamsamani did [250], but going from weak ∞-categories to weak n-categories. Thus an ∞-category $\mathcal{A}$ would lead to a compatible system of n-categories $\tau_{\leq n}(\mathcal{A})$ for all n.

This suggests a different kind of definition: it appears that one should get the right theory by taking a homotopy limit of the theories of n-categories. Jacob Lurie had mentioned something like this in correspondence some time ago. Given the compatibility of the Rezk–Barwick theory with homotopy limits shown by Bergner [46, 47], that might be specially adapted to this task.

One might alternatively be able to view the theory of ∞-categories as some kind of first "fixed point" of the operation $\mathcal{M} \to \mathcal{PC}(\mathcal{M})$, which we will construct in the main chapters.

We will leave these considerations on a speculative level for now, hoping only that the techniques to be developed in the main part of the book will be useful in attacking the problem of ∞-categories later.

4

Operadic approaches

One of the main directions towards the theory of higher categories is comprised of a number of *operadic approches*, which are definitions of higher categories based on Peter May's notion of "operad." This contrasts with the simplicial directions, which we review in the next chapter, but which will form the basis for most of our discussion in the book. This dichotomy is not surprising, given that operads and simplicial objects are the two main ways of doing delooping machines in algebraic topology. The operadic approaches are not the main subject of this book, so our presentation will be succinct and mostly designed to inform the reader of what is out there.

Tom Leinster's book [184] has a very complete discussion of the relationship between operads and higher categories. His paper [183] gives a brief but detailed exposition of numerous different definitions of higher categories, including several in the operadic direction, and that was the first appearence in print of some definitions, such as Trimble's for example.

4.1 May's delooping machine

We start by recalling Peter May's delooping machine [203, 204]. An *operad* is a collection of sets $O(n)$, thought of as the "set of n-ary operations," together with some maps which are thought of as the result of substitutions:

$$\psi : O(m) \times O(k_1) \times \cdots \times O(k_m) \to O(k_1 + \cdots + k_m),$$

for any uplets of natural numbers $(m; k_1, \ldots, k_m)$. If we think of an element $u \in O(n)$ as representing a function $u(x_1, \ldots, x_n)$, then $\psi(u; v_1, \ldots, v_m)$ is the function of $k_1 + \cdots + k_m$ variables

$$u(v_1(x_1, \ldots, x_{k_1}), v_2(x_{k_1+1}, \ldots, x_{k_{k_1+k_2}}), \ldots,$$
$$v_m(x_{k_1+\cdots+k_{m-1}+1}, \ldots, x_{k_1+\cdots+k_m})).$$

The substitution operation ψ is required to satisfy the appropriate axioms *loc. cit.* [203].

A *topological operad* is the same, but where the $O(n)$ are topological spaces; it is said to be *contractible* if each $O(n)$ is contractible. More generally, we can consider the notion of operad in any category admitting finite products (or in fact in any symmetric monoidal category).

There is a notion of *action of an operad O on a set X*, which means an association to each $u \in O(n)$ of an actual n-ary function $X \times \cdots \times X \to X$ such that the substitution functions ψ map to function substitution as described above.

If X is a space and O is a topological operad, we can require that the action consist of continuous functions $O(n) \times X^n \to X$, or, more generally, if O is an operad in $\mathcal{M}$ then an action of O on $X \in \mathcal{M}$ is a collection of morphisms $O(n) \times X^n \to X$ satisfying the appropriate compatibility conditions.

A *delooping structure* on a space X is an action of a contractible operad on it.

The typical example is that of the *little intervals operad*: here $O(n)$ is the space of inclusions of n consecutive intervals into the given interval $[0, 1]$, and substitution is given by pasting in. A variant is used in Section 4.4 below.

4.2 Baez–Dolan's definition

In Baez and Dolan's approach, the notion of operad is first and foremost used to determine the shapes of higher-dimensional cells. They introduce a category of *opetopes* and a notion of *opetopic set*, which, like the case of simplicial sets, just means a presheaf on the category of opetopes. They then impose filler conditions. Their scheme of filler conditions is inductive on the dimension of the opetopes, but is rather intricate. We describe the category of opetopes by drawing some of the standard pictures, and then give an informal discussion of the filler conditions. In addition to Baez and Dolan's original papers [8, 11], Leinster's paper [183] was one of our main sources. Readers may also consult Cheng [72], Kock *et al.* [174] and Zawadowski [267] for other approaches to defining and calculating with opetopes.

An n-dimensional opetope should be thought of as a roughly globular n-dimensional object, with an output face which is an $(n-1)$-dimensional opetope, and an input which is a *pasting diagram* of $(n-1)$-dimensional opetopes. To paste opetopes together, match up the output faces with the different pieces of the input faces. Pasting diagrams were introduced by Johnson [154].

The only 0-dimensional opetope is a point. The only 1-dimensional opetope has as input and output a single point, so it is a single arrow:

A pasting diagram of 1-dimensional opetopes can therefore be composed of several arrows joined head to tail:

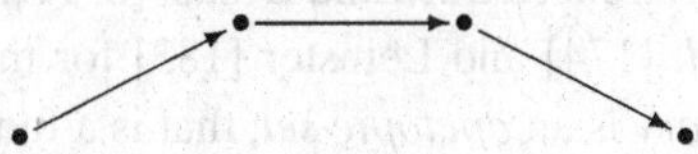

A 2-dimensional opetope can then have such a pasting diagram as input, with a single 1-dimensional opetope or arrow as output:

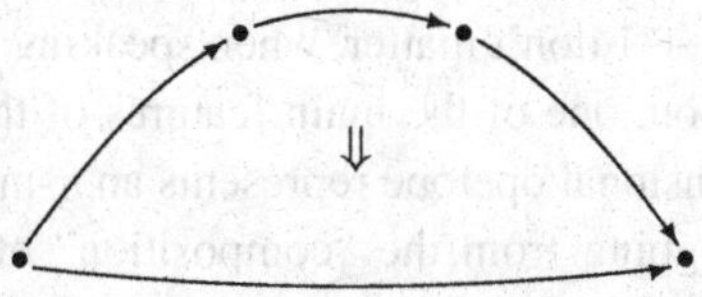

A pasting diagram of 2-dimensional opetopes would arise if we add on some other opetopes, with the output edges attached to the input edges. This can be done recursively. A picture where we add three more opetopes on the three inputs, with two, one and three input edges respectively; then a fourth one on the second input edge of the first new one, would look like this:

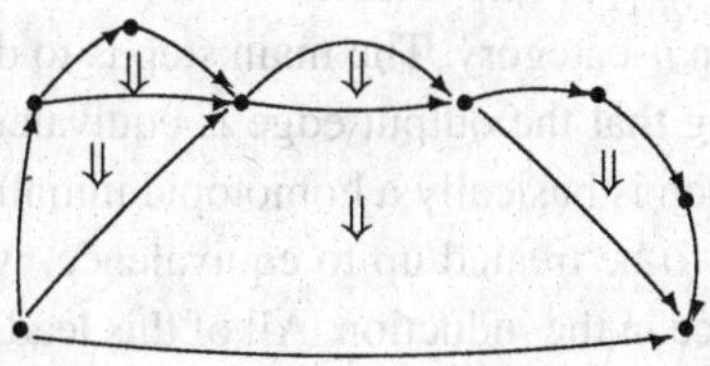

Now, a 3-dimensional opetope could have the above pasting diagram as input; in that case, the output opetope is supposed to have the shape of the boundary of the pasting diagram:

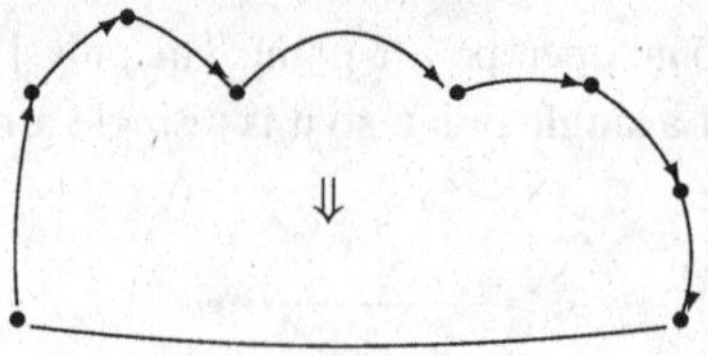

That is to say, it is a 2-dimensional opetope with seven input edges. We don't draw the 3-dimensional opetope; it should look like a "cushion."

Baez and Dolan define in this way the *category of opetopes* OTP. The reader is referred to the main references Baez and Dolan, [8, 11], Hermida *et al.* [142], Cheng [72], Kock *et al.* [174] and Leinster [183] for the precise definitions. A *Baez–Dolan n-category* is an *opetopic set*, that is a functor

$$\mathcal{A} : \mathrm{OTP}^o \to \mathrm{SET},$$

which is required to satisfy some conditions. Refer again to *loc. cit.* [8, 11, 72, 142, 183] for the precise statements of these conditions. See also Zawadowski [267] for one of the most recent treatments.

Included in the conditions are things having to do with truncation so that the cells of dimension $> n + 1$ don't matter, when speaking of an n-category.

Beyond this truncation, one of the main features of the Baez–Dolan viewpoint is that an n-dimensional opetope represents an n-morphism which is not necessarily invertible, going from the "composition" of the opetopes in the input face, to that of the output face. In particular, the output face of an opetope is not necessarily the same as the composition of the input faces; this distinguishes their setup from the Segal-style picture we are mostly considering in the present book, in which a simplex represents a composition with the outer edge being equivalent to the composition of the principal edges.

The idea that opetopes represent arbitrary morphisms from the composition of the inputs, to the output, makes it somewhat complicated to collect together and write down all of the appropriate conditions which an opetopic set should satisfy in order to be an n-category. The main step is to designate certain cells as "universal," meaning that the output edge is equivalent to the composition of the input edges, which is basically a homotopic initiality property. The necessary uniqueness has to be treated up to equivalence, whence the need for a definition of equivalence in the induction. All of this leads to Baez and Dolan's notions of *niches*, *balanced cells* and so forth.

As a rough approximation, we can say that these conditions are a form of horn-filler conditions, generalizing the restricted Kan condition of Boardman and Vogt [50] used in Joyal's definition of quasicategory [157] but adapted to the opetopic context.

To close out this section, here are a few thoughts on the possible relationship between this theory and the many other theories of n-categories. For one thing, the fact that the opetopic cells represent explicitly the n-morphisms, which are not necessarily invertible, would seem to render this theory particularly well adapted to looking at things like lax functors (what Bénabou [34] calls "morphisms" as opposed to "homomorphisms").

On the other hand, for a comparison with other theories, it would be interesting to investigate functors $\mathbf{F} : \mathrm{OTP} \to \mathcal{P}$ where $\mathcal{P}$ is a model category of "n-categories" (for example, the $\mathcal{P} = \mathcal{PC}^n(\mathrm{SET})$ which we are going to be constructing in the rest of the book). Given such a functor, if $\mathcal{B} \in \mathcal{P}$ is a fibrant object, we would obtain an opetopic set $\mathbf{F}^*(\mathcal{B})$, the functor sending an opetope O to $\mathrm{HOM}(\mathbf{F}(O), \mathcal{B})$; and conversely, taking the left adjoint of $\mathbf{F}^*$, given an opetopic set $\mathcal{A}$ we could construct its realization $\mathbf{F}_!(\mathcal{A})$ in $\mathcal{P}$. Under the right hypotheses, one hopes, these should set up a Quillen adjunction between a model category of opetopic sets, and the other model category $\mathcal{P}$.

If both of the above remarks could be realized, it would lead to the introduction of powerful new techniques for treating lax functors in any of the other theories of n-categories.

4.3 Batanin's definition

Batanin's definition is certainly the closest to Grothendieck's original vision. Recall the passage that we have quoted from "Pursuing stacks" on page 9 in Chapter 1 above, saying that whenever two morphisms which are naturally obtained as some kind of composition have the same source and target, there should be a homotopy between them at one level up. Batanin's definition puts this into place, by carefully studying the notion of possible composition of arrows in a higher category. It was recently pointed out by Maltsiniotis [199] that Grothendieck had in fact given a definition of higher groupoid and [200] that a somewhat minor modification of that approach yields a definition of higher category which is very close to Batanin's. Ara explains a version of this approach in his thesis [5].

Tom Leinster [183, 184] and Eugenia Cheng [74] have refined Batanin's original work, and our discussion will be informed by their expositions, to which the reader should refer for more details. Leinster has also introduced some related definitions of weak n-category based on the notion of multicategory, again for this we refer to his book [184].

One of Batanin's innovations was to introduce a notion of operad adapted to higher categories, based on the notion of *globular set*, a presheaf on the category $\mathbb{G}$ which has objects $\mathbf{g}_i$ for each i, and maps $s^!, t^! : \mathbf{g}_i \to \mathbf{g}_{i+1}$. The objects $\mathbf{g}_i$ are supposed to represent globular pictures of i-morphisms, for example $\mathbf{g}_2$ may be pictured as

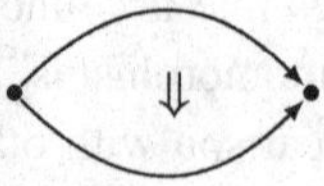

and the maps $s^!, t^!$ are viewed as inclusions of smaller-dimensional globules on the boundary. Thus, the inclusion maps are subject to the relations

$$s^! s^! = t^! s^!, \quad t^! s^! = t^! t^!.$$

Dually, a globular set is a collection of sets $\mathcal{A}_i := \mathcal{A}(\mathbf{g}_i)$ with source and target maps

$$\cdots \mathcal{A}_i \underset{t}{\overset{s}{\rightrightarrows}} \mathcal{A}_{i-1} \cdots \mathcal{A}_1 \underset{t}{\overset{s}{\rightrightarrows}} \mathcal{A}_0,$$

subject to the relations $s \circ s = s \circ t$ and $s \circ t = t \circ t$.

This definition differs slightly from the one which was suggested in Chapter 2; there we looked at what should most precisely be called a *reflexive globular set* having degeneracy maps going back in the other direction $i : \mathcal{A}_i \to \mathcal{A}_{i+1}$. For the present purposes, we consider globular sets without degeneracy maps.

Batanin's idea is to use the notion of globular set to generate the appropriate kind of "collection" for a notion of higher operad. One can think of an operad as specifying a collection of operations of each possible arity; and an arity is a possible configuration of the collection of inputs. For Batanin's globular operads, an input configuration is represented by a *globular pasting diagram* P, and the family of operations of arity P should itself form a globular set. A globular pasting diagram in dimension n is an n-cell in the free strict ∞-category generated by a single non-identity cell in each dimension.

Batanin gives an explicit description of the n-cells in this free ∞-category, using planar trees. The nodes of the trees come from the source and target maps between stages in a globular set, which is not to be confused with the occurence of trees in parametrizations of elements of a free nonassociative algebra (where the nodes correspond to parenthetizations). For the free ∞-category involved here, it is strictly associative and going up a level in the tree corresponds rather to going from i-morphisms to $(i + 1)$-morphisms.

As Cheng describes in a detailed example [74], a globular pasting diagram is really just a picture of how one might compose together various i-morphisms.

Given any globular set $\mathcal{A}$, and a globular pasting diagram P, we get a set denoted $\mathcal{A}_P$ consisting of all the ways of filling in P with labels from the globular set $\mathcal{A}$, consistent with sources and targets.

Now, a globular operad $\mathcal{B}$ consists of specifying, for each globular pasting diagram P, a globular set $\mathcal{B}(P)$ of operations of arity P. This globular set consists in particular of sets $\mathcal{B}(P)_n$ for each n called the *set of P-ary operations of level n*, together with sources and targets

$$\mathcal{B}(P)_{n+1} \overset{s}{\underset{t}{\rightrightarrows}} \mathcal{B}(P)_n$$

satisfying the globularity relations. An *action* of $\mathcal{B}$ on a globular set $\mathcal{A}$ consists of specifying for each globular pasting diagram P of degree n, a map of sets

$$\mathcal{B}(P)_n \times \mathcal{A}_P \to \mathcal{A}_n$$

such that the source diagram

$$\begin{array}{ccc} \mathcal{B}(P)_n \times \mathcal{A}_P & \longrightarrow & \mathcal{A}_n \\ \Big\downarrow s & & \Big\downarrow s \\ \mathcal{B}(sP)_{n-1} \times \mathcal{A}_{sP} & \longrightarrow & \mathcal{A}_{n-1} \end{array}$$

commutes and similarly for the target diagram. Here sP and tP are the source and target of the pasting diagram P, which are pasting diagrams of degree $(n-1)$. Of course we need to describe an additional operadic structure on $\mathcal{B}$ and the action should be compatible with this too, but it is easier to first consider what data an algebra should have.

To describe what an operadic structure should mean, notice that there is an operation of substitution of globular pasting diagrams, indeed the free ∞-category on one cell in each dimension can be denoted $\mathcal{GP}$, and if $P \in \mathcal{GP}_m$ is a globular pasting diagram in degree m then we get a map of globular sets

$$\mathcal{GP}_P \to \mathcal{GP},$$

that is to say that given a labeling of the cells of P where the labels L_j are themselves globular pasting diagrams, we can substitute the labels into the cells and obtain a big resulting globular pasting diagram.

Now the operadic structure, which is in addition to the structure of $\mathcal{B}$ described above, should say that for any element of $\mathcal{GP}_P$ consisting of labels denoted $L_j \in \mathcal{GP}_{n_j}$ for the cells of P and yielding an output pasting diagram S, if we are given elements of the $\mathcal{B}(L_j)$ plus an element of $\mathcal{B}(P)$ then there should be a big output element of $\mathcal{B}(S)$. This needs to be compatible with source and target operations, as well as compatible with iteration in the style of usual operads.

Batanin describes explicitly the combinatorics of this using the identification between pasting diagrams and trees, whereas Leinster takes a more abstract approach using monads and multicategories. We refer the reader to the references for further details.

The main point is that this discussion establishes a language in which to say that the system of coherencies should satisfy a globular contractibility condition. Observe that contractibility is a very easily defined property of a globular set, and it doesn't depend on any kind of composition law: a globular set $\mathcal{A}$ is contractible if it is nonempty, and if for any two $f, g \in \mathcal{A}_m$ with $s(f) = s(g)$ and $t(f) = t(g)$, there exists an $h \in \mathcal{A}_{m+1}$ with $s(h) = f$ and $t(h) = g$.

Note that this definition is really only appropriate if we are working with globular sets which are truncated at some level (i.e. trivial above a certain degree n), or ones which are supposed to represent (∞, n)-categories, that is ones in which all i-morphisms are declared invertible for $i > n$. For the purposes of Batanin's definition of n-categories, this is the case.

In the case of a globular operad $\mathcal{B}$, the globular sets vary as a function of the pasting diagram P. So we can adapt the notion of contractibility, following Leinster [184, chapter 9]. Suppose given a pasting diagram P and an element $u \in \mathcal{B}(P)_m$. Then $s(u) \in \mathcal{B}(sP)_{m-1}$ and $t(u) \in \mathcal{B}(tP)_{m-1}$. Say that $\mathcal{B}$ is *contractible* if, for any pasting diagram P and any $v \in \mathcal{B}(sP)_{m-1}$ and $w \in \mathcal{B}(tP)_{m-1}$ such that $s(v) = s(w)$ and $t(v) = t(w)$ (i.e. v and w are "parallel"), there exists $u \in \mathcal{B}(P)_m$ with $s(u) = v$ and $t(u) = w$. A *contraction* is a choice of such u for all parallel (v, w). See *loc. cit.* [184] for what to say when $m \geq n$.

As Leinster points out [184, 10.1.6], Batanin originally used a simpler condition of contractibility which Leinster calls *coherence*, but then included also a condition of the existence of a system of compositions, and in fact he generally means both when he says "contractible." The conjunction of coherence and existence of a system of compositions is probably essentially the same as Leinster's contractibility condition; this distinction creates an *a priori* difference between Batanin's original definition and Leinster's version. Refer to *loc. cit.* [184] for a more detailed discussion.

Now, a Batanin n-category is a globular set $\mathcal{A}$ provided with an action of a globular operad $\mathcal{B}$, such that $\mathcal{B}$ is contractible (coherent with a system of compositions). Batanin constructs a *universal* contractible globular operad, but it can also be convenient to work with other contractible globular operads, as in Cheng's comparison with Trimble's work which we discuss next.

The elements of $\mathcal{B}(P)_n$ are the "natural operations taking a collection of morphisms of various degrees, and combining together to get an n-morphism." Contractibility of $\mathcal{B}$ really puts into effect Grothendieck's dictum that, given two natural operations f and g with the same source and target, there should be an operation h one level higher whose source is f and whose target is g. So, Batanin's definition is the closest to what Grothendieck was asking for.

4.4 Trimble's definition and Cheng's comparison

Trimble's definition of higher category has acquired a central role because of Cheng's recent work [74] comparing it to Batanin's definition. This has also been taken up by Batanin *et al.* [30]. Trimble's framework has the advantage that it is iterative. In the future it should be possible to establish a comparison with the iterative Segal approach we are discussing in the rest of the book. Such a comparison result would be very interesting, but we don't discuss it here. Instead we just give the basic outlines of Trimble's approach and state Cheng's comparison theorem. We are following very closely her article [74].

Consider the following operad O^T in TOP:

$$O^T(n) \subset C^0([0,1],[0,n])$$

is the subset of endpoint-preserving continuous maps. The operad structure ψ^T is given by

$$\psi(f; g_1, \ldots, g_m)(t) := g_j(f(t) - (j-1)), \quad f(t) \in [j-1, j] \subset [0, m].$$

The spaces $O^T(n)$ are contractible. This topological operad is particularly adapted to loop spaces and path spaces. If $Z \in$ TOP and $x, y \in Z$, let $\mathrm{Path}^{x,y}(Z)$ denote the space of paths $\gamma : [0,1] \to Z$ with $\gamma(0) = x$ and $\gamma(1) = y$. For any sequence of points $x_0, \ldots, x_m \in X$ we have a "substitution" map

$$O^T(m) \times \mathrm{Path}^{x_0,x_1}(Z) \times \cdots \times \mathrm{Path}^{x_{n-1},x_n}(Z),$$

and these are compatible with the operad structure. In particular, when the points are all the same, this gives an action of O^T on the loop space $\Omega^x(Z)$.

A *Trimble topological category* consists of a set X "of objects," together with a collection of spaces $\mathcal{A}(x, y)$ for any $x, y \in X$, and collection of maps

$$\phi_{x.} : O^T(n) \times \mathcal{A}(x_0, x_1) \times \cdots \times \mathcal{A}(x_{n-1}, x_n) \to \mathcal{A}(x_0, x_n)$$

for any sequence $(x_0, \ldots, x_n)$ in X. These should satisfy a compatibility condition with the operad structure ψ^T for O^T: given a sequence $x_0, \ldots, x_m$ and sequences $y^i_0, \ldots, y^i_{k_i}$ with $y^i_0 = x_{i-1}$ and $y^i_{k_i} = x_i$,

$$\phi\left(\psi^T(f; g_1, \ldots, g_m); u^1_1, \ldots, u^1_{k_1}, \ldots, u^m_1, \ldots, u^m_{k_m}\right)$$
$$= \phi\left(f; \phi\left(g_1; u^1_1, \ldots, u^1_{k_1}\right), \ldots, \phi\left(g_1 m; u^m_1, \ldots, u^m_{k_m}\right)\right).$$

Similarly, for any category $\mathcal{M}$ admitting finite products, and any operad (O, ψ) in $\mathcal{M}$, we can define a notion of $(\mathcal{M}, O, \psi)$-category; this consists of a set X of objects, together with a collection of $\mathcal{A}(x, y) \in \mathcal{M}$ for any $x, y \in X$, and a collection of maps ϕ as above satisfying the same compatibility condition.

Suppose we are given a category $\mathcal{M}$ with finite products, and a functor $\Pi : \text{TOP} \to \mathcal{M}$, then we obtain an operad $\Pi(O^T)$ in $\mathcal{M}$. We get the notion of $(\mathcal{M}, \Pi(O^T), \Pi(\psi^T))$-category. If contractibility makes sense in $\mathcal{M}$ and $\Pi(O^T(n))$ is contractible, then this is a generalization due to Cheng [74], of Trimble's notion of higher category enriched in $\mathcal{M}$.

Trimble's original definition, which first appeared publicly in Leinster's compendium [183], included an inductive construction of the Poincaré n-groupoid functor Π_n. He defines inductively a sequence of categories which Cheng denotes by $\mathcal{V}_n$, starting with $\mathcal{V}_0 = \text{SET}$; together with product-compatible Poincaré n-groupoid functors $\Pi_n : \text{TOP} \to \mathcal{V}_n$. The inductive definition is that $\mathcal{V}_{n+1}$ is the category of $(\mathcal{V}_n, \Pi_n(O^T), \Pi_n(\psi^T))$-categories, and

$$\Pi_{n+1}(Z) = (X, \mathcal{A}, \phi),$$

where $X := Z^{\mathbf{disc}}$ is the discrete set of points of Z, where $\mathcal{A}$ is defined by $\mathcal{A}(x, y) := \Pi_n(\text{Path}^{x,y}(Z))$, and ϕ is defined using the action described above of O^T on the path spaces. See Cheng [74] for further details, as well as for the generalization to the case where an arbitrary contractible operad P_n replaces $\Pi_n(O^T)$.

Cheng goes on to compare this family of definitions, with Batanin's definition: she shows how to combine the $P_0, P_1, \ldots, P_{n-1}$ together to form a contractible globular operad $Q^{(n)}$ such that the category of globular algebras of $Q^{(n)}$ is $\mathcal{V}_n$. This expresses $\mathcal{V}_n$ as a category of Batanin n-categories for this particular choice of contractible globular operad. The reader is refered to *loc. cit.* [74], as well as to Berger [36], Batanin *et al.* [30] and Cisinski [78] for related aspects of this kind of comparison.

4.5 Weak units

In the course of investigating the nonrealization of homotopy 3-types by strict 3-groupoids, the main obstruction seemed to be the strict unit condition in a strict n-category. This is one of the main aspects which allows the Eckmann–Hilton argument to work. That was explained to me by Georges Maltsiniotis and Alain Bruguières, but has of course been well known for a long time. This led to the conjecture [238] that maybe it would be sufficient to keep the strict associativity of composition, but to weaken the unit condition.

Recall from homotopy theory (cf. Lewis [187]) the yoga that it suffices to weaken any one of the principal structures involved. Most weak notions of n-category involve a weakening of the associativity, or eventually of the Godement interchange conditions.

O. Leroy [186] and apparently, independently, Joyal and Tierney [161] were the first to do this in the context of 3-types. See also Gordon *et al.* [123] and Berger [35] for weak 3-categories and 3-types. Baues [32] showed that 3-types correspond to *quadratic modules* (a generalization of the notion of crossed complex). Then come the models for weak higher categories which we are considering in the rest of the book.

It seems likely that the Kapranov–Voevodsky arguments [168] would show that one could instead weaken the condition of being *unital*, that is having identities, and keep associativity and Godement. We give a proposed definition of what this would mean and then state two conjectures.

This can be motivated by looking at the *Moore loop space* $\Omega^x_M(X)$ of a space X based at $x \in X$, cited by Kapranov and Voevodsky [168] as a motivation for their construction. Recall that $\Omega^x_M(X)$ is the space of *pairs* (r, γ) where r is a real number $r \geq 0$ and $\gamma = [0, r] \to X$ is a path starting and ending at x. This has the advantage of being a strictly associative monoid. On the other side of the coin, the "length" function

$$\ell : \Omega^x_M(X) \to [0, \infty) \subset \mathbb{R}$$

has a special behavior over $r = 0$. Note that over the open half-line $(0, \infty)$ the length function ℓ is a fibration (even a fiber-space) with fiber homeomorphic to the usual loop space. However, the fiber over $r = 0$ consists of a single point, the constant path $[0, 0] \to X$ based at x. This additional point (which is the unit element of the monoid $\Omega^x_M(X)$) doesn't affect the topology of Ω^x_M (at least if X is locally contractible at x) because it is glued in as a limit of paths which are more and more concentrated in a neighborhood of x. However, the map ℓ is no longer a fibration over a neighborhood of $r = 0$. This is a bit of a problem because Ω^x_M is not compatible with cartesian products of the space

X; in order to obtain a compatibility one has to take the fiber product over $\mathbb{R}$ via the length function:

$$\Omega_M^{(x,y)}(X \times Y) = \Omega_M^x(X) \times_{\mathbb{R}} \Omega_M^y(Y),$$

and the fact that ℓ is not a fibration could end up causing a problem in an attempt to iteratively apply a construction like the Moore loop-space.

Things seem to get better if we restrict to

$$\Omega_{M'}^x(X) := \ell^{-1}((0, \infty)) \subset \Omega_M^x(X),$$

but this associative monoid no longer has a strict unit. Even so, the constant path of any positive length gives a weak unit.

A motivation coming from a different direction was an observation made by Tamsamani early in the course of his thesis work. He was trying to define a strict 3-category 2CAT whose objects would be the strict 2-categories and whose morphisms would be the weak 2-functors between 2-categories (plus notions of weak natural transformations and 2-natural transformations). At some point he came to the conclusion that one could adequately define 2CAT as a strict 3-category except that he couldn't get strict identities. Because of this problem we abandonned the idea and looked toward weakly associative n-categories. In retrospect it would be interesting to pursue Tamsamani's construction of a strict 2CAT, but with only weak identities.

In my preprint [238] I introduced a preliminary definition of *weakly unital strict n-category* (called "snucategory" there), including a notion of cartesian product. The proposed definition went as follows. Suppose we know what these are for $(n-1)$. Then a weakly unital strict n-category $\mathcal{C}$ consists of a set $\mathrm{Ob}(\mathcal{C})$ of objects together with, for every pair of objects $x, y \in \mathrm{Ob}(\mathcal{C})$ a weakly unital strict $(n-1)$-category $\mathcal{C}(x, y)$ and composition morphisms

$$\mathcal{C}(x, y) \times \mathcal{C}(y, z) \to \mathcal{C}(x, z),$$

which are strictly associative, such that a *weak unital condition* holds. We now explain this condition. An element $e_x \in \mathcal{C}(x, x)$ is called a *weak identity* if:

- composition with e induces equivalences of weakly unital strict $(n-1)$-categories

$$\mathcal{C}(x, y) \to \mathcal{C}(x, y), \quad \mathcal{C}(y, x) \to \mathcal{C}(y, x);$$

- and if $e \cdot e$ is equivalent to e. This condition can furthermore be completed to the fuller collection of coherence conditions introduced by Kock [172].

In order to complete the recursive definition we must define the notion of when a morphism of weakly unital strict n-categories is an equivalence, and we must define what it means for two objects to be equivalent. A morphism is said to be an equivalence if the induced morphisms on the $\mathcal{C}(x, y)$ are equivalences of weakly unital strict $(n-1)$-categories and if it is essentially surjective on objects: each object in the target is equivalent to the image of an object. It thus remains just to be seen what internal equivalence of objects means. For this we introduce the *truncations* $\tau_{\leq i}\mathcal{C}$ of a weakly unital strict n-category $\mathcal{C}$. Again this is done in the same way as usual: $\tau_{\leq i}\mathcal{C}$ is the weakly unital strict i-category with the same objects as $\mathcal{C}$ and whose morphism $(i-1)$-categories are the truncations

$$\text{Hom}_{\tau_{\leq i}\mathcal{C}}(x, y) := \tau_{\leq i-1}\mathcal{C}(x, y).$$

This works for $i \geq 1$ by recurrence, and for $i = 0$ we define the truncation to be the set of isomorphism classes in $\tau_{\leq 1}\mathcal{C}$. Note that truncation is compatible with cartesian product (cartesian products are defined in the obvious way) and takes equivalences to equivalences. These statements used recursively allow us to show that the truncations themselves satisfy the weak unary condition. Finally, we say that two objects are equivalent if they map to the same thing in $\tau_{\leq 0}\mathcal{C}$.

Proceeding in the same way as in Chapter 2, we can define the notion of weakly unital strict n-groupoid:

Conjecture 4.5.1 *There are functors Π_n and $\Re$ between the categories of weakly unital strict n-groupoids and n-truncated spaces (going in the usual directions) together with adjunction morphisms inducing an equivalence between the localization of weakly unital strict n-groupoids by equivalences, and n-truncated spaces by weak equivalences.*

Joachim Kock [172, 173] has developed the full collection of higher coherence relations entering into the notion of weakly unitary strict n-category, beyond just asking that $e \sim e \cdot e$. We refer the reader there for his definition, using a "fat" version of Δ, which, for the problem of fully understanding the weak unit condition, clearly supersedes the preliminary version described above. One can nonetheless ask whether these higher coherence conditions need to be assumed, or whether they might be natural consequences of the very weak condition $e \sim e \cdot e$. Some hints may be found in Joyal and Kock [160]. Working with strictly associative weakly unitary monoidal strict 2-categories, they show that given I such that $I \otimes -$ and $- \otimes I$ are equivalences, and $\alpha : I \cdot I \sim I$, there is a canonical associator 2-cell which automatically satisfies the higher coherence relations such as the pentagon axiom; the space of such weak units is contractible, they are compatible with morphisms, and this data corresponds to what can be distilled from Gordon *et al.* [123]. These results in the 3-dimensional case suggest that investigating the precise role of Kock's higher coherence conditions is an interesting question requiring further study.

In other recent work on this kind of question, Hirsh and Millès [146] discuss homotopy coherence for weak units and its relationship with Koszul duality in the setting of A_∞-algebras.

Joyal and Kock [159] have proven Conjecture 4.5.1 for the 1-connected case when $n = 3$. For general n, one could hope to apply the argument of Kapranov and Voevodsky [168].

These results concern the case of groupoids; however, we might also expect that weakly unital strict n-categories serve to model all weak n-categories:

Conjecture 4.5.2 *The localization of the category of weakly unital strict n-categories by equivalences is equivalent to the localizations of the categories of weak n-categories of Tamsamani and/or Baez–Dolan and/or Batanin by equivalences.*

While we're discussing the subject of unitality conditions, the following remark is in order. The role of strict unitality conditions in the interchange rule or Godement relations, and the consequent nonrealization of homotopy types with nontrivial Whitehead bracket, suggests that we need to take some care about this point in the general argument which will be developed in Parts III and IV. It turns out that, in order to insure a good cartesian property, our Segal-style weakly enriched categories should nonetheless be endowed with strict units in a certain sense. These correspond to the degeneracies in the category of simplices Δ, and are important for the Eilenberg–Zilber argument, which yields the cartesian property. They don't correspond to full strict units in the maximal possible way, because the composition operation will not even be well defined; that is why we will be able to impose the unitality condition in Part III, without running up against the problems identified in the second half of Chapter 2.

4.6 Other notions

The theory of n-categories is an essentially *globular* theory. This means that an i-morphism has a single $(i-1)$-morphism as source, and a single one as target. This basic shape can be relaxed in many ways. For example, Leinster and others have investigated notions of *multicategory* where the input is a collection of objects rather than just one object. This is somewhat related to the opetopic shapes introduced by Baez and Dolan.

Another way of relaxing the globular shape is to iterate the internal category construction. Brown and Loday constructed the first algebraic representation for homotopy n-types, with the notion of Cat^n*-group*. Let $Cat^1(\text{GP})$ denote the category of internal categories in the category GP, then $Cat^{n+1}(\text{GP})$ is the category of internal categories in $Cat^n(\text{GP})$. There is a natural realization functor from $Cat^n(\text{GP})$ to homotopy n-types, and Brown and Loday prove that all homotopy n-types are realized.

Paoli [210] has recently refined this model to go back in the globular direction, by introducing a notion of *special* Cat^n*-group*. If we think of an internal category as being a pair of objects connected by several morphisms, then an internal n-fold category may be seen as a collection of objects arranged at the vertices of an n-cube. The speciality condition requires that certain faces of the cube be contractible.

The speciality condition is a sort of weak globularity condition. An internal n-fold category is not in and of itself a globular object, because the object of objects may be non-trivial. A strict globular condition would have the object of objects be a discrete set; the speciality condition requires only that it be a disjoint union of contractible Cat^{n-1}-groups.

Paoli shows that the special Cat^n-groups model homotopy n-types, and she relates this model to Tamsamani's model. This yields a semistrictification result saying that we can have a strictly associative composition at any one stage of Tamsamani's model. An alternative proof of this result for semistrictification at the last stage, may be obtained using the fact that Segal categories are equivalent to strict simplicial categories.

Paoli's model relates to Lewis's principle [187] cited above, in an interesting way: in a Cat^n-group all the structures–associativity, units, inverses, interchange–are strict; the special Cat^n-groups weaken instead the globularity condition itself.

Several authors have considered the *Gray tensor product* and its applications to definitions of higher categories. See, for example, Leroy [186], Crans [86], Berger [35], Gurski [134], Batanin *et al.* [30] and Lack [178].

Penon's definition [212] is completely algebraic in the sense that a weak n-category is an algebra over a monad. For a rapid description of the monad, one can refer to Leinster [183, pp 14–17], see also Batanin [28], Cheng and Lauda [75], Cheng and Makkai [76] and Futia [116]. Penon [212] introduces a category $\mathcal{Q}$ consisting of arrows of "ω-magmas" $M \to S$, where S is a strict ω-category, together with a contractive structure on π. The monad is adjoint to the functor "underlying globular set of $\mathcal{A}^\sharp$." If $\mathcal{A}$ is a globular set then it goes to an element of $\mathcal{Q}$ with S being the free strict ω-category generated by $\mathcal{A}$. Note that this free category is the set of globular pasting diagrams as in Section 4.3, and M may be viewed as some sort of family of elements over S with a contractible structure. In this sense, Penon's definition uses objects of the same sort as Batanin's definition – indeed Batanin [28] has made a more precise comparison. Penon's definition uses globular sets with identities ("reflexive globular sets") whereas Batanin's were without them, so Batanin [28] proposes a modified version of Penon's definition with nonreflexive globular sets. Cheng and Makkai [76] have pointed out that it is better to use the nonreflexive version, since the reflexive version doesn't lead to all the objects one would want, essentially because of the Eckmann–Hilton argument. Futia [116] proposes a generalized family of Penon-style definitions. In these definitions, one could say that globally the goal is to be able to parametrize higher compositions sorted according to their shapes which are globular pasting diagrams.

5
Simplicial approaches

There are a number of approaches to weak higher categories based on the category of simplices Δ, including the Segal approach and its iterations, which are the main subject of our book. We also discuss several other approaches, which concern first and foremost the theory of $(\infty, 1)$-categories.

5.1 Strict simplicial categories

A *simplicial category* is a $\mathcal{K}$-enriched category. It has a set of objects $\mathrm{Ob}(\mathcal{A})$, and for each pair $x, y \in \mathrm{Ob}(\mathcal{A})$ a simplicial set $\mathcal{A}(x, y)$ thought of as the "space of morphisms" from x to y. The composition maps are morphisms of simplicial sets $\mathcal{A}(x, y) \times \mathcal{A}(y, z) \to \mathcal{A}(x, z)$ satisfying the associativity condition strictly, that is for any x, y, z, w the diagram of simplicial sets

$$\begin{array}{ccc} \mathcal{A}(x, y) \times \mathcal{A}(y, z) \times \mathcal{A}(z, w) & \to & \mathcal{A}(x, y) \times \mathcal{A}(y, w) \\ \downarrow & & \downarrow \\ \mathcal{A}(x, z) \times \mathcal{A}(z, w) & \longrightarrow & \mathcal{A}(x, w) \end{array}$$

commutes. The identities of $\mathcal{A}$ are points (i.e. vertices) of the simplicial sets $1_x \in \mathcal{A}(x, x)_0$ satisfying left and right identity conditions which are equalities of maps. Note that 1_x provides a map of simplicial sets $* \to \mathcal{A}(x, x)$ and we require that the compositions

$$* \times \mathcal{A}(x, y) \to \mathcal{A}(x, x) \times \mathcal{A}(x, y) \to \mathcal{A}(x, y),$$

$$\mathcal{A}(x, y) \times * \to \mathcal{A}(x, y) \times \mathcal{A}(y, y) \to \mathcal{A}(x, y)$$

should both be equal to the identity map of $\mathcal{A}(x, y)$.

A *functor* of simplicial categories $f : \mathcal{A} \to \mathcal{B}$ consists of a map $f : Ob(\mathcal{A}) \to \mathrm{Ob}(\mathcal{B})$ and for each $x, y \in \mathrm{Ob}(\mathcal{A})$, a map of simplicial sets $f_{x,y} : \mathcal{A}(x, y) \to \mathcal{B}(f(x), f(y))$, compatible with the composition maps and identities in an obvious way. In keeping with our general notation for enriched categories, the category of simplicial categories is denoted $\mathrm{CAT}(\mathcal{K})$.

Given a simplicial category $\mathcal{A}$, we define its *truncation* $\tau_{\leq 1}(\mathcal{A})$ to be the category whose set of objects is the same as $\mathrm{Ob}(\mathcal{A})$, but for any $x, y \in \mathrm{Ob}(\mathcal{A})$

$$\tau_{\leq 1}(\mathcal{A})(x, y) := \pi_0(\mathcal{A}(x, y)).$$

The composition maps and identities for $\mathcal{A}$ define composition maps and identities for $\tau_{\leq 1}(\mathcal{A})$, and we obtain a functor

$$\tau_{\leq 1} : \mathrm{CAT}(\mathcal{K}) \to \mathrm{CAT}.$$

A functor $f : \mathcal{A} \to \mathcal{B}$ between simplicial categories is said to be *fully faithful* if for every $x, y \in \mathrm{Ob}(\mathcal{A})$ the map $f_{x,y} : \mathcal{A}(x, y) \to \mathcal{B}(f(x), f(y))$ is a weak equivalence of simplicial sets, in other words a weak equivalence in the standard model structure of $\mathcal{K}$. A functor f is said to be *essentially surjective* if the functor $\tau_{\leq 1}(f)$ between usual 1-categories is essentially surjective, in other words it induces a surjection on sets of isomorphism classes $\mathbf{Iso}\tau_{\leq 1}(\mathcal{A}) \twoheadrightarrow \mathbf{Iso}\tau_{\leq 1}(\mathcal{B})$. A functor $f : \mathcal{A} \to \mathcal{B}$ is said to be a *Dwyer–Kan equivalence* between simplicial categories, if it is fully faithful and essentially surjective. In this case, $\tau_{\leq 1}(f)$ is also an equivalence of categories, in particular it is bijective on sets of isomorphism classes.

Given a simplicial category $\mathcal{A}$, its *underlying category* is the category with objects $\mathrm{Ob}(\mathcal{A})$, but the morphisms from x to y are the set of points or vertices of the simplicial set $\mathcal{A}(x, y)$. This is not to be confused with $\tau_{\leq 1}(\mathcal{A})$, but there is a natural projection functor from the underlying category to the truncated category. An "arrow" in $\mathcal{A}$ from x to y means a map in this underlying category; such an arrow is said to be an *internal equivalence* if it projects to an isomorphism in $\tau_{\leq 1}(\mathcal{A})$. In these terms, the essential surjectivity condition for a functor $f : \mathcal{A} \to \mathcal{B}$ may be rephrased as saying that any object of $\mathcal{B}$ is internally equivalent to the image of an object of $\mathcal{A}$.

Dwyer *et al.* [99] and Bergner [39] have constructed a model category structure on $\mathrm{CAT}(\mathcal{K})$ such that the weak equivalences are the Dwyer–Kan equivalences, and the fibrations are the functors $f : \mathcal{A} \to \mathcal{B}$ such that each $f_{x,y}$ is a fibration of simplicial sets, and furthermore f satisfies an additional lifting condition which basically says that an internal equivalence in $\mathcal{B}$ should lift to $\mathcal{A}$ if one of its endpoints lifts.

It is interesting to note that Dwyer and Kan [101, 102, 103] started first by constructing a model structure on $\mathrm{CAT}(X, \mathcal{K})$, the category of simplicial categories with a fixed set of objects X. We will also adopt this route, following a suggestion by Clark Barwick.

Simplicial categories appear in an important way in homotopy theory. Quillen defined the notion of *simplicial model category*, and if $\mathcal{N}$ is a simplicial model category then we obtain a simplicial category $\mathcal{N}_{cf}^{\mathrm{spl}}$ of fibrant and cofibrant objects, such that its truncation $\tau_{\leq 1}\left(\mathcal{N}_{cf}^{\mathrm{spl}}\right) \cong \mathrm{ho}(\mathcal{N})$ is the homotopy category of $\mathcal{N}$. Dwyer and Kan then developed the theory of *simplicial localization*, which gives a good simplicial category even when $\mathcal{N}$ doesn't have a simplicial model structure. If $\mathcal{C}$ is any category, and if we are given a subcollection of arrows $\Sigma \subset \mathrm{ARR}(\mathcal{C})$, then Dwyer and Kan define a simplicial category $L(\mathcal{C}, \Sigma)$ whose truncation is the classical Gabriel–Zisman localization: $\tau_{\leq 1}(L(\mathcal{C}, \Sigma)) \cong \Sigma^{-1}(\mathcal{C})$. In the case where $\mathcal{N}$ is a simplicial model category, then the two options $L(\mathcal{N}, \mathcal{N}_w)$ (where $\mathcal{N}_w$ denotes the class of weak equivalences) and $\mathcal{N}_{cf}^{\mathrm{spl}}$, are Dwyer–Kan equivalent as simplicial categories [102, 103].

Even though simplicial categories have strictly associative composition, they are weaker than strict n-categories in the sense that the higher categorical structure is encoded by the simplicial morphism sets rather than by strict $(n-1)$-categories. Hence, the need for weak composition described in the previous chapter is not contradicatory with the fact that strict simplicial categories model all $(\infty, 1)$-categories. For the weaker versions to be discussed next, one can rectify back to a strict simplicial category, as was originally shown by Dwyer *et al.* [104] and Schwänzl and Vogt [224], then extended to Quillen equivalences between the corresponding model structures by Bergner [40].

5.2 Segal's delooping machine

The best-known version of Segal's theory is his notion of an infinite delooping machine or Γ-space. Grothendieck mentioned some correspondence from Larry Breen in 1975 concerning this idea:

Dear Larry,

…The construction which you propose for the notion of a non-strict n-category, and of the nerve functor, has certainly the merit of existing, and of being a first precise approach…

Otherwise, not having understood the idea of Segal in your last letter…

In the first letter of 1983, Grothendieck also mentioned the notion of a multisimplicial nerve of a (strict) n-category. So it would seem that the idea of applying Segal's delooping machine, much as was done by Tamsamani [250], was present in some sense at the time.

The starting point is Segal's 1-delooping machine. Recall from topology that for a pointed space (X, x) the *loop space* ΩX is the space of pointed loops $(S^1, 0) \to (X, x)$. These can be composed by reparametrizing the loops in a well-known way, although the resulting composition is only associative and unital up to homotopy. It is possible to replace ΩX by a topological group, for example with Quillen's realization of homotopy types as coming from simplicial groups. See Ellis [110] for a finiteness property of these realizations. A classifying space construction usually denoted $B(\cdot)$ allows one to get back to the original space:

$$B(\Omega X) \sim X.$$

A popular question in topology in the 1960s was how to define various types of structure on spaces homotopy equivalent to ΩX, which would be weaker than the strong structure of topological group, but which would include sufficiently much homotopical data to let us get back to X by a classifying space construction $B(\cdot)$. Such a kind of structure was known as a "delooping machine." There were a number of examples including A_∞-algebras (in the linearized case), PROPs, operads, and the one which we will be considering: Segal's simplicial delooping machine.

Any of these delooping machines should lead to one or several notions of higher category, indeed this has been the case as we shall discuss elsewhere.

Let Δ denote the category of simplices whose objects are denoted m for positive integers m, and where the morphisms $p \to m$ are the (not-necessarily strictly) order-preserving maps

$$\{0, 1, \ldots, p\} \to \{0, 1, \ldots, m\}.$$

A morphism $1 \to m$ sending 0 to $i - 1$ and 1 to i is called a *principal edge* of m. A morphism which is not injective is called a *degeneracy*.

A simplicial set $\mathcal{A} : \Delta^o \to \text{SET}$ such that $\mathcal{A}_0 = *$ and such that the *Segal maps* obtained by the principal edges $(01), (12), \ldots, ((n-1)n) \subset (0123\cdots n) = [n]$

$$\mathcal{A}_m \to \mathcal{A}_1 \times \cdots \times \mathcal{A}_1$$

are isomorphisms of sets, corresponds to a structure of a monoid on the set $\mathcal{A}_1$. Indeed, the diagram

$$\begin{array}{ccc} \mathcal{A}_2 & \xrightarrow{\cong} & \mathcal{A}_1 \times \mathcal{A}_1 \\ \downarrow & & \\ \mathcal{A}_1 & & \end{array}$$

where the horizontal map is the Segal map and the vertical map is given by the third edge $(02) \subset (012)$, provides a composition law $\mathcal{A}_1 \times \mathcal{A}_1 \to \mathcal{A}_1$. The degeneracy map $\mathcal{A}_0 \to \mathcal{A}_1$ provides a unit–proved using the degeneracy maps for $\mathcal{A}_2$–and consideration of $\mathcal{A}_3$ gives the proof of associativity.

In Segal's 1-delooping theory, this characterization of monoids is weakened by replacing the condition of isomorphism by the condition of weak homotopy equivalences (i.e. maps inducing isomorphisms on the π_i). Thus, a loop space is defined to be a simplicial space

$$\mathcal{A}. : \Delta^o \to \text{TOP},$$

such that $\mathcal{A}_0$ is a single point, such that the Segal maps, again using the principal edges $(01), (12), \ldots, ((n-1)n) \subset (0123\cdots n) = [n]$

$$\mathcal{A}_m \to \mathcal{A}_1 \times \cdots \times \mathcal{A}_1$$

are weak homotopy equivalences, and which is *grouplike* in that the monoid which results when we compose

$$\pi_0 \circ \mathcal{A}. : \Delta^o \to \text{TOP} \to \text{SET}$$

should be a group. Segal [228] explains how to deloop such an object: if TOP is replaced by the category of simplicial sets then the structure $\mathcal{A}.$ is a bisimplicial set, and its delooping $B(\mathcal{A}.)$ is just the diagonal realization. This approach to delooping will be discussed some more in Section 15.3.

As was well known at the time, the characterization of monoids generalizes to give a characterization of the nerve of a category in terms of Segal maps being isomorphisms. Indeed, a monoid can be viewed as a category with a single object, and a small change in the definition makes it apply to the case of categories with an arbitrary set of objects: a simplicial set

$$\mathcal{A}. : \Delta^o \to \text{SET}$$

is the nerve of a 1-category if and only if the Segal maps made using fiber products are isomorphisms

$$\mathcal{A}_m \overset{\cong}{\to} \mathcal{A}_1 \times_{\mathcal{A}_0} \cdots \times_{\mathcal{A}_0} \mathcal{A}_1.$$

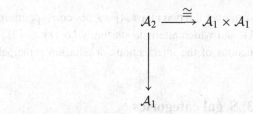

Here the fiber products are taken over the two maps $\mathcal{A}_1 \to \mathcal{A}_0$ corresponding to $(0) \subset (01)$ and $(1) \subset (01)$, and which alternate starting with $(1) \subset (01)$. These correspond to the inclusions of the intersections of adjacent principal edges.

5.3 Segal categories

A *Segal precategory* is a bisimplicial set

$$\mathcal{A} = \{\mathcal{A}_{p,k}, \quad p, k \in \Delta\},$$

in other words a functor $\mathcal{A} : \Delta^o \times \Delta^o \to \text{SET}$ satisfying the *globular condition* that the simplicial set $k \mapsto \mathcal{A}_{0,k}$ is constant equal to a set which we denote by $\mathcal{A}_0$ (called the set of *objects*).

If $\mathcal{A}$ is a Segal precategory then for $p \geq 1$ we obtain a simplicial set

$$k \mapsto \mathcal{A}_{p,k},$$

which we denote by $\mathcal{A}_{p/}$. This yields a simplicial collection of simplicial sets, or a functor $\Delta^o \to \mathcal{K}$ to the Kan–Quillen model category $\mathcal{K}$ of simplicial sets. One could instead look at *simplicial spaces*, that is functors $\Delta^o \to \text{TOP}$ such that $\mathcal{A}_0$ is a discrete space thought of as a set. This gives an equivalent theory, although there are degeneracy problems which apparently need to be treated in an appendix in that case, see May and Thomason [205] and Thomason [253]. We often use the "simplicial space" point of view when speaking informally, as it is more intuitively compelling; however, we don't want to get into details of defining a model structure on TOP, and instead use $\mathcal{K}$ for technical statements.

For each $m \geq 2$ there is a morphism of simplicial sets whose components are given by the principal edges of m, which we call the *Segal map*:

$$\mathcal{A}_{m/} \to \mathcal{A}_{1/} \times_{\mathcal{A}_0} \cdots \times_{\mathcal{A}_0} \mathcal{A}_{1/}.$$

The morphisms in the fiber product $\mathcal{A}_{1/} \to \mathcal{A}_0$ are, alternatively, the inclusions $0 \to 1$ sending 0 to the object 1, or to the object 0.

We would like to think of the inverse image $\mathcal{A}_{1/}(x, y)$ of a pair $(x, y) \in \mathcal{A}_0 \times \mathcal{A}_0$ by the two maps $\mathcal{A}_{1/} \to \mathcal{A}_0$ referred to above, as the *simplicial set of maps from x to y*.

We say that a Segal precategory $\mathcal{A}$ is a *Segal category* if it satisfies the "Segal conditions" saying that for all $m \geq 2$ the Segal maps

$$\mathcal{A}_{m/} \to \mathcal{A}_{1/} \times_{\mathcal{A}_0} \cdots \times_{\mathcal{A}_0} \mathcal{A}_{1/}$$

are weak equivalences of simplicial sets. This notion was introduced by Dwyer *et al.* [104] and Schwänzl and Vogt [224].

Given a strict simplicial category $\mathcal{A}$, we obtain a corresponding Segal precategory by setting

$$\mathcal{A}_{n/} := \coprod_{(x_0,\ldots,x_n)\in \mathrm{Ob}(\mathcal{A})^{n+1}} \mathcal{A}(x_0, x_1) \times \cdots \times \mathcal{A}(x_{n-1}, x_n).$$

This is a Segal category, because the Segal maps are *isomorphisms*. In the other direction, a Segal category such that the Segal maps are isomorphisms comes from a unique strict simplicial category.

The "generators and relations" operation introduced in Chapter 14 is a way of starting with a Segal precategory and enforcing the condition of becoming a Segal category, by forcing the condition of weak equivalence on the Segal maps. As a general matter we will call operations of this type $\mathcal{A} \mapsto \mathbf{Seg}(\mathcal{A})$.

Suppose $\mathcal{A}$ is a Segal category. Then the simplicial set $p \mapsto \pi_0(\mathcal{A}_{p/})$ is the nerve of a category which we call $\tau_{\leq 1}\mathcal{A}$. We say that $\mathcal{A}$ is a *Segal groupoid* if $\tau_{\leq 1}\mathcal{A}$ is a groupoid. This means that the 1-morphisms of $\mathcal{A}$ are invertible up to equivalence.

In fact we can make the same definition even for a Segal precategory $\mathcal{A}$: we define $\tau_{\leq 1}\mathcal{A}$ to be the simplicial set $p \mapsto \pi_0(\mathcal{A}_{p/})$.

We can now describe exactly the situation envisaged by Segal [228] as reported also by Adams [3]: a Segal category $\mathcal{A}$ with only one object, $\mathcal{A}_0 = *$. We call this a *Segal monoid*. If $\mathcal{A}$ is a groupoid then the homotopy theorists' terminology is to say that it is *grouplike*.

5.3.1 Equivalences of Segal categories

The basic intuition is to think of Segal categories as the natural weak version of the notion of topological (i.e. TOP-enriched) category. One of the main concepts in category theory is that of a functor which is an "equivalence of categories." This may be generalized to Segal categories. For simplicial (i.e. strictly $\mathcal{K}$-enriched) categories, this notion is due to Dwyer and Kan, and is often called *DK-equivalence*. The same thing in the context of n-categories is well known (see Kapranov and Voevodsky [168] for example); in the weak case it is described in Tamsamani's paper [250].

We say that a morphism $f : \mathcal{A} \to \mathcal{B}$ of Segal categories is an *equivalence* if it is *fully faithful*, meaning that for $x, y \in \mathcal{A}_0$ the map

$$\mathcal{A}_{1/}(x, y) \to \mathcal{B}_{1/}(f(x), f(y))$$

is a weak equivalence of simplicial sets; and *essentially surjective*, meaning that the induced functor of categories

$$\tau_{\leq 1}(\mathcal{A}) \xrightarrow{\tau_{\leq 1} f} \tau_{\leq 1}(\mathcal{B})$$

is surjective on isomorphism classes of objects. Note that this induced functor $\tau_{\leq 1} f$ will be an equivalence of categories as a consequence of the fully faithful condition.

The homotopy theory that we are interested in is that of the category of Segal categories modulo the above notion of equivalence. In particular, when we search for the "right answer" to a question, it is only up to the above type of equivalence. Of course when dealing with Segal categories having only one object (as will actually be the case in what follows) then the essentially surjective condition is vacuous and the fully faithful condition just amounts to equivalence on the level of the "underlying space" $\mathcal{A}_{1/}$.

In order to have an appropriately reasonable point of view on the homotopy theory of Segal categories, one should look at the model structure (which is one of our main goals, specialized to the model category $\mathcal{K}$, see Chapter 15): the right notion of weak morphism from $\mathcal{A}$ to $\mathcal{B}$ is that of a morphism from $\mathcal{A}$ to $\mathcal{B}'$ where $\mathcal{B} \hookrightarrow \mathcal{B}'$ is a fibrant replacement of $\mathcal{B}$.

5.3.2 Segal's theorem

We define the *realization* of a Segal category $\mathcal{A}$ to be the space $|\mathcal{A}|$ which is the realization of the bisimplicial set $\mathcal{A}$. Suppose $\mathcal{A}_0 = *$. Then we have a morphism

$$|\mathcal{A}_{1/}| \times [0, 1] \to |\mathcal{A}|,$$

giving a morphism

$$|\mathcal{A}_{1/}| \to \Omega|\mathcal{A}|.$$

The notation $|\mathcal{A}_{1/}|$ means the realization of the simplicial set $\mathcal{A}_{1/}$ and $\Omega|\mathcal{A}|$ is the loop space based at the basepoint $* = \mathcal{A}_0$.

Theorem 5.3.1 (G. Segal [228], proposition 1.5) *Suppose $\mathcal{A}$ is a Segal groupoid with one object. Then the morphism*

$$|\mathcal{A}_{1/}| \to \Omega|\mathcal{A}|$$

is a weak equivalence of spaces.

Refer to Segal's paper, or May [204, 8.7], for a proof. Tamsamani noted that the same works in the case of many objects, and indeed this was a key step in his proof of the topological realization theorem for n-categories.

Corollary 5.3.2 *Suppose $\mathcal{A}$ is a Segal groupoid. Then the morphism*

$$|\mathcal{A}_{1/}| \to \Omega|\mathcal{A}|$$

is a weak equivalence in $\mathcal{K}$.

Proof This follows from Segal's theorem by noting that the reduction from a connected many-object Segal groupoid, to a one-object Segal groupoid, doesn't change the homotopy type. See Tamsamani [250] for more details. This corollary will be discussed further, restated, and iterated, in Sections 15.2 and 20.4. □

In order to do these things inside the world of simplicial spaces, the additional cofibrancy conditions in TOP would necessitate a discussion of "whiskering" as is standard in delooping and classifying space constructions (cf. Segal [228], May [203], May and Thomason [205] and Thomason [253]). This is why we have replaced "spaces" by "simplicial sets" in the above discussion, and corresponds also to our use of Reedy model structures in the main chapters.

5.3.3 $(\infty, 1)$-categories

Simplicial categories and Segal categories are two models for what Lurie calls the notion of $(\infty, 1)$-*category*, meaning ∞-categories where the i-morphisms are invertible (analogous to being an inner equivalence) for $i \geq 2$. The Dwyer–Kan simplicial localization may be viewed as the localization in the $(\infty, 2)$-category of $(\infty, 1)$-categories. Part of our goal in this book is to develop an algebraic formalism that is useful for looking at these situations, as well as their iterative counterparts for (∞, n)-categories.

5.3.4 Strictification and Bergner's comparison result

The two models of $(\infty, 1)$-categories discussed above all furnish essentially the same homotopy theory. Such a rectification result was known very early for homotopy monoid structures. For Segal categories, the first result of this kind was due to Dwyer *et al.* [104], who showed how to rectify a Segal category into a strict simplicial category. Similarly, Schwänzl and Vogt [224] showed the same thing in their paper introducing Segal categories. The full homotopy equivalence result, stating that the rectification operation is a Quillen equivalence between model structures, was shown by Bergner [40] at the same time as she constructed the model structures in question. The model structure for

Segal categories is the special case $\mathcal{M} = \mathcal{K}$ of the global construction we are doing in the present book.

With respect to the models we are going to discuss next, Bergner also gave a Quillen equivalence with Rezk's model category of complete Segal spaces, and the comparison can be extended to quasicategories too, as shown by Joyal and Tierney [162], and also by Lurie [190] and Dugger and Spivak [96]. A comparison with quasicategories is also available for the new model of "relative categories" of Barwick and Kan [21, 22].

5.3.5 Enrichment over monoidal structures

Following Leinster [185] and a question of Bergner, Bacard [6] proposes a Segal-type method for weak enrichment over a monoidal model category $(\mathcal{M}, \otimes)$. The theory we consider in this book corresponds to the case when $\otimes = \times$ is the cartesian product. Bacard's theory also extends to more general situations where the base of enrichment is a 2-category with several objects.

5.3.6 Iteration

The passage from the model category $\mathcal{K}$ of simplicial sets, to some kind of $\mathcal{K}$-enriched categories, is something which we would like to iterate in order to obtain the theory of (∞, n)-categories. This iteration is the main subject of the book.

One first attempt would be to iterate the process $\mathcal{K} \mapsto \text{CAT}(\mathcal{K})$ of taking strictly enriched categories. A subtle point is that simplicial categories don't behave well under cartesian products: the Dwyer–Kan–Bergner model structure on $\text{CAT}(\mathcal{K})$ is not cartesian (cartesian model categories are defined in Section 7.7) because a product of two cofibrant objects is no longer cofibrant. Thus, if we try to continue by looking at $\text{CAT}(\text{CAT}(\mathcal{K}))$ the resulting theory doesn't have the right properties. This hooks up with what we have seen in Chapter 2, that iterating the strict enriched category construction doesn't lead to enough objects.

If we use Segal's method, on the other hand, one can iterate the construction with a better effect. This leads to Tamsamani's iterative definition of n-categories. See Chapter 6. It is also related to Dunn's theory [97] of iterated n-fold Segal delooping machines, and it will undoubtedly be profitable to compare Tamsamani [250] and Dunn [97].

Some further discussion of the iterative versions of Segal's theorem, relating homotopy types with Segal n-groupoids, will be done in Chapters 15 and 20.

The next two sections will be devoted to brief descriptions of two other major points of view on $(\infty, 1)$-categories. After that, we discuss the comparison between the various theories.

5.4 Rezk categories

Rezk has given a different way of using the Segal maps to specify an $(\infty, 1)$-categorical structure. Barwick showed how to iterate this construction, and this iteration has now also been taken up by Lurie and Rezk. Their iteration is philosophically similar to what we are doing in the main part of this book. In the present section we discuss Rezk's original case, which he called "complete Segal spaces." These objects enter into Bergner's three-way comparison [40].

It will be convenient to start our discussion by refering to a generic notion of $(\infty, 1)$-category, which could be concretized by simplicial categories, or Segal categories. Recall that an $(\infty, 1)$-category $\mathcal{A}$ has an $(\infty, 0)$-category or ∞-groupoid, as its *interior* denoted $\mathcal{A}^{\rm int}$. The interior is the universal ∞-groupoid mapping to $\mathcal{A}$. For any $x, y \in \mathrm{Ob}(\mathcal{A})$, the mapping space is the subspace $\mathcal{A}^{\rm int}(x, y) \subset \mathcal{A}(x, y)$ union of all the connected components corresponding to maps which are invertible up to equivalence. By Segal's theorem (which will be discussed further in Chapter 15), this corresponds to a space which we can denote by $|\mathcal{A}^{\rm int}|$. It is the "moduli space of objects of $\mathcal{A}$ up to equivalence": there is a separate connected component for each equivalence class of objects. The vertices coming from the 0-simplices correspond to the original objects $\mathrm{Ob}(\mathcal{A})$, and within a connected component the space of paths from one vertex to another, is the space $\mathcal{A}^{\rm int}(x, y)$ of equivalences between the corresponding objects.

In Rezk's theory, our $(\infty, 1)$-category is represented by a simplicial space $\mathcal{A}^R$ with $\mathcal{A}^R_0 = |\mathcal{A}^{\rm int}|$ in degree zero. The *homotopy fiber* of the map

$$\mathcal{A}^R_1 \to \mathcal{A}^R_0 \times \mathcal{A}^R_0$$

over a point (x, y) is (canonically equivalent to) the space of morphisms $\mathcal{A}(x, y)$. The categorical structure is defined by imposing a Segal condition on homotopy fiber products: for any n, there is a version of the Segal map going to the homotopy fiber product

$$\mathcal{A}^R_n \to \mathcal{A}^R_1 \times^h_{\mathcal{A}^R_0} \mathcal{A}^R_1 \times^h_{\mathcal{A}^R_0} \cdots \times^h_{\mathcal{A}^R_0} \mathcal{A}^R_1,$$

and this is required to be a weak equivalence. A *complete Segal space* is a simplicial space satisfying these Segal conditions, and also the *completeness* condition which corresponds to the requirement $\mathcal{A}^R_0 = |\mathcal{A}^{\rm int}|$. We had formulated that requirement by first considering a generic theory of $(\infty,1)$-categories. Internally to Rezk's theory, the completeness condition says that $\mathcal{A}_1^{R,\mathrm{int}} \to \mathcal{A}_0^R \times \mathcal{A}_0^R$ should be equivalent to the path space fibration, where $\mathcal{A}_1^{R,\mathrm{int}} \subset \mathcal{A}_1^R$ denotes the union of connected components corresponding to morphisms which are invertible up to equivalence. This condition is shown to be equivalent to a more abstract condition useful for manipulating the model structure, see Rezk [218, 6.4], Bergner [40, 3.7], and Joyal and Tierney [162, section 4].

Rezk's theory is a little bit more complicated in its initial stages than the theory of Segal categories. The Segal maps go to a homotopy fiber product, which nevertheless can be assumed to be a plain fiber product by imposing a Reedy fibrant condition on the simplicial space, for example. Since the set of objects is not really too well defined, the kind of reasoning which we are considering here (and which was also followed by Dwyer and Kan in their series of papers), breaking up the problem into first a problem for higher categories with a fixed set of objects, then varying the set of objects, is less available.

On the other hand, Rezk's theory has the advantage that $\mathcal{A}^R$ is a canonical model for $\mathcal{A}$ up to levelwise homotopy equivalence in the category of diagrams $\Delta^o \to \textsc{Top}$. Thus, a map $\mathcal{A}^R \to \mathcal{B}^R$ of complete Segal spaces is an equivalence if and only if each $\mathcal{A}_n^R \to \mathcal{B}_n^R$ is a weak equivalence of spaces (and it suffices to check $n = 0$ and $n = 1$ because of the Segal conditions). This contrasts with the case of Segal categories, where the set of objects $\mathcal{A}_0 = \mathrm{Ob}(\mathcal{A})$ is not invariant under equivalences of categories.

As Bergner has pointed out [46, 47], the canonical nature of the spaces involved makes Rezk's theory particularly amenable to calculating limits. For example, if

$$\mathcal{A}^R \to \mathcal{B}^R \leftarrow \mathcal{C}^R$$

are two arrows between complete Segal spaces, then the levelwise homotopy fiber product

$$\mathcal{U}_n^R := \mathcal{A}_n^R \times^h_{\mathcal{B}_n^R} \mathcal{C}_n^R$$

is again a complete Segal space, and it is the right homotopy fiber product in the world of $(\infty, 1)$-categories.

This again contrasts with the case of Segal categories, or indeed even usual 1-categories. For example, letting $\mathbb{I}$ denote the 1-category with two isomorphic objects υ_0 and υ_1, the inclusion maps

$$\{\upsilon_0\} \to \mathbb{I} \leftarrow \{\upsilon_1\}$$

are equivalences of categories, so the homotopy fiber product in any reasonable model structure for 1-categories, should also be equivalent to a discrete singleton category. However, the fiber product of categories, or of simplicial sets (the nerves) is empty. In Rezk's theory, the degree zero space $\mathbb{I}_0^R$ will again be contractible, since there is only a single equivalence class of objects of $\mathbb{I}$ and they have no nontrivial automorphisms.

As usual, for treating the technical aspects of the theory it is better to look at bisimplicial sets rather than simplicial spaces. Rezk [218] constructs a model structure on the category of bisimplicial sets, such that the fibrant objects are complete Segal spaces which are Reedy fibrant as Δ^o-diagrams and levelwise fibrant. Bergner [40] considers further this theory and shows the equivalence with simplicial categories and Segal categories.

Barwick has suggested to iterate this construction to a Rezk-style theory of (∞, n) categories for all n, and Rezk [219] has taken this up more recently. He shows that the resulting model categories are cartesian, in particular this gives a construction of the $(\infty, n+1)$-category of (∞, n)-categories. Various different versions of Rezk's iterative model are discussed in Bergner's recent paper [48] comparing the iterative approaches to (∞, n)-categories.

Barwick and Kan [21] propose to model $(\infty, 1)$-categories as pairs (C, W) consisting of a category and a subcategory. The idea is that such a pair corresponds to the $(\infty, 1)$-category $L(C, W)$ obtained by Dwyer–Kan localization. They construct a model structure on the category of pairs, Quillen equivalent to Rezk's model category. Most recently [23] they propose a model for (∞, n)-categories in the same spirit.

5.5 Quasicategories

Joyal and Lurie have developed extensively the theory of *quasicategories*. These first appeared in the book of Boardman and Vogt [50] under the name "restricted Kan complexes." An important example appeared in work of Cordier and Porter [81, 82, 83].

A good place to start is to recall Kan's original horn-filling conditions for the category of simplicial sets $\mathcal{K}$. As $\mathcal{K}$ is a category of diagrams $\Delta^o \to$ SET, we have in particular the *representable diagrams* which we shall denote $R(n)$, defined by $R(n)_m := \Delta([m], [n])$. This is the "standard n-simplex," classically denoted by $R(n) = \Delta[n]$. For our purposes this classical notation would seem to risk some confusion with too many symbols Δ around, so we call it $R(n)$ instead. Now, $R(n)$ has a standard simplicial subset denoted $\partial R(n)$, which is the "boundary." It can be defined as the $(n-1)$-skeleton of $R(n)$, or as the union of the $(n-1)$-dimensional faces of $R(n)$. The faces are indexed by $0 \leq k \leq n$; in terms of linearly ordered sets, the k-th face corresponds to the linearly ordered subset of $[n]$ obtained by crossing out the k-th element. Now, the k-th *horn* $\Lambda(n,k)$ is the subset of $\partial R(n)$ which is the union of all the $(n-1)$-dimensional faces except the k-th one.

If $\mathcal{X} \in \mathcal{K}$ is a simplicial set, then the universal property of the representable $R(n)$ says that $\mathcal{X}_n = \textsc{Hom}_{\mathcal{K}}(R(n), \mathcal{X})$. Kan's *horn-filling condition* says that any map $\Lambda(n,k) \to \mathcal{X}$ extends to a map $R(n) \to \mathcal{X}$. The simplicial sets $\mathcal{X}$ satisfying this condition are the fibrant objects of the model structure on $\mathcal{K}$.

Boardman and Vogt introduced the *restricted Kan condition*, satisfied by a simplicial set X whenever any map $\Lambda(n,k) \to \mathcal{X}$ extends to $R(n) \to \mathcal{X}$, for each $0 < k < n$. In other words, they consider only the horns obtained by taking out any except for the first and last faces.

This condition corresponds to keeping a directionality of the 1-cells in $\mathcal{X}$. This may be seen most clearly by looking at the case $n = 2$. A 2-cell may be drawn as

$R(2)$:

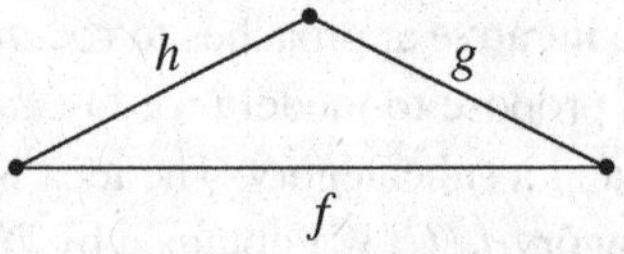

where h, g and f are the 1-cells corresponding to edges (01), (12) and (02) respectively. Such a 2-cell is thought of as the relation $f = gh$. In the usual Kan condition, there are three horns which need to be filled:

$\Lambda(2,0)$:

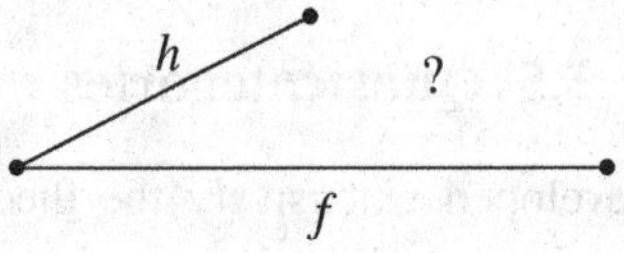

$\Lambda(2,1)$:

$\Lambda(2,2)$:

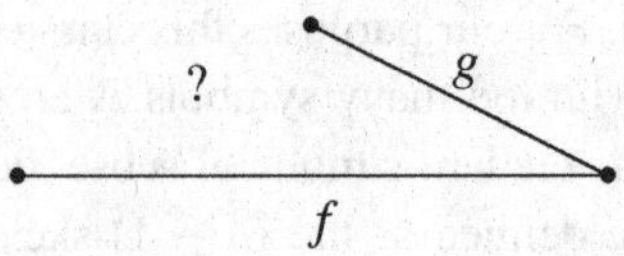

However, in the restricted Kan condition, only the middle horn $\Lambda(2,1)$ is required to be filled. This corresponds to saying that for any composable arrows g and h, there is a composition $f = gh$. On the other hand, filling the horn $\Lambda(2,0)$ would correspond to saying that given f and h there is g such that $f = gh$, which essentially means we look for $g = fh^{-1}$; and filling $\Lambda(2,2)$ would correspond to saying that given f and g there is h such that $f = gh$, that is $h = g^{-1}f$.

When we look at things in this way, it is clear that the full Kan condition corresponds to imposing, in addition to the categorical composition of arrows, some kind of groupoid condition of existence of inverses. It isn't surprising, then, that Kan complexes correspond to ∞-groupoids.

Following through this philosophy has led Joyal to the theory of quasicategories, which are simplicial sets satisfying the restricted Kan condition, but viewed as $(\infty, 1)$-categories with arrows which are not necessarily invertible.

Making the translation from restricted Kan simplicial sets to $(\infty, 1)$-categories is not altogether trivial, most notably for any two vertices x, y of a quasicategory $\mathcal{X}$ we need to define the *simplicial mapping space* $\mathcal{X}(x, y)$; one possibility is to say that it is the Kan simplicial set

$$k \mapsto \mathrm{HOM}^{x,y}(R(1) \times R(k), \mathcal{X})$$

where the superscript indicates maps sending $0 \times R(k)$ to x and $1 \times R(k)$ to y. See Dugger and Spivak [96] for more on the simplicial mapping spaces of a quasicategory. There is also a way of describing directly a simplicial category which is the rectification of the corresponding $(\infty, 1)$-category; see Riehl [220] for a detailed discussion. Nichols-Barrer [209] provides a self-contained treatment of the theory.

Joyal constructs a model category structure whose underlying category is that of simplicial sets, in for which the fibrant objects are exactly those satisfying the restricted Kan condition. The passage from a general simplicial set to its fibrant replacement, done by enforcing the restricted Kan horn-filling conditions using the small object argument, is a version of the "calculus of generators and relations" and very similar to what we will be discussing in Chapters 14 and 15 for the case of Segal categories.

In the various kinds of simplicial objects which we now have representing $(\infty, 1)$-categories with weak composition, we can see a trade-off between information content and simplicity. The simplest model is that of quasicategories, which are just simplicial sets satisfying a very classical horn-filling condition; but in this case it isn't easy to get back some of the main pieces of information in an $(\infty, 1)$-category, such as the simplicial mapping sets. At the other end, in Rezk's complete Segal spaces, the full information of the

∞-groupoid interior is contained within the object, to the extent that the homotopy type of the Δ^o-diagram is an invariant of the $(\infty, 1)$-category up to equivalence; on the other hand, the initial steps of the theory are more complicated. The theory of Segal categories fits in between: a Segal category has more information readily at hand than a quasicategory, but less than a complete Segal space; and the initial theory is more complicated than for quasicategories but less than for complete Segal spaces.

Apparently, the quasicategory model can also be iterated to give a model for (∞, n)-categories. This was suggested by Joyal in his original paper, and Barwick and Schommer-Pries are working on it using the technique of operator categories.

5.6 Going between Segal categories and n-categories

We mention briefly the relationship between the notions of Segal category and n-category. Tamsamani's definition of n-category is recursive. The basic idea is to use the same definition as above for Segal category, but where the $\mathcal{A}_{p/}$ are themselves $(n-1)$-categories. The appropriate condition on the Segal maps is the condition of equivalence of $(n-1)$-categories, which in turn is defined (inductively) in the same way as the notion of equivalence of Segal categories explained above.

Tamsamani [250] showed, as we shall discuss further in Section 20.4, that the homotopy category of n-groupoids is the same as that of n-truncated spaces. The two relevant functors are the realization and Poincaré n-groupoid Π_n functors. Applying this to the $(n-1)$-categories $\mathcal{A}_{p/}$ we obtain the following relationship. An n-category $\mathcal{A}$ is said to be $^{>}1$-*groupic* (notation introduced in my preprint [235]) if the $\mathcal{A}_{p/}$ are $(n-1)$-groupoids. In this case, replacing the $\mathcal{A}_{p/}$ by their realizations $|\mathcal{A}_{p/}|$ we obtain a simplicial space which satisfies the Segal condition. Conversely, if $\mathcal{A}_{p/}$ are spaces or simplicial sets then replacing them by their $\Pi_{n-1}(\mathcal{A}_{p/})$ we obtain a simplicial collection of $(n-1)$-categories, again satisfying the Segal condition. These constructions are not quite inverses because

$$|\Pi_{n-1}(\mathcal{A}_{p/})| = \tau_{\leq n-1}(\mathcal{A}_{p/})$$

is the Postnikov truncation. If we think (heuristically) of setting $n = \infty$ then we get inverse constructions. Thus – in a sense which I will not currently make more precise than the above discussion and the next subsection – one can say that Segal categories are the same thing as $^{>}1$-groupic ∞-categories.

The passage from simplicial sets to Segal categories is the same as the inductive passage from $(n-1)$-categories to n-categories. In my preprint [234] was introduced the notion of *n-precat*, the analogue of the above Segal precat. Noticing that the results and arguments in *loc. cit.* [234] are basically organized into one gigantic inductive step passing from $(n-1)$-precats to n-precats, the same step applied only once works to give the analogous results in the passage from simplicial sets to Segal precats.

The notion of Segal category thus presents, from a technical point of view, an aspect of a "baby" version of the notion of n-category. On the other hand, it allows a first introduction of homotopy going all the way up to ∞ (i.e. it allows us to avoid the n-truncation inherent in the notion of n-category).

One can easily imagine combining the two into a notion of "Segal n-category," which is an n-simplicial simplicial set satisfying the globular condition at each stage. It is interesting and historically important to note that the notion of Segal n-category with only one i-morphism for each $i \leq n$, is the same thing as the notion of *n-fold delooping machine*. This translation comes out of Dunn [97], which apparently dates essentially back to 1984. In retrospect it is not too hard to see how to go from Dunn's notion of E_n-machine, to Tamsamani's notion of n-category, simply by relaxing the conditions of having only one object. Metaphorically, n-fold delooping machines correspond to the Whitehead tower, whereas n-groupoids correspond to the Postnikov tower.

There are other proposals for simplicial models for n-categories which we haven't been able to discuss. For example, Street [244, 245] proposed a model based on simplicial sets with certain distinguished simplicial subsets, which he calls "thin subcomplexes," and Fiedorowicz and Vogt [111, 112] apply an n-fold iteration of the notion of moniodal category to study iterated loop spaces.

6
Weak enrichment over a cartesian model category: an introduction

To close out the first part of the book, we describe in this chapter the basic outlines of the theory which will occupy the rest of the work. The basic idea, already considered in Pellissier's thesis, is to abstract Tamsamani's iteration process to obtain a theory of $\mathcal{M}$-enriched categories, weak in Segal's sense, for a model category $\mathcal{M}$.

6.1 Simplicial objects in $\mathcal{M}$

The original definitions of Segal category, Tamsamani n-category and Pellissier's enriched categories took as basic object a functor

$$\mathcal{A} : \Delta^o \to \mathcal{M}.$$

The first condition is that the image $\mathcal{A}_0$ of $[0] \in \Delta$ should be a "discrete object," that is the image of a set under the natural inclusion SET $\to \mathcal{M}$ which sends a set X to the colimit of $*$ over the discrete category corresponding to X. This version of the theory therefore requires, at least, some axioms saying that the functor SET $\to \mathcal{M}$ is fully faithful and compatible with disjoint unions. Thus $\mathcal{A}_0$ may be viewed as a set and the expression $x \in \mathcal{A}_0$ means that x is an element of the corresponding set, equivalently $x : * \to \mathcal{A}_0$. The higher elements of the simplicial object will be denoted $\mathcal{A}_{m/}$.

Then, the Segal category condition says that the Segal maps

$$\mathcal{A}_{m/} \to \mathcal{A}_{1/} \times_{\mathcal{A}_0} \cdots \times_{\mathcal{A}_0} \mathcal{A}_{1/}$$

are supposed to be weak equivalences.

The pair of structural maps $(\partial_0, \partial_1) : \mathcal{A}_{1/} \to \mathcal{A}_0 \times \mathcal{A}_0$ serves to decompose

$$\mathcal{A}_{1/} = \coprod_{x,y \in \mathcal{A}_0} \mathcal{A}(x, y),$$

where $\mathcal{A}(x, y)$ is the inverse image of $(x, y) \in \mathcal{A}_0 \times \mathcal{A}_0$, or more precisely

$$\mathcal{A}(x, y) := \mathcal{A}_{1/} \times_{\mathcal{A}_0 \times \mathcal{A}_0} *,$$

with the right map of the fiber product being given by $(x, y) : * \to \mathcal{A}_0 \times \mathcal{A}_0$. We can similarly decompose

$$\mathcal{A}_{m/} = \coprod_{(x_0,\ldots,x_m)\in \mathcal{A}_0^{m+1}} \mathcal{A}(x_0, \ldots, x_m),$$

and the Segal condition may be expressed equivalently as saying that

$$\mathcal{A}(x_0, \ldots, x_m) \to \mathcal{A}(x_0, x_1) \times \cdots \times \mathcal{A}(x_{m-1}, x_m)$$

is a weak equivalence in $\mathcal{M}$.

6.2 Diagrams over Δ_X

Upon closer inspection, most of the arguments about $\mathcal{M}$-Segal categories can really be phrased in terms of the objects $\mathcal{A}(x_0, \ldots, x_m)$; and in these terms, the Segal condition involves only a product rather than a fiber product. So, it is natural and useful to consider the objects $\mathcal{A}(x_0, \ldots, x_m)$ as the primary objects of study rather than the $\mathcal{A}_{m/}$. This economizes some hypotheses and arguments about discrete objects and fiber products.

This point of view, folklore for perhaps some time, has been introduced in the literature by Lurie [192]. For any set X, define the category Δ_X whose objects are finite linearly ordered sets decorated by elements of X, that is to say an object of Δ_X is an ordered set $[m] \in \Delta$ plus a map of sets $x. : [m] \to X$. This pair will be denoted $(x_0, \ldots, x_m)$, that is it is an $(m + 1)$-tuple of elements of X. The morphisms in the category Δ_X are the morphisms of Δ which respect the decoration, so for example the three injective morphisms $[1] \to [2]$ yield morphisms of the form

$$(x_0, x_1) \to (x_0, x_1, x_2), \quad (x_1, x_2) \to (x_0, x_1, x_2), \quad (x_0, x_2) \to (x_0, x_1, x_2),$$

whereas the three degenerate $[1] \to [2]$ yield

$$(x_0, x_0) \to (x_0, x_1, x_2), \quad (x_1, x_1) \to (x_0, x_1, x_2), \quad (x_2, x_2) \to (x_0, x_1, x_2).$$

Now, an $\mathcal{M}$-Segal category will be a pair $(X, \mathcal{A})$, where X is a set, called the *set of objects*, and $\mathcal{A} : \Delta_X^o \to \mathcal{M}$ is a functor denoted by

$$([m], x.) = (x_0, \ldots, x_m) \mapsto \mathcal{A}(x_0, \ldots, x_m),$$

or just $(x_0, \ldots, x_m) \mapsto \mathcal{A}(x_0, \ldots, x_m)$ if there is no danger of confusion, such that the Segal maps

$$\mathcal{A}(x_0, \ldots, x_m) \to \mathcal{A}(x_0, x_1) \times \cdots \times \mathcal{A}(x_{m-1}, x_m)$$

are weak equivalences in $\mathcal{M}$. At $m = 0$ the Segal condition says that $\mathcal{A}_0(x_0) \to *$ is a weak equivalence. This is a sort of weak unitality condition, but for our purposes it is generally speaking better to impose the *strict unitality condition* that $\mathcal{A}_0(x_0) = *$ for any x_0. This condition becomes essential when we consider cartesian products.

At $m = 1$, the morphism space between two elements $x, y \in X$ is $\mathcal{A}(x, y)$. At $m = 2$ the usual diagram using the three standard morphisms, serves to define the composition operation in a weak sense:

$$\mathcal{A}(x, y) \times \mathcal{A}(y, z) \leftarrow \mathcal{A}_2(x, y, z) \to \mathcal{A}(x, z) \qquad (6.2.1)$$

with the leftward arrow being a weak equivalence in $\mathcal{M}$. For higher values of m we get the higher homotopy coherence conditions starting with associativity at $m = 3$.

6.3 Hypotheses on $\mathcal{M}$

In the weak enrichment, the composition operation is given by a diagram of the form (6.2.1) above, using the usual cartesian product $\times$ in $\mathcal{M}$ and where the leftward arrow is the Segal map which is required to be a weak equivalence.

Therefore, the main condition which we need to impose upon $\mathcal{M}$ is that it be *cartesian*, that is to say a monoidal model category whose monoidal operation is the cartesian product. This insures that cartesian product is compatible with the cofibrations and weak equivalences. Monoidal model categories have been considered by many authors, see Hovey [149] for example, and the cartesian theory is a special case. This condition will be discussed in Section 7.7.

For convenience we also impose the conditions that $\mathcal{M}$ be *left proper*, and *tractable*. Tractability is Barwick's slight modification of J. Smith's notion of *combinatorial model category*. A combinatorial model category is a cofibrantly generated one whose underlying category is locally presentable–locally presentable categories are the most appropriate environment for using the small object argument, one of our staples. Barwick's tractability adds the condition that the domains of the generating cofibrations and trivial cofibrations, be themselves cofibrant objects. This is useful at some technical places in the small object argument. Our discussion of these topics is put together in Chapters 8 and 7.

In Section 10.7 we consider some additional hypotheses on $\mathcal{M}$ saying that disjoint unions behave like we think they do; if $\mathcal{M}$ satisfies these hypotheses then the discrete-set objects in $\mathcal{M}$ work well, and we can use the notation $\mathcal{A}_{m/}$ for the disjoint union of the $\mathcal{A}(x_0, \ldots, x_m)$. This reduces to consideration of simplicial objects in $\mathcal{M}$ rather than functors from Δ_X^o. If $\mathcal{M}$ is a category of presheaves over a connected category Φ then it satisfies the additional hypotheses, and the category of $\mathcal{M}$-precategories discussed next will again be a category of presheaves (over a quotient of $\Delta \times \Phi$). The fact that iteratively we stay within the world of presheaf categories is convenient if one wants to think of the small object argument in a simplified way.

6.4 Precategories

A tractable left proper cartesian model category $\mathcal{M}$ is fixed. For the original case of n-categories, $\mathcal{M}$ would be the model category for $(n-1)$-categories constructed according to the inductive hypothesis. In order for the induction to work, the main goal is to construct from $\mathcal{M}$ a new model category, whose objects represent up to homotopy the $\mathcal{M}$-enriched Segal categories, and which satisfies the same hypotheses of tractability, left properness, and the cartesian condition.

If $\mathcal{M}$ satisfies the additional hypotheses on disjoint unions, then an $\mathcal{M}$-enriched category is a functor $\mathcal{A} : \Delta^o \to \mathcal{M}$ such that $\mathcal{A}_0$ is a discrete set also called $\mathrm{Ob}(\mathcal{A})$, and such that the Segal maps

$$\mathcal{A}_{n/} \to \mathcal{A}_{1/} \times_{\mathcal{A}_0} \cdots \times_{\mathcal{A}_0} \mathcal{A}_{1/}$$

are weak equivalences in $\mathcal{M}$. (The slash notation indicates that $\mathcal{A}_{m/}$ is an object of the category $\mathcal{M}$, which will typically itself be the category of $(n-1)$-precategories, see Sections 10.7 and 20.2.)

However, looking at the category of all such functors, the Segal condition is not preserved by limits or colimits of diagrams. It would be preserved by homotopy limits, but not even by homotopy colimits, and indeed the problem of taking a homotopy colimit of diagrams and then imposing the Segal condition is our main technical difficulty.

So, in order to obtain a model category structure, we have to leave out the Segal condition. This leads to the basic notion of *$\mathcal{M}$-precategory*. Our utilisation of the word "precategory" is similar to but not the same as that of Janelidze [152]. The reader may refer to the introduction of my preprint [234] for a discussion of this notion in the original n-categorical context.

If $\mathcal{M}$ satisfies the additional condition about disjoint unions, then an $\mathcal{M}$-precategory may be defined as a functor $\mathcal{A} : \Delta^o \to \mathcal{M}$ such that $\mathcal{A}_0$ is a discrete set, that is to say a disjoint sum of copies of the coinitial object $* \in \mathcal{M}$.

In the more general case, an $\mathcal{M}$-enriched precategory is a pair $(X, \mathcal{A})$ where X is a set (often denoted by $\mathrm{Ob}(\mathcal{A})$), and $\mathcal{A} : \Delta^o_X \to \mathcal{M}$ is a functor satisfying the *unitality condition* that $\mathcal{A}(x_0) = *$ for any sequence of length zero (i.e. having only a single element).

In either of these situations, the category $\mathcal{PC}(\mathcal{M})$ of such diagrams is closed under limits and colimits, and furthermore if $\mathcal{M}$ is locally presentable then $\mathcal{PC}(\mathcal{M})$ is locally presentable too. The category $\mathcal{PC}(\mathcal{M})$ will thus serve as a suitable substrate for our model structure. To make up for the fact that we left out the Segal conditions for general objects of $\mathcal{PC}(\mathcal{M})$, one of the goals of the model structure to be constructed will be to make sure that the fibrant objects do satisfy the Segal conditions.

An additional benefit of the notation Δ_X is that it allows us to break down the argument into two pieces, a suggestion of Clark Barwick [16]. Indeed, we obtain two different categories, $\mathcal{PC}(X, \mathcal{M})$ and $\mathcal{PC}(\mathcal{M})$. The first consists of all $\mathcal{M}$-precategories with a fixed set of objects X. It is just the full subcategory of the diagram category $\mathrm{FUNC}\left(\Delta^o_X, \mathcal{M}\right)$ consisting of diagrams satisfying the strict unitality condition $\mathcal{A}_0(x_0) = *$. The study of $\mathcal{PC}(X, \mathcal{M})$, considered first, is therefore almost the same as the study of the category of $\mathcal{M}$-valued diagrams on a fixed category Δ^o_X.

The category $\mathcal{PC}(\mathcal{M})$ is obtained by letting X vary, with a natural definition of morphism $(X, \mathcal{A}) \to (Y, \mathcal{B})$. Once everything is well under way and the objects of $\mathcal{PC}(\mathcal{M})$ become our main objects of study, then we will drop the set X from the notation: an object of $\mathcal{PC}(\mathcal{M})$ will be denoted $\mathcal{A}$ and its set of objects by $\mathrm{Ob}(\mathcal{A})$, but with the same letter for the functor $\mathcal{A} : \Delta^o_{\mathrm{Ob}(\mathcal{A})} \to \mathcal{M}$, in other words $\mathcal{A}$ denotes $(\mathrm{Ob}(\mathcal{A}), \mathcal{A})$.

6.5 Unitality

The strict unitality condition says that $\mathcal{A}(x_0) = *$. The reason for imposing this condition, aside from its convenience, is that it is needed to obtain the cartesian condition on the model category of $\mathcal{M}$-precategories. Indeed, if we don't impose the unitality condition, then the precategories must be allowed to have $\mathcal{A}(x_0)$ arbitrary, even those would be forced to be contractible by the Segal condition of length 0. A product with a non-unital precategory, such that $\mathcal{B}(x_0, \ldots, x_m) = \emptyset$ for all sequences of objects, is not compatible with weak equivalences (as will be discussed in Section 17.3.1).

Given the fact that the Eckmann–Hilton argument rules out a number of different approaches to higher categories, as we have seen in Chapter 2 but also as in Cheng and Makkai's remark [76] on Penon's original definition, we should justify why our version of the unitality condition which says that $\mathcal{A}(x_0) = *$ doesn't also lead to an interchange rule or the Eckmann–Hilton argument.

The point is that in the Segal-style definitions, the composition is not a well-defined operation. So, even if there exist cells which are supposed to be the "identities," there is not a single well-defined composition with the identity. The degeneracies provide 2-cells which say that, for an i-morphism f, some possible composition of the form $1_{t(f)} \circ f$ or $f \circ 1_{s(f)}$ will be equal to f, but these choices (which we call respectively the left and right degeneracies) are not the only possible ones. The main step of the Eckmann–Hilton argument going from

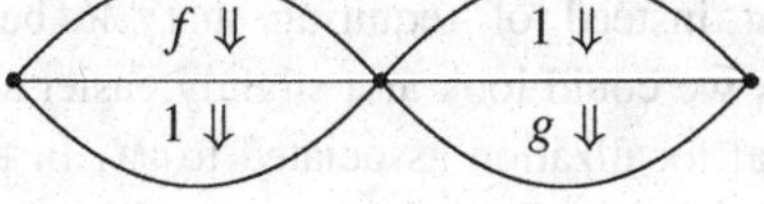

to

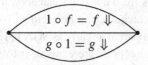

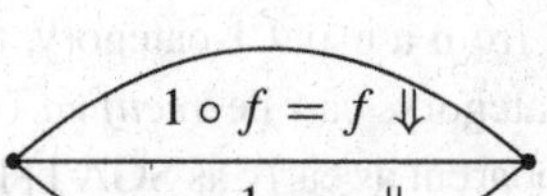

involves glueing the left degeneracy for f on top to the right degeneracy for g on the bottom, generating an amalgamated sum of cells which doesn't fit into any canonical global composition operation for the four 2-morphisms at once. And similarly for the step involving vertical compositions. The information on composition with units, which comes from the unitality condition and the degeneracies of Δ, is luckily not enough to make the Eckmann–Hilton argument work. Because we're close to the borderline here, it is clear that some care should be taken to verify everything related to the unitality condition in the technical parts of our construction.

Unitality will therefore be considered in the context of more general up-to-homotopy finite product theories, in Chapter 11.

6.6 Rectification of Δ_X-diagrams

The reader should now be asking the following question: wouldn't it be better to consider $\mathcal{M}$ as some kind of higher category, and to look at *weak* functors $\Delta^o_X \to \mathcal{M}$? This would certainly seem like the most natural thing to do. Unfortunately, this idea leads to "bootstrapping" problems both philosophical as well as practical. On the philosophical level, the really good version of $\mathcal{M}$ as a higher category, is to think of $\mathcal{M}$ as being enriched over itself. We exploit this point of view starting from Section 20.5, where $\mathcal{M}$ considered as an $\mathcal{M}$-enriched category is called $\mathbf{Enr}(\mathcal{M})$. However, if we are looking to define a notion of $\mathcal{M}$-enriched category, then we shouldn't start with something which is itself an $\mathcal{M}$-enriched category. One can imagine getting around this problem by noting that $\mathcal{M}$, considered as a category enriched over itself, is actually strictly associative; however, for looking at functors to $\mathcal{M}$ we need to go to a weaker model, and we end up basically having at least to pass through the notion of strict functors $\Delta_X \to \mathcal{M}$. One could alternatively say that instead of requiring that $\mathcal{M}$ be considered as an $\mathcal{M}$-enriched category, we could look at a slightly easier structure such as the Dwyer–Kan simplicial localization associated to $\mathcal{M}$. In this case, we would need a theory of weak functors from Δ^o_X to a simplicial category. This theory has already been done by Bergner [39, 40], so it would be possible to go that route. However, it would seem to lead to many notational and mathematical difficulties.

Luckily, we don't need to worry about this issue. It is well known that any kind of weak functors from a usual 1-category, to a higher category such as comes from a model category, can be *rectified* (or "strictified") to actual 1-functors. This became apparent as early as SGA I [131], where Grothendieck pointed out that fibered categories are equivalent to strictly cartesian fibered categories. Since then it has been well known to homotopy theorists working on diagram categories, and indeed the various model structures on the category of diagrams $\mathrm{FUNC}(\Delta^o_X, \mathcal{M})$ serve to provide model categories whose corresponding higher categories, in whatever sense one would like, are equivalent to the higher category of weak functors. In the context of diagrams towards Segal categories, an argument is given by Hirschowitz and myself [145].

Due to the philosophical bootstrapping problem mentioned above, I don't see any way of making the argument given in the previous paragraph into anything other than the heuristic consideration that it is. But, taking it as a basic principle, we shall stick to the notion of a usual functor $\Delta^o_X \to \mathcal{M}$ as being the underlying object of study.

6.7 Enforcing the Segal condition

We left out the Segal condition in order to get a good locally presentable category $\mathcal{PC}(\mathcal{M})$ of $\mathcal{M}$-precategories. The Segal condition should then be built into the model structure, for example it is supposed to be satisfied by the fibrant objects. This guides our construction of the model structure: a fibrant replacement should impose or "force" the Segal condition, and such a process in turn tells us how to define the notion of weak equivalence.

To understand this, one should view an $\mathcal{M}$-enriched precategory as being a prescription for constructing an $\mathcal{M}$-enriched category by a collection of "generators and relations." The notion of precategory was made necessary by the need for colimits, so one should think of a precategory as being a colimit of smaller pieces. The associated $\mathcal{M}$-enriched category should then be seen as the homotopy colimit of the same pieces, in the model category we are looking for. That is to say it is an object specified by generators and relations. This will be explained in some detail for the case of 1-categories in Section 14.8.

The *calculus of generators and relations* is the process whereby $\mathcal{A}$ may be replaced with $\mathbf{Seg}(\mathcal{A})$ which is, in a homotopical sense, the minimal object satisfying the Segal conditions with a map $\mathcal{A} \to \mathbf{Seg}(\mathcal{A})$. The minimal object is obtained by a sequence of pushouts. For example, one of the first steps supplies the composition of elements $f \in \mathcal{A}(x, y)$ and $g \in \mathcal{A}(y, z)$ by formally adding an element h to $\mathcal{A}(x, y, z)$, with the relation that h projects to f and g. In practice, this will be done in more abstract terms by enforcing the Segal conditions using the small object argument. In order to find the model category, we should define and investigate closely this process of generating an $\mathcal{M}$-enriched category.

The construction $\mathbf{Seg}(\mathcal{A})$ doesn't in itself change the set of objects of $\mathcal{A}$, so we can look at it in the smaller category $\mathcal{PC}(X, \mathcal{M})$. There, it can be considered as a case of left Bousfield localization. This way of breaking up the procedure was suggested by Barwick. Luckily, the left Bousfield localization which occurs here has a particular form which we call "direct," in which the the weak equivalences may be characterized explicitly, and we develop that theory with general notations in Chapter 9. Going to a more general situation helps to clarify and simplify notations at each stage; it isn't clear that these discussions would have significant other applications although that cannot be ruled out. Continuing in this way, we discuss in Chapter 11 the application of direct left Bousfield localization to algebraic theories in diagram categories. This formalizes the idea of requiring certain cartesian product maps to be weak equivalences, with the objective of applying it to the Segal maps. Here we refer implicitly to the theories of sketches and algebraic theories.

Then, in Chapters 12 and 14 we apply the preceding general discussions to the case of $\mathcal{M}$-enriched precategories, and define the operation $\mathcal{A} \mapsto \mathbf{Seg}(\mathcal{A})$ which to an $\mathcal{M}$ enriched precategory $\mathcal{A}$ associates an $\mathcal{M}$-enriched category, i.e. a precategory satisfying the Segal condition. As a rough approximation the idea is to "force" the Segal condition in a minimal way, an operation that can be accomplished using a series of pushouts along standard cofibrations.

The passage from precategories to Segal $\mathcal{M}$-categories is inspired by the workings of the theory of simplicial presheaves as developed by Joyal [156] and Jardine [153]. Whereas their ultimate objects of interest were simplicial presheaves satisfying a descent condition, it was most convenient to consider all simplicial presheaves and impose a model structure such that the fibrant objects will satisfy descent. The Segal condition is very close to a descent condition, as has been remarked by Berger [36].

As in Jardine [153], we are tempted to use the *injective model structure* for diagrams, defining the cofibrations to be all maps of diagrams $\mathcal{A} \to \mathcal{B}$ which induce cofibrations at each stage $\mathcal{A}_n \to \mathcal{B}_n$. It turns out that a slightly better alternative is to use a Reedy definition of cofibration, see Chapter 13. The Reedy structure exists in greater generality, and yields the important property that degeneracy maps such as $* \to \mathcal{A}(x, x)$ are cofibrations. If $\mathcal{M}$ is itself a presheaf category with monomorphisms as cofibrations, then they coincide. This case is sufficient for our iterative application of the construction, and the reader is invited to think of injective cofibrations rather than Reedy ones for simplicity. It can also be helpful to maintain a parallel *projective model structure* where the cofibrations are generated by elementary cofibrations as originally done by Bousfield. However, the projective structure is not appropriate for the iteration step: it will not generally give back a cartesian model category.

Once we have a construction $\mathcal{A} \mapsto \mathbf{Seg}(\mathcal{A})$ which enforces the Segal condition, a map $\mathcal{A} \to \mathcal{B}$ is said to be a "weak equivalence" if $\mathbf{Seg}(\mathcal{A}) \to \mathbf{Seg}(\mathcal{B})$ satisfies the usual conditions for being an equivalence of enriched categories, essential surjectivity and full faithfulness. A map of $\mathcal{M}$-enriched categories $f : \mathcal{A} \to \mathcal{B}$ is *fully faithful* if, for any two objects $x, y \in \mathrm{Ob}(\mathcal{A})$ the map $\mathcal{A}(x, y) \to \mathcal{B}(f(x), f(y))$ is a weak equivalence in $\mathcal{M}$. Taking a homotopy class projection $\pi_0 : \mathcal{M} \to \mathrm{SET}$ gives a truncation operation $\tau_{\leq 1}$ from $\mathcal{M}$-enriched categories to 1-categories, and we say that $f : \mathcal{A} \to \mathcal{B}$ is *essentially surjective* if $\tau_{\leq 1}(f) : \tau_{\leq 1}(\mathcal{A}) \to \tau_{\leq 1}(\mathcal{B})$ is an essentially surjective map of 1-categories, i.e. it is surjective on isomorphism classes. The isomorphism classes of $\tau_{\leq 1}(\mathcal{A})$ should be thought of as the "equivalence classes" of objects of $\mathcal{A}$. Putting these together, we say that a map $f : \mathcal{A} \to \mathcal{B}$ is an *equivalence of $\mathcal{M}$-enriched categories* if it is fully faithful and essentially surjective.

Now, we say that a map $f : \mathcal{A} \to \mathcal{B}$ of $\mathcal{M}$-enriched precategories, is a *weak equivalence* if the corresponding map between the $\mathcal{M}$-enriched categories obtained by generators and relations $\mathbf{Seg}(f) : \mathbf{Seg}(\mathcal{A}) \to \mathbf{Seg}(\mathcal{B})$ is an equivalence in the above sense. With this definition and any one of the classes of cofibrations briefly referred to above and considered in detail in Chapter 13, the specification of the model structure is completed by defining the fibrations to be the morphisms satisfying right lifting with respect to trivial cofibrations, i.e. cofibrations which are weak equivalences.

6.8 Products, intervals and the model structure

The introduction of $\mathcal{M}$-precategories together with the operation **Seg** allows us to define *pushouts of weakly $\mathcal{M}$-enriched categories*: if $\mathcal{A} \to \mathcal{B}$ and $\mathcal{A} \to \mathcal{C}$ are morphisms of weak $\mathcal{M}$-enriched categories, then the pushout of diagrams $\Delta^o \to \mathcal{M}$ gives an $\mathcal{M}$-precategory $\mathcal{B} \cup^{\mathcal{A}} \mathcal{C}$. The associated pushout in the world of weakly $\mathcal{M}$-enriched categories is supposed to be $\mathbf{Seg}(\mathcal{B} \cup^{\mathcal{A}} \mathcal{C})$. Proving that the collections of maps we have defined above, really do define a model category, may be viewed as showing that this pushout operation behaves well. As came out pretty clearly in Jardine's construction [153] but was formalized in *Smith's recognition principle* as explained by Beke [33], Dugger [95] and Barwick [18], the key step is to prove that pushout by a trivial cofibration is again a trivial cofibration.

Before getting to the proof of this property, one has to calculate something somewhere, which is what is done in Chapters 16 and 17 leading up to the theorem that the calculus of generators and relations is compatible with cartesian products:

$$\mathbf{Seg}(\mathcal{A}) \times \mathbf{Seg}(\mathcal{B}) \to \mathbf{Seg}(\mathcal{A} \times \mathcal{B})$$

is a weak equivalence. Our proof of this compatibility really starts in Chapter 16 about free ordered $\mathcal{M}$-enriched categories. These may be used as basic building blocks for the generators defining the model structure, so it suffices to check the product condition on them, which is then done in Chapter 17.

The compatibility between **Seg** and cartesian products leads to what will be the main part of the cartesian property for the model category which is being constructed. This is a categorical analogue of the Eilenberg–Zilber theorem for simplicial sets. It wouldn't be true if we hadn't kept the degeneracy maps in Δ, and the strict unitality condition seems to be essential too.

From this result on cartesian products, a trick lets us conclude the main result for constructing the model structure via a Smith-type recognition theorem: that trivial cofibrations are preserved by pushout. For that trick, one requires also a good notion of interval, which was the subject of Pellissier's correction [211] to an error of mine [234]. Although Pellissier discussed only the case of Segal categories enriched over the model category $\mathcal{K}$ of simplicial sets, his construction transfers to $\mathcal{PC}(\mathcal{M})$ by functoriality using a functor $\mathcal{K} \rightarrow \mathcal{M}$. A somewhat similar correction was made by Bergner in her construction [39] of the model category structure for simplicial categories originally suggested by Dwyer and Kan.

The construction of a natural "interval category" is described in Chapter 18. It is a sort of versal replacement for the simple category $0 \leftrightarrow 1$ with two isomorphic objects. This is the point where Pellissier's correction [211] of *loc. cit.* [234] comes in, and in order to make the process fully iterative we just have to point out that an interval for the case of the standard model category $\mathcal{M} = \mathcal{K}$ of simplicial sets, leads by functoriality to an interval for any other $\mathcal{M}$. The good version of the versal interval constructed by Pellissier [211] is similar to the interval object for dg-categories subsequently introduced by Drinfeld [93]. We modify slightly Pellissier's construction, but one could use his original one too.

Once all of these ingredients are in place, we can construct the model structure in Chapter 19, a model category structure on $\mathcal{PC}(\mathcal{M})$ which again satisfies all of the hypotheses which were required of $\mathcal{M}$, so the process can be iterated.

Starting with the trivial model category structure on SET, the n-th iterate $\mathcal{PC}^n(\text{SET})$ is the model category structure for n-precategories as considered in *loc. cit.* [234]. If instead we start with the Kan–Quillen model category $\mathcal{K}$ of simplicial sets, then $\mathcal{PC}^n(\mathcal{K})$ is the model category of Segal n-precategories which was used in the work of Hirschowitz and myself [145] about n-stacks. We discuss these iterations for weak n-categories (which are Tamsamani's *n-nerves*) and Segal n-categories, together with a few variants where the initializing category of sets is replaced by a category of graphs or other things, in Chapter 20.

The internal $\underline{\text{HOM}}$ operation then leads to a category enriched over our new model category, which in the iterative scheme for n-categories gives a construction of the $(n+1)$-category nCAT.

The last part of the book is dedicated to considering how to write in this language some basic elements of higher category theory, such as inverting morphisms, and limits and colimits. We construct limits and colimits in the $(n+1)$-category nCAT of n-categories. The last chapter is about the proof of the Breen–Baez–Dolan stabilization hypothesis, relating the behavior of the theories of n-categories for different values of n.

PART II

Categorical preliminaries

7
Model categories

To start off the categorical prerequisites of the theory, this chapter reviews some of the basic elements of Quillen's theory of model categories and modern variants. We cover the main definitions and what can be done on a formal level. A more thorough discussion of the process of adding on infinitely many cells to form a cell complex, and the resulting constructions which are grouped around what is usually known as the "small object argument," are left for the next chapter. These are the places which use the notion of *locally presentable category* in a fundamental way, so that treatment will start off the next chapter. For a reference to the distinction between large and small categories, the reader is therefore asked to skip ahead to the beginning of the next chapter.

Near the end of the present chapter, the notion of *cartesian model category* is introduced. This is crucial to the subsequent development of an iterative theory of higher categories, as has been discussed in the first part of the book. We show here how to make use of the cartesian property to gain an internal <u>Hom</u>, a construction which eventually leads to the $(n+1)$-category nCAT of n-categories.

A few words are in order about why the theory of Quillen model categories is useful for treating higher categories. It seems clear that Grothendieck was aware of the applicability of the theory of model categories to these questions, at least on an intuitive level, because "Pursuing stacks" [132] started out as a series of letters to Quillen.

We have seen that iterating the theory of strictly enriched categories doesn't give an adequate theory. Thus, the notion of "equivalence" has to be taken into account somewhere in the theory. This inexorably calls into play some kind of homotopy theory, or what is essentially equivalent, some kind of localization problem. More particularly, the third chapter of this part, about left Bousfield localization, is specifically designed to yield the process $\mathcal{A} \mapsto \mathbf{Seg}(\mathcal{A})$, whereby a precategory is replaced by its associated Segal category. In the current state of the art, the most complete array of available methods for treating these kinds of questions centers around the theory of model categories. So, while the utilization of model categories is not set in stone–one can imagine working up to a fully fledged machine for dealing with homotopy theory based on any one of the many available concrete theories of $(\infty, 1)$-categories such as Lurie [190] and Joyal [157] did with quasicategories–for now it is the most practical way to approach the problem. And in any case, most of the main steps will undoubtedly persist whatever underlying technical tools are used.

On a more precise level, one of our aims is to construct the $(n+1)$-category nCAT of all n-categories. Its mapping objects will be represented by the internal $\underline{\text{HOM}}$ construction:

$$n\text{CAT}(\mathcal{A}, \mathcal{B}) = \underline{\text{HOM}}(\mathcal{A}, \mathcal{B}).$$

We need a way of insuring that this is the "homotopically correct" n-category of morphisms from $\mathcal{A}$ to $\mathcal{B}$, for example it should be invariant under equivalences in $\mathcal{A}$ and $\mathcal{B}$. The theory of model categories provides precisely a framework for doing that in the context of the usual (not internal) HOM: the idea is that if $\mathcal{A}$ is cofibrant, and $\mathcal{B}$ is fibrant, then $\text{HOM}(\mathcal{A}, \mathcal{B})$ is the right thing. The cartesian model condition allows this to be upgraded to the internal $\underline{\text{HOM}}(\mathcal{A}, \mathcal{B})$.

7.1 Lifting properties

Before getting started, recall the following terminology. Suppose $A \subset \text{ARR}(\mathcal{M})$ is a class of morphisms. We say that a morphism $f : X \to Y$ *satisfies the right lifting property with respect to* A if, for any commutative diagram

$$\begin{array}{ccc} U & \xrightarrow{a} & X \\ {\scriptstyle u}\downarrow & & \downarrow{\scriptstyle f} \\ V & \xrightarrow{b} & Y \end{array}$$

such that $u \in A$, there exists a lifting $s : V \to X$ such that $fs = b$ and $su = a$. We say that f *satisfies the left lifting property with respect to* A if, for any commutative diagram

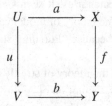

$$\begin{array}{ccc} X & \xrightarrow{a} & U \\ {\scriptstyle f}\downarrow & & \downarrow{\scriptstyle v} \\ Y & \xrightarrow{b} & V \end{array}$$

such that $v \in A$, there exists a lifting $s : Y \to U$ such that $vs = b$ and $sf = a$.

Refer to Subsection 8.0.1 at the start of the next chapter for our conventions on universes and the terminology of large and small categories.

7.2 Quillen's axioms

Start by recalling Quillen's definition[1] of a *model category* [215]. This is a category $\mathcal{M}$ provided with three classes of morphisms, the "weak equivalences," the "cofibrations" and the "fibrations." The intersection of the classes of cofibrations and weak equivalences is called the class of "trivial cofibrations," and similarly the intersection of the classes of fibrations and weak equivalences is called the class of "trivial fibrations." Quillen asks that these should satisfy the following axioms.

(CM1)–The category $\mathcal{M}$ should be closed under finite limits and colimits. Following modern tradition, we really require that it be closed under all small limits and colimits.

(CM2)–The class of weak equivalences satisfies *3 for 2*: given a composable sequence of arrows in $\mathcal{M}$

$$X \xrightarrow{f} Y \xrightarrow{g} Z,$$

if any two of f, g and gf are weak equivalences, then so is the third one.

(CM3)–The classes of cofibrations, fibrations and weak equivalences should be closed under retracts: given a diagram

$$\begin{array}{ccccc} A & \xrightarrow{i} & X & \xrightarrow{r} & A \\ \downarrow & & \downarrow & & \downarrow \\ B & \xrightarrow{j} & Y & \xrightarrow{s} & B \end{array}$$

[1] As is now usual, we use the terminology "model category" for what Quillen originally called a "closed model category"; since the nonclosed case doesn't appear, the adjective "closed" is dropped for brevity.

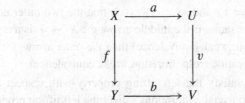

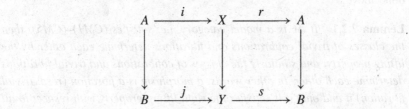

such that $ri = 1_A$ and $sj = 1_B$, such that the two outer downward arrows are the same, if the middle arrow $g : X \to Y$ is a cofibration (resp. fibration, weak equivalence) then the outer arrow $f : A \to B$ is also a cofibration (resp. fibration, weak equivalence).

(CM4)–Cofibrations satisfy the left lifting property with respect to trivial fibrations, and trivial cofibrations satisfy the left lifting property with respect to fibrations.

(CM5)–If $f : X \to Y$ is any morphism, then there exist factorizations (i) and (ii) of f as the composition $X \overset{g}{\to} Z \overset{p}{\to} Y$ such that:

(i) g is a cofibration and p is a trivial fibration;

(ii) g is a trivial cofibration and p is a fibration.

It follows from these axioms that any two of the classes of cofibrations, fibrations and weak equivalences, determine the third. For example a morphism is a fibration (resp. trivial fibration) if and only if it satisfies right lifting with respect to any trivial cofibration (resp. cofibration), and dually. A morphism is a weak equivalence if and only if it factors as a composition of a trivial fibration and a trivial cofibration, these classes being determined from the classes of cofibrations and fibrations respectively by the lifting property. Each of the three classes contains any isomorphism.

We point out [234] that the diagram included in Quillen's monograph [215] for the definition of "retract" is visibly wrong, so his notion of "retract" is not well defined. There could be two reasonable interpretations of this condition. For condition (CM2) we have adopted the weak interpretation. The stronger interpretation would have the arrows on the bottom row going in the opposite direction. If f is a strong retract of g then it is also a weak retract of g. Hence, closure under retracts as we require in (CM2) also implies closure under strong retracts. This choice coincides with what was said by Dwyer and Spalinski [105, 2.6]. Similarly, Hinich [143] uses the retract condition stated as we have done above.

Lemma 7.2.1 *If $\mathcal{M}$ is a model category, i.e. satisfies (CM1)–(CM5), then the classes of trivial cofibrations and fibrations determine each other by the lifting property; and similarly the classes of cofibrations and trivial fibrations determine each other. In other words, a morphism is a fibration (resp. trivial fibration) if and only if it satisfies the right lifting property with respect to all trivial cofibrations (resp. all cofibrations). And a morphism is a cofibration (resp. trivial cofibration) if and only if it satisfies the left lifting property with respect to all trivial fibrations (resp. all fibrations).*

Proof There are four things to prove. Consider for example the statement that a morphism is a fibration if and only if it satisfies the right lifting property with respect to trivial cofibrations; the other three arguments are identical. If f is a fibration then by (CM4) it satisfies the right lifting property with respect to all trivial cofibrations. Suppose on the other hand that f is a morphism which satisfies the right lifting property with respect to all trivial cofibrations. Using (CM5), factor $f = pg$,

$$X \xrightarrow{g} Z \xrightarrow{p} Y,$$

where g is a trivial cofibration and p is a fibration. Apply the right lifting property being assumed for f, to the diagram

$$\begin{array}{ccc} X & \xrightarrow{=} & X \\ {\scriptstyle g}\downarrow & & \downarrow{\scriptstyle f} \\ Z & \xrightarrow{p} & Y \end{array}$$

to get a morphism $s : Z \to X$ such that $sg = 1_X$ and $fs = p$. Putting s into the diagram

$$\begin{array}{ccccc} X & \xrightarrow{g} & Z & \xrightarrow{s} & X \\ \downarrow & & \downarrow & & \downarrow \\ Y & \xrightarrow{=} & Y & \xrightarrow{=} & Y \end{array}$$

now says that f is a retract of the fibration $p : Z \to Y$, so by (CM3) f is a fibration. □

7.2.1 Quillen adjunctions

Suppose $\mathcal{M}$ and $\mathcal{N}$ are model categories. A *Quillen adjunction* from $\mathcal{M}$ to $\mathcal{N}$ is an adjoint pair of functors $L : \mathcal{M} \to \mathcal{N}$ and $R : \mathcal{N} \to \mathcal{M}$, with L left adjoint and R right adjoint, such that:

(QA1)–L sends cofibrations to cofibrations and trivial cofibrations to trivial cofibrations;

(QA2)–R sends fibrations to fibrations and trivial fibrations to trivial fibrations.

Either one of these conditions implies the other, by the adjunction formula.

If $L : \mathcal{M} \to \mathcal{N}$ is a functor admitting a right adjoint R such that (L, R) is a Quillen adjunction, we say that L is a *left Quillen functor*, similarly the right adjoint R is a *right Quillen functor*. We say that L or R or (L, R) is a *Quillen equivalence* if a map $L(x) \to y$ is a weak equivalence if and only if the corresponding map $x \to R(y)$ is a weak equivalence.

7.3 Left properness

Recall that a model category $\mathcal{M}$ is *left proper* if, in any pushout square

$$\begin{array}{ccc} X & \longrightarrow & Y \\ \downarrow & & \downarrow \\ Z & \longrightarrow & W \end{array}$$

such that $X \to Y$ is a cofibration and $X \to Z$ is a weak equivalence, then $Y \to W$ is also a weak equivalence.

Lemma 7.3.1 *Suppose $\mathcal{M}$ is a left proper model category, and suppose we are given a diagram*

$$\begin{array}{ccc} X & \longrightarrow & Y \\ \downarrow & \searrow & \downarrow \\ Z & & V \end{array}$$

such that $Y \to V$ is a weak equivalence. Suppose either that $X \to Z$ is a cofibration, or that both maps $X \to Y$ and $X \to V$ are cofibrations. Then the map

$$Z \cup^X Y \to Z \cup^X V$$

is a weak equivalence.

Proof The case where $X \to Z$ is a cofibration is a straightforward application of the definition of left properness. Suppose that $X \to Y$ and $X \to V$ are cofibrations. Choose a factorization of the map $X \to Z$ into $X \xrightarrow{c} W \xrightarrow{f} Z$ such that c is a cofibration and f is a trivial fibration. The map

$$W \cup^X Y \to (W \cup^X Y) \cup^Y V = W \cup^X V$$

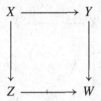

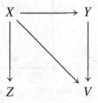

is the pushout of the weak equivalence $Y \to V$ along the cofibration $Y \to W \cup^X Y$, so by the left properness condition, it is a weak equivalence. On the other hand, the morphisms $W \cup^X Y \to Z \cup^X Y$ and $W \cup^X V \to Z \cup^X V$ are pushouts of the weak equivalence $W \to Z$ along the cofibrations $W \to W \cup^X Y$ and $W \to W \cup^X V$ respectively. Again by left properness, these maps are weak equivalences. Hence, in the diagram

$$\begin{array}{ccc} W \cup^X Y & \longrightarrow & W \cup^X V \\ \downarrow & \searrow & \downarrow \\ Z \cup^X Y & & Z \cup^X V \end{array}$$

the top arrow and both vertical arrows are weak equivalences; hence the bottom arrow is a weak equivalence as desired. □

The following corollary is known as "Reedy's lemma" [217]:

Corollary 7.3.2 (Reedy's lemma) *Suppose $\mathcal{M}$ is a left proper model category and suppose given a diagram*

$$\begin{array}{ccccc} X & \longleftarrow & Y & \longrightarrow & Z \\ \downarrow & & \downarrow & & \downarrow \\ A & \longleftarrow & B & \longrightarrow & C \end{array}$$

such that the vertical arrows are weak equivalences, and the left horizontal arrows are cofibrations. Then the map

$$X \cup^Y Z \to A \cup^B C$$

is a weak equivalence.

Proof Applying the previous lemma on both sides gives the required statement in the case $Y = B$. In general let $A' := X \cup^Y B$. The map $X \to A'$ is a weak equivalence by left properness, so $A' \to A$ is a weak equivalence by 3 for 2. The map $B \to A'$ is a cofibration. Applying the case where the middle map is an isomorphism gives the statement that

$$A' \cup^B C \to A \cup^B C$$

is a weak equivalence. That case, or really the previous lemma, also implies that the map

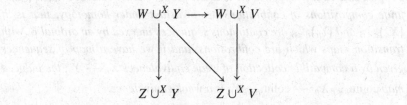

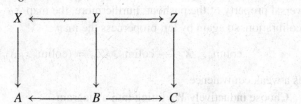

$$X \cup^Y Z \to X \cup^Y C = (X \cup^Y B) \cup^B C = A' \cup^B C$$

is a weak equivalence. Putting these together gives the required statement. □

We sometimes need an analogous invariance property for transfinite compositions:

Proposition 7.3.3 *Suppose $\mathcal{M}$ is a left proper model category. Then transfinite compositions of cofibrations are invariant under homotopy; that is, if $\{X_n\}_{n<\beta}$ and $\{Y_n\}_{n<\beta}$ are continuous sequences indexed by an ordinal β, with transition maps which are cofibrations, and if we have a map of sequences given by a compatible collection of weak equivalences $X_n \xrightarrow{g_n} Y_n$, the induced map $\mathrm{colim}_{n<\beta} X_n \xrightarrow{g_\beta} \mathrm{colim}_{n<\beta} Y_n$ is a weak equivalence.*

Proof The proof is by induction on the ordinal β; we may assume it is known for all sequences indexed by ordinals $\alpha < \beta$. Suppose given continuous sequences $\{X_n\}_{n<\beta}$ and $\{Y_n\}_{n<\beta}$ with transition maps which are cofibrations, and $X_n \xrightarrow{g_n} Y_n$ compatible with the transition maps of the sequences. Put $Z_n := X_n \cup^{X_0} Y_0$. By left properness, the maps $X_n \to Z_n$ are weak equivalences, hence by 3 for 2 the same holds for the maps $Z_n \to Y_n$ induced by the universal property of the pushout. Furthermore, the map $X_0 \to \mathrm{colim}_{n<\beta} X_n$ is a cofibration, so again by left properness the map

$$\mathrm{colim}_{n<\beta} X_n \to \mathrm{colim}_{n<\beta} Z_n = (\mathrm{colim}_{n<\beta} X_n) \cup^{X_0} Y_0$$

is a weak equivalence.

Choose inductively W_n fitting into a diagram

$$\begin{array}{ccccccccc}
Z_0 & \longrightarrow & Z_1 & \longrightarrow & \cdots & \longrightarrow & Z_n & \longrightarrow & \cdots \\
\| & & \downarrow{\scriptstyle i_1} & & & & \downarrow{\scriptstyle i_n} & & \\
W_0 & \longrightarrow & W_1 & \longrightarrow & \cdots & \longrightarrow & W_n & \longrightarrow & \cdots \\
\| & & \downarrow{\scriptstyle p_1} & & & & \downarrow{\scriptstyle p_n} & & \\
Y_0 & \longrightarrow & Y_1 & \longrightarrow & \cdots & \longrightarrow & Y_n & \longrightarrow & \cdots
\end{array}$$

such that i_n are trivial cofibrations and p_n are trivial fibrations, for any non-limiting ordinal $n < \beta$. Furthermore, assume that once given the choice up to n, the next step is obtained by applying the factorization property to get

$$W_n \cup^{Z_n} Z_{n+1} \to W_{n+1} \xrightarrow{p_{n+1}} Y_n,$$

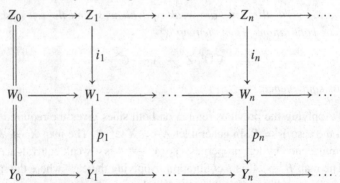

a cofibration followed by a trivial fibration. Using left properness, the map $Z_{n+1} \to W_n \cup^{Z_n} Z_{n+1}$ is a weak equivalence so the composed map above is a weak equivalence, meaning that the left map in the factorization is a trivial cofibration. This in turn says that i_{n+1} is a trivial cofibration.

If n is a limit ordinal, choose $W_n = \lim_{j<n} W_j$, so that $W.$ is continuous. Note that the map $Z_n \to W_n$ will still satisfy the left lifting property against any fibration, as may be seen using induction and the factorization used to choose the i_{n+1}. Therefore $Z_n \to W_n$ is still a trivial cofibration when n is a limit ordinal. Similarly, once we are done, the map $\mathrm{colim}_{n<\beta} Z_n \to \mathrm{colim}_{n<\beta} W_n$ will be a trivial cofibration. On the other hand, for a limiting ordinal $n < \beta$ the induction hypothesis implies that the map p_n is still a weak equivalence, allowing the inductive choice to continue. Note furthermore that the transition maps of $W.$, as chosen above, are cofibrations.

The trivial fibration property at successor ordinals, and continuity at limit ordinals, plus the fact that the transition maps of $Y.$ are cofibrations, allow us to choose a sequence of maps $Y_n \xrightarrow{s_n} W_n$ which are sections of p_n and commute with the transition maps. The map $p_\beta : \mathrm{colim}_{n<\beta} W_n \to \mathrm{colim}_{n<\beta} Y_n$ therefore admits a section s_β given by the colimit of the s_n. This shows that p_β admits a right inverse.

We have now proven that our original map g_β admits a right inverse in the homotopy category. However, s_β comes from a system satisfying the same hypotheses, so s_β also admits a right inverse in the homotopy category; but it has by construction a left inverse, so s_β is a weak equivalence; then its left inverse p_β is a weak equivalence, which implies that g_β is one too. □

7.4 The Kan–Quillen model category of simplicial sets

The most important model category is the category of simplicial sets with the structure of Quillen model category originally defined by Kan [163, 164, 165, 166]. We denote this model category by $\mathcal{K}$. As a category, it is the category of functors $\Delta^o \to \mathrm{SET}_{\mathbb{U}}$, where $\mathbb{U}$ denotes our main chosen universe. The cofibrations are just the monomorphisms. The fibrations are the maps satisfying the Kan lifting condition for all horns over standard simplices. Modern proofs that these classes of maps provide a model structure are given by Goerss and Jardine [122] and Hovey [149] for example.

7.4.1 Generating sets

The model category $\mathcal{K}$ provides a good first example on which to illustrate the notion of generating sets. Let $h(n)$ denote the standard simplex of dimension

n, which as a diagram $\Delta^o \to \text{SET}_{\mathbb{U}}$ is just the functor represented by $[n]$. Let $\partial h(n)$ denote the boundary, defined by the condition that $(\partial h(n))_k$ is the subset of simplices in $h(n)_k$ which factor through some principal face $h(n-1) \subset h(n)$. Let $I \subset \text{ARR}(\mathcal{K})$ be the set consisting of all inclusions of boundaries $\partial h(n) \hookrightarrow h(n)$.

The set I generates the class of cofibrations in the following sense: let **cell**(I) denote the class of maps obtained by transfinite compositions of pushouts along arrows in I. Let **inj**(I) denote the class of maps which satisfy the right lifting property with respect to **cell**(I), or equivalently with respect to just I itself. It is equal to the class of trivial fibrations. The class of cofibrations is equal to **cof**(I), the class of maps which satisfy left lifting with respect to **inj**(I). In fact, in this case it turns out that **cof**(I) = **cell**(I) (see Hovey [149, 3.2.2] for example). That property doesn't usually hold in general, but on the other hand Lurie [190] shows that it can sometimes be attained by widening the generating sets, as we shall discuss in Theorem 8.6.1.

The *k-th horn* $\partial\langle k\rangle h(n)$ is the subobject spanned by all principal faces except the k-th one. Let $J \subset \text{ARR}(\mathcal{K})$ denote the set consisting of all inclusions of the form $\partial\langle k\rangle h(n) \hookrightarrow h(n)$. As before, **cell**$(J)$ is defined to be the class obtained by transfinite compositions of pushouts along arrows in J. The *Kan fibrations* are the maps in **inj**(J), those which satisfy right lifting with respect to **cell**(J) or equivalently J. Then **cof**(J), the class of maps satisfying left lifting with respect to **inj**(J), is the class of trivial cofibrations. These may also be characterized as the retracts of maps in **cell**(J), and for the model category of simplicial sets they are often called "anodyne extensions."

Kan's fibrant replacement functor $\mathbf{Ex}^\infty$ is a concrete version of what we get upon applying the "small object argument" to J: given a simplicial set X, just add in infinitely many times all possible pushouts by elements of J. This operation forms the basis for most constructions in the theory of model categories. A good understanding requires careful consideration of the class of cell complexes **cell**(J), the subject of the next chapter.

Following this basic example, the process of specifying "generating sets" for the cofibrations and trivial cofibrations, via the small object argument, has classically been one of the most useful and common ways of constructing model catgories. They are called "cofibrantly generated model categories." The next chapter is devoted to this theory, because of its importance for all that follows. For expositional purposes, some of the following basic facts about model categories will refer to this notion, so we introduce the terminology here.

In the context of a general category $\mathcal{M}$, if $I \subset \text{ARR}(\mathcal{M})$ is a small subset of arrows, then **cof**(I) is defined to be the smallest class of arrows containing I and closed under retracts, pushouts and transfinite composition. Dually,

inj(I) is defined to be the largest class of arrows which satisfy the right lifting property with respect to arrows in I. These come from **cell**(I) as described above.

A pair of subsets I, J constitutes a *pair of generating sets* for a model category structure on $\mathcal{M}$, if **cof**(I) is the class of cofibrations, **cof**(J) is the class of trivial cofibrations, **inj**(J) is the class of fibrations, and **inj**(I) is the class of trivial fibrations. If there exists a pair of generating sets, then the model structure is said to be *cofibrantly generated.*

If furthermore $\mathcal{M}$ is a locally presentable category (a property to be discussed at the beginning of the next chapter) then the model structure is said to be *combinatorial.* In a combinatorial model category, if there exists a pair of generating sets I and J, all of whose arrows have domains which are themselves cofibrant objects, the model structure is said to be *tractable.* The reader is requested to consult the next chapter, in order to follow the few technical steps which refer to these notions in some places of the present one.

The Kan–Quillen model category of simplicial sets $\mathcal{K}$, being the source of most of our intuition on these matters, satisfies all of the various properties to be considered: it is tractable (hence combinatorial and cofibrantly generated), left proper, cartesian, etc.

7.5 Homotopy liftings and extensions

For $X \in \mathcal{M}$ a *cylinder object* is a diagram

$$X \cup^{\emptyset} X \xrightarrow{i_0 \cup i_1} C \xrightarrow{p} X$$

such that $i_0 \cup i_1$ is a cofibration and p is a trivial fibration. The existence is guaranteed by axiom (CM5)(i). Similarly, if $X \to Y$ is a cofibration there exists a *relative cylinder object*, which is a diagram

$$Y \cup^{X} Y \xrightarrow{i_0 \cup i_1} C \xrightarrow{p} Y$$

again with $i_0 \cup i_1$ a cofibration and p a trivial fibration. Quillen shows that if X is cofibrant and A fibrant, then two maps $f_0, f_1 : X \to A$ are homotopic, that is project to the same map in ho$(\mathcal{M})$, if and only if there is a cylinder object and a map (or for any cylinder object there is a map) $C \to A$ inducing the two given maps. Given a cofibration $X \to Y$ and two maps $f_0, f_1 : Y \to A$ which agree when restricted to X, we say they are *homotopic relative to* X if there is a relative cylinder object and a map $C \to A$ inducing the two maps.

Weak equivalences between fibrant objects can be characterized by a homotopy lifting property, being careful to look at homotopies relative to the subobject of the cofibration. This is a classical fact from the theory of model categories but, as it is a technical step needed in Chapter 11, the proof is included here for completeness.

Lemma 7.5.1 *Suppose $g : X \to Y$ is a map between fibrant objects in a tractable model category $\mathcal{M}$, such that for any diagram*

$$\begin{array}{ccc} U & \xrightarrow{u} & X \\ {\scriptstyle f}\downarrow & & \downarrow{\scriptstyle g} \\ V & \xrightarrow{v} & Y \end{array}$$

where f is a generating cofibration, there exists a lifting $r : V \to X$ such that $rf = u$, together with a relative cylinder object

$$V \cup^U V \xrightarrow{j_0 \cup j_1} IV \xrightarrow{q} V$$

with $j_0 \cup j_1$ a cofibration and q a weak equivalence, and a map $h : IV \to Y$ restricting to $gu = vf$ on U, such that $hj_0 = gr$ and $hj_1 = v$. Then g is a weak equivalence.

Proof Factor $X \xrightarrow{k} X' \xrightarrow{g'} Y$ where k is a trivial cofibration and g' is a fibration. Since X is assumed fibrant there is a retraction s from X' to X with $sk = 1_X$. We show that g' is a trivial fibration by the lifting property. Suppose given a diagram

$$\begin{array}{ccc} U & \xrightarrow{u'} & X' \\ {\scriptstyle f}\downarrow & & \downarrow{\scriptstyle g'} \\ V & \xrightarrow{v'} & Y \end{array}$$

with f a generating cofibration. Put $u := su' : U \to X$; there is a homotopy between ku and u', given by a cylinder object

$$U \cup^{\emptyset} U \xrightarrow{i_0 \cup i_1} IU \xrightarrow{p} U$$

with a map $z : IU \to X'$ with $zi_0 = ku$ and $zi_1 = u'$. Choose a compatible cylinder object for V, fitting into a diagram

$$
\begin{array}{ccccc}
U \cup^{\emptyset} U & \xrightarrow{i_0 \cup i_1} & IU & \xrightarrow{p} & U \\
{\scriptstyle f}\downarrow & & {\scriptstyle If}\downarrow & & \downarrow{\scriptstyle f} \\
V \cup^{\emptyset} V & \xrightarrow{l_0 \cup l_1} & I'V & \xrightarrow{q} & V.
\end{array}
$$

Extend the map $IU \cup^{i_1,U,f} V \to Y$ given by $g'z \cup v'$, to a map $t : I'V \to Y$ (this is an extension along a trivial cofibration, for maps to the fibrant object Y). Restricting along l_0 now gives a map $v : V \to Y$ such that $vf = g'ku = gu$. We obtain a diagram

$$
\begin{array}{ccc}
U & \xrightarrow{u} & X \\
{\scriptstyle f}\downarrow & & \downarrow{\scriptstyle g} \\
V & \xrightarrow{v} & Y.
\end{array}
$$

By hypothesis there is a lifting $r : V \to X$ such that $rf = u$ and gr is homotopic to v relative to U. This gives a lifting $kr : V \to X'$ such that $krf = ku$ and $g'kr$ is homotopic to v relative to U. Putting this back together with the previous homotopy, we get the top map in the diagram

$$
\begin{array}{ccc}
IU \cup^{i_0,U,f} V & \xrightarrow{z \cup kr} & X' \\
{\scriptstyle If \cup l_0}\downarrow & & \downarrow{\scriptstyle g'} \\
I'V & \xrightarrow{t} & Y
\end{array}
$$

where the left map is a trivial cofibration and g' is a fibration. This diagram doesn't commute; however, by the hypothesis of the lemma (and using the notations from there) there is a homotopy relative to IU making it commute. Adding this on and shifting the map from the component V on the upper left, to the other side of the new cylinder object IV, gives the commutative diagram

$$
\begin{array}{ccc}
IU \cup^{i_0,U,f} V & \xrightarrow{z \cup kr} & X' \\
{\scriptstyle If \cup j_1}\downarrow & & \downarrow{\scriptstyle g'} \\
I'V \cup^{l_0,V,j_0} IV & \xrightarrow{t \cup h} & Y
\end{array}
$$

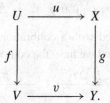

where the left map remains a trivial cofibration. Hence there exists a lifting $I'V \cup^{l_0,V,j_0} IV \to X'$ which, when restricted along l_1, gives the desired lifting to show that g' is a trivial fibration. This completes the proof. □

A similar consideration holds for extensions:

Lemma 7.5.2 *Suppose given two cofibrations $X \stackrel{a}{\to} Y \stackrel{b}{\to} Z$ and maps $Y \stackrel{f}{\to} A$ and $Z \stackrel{g}{\to} A$ to a fibrant object A, such that $fa = gba$. Suppose that gb is homotopic to f relative to X. Then there exists a map $Z \stackrel{g'}{\to} A$ such that $g'b = f$.*

Proof Let C be a relative cylinder object for $X \to Y$. Choose a factorization

$$Z \cup^{Y,i_0} C \cup^{Y,i_1} Z \stackrel{d}{\to} D \stackrel{q}{\to} Z$$

of a trivial fibration composed with a cofibration. Then D is also a relative cylinder object for $X \to Z$ and we have the compatibility diagram

$$\begin{array}{ccccc}
Y \cup^X Y & \xrightarrow{i_0 \cup i_1} & C & \xrightarrow{p} & Y \\
\downarrow{\scriptstyle b \cup b} & & \downarrow{\scriptstyle c} & & \downarrow{\scriptstyle b} \\
Z \cup^X Z & \xrightarrow{j_0 \cup j_1} & D & \xrightarrow{q} & Z.
\end{array}$$

Choose a map $C \stackrel{h}{\to} A$ such that $hi_0 = f$ and $hi_1 = gb$. We get a map

$$C \cup^{Y,i_1} Z \stackrel{h \cup g}{\longrightarrow} A,$$

but the cofibration $Y \stackrel{i_1}{\to} C$ is a weak equivalence, so the pushout along i_1 is a trivial cofibration $Z \to C \cup^{Y,i_1} Z$. Using the fact that d was chosen to be a cofibration at the start, it follows that

$$C \cup^{Y,i_1} Z \stackrel{d'}{\to} D$$

is a trivial cofibration. The fibrant condition for A now allows us to extend the previous map $h \cup g$ along d', so we get a map $D \stackrel{h'}{\to} A$ extending h. The restriction $g' := h' j_0 : Z \to A$ provides the required map. □

7.6 Model structures on diagram categories

Suppose Φ is a small category and $\mathcal{M}$ a category. Consider the *diagram category* $\text{FUNC}(\Phi, \mathcal{M})$ of functors $\Phi \to \mathcal{M}$. If $\mathcal{M}$ is complete (resp. cocomplete) then so is $\text{FUNC}(\Phi, \mathcal{M})$ and limits (resp. colimits) of diagrams are computed levelwise, that is over each object of Φ.

7.6.1 Some adjunctions

Suppose $f : \Phi \to \Psi$ is a functor between small categories. Given any diagram $A : \Psi \to \mathcal{M}$ then the composition $A \circ f$ is a diagram $\Phi \to \mathcal{M}$, which will also be denoted $f^*(A)$. This gives a functor $f^* : \mathrm{FUNC}(\Psi, \mathcal{M}) \to \mathrm{FUNC}(\Phi, \mathcal{M})$. If $\mathcal{M}$ is complete (resp. cocomplete) then the functor f^* preserves limits (resp. colimits) since they are computed levelwise in both $\mathrm{FUNC}(\Phi, \mathcal{M})$ and $\mathrm{FUNC}(\Psi, \mathcal{M})$.

We consider the left and right adjoints of f^*. These are left and right Kan extensions, see Mac Lane [196, X.3]. They were used with A. Hirschowitz in our preprint [145, chapter 4], see also Lurie [190, A.2.8.7].

If $\mathcal{M}$ is cocomplete then we can construct a left adjoint

$$f_! : \mathrm{FUNC}(\Phi, \mathcal{M}) \to \mathrm{FUNC}(\Psi, \mathcal{M})$$

as follows. For any object $y \in \Psi$ consider the category f/y of pairs (x, a) where $x \in \Phi$ and $a : f(x) \to y$ is an arrow in Ψ. There is a forgetful functor $r_{f,y} : f/y \to \Phi$ sending (x, a) to x.

Suppose $A \in \mathrm{FUNC}(\Phi, \mathcal{M})$. Put $f_!(A)(y) := \mathrm{colim}_{f/y} r^*_{f,y}(A)$. Suppose $g : y \to y'$ is an arrow. Then we obtain a functor $c_g : f/y \to f/y'$ sending (x, a) to (x, ga). Furthermore this commutes with the forgetful functors in the sense that $r_{f,y'} \circ c_g = r_{f,y}$. Thus $r^*_{f,y}(A) = c^*_g\big(r^*_{f,y'}(A)\big)$. Applying the above remark about colimits, we get a natural map

$$f_!(A)(y) := \mathrm{colim}_{f/y} c^*_g\left(r^*_{f,y'}(A)\right) \to \mathrm{colim}_{f/y'} r^*_{f,y'}(A) =: f_!(A)(y').$$

Using the last part of the paragraph about colimits above, we see that this collection of maps turns $f_!(A)$ into a functor from Ψ to $\mathcal{M}$, that is an object in $\mathrm{FUNC}(\Psi, \mathcal{M})$. The construction is functorial in A so it defines a functor

$$f_! : \mathrm{FUNC}(\Phi, \mathcal{M}) \to \mathrm{FUNC}(\Psi, \mathcal{M}).$$

The structural maps for the colimit defining $f_!(A)(y)$ are maps of the form $A(x) \to f_!(A)(y)$ for any $a : f(x) \to y$. In particular, when $y = f(x)$ and a is the identity we get maps $A(x) \to f_!(A)(f(x)) = f^* f_!(A)(x)$. This is a natural transformation from the identity on $\mathrm{FUNC}(\Phi, \mathcal{M})$ to $f^* f_!$.

On the other hand, suppose $A = f^*(B)$ for $B \in \mathrm{FUNC}(\Psi, \mathcal{M})$. Then, for any $(x, a) \in f/y$ we get a map

$$r^*_{f,y}(f^*(B))(x, a) = B(f(x)) \xrightarrow{B(a)} B(y).$$

This gives a map from $r^*_{f,y}(f^*(B))$ to the constant diagram with values $B(y)$, hence a map on the colimit

$$f_! f^*(B)(y) \mathrm{colim}_{f/y} r^*_{f,y}(f^*(B)) \to B(y).$$

It is functorial in y and B so it gives a natural transformation from $f_! f^*$ to the identity.

Lemma 7.6.1 *Supposing that $\mathcal{M}$ is cocomplete and with these natural transformations, $f_!$ becomes left adjoint to f^* and one has the formula*

$$f_!(A)(y) = \mathrm{colim}_{f/y} r^*_{f,y}(A).$$

Proof Suppose $A \in \mathrm{FUNC}(\Phi, \mathcal{M})$ and $B \in \mathrm{FUNC}(\Psi, \mathcal{M})$. To give a map $A \to f^*(B)$ consists of giving, for each $x \in \Phi$, a map $A(x) \to B(f(x))$. This is equivalent to giving, for each $y \in \Psi$ and $(x, a) \in f/y$, a map $A(x) \to B(y)$ subject to some naturality constraints as x, a, y vary. This in turn is the same as giving a map of f/y-diagrams from $r^*_{f,y}(A)$ to the constant diagram with values $B(y)$, which in turn is the same as giving a map from $f_!(A) = \mathrm{colim}_{f/y} r^*_{f,y}(A)$ to B. It is left to the reader to verify that these identifications are the same as the ones given by the above-defined adjunction maps. □

Lemma 7.6.2 *Suppose that $\mathcal{M}$ is complete. Then f^* has a right adjoint denoted f_*, given by the formula*

$$f_*(A)(y) = \lim_{f/y} s^*_{f,y}(A),$$

where f/y is the category of pairs (z, u), where $z \in \Phi$ and $f(z) \xrightarrow{u} y$ is an arrow in Ψ, and $s_{f,y} : f/y \to \Phi$ is the forgetful functor.

Proof Apply the previous lemma to the functor $f^o : \Phi^o \to \Psi^o$ for diagrams in the opposite category $\mathcal{M}^o$. □

7.6.2 Injective and projective diagram structures

Suppose Φ is a small category, and $\mathcal{M}$ is a model category. There two main types of potential model structures on the category of functors $\mathrm{FUNC}(\Phi, \mathcal{M})$, usually known as the *projective* and *injective* model structures. In both model structures, the weak equivalences are defined to be the *levelwise* ones; in other words $A \to B$ is a weak equivalence if and only if $A(x) \to B(x)$ is a weak equivalence for each $x \in \Phi$.

In the projective structure, the fibrations are defined as the levelwise ones. The cofibrations are then defined to be the maps which satisfy the left lifting property with respect to levelwise trivial fibrations. In most cases, one can give an explicit generating set for the projective cofibrations.

Dually, in the injective structure the cofibrations are defined as the levelwise ones, and the fibrations are defined as those maps having the right lifting property with respect to levelwise trivial cofibrations. Usually the fibrations don't have an easy description.

Once we have defined the classes of weak equivalences, cofibrations and fibrations, the question is whether these classes form a model structure on $\text{FUNC}(\Phi, \mathcal{M})$. This is not known in general, but when $\mathcal{M}$ has further properties of being generated by a small set of morphisms, in a way which will be discussed in the next chapter, then the projective and injective model structures exists. We state some versions of the existence theorems here; the notion of "combinatorial" model category will be considered in further detail in the next chapter.

Theorem 7.6.3 *Suppose Φ is a small category, and $\mathcal{M}$ is a combinatorial model category (cf Definition 8.4.2). Then* $\text{FUNC}(\Phi, \mathcal{M})$ *has two structures of combinatorial model category, the projective and the injective ones, with classes of morphisms described above. If $\mathcal{M}$ is left proper then so are the injective and projective diagram structures.*

These model structures have been built up successively by many authors, with a long history going back to Bousfield and Kan and others, more recently considered by Hirschhorn [144], Smith (see Beke [33], Dugger [95]), Blander [49] and Barwick [18]. See Barwick [18] and Lurie [190, proposition A.2.8.2] for the proof in the full generality stated here. Left properness may be verified levelwise. The main difficulty is in the construction of the injective model structure, to get a generating set for the cofibrations. This is done using Lurie's theorem, which will be reviewed as Theorem 8.6.2, before restating the theorem as a part of Corollary 8.6.4 in the next chapter. For most special cases including many of the cases of interest to us, the reader may refer to any of a number of prior references.

Barwick [18] generalizes this result to a relative situation, using a generalization of Lurie's theorem. A *left Quillen presheaf* is a presheaf of model categories over a base category, such that the transition functors are left Quillen functors. The *category of sections* is the category of sections of the associated fibered category. These were considered by Hirschowitz and myself [145]. Barwick [18] has the following theorem, which for the injective structure uses Lurie's techniques to be discussed in Theorems 8.6.1 and 8.6.2. The notion of left Quillen presheaf will not be used later, since the case we need is Theorem 7.6.3, which corresponds to a constant left Quillen presheaf with values $\mathcal{M}$. For informational purposes, here is Barwick's more general relative statement:

Theorem 7.6.4 *The category of sections of a left Quillen presheaf, whose values are combinatorial model categories, has an "injective" and a "projective" combinatorial model structure.*

In Chapter 11 below, we consider a variant of diagrams called *unital diagram categories*. Given $\Phi_0 \subset \mathrm{Ob}(\Phi)$ and a single model category $\mathcal{M}$, we could define a presheaf of model categories by setting $\mathcal{M}_x := \mathcal{M}$ if $x \notin \Phi_0$, with $\mathcal{M}_x := \{*\}$ for $x \in \Phi_0$. The unital diagram category $\mathrm{FUNC}(\Phi/\Phi_0, \mathcal{M})$ is the category of sections of the associated fibered category. This presheaf of categories is usually not a left Quillen presheaf so the model structure to be discussed in Corollary 8.6.4 as well as Propositions 11.4.3 and 11.4.2 later, can't be viewed as a direct corollary of 7.6.4, even though the techniques of proof are the same.

7.6.3 Reedy diagram structures

Another important type of model structure on certain diagram categories is the *Reedy* model structure, which is defined when Φ has a structure of "Reedy category." This lies in between the projective and the injective structures, and indeed a closely related variant will be provide an important class of cofibrations in Chapter 13 below. Some references for this discussion are Reedy [217], Bousfield and Kan [58], Hirschhorn [144], Dwyer *et al.* [99], Dwyer and Kan [103], Goerss and Jardine [122], Barwick [19] and Hirschowitz and myself [145, chapter 17].

A *Reedy category* is a category Φ provided with two subcategories on the same object set, called the *direct subcategory* Φ^d and the *inverse subcategory* Φ^i and a function *degree* from the set of objects to an ordinal (usually ω), such that the non-identity direct maps strictly increase the degree, the non-identity inverse maps strictly decrease the degree, and any morphism f factors uniquely as $f^d f^i$ where f^d is direct and f^i is inverse. The degree of the middle object in this factorization is $\leq$ the degrees of the source and target of f, and in case of equality f is either direct or inverse (or both in which case it is the identity).

The basic example is Δ itself. The direct subcategory consists of the injective arrows and the inverse subcategory, of the surjective arrows. The opposite category of a Reedy category is again Reedy, with the two subcategories interchanged; so for example the direct subcategory of Δ^o consists of the opposites of surjective arrows, and the inverse subcategory consists of the opposites of injective arrows.

If Φ is a Reedy category, for $y \in \Phi$ define $\mathrm{Latch}(y)$ to be Φ^d/y minus $\{y\}$ and $\mathrm{Match}(y)$ to be y/Φ^i minus $\{y\}$.

Suppose $A : \Phi \to \mathcal{M}$ is a diagram. The *latching and matching objects* at $y \in \Phi$ are defined to be

$$\mathrm{latch}(A, y) := \mathrm{colim} A|_{\mathrm{Latch}(y)},$$

$$\mathrm{match}(A, y) := \lim A|_{\mathrm{Match}(y)}.$$

A diagram $A \in \mathrm{FUNC}(\Phi, \mathcal{M})$ is *Reedy cofibrant* if the morphisms

$$\mathrm{latch}(A, y) \to A(y)$$

are cofibrations in $\mathcal{M}$, and *Reedy fibrant* if the morphisms

$$A(y) \to \mathrm{match}(A, y)$$

are fibrations in $\mathcal{M}$. A morphism $A \xrightarrow{f} B$ in $\mathrm{FUNC}(\Phi, \mathcal{M})$ is said to be a *Reedy cofibration* if the relative latching maps

$$\mathrm{latch}(f, y) := \mathrm{latch}(B, y) \cup^{\mathrm{latch}(A,y)} A(y) \to B(y)$$

are cofibrations, and a *Reedy fibration* if the relative matching maps

$$A(y) \to \mathrm{match}(A, y) \times_{\mathrm{match}(B,y)} B(y) =: \mathrm{match}(f, y)$$

are fibrations.

Cofibrant generation properties such as "combinatorial" or "tractable" (to be discussed in the next chapter) are not needed in order to get a Reedy model structure on the diagram category. However, these properties are preserved, statements we have included here for completeness.

Proposition 7.6.5 *Suppose Φ is a Reedy category. The category of diagrams* $\mathrm{FUNC}(\Phi, \mathcal{M})$ *provided with the levelwise weak equivalences, and the above classes of cofibrations and fibrations, is a model category, fitting in the middle of a sequence of left Quillen functors*

$$\mathrm{FUNC}_{\mathrm{proj}}(\Phi, \mathcal{M}) \to \mathrm{FUNC}_{\mathrm{Reedy}}(\Phi, \mathcal{M}) \to \mathrm{FUNC}_{\mathrm{inj}}(\Phi, \mathcal{M}).$$

The Reedy structure is combinatorial (resp. tractable, left proper) whenever $\mathcal{M}$ is.

Proof See the references *loc. cit.* [19, 58, 99, 103, 122, 144, 217]. In particular, inheritance of the combinatorial or tractable properties is shown by Barwick [19, lemmas 3.10, 3.11]. The left properness condition may be verified levelwise, since Reedy cofibrations are injective ones *loc. cit.* [19, lemma 3.1]. □

7.7 Cartesian model categories

The notion of *cartesian model category* plays two important roles. First of all, the Segal conditions involve cartesian products, so it is important that they behave well homotopically in the model category $\mathcal{M}$ used as input. The second application is the fact that a cartesian model category $\mathcal{P}$ leads to a $\mathcal{P}$-enriched category $\mathbf{Enr}(\mathcal{P})$ obtained by looking at its fibrant and cofibrant objects and using the internal $\underline{\text{HOM}}$ to define morphism objects. Once we will have constructed by induction the cartesian model structure on $\mathcal{P} = \mathcal{PC}^n(\text{SET})$, then $nCAT = \mathbf{Enr}(\mathcal{P})$ will be the $(n+1)$-category of n-categories.

So, the cartesian condition is one of the main hypotheses but also one of the main properties which we would like to prove for our construction of a model category $\mathcal{PC}(\mathcal{M})$ of $\mathcal{M}$-enriched precategories. This compatibility with products will be shown in Chapter 17, furthermore it will provide a useful trick to help establishing the model structure.

By "cartesian model category," we mean a symmetric monoidal model category in the sense of Hovey [149] for the monoidal operation given by cartesian product. We add a condition about commutation of cartesian products with colimits, in order to be able to get an internal $\underline{\text{HOM}}$. Rezk [219] considers this notion in his recent paper.

Recall that $\emptyset$ denotes the initial object and $*$ the coinitial object of $\mathcal{M}$.

Definition 7.7.1 Say that a combinatorial model category $\mathcal{M}$ is *cartesian* if:

(DCL)–the cartesian product preserves colimits: if $\{A_i\}_{i\in\alpha}$ and $\{B_j\}_{j\in\beta}$ are diagrams, then

$$\text{colim}_{\alpha\times\beta}(A_i \times B_j) = (\text{colim}_\alpha A_i) \times (\text{colim}_\beta B_j);$$

(AST)–the object $*$ is cofibrant, i.e. the map $\emptyset \to *$ is a cofibration;

(PROD)–for any cofibrations $A \to B$ and $C \to D$ the map

$$A \times D \cup^{A\times C} B \times C \to B \times D \tag{7.7.1}$$

is a cofibration; if in addition at least one of $A \to B$ or $C \to D$ is a trivial cofibration then (7.7.1) is a trivial cofibration.

See the definition of monoidal model category given for example by Hovey [148] and Shipley and Schwede [229]. Axiom (AST) says that the unit object for the cartesian product is cofibrant, which is stronger than the unit axiom of Shipley and Schwede [229]. Note that since $\mathcal{M}$ is combinatorial, the factorizations can be chosen functorially so as to approach Hovey's definition.

We record now some first consequences of this definition.

Lemma 7.7.2 *Condition (DCL) implies that for any object X the natural map $\emptyset \to X \times \emptyset$ is an isomorphism. In turn this implies that the object $\emptyset$ is a strict initial object, which is to say that it is really empty: if $X \to \emptyset$ is any morphism, then $X \cong \emptyset$.*

Proof Note that $\emptyset$ is the colimit of the empty diagram, that is $\emptyset = \mathrm{colim}_{\emptyset\mathrm{cat}} F$ where $\emptyset$cat is the empty category and $F : \emptyset\mathrm{cat} \to \mathcal{M}$ is the unique functor. Compatibility of products and colimits (DCL) therefore implies that for any $X = \mathrm{colim}_1 X$ (here 1 is the one-object category) we have

$$X \times \emptyset = \mathrm{colim}_1 X \times \mathrm{colim}_{\emptyset\mathrm{cat}} F = \mathrm{colim}_{1\times\emptyset\mathrm{cat}}(X \times F) = \emptyset,$$

the last equality coming from the fact that $1 \times \emptyset\mathrm{cat} = \emptyset\mathrm{cat}$ is again the empty category.

Suppose now that $f : X \to \emptyset$ is a morphism. Letting $e : \emptyset \to X$ denote the unique morphism we get $fe = 1_\emptyset$. On the other hand, the morphism $(1_X, f) : X \to X \times \emptyset$ factors through $(e, 1_\emptyset) : \emptyset \xrightarrow{\cong} X \times \emptyset$. Projecting to the second factor shows that the factorization map is $f : X \to \emptyset$, and projecting to the first factor then shows that $1_X = ef$. Thus f is an isomorphism inverse to e. □

Using the last statement of the previous lemma leads to the following observation: if there exists a morphism $* \to \emptyset$ then $\mathcal{M}$ is equivalent to the trivial category with one object and one morphism. Indeed from the lemma, any object is isomorphic to $\emptyset$, but its universal property says that there is only a single map $\emptyset \to \emptyset$ so $\mathcal{M} \cong *$. Such a trivial model category can come up as a sort of very first initial condition in our iterative construction, but apart from this case the reader may safely imagine that $\emptyset$ has no points.

Lemma 7.7.3 *If $\mathcal{M}$ satisfies (PROD) and (DCL) then, for any pair of weak equivalences $A \xrightarrow{f} B$ and $C \xrightarrow{g} D$, the resulting product map $A\times C \xrightarrow{(f,g)} B\times D$ is a weak equivalence.*

Proof Suppose first that f and g are trivial cofibrations between cofibrant objects. Apply (PROD) to f and $\emptyset \to C$, then to $\emptyset \to B$ and g, and use (DCL) via the result of Lemma 7.7.2. These give that both maps

$$A \times C \to B \times C \to B \times D$$

are trivial cofibrations; hence their composition is a weak equivalence.

Suppose f is a cofibration between cofibrant objects, and C is any object. Choose a replacement $C' \xrightarrow{p} C$, where C' is cofibrant and p is a trivial fibration. By the first paragraph, $A \times C' \to B \times C'$ is a weak equivalence. On the other hand, $A \times C' \to A \times C$ and $B \times C' \to B \times C$ are trivial fibrations as can be seen directly from the lifting property. Writing a commutative square and using 3 for 2 we get that the map $A \times C \to B \times C$ is a weak equivalence.

Suppose now that f is an arbitrary weak equivalence and C is any object. Choose a cofibrant replacement $A' \to A$, then complete to a square

$$\begin{array}{ccc} A' & \longrightarrow & B' \\ \downarrow & & \downarrow \\ A & \longrightarrow & B \end{array}$$

such that $B' \to B$ is a cofibrant replacement, and $A' \to B'$ is a trivial cofibration. The vertical maps are trivial fibrations so their products with C remain trivial fibrations, and by the previous paragraph the map $A' \times C \to B' \times C$ is a weak equivalence. By 3 for 2, the map $A \times C \to B \times C$ is a weak equivalence.

Similarly $B \times C \to B \times D$ is a weak equivalence, and composing we get the statement of the lemma. □

Lemma 7.7.4 *Suppose $\mathcal{M}$ is a cartesian model category. Then if A and B are cofibrant, so is $A \times B$. If $A \to B$ and $C \to D$ are cofibrations (resp. trivial cofibrations) between cofibrant objects then $A \times C \to B \times D$ is a cofibration (resp. trivial cofibration).*

Proof Follow the argument in the first paragraph of the proof of the previous lemma. □

Condition (DCL) holds for a wide variety of categories, notably any presheaf category.

Lemma 7.7.5 *If $\mathcal{M} = \text{PRESH}(\Phi)$ is the category of presheaves of sets over a small category Φ, then it satisfies condition (DCL).*

Proof Products and colimits are computed levelwise, and these properties hold in SET. □

7.8 Internal *Hom*

In this section, we explain how compatibility with products induces an internal HOM, starting with some purely categorical considerations.

Suppose $\mathcal{M}$ is a locally presentable category. Condition (DCL) says that *cartesian product distributes over colimits* , which can be expressed more precisely as saying that any small diagram $A : \eta \to \mathcal{M}$ and any object $B \in \mathcal{M}$ the natural map

$$\text{colim}_{i \in \eta}(A_i \times B) \to (\text{colim}_{i \in \eta} A_i) \times B$$

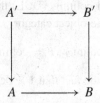

is an isomorphism. If this is the case, then for any pair of diagrams $A : \eta \to \mathcal{M}$ and $B : \zeta \to \mathcal{M}$, the natural map

$$\text{colim}_{(i,j)\in\eta\times\zeta}(A_i \times B_j) \to (\text{colim}_{i\in\eta} A_i) \times (\text{colim}_{j\in\zeta} B_j)$$

is an isomorphism, so these two ways of stating the condition are equivalent.

We say that $\mathcal{M}$ *admits an internal* $\underline{\text{HOM}}$ if, for any $A, B \in \mathcal{M}$ the functor $E \mapsto \text{HOM}_{\mathcal{M}}(A \times E, B)$ is representable by an object $\underline{\text{HOM}}(A, B)$ contravariantly functorial in A and covariantly functorial in B together with a natural transformation $\underline{\text{HOM}}(A, B) \times A \to B$. That is to say that a map $E \to \underline{\text{HOM}}(A, B)$ is the same thing as a map $A \times E \to B$.

Proposition 7.8.1 *Suppose $\mathcal{M}$ is a locally presentable category such that cartesian product distributes over colimits. Then $\mathcal{M}$ admits an internal* $\underline{\text{HOM}}$.

Proof See Adámek *et al.* [1, theorem 27.4], applying the fact that locally presentable categories are co-wellpowered (Adámek and Rosický [2, remark 1.56(3)]). □

Corollary 7.8.2 *Suppose Φ is a small category. Then the category of presheaves of sets* $\text{PRESH}(\Phi) = \text{FUNC}(\Phi^o, \text{SET})$ *admits an internal* $\underline{\text{HOM}}$. *This may be calculated as follows: if A, B are presheaves of sets on Φ then* $\underline{\text{HOM}}(A, B)$ *is the presheaf which to $x \in \Phi$ associates the set* $\text{HOM}_{\text{PRESH}(\Phi/x)}(A|_{\Phi/x}, A|_{\Phi/x})$.

Proof Note that $\text{PRESH}(\Phi)$ is locally presentable. Cartesian products and colimits are calculated levelwise, so cartesian product distributes over colimits since the same is true for the category SET. The explicit description of $\underline{\text{HOM}}(A, B)$ is classical. □

Corollary 7.8.3 *Suppose $\mathcal{M}$ is a locally presentable category such that cartesian product distributes over colimits, and suppose Φ is a small category. Then cartesian product distributes over colimits in* $\text{FUNC}(\Phi^o, \mathcal{M})$ *and this category admits an internal* $\underline{\text{HOM}}$.

Proof Again, cartesian product and colimits are calculated levelwise in $\text{FUNC}(\Phi^o, \mathcal{M})$. □

Turn now to the application of this notion for model categories. Suppose $\mathcal{P}$ is a tractable left proper cartesian model category. In practice, $\mathcal{P}$ will be the model category $\mathcal{PC}(\mathcal{M})$ of $\mathcal{M}$-enriched precategories which we are going to construct, starting with a tractable left proper cartesian model category $\mathcal{M}$. This is why we use the notation $\mathcal{P}$ rather than $\mathcal{M}$ in the present discussion.

The underlying category of $\mathcal{P}$ is locally presentable, so commutation of cartesian product with colimits yields an internal $\underline{\text{HOM}}$ (Proposition 7.8.1). For any A, B this is an object $\underline{\text{HOM}}(A, B)$ together with a map $\underline{\text{HOM}}(A, B) \times A \to B$ such that, for any $E \in \mathcal{M}$, to give a map $E \to \underline{\text{HOM}}(A, B)$ is the same as to give a map $E \times A \to B$.

The cartesian condition is designed exactly so that the internal $\underline{\text{HOM}}$ will be compatible with the model structure:

Theorem 7.8.4 *Suppose $\mathcal{P}$ is a tractable cartesian combinatorial model category. Then:*

(a) internal $\underline{\text{HOM}}(A, B)$ exists for any $A, B \in \mathcal{P}$, and takes pushouts in the first variable or fiber products in the second variable, to fiber products. For example, given morphisms $A \to B$ and $A \to C$ and any D, we have

$$\underline{\text{HOM}}(B \cup^A C, D) = \underline{\text{HOM}}(B, D) \times_{\underline{\text{HOM}}(A,D)} \underline{\text{HOM}}(C, D);$$

(b) if A is cofibrant and B is fibrant, then $\underline{\text{HOM}}(A, B)$ is fibrant;

(c) if $A' \to A$ is a cofibration (resp. trivial cofibration) and B is fibrant, then the induced map $\underline{\text{HOM}}(A, B) \to \underline{\text{HOM}}(A', B)$ is a fibration (resp. trivial fibration);

(d) if A is cofibrant and $B \to B'$ is a fibration (resp. trivial fibration), then the induced map $\underline{\text{HOM}}(A, B) \to \underline{\text{HOM}}(A, B')$ is a fibration (resp. trivial fibration);

(e) if $A' \to A$ (resp. $B \to B'$) is a weak equivalence between cofibrant (resp. fibrant) objects then the induced map $\underline{\text{HOM}}(A, B) \to \underline{\text{HOM}}(A', B')$ is a weak equivalence.

Proof Part (a) comes from Proposition 7.8.1, and the fiber product formulae come from the adjunction definition of $\underline{\text{HOM}}$ and the fact that cartesian product preserves colimits.

For (c), suppose $A \xrightarrow{f} A'$ is a trivial cofibration and B is fibrant. Then for any cofibration $E \xrightarrow{h} F$, the map

$$A \times F \cup^{A \times E} A' \times E \to A' \times F$$

is a trivial cofibration. Hence B satisfies the right lifting property with respect to it. This translates, by the adjunction property of $\underline{\text{HOM}}$, to the right lifting property for

$$f^* : \underline{\text{HOM}}(A', B) \to \underline{\text{HOM}}(A, B)$$

along h. This shows that f^* is a trivial fibration. Similarly if f was a cofibration then f^* is a fibration.

For (d), suppose A is cofibrant and $B \xrightarrow{f} B'$ is a fibration; then if $E \xrightarrow{h} F$ is a trivial cofibration, the product $A \times E \xrightarrow{1_A \times h} A \times F$ is also a trivial cofibration. Since f is fibrant, it satisfies right lifting with respect to this product map $1_A \times h$, which is equivalent to the right lifting property for $\underline{\text{HOM}}(A, f)$ with respect to h by the adjunction property of $\underline{\text{HOM}}$. Thus $\underline{\text{HOM}}(A, B) \to \underline{\text{HOM}}(A, B')$ is fibrant. Applied to $B' = *$ this says that if B is fibrant and A cofibrant then $\underline{\text{HOM}}(A, B)$ is fibrant, giving (b). For the other part of (d), note that similarly a trivial fibration is tranformed to a trivial fibration.

For (e), any weak equivalence between cofibrant objects $A \xrightarrow{f} A'$ can be decomposed as $f = pi$, where i is a trivial cofibration from A to some A' and p is a trivial fibration from A'' to A'; in turn p admits a section $s : A' \to A''$ with $ps = 1_{A'}$. Now $\underline{\text{HOM}}$ transforms (contravariantly) i to i^*, which is a trivial fibration, in particular invertible in the homotopy category. Thus, in $\text{ho}(\mathscr{P})$ we have

$$\begin{aligned}\text{ho}(s^*)\text{ho}(i^*)^{-1}\text{ho}(f^*) &= \text{ho}(s^*)\text{ho}(i^*)^{-1}\text{ho}(i^*)\text{ho}(p^*)\\ &= \text{ho}(s^*)\text{ho}(p^*) = \text{ho}((ps)^*) = 1,\end{aligned}$$

which says that $\text{ho}(f^*)$ admits a left inverse. On the other hand, s is also a weak equivalence between cofibrant objects so $\text{ho}(s^*)$ also admits a left inverse, whereas from the above formula it admits a right inverse too. Thus $\text{ho}(s^*)$ is invertible, which then implies that $\text{ho}(f^*)$ is invertible and f^* is a weak equivalence. This was just a contravariant version of the standard argument which appears elsewhere.

If A is cofibrant and $E \xrightarrow{h} F$ is a cofibration then $A \times E \to A \times F$ is again a cofibration; it follows as usual from the adjunction property that a trivial fibration $B \to B'$ induces a trivial fibration

$$\underline{\text{HOM}}(A, B) \to \underline{\text{HOM}}(A, B'),$$

and then repeating an argument similar to the previous one (but covariantly this time) yields that if $B \to B'$ is a weak equivalence between fibrant objects, then $\underline{\text{HOM}}(A, B) \to \underline{\text{HOM}}(A, B')$ is a weak equivalence. □

7.9 Enriched categories

Enriched categories have been familiar objects for quite a while, see Kelly's book [171]. Our overall goal is to discuss a homotopical analogue of categorical enrichment which is susceptible of being iterated. For this, we will make use of the notion of cartesian model category.

The notion of enriched 1-category will provide an important intermediate step of our argument: a weakly enriched category over a model category $\mathcal{M}$ gives rise to a ho($\mathcal{M}$)-enriched category in the classical sense, and this construction is conservative for weak equivalences. This is used notably for Proposition 12.5.4. So, it is worthwile to start by considering enrichment for plain 1-categories.

Suppose $\mathcal{E}$ is a category admitting finite cartesian products. This includes existence of the coinitial object $*$ which is the empty cartesian product.

If X is a set, then a *strict $\mathcal{E}$-enriched category on object set X* is a collection of objects $A(x, y) \in \mathcal{E}$ for $x, y \in X$, together with morphisms $A(x, y) \times A(y, z) \to A(x, z)$ and $* \to A(x, x)$, such that for any x, y the composed map

$$* \times A(x, y) \to A(x, x) \times A(x, y) \to A(x, y)$$

is the identity; the composed map

$$A(x, y) \times * \to A(x, y) \times A(y, y) \to A(x, y)$$

is the identity; and for any x, y, z, w the diagram

$$\begin{array}{ccc} A(x, y) \times A(y, z) \times A(z, w) & \to & A(x, y) \times A(y, w) \\ \downarrow & & \downarrow \\ A(x, z) \times A(z, w) & \longrightarrow & A(x, w) \end{array}$$

commutes.

An *strict $\mathcal{E}$-enriched category* is a pair (X, A) as above, but often this will be denoted just by A with $X = \mathrm{Ob}(A)$. A functor between $\mathcal{E}$-enriched categories $f : A \to B$ consists of a map of sets $f : \mathrm{Ob}(A) \to \mathrm{Ob}(B)$, and for each $x, y \in \mathrm{Ob}(A)$ a morphism $f_{x,y} : A(x, y) \to B(f(x), f(y))$ in $\mathcal{E}$, such that the diagrams

$$\begin{array}{ccc} * & \longrightarrow & A(x, x) \\ \downarrow & & \downarrow \\ * & \longrightarrow & B(f(x), f(x)) \end{array}$$

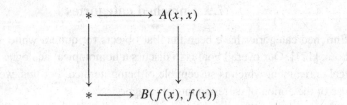

and

$$\begin{array}{ccc} A(x,y)\times A(y,z) & \longrightarrow & A(x,z) \\ \downarrow & & \downarrow \\ B(f(x),f(y))\times B(f(y),f(z)) & \rightarrow & B(f(x),f(z)) \end{array}$$

commute.

Let $\text{CAT}(\mathcal{E})$ denote the category of strict $\mathcal{E}$-enriched categories. It admits cartesian products too: if A and B are strict $\mathcal{E}$-enriched categories then

$$\text{Ob}(A \times B) = \text{Ob}(A) \times \text{Ob}(B),$$

and for (x, x') and (y, y') in $\text{Ob}(A) \times \text{Ob}(B)$ we have

$$(A \times B)((x, x'), (y, y')) = A(x, y) \times B(x', y').$$

Hence this construction can be iterated and we can obtain $\text{CAT}^n(\mathcal{E})$, the category of strict n-categories enriched in $\mathcal{E}$ at the top level.

Starting with $\mathcal{E} = \text{SET}$ yields $\text{CAT}(\text{SET}) = \text{CAT}$, and $\text{CAT}^n(\text{SET})$ is the category of strictly associative and strictly unital n-categories. These objects have been studied a great deal. However, as we have seen in Chapter 2, these objects do not have a sufficiently rich homotopy theory, in particular the groupoid objects therein do not model homotopy types in any reasonable way. This obervation is the motivation for considering weakly associative objects as in the remainder of this work.

If $\varphi : \mathcal{E} \to \mathcal{E}'$ is a functor compatible with cartesian products, applying it to the morphism objects of a strict $\mathcal{E}$-enriched category A gives a strict $\mathcal{E}'$-enriched category $\text{CAT}(\varphi)(A)$ with the same set of objects, and morphism objects defined by

$$\text{CAT}(\varphi)(A)(x, y) := \varphi(A(x, y)).$$

We apply this in particular to the functor $\tau_{\leq 0} : \mathcal{E} \to \text{SET}$ defined by $\tau_{\leq 0}(E) := \text{HOM}_{\mathcal{E}}(*, E)$. Define $\tau_{\leq 1} := \text{CAT}(\tau_{\leq 0})$, i.e.

$$\tau_{\leq 1}A(x, y) = \text{HOM}_{\mathcal{E}}(*, A(x, y)).$$

With this, $\tau_{\leq 1}A \in \text{CAT}(\text{SET})$ is a usual category. Let $\mathbf{Iso}\tau_{\leq 1}A$ denote its set of isomorphism classes. A functor $A \to B$ of strict $\mathcal{E}$-enriched categories is said to be *essentially surjective* if the induced map

$$\mathbf{Iso}\tau_{\leq 1}A \to \mathbf{Iso}\tau_{\leq 1}B$$

is surjective. A functor $f : A \to B$ of strict $\mathcal{E}$-enriched categories is said to be *fully faithful* if, for each $x, y \in \mathrm{Ob}(A)$, the morphism

$$f_{x,y} : A(x, y) \to B(f(x), f(y))$$

is an isomorphism in $\mathcal{E}$. A functor is an *equivalence of strict $\mathcal{E}$-enriched categories* if it is essentially surjective and fully faithful.

These definitions are useful because, as we shall see in Chapter 12, a morphism of weak $\mathcal{M}$-enriched precategories is a global weak equivalence if and only if the associated morphism of strict $\mathrm{ho}(\mathcal{M})$-enriched categories is an equivalence in the present sense. So, we can already obtain versions of some of the main closure properties: closure under retracts and 3 for 2.

Lemma 7.9.1 *Suppose $\varphi : \mathcal{E} \to \mathcal{E}'$ is a functor commuting with cartesian products. Suppose $f : A \to B$ is an equivalence of strict $\mathcal{E}$-enriched categories. Then*

$$\mathrm{CAT}(\varphi)(f) : \mathrm{CAT}(\varphi)(A) \to \mathrm{CAT}(\varphi)(B)$$

is an equivalence of strict $\mathcal{E}'$-enriched categories. This applies in particular to the functor $\varphi = \tau_{\leq 0}$, to conclude that $\tau_{\leq 1}(f) : \tau_{\leq 1}(A) \to \tau_{\leq 1}(B)$ is an equivalence of categories; hence f induces an isomorphism of sets $\mathbf{Iso}\tau_{\leq 1}A \cong \mathbf{Iso}\tau_{\leq 1}B$.

Proof The functor φ sends $*_{\mathcal{E}}$ to $*_{\mathcal{E}'}$ so it induces a map

$$\mathrm{HOM}_{\mathcal{E}}(*, A) \to \mathrm{HOM}_{\mathcal{E}'}(*, \varphi(A)).$$

This natural transformation induces a natural transformation of functors

$$\tau_{\leq 1,\varphi} : \tau_{\leq 1,\mathcal{E}} \to \tau_{\leq 1,\mathcal{E}'} \circ \mathrm{CAT}(\varphi)$$

from $\mathrm{CAT}(\mathcal{E})$ to CAT, which is the identity on underlying sets of objects. Therefore the resulting natural transformation

$$\mathbf{Iso}\tau_{\leq 1,\varphi}(A) : \mathbf{Iso}\tau_{\leq 1,\mathcal{E}}(A) \to \mathbf{Iso}\tau_{\leq 1,\mathcal{E}'}(\mathrm{CAT}(\varphi)(A))$$

is surjective. It follows that if $f : A \to B$ is essentially surjective in $\mathrm{CAT}(\mathcal{E})$ then $\mathrm{CAT}(\varphi)(f) : \mathrm{CAT}(\varphi)(A) \to \mathrm{CAT}(\varphi)(B)$ is also essentially surjective. On the other hand, φ takes isomorphisms to isomorphisms, so if f is fully faithful then, for any $x, y \in \mathrm{Ob}(\mathrm{CAT}(\varphi)(A)) = \mathrm{Ob}(A)$, the map

$$\mathrm{CAT}(\varphi)(f)_{x,y} = \varphi(f_{x,y}) : \varphi(A(x, y)) \to \varphi(B(f(x), f(y)))$$

is an isomorphism in $\mathcal{E}'$. This shows that if f is an equivalence in $\mathrm{CAT}(\mathcal{E})$ then $\mathrm{CAT}(\varphi)(f)$ is an equivalence in $\mathrm{CAT}(\varphi)(\mathcal{E}')$.

For the second part of the statement, apply this to $\varphi := \tau_{\leq 0}$, which preserves products. □

Lemma 7.9.2 *In any category $\mathcal{E}$, the class of isomorphisms is closed under retracts and satisfies 3 for 2.*

Proof It is easy to see for the category of sets. A retract of objects in $\mathcal{E}$ gives a retract of representable functors to sets; so if we have a retract of an isomorphism then the resulting retracted natural transformation is a natural isomorphism, and a morphism in $\mathcal{E}$ which induces an isomorphism between representable functors, is an isomorphism.

The 3 for 2 property is easy to see using the inverses of the isomorphisms in question. □

Theorem 7.9.3 *The notion of equivalence of strict $\mathcal{E}$-enriched categories is closed under retracts and satisfies 3 for 2. If $A \xrightarrow{f} B$ and $B \xrightarrow{g} A$ are functors between strict $\mathcal{E}$-enriched categories such that fg and gf are equivalences, then f and g are equivalences.*

Proof Suppose $f : A \to B$ is a retract of an equivalence of strict $\mathcal{E}$-enriched precategories, by a commutative diagram

$$\begin{array}{ccccc} A & \longrightarrow & U & \longrightarrow & A \\ {\scriptstyle f}\downarrow & & \downarrow & & \downarrow{\scriptstyle f} \\ B & \longrightarrow & V & \longrightarrow & B \end{array}$$

such that the horizontal compositions are the identity and the middle vertical arrow is an equivalence. We get a corresponding diagram of sets

$$\begin{array}{ccccc} \mathbf{Iso}\tau_{\leq 1}(A) & \to & \mathbf{Iso}\tau_{\leq 1}(U) & \to & \mathbf{Iso}\tau_{\leq 1}(A) \\ \downarrow & & \downarrow & & \downarrow \\ \mathbf{Iso}\tau_{\leq 1}(B) & \to & \mathbf{Iso}\tau_{\leq 1}(V) & \to & \mathbf{Iso}\tau_{\leq 1}(B) \end{array}$$

where the middle vertical arrow is an isomorphism by Lemma 7.9.1. It follows that $\mathbf{Iso}\tau_{\leq 1}(A) \to \mathbf{Iso}\tau_{\leq 1}(B)$ is an isomorphism, in particular f is essentially surjective.

If $x_0, x_1 \in \mathrm{Ob}(A)$ is a pair of objects, denote the image objects in U, V and B respectively by u_i, v_i and y_i. Then we get a commutative diagram

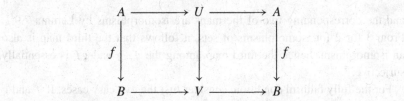

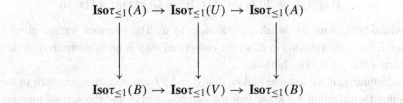

$$\begin{array}{ccccc} A(x_0,x_1) & \rightarrow & U(u_0,u_1) & \rightarrow & A(x_0,x_1) \\ \downarrow & & \downarrow & & \downarrow \\ B(y_0,y_1) & \rightarrow & V(v_0,v_1) & \rightarrow & B(y_0,y_1) \end{array}$$

in which the horizontal compositions are the identity and the middle vertical map is an equivalence. The class of isomorphisms in $\mathcal{E}$ is closed under retracts, by Lemma 7.9.2. It follows that $A(x_0, x_1) \to B(y_0, y_1)$ is an isomorphism in $\mathcal{E}$. This proves that f is fully faithful, completing the proof that the class of equivalences between strict $\mathcal{E}$-enriched categories is closed under retracts.

Turn to the proof of the 3 for 2 property, which says that if

$$A \xrightarrow{f} B \xrightarrow{g} C$$

is a composable pair of morphisms, and if any two of f, g and gf are equivalences, then the third one is too.

Suppose that some two of f, g and gf are equivalences of strict $\mathcal{E}$-enriched categories. Applying the truncation functor gives a composable pair of morphisms of sets

$$\mathbf{Iso}\tau_{\leq 1}(A) \to \mathbf{Iso}\tau_{\leq 1}(B) \to \mathbf{Iso}\tau_{\leq 1}(C)$$

and the corresponding two of the maps are isomorphisms by Lemma 7.9.1. From 3 for 2 for isomorphisms of sets, it follows that the third map is also an isomorphism, hence the third map among the f, g and gf is essentially surjective.

For the fully faithful condition, consider first the two easy cases. If f and g are global weak equivalences, then for any pair of objects $x_0, x_1 \in \mathrm{Ob}(A)$ we have a factorization

$$A(x_0, x_1) \to B(f(x_0), f(x_1)) \to C(gf(x_0), gf(x_1)),$$

where both maps are weak equivalences in $\mathcal{E}$. The previous lemma gives 3 for 2 for isomorphisms in $\mathcal{E}$, so the composed map is an isomorphism, which shows that gf is fully faithful.

Similarly, if we assume known that g and gf are equivalences, then in the same factorization we know that the composed map and the second map are isomorphisms in $\mathcal{E}$, so again by Lemma 7.9.2 it follows that the first map is a weak equivalence, showing that f is fully faithful.

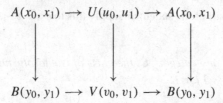

More work is needed for the third case: with the assumption that f and gf are equivalences, to show that g is an equivalence. Applying Lemma 7.9.1 we get that $\mathbf{Iso}\tau_{\leq 1}(f)$ and $\mathbf{Iso}\tau_{\leq 1}(g) \circ \mathbf{Iso}\tau_{\leq 1}(f)$ are isomorphisms of sets, so $\mathbf{Iso}\tau_{\leq 1}(g)$ is an isomorphism. The problem is to show the fully faithful condition. Suppose $x, y \in \mathrm{Ob}(B)$. Choose $x', y' \in \mathrm{Ob}(A)$ and isomorphisms in $\tau_{\leq 1}(B)$, $u \in \tau_{\leq 1}(B)(f(x'), x)$ and $v \in \tau_{\leq 1}(B)(f(y'), y)$ plus their inverses denoted u^{-1} and v^{-1}. These are really maps $u : * \to B(f(x'), x)$ and $v : * \to B(f(y'), y)$, and similarly for u^{-1} and v^{-1}. The composition map

$$B(f(x'), x) \times B(x, y) \times B(y, f(y')) \to B(f(x'), f(y')),$$

composed with

$$u \times 1 \times v^{-1} : * \times B(x, y) \times * \to B(f(x'), x) \times B(x, y) \times B(y, f(y')),$$

gives

$$B(x, y) \to B(f(x'), f(y')).$$

Using u^{-1} and v gives $B(f(x'), f(y')) \to B(x, y)$, and the associativity axiom, definition of inverses and unit axioms combine to say that these are inverse isomorphisms of objects in $\mathcal{E}$. The action of g on morphism objects respects all of these operations so we get a diagram

$$\begin{array}{ccc} B(x, y) & \longrightarrow & C(g(x), g(y)) \\ \downarrow & & \downarrow \\ B(f(x'), f(y')) & \to & C(gf(x'), gf(y')). \end{array}$$

The first vertical map is an isomorphism as described above. The second vertical map is an isomorphism for the same reason applied to C and noting that $g(u)$ and $g(v)$ are isomorphisms in $\tau_{\leq 1}(C)$. The bottom map is an isomorphism by Lemma 7.9.2 and because of the hypotheses that f and gf are fully faithful. Therefore the top map is an isomorphism, showing that g is fully faithful.

For the last part, suppose given functors between strict $\mathcal{E}$-enriched categories $A \xrightarrow{f} B$ and $B \xrightarrow{g} A$ such that fg and gf are equivalences. On the level of sets $\tau_{\leq 0}$ we get a pair of maps whose compositions in both directions are isomorphisms, it follows that $\tau_{\leq 0}(f)$ and $\tau_{\leq 0}(g)$ are isomorphisms. It also easily follows that for any objects $x, y \in \mathrm{Ob}(A)$, the map

$$B(f(x), f(y)) \to A(gf(x), gf(y))$$

has both a left and right inverse, so it is invertible. But since any object of B is isomorphic to some $f(x)$, an argument similar to the previous one shows that $B(u, v) \to A(g(u), g(v))$ is an isomorphism for any $u, v \in \mathrm{Ob}(B)$. Doing the same in the other direction we see that both f and g are equivalences. □

7.9.1 Interpretation of enriched categories as functors $\Delta^o_X \to \mathcal{E}$

Lurie [190] has used an important variant on the notion of nerve of a category (undoubtedly well known in the 1-categorical context). This makes the usual nerve construction apply to an enriched category, without needing to assume anything further about $\mathcal{E}$. In this point of view, the set of objects is singled out as a set while the morphism objects of a strict $\mathcal{E}$-enriched category are considered as objects of $\mathcal{E}$. The nerve is then neither a functor from Δ^o to SET, nor a functor to $\mathcal{E}$, but rather a mixture of the two. We will adopt this point of view when defining $\mathcal{M}$-enriched precategories later on. It has the advantage of allowing us to avoid consideration of disjoint unions, sidestepping some of the difficulties met by Pellissier [211].

If X is a set, define the category Δ_X to consist of all sequences of elements of X denoted $(x_0, \ldots, x_n)$ with $n \geq 0$. One should think of such a sequence as a decoration of the basic object $[n] \in \Delta$. The morphisms of Δ_X are defined in an obvious way generalizing the morphisms of Δ, by just requiring compatibility of the decorations on the source and target. This will be discussed further in Chapter 10.

If A is a strict $\mathcal{E}$-enriched category, let $X := \mathrm{Ob}(A)$. The *nerve* of A is the functor $\Delta^o_X \to \mathcal{E}$, denoted also by A, defined by

$$A(x_0, \ldots, x_n) := A(x_0, x_1) \times \cdots \times A(x_{n-1}, x_n).$$

Here the notations $A(x, y)$ used on the right-hand side are those of the enriched category, but after having made the definition they are seen to be the same as the notations for the nerve, so there is no contradiction in this notational shortcut. The transition maps for the functor $A : \Delta^o_X \to \mathcal{E}$ are obtained using the composition maps of A, for example

$$A(x_0, x_1, x_2) = A(x_0, x_1) \times A(x_1, x_2) \to A(x_0, x_2)$$

in the main case that was discussed in the introduction.

Theorem 7.9.4 *The category* CAT$(\mathcal{E})$ *of strict* $\mathcal{E}$*-enriched categories becomes equivalent, via the above construction, to the category of pairs* (X, A) *where* X *is a set and* $A : \Delta^o_X \to \mathcal{E}$ *is a functor satisfying the* Segal condition *that for any sequence of elements* $x_0, \ldots, x_n$ *the morphism*

$$A(x_0, \ldots, x_n) \to A(x_0, x_1) \times \cdots \times A(x_{n-1}, x_n)$$

is an isomorphism in $\mathcal{E}$. *The straightforward definition of morphisms between pairs* $(X, A) \to (Y, B)$ *is discussed further in Chapter 10 below.*

Understanding this theorem is crucial to understanding the weakly enriched version which is the object of this book. It is left as an exercise for the reader.

7.9.2 The enriched category associated to a cartesian model category

Keep the a tractable left proper cartesian model category $\mathscr{P}$. Then we obtain a strict $\mathscr{P}$-enriched category of cofibrant and fibrant objects, denoted $\mathbf{Enr}(\mathscr{P})$ defined as follows. It is in the next higher universe level. The object class of $\mathbf{Enr}(\mathscr{P})$ is defined to be $\mathrm{Ob}(\mathscr{P}_{cf})$, the class of cofibrant and fibrant objects of $\mathscr{P}$. For any two such objects A and B, put

$$\mathbf{Enr}(\mathscr{P})(A, B) := \underline{\mathrm{HOM}}(A, B) \in \mathscr{P}.$$

The structural morphisms for the enriched category structure come from the standard morphisms for the internal $\underline{\mathrm{HOM}}$. These satisfy the strict associativity and identity constraints. We get a structure of strict $\mathscr{P}$-enriched category.

Although $\mathbf{Enr}(\mathscr{P})$ is strictly associative as a $\mathscr{P}$-enriched category, we can also consider it as a weakly $\mathscr{P}$-enriched category, that is

$$\mathbf{Enr}(\mathscr{P}) \in \mathscr{PC}(\mathscr{P})$$

in the notation of Chapter 10.

An early example using an internal $\underline{\mathrm{HOM}}$ to obtain a higher categorical structure was the notion of "enhanced triangulated category" of Bondal and Kapranov [51].

The morphism objects of $\mathbf{Enr}(\mathscr{P})$ are by definition $\underline{\mathrm{HOM}}(A, B)$ for A and B cofibrant and fibrant, in particular they are fibrant. If we want $\mathbf{Enr}(\mathscr{P})$ to be enriched over $\mathscr{P}_{cf}$ we need to include an additional hypothesis, such as supposing that all objects are cofibrant, in order to ensure that the $\underline{\mathrm{HOM}}(A, B)$ are cofibrant. I would like to thank G. Maltsiniotis for pointing this out.

8

Cell complexes in locally presentable categories

The theory of model categories is most effective when used in conjunction with the *small object argument*. In a nutshell, this says that by adding in enough copies of pushouts along arrows from a fixed set, we can ensure the right lifting property with respect to those arrows. This kind of discussion usually makes essential use of arguments about cardinality, and the most convenient categorical setting in which to do that is the notion of *locally presentable category*.

We therefore start with a review of that part of category theory. Our discussion is based in large part on the book of Adamek and Rosický [2] about locally presentable and accessible categories. Refer there for historical remarks about these notions. The applicability of this theory, in its abstract form, to model categories came out with J. Smith's notion of *combinatorial model category* [239] (see Beke [33], Dugger [95] and Rosický [223]), slightly modified by Barwick with his notion of *tractable model category* [18]. In turn, these authors were formalizing arguments which, for the basic cases derived from the category of simplicial sets, were due to Quillen [215], Bousfield and Kan [58], Jardine [153], Hirschhorn [144] and others.

The idea of "adding on many copies of something" translates into the notion of *cell complex*, which has long played an important role in algebraic topology. From the basic vision of a triangulated manifold obtained by successively adding on simplices, through the notion of CW-complex, the modern point of view allows us to see any transfinite composition of pushouts as being a cell complex in a generalized sense. Hirschhorn [144] has formalized this use of cell complexes for the small object argument and left Bousfield localization, in the context of a general locally presentable category. However, he used an additional assumption of a monomorphism property of elements of the generating set of arrows $I \subset \mathrm{ARR}(\mathcal{M})$, encoded in his notion of *cellular model category*.

For a general discussion, we would like to avoid this hypothesis. Indeed, one of the main examples which we can use to start out our induction is the model category on SET, which as Hirschhorn pointed out is not cellular. It turns out that a somewhat more abstract approach to cell complexes works pretty well. One can note, on the other hand, that if we start the induction with the model category $\mathcal{K}$ of simplicial sets, Hirschhorn's cellularity hypothesis is preserved and the reader may, in that way, avoid the more arcane level of our treatment.

Our discussion covers much the same materiel as Lurie in the appendix to his book [190]. Lurie introduces a notion of "tree" generalizing the standard transfinite cell-addition process. The basic idea is that once we have attached a certain number of cells, the next cell is attached along a κ-presentable subcomplex, but this information is lost under the usual indexation by an ordinal. In our discussion, just to be different, we'll stick to the standard ordinal presentation, but introduce a category of "inclusions of cell complexes," and show that the category of κ-small inclusions of cell complexes into a given one, is κ-filtered. Roughly speaking, an inclusion of cell complexes corresponds to a downward-closed subset of a tree. We sketch a proof of Lurie's theorem [190, proposition A.1.5.12] that cofibrations are cell complexes over κ-small cofibrations, rather than just retracts of such.

A main application of this result is to construct the generating set for injective cofibrations. Again we give a brief account of a proof of the main technical result in the present chapter (Section 8.6), although the reader can also refer to Lurie [190] and Barwick [18].

This discussion prepares the way for the notion of "pseudo-generating set" introduced in Section 8.7, based on Smith's recognition principle as reported by Beke [33] and Barwick [18] and which we review in Section 8.5. Smith's recognition principle has, as one of its main hypotheses, the accessibility of the class of weak equivalences. The notion of pseudo-generating set is designed to give a statement which encodes the accessibility argument in an algebraic condition. The advantage is that the notion of accessibility then no longer appears in the statement, so we can use that in later chapters to construct model categories without needing to discuss the notion of accessibility anymore.

So, in a certain sense what we are doing here is to evacuate some of the more technical details in the theory of model categories, towards the first part of this chapter. We hope that this will be helpful to the reader who wishes to avoid this kind of discussion: if willing to take for granted the recognition principle, which will be stated in the last section as Theorem 8.7.3, and the existence of model structures on diagram categories, the reader may largely skip over the most technical parts of this chapter.

8.0.1 Universes and set theory

In order to avoid repetitive language, we often apply the following conventions about universes. We assume given at least two universes $\mathbb{U} \in \mathbb{V}$. Recall that these are sets which themselves provide models for ZFC set theory, also $\mathbb{U} \subset \mathbb{V}$. A *category* will mean a category object in $\mathbb{V}$. An example is the category $\mathrm{SET}_{\mathbb{U}}$ of sets in $\mathbb{U}$. A *small category* will be a category object in $\mathbb{U}$, which is also one in $\mathbb{V}$. Often a category $\mathcal{C}$ will have *small morphism sets*, that is for any $x, y \in \mathrm{Ob}(\mathcal{C})$ the set $\mathrm{HOM}_{\mathcal{C}}(x, y) \in \mathbb{V}$ is isomorphic to a set in $\mathbb{U}$. Depending on context, the word "category" can mean "small category," or sometimes "category with small morphism sets." However, when we need to consider categories outside of $\mathbb{V}$ this will be explicitly mentioned.

For the theory of ordinals and cardinals, the reader may refer to Krivine [177] for example. Recall that an *ordinal* is a set a, such that if $x \in y$ and $y \in a$ then $x \in a$; and such that a is well ordered by the strict relation $x < y \Leftrightarrow x \in y$ for $x, y \in a$. For the corresponding nonstrict order relation we then have $x \leq y \Leftrightarrow x \subset y$, and for any $x \in a$ the successor of x is $x \cup \{x\}$. A limit ordinal is one which is not a successor, meaning that it is the least upper bound of its set of strictly smaller ordinals.

A *cardinal* is an ordinal a with the property that for any $b \in a$, b is not isomorphic to a. Any set x has a unique *cardinality* $|x|$ which is a cardinal such that $x \cong |x|$.

An *ordinal (resp. cardinal) of* $\mathbb{U}$ is an ordinal (resp. cardinal) which is an element of $\mathbb{U}$. These are the ordinals (resp. cardinals) for the model of set theory given by $\mathbb{U}$. In particular, for any $x \in \mathbb{U}$ we have $|x| \in \mathbb{U}$.

We say that an ordinal α is *approached by a sequence of cardinality* λ if there is a subset $x \subset \alpha$ with $|x| = \lambda$, such that α is the least upper bound of x. A cardinal κ is *regular* if it is not approached by any sequence of cardinality $< \kappa$. The first regular cardinal is ω.

8.1 Locally presentable categories

Fix a regular cardinal κ. A category Φ is said to be *κ-filtered* if for any collection of $< \kappa$ objects $X_i \in \Phi$, there exists an object Y and morphisms $X_i \to Y$; and for any pair of objects X and Y, and any collection of $< \kappa$ morphisms $f_i : X \to Y$ there exists a morphism $g : Y \to Z$ such that all the gf_i are equal. Note that taking an empty set of objects in the first condition implies that Φ is nonempty. A *κ-filtered colimit* is a colimit over a κ-filtered index category. See Adámek and Rosický [2, remark 1.21].

Let $\mathcal{C}$ be a category. We assume that $\mathcal{C}$ admits κ-filtered colimits. Then, say that an object $X \in \mathcal{C}$ is *κ-presentable* if, for any κ-filtered colimit $\mathrm{colim}_{i\in\Phi} Y_i = Z$, the map

$$\mathrm{colim}_{i\in\Phi}\mathrm{HOM}_{\mathcal{C}}(X, Y_i) \to \mathrm{HOM}_{\mathcal{C}}(X, Z)$$

is an isomorphism of sets. An object $Z \in \mathcal{C}$ is said to be *κ-accessible* if it can be expressed as a κ-filtered colimit of κ-presentable objects.

Definition 8.1.1 Let κ be a regular cardinal. A category $\mathcal{C}$ with small morphism sets is called *κ-accessible* if:

(1) $\mathcal{C}$ admits κ-filtered colimits;
(2) the full subcategory of κ-presentable objects is equivalent to a small category;
(3) every object of $\mathcal{C}$ is κ-accessible.

Furthermore, if $\mathcal{C}$ admits all small colimits then we say that $\mathcal{C}$ is *locally κ-presentable*. A category which is locally κ-presentable for some regular cardinal κ is called *locally presentable*.

For the countable cardinal $\kappa = \omega$, the terminology "locally finitely presentable" is interchangeable with "locally ω-presentable." The first and most basic example is the category SET of small sets, which is locally finitely presentable. There the κ-presentable objects are the sets of cardinality $< \kappa$; for $\kappa = \omega$ the finitely presentable objects are the finite sets. Any set is an ω-filtered colimit of its finite subsets, so all sets are ω-accessible.

Theorem 8.1.2 *Suppose $\mathcal{C}$ is locally presentable. Then it is complete, i.e. it admits small limits too. Each object has only a small set of subobjects up to isomorphism. All κ-filtered colimits commute with κ-small limits (i.e. limits over categories of cardinality $< \kappa$). For any $X \in \mathcal{C}$, the subcategory $\mathcal{C}_\kappa/X$ of κ-presentable objects of $\mathcal{C}$ over X, is κ-filtered and X is canonically the colimit of the forgetful functor on $\mathcal{C}_\kappa/X$.*

If $\mathcal{C}$ is locally κ-presentable then for any regular cardinal $\kappa' > \kappa$ it is locally κ'-presentable too.

Proof See Makkai and Paré [197], or Adámek and Rosický [2, proposition 1.22, corollary 1.28, remark 1.56 and proposition 1.59]. For the last sentence see *loc. cit.* [2, page 22]. □

Lemma 8.1.3 *Suppose Ψ is a small category, and $\mathcal{C}$ is locally κ-presentable. Then the category* FUNC($\Psi, \mathcal{C}$) *of diagrams from Ψ to $\mathcal{C}$, is locally κ-presentable. The κ-presentable diagrams in* FUNC($\Psi, \mathcal{C}$) *are exactly the functors $F : \Phi \to \mathcal{C}$ such that $F(a)$ is κ-presentable in $\mathcal{C}$ for every $a \in \Phi$.*

Proof See Makkai and Paré [197], or Adámek and Rosický [2, corollary 1.54]. □

Corollary 8.1.4 *Suppose Ψ is a small category, then the category* $\mathrm{PRESH}(\Psi) = \mathrm{SET}^{\Psi^o}$ *of presheaves of sets on Ψ, is locally finitely presentable. When necessary the universe in which the values are supposed to lie, is indicated by a subscript.*

Proof As pointed out above, the category of sets is locally finitely presentable. □

If $\mathcal{C}$ is a category, let $\mathrm{ARR}(\mathcal{C})$ be the category of arrows of $\mathcal{C}$, whose objects are the diagrams of shape $X \xrightarrow{f} Y$ in $\mathcal{C}$. The morphisms in $\mathrm{ARR}(\mathcal{C})$ from $X \xrightarrow{f} Y$ to $X' \xrightarrow{f'} Y'$ are the commutative squares

$$
\begin{array}{ccc}
X & \longrightarrow & X' \\
{\scriptstyle f}\downarrow & & \downarrow{\scriptstyle f'} \\
Y & \longrightarrow & Y'
\end{array}
$$

Corollary 8.1.5 *Suppose $\mathcal{C}$ is a locally κ-presentable category. Then* $\mathrm{ARR}(\mathcal{C})$ *is locally κ-presentable.*

Proof Indeed, $\mathrm{ARR}(\mathcal{C}) = \mathrm{FUNC}(\mathcal{E}, \mathcal{C})$, where $\mathcal{E}$ is the category with two objects $0, 1$ and a single morphism $0 \to 1$ besides the identities. Therefore 8.1.3 applies. □

In a similar way, any category of commutative diagrams of a given shape in a locally presentable category, will again be locally presentable.

A functor $q : \alpha \to \beta$ is *cofinal* if:

- for any object $i \in \beta$ there exists $j \in \alpha$ and an arrow $i \to q(j)$;
- for any pair of arrows $i \to q(j)$ and $i \to q(j')$ in β there are arrows $j \to j''$ and $j' \to j''$ in α such that the diagram

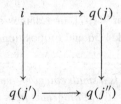

$$
\begin{array}{ccc}
i & \longrightarrow & q(j) \\
\downarrow & & \downarrow \\
q(j') & \longrightarrow & q(j'')
\end{array}
$$

commutes (see Adámek and Rosický [2, 0.11]; note, however, that some authors call this "final").

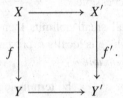

Recall that if $q : \alpha \to \beta$ is a cofinal functor, then it induces an equivalence between the theory of colimits indexed by α and the theory of colimits indexed by β, see *loc. cit.* [2, page 4].

A basic and motivating example of a locally presentable category is when Φ^o is a site, and $\mathcal{C} \subset \text{PRESH}_{\mathbb{U}}(\Phi)$ is the subcategory of sheaves. In this case the identity inclusion H^* has as left adjoint $T = H_!$ the sheafification functor. This motivates the following characterization:

Proposition 8.1.6 *A $\mathbb{V}$-category $\mathcal{C}$ with $\mathbb{U}$-small morphism sets is locally presentable if it has $\mathbb{U}$-small limits and colimits, and if there exists a $\mathbb{U}$-small category Φ and an adjunction*

$$H_! : \text{PRESH}_{\mathbb{U}}(\Phi) \leftrightarrow \mathcal{C} : H^*,$$

and a regular cardinal $\kappa \in \mathbb{U}$ such that $H_! H^*$ is the identity, and the composition $T := H^* H_! : \text{PRESH}_{\mathbb{U}}(\Phi) \to \text{PRESH}_{\mathbb{U}}(\Phi)$ commutes with κ-filtered colimits. Or equivalently, that H^* itself preserves κ-filtered colimits.

Proof This rephrases *loc. cit.* [2, theorem 1.46], using an adjoint pair of functors such that one composition is the identity, instead of a full reflective subcategory. □

A locally κ-presentable category $\mathcal{C}$ will in general have the property that $|\text{HOM}(X, Y)| > \kappa$ for two κ-presentable objects. For example, $\text{PRESH}_{\mathbb{U}}(\Phi)$ is locally κ-presentable whatever the size of Φ, but if $|\Phi| > \kappa$ then there can be $> \kappa$ morphisms between κ-presentable objects. This is rectified by taking κ big enough, but the bound has to be exponential in κ because we look at maps from objects of size $< \kappa$.

Corollary 8.1.7 *Suppose $\mathcal{C}$ is a locally presentable category. There is a regular cardinal κ such that $\mathcal{C}$ is locally κ-presentable, such that for any regular cardinals $\lambda, \mu > \kappa$ and objects X, Y such that X is λ-presentable and Y is μ-presentable, the set of morphisms $\text{HOM}_{\mathcal{C}}(X, Y)$ has size $< \mu^{\lambda}$.*

Proof Suppose $\mathcal{C}$ is κ_0-presentable to begin with. Use the characterisation of the previous proposition, and choose a new regular cardinal $\kappa_1 > \sup(\kappa_0, |\Phi|)$. The total cardinality of a presheaf $A \in \text{PRESH}_{\mathbb{U}}(\Phi)$ is the sum of the cardinalities of the values $A(x)$ for $x \in \Phi$. For any $\lambda \geq \kappa_1$ the λ-presentable objects of $\mathcal{C}$ are those of the form $H_!(A)$ for presheaves $A : \Phi \to \text{SET}_{\mathbb{U}}$ of total cardinality $< \lambda$ (see Adámek and Rosický [2, example 1.31]). Choose a regular cardinal $\kappa \geq \kappa_1$ so that the total cardinality of $H^* H_!(B)$ is $< \kappa$ for any presheaf B of total cardinality $< \kappa_1$.

Now suppose $\lambda, \mu > \kappa$. By *loc. cit.* [2, remark 1.30(2)], the λ-presentable objects of $\mathcal{C}$ are κ_1-filtered colimits of size $< \lambda$, of κ_1-presentable objects (and the same for μ). Suppose X and Y are λ-presentable and μ-presentable objects respectively. Write $X = \mathrm{colim}_{i\in I} H_!(A_i)$ (resp. $Y = \mathrm{colim}_{j\in J} H_!(B_j)$), where I (resp. J) is a κ_1-filtered category of size $< \lambda$ (resp. of size $< \mu$) and A_i and B_j are presheaves of total cardinality $< \kappa_1$. Then $H^* H_!(B_j)$ has total cardinality $< \kappa$. Now

$$\mathrm{HOM}_{\mathcal{C}}(X, Y) = \lim_{i\in I} \left(\mathrm{colim}_{j\in J}\mathrm{HOM}_{\mathcal{C}}(A_i, H^* H_!(B_j))\right),$$

which has size $< \mu^{\lambda}$.

One should also be able to prove this using the characterization of locally presentable categories as categories of models of limit theories *loc. cit.* [2, theorem 5.30]. □

8.1.1 Miscellany about limits and colimits

Here is an elementary observation about limits, useful in the construction of adjoints:

Lemma 8.1.8 *Suppose $p : \alpha \to \beta$ is a functor between small categories, and suppose $A : \beta \to \mathcal{M}$ is a diagram with values in a cocomplete category $\mathcal{M}$. Then there is a natural map* $\mathrm{colim}_\alpha p^*(A) \to \mathrm{colim}_\beta A$*, satisfying compatibility conditions in case of compositions of functors.*

Proof There is a tautological natural transformation of β-diagrams from A to the constant diagram with values $\mathrm{colim}_\beta A$, and the pullback of this natural transformation to α is a natural transformation of α-diagrams from $p^*(A)$ to the constant diagram with values $\mathrm{colim}_\beta A$, which gives the map $\mathrm{colim}_\alpha p^*(A) \to \mathrm{colim}_\beta A$ in question. If $q : \delta \to \alpha$ is another functor then the composition of the maps for p and for q

$$\mathrm{colim}_\alpha q^*(p^*(A)) \to \mathrm{colim}_\alpha p^*(A) \to \mathrm{colim}_\beta A$$

is the natural map for pq. Similarly, if p is the identity functor then the associated natural map is the identity. □

The following lemma will be useful in dealing with unitality conditions in Chapter 11:

Lemma 8.1.9 *Suppose $\mathcal{M}$ is a category with coinitial object $*$, and suppose α is a nonempty connected small category (that is, a category whose nerve is a connected simplicial set). Then the colimit of the constant functor $C. : \alpha \to \mathcal{M}$, defined by $C_i = *$ for all $i \in \alpha$, exists and is equal to $*$.*

Proof There is a unique compatible system of morphisms $\phi_i : C_i = * \to *$. We claim that this makes $*$ into a colimit of $C.$. Suppose $U \in \mathcal{M}$ and $\psi_i : C_i \to U$ is a compatible system of morphisms. Pick $i_0 \in \alpha$ and use $f := \psi_{i_0} : C_{i_0} = * \to U$. We claim that for any $j \in \alpha$ the composition $f\phi_j$ is equal to ψ_j. Let α' be the subset of objects of α for which this is true, nonempty since it contains i_0. If $j' \in \alpha'$ and if $g : j \to j'$ is an arrow of α then

$$f\phi_j = f\phi_{j'}C_g = \psi_{j'}C_g = \psi_j,$$

so $j \in \alpha'$. Suppose $j' \in \alpha'$ and if $g : j' \to j$ is an arrow of α. Then

$$f\phi_j C_g = f\phi_{j'} = \psi_{j'} = \psi_j C_g,$$

but C_g is an isomorphism so $f\phi_j = \psi_j$ and $j \in \alpha'$. These two steps imply inductively, using connectedness of α, that $\alpha' = \mathrm{Ob}(\alpha)$. Thus $f\phi. = \psi.$. Clearly f is unique, so we get the required universal property. □

8.2 The small object argument

For the basic treatment we follow Hirschhorn [144]. Fix a locally presentable category $\mathcal{M}$.

If α is an ordinal, denote by $[\alpha]$ the set $\alpha + 1$, that is the set of all ordinals $j \leq \alpha$. This notation extends the usual notation $[n]$ used in designating objects of the category of simplices Δ. Note that by definition $[\alpha]$ is again an ordinal. We can write interchangeably $i \leq \alpha$ or $i \in [\alpha]$.

A *sequence* is a pair $(\beta, X.)$, where β is an ordinal and $X : [\beta] \to \mathcal{M}$ is a functor. We usually denote this by the collection of objects X_i for $i \leq \beta$, with morphisms $\phi_{ij} : X_i \to X_j$ whenever $i < j \leq \beta$. A sequence is *continuous* if for any $j \leq \beta$ such that j is a limit ordinal, the map

$$\mathrm{colim}_{i<j} X_i \to X_j$$

is an isomorphism. A sequence gives in particular a morphism $X_0 \to X_\beta$.

Suppose we are given a set of arrows $I \subset \mathrm{ARR}(\mathcal{M})$. An *$I$-cell complex* is a continuous sequence $X : [\beta] \to \mathcal{M}$ together with the data, for each $i < \beta$, of $f_i \in I$ and a pushout diagram

$$\begin{array}{ccc} U_i & \xrightarrow{f_i} & V_i \\ {\scriptstyle u_i}\downarrow & & \downarrow{\scriptstyle v_i} \\ X_i & \longrightarrow & X_{i+1} \end{array} \qquad (8.2.1)$$

inducing an isomorphism $X_i \cup^{U_i} V_i \cong X_{i+1}$. In general, the choice of data $\{(f_i, u_i, v_i)\}_{i\leq\beta}$ is not uniquely determined by the choice of sequence $X.$ although by abuse of notation we usually say just that $\{X_i\}_{i\leq\beta}$ is a cell complex.

The *length* of a cell complex or of its underlying sequence $(\beta, X.)$, is by definition the ordinal β.

Denote by $\mathbf{cell}(I) \subset \mathrm{ARR}(\mathcal{M})$ the class of arrows $X \xrightarrow{f} Y$ in $\mathcal{M}$ such that there exists an I-cell complex $(\beta, X.)$ with $X_0 = X$ and $X_\beta = Y$, with f being the transition map from X_0 to X_β. The class $\mathbf{cell}(I)$ defined this way is clearly closed under compositions.

Let $\mathbf{inj}(I) \subset \mathrm{ARR}(\mathcal{M})$ be the class of maps which satisfy the right lifting property with respect to $\mathbf{cell}(I)$, and let $\mathbf{cof}(I) \subset \mathrm{ARR}(\mathcal{M})$ be the class of maps which satisfy the left lifting property with respect to $\mathbf{inj}(I)$. Note that $\mathbf{cell}(I) \subset \mathbf{cof}(I)$. The famous *small object argument*, as it applies to locally presentable categories, can be summed up in the following theorem:

Theorem 8.2.1 *Suppose $\mathcal{M}$ is a locally presentable category, and $I \subset \mathrm{ARR}(\mathcal{M})$ is a small subset of morphisms. Then:*

- *any morphism $f : X \to Y$ admits a factorization $f = pg$ where $X \xrightarrow{g} Z \xrightarrow{p} Y$, such that $g \in \mathbf{cell}(I)$ and $p \in \mathbf{inj}(I)$;*
- *one may choose a factorization which is functorial in f;*
- *the class $\mathbf{cof}(I)$ is closed under retracts and is equal to the class of morphisms $f : X \to Y$ such that Y is a retract of some $g : X \to Z$ in $\mathbf{cell}(I)$, in the category $X/\mathcal{M}$ of objects under X;*
- *the class $\mathbf{inj}(I)$ is also equal to the class of morphisms which satisfy the right lifting property with respect to $\mathbf{cof}(I)$.*

Proof Refer to the numerous discussions in the literature. □

We record, in the next few statements, some basic facts about lifting properties.

Lemma 8.2.2 *Suppose $\mathcal{M}$ is a category and $\mathcal{A}$ is a class of arrows in $\mathcal{M}$. Then the class $\mathcal{F}$ of morphisms in $\mathcal{M}$ which satisfy the right lifting property with respect to all morphisms of $\mathcal{A}$, is closed under retracts. Similarly, if $\mathcal{B}$ is a class of arrows then the class $\mathcal{G}$ of morphisms which satisfy the left lifting property with respect to all morphisms of $\mathcal{B}$, is closed under retracts.*

Proof Suppose $f : X \to Y$ is in $\mathcal{F}$, and $g : A \to B$ is a retract of f with a retract diagram

$$\begin{CD} A @>i>> X @>r>> A \\ @VVV @VVV @VVV \\ B @>j>> Y @>s>> B. \end{CD}$$

Suppose

$$\begin{CD} U @>a>> A \\ @VuVV @VVgV \\ V @>b>> B \end{CD}$$

is a diagram with $u \in \mathcal{A}$. Compose with the left square of the retract diagram to get

$$\begin{CD} U @>ia>> X \\ @VuVV @VVfV \\ V @>jb>> Y. \end{CD}$$

The lifting property for f says that there exists $t : V \to X$ with $tu = ia$ and $ft = jb$. Composing with r gives a lifting $rt : V \to A$ such that $rtu = ria = a$ and $grt = sft = sjb = b$, so rt is a lifting for the original diagram to g.

The proof for $\mathcal{G}$ is similar. □

Corollary 8.2.3 *Suppose I is a set of morphisms in a locally presentable category $\mathcal{M}$. Then the class* **inj**(I) *is closed under retracts.*

Proof The class **inj**(I) is defined by the right lifting property with respect to **cell**(I). □

Lemma 8.2.4 *Suppose $\mathcal{M}$ is a complete and cocomplete category, and $\mathcal{A}$ is a class of arrows in $\mathcal{M}$. Then the class $\mathcal{F}$ of morphisms in $\mathcal{M}$ which satisfy the right lifting property with respect to all morphisms of $\mathcal{A}$, is closed under fiber products and sequential limits. Similarly, if $\mathcal{B}$ is a class of arrows then the class $\mathcal{G}$ of morphisms which satisfy the left lifting property with respect to all morphisms of $\mathcal{B}$, is closed under pushouts and transfinite composition.*

Proof Left to the reader. □

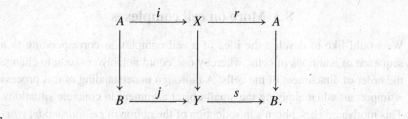

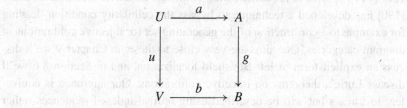

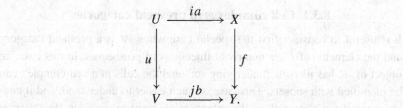

8.3 More on cell complexes

We would like to develop the idea of a cell complex as corresponding to a sequence of additions of cells, whereby one could notably envision to change the order of attachment of the cells. A thorough understanding of this process is important when applying the small object argument in concrete situations. This motivated Hirschhorn's introduction of the notion of *cellular model category* [144], applied to the construction of the left Bousfield localization. Lurie [190] has developed a technique to bypass the cellularity condition, leading for example to a construction of the generating set for injective cofibrations in diagram categories. Our aims are very close to these: in Chapter 9 we'll discuss an explicit form of left Bousfield localization, and in Section 8.6 we'll discuss Lurie's theorems on injective cofibrations. Our approach is equivalent to Lurie's but will be described using ordinal-indexed sequences rather than trees.

8.3.1 Cell complexes in presheaf categories

It is useful to consider first the special case when $\mathcal{M}$ is a presheaf category and the elements of I are nontrivial injections of presheaves. In this case, an object of $\mathcal{M}$ has its own "underlying set" and the cells in a cell complex can be identified with subsets. This case is much easier to understand, and it covers many if not almost all of the examples we want to consider. We describe the notion of inclusion of cell complexes in this special case. It is sufficient for most applications, so the present section is an option replacing the more general discussion in the next subsections.

Suppose Ψ is a small category, and $\mathcal{M} = \text{PRESH}(\Psi) = \text{FUNC}(\Psi^o, \text{SET})$. For $A \in \mathcal{M}$ define its *underlying set* to be

$$\mathcal{U}(A) := \coprod_{x \in \Psi} A(x).$$

This construction gives a functor $\mathcal{U} : \mathcal{M} \to \text{SET}$ which is faithful and compatible with colimits. A morphism in $\mathcal{M}$ is a monomorphism if and only if it goes to an injection of underlying sets.

Suppose given a set of morphisms I in $\mathcal{M}$, which we suppose to be monomorphisms but not isomorphisms. Then, given an I-cell complex $\{X_i\}_{i \leq \beta}$, the set $\mathcal{U}(X_\beta) - \mathcal{U}(X_0)$ is partitioned into nonempty subsets which we call the *cells*, and which are indexed by the successor ordinals $i + 1 \leq \beta$. The cell C_{i+1} is by definition the complement $\mathcal{U}(X_{i+1}) - \mathcal{U}(X_i)$. Because of the assumption that elements of I are monomorphisms, all of the transition maps in the cell complex are monomorphisms.

Associated to a cell C_{i+1} is its *attaching diagram* (8.2.1) as above. With those notations, in our case the map $u_i : U_i \to X_i$ is uniquely determined by the composed map $U_i \to X_\beta$.

If $\{Y_j\}_{j\leq\alpha}$ is another cell complex, an *inclusion of cell complexes* consists of morphisms $Y_0 \to X_0$ and $Y_\alpha \to X_\beta$ such that the morphism

$$\mathcal{U}(Y_\alpha) - \mathcal{U}(Y_0) \to \mathcal{U}(X_\beta) - \mathcal{U}(X_0)$$

is injective, respects the partitions and the order on the pieces, induces an isomorphism from each cell for $Y.$ to one of the cells for $X.$, and such that the attaching maps for these corresponding cells are the same, where the attaching maps are considered as maps into Y_α and X_β respectively. The ordering of the cells in $Y.$ is determined by that of the larger containing cell complex $X.$, which means that the essential information consists of specifying the subset of cells of $X.$ which are in $Y.$ rather than allowing the choice of a new and different order too.

A presheaf category will satisfy the cellularity hypotheses used by Hirschhorn [144], so in this case we can refer there. The reader may note that, starting with a presheaf category $\mathcal{M}$, our construction of the model category $\mathcal{PC}(X, \mathcal{M})$ keeps us within the realm of presheaf categories, and this remark can be applied iteratively. So, for many of the main examples to be envisioned, the case of cell complexes in a presheaf category is sufficient.

8.3.2 Inclusions of cell complexes

In spite of the previous remark, it seems like a good idea to search for the highest convenient level of generality. Thus, we turn now to cell complexes in the general situation where $\mathcal{M}$ is a locally presentable category and $I \subset \mathrm{ARR}(\mathcal{M})$ is a subset of morphisms.

In order to develop a theory allowing us to interchange the order of cell attachment, we first define a notion of inclusion of cell complexes. After Hirschhorn's cellular model categories, the general case was treated by Lurie in Appendix 1 to his book on higher topos theory [190]. We give a somewhat different discussion which is certainly less streamlined than Lurie's, but we hope it will help the reader to gain an intuitive picture of what is going on.

Consider a strictly increasing map of ordinals $q : [\alpha] \to [\beta]$. Define the map $q^- : [\alpha] \to [\beta]$ by

$$q^-(i) := \inf\{k, \ \forall j < i, q(j) < k\}.$$

Thus $q^-(0) = 0$ and for $i > 0$, $q^-(i) = \sup_{j<i}(q(j) + 1)$. For successor ordinals, $q^-(i + 1) = q(i) + 1$, whereas if i is a limit ordinal then $q^-(i)$ is

the limit of $q(j)$ for $j < i$. It follows that q^- is strictly increasing, and also continuous, which is to say that if i is a limit ordinal then $q^-(i)$ is the limit of $q^-(j)$ for $j < i$. Furthermore $q^-(i) \leq q(i)$. One can see from the above characterization that the intervals

$$[q^-(i), q(i)] := \{j \in [\beta], \ \ q^-(i) \leq j \leq q(i)\}$$

are disjoint and cover $[\beta]$, which is another way of understanding why to introduce q^-.

The basic idea of the following definition is that a cell complex is a transfinite sequence of pushouts along specified morphisms in I, with these specifications forming part of the data of the cell complex; in particular, a cell complex consists of more than just the resulting morphism $X_0 \to X_\beta$. An "inclusion of cell complexes" means a map compatible with the sequences of pushouts, via a map of ordinals which should make the labels in I correspond.

Here is the precise definition. Suppose we are given two cell complexes $(\alpha, X.) = (\alpha, X_i, \phi_{ij}, f_i, u_i, v_i)$ and $(\beta, Y.) = (\beta, Y_i, \psi_{ij}, g_i, r_i, s_i)$. Here f_i, u_i and v_i are the structural maps for the first cell complex with the previous notations, and $g_i : R_i \to S_i$, in I, together with $r_i : R_i \to Y_i$ and $s_i : S_i \to Y_{i+1}$ are the structural maps for $(\beta, Y.)$.

An *inclusion of cell complexes* from $(\alpha, X.)$ to $(\beta, Y.)$ is a pair consisting of a strictly increasing map $q : \alpha \to \beta$, and a collection of morphisms $\xi_i : X_i \to Y_{q^-(i)}$, subject to some conditions. Before explaining the conditions, extend the map to $q : [\alpha] \to [\beta]$ by setting $q(\alpha) := \beta$ and notice that the maps ξ_i induce maps denoted $\xi_{i,+} : X_i \to Y_{q(i)}$, defined by $\xi_{i,+} := \psi_{q^-(i)q(i)} \circ \xi_i$ because $q^-(i) \leq q(i)$. This includes $\xi_{\alpha,+} : X_\alpha \to Y_\beta$. In view of the relation $q^-(i+1) = q(i) + 1$, we have $\xi_{i+1} : X_{i+1} \to Y_{q(i)+1}$. The conditions are as follows:

(i) first of all that $f_i = g_{q(i)}$ as elements of the set I;

(ii) second, if i is a limit ordinal then ξ_i is the colimit of the ξ_j for $j < i$, going from $X_i = \operatorname{colim}_{j<i} X_j$ to $Y_{q^-(i)} = \operatorname{colim}_{j<i} Y_{q^-(j)}$;

(iii) and third, for any $i < \alpha$ the composition

$$\begin{CD}
U_i @>{u_i}>> X_i @>{\xi_{i,+}}>> Y_{q(i)} \\
@V{f_i}VV @VVV @VVV \\
V_i @>{v_i}>> X_{i+1} @>{\xi_{i+1}}>> Y_{q(i)+1}
\end{CD}$$

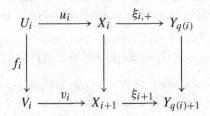

is equal to the diagram

$$\begin{array}{ccc} R_{q(i)} & \xrightarrow{r_{q(i)}} & Y_{q(i)} \\ {\scriptstyle g_{q(i)}}\downarrow & & \downarrow \\ S_{q(i)} & \xrightarrow{s_{q(i)}} & Y_{q(i)+1} \end{array}$$

for the complex $Y.$.

Notice that the initial map $\xi_0 : X_0 \to Y_0$ can be arbitrary. However, the choice of remaining maps satisfies the rigidity requirement that one would expect of a map between cell complexes, as will be pointed out in Lemma 8.3.3 and Corollary 8.3.4 below.

Suppose $(\gamma, Z.)$ is a third cell complex and $(p, \zeta.)$ is a map of cell complexes from $(\beta, Y.)$ to $(\gamma, Z.)$. We define the *composition* $(p, \zeta.) \circ (q, \xi.)$ to be the map of cell complexes $(h, \eta.)$ given as follows. First of all, $h := p \circ q : [\alpha] \to [\gamma]$. Notice that $h^-(i) = p^-(q^-(i))$, as can be seen from the definitions of p^- and q^-. Thus it makes sense to define η_i to be the composition

$$X_i \xrightarrow{\xi_i} Y_{q^-(i)} \xrightarrow{\zeta_{q^-(i)}} Z_{h^-(i)} = Z_{p^-(q^-(i))}.$$

This collection of compositions satisfies the three conditions for being a morphism of cell complexes.

Let $\text{CELL}(\mathcal{M}; I)$ denote the category whose objects are I-cell complexes in $\mathcal{M}$ and whose morphisms are the inclusions of cell complexes. This is provided with functors $\mathbf{s}$ and $\mathbf{t}$ to $\mathcal{M}$ defined by $\mathbf{s}(\alpha, X.) := X_0$ and $\mathbf{t}(\alpha, X.) := X_\alpha$. If $(q, \xi.)$ is an inclusion from $(\alpha, X.)$ to $(\beta, Y.)$ then $\xi_{\alpha,+} : X_\alpha \to Y_{q(\alpha)}$ and we can compose with the transition map for $Y.$ to get a map to Y_β. This defines the functoriality maps for $\mathbf{t}$. The collection of maps $X_0 \to X_\alpha$ provides a natural transformation from $\mathbf{s}$ to $\mathbf{t}$ or equivalently a functor $\mathbf{a} : \text{CELL}(\mathcal{M}; I) \to \text{ARR}(\mathcal{M})$ whose compositions with the source and target functors are $\mathbf{s}$ and $\mathbf{t}$ respectively.

Recall that $\mathbf{cell}(I)$ denotes the class of arrows in the essential image of $\mathbf{a}$, in other words it is the class of morphisms $f : X \to Y$ in $\mathcal{M}$ such that there exists $(\alpha, X.) \in \text{CELL}(\mathcal{M}; I)$ with $X_0 \to X_\alpha$ isomorphic to f in $\text{ARR}(\mathcal{M})$.

Lemma 8.3.1 *If $(\alpha, X.)$ is a cell complex and $X_0 \to Y_0$ is a morphism, then define $Y_i := X_i \cup^{X_0} Y_0$. Using the induced maps for attaching data, we get a cell complex $(\alpha, Y.)$. In particular, if*

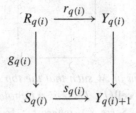

$$\begin{array}{ccc} X & \longrightarrow & Y \\ \Big\downarrow & & \Big\downarrow \\ Z & \longrightarrow & W \end{array}$$

is a cocartesian diagram in $\mathcal{M}$ such that the top arrow is in $\mathbf{cell}(I)$*, then the bottom arrow is also in* $\mathbf{cell}(I)$*. There is a tautological inclusion of cell complexes $(t, \phi) : (\alpha, X.) \to (\alpha, Y.)$, where $t : [\alpha] \to [\alpha]$ is the identity and $\phi_i : X_i \to Y_i$ is the tautological inclusion.*

Proof With the notations used in the definition of cell complex, given that $X_i \cup^{U_i} V_i \cong X_{i+1}$ we get

$$Y_{i+1} = Y_0 \cup^{X_0} X_{i+1} = Y_0 \cup^{X_0} (X_i \cup^{U_i} V_i) = (Y_0 \cup^{X_0} X_i) \cup^{U_i} V_i = Y_i \cup^{U_i} V_i.$$

Similarly $Y.$ satisfies the continuity condition at limit ordinals. □

Lemma 8.3.2 *Given a continuous sequence indexed by an ordinal such that the transition maps $X_i \to X_{i+1}$ are in* $\mathbf{cell}(I)$*, the composition $X_0 \to$ $\mathrm{colim}_i X_i$ is again in* $\mathbf{cell}(I)$*.*

Proof Just combine together the sequences. □

Inclusions of cell complexes enjoy some rigidity properties because of the fact that the map of labels is included as part of the data:

Lemma 8.3.3 *Suppose $q : [\alpha] \to [\beta]$ is a strictly increasing map, and suppose that we are given two cell complexes $(\alpha, X.)$ and $(\beta, Y.)$. For a given initial map $\xi_0 : X_0 \to Y_0$, there is at most one inclusion of cell complexes*

$$(q, \xi.) : (\alpha, X.) \to (\beta, Y.),$$

based on q and starting with ξ_0. If q and ξ_0 are isomorphisms, and if $(q, \xi.)$ exists, then it is an isomorphism.

Proof Suppose $(q, \xi'.)$ is another inclusion of cell complexes with $\xi'_0 = \xi_0$. We prove by induction that $\xi'_j = \xi_j$ for all $j \leq \alpha$. Suppose it is known for all $i < j$. If j is a limit ordinal then the universal property of the colimit $X_j = \mathrm{colim}_{i<j} X_i$ means that the map

$$\xi_j : X_j \to Y_{q^-(j)}$$

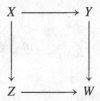

is determined by the ξ_i for $i < j$, and the same is true of ξ'_j. The inductive hypothesis that $\xi'_i = \xi_i$ for $i < j$ therefore implies $\xi'_j = \xi_j$. This treats the case of limit ordinals; the other possible case is $j = i + 1$, which is now assumed. Then the map $\xi_j = \xi_{i+1}$ fits into the diagram

$$\begin{array}{ccccc} R_i & \xrightarrow{r_i} & X_i & \xrightarrow{\xi_{i,+}} & Y_{q(i)} \\ \downarrow g_i & & \downarrow & & \downarrow \\ S_i & \xrightarrow{s_i} & X_{i+1} & \xrightarrow{\xi_{i+1}} & Y_{q(i)+1} \end{array}$$

where the horizontal compositions are given as the attaching maps for $(\beta, Y.)$. On the other hand, X_{i+1} is a pushout in the left square, that is the left square is cocartesian. Thus ξ_{i+1} is uniquely determined by $\xi_{i,+}$ and the attaching map $S_i = V_{q(i)} \xrightarrow{v_{q(i)}} Y_{q(i)+1}$, which identifies with $\xi_{i+1} \circ s_i$. The same determination holds for ξ'_{i+1}, but $\xi_{i,+}$ is determined by ξ_i so $\xi'_{i,+} = \xi_{i,+}$, therefore $\xi'_{i+1} = \xi_{i+1}$. This completes the induction step.

Suppose q and ξ_0 are isomorphisms. Then $\alpha = \beta$ and $q^-(i) = q(i) = i$ for all $i \leq \alpha$, and a similar induction shows that the ξ_i are all isomorphisms. □

An application of the rigidity property will help with the uniqueness part of the main accessibility result:

Corollary 8.3.4 *Suppose $(\alpha, X.)$ is a cell complex, and suppose*

$$(p, \eta.) : (\beta, Y.) \to (\gamma, Z.)$$

is an inclusion of cell complexes. Suppose given a set A and a family of inclusions of cell complexes

$$(q(a), \xi(a)) : (\alpha, X.) \to (\beta, Y.)$$

indexed by $a \in A$, such that the compositions $(p, \eta.) \circ (q(a), \xi(a))$ are all the same. Suppose furthermore that the maps $X_0 \xrightarrow{\xi_0(a)} Y_0$ are all the same. Then the $(q(a), \xi(a))$ are all the same.

Proof Since p is strictly increasing, it is injective. Thus, the condition that the $p \circ q(a)$ are all the same implies that the $q(a)$ are all the same; the previous lemma immediately says that the $(q(a), \xi(a))$ are all the same. □

It is useful to have a different representation of an inclusion of cell complexes $(q, \xi.) : (\alpha, X.) \to (\beta, Y.)$. Define a family of objects denoted X^j and indexed by $j \in [\beta]$ as follows: let

$$X^j := X_i, \quad q^-(i) \leq j \leq q(i).$$

Note that for any $j \in [\beta]$ there exists a unique $i \in [\alpha]$ such that $q^{-}(i) \leq j \leq q(i)$ (see one of the first properties of q^{-} above). We have maps $X^j \to Y_j$, and compatible transition maps $X^j \to X^k$ for any $j \leq k$. For exponents in the same sub-interval $q^{-}(i) \leq j \leq k \leq q(i)$, the transition maps are the identity.

In these terms, the cell attaching data consists of a subset $c(X^{\cdot}) \subset \beta$, defined as the set of all $q(j)$ for $j \in \alpha$, which we think of as the *subset of cells in* $X^{\cdot}$; plus, for each $j \in c(X^{\cdot})$ two maps $u^j(X^{\cdot})$ and $v^j(X^{\cdot})$ fitting into a diagram

$$
\begin{array}{ccccc}
U_j & \xrightarrow{u^j(X^{\cdot})} & X^j & \longrightarrow & Y_j \\
{\scriptstyle g_j}\downarrow & & \downarrow & & \downarrow \\
V_j & \xrightarrow{v^j(X^{\cdot})} & X^{j+1} & \longrightarrow & Y_{j+1}
\end{array}
$$

where $g_j : U_j \to V_j$ is the cell attached to $Y_{\cdot}$ at $j \in \beta$ and the horizontal compositions are the attaching maps $u_j : U_j \to Y_j$ and $v_j : V_j \to Y_{j+1}$ for $Y_{\cdot}$. For any $j \in c(X^{\cdot})$ the above diagram has a cocartesian left square. The composed outer square is cocartesian, being the attaching diagram for $Y_j \to Y_{j+1}$, which implies that the right square is cocartesian also. On the other hand, for $j \notin c(X)$ the map $X^j \to X^{j+1}$ is an isomorphism. In the new notation note that the last element is $X^{\beta} = X_{\alpha}$. The maps $u^j(X^{\cdot})$ and $v^j(X^{\cdot})$ are contained in the data of the cell complex $(\alpha, X_{\cdot})$ with appropriate renumbering via $c(X^{\cdot}) \cong \alpha$.

The collection of data described above is the same as the inclusion of cell complexes. Furthermore, compositions of inclusions of cell complexes can be understood in this notation: if $X^{\cdot}$ is a subcomplex of $Y_{\cdot}$ then the indexing ordinal α for X is isomorphic to the subset $c(X^{\cdot}) \subset \beta$; a subcomplex $Z^{\cdot}$ of $X^{\cdot}$ then consists of a subset

$$c(Z^{\cdot}) \subset \alpha \cong c(X^{\cdot}) \subset \beta,$$

and the composed inclusion of cell complexes from $Z^{\cdot}$ to $Y_{\cdot}$ corresponds to the subset $c(Z^{\cdot}) \subset \beta$ obtained using transport of structure along the isomorphism in the middle.

The category of ordinals with strictly increasing maps isn't closed under sequential transfinite colimits. However, the category of ordinals mapping to a given fixed ordinal does admit sequential colimits. Therefore, the same is true of our notion of inclusion of cell complexes: whereas $\text{CELL}(\mathcal{M}; I)$ is not itself closed under sequential colimits, if we look at cell complexes included into a given $(\beta, Z_{\cdot})$, then we can take sequential colimits.

Proposition 8.3.5 *Suppose $(\beta, Z^\cdot) \in$ CELL$(\mathcal{M}; I)$ is a cell complex. Then the category* CELL$(\mathcal{M}; I)/(\beta, Z^\cdot)$ *is closed under filtered colimits: if $(\alpha(k), X^\cdot(k))$ is a family of cell complexes indexed by a filtered category $k \in \zeta$ such that the transition maps are inclusions of cell complexes, all provided with compatible inclusions $(\alpha(k), X^\cdot(k)) \to (\beta, Z^\cdot)$, then there is a colimit cell complex $(\alpha(\zeta), Y^\cdot)$ again mapping to $(\beta, Z^\cdot)$ and such that $Y_0 = \mathrm{colim}_{k\in\zeta} X_0(k)$ and $Y_{\alpha(\zeta)} = \mathrm{colim}_{k\in\zeta} X_{\beta(k)}(k)$.*

Proof Use the alternative description of an inclusion of cell complexes to $(\beta, Z^\cdot)$. We obtain a family $X^j(k)$ doubly indexed by $j \in [\beta]$ and $k \in \zeta$. Put

$$Y^j := \mathrm{colim}_{k\in\zeta} X^j(k).$$

The maps $Y^j \to Z^j$ are given by the universal property of the colimit. The subset of cells is defined as $c(Y^\cdot) := \bigcup_{k\in\zeta} c(X^\cdot(k))$. If $j \in c(Y^\cdot)$, choose $k_j \in \zeta$ such that $j \in c(X^\cdot(k_j))$. Then let k_j/ζ be the category of arrows $k_j \to m$ in ζ. The functor $k_j/\zeta_j \to \zeta$ is cofinal, so

$$Y^j = \mathrm{colim}_{m\in k_j/\zeta} X^j(m).$$

For each $k_j \to m$ we have the attaching maps

$$U_j \to X^j(k_j) \to X^j(m), \quad V_j \to X^{j+1}(k_j) \to X^{j+1}(m),$$

which yield attaching maps

$$U_j \to Y^j, \quad V_j \to Y^{j+1}.$$

If $j \notin c(Y^\cdot)$ then for any $k \in \zeta$ the map $X^j(k) \to X^{j+1}(k)$ is an isomorphism, so the map $Y^j \to Y^{j+1}$ is an isomorphism. If j is a limit ordinal then

$$X^j(k) = \mathrm{colim}_{i<j} X^i(k)$$

so passing to the double colimit, we have $Y^j = \mathrm{colim}_{i<j} Y^i$. This gives $Y^\cdot$ the structure of cell complex with cell inclusion to $Z^\cdot$ according to our second description above.

This provides a colimit in the category CELL$(\mathcal{M}; I)/(\beta, Z^\cdot)$. Indeed, if $(\beta', Z'^\cdot)$ is an object of the category provided with inclusions from all the $(X^\cdot(k))$, then we can understand the above construction as corresponding to the same construction in CELL$(\mathcal{M}; I)/(\beta', Z'^\cdot)$, which gives the inclusion of cell complexes from $Y^\cdot$ to $(\beta', Z'^\cdot)$. □

8.3.3 Cutoffs

Suppose $(\beta, Z^\cdot)$ is a cell complex, and $\beta' \leq \beta$. Define the *cutoff* $C_{\beta'}(\beta, Z^\cdot)$ to be the cell complex consisting of ordinal β' and family of Z_i for $i \leq \beta' \leq \beta$ with the same structural data restricted to the subset of values of i. If

$$(q, \xi.) : (\alpha, X.) \hookrightarrow (\beta, Z.)$$

is an inclusion of cell complexes, define its *relative cutoff*

$$C_{\beta'}(q, \xi.) : C_{\alpha'}(\alpha, X.) \to C_{\beta'}(\beta, Z.)$$

as follows. Using the strictly increasing map $q : \alpha \to \beta$, put $\alpha' := \sup\{i \leq \alpha, q(i) \leq \beta'\}$. Define a new map q' equal to q on α, but extended to $[\alpha]$ by setting $q'(\alpha) := \beta'$. In general this is different from $q(\alpha)$.

Define the relative cutoff inclusion of cell complexes to be given by the restricted map $q' : [\alpha'] \to [\beta']$ and the family of maps ξ_i and associated cell identification data for $X.$ restricted to $i \leq \alpha'$.

In terms of the alternative notation for inclusions of cell complexes, we denote the cutoff of a subcomplex $X^{\cdot}$ by just $C_{\beta'}(X^{\cdot})$. In these terms, the subset of cells is just the intersection

$$c(C_{\beta'}(X^{\cdot})) = c(X^{\cdot}) \cap \beta',$$

and the attaching data are the same as those of $X^{\cdot}$.

8.3.4 The filtered property for subcomplexes

In the context of our next main accessibility result for cell complexes, we would like to consider the *join* of a set of inclusions of cell complexes. Suppose $(\beta, Z.) = (\beta, Z., f., u., v.)$ is a cell complex, J is a set, and $(q^j, \xi.(j)) : (\alpha^j, X.(j)) \hookrightarrow (\beta, Z.)$ is a family of inclusions of cell complexes. The "join" should be a cell complex sitting in the middle

$$(\alpha^j, X.(j)) \hookrightarrow (\varphi, Y.) \hookrightarrow (\beta, Z.).$$

The set of cells $c(Y^{\cdot})$ should be the union of the sets of cells $c(X.(j)) \subset \beta$. In order to start off the process, we need to make a choice of the place to start Y_0, which should fit into factorizations

$$X_0^j \to Y_0 \to Z_0,$$

for all j. From there, one can proceed to attach the required cells without problem, provided $\mathcal{M}$ is a *cellular model category* in the terminology of Hirschhorn [144]. This condition basically means that the arrows in I are monomorphisms enjoying good properties; for example, a presheaf category in which the cofibrations are contained in the injections, is cellular.

In the general case, we run into the problem that the cell attaching maps needed to construct $Y^{\cdot}$ are not uniquely determined by those of $Z.$. So, we proceed differently. The general situation has been treated by Lurie [190, Appendix]. Between the cellular case treated by Hirschhorn [144] and the general treatment of Lurie's [190], our present discussion is undoubtedly superfluous, but is included for completeness.

Suppose $\mathcal{M}$ is locally κ-presentable, that the elements of I are arrows whose source and target are κ-presentable, and that I has $< \kappa$ elements. Let CELL $(\mathcal{M}; I)_\kappa$ denote the full subcategory of cell complexes $(\alpha, X.)$ such that $|\alpha| < \kappa$ and X_0 is κ-presentable. The following theorem lets us replace the general notion of cofibration by a cell complex. This was done by Lurie [190, A.1.5.12], so the reader could refer there instead.

Theorem 8.3.6 *With the above hypotheses, suppose given a cell complex $(\beta, Z.) \in \text{CELL}(\mathcal{M}; I)$. Then the category $\text{CELL}(\mathcal{M}; I)_\kappa/(\beta, Z.)$ is κ-filtered. Furthermore $(\beta, Z.)$ is the colimit in $\text{CELL}(\mathcal{M}; I)$ of the tautological functor*

$$\text{CELL}(\mathcal{M}; I)_\kappa/(\beta, Z.) \to \text{CELL}(\mathcal{M}; I),$$

and the arrow $Z_0 \to Z_\beta$ is the colimit in $\text{ARR}(\mathcal{M})$ of the composition of the tautological functor with $\text{CELL}(\mathcal{M}; I) \to \text{ARR}(\mathcal{M})$.

Proof The objects of $\text{CELL}(\mathcal{M}; I)_\kappa/(\beta, Z.)$ are inclusions of cell complexes $(\alpha, X.) \hookrightarrow (\beta, Z.)$ such that $|\alpha| < \kappa$ and such that X_0 is κ-presentable.

We first show the uniqueness half of the κ-filtered property. Suppose

$$(p, \eta.) : (\alpha, X.) \to (\beta, Z.)$$

and

$$(p', \eta'_.) : (\alpha', X'_.) \to (\beta, Z.)$$

are two objects of $\text{CELL}(\mathcal{M}; I)_\kappa/(\beta, Z.)$, and suppose given a set A of cardinality $|A| < \kappa$ indexing a family of morphisms from one to the other in $\text{CELL}(\mathcal{M}; I)_\kappa/(\beta, Z.)$, that is to say a family of inclusions of cell complexes

$$(q(a), \xi.(a)) : (\alpha, X.) \to (\alpha', X'_.)$$

such that $(p', \eta'_.) \circ (q(a), \xi.(a)) = (p, \eta.)$. Injectivity of p' implies that the $q(a)$ are all the same. On the other hand we get the family of maps

$$\xi_0(a) : X_0 \to X'_0$$

such that $\eta'_0 \circ \xi_0(a) = \eta_0$. Recall that X_0 and X'_0 are required to be κ-presentable. The same uniqueness part of the property that $\mathcal{M}_\kappa/Z_0$ is κ-filtered, says that the map $X'_0 \to Z_0$ factors as $X'_0 \stackrel{\phi_0}{\to} Y_0 \to Z_0$ through a κ-presentable object Y_0, such that all of the maps

$$\phi_0 \circ \xi_0(a) : X_0 \to Y_0$$

are the same. As in Lemma 8.3.1, define a new cell complex $(\alpha', Y.)$ by setting

$$Y_i := Y_0 \cup^{X'_0} X'_i,$$

and keeping the same attaching data as for $X'_.$, and with the tautological inclusion of cell complexes $(t, \phi.)$ from $(\alpha', X'_.)$ to $(\alpha', Y.)$. The condition on the choice of Y_0 together with the fact that the $q(a)$ are all the same, provide the hypotheses required in order to apply Corollary 8.3.4 to conclude that all the maps of cell complexes

$$(t, \phi.) \circ (q(a), \xi.(a)) : (\alpha, X.) \to (\alpha', Y.)$$

are the same.

The elements Y_i are κ-presentable, so $(\alpha', Y.) \in \text{CELL}(\mathcal{M}; I)_\kappa$. The map of cell complexes $(p', \eta'_.)$ extends, using the pushout expressions for Y_i and the maps $Y_0 \to Z_i$, to a map of cell complexes

$$(p', \zeta.) : (\alpha', Y.) \to (\beta, Z.).$$

Via this map, we may see $(\alpha', Y.)$ as an element of $\text{CELL}(\mathcal{M}; I)_\kappa/(\beta, Z.)$.

Note that

$$(p', \zeta.) \circ (t, \phi.) = (p', \eta'_.),$$

so $(t, \phi.)$ may be viewed as a map in $\text{CELL}(\mathcal{M}; I)_\kappa/(\beta, Z.)$. This gives exactly a map there whose compositions with the $(q(a), \xi.(a))$ are all the same, serving to prove the uniqueness half of the κ-filtered property.

For the remainder of the theorem, we prove the full statement by induction on the length β of the cell complex Z. If $\beta = 0$ there is nothing to prove. Hence, we may assume that the statement of the theorem is known for all cell complexes $(\beta', Z'_.)$ with $\beta' < \beta$.

The main step is to prove that $\text{CELL}(\mathcal{M}; I)_\kappa/(\beta, Z.)$ is κ-filtered. Suppose that $X^{\cdot}(a)$ is a collection of subcomplexes of $(\beta, Z.)$ which are κ-small, that is with X^0 being κ-presentable and $|c(X^{\cdot})| < \kappa$, and indexed by a set $a \in A$ with $|A| < \kappa$. We have to show that they all map to a single κ-small subcomplex $Y^{\cdot}$.

For any $\beta' < \beta$, consider the collection of cutoff complexes (using the alternate notation at the end of Section 8.3.3) $C_{\beta'}(X^{\cdot}(a))$ mapping by inclusions of cell complexes to $C_{\beta'}(\beta, Z.)$. By the induction hypothesis, there is a single κ-small subcomplex $Y^{\cdot}(\beta')$ of $C_{\beta'}(\beta, Z.)$, such that all of the $C_{\beta'}(X^{\cdot}(a))$ map to $Y^{\cdot}(\beta')$.

Assume that β is a limit ordinal approached by a sequence of $\beta' < \beta$ of cardinality $< \kappa$. Then, by transfinite induction over this sequence we may assume that the choice of $Y^{\cdot}(\beta')$ is provided with transition inclusion maps from $Y^{\cdot}(\beta')$ to $Y^{\cdot}(\beta'')$ whenever $\beta' \leq \beta''$ are two members of the sequence. Furthermore, we may take the colimit of all of the $Y^0(\beta')$ and use this as starting element (it is a colimit of length $< \kappa$ so it remains κ-presentable). Thus we may assume that the $Y^0(\beta')$ are all the same, so that Lemma 8.3.5 applies. Set $Y^{\cdot} := \mathrm{colim}_{\beta'} Y^{\cdot}(\beta')$. There are maps from $C_{\beta'}(X^{\cdot}(a))$ to $Y^{\cdot}$. The different maps from the $X^0(a)$ to Y^0, depending on β', compose to the same map into Z^0. Since $X^0(a)$ is κ-presentable, and the κ-presentable objects mapping to Z^0 form a κ-filtered category, see Adámek and Rosický [2, proposition 1.22], and also since both A and the sequence of β' have size $< \kappa$, we may choose W fitting into $Y^0 \to W^0 \to Z^0$ but with W still being κ-presentable. Then replace Y^j by $W \cup^{Y^0} Y^j$. This way, all of the maps $X^0(a) \to Y^0$ will be the same independent of β'. Then a transfinite induction argument on the cutoff level β' shows that the maps $C_{\beta'}(X^{\cdot}(a)) \to Y^{\cdot}$ fit together in the colimit as $\beta' \to \beta$, to a collection of maps $X^{\cdot}(a) \to Y^{\cdot}$. This completes the proof of the filtered property when β is a limit ordinal approached by a sequence of cardinality $< \kappa$.

If on the other hand β is a limit ordinal which is not approached by any sequence of cardinality $< \kappa$, then there is some $\beta' < \beta$ such that all of the cells in $c(X^{\cdot}(a))$ are at $j < \beta'$ for all $a \in A$. Therefore $C_{\beta'}(X^{\cdot}(a)) = X^{\cdot}(a)$, and these are cell complexes included in $C_{\beta'}(\beta, Z.)$. The inductive hypothesis says that they all map to the same subcomplex $Y^{\cdot}$ of $C_{\beta'}(\beta, Z.)$, and this $Y^{\cdot}$ serves to show the filtered property for the family of cell complex inclusions $X^{\cdot}(a) \to (\beta, Z.)$.

This leaves us with the case when β is a successor ordinal: $\beta = \eta + 1$. It is in some sense the main case where we need some work. This extra work is occasionned by the goal of working without a monomorphism axiom, such as was used by Hirschhorn [144] for the elements of I. The basic problem is that when we try to add in the last cell, the attaching maps may be ill defined because of various different collections of cell attachments up until then. Hence, we need to backtrack and add some more cells so as to stabilize the collection of attaching maps. For this we use the full inductive hypothesis about the colimit over our filtered category, which applies to the cell complex $C_\eta(\beta, Z.)$ of length one less. This says that Z_η is a κ-filtered colimit of things obtained by inclusions of κ-small cell complexes.

To set things up more carefully, write $A = A' \cup A''$, where A' consists of those a such that $X^{\cdot}(a)$ involves the last cell, that is to say $\eta \in c(X^{\cdot}(a))$; and A'' consists of those a for which $\eta \notin c(X^{\cdot}(a))$. One could suppose that A'' consists of a single element, indeed the $X^{\cdot}(a)$ which don't involve the last cell are all subcomplexes of a single subcomplex of $C_\eta(\beta, Z.)$ by the inductive hypothesis, and we could take this one as the single subcomplex indexed by A''.

Now for each $a \in A'$ we have $X^\eta(a) \to Z_\eta$. Furthermore, we have attaching maps

$$u_\eta(X^\cdot(a)) : U_\eta \to X^\eta(a), \quad v_\eta(X^\cdot(a)) : V_\eta \to X^\beta(a) \cong X^\eta(a) \cup^{U_\eta} V_\eta.$$

These lift the attaching maps $u_\eta : U_\eta \to Z_\eta$ and $v_\eta : V_\eta \to Z_\beta$. By the inductive hypothesis, there is a single κ-small cell complex $Y^\cdot$ mapping by a cell complex inclusion to $C_\eta(\beta, Z.)$, such that all of the $C_\eta(X^\cdot(a))$ for $a \in A'$, and all of the $X^\cdot(a)$ for $a \in A''$, map to $Y^\cdot$. Choose such maps for each a, denoted $\psi^\cdot(a)$. Then for any $a \in A'$ we get a composed attaching map

$$u_\eta(a) := \psi^\eta(a) \circ u_\eta(X^\cdot(a)) : U_\eta \to Y^\eta.$$

The composition of these maps with $Y^\eta \to Z_\eta$ are all the same, they are the attaching map for the original complex Z..

The inductive hypothesis tells us that Z_η is a κ-filtered colimit of W^η as $W^\cdot$ runs over the κ-filtered category of κ-small cell complexes with inclusions to $C_\eta(\beta, Z.)$. The category of objects $W^\cdot$ under $Y^\cdot$ is cofinal in here, so we can view this colimit as being over factorizations $Y^\cdot \to W^\cdot \to C_\eta(\beta, Z.)$.

As, on the other hand, the category of κ-presentable objects mapping to Z_η is κ-filtered, there is a factorization $Y^\eta \to T \to Z_\eta$ with T being κ-presentable, such that all of the above maps $u_\eta(a)$ compose into the same map $U_\eta \to T$. Because Z_η is a κ-filtered colimit of the W^η, the κ-presentable property of T tells us that there is a factorization $T \to W^\eta \to Z_\eta$ for some $Y^\cdot \to W^\cdot \to C_\eta(\beta, Z.)$. Thus, all of the composed maps

$$U_\eta \xrightarrow{u_\eta(a)} Y^\eta \to W^\eta$$

are the same. We may now use this unique map as attaching map to add on the last cell, to create a cell complex $\tilde{W}^\cdot$ with an inclusion to $(\beta, Z.)$, whose cutoff at η is $C_\eta(\tilde{W}^\cdot) = W^\cdot$. The identity of all of the composed attaching maps provides us with inclusions of cell complexes $X^\cdot(a) \to \tilde{W}^\cdot$ for all $a \in A$ (the cell attachment provides these inclusions for $a \in A'$ and they are automatic for $a \in A''$). This finishes the proof of the κ-filtered property for $\text{CELL}(\mathcal{M}; I)_\kappa/(\beta, Z.)$.

To complete the proof of the inductive step we just have to note that the colimit of the κ-small cell complex inclusions, which we now know is κ-filtered and hence exists by Lemma 8.3.5, is equal to the full $(\beta, Z.)$. For this, note first of all that the colimit of all of the X^0 will be Z^0 by the locally κ-presentable property of $\mathcal{M}$. To conclude, it suffices to note that every $j \in \beta$ is a cell in at least one of the κ-small subcomplexes $X^\cdot$. If β is a limit ordinal, apply the inductive hypothesis saying that we know the statement of the theorem for any $\beta' < \beta$, and note that for any $j < \beta$ we can choose $j < \beta' < \beta$. If $\beta = \eta + 1$

is a successor ordinal, the inductive hypothesis gives the required statement for any $j < \eta$ so we may assume $j = \eta$. Then, argue as above. We know that $C_\eta(\beta, Z.)$ is a κ-filtered colimit of its κ-small subcomplexes, and as before it follows that there will be a single κ-small subcomplex $W^\cdot$ such that the attaching map $U_\eta \to Z_\eta$ factors through $U_\eta \to W^\eta$. The complex $\tilde{W}^\cdot$ obtained by attaching the last cell $U_\eta \to V_\eta$ onto $W^\cdot$ satisfies the current requirement that $j = \eta$ be an element of $c(\tilde{W}^\cdot)$.

This completes the inductive proof of the theorem. □

We now restate the result of the theorem in its simplified form which will be used later for the pseudo-generating set construction of a model category structure in Section 8.7:

Corollary 8.3.7 *Suppose $f : X \to Y$ is a morphism in* **cell**(I). *Then we can express f as a κ-filtered colimit of arrows $f_i : X_i \to Y_i$, in particular $X = \mathrm{colim}_i X_i$ and $Y = \mathrm{colim}_i Y_i$, such that X_i, Y_i are κ-presentable and $f_i \in$* **cell**(I) *are cell complexes of length $< \kappa$.*

Proof There is a cell complex $(\beta, Z.) \in \mathrm{CELL}(\mathcal{M}; I)$ such that f is equal to the map $Z_0 \to Z_\beta$, by the definition of **cell**(I). The theorem then says exactly that f is a κ-filtered colimit of morphisms coming from cell complexes in $\mathrm{CELL}(\mathcal{M}; I)_\kappa$. □

We now describe how to fine-tune a map between colimits so that it comes from a levelwise map of diagrams. Let $\mathcal{M}$ be a locally κ-presentable category. Suppose α is a κ-filtered category, and $F : \alpha \to \mathcal{M}$ is a diagram such that each F_i is κ-presentable. Suppose β is another κ-filtered category and $G : \beta \to \mathcal{M}$ is another diagram. Suppose given a map

$$f : \mathrm{colim}_{i\in\alpha} F_i \to \mathrm{colim}_{j\in\beta} G_j.$$

Assume that the F_i are κ-presentable objects.

Lemma 8.3.8 *In the above situation, there are a κ-filtered category ψ and cofinal functors $p : \psi \to \alpha$ and $q : \psi \to \beta$ and a natural collection of maps $f_k : F_{p(k)} \to G_{q(k)}$ depending on $k \in \psi$, i.e. a natural transformation $p^*F \to q^*(G)$, such that the composition*

$$\mathrm{colim}_{i\in\alpha} F_i = \mathrm{colim}_{k\in\psi} F_{p(k)} \xrightarrow{\mathrm{colim} f_k} \mathrm{colim}_{k\in\psi} G_{q(k)} = \mathrm{colim}_{j\in\beta} G_j$$

is equal to f.

Proof Let ψ be the category of triples (i, j, u), where $i \in \alpha$, $j \in \beta$ and $u : F_i \to G_j$ is a morphism such that the diagram

$$\begin{array}{ccc} F_i & \longrightarrow & \mathrm{colim}_{i\in\alpha} F_i \\ \downarrow & & \downarrow \\ G_j & \longrightarrow & \mathrm{colim}_{j\in\beta} G_j \end{array}$$

commutes. Since the F_i are κ-presentable, any i is part of a triple $(i, j, u) \in \psi$. Any j sufficiently far out in β is also part of a triple, and indeed ψ is κ-filtered, and the forgetful functors $\psi \to \alpha$ and $\psi \to \beta$ are cofinal. The third variable u provides the desired natural transformation $\{f_k\}$ such that the diagram

$$\begin{array}{ccc} \mathrm{colim}_{k\in\psi} F_{p(k)} & \stackrel{\cong}{\rightarrow} & \mathrm{colim}_{i\in\alpha} F_i \\ {\scriptstyle \mathrm{colim}\, f_k}\downarrow & & \downarrow {\scriptstyle f} \\ \mathrm{colim}_{k\in\psi} G_{q(k)} & \stackrel{\cong}{\rightarrow} & \mathrm{colim}_{j\in\beta} G_j \end{array}$$

commutes. □

The following proposition represents the idea that, given an inclusion of cell complexes, we can rearrange things so that the subcomplex comes first and then the rest of the complex is added on later.

Proposition 8.3.9 *If $(\alpha, X.) \hookrightarrow (\beta, Y.)$ is an inclusion of cell complexes, then there is another inclusion of cell complexes $(\delta, Z.) \hookrightarrow (\beta, Y.)$ with $Z_0 = X_\alpha \cup^{X_0} Y_0$ and $Z_\delta = Y_\beta$, such that β is the disjoint union of $c(X^{\cdot})$ and $c(Y^{\cdot})$.*

Proof Left to the reader. □

8.4 Cofibrantly generated, combinatorial and tractable model categories

There is an important class of model categories that are particularly easy to work with, and which contains many if not most of the examples currently considered as important. Useful examples of model categories which are not cofibrantly generated will often be Quillen equivalent to cofibrantly generated ones. On the other hand, this notion is very helpful for a number of the operations we need, such as taking the model category of $\mathcal{M}$-diagrams, and left Bousfield localization. The reader is referred to Hirschhorn's excellent book [144] for a full explanation of everything concerning cofibrantly generated model categories.

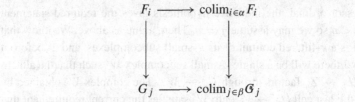

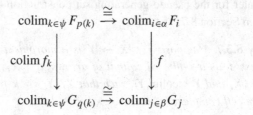

Suppose $J \subset \mathrm{Mor}(\mathcal{M})$ is a small subset of morphisms, then we define classes of morphisms $\mathbf{cell}(J)$, $\mathbf{inj}(J)$ and $\mathbf{cof}(J)$. The class $\mathbf{cell}(J)$ is the smallest subclass of morphisms containing J and closed under pushout (i.e. if $X \to Y$ is in $\mathbf{cell}(J)$ and $X \to Z$ is any morphism then $Z \to Z \cup^X Y$ is in $\mathbf{cell}(J)$) and transfinite composition. Then $\mathbf{inj}(J)$ is the class of those morphisms which satisfy the right lifting property with respect to all morphisms in $\mathbf{cell}(J)$, and finally $\mathbf{inj}(J)$ is the class of those morphisms which satisfy the left lifting property with respect to all morphisms in $\mathbf{cell}(J)$. One shows that $\mathbf{cof}(J)$ consists of arrows $X \to Y$ which are retracts, in the category of objects under X, of elements of $\mathbf{cell}(J)$. This condition means that there should exist $X \to Z$ in $\mathbf{cell}(J)$ and maps $Y \to Z \to Y$ compatible with the maps from X and composing to the identity of Y. The notion of cell complex will be discussed further in Sections 8.3.2.

Recall that a $\mathbb{U}$-model category $\mathcal{M}$ is *cofibrantly generated* if it is closed under $\mathbb{U}$-small limits and colimits, and if there are $\mathbb{U}$-small subsets of arrows $I, J \subset \mathbf{Arr}(\mathcal{M})$ satisfying the following properties:

(CG1) – the sources and targets of arrows in I and J are small in $\mathcal{M}$;

(CG2a) – the arrows in I are cofibrations;

(CG2b) – the trivial fibrations of $\mathcal{M}$ are the morphisms satisfying right lifting with respect to arrows in I;

(CG3a) – the arrows in J are trivial cofibrations;

(CG3b) – the fibrations of $\mathcal{M}$ are the morphisms satisfying right lifting with respect to arrows in J.

As a matter of notation, we say that $(\mathcal{M}, I, J)$ *is a cofibrantly generated model category*, if $\mathcal{M}$ is a model category and I and J are subsets of arrows satisfying the above axioms.

This notation is convenient in that I and J determine the model category structure, indeed I determines the class of trivial fibrations by (CG2b) and J determines the class of fibrations by (CG3b). These in turn determine respectively the classes of cofibrations and trivial cofibrations by the saturated lifting properties. Then the class of weak equivalences is determined by either one of the factorization properties.

Given a triple $(\mathcal{M}, I, J)$ consisting of a category admitting $\mathbb{U}$-small limits and colimits, and two subsets of arrows, we can try to define a model structure following the recipe of the previous paragraph. If these classes of (trivial) cofibrations, (trivial) fibrations and weak equivalences do in fact form a model structure, and if (CG1) is satisfied, then $(\mathcal{M}, I, J)$ is a cofibrantly generated model category.

See Hirschhorn [144], Goerss and Jardine [122], Dwyer and Spalisnki [105] and other references for discussions of various recognition properties telling when a triple $(\mathcal{M}, I, J)$ yields a cofibrantly generated model category. For example, the following statement will be useful:

Proposition 8.4.1 *Suppose $(\mathcal{M}, I, J)$ is a triple consisting of a locally presentable category and two small sets of morphisms. Start with the three classes of cofibrations, fibrations and weak equivalences defined from I and J as above; then define trivial cofibrations as the intersection of cofibrations and weak equivalences and similarly for trivial fibrations. If I and J satisfy properties (CG1) – (CG3b); and if furthermore we know that weak equivalences are closed under retracts and satisfy 3 for 2, then it is a cofibrantly generated model category.*

Proof Indeed, (CM5) comes from the small object argument, (CM4) comes from the hypotheses (CG2b) and (CG3b), (CM3) is supposed for weak equivalences and follows for **inj**(J) and **cof**(I), and (CM1) and (CM2) are supposed. □

It is most convenient to combine the notion of cofibrantly generated model category with a good category-theoretical condition on $\mathcal{M}$ which guarantees the small object argument. As was observed by J. Smith and reported in several papers by Beke [33], Dugger [95] and, more recently, Rosický [223], the appropriate condition is to require that $\mathcal{M}$ be locally presentable:

Definition 8.4.2 A *combinatorial model category* is a cofibrantly generated model category which is also locally presentable. In this case all objects are small in $\mathcal{M}$ so (CG1) is automatic.

Barwick [18] then refined this by requiring that the domains of generating cofibrations and trivial cofibrations also be cofibrant:

Definition 8.4.3 A combinatorial model category is *tractable* if in addition for the given generating sets I and J:

(TR) – the sources of arrows in I and J are cofibrant.

This condition simplifies considerably the application of the small object argument, in particular it will be important for our discussion of direct left Bousfield localization in Chapter 9. That explains why tractability will be one of the main hypotheses for our model category $\mathcal{M}$ retained throughout the next parts of the book.

As we shall discuss below, diagrams with values in a combinatorial model category have injective and projective model structures which are also combinatorial. In the projective structure, originally constructed by Bousfield and Kan [58], the fibrations are defined as the levelwise ones. The cofibrations are generated by the $i_{x,!}(f)$ where $i_x : \{x\} \rightarrow \Phi$ is inclusion of a single object, and f runs through a generating set of cofibrations for $\mathcal{M}$. It follows that if the target category is tractable, then the projective model structure on diagrams is also tractable. This is no longer necessarily true for the injective model structure, and even the cofibrant generation property is not easy to show: it will be discussed in Theorem 8.6.2 below.

If all objects of $\mathcal{M}$ are cofibrant (a condition preserved when we pass to injective diagram categories) then the tractable and combinatorial conditions are equivalent. This condition holds for most of the model categories to be envisioned in practice, obtained by interation of our basic construction. In that case, one can say that the injective diagram categories are tractable, so these become available and it will be worthwhile to devote Section 8.6 to the construction of these structures. For the reader wanting to keep to a greater level of generality, the injective structure can be replaced by the Reedy diagram structure, as will appear in many of the statements in later parts of the book.

8.5 Smith's recognition principle

The basic step in the construction by Joyal [156] and Jardine [153] of the model structure for simplicial presheaves, was to show that trivial cofibrations are closed under pushout. This has been the basis for a number of "recognition principles" allowing one to deduce the existence of a model structure by starting with a distinguished subset of properties. One of the most useful of these is due to J. Smith, and has been reported by Beke [33] and Barwick [20]. It will form the basis for our construction via pseudo-generating sets in Section 8.7, but also provides a good way of approaching the classical projective and injective model structures on diagram categories.

Definition 8.5.1 Suppose $\mathcal{M}$ is a locally κ-presentable category for some regular cardinal κ, and $\mathcal{F} \subset \text{ARR}(\mathcal{M})$ is a subclass of arrows. Let $\mathcal{F}_\kappa$ denote the subclass of $\mathcal{F}$ consisting of those $X \xrightarrow{f} Y$ in $\mathcal{F}$ such that X and Y are κ-presentable. We say that $\mathcal{F}$ is *κ-accessible* if every $X \xrightarrow{f} Y$ in $\mathcal{F}$ may be represented as a κ-filtered colimit of $X_i \xrightarrow{f_i} Y_i$ in $\mathcal{F}_\kappa$. More precisely that colimit expression means that the f_i form a functor defined on objects i in a κ-filtered category α, with transition diagrams

$$\begin{array}{ccc} X_i & \longrightarrow & X_j \\ {\scriptstyle f_i}\downarrow & & \downarrow{\scriptstyle f_j} \\ Y_i & \longrightarrow & Y_j \end{array}$$

whenever $i \to j$ in α, with $X = \mathrm{colim}_{i\in\alpha} X_i$, $Y = \mathrm{colim}_{i\in\alpha} Y_i$ and f is the colimit of the f_j.

We now describe the setup for Smith's recognition theorem, see Barwick [20, proposition 2.2]. Suppose now we are given a locally presentable category $\mathcal{M}$, with a class of arrows called weak equivalences, and a set of arrows $I \subset \mathrm{ARR}(\mathcal{M})$. Assume the following hypotheses:

(SR1)–the class of weak equivalences is κ-accessible, for a regular cardinal κ such that $\mathcal{M}$ is locally κ-presentable (Definition 8.5.1);

(SR2)–the class of weak equivalences is closed under retracts and satisfies 3 for 2;

(SR3)–any morphism in **inj**(I) is a weak equivalence;

(SR4)–the class of trivial cofibrations, defined to be the intersection of **cof**(I) and the weak equivalences, is closed under pushout and transfinite composition;

(SR5)–if $X \stackrel{f}{\to} Y$ is in I then $\emptyset \to X$ is in **cof**(I).

A morphism in **cof**(I) is called a cofibration; and if it is furthermore a weak equivalence, it is called a trivial cofibration. The morphisms that satisfy the right lifting property with respect to the trivial cofibrations are called fibrations; and the fibrations that are weak equivalences are called trivial fibrations.

The goal, to be completed in Theorem 8.5.6 below, is to show that these classes define a tractable model category structure on $\mathcal{M}$. The tractability, which uses hypothesis (SR5) saying that the elements of I have cofibrant domains, is Barwick's Corollary 2.7 [20]; Smith's original result didn't use this hypothesis, and gives a combinatorial model category instead *loc. cit.* [20]. Before getting to this main theorem, we give some preliminary definitions and lemmas.

A *cofibrant replacement* of an object $X \in \mathcal{M}$ is a morphism $p : X' \to X$ such that $p \in$ **inj**(I) and $\emptyset \to X'$ is in **cell**(I). This exists by the small object argument for I, and by Axiom (SR3) above, it follows that $p \in$ **inj**(I) is a weak equivalence.

Choose a regular cardinal κ such that $\mathcal{M}$ is locally κ-presentable, and the class of weak equivalences is κ-accessible (Definition 8.5.1). The first step is to define a set J of trivial cofibrations with cofibrant domains which will serve as the generating set for the trivial cofibrations.

Recall that the category $\mathcal{M}$ is locally κ-presentable. Choose a small set N_1 of representatives for the isomorphism classes of arrows $f : X \to Y$ which are weak equivalences between κ-presentable objects. Since the isomorphism classes of κ-presentable objects form a small set (axiom (2) of Definition 8.1.1 for the locally κ-presentable category $\mathcal{M}$), we can choose N_1 as a small set. For each $f \in N_1$, choose a cofibrant replacement $p : X' \to X$, let $f' : X' \to Y$ be the composition $f' = fp$, and choose a factorization $f' = gh$ where $h : X' \to Z$ is in $\mathbf{cell}(I)$ and $g : Z \to Y$ is in $\mathbf{inj}(I)$, in particular g is a trivial fibration. Let J be the set of all cofibrations h obtained in this way. Using the facts that g and p are weak equivalences, and the hypothesis that f is a weak equivalence, we get that f' and then h are weak equivalences by (SR2). Thus the elements of J are trivial cofibrations with cofibrant domains. We have to show that a map in $\mathbf{inj}(J)$ is a fibration, that is that it satisfies lifting with respect to any trivial cofibration. Or, equivalently, to show that any trivial cofibration is in $\mathbf{cof}(J)$.

Lemma 8.5.2 *The elements of* $\mathbf{cof}(J)$ *are trivial cofibrations.*

Proof Trivial cofibrations are closed under pushouts and transfinite compositions (SR4), hence the elements of $\mathbf{cell}(J)$ are trivial cofibrations. The class $\mathbf{cof}(J)$ is the closure of $\mathbf{cell}(J)$ under retracts; by definition the class $\mathbf{cof}(I)$ of cofibrations is closed under retracts, and by (SR2) the class of weak equivalences is closed under retracts, so elements of $\mathbf{cof}(J)$ are trivial cofibrations. □

Lemma 8.5.3 *Our set of arrows J generates the trivial cofibrations, in other words the class* $\mathbf{cof}(J)$ *equals the class of all trivial cofibrations. In particular,* $\mathbf{inj}(J)$ *equals the class of fibrations.*

Proof Suppose we are given a trivial cofibration $f : X \to Y$. For now suppose also that f is in $\mathbf{cell}(I)$.

Choose a cellular expression for f, that is a sequence $\{X_n\}_{n\leq\beta}$ indexed by an ordinal β with $X_0 = X$, $X_\beta = Y$, $X_m = \mathrm{colim}_{n<m} X_m$ when m is a limit ordinal, and $X_n \to X_{n+1}$ obtained by pushout along an element of I.

We will choose a sequence of diagrams $X_n \to Z_n \to Y$, where $\{Z_n\}_{n\leq\beta}$ is a transfinite sequence, such that $X \to Z_n$ is in $\mathbf{cell}(J)$ in particular a trivial cofibration; $Z_n \to Z_{n+1}$ is pushout along an element of J, and again $Z_m = \mathrm{colim}_{n<m} Z_n$ when m is a limit ordinal. The transition maps in the sequence $\{Z_n\}_{n\leq\beta}$ are supposed to be compatible with those of the sequence $\{X_n\}_{n\leq\beta}$ and with the maps to Y.

For the induction step at a limit ordinal m, just let Z_m be the colimit of the Z_n for $n < m$.

Suppose n is a successor ordinal with the Z_i for $i \leq n-1$ given, and we want to choose Z_n. Consider the map $w_n : U_n \to V_n$ in I such that $X_{n-1} \to X_n$ is pushout along w_n; we have a commutative diagram

$$\begin{CD} U_n @>>> X_{n-1} @>>> Z_{n-1} \\ @VVV @VVV @VVV \\ V_n @>>> X_n @>>> Y \end{CD}$$

where the first square is a pushout. The map $X \to Z_{n-1}$ is a weak equivalence by the inductive hypothesis, and $X \to Y$ is a weak equivalence by assumption, so by (SR2), the map $Z_{n-1} \to Y$ is a weak equivalence. On the other hand, $w_n : U_n \to V_n$ is in I, so by the choice of κ it is κ-presentable in $\mathrm{ARR}(\mathcal{M})$. By Proposition 8.7.2 the map $Z_{n-1} \to Y$ is a κ-filtered colimit of κ-presentable arrows in $\mathrm{ARR}(\mathcal{M})$ which are weak equivalences–that is, elements of N_1. Since w_n is κ-presentable it factors through one of them. Therefore, there exists a factorization

$$\begin{CD} U_n @>a>> A @>z>> Z_{n-1} \\ @VVV @VVV @VVV \\ V_n @>b>> B @>y>> Y \end{CD}$$

with the middle vertical arrow in N_1. Consider the choice of cofibrant replacement $A' \to A$ used for the definition of J and, since U_n is cofibrant (SR5), we can choose a lifting $U_n \to A'$. In particular, we may replace A by A' in the previous diagram. Consider the choice of factorization $A' \xrightarrow{u} C \xrightarrow{v} B$ with $u \in J$ and $v \in \mathbf{inj}(I)$, and choose a lifting $c : V_n \to C$ with $vc = b$ and c restricting to the given map on U_n. This gives a factorization

$$\begin{CD} U_n @>a>> A' @>>> Z_{n-1} \\ @VVV @VVV @VVV \\ V_n @>c>> C @>yv>> Y \end{CD}$$

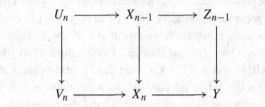

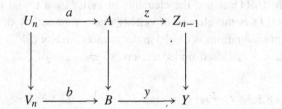

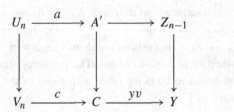

where the middle vertical arrow is in J. Let $Z_n := Z_{n-1} \cup^{A'} C$ be the pushout along this arrow. Then $X \to Z_n$ is again in **cell**(J), in particular it is a trivial cofibration (Lemma 8.5.2). From the second square in the above diagram we get a map $Z_n \to Y$, which is compatible with the map on Z_{n-1}. But, given that $X_n = X_{n-1} \cup^{U_n} V_n$, we get a map $X_n \to Z_n$, which is compatible with given map on X_{n-1}. This completes the inductive step, giving the construction of the sequence $\{Z_n\}_{n\leq\beta}$ as desired.

Letting $Z := Z_\beta$ we get a map $g : X \to Z$ in **cell**(J) with projection $p : Z \to Y$ and a splitting $s : Y \to Z$, $ps = 1$ so that f is a retract of g in objects under X. This shows $f \in$ **cof**(J).

We have shown that a map f in **cell**(I) which is a weak equivalence, is in **cof**(J). Suppose $f \in$ **cof**(I) is a weak equivalence. Choose a factorization $f = gh$ with $X \xrightarrow{h} V \xrightarrow{g} Y$, where $h \in$ **cell**(I) and $g \in$ **inj**(I). Then g is a weak equivalence from the definition (PG), so h is a weak equivalence (SR2). We have shown above that $h \in$ **cof**(J). On the other hand, since $f \in$ **cof**(I) it satisfies lifting with respect to h, so there is a section $s : Y \to Z$, making f into a retract of h in objects under X. In general the cofibrations for a set of arrows forms a class closed under retracts, and f is a retract of $h \in$ **cof**(J) so $f \in$ **cof**(J).

This shows that **cof**(J) equals the class of trivial cofibrations. Given a map $g \in$ **inj**(J) it satisfies lifting with respect to **cell**(J), and any trivial cofibration is a retract of something in **cell**(J). It follows that g satisfies lifting with respect to any trivial cofibration, so g is a fibration. Conversely a fibration is in **inj**(J), as follows from Lemma 8.5.2. □

Corollary 8.5.4 *Any map $f : X \to Y$ in $\mathcal{M}$ factors as $f = gh$ with $X \xrightarrow{h} Z \xrightarrow{g} Y$, where h is a trivial cofibration (which can be assumed even in* **cell**(J)*) and g is a fibration.*

Proof This is just the factorization into $h \in$ **cell**(J) and $g \in$ **inj**(J) for the set J, given by the small object argument. Note that h is a trivial cofibration by Lemma 8.5.2, and g is a fibration by Lemma 8.5.3. □

Lemma 8.5.5 *The class of trivial fibrations, defined as the intersection of the classes of fibrations and weak equivalences, is equal to* **inj**(I).

Proof An element of **inj**(I) satisfies lifting with respect to all cofibrations, in particular with respect to trivial ones, so **inj**(I) is contained in the class of fibrations. It is contained in the class of weak equivalences by the definition (PG), which shows that it is contained in the class of trivial fibrations.

Suppose $f : X \to Y$ is a trivial fibration. Applying Lemma 8.5.3, this means that $f \in \mathbf{inj}(J)$ and f is a weak equivalence. Use the small object argument to choose a factorization of f as the composition of

$$X \xrightarrow{g} Z \xrightarrow{p} Y,$$

such that $g \in \mathbf{cell}(I)$ and p is in $\mathbf{inj}(I)$. It follows from (SR2) and (SR3) that p is a weak equivalence. By 3 for 2, the map g is also a weak equivalence, so it is a trivial cofibration. Thus, the condition that f be a fibration implies that it satisfies right lifting with respect to g. Use this lifting property on the square with the identity 1_X along the top, f on the right, g on the left and p on the bottom; hence there is a map $t : Z \to X$ such that $tg = 1_X$ and $ft = p$. This presents f as a retract of p, but $p \in \mathbf{inj}(I)$ and $\mathbf{inj}(I)$ is closed under retracts (see Lemma 8.2.2), so $f \in \mathbf{inj}(I)$. □

Theorem 8.5.6 (Smith) *Given a locally presentable category $\mathcal{M}$, a class of "weak equivalences," and a small set of morphisms $I \subset \text{ARR}(\mathcal{M})$ satisfying axioms (SR1) – (SR5), define the class* **cof**(I) *of cofibrations, and the class of fibrations by the right lifting property with respect to the cofibrations which are weak equivalences. These classes form a tractable model category. In the absence of (SR5), they still form a combinatorial model category.*

Proof Axioms (CM1) and (CM2) are given by the hypotheses. We have constructed a set $J \subset \text{ARR}(\mathcal{M})$ of arrows with cofibrant domains, generating the class of trivial cofibrations. Since the cofibrations are given as $\mathbf{cof}(I)$ and the fibrations are given as $\mathbf{inj}(J)$, both classes are closed under retracts. Closure of weak equivalences under retracts is (SR3), giving (CM3). By Lemma 8.5.5, the trivial fibrations are $\mathbf{inj}(I)$, which gives one half of (CM4); the other half comes from the definition of fibrations. The factorizations (CM5) come from the small object argument applied to I and J, in view of Lemma 8.5.3, Corollary 8.5.4 and Lemma 8.5.5. This proves that $\mathcal{M}$ is a model category; similarly the axioms (CG1) – (CG3b) for cofibrant generation are immediate from the above facts. Since $\mathcal{M}$ is by hypothesis locally presentable, it is a combinatorial model category. Furthermore, by hypothesis (SR5) for I and by construction for J, the domains of arrows in both generating sets are cofibrant, so $\mathcal{M}$ is tractable. Note that if we don't assume (SR5), then an easier construction of J yields a combinatorial but not tractable model category, see Barwick [20] and Beke [33]. □

In the next section, we'll see how this applies to give the injective model structure on diagram categories. Then in Section 8.7, a further refinement of this recognition principle will allow the replacement of the hypothesis of accessibility of weak equivalences by a more algebraic hypothesis on how the weak equivalences are generated by a "pseudo-generating set."

8.6 Injective cofibrations in diagram categories

We turn now to two of the main results from the appendix to Lurie's book [190]; these are notably refered to by Barwick [18], who gives a general discussion of the injective model structure on section categories of left Quillen presheaves.

These results allow us to construct injective model categories in what follows, and we need to understand something about the proof in order to apply it to the case of unital diagram categories where we require that some values are equal to $*$. While it is possible to skip this section, refering to Lurie [190] for the results, and imagining their extension to the unital case, Lurie's technique motivates the similar considerations which we will use in the section on pseudo-generating sets. One can also note that the arguments can be made more concrete in the case (due originally to Heller [141]) where $\mathcal{M}$ is a presheaf category and the cofibrations are monomorphisms.

Fix a locally κ-presentable category $\mathcal{M}$ and a subset I of morphisms. Let $\mathbf{cof}(I)$ denote the class of I-cofibrations, that is morphisms which are retracts of morphisms in $\mathbf{cell}(I)$. Let $\mathbf{cof}(I)_\kappa$ denote a set of representatives for the isomorphism classes of morphisms $A \to B$ in $\mathbf{cof}(I)$ such that A and B are κ-presentable. Lurie's first theorem [190, A.1.5.12] is:

Theorem 8.6.1 *In the above situation,*

$$\mathbf{cell}(\mathbf{cof}(I)_\kappa) = \mathbf{cof}(\mathbf{cof}(I)_\kappa) = \mathbf{cof}(I),$$

that is to say any I-cofibration can be expressed as a cell complex whose attaching maps are taken from the set $\mathbf{cof}(I)_\kappa$.

Proof The inclusions $\mathbf{cell}(\mathbf{cof}(I)_\kappa) \subset \mathbf{cof}(\mathbf{cof}(I)_\kappa) \subset \mathbf{cof}(I)$ are immediate, we need to show that $\mathbf{cof}(I) \subset \mathbf{cell}(\mathbf{cof}(I)_\kappa)$. Suppose $w : A \to V$ is in $\mathbf{cof}(I)$. Using the small object argument, we can choose a diagram

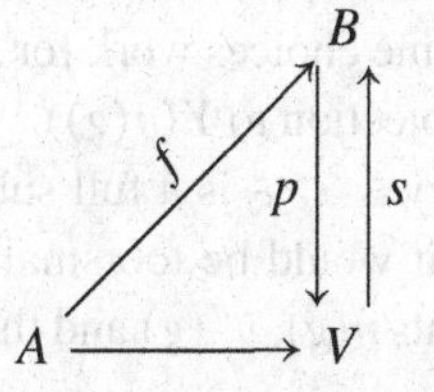

with a map $A \xrightarrow{f} B$ in **cell**(I), a projection $B \xrightarrow{p} V$ in **inj**(I), and a section $V \xrightarrow{s} B$ compatible with the map from A, such that $ps = 1_V$. Let $\pi_B := sp : B \to B$ be the idempotent $\pi_B^2 = \pi_B$, compatible with the identity map of A.

For the transfinite induction step, it will be convenient to consider more generally the case when π_B is compatible with π_A, which can be a nontrivial idempotent on A.

The idea is to find κ-presentable cell complexes $f_i : A_i \to B_i$ included into f, together with definitions of $\pi_{B,i}$ and $\pi_{A,i}$ on the source and target. Note that, given an idempotent $\pi_{B,i} : B_i \to B_i$ we can let V_i be the colimit of the diagram

$$B_i \xrightarrow{\pi_{B,i}} B_i \xrightarrow{\pi_{B,i}} B_i \xrightarrow{\pi_{B,i}} \cdots,$$

then we have a projection $p_i : B_i \to V_i$ and the system of maps π_i on the colimit gives $s_i : V_i \to B_i$ with $p_i s_i = 1_{U_i}$. Similarly let U_i be the image of $\pi_{A,i}$ defined in the same way, so if $A_i \to B_i$ are in **cell**(I) then their retracts $U_i \to V_i$ are in **cof**(I) since **cof**(I) is closed under retracts.

In order to construct the f_i, proceed as follows. Let

$$\xi := \text{CELL}(\mathcal{M}; I)_\kappa / f$$

denote the category of inclusions of cell complexes over f. By Theorem 8.3.6, ξ is κ-filtered. For $g \in \xi$ let $h(g) : X(g) \to Y(g)$ denote the corresponding cell complex, mapping to f by an inclusion of cell complexes. We have $A = \text{colim}_{g \in \xi} X(g)$, $B = \text{colim}_{g \in \xi} Y(g)$, and f is the colimit of the $g \in \xi$. Let $x(g) : X(g) \to A$ and $y(g) : Y(g) \to B$ denote the maps to the colimits. On the other hand, $X(g)$ and $Y(g)$ are κ-presentable. In particular, the map

$$\pi_B y(g) : Y(g) \to B$$

has to factor through a map $\pi_B(g) : Y(g) \to Y(n(g))$ for some $g \to n(g)$ in ξ which is a function of g. Furthermore for any $v : g \to h$ in ξ there is $n(g), n(h) \to n'(v)$ such that the projections of $\pi_B(g)$ and $\pi(h)$ into $Y(n'(v))$ are the compatible along $Y(v)$, and finally there is $n(n(g)) \to n''(g)$ such that $\pi_B(n(g))\pi(g) = \pi_B(g)$ after projecting into $Y(n''(g))$. Choose similarly $\pi_A(g)$ and assume that the same choices work for $\pi_A(g)$, and furthermore that $\pi_B(g)h(g) = \pi_A(g)$ after projection to $Y(n(g))$.

A *good filtered subcategory* $\xi_i \subset \xi$ is a full subcategory with $< \kappa$ objects, filtered (note, however,.that it would be too small to be κ-filtered), and such that for any $g \in \xi_i$ the elements $n(g)$, $n''(g)$ and their arrows are also in ξ_i; and for an arrow v in ξ_i the object $n'(v)$ with its arrows is also in ξ_i. If ξ_i is a good filtered subcategory, then

$$B_i := \mathrm{colim}_{g\in\xi_i} Y(g)$$

has an endomorphism $\pi_{B,i} : B_i \to B_i$ defined by using the $\pi_B(g)$. Similarly $\pi_{A,i} : A_i \to A_i$ is defined using the $\pi_A(g)$; these are compatible and $\pi_i^2 = \pi_i$. By Proposition 8.3.5, f_i is in **cell**(I), so the images U_i (resp. V_i) of the idempotents $\pi_{B,i}$ (resp. $\pi_{A,i}$) as defined above, have an I-cofibration $w_i : U_i \to V_i$. Note that $w_i \in \mathbf{cof}(I)_\kappa$.

Recall that by a *cell* of f we mean an element of the ordinal indexing the cell attachements. An inclusion of cell complexes generates a corresponding subset of cells, although in the general case we are currently considering, the specification of the subset of cells is not sufficient to specify the inclusion of cell complexes, because our attaching maps are not necessarily monomorphisms.

For any given cell of f, we can choose a good filtered subcategory such that the inclusion of cell complexes from B_i to B, contains the given cell. Hence $A \cup^{A_i} B_i$ is a pushout of A along an element $w_i \in \mathbf{cof}(I)_\kappa$ containing the given cell.

Start now with our given situation $w : A \to V$ being a retract of $f : A \to B$. We construct by transfinite induction, a sequence of inclusions of cell complexes $A \to C_j \to B$ together with definitions of the idempotent $\pi_{C,j}$ on C_j, compatible with 1_A and π_B. Start with $C_0 = 1$. Assume C_j is chosen for $j < j_0$. If j_0 is a limit ordinal then let $C_j := \mathrm{colim}_{j<j_0} C_j$. If not, $j_0 = k+1$ and we are in the general situation envisioned above: C_k has its idempotent $\pi_{C,k}$ compatible with π_B via the cell complex $b_k : C_k \to B$. Fix the smallest cell of B not contained in the subset of cells of C_k, and choose according to the previous procedure applied to the map b_k, a κ-presentable cell complex $w_j : A_j \to B_j$ with an inclusion of cell complexes from w_i to $(b_k : C_k \to B)$, and with compatible idempotents $\pi_{A,j}$ and $\pi_{B,j}$. Set

$$C_j = C_{k+1} := C_k \cup^{A_j} B_j,$$

with its induced idempotent which will be called $\pi_{C,j}$.

The process stops when there are no longer any cells of B but not in C_k, that is to say $C_k = B$. Now let $W_0 := A$ and let W_j be the image of the idempotent $\pi_{C,j}$ defined as

$$W_j := \mathrm{colim}\left(C_j \xrightarrow{\pi_{C,j}} C_j \xrightarrow{\pi_{C,j}} \cdots\right).$$

If U_j and V_j denote the images of $\pi_{A,j}$ and $\pi_{B,j}$ respectively, we have $U_j \to V_j \in \mathbf{cof}(I)_\kappa$. Note that

$$W_j = W_k \cup^{U_j} V_j$$

by commutation of colimits. Similarly if j is a limit ordinal then $W_j = \mathrm{colim}_{i<j} W_i$. The sequence $\{W_j\}$ is an expresssion of $f : A \to V$ as an element of $\mathbf{cell}(\mathbf{cof}(I)_\kappa)$. □

Lurie's second theorem is the application to diagram categories. Keeping the previous notations, suppose Φ is a κ-small category (i.e. its object and morphism sets have cardinality $< \kappa$). Within the category $\text{FUNC}(\Phi, \mathcal{M})$ say that a map $A \to B$ is an *injective cofibration* if $A(x) \to B(x)$ is in $\mathbf{cof}(I)$ for any $x \in \mathrm{Ob}(\Phi)$. This class of morphisms is denoted $\text{FUNC}(\Phi, \mathbf{cof}(I))$. Let $\text{FUNC}(\Phi, \mathbf{cof}(I))_\kappa$ denote a set of representatives for isomorphism classes of injective cofibrations between κ-presentable objects, which is also the set of morphisms $A \to B$ such that each $A(x) \to B(x)$ is in $\mathbf{cof}(I)_\kappa$. The following statement [190, A.2.8.3] has an argument similar to the previous one.

Theorem 8.6.2 *In the above situation,*

$$\text{FUNC}(\Phi, \mathbf{cof}(I)) = \mathbf{cof}(\text{FUNC}(\Phi, \mathbf{cof}(I))_\kappa),$$

i.e. the set $\text{FUNC}(\Phi, \mathbf{cof}(I))_\kappa$ *generates the class of injective cofibrations.*

Proof In light of the previous theorem, we may assume that $\mathbf{cof}(I) = \mathbf{cell}(I)$. Suppose $f : A \to B$ is in $\text{FUNC}(\Phi, \mathbf{cof}(I))$, then each $f(x)$ can be given a structure of I-cell complex. Make such a choice for each $x \in \mathrm{Ob}(\Phi)$, although these choices are not compatible with the structure of diagram.

Let $\xi(x)$ be the κ-filtered category of κ-presentable inclusions of cell complexes into $f(x)$ (see Theorem 8.3.6). For $i \in \xi(x)$ we have a κ-presentable cell complex $f(x,i) : A(x,i) \to B(x,i)$ with an inclusion to f, and $B(x) = \mathrm{colim}_{i \in \xi(x)} B(x,i)$. If $i \in \xi(x)$ and $\varphi : x \to y$ is an arrow in Φ then there is $j \in \xi(y)$ plus a lifting to a map $B(x,i) \to B(y,j)$ compatible with $B(\varphi)$. We can assume that the same j works for any φ; for x, y, z and $i \in \xi(x)$ we can choose $k \in \xi(z)$ such that the compositions of any transition maps from $B(x,i)$ to $B(y,j)$ then to $B(z,k)$ satisfy the product rule. Continuing in this way, we obtain a collection of operations defined on $\bigcup_x \xi(x)$ together with lifting data, such that if $\xi'(x) \subset \xi(x)$ is a collection of filtered subcategories preserved by these operations, then the collection of $B'(x) := \mathrm{colim}_{i \in \xi'(x)} B_i$ is given a structure of diagram B'. Similarly for A' and the map $A' \to B'$ is an injective cofibration. We can choose $\xi'(x)$ with cardinality $< \kappa$, so $A' \to B'$ is a κ-presentable injective cofibration, and in such a way that $B'(x)$ contains any given cell of some $B(x_0)$.

Now applying the same inductive argument as at the end of the previous theorem, which we don't repeat, gives an expression of $A \to B$ as a transfinite composition of pushouts along such $A' \to B'$. □

The version needed later is a variant in which the diagram category is replaced by the *unital diagram category* denoted $\text{FUNC}(\Phi/\Phi_0, \mathcal{M})$, discussed in more detail in Chapter 11. Here Φ_0 is a subset of objects of Φ and $A \in \text{FUNC}(\Phi/\Phi_0, \mathcal{M})$ means that $A : \Phi \to \mathcal{M}$ is a diagram such that $A(x) \cong *$ is the coinitial object of $\mathcal{M}$, for all $x \in \Phi_0$. As before, if I is a fixed set of maps in $\mathcal{M}$, a map $A \to B$ in this category is an *injective cofibration* if each $A(x) \to B(x)$ is in $\mathbf{cof}(I)$. Note that isomorphisms are always contained in $\mathbf{cof}(I)$, so this condition holds automatically whenever $x \in \Phi_0$.

Theorem 8.6.3 *The class* $\text{FUNC}(\Phi/\Phi_0, \mathbf{cof}(I))$ *of injective cofibrations in the unital diagram category, equals* $\mathbf{cof}(\text{FUNC}(\Phi/\Phi_0, \mathbf{cof}(I))_\kappa)$, *i.e. it is generated by the set* $\text{FUNC}(\Phi, \mathbf{cof}(I))_\kappa$ *of representatives for isomorphism classes of* κ*-presentable injective cofibrations.*

Proof Follow the same procedure as in the previous theorem, noting that all colimits involved are taken over connected categories, and by Lemma 8.1.9 the unitality condition is preserved by such colimits. The κ-presentable objects of $\text{FUNC}(\Phi/\Phi_0, \mathcal{M})$ are the unital diagrams A which are κ-presentable as diagrams, or equivalently such that each $A(x)$ is κ-presentable in $\mathcal{M}$. So, the process described in the proof of the previous theorem leads to an expression of an arbitrary injective cofibration as a transfinite composition of pushouts along κ-presentable ones, and all of the objects intervening here remain unital. □

Corollary 8.6.4 *If* $\mathcal{M}$ *is a tractable model category, then for any small category* Φ, *the injective model structure* $\text{FUNC}_{\text{inj}}(\Phi, \mathcal{M})$ *is combinatorial. Similarly if* $\Phi_0 \subset \Phi$ *is a full subcategory, the injective model structure on the unital diagram category* $\text{FUNC}_{\text{inj}}(\Phi/\Phi_0, \mathcal{M})$ *is combinatorial.*

Proof This follows from Smith's recognition principle (Theorem 8.5.6) using Theorems 8.6.2 and 8.6.3 above to provide the generating sets. See Barwick [18] for details on the proof of accessibility of the class of levelwise weak equivalences. □

The generating set constructed in Theorem 8.6.2 consists of arrows which don't necessarily have cofibrant domains. Here is an example to show that this is unavoidable. Let $\mathcal{M}$ be the category of diagrams of sets $A \to B \leftarrow C$ denoted (A, B, C) for short. Consider the set I consisting of the two arrows

$$(\emptyset, \emptyset, \emptyset) \to (*, *, \emptyset)$$

$$(*, *, \emptyset) \to (*, *, *).$$

An I-cofibrant object is a diagram of the form

$$A \xrightarrow{\cong} B \hookleftarrow C.$$

If a map

$$(A, B, C) \to (A', B', C')$$

is in **cell**(I), then there is a canonical expression

$$(A', B', C') = (A, B, C \sqcup U) \sqcup (V, V, W),$$

where (V, V, W) is an I-cofibrant object and U is a set provided with a map $U \to B$ such that there exists a factorization $U \to A \to B$ (but the factorization is not part of the data). The class of maps having such an expression is closed under retracts, so it is also equal to **cof**(I).

In an I-cofibration between I-cofibrant objects, the map $A \to B$ is also an isomorphism so the factorization in question is unique, hence compatible with any kind of group action. This is the phenomenon which leads to a problem when we go to equivariant objects.

Let G be a nontrivial group (such as $\mathbb{Z}/2\mathbb{Z}$) and let Φ be the category with one object having G as group of endomorphisms. A diagram $\Phi \to \mathcal{M}$ is the same thing as a diagram $A \to B \leftarrow C$ of G-sets. Consider the map

$$(G, *, \emptyset) \xrightarrow{f} (G, *, *).$$

If $(G, *, \emptyset) \to (A, B, C)$ is any composition of pushouts along the I-cofibrations between I-cofibrant G-equivariant objects, then one can see from the previous description that the subobject of C which maps to the image of $* \subset B$ has to be free as a G-set. This latter condition is preserved under retracts, but is not satisfied by f (since G is assumed to be a nontrivial group so $*$ is certainly not a free G-set). Therefore f is not the retract of a map in **cell**(I'), for any set I' of G-equivariant I-cofibrations between I-cofibrant G-equivariant objects. This shows that the levelwise I-cofibrations in $\text{FUNC}(\Phi, \mathcal{M})$ are not generated in a tractable way.

The above example can be made into a model category by adopting a trivial "coarse" model structure on $\mathcal{M}$ in which all morphisms are weak equivalences. The cofibrations and trivial cofibrations are **cof**(I), whereas the fibrations and trivial fibrations are **inj**(I). This model category is tractable, but the injective model structure $\text{FUNC}_{\text{inj}}(\Phi, \mathcal{M})$ of the previous example is combinatorial but not tractable.

Luckily, most of the model structures we shall meet along the way by iterating our basic constructions do not exhibit this somewhat pathological behavior. Indeed, in these structures all objects are cofibrant and we have the following easy corollary:

Corollary 8.6.5 *If $\mathcal{M}$ is a combinatorial model category in which all objects are cofibrant (and is therefore tractable), then for any small category Φ, the injective model structure* $\text{FUNC}_{\text{inj}}(\Phi, \mathcal{M})$ *is also tractable. Similarly, if $\Phi_0 \subset \Phi$ is a full subcategory, the injective model structure on the unital diagram category* $\text{FUNC}_{\text{inj}}(\Phi/\Phi_0, \mathcal{M})$ *is tractable too.*

Proof If all objects of $\mathcal{M}$ are cofibrant then for any small category Φ (resp. $\Phi_0 \subset \Phi$) all objects of $\text{FUNC}_{\text{inj}}(\Phi, \mathcal{M})$ (resp. $\text{FUNC}_{\text{inj}}(\Phi/\Phi_0, \mathcal{M})$) are cofibrant. So the injective model structure, which is combinatorial by the previous corollary, is automatically tractable. □

8.7 Pseudo-generating sets

The cofibrant generating sets for a cofibrantly generated model category do not often have as simple and geometric a meaning as for the original case of simplicial sets $\mathcal{K}$. This problem tends to occur particularly for the generating set for trivial cofibrations, which is often obtained by an abstract accessibility argument leading to a set containing all trivial cofibrations up to a given cardinality. In this section, we explore a way of defining a cofibrantly generated model category using sets I and K. The first will be the generating set for cofibrations; but the set K will only generate the weak equivalences in a roundabout way, thus the terminology "pseudo-generating sets." The construction of a cofibrantly generated model structure in Theorem 8.7.3 uses exactly the argument where we throw in everything up to a given cardinality. Once we have the statement of Theorem 8.7.3 we will be able to apply it in later chapters without having to come back to this cardinality argument. The reader hoping to avoid too much theory of model categories could therefore skip this section and just take Theorem 8.7.3 as a "black box" for constructing model structures. The pseudo-generating sets I and K used later will have some geometric meaning and hence be motivated outside of the technicalities of model category theory.

Suppose we are given a locally presentable category $\mathcal{M}$, and two sets of morphisms $I, K \subset \text{ARR}(\mathcal{M})$. Here both I and K are assumed to be small sets. Say that a morphism $f : A \to B$ is a *weak equivalence* if and only if (PG) there exists a diagram

$$\begin{array}{ccc} A & \longrightarrow & A' \\ {\scriptstyle f}\downarrow & & \downarrow{\scriptstyle f'} \\ B & \longrightarrow & B' \end{array}$$

such that the horizontal morphisms are in **cell**(K), and the morphism f' is in **inj**(I). Note in particular that any morphism in **inj**(I) is a weak equivalence.

Define the class of cofibrations to be **cof**(I), the trivial cofibrations to be the cofibrations which are weak equivalences, and the fibrations to be the morphisms satisfying right lifting with respect to trivial cofibrations. As usual a cofibrant object means an object X such that the morphism $\emptyset \to X$ is a cofibration.

Suppose the following axioms:

(PGM1)–the category $\mathcal{M}$ is locally presentable, and I and K are small sets of morphisms;

(PGM2)–the domains of arrows in I and K are cofibrant, and $K \subset$ **cof**(I);

(PGM3)–the class of weak equivalences is closed under retracts;

(PGM4)–the class of weak equivalences satisfies 3 for 2;

(PGM5)–the class of trivial cofibrations is closed under pushouts;

(PGM6)–the class of trivial cofibrations is closed under transfinite composition.

Using these axioms, we would like to show that these classes define a cofibrantly generated, and indeed tractable, model category structure on $\mathcal{M}$. Note that the class of trivial fibrations is defined as the intersection of the fibrations and the weak equivalences; we don't know a priori that this is the same as **inj**(I)–that will have to be proven as a consequence of the axioms. One can say, however, that **inj**(I) is contained in the class of trivial fibrations, indeed an element of **inj**(I) satisfies right lifting with respect to **cof**(I) so it is a fibration, and from the definition (PG) it is a weak equivalence.

One important preliminary result is Corollary 8.3.7 from the previous chapter, saying that any morphism in **cell**(I) is a κ-filtered colimit of κ-presentable cell complexes. The other main ingredient is the following observation, which is a sort of accessibility property for morphisms in **inj**(I):

Lemma 8.7.1 *We suppose given three regular cardinals $\mu < \lambda < \kappa$ such that $\mathcal{M}$ is locally λ-presentable (hence also locally κ-presentable), such that $|I| < \kappa$, and such that the sources and targets of arrows in I are μ-presentable. We assume that $2^{\mu} < \kappa$.*

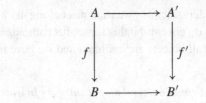

Suppose $f : X \to Y$ *is in* $\mathbf{inj}(I)$*, and suppose it can be expressed* $f = \operatorname{colim}_{i \in \alpha} f_i$*, where* $f_i : X_i \to Y_i$ *are arrows between* κ*-presentable objects and* α *is* κ*-filtered. Then there is a collection of* λ*-filtered categories* β_j *of size* $|\beta_j| < \kappa$*, together with functors* $q_j : \beta_j \subset \alpha$*, all indexed by a* κ*-filtered poset* $j \in \psi$*, such that for any* j*,* $f(\beta_j) := \operatorname{colim}_{i \in \beta_j} f_i$ *is in* $\mathbf{inj}(I)$*, and* $f = \operatorname{colim}_{j \in \psi} f(\beta_j)$.

Proof The first step is to say the following. For each $i \in \alpha$ there is $t(i) \in \alpha$ with an arrow $i \to t(i)$, such that for any diagram

$$\begin{array}{ccc} U & \longrightarrow & X_i \\ {\scriptstyle u}\downarrow & & \downarrow \\ V & \longrightarrow & Y_i \end{array}$$

with $u \in I$, there exists a lifting $V \to X_{t(i)}$ so that the two triangles in

$$\begin{array}{ccc} U & \longrightarrow & X_{t(i)} \\ \downarrow & \nearrow & \downarrow \\ V & \longrightarrow & Y_{t(i)} \end{array}$$

commute. Here the horizontal maps are the compositions of the previous ones, with the transition maps $X_i \to X_{t(i)}$ and $Y_i \to Y_{t(i)}$.

To prove this, note that there exist liftings $V \to X$ for any diagram. However, the X_i and Y_i are κ-presentable, the arrows in I are μ-presentable, and $|I| < \kappa$. Thus, the cardinality of the set of diagrams we need to consider is $< \kappa^{\mu} \leq \kappa$, by Corollary 8.1.7. Since again U is κ-presentable, there is some $t \in \alpha$ which works for each diagram; but since α is κ-filtered we can choose a single $t(i)$ for all the diagrams at a given value of i. This completes the proof of the first step.

Now we can exhaust α by a family of λ-filtered subcategories β_j with $|\beta_j| < \kappa$, indexed by $j \in \psi$ where ψ is a κ-filtered partially ordered set. We can do it in such a way that for any $i \in \beta_j$ we also have $t(i) \in \beta_j$. The exhaustion condition means that for any $i \in \alpha$ there is $j \in \psi$ and $k \in \beta_j$ with $i \to k$, and it implies that $f = \operatorname{colim}_{j \in \psi} f(\beta_j)$.

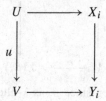

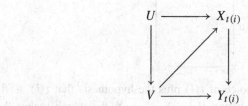

Now put

$$\begin{array}{rcl} X(\beta_j) & := & \operatorname{colim}_{i\in\beta_j} X_i \\ & & \Big\downarrow f(\beta_j) := \operatorname{colim}_{i\in\beta_j} f_i \\ Y(\beta_j) & := & \operatorname{colim}_{i\in\beta_j} Y_i . \end{array}$$

We claim that $f(\beta_j) \in \mathbf{inj}(I)$. If

$$\begin{array}{ccc} U & \longrightarrow & X(\beta_j) \\ {\scriptstyle u}\Big\downarrow & & \Big\downarrow \\ V & \longrightarrow & Y(\beta_j) \end{array}$$

is a diagram with $u \in I$, then since β_j is λ-filtered and u is λ-small, there exists a lifting to a diagram of the form

$$\begin{array}{ccc} U & \longrightarrow & X_i \\ {\scriptstyle u}\Big\downarrow & & \Big\downarrow \\ V & \longrightarrow & Y_i \end{array}$$

for some $i \in \beta_j$. By our choice of $t(i)$ plus the hypothesis that $t(i) \in \beta_j$ whenever $i \in \beta_j$, we get a lifting $V \to X_{t(i)} \to X(\beta_j)$ which makes the triangles in

$$\begin{array}{ccc} U & \longrightarrow & X(\beta_j) \\ {\scriptstyle u}\Big\downarrow & \nearrow & \Big\downarrow \\ V & \longrightarrow & Y(\beta_j) \end{array}$$

commute. This shows that $f(\beta_j) \in \mathbf{inj}(I)$. □

The following proposition says that the class of weak equivalences is accessible (Definition 8.5.1). In this argument, we are following Barwick [18].

Proposition 8.7.2 *Suppose $f : X \to Y$ is a weak equivalence. Then f can be expressed as a κ-filtered colimit of arrows f_i which are weak equivalences between κ-presentable objects.*

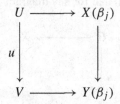

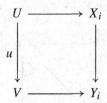

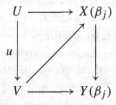

Proof The category $\mathrm{ARR}(\mathcal{M})$ is locally κ-presentable, so we can write the arrow $f : X \to Y$ as a κ-filtered colimit of arrows $\mathrm{colim}_{i\in\alpha} f_i$, such that $f_i : X_i \to Y_i$ is κ-presentable in $\mathrm{ARR}(\mathcal{M})$. In particular, $X = \mathrm{colim}_{i\in\alpha} X_i$ and $Y = \mathrm{colim}_{i\in\alpha} Y_i$ are expressed as κ-filtered colimits of κ-presentable objects in $\mathcal{M}$ (see Lemma 8.1.3).

By the definition of weak equivalence depending on I and K, there exists a diagram

$$\begin{array}{ccc} X & \xrightarrow{a} & A \\ {\scriptstyle f}\downarrow & & \downarrow{\scriptstyle g} \\ Y & \xrightarrow{b} & B \end{array}$$

such that a and b are in $\mathbf{cell}(K)$, and such that $g \in \mathbf{inj}(I)$.

Apply Corollary 8.3.7 to a and b. Thus

$$A = \mathrm{colim}_{i\in\alpha} A_i, \quad B = \mathrm{colim}_{i\in\alpha} B_i,$$

such that a and b are the colimits of systems $a_i : X_i \to A_i$ and $b_i : Y_i \to B_i$, with A_i and B_i being κ-presentable, and $a_i, b_i \in \mathbf{cell}(K)$. For each i we have a diagram

$$\begin{array}{ccc} X_i & \xrightarrow{a_i} & A_i \\ {\scriptstyle f_i}\downarrow & & \downarrow{\scriptstyle g_i} \\ Y_i & \xrightarrow{b_i} & B = \mathrm{colim}_{j\in\alpha} B_j. \end{array}$$

This gives a collection of maps, natural in $i \in \alpha$,

$$Y_i \cup^{X_i} A_i \to B,$$

with the sources being κ-presentable. By Lemma 8.3.8, there is a functor $q : \alpha \to \alpha$ together with factorizations

$$Y_i \cup^{X_i} A_i \to B_{q(i)} \to B,$$

natural in i.

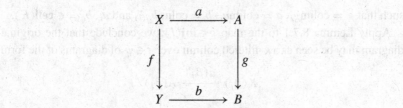

This gives a functor of diagrams depending on $i \in \alpha$

$$\begin{array}{ccc} X_i & \xrightarrow{a_i} & A_i \\ {\scriptstyle f_i}\downarrow & & \downarrow{\scriptstyle g_i} \\ Y_{q(i)} & \xrightarrow{b_{q(i)}} & B_{q(i)} \end{array}$$

such that $g = \mathrm{colim} g_i$, $a = \mathrm{colim} a_i$, $b = \mathrm{colim} b_{q(i)}$ and $a_i, b_{q(i)} \in \mathbf{cell}(K)$.

Apply Lemma 8.7.1 to the map $g \in \mathbf{inj}(I)$; we conclude that the original diagram may be seen as a κ-filtered colimit over $j \in \psi$ of diagrams of the form

$$\begin{array}{ccc} X(\beta_j) & \xrightarrow{a(\beta_j)} & A(\beta_j) \\ {\scriptstyle f(\beta_j)}\downarrow & & \downarrow{\scriptstyle g(\beta_j)} \\ Y(\beta_j) & \xrightarrow{b(\beta_j)} & B(\beta_j) \end{array}$$

where $X(\beta_j) = \mathrm{colim}_{i \in \beta_j} X_i$ etc., and such that the maps $g(\beta_j)$ are in $\mathbf{inj}(I)$. The maps $a(\beta_j)$ and $b(\beta_j)$ are still in $\mathbf{cell}(K)$, so this shows that $f(\beta_j) : X(\beta_j) \to Y(\beta_j)$ are weak equivalences. The objects $X(\beta_j)$ and $Y(\beta_j)$ are κ-presentable, and $f = \mathrm{colim}_{j \in \psi} f(\beta_j)$, which completes the proof of the proposition. □

Smith's recognition theorem (Beke [33], Barwick [18]), which we have treated as Theorem 8.5.6, now applies. For the reader's convenience we repeat the definition of weak equivalence (PG) and the axioms equivalent to (PGM1)–(PGM6). The present statement sums up the main result from the first two chapters, which will be used later on:

Theorem 8.7.3 *Suppose $\mathcal{M}$ is a locally presentable category, and $I \subset$ $\mathrm{ARR}(\mathcal{M})$ and $K \subset \mathbf{cof}(I)$ are sets of morphisms. Say that a morphism $f : A \to B$ is a* weak equivalence *if and only if there exists a diagram*

$$\begin{array}{ccc} A & \longrightarrow & A' \\ {\scriptstyle f}\downarrow & & \downarrow{\scriptstyle f'} \\ B & \longrightarrow & B' \end{array}$$

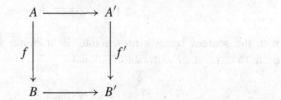

such that the horizontal morphisms are in $\mathbf{cell}(K)$, and f' is in $\mathbf{inj}(I)$. Define the class of cofibrations to be $\mathbf{cof}(I)$. Define the trivial cofibrations to be the *cofibrations which are weak equivalences, and the fibrations to be the morphisms satisfying right lifting with respect to trivial cofibrations. Suppose:*

- *the domains of arrows in I and K are cofibrant;*
- *the class of weak equivalences is closed under retracts, and satisfies 3 for 2;*
- *the class of trivial cofibrations is closed under pushouts and transfinite composition.*

Then the class of trivial fibrations is exactly $\mathbf{inj}(I)$*, and* $\mathcal{M}$ *with the given classes is a cofibrantly generated and indeed tractable model category. We say that I and K are* pseudo-generating sets *for the model structure.*

Proof Note that the class of weak equivalences is the one defined by condition (PG) above, and the hypotheses stated in the theorem are equivalent to the system of axioms (PGM1) – (PGM6). Let J be the set of morphisms given for Lemma 8.5.3, which says that $\mathbf{inj}(J)$ is the class of fibrations and $\mathbf{cof}(J)$ is the class of trivial cofibrations. By Lemma 8.5.5, $\mathbf{inj}(I)$ is the class of trivial fibrations.

Verify first the axioms for a model category.

Axiom (CM1) comes from the fact that $\mathcal{M}$ is locally presentable.

Axiom (CM2) is a hypothesis.

Axiom (CM3) for weak equivalences is a hypothesis. In a locally presentable category, for a given subset of morphisms I, the class $\mathbf{cof}(I)$ is closed under retracts. This gives (CM3) for the cofibrations. The class of fibrations is defined to be the class of morphisms which satisfy the right lifting property with respect to the trivial cofibrations. By Lemma 8.2.2, the class of fibrations is closed under retracts to give (CM3).

The fibrations are defined to be the maps satisfying the right lifting property with respect to trivial cofibrations. Therefore, the trivial cofibrations satisfy the left lifting property with respect to any fibrations. This is one half of (CM4). For the other half, Lemma 8.5.5 says that the class of trivial fibrations is equal to $\mathbf{inj}(I)$, but the class of cofibrations is equal to $\mathbf{cof}(I)$ so the cofibrations satisfy the left lifting property with respect to trivial fibrations.

For (CM5), suppose $f : X \to Y$ is any morphism. It can be factored by the small object argument in two ways, as

$$X \xrightarrow{g} Z \xrightarrow{p} Y,$$

with $g \in \mathbf{cell}(J) \subset \mathbf{cof}(J)$ and $p \in \mathbf{inj}(J)$, or as

$$X \xrightarrow{g'} Z' \xrightarrow{p'} Y,$$

with $g' \in \mathbf{cell}(I) \subset \mathbf{cof}(I)$ and $p' \in \mathbf{inj}(I)$. In the first, g is a trivial cofibration and p is a fibration by Lemma 8.5.3; in the second g' is a cofibration in view of the definition of cofibrations, and p' is a trivial fibration by Lemma 8.5.5. This proves the two parts of (CM5) and completes the proof of the Quillen model structure.

Next we note that the model structure is cofibrantly generated, indeed we have exhibited the required generating sets I and J. Axiom (CG1) is automatic since $\mathcal{M}$ is locally presentable (all objects are small); axiom (CG2a) is the definition of cofibrations and (CG2b) is Lemma 8.5.5; and axioms (CG3a) and (CG3b) are given by Lemma 8.5.3.

As $\mathcal{M}$ is locally presentable, the model structure is combinatorial. Furthermore it is tractable: indeed we have required in (PGM2) that the domains of arrows in I are cofibrant, and the arrows in J have cofibrant domains by construction. □

It should perhaps be stressed that the main work was done in Smith's recognition theorem 8.5.6, and then the discussion of cell complexes inspired by Lurie's theorems, and the proof of Theorem 8.3.6 – via Corollary 8.3.7 used in the proof of Proposition 8.7.2 saying that weak equivalences are accessible. The advantage of the statement of Theorem 8.7.3 is that it makes no reference to accessibility (other than the hypothesis that $\mathcal{M}$ be locally presentable), so when we apply it later on we don't need to do any more work with cardinals and presentability.

It might be possible to relax the hypothesis that the domains of arrows in K be cofibrant, using the argument of Hovey and Barwick [20, corollary 2.7]. It is left to the reader to explore this possibility.

We can get some further information on trivial cofibrations, fibrant objects, and fibrations between fibrant objects. This result will, as it is applied successively in later chapters, eventually turn into the version for constant object set of Bergner's result characterizing fibrant Segal categories [42].

Proposition 8.7.4 *In the situation of Theorem 8.7.3, suppose $X \xrightarrow{f} Y$ is a cofibration (i.e. in $\mathbf{cof}(I)$). Then f is a trivial cofibration if and only if there exists a diagram*

$$\begin{array}{ccc} X & \xrightarrow{\quad a \quad} & A \\ {\scriptstyle f}\Big\downarrow & & {\scriptstyle s}\Big\uparrow\Big\downarrow{\scriptstyle g} \\ Y & \xrightarrow{\quad b \quad} & B \end{array}$$

such that $ga = bf$, $sbf = a$, $gs = 1_B$, and $a, b \in \mathbf{cell}(K)$.

An object $U \in \mathcal{M}$ is fibrant if and only if $U \in \mathbf{inj}(K)$. If this is the case, then a morphism $W \xrightarrow{p} U$ is a fibration if and only if $p \in \mathbf{inj}(K)$.

Proof If f is a trivial cofibration, then it fits into a diagram with a two maps $a, b \in \mathbf{cell}(K)$ and a morphism $g \in \mathbf{inj}(I)$ by the definition of weak equivalences (PG). Since bf is a cofibration, there exists a lifting s with $gs = 1_B$ and $sbf = a$. This is a diagram of the required form.

Suppose f is a cofibration such that there exists a diagram as above (but in this case g is no longer assumed in $\mathbf{inj}(I)$). Then bf is a retract of a which is a weak equivalence. By closure of weak equivalences under retracts and then 3 for 2, it follows that f is a weak equivalence hence a trivial cofibration. This proves the first statement.

If U is fibrant then it is clearly K-injective since the elements of K are trivial cofibrations. Suppose $U \in \mathbf{inj}(K)$. If $X \xrightarrow{f} Y$ is a trivial cofibration, there exists a diagram as in the first part. For any map $X \to U$ we can extend it to a map $A \to U$ and composing with sb gives the required extension to $Y \to A$. This shows that U is fibrant.

Suppose now that U is fibrant and $W \xrightarrow{p} U$ is a map. If p is a fibration then clearly it is in $\mathbf{inj}(K)$. On the other hand, suppose $p \in \mathbf{inj}(K)$, and

$$\begin{array}{ccc} X & \xrightarrow{w} & W \\ {\scriptstyle f}\downarrow & & \downarrow{\scriptstyle p} \\ Y & \xrightarrow{u} & U \end{array}$$

is a diagram with f being a trivial cofibration. Choose a diagram as in the first part of the proposition for f. Since U is assumed fibrant, we can extend the bottom map to a map $B \xrightarrow{t} U$ with $tb = u$ and which composes with g to give a diagram

$$\begin{array}{ccc} X & \xrightarrow{w} & W \\ {\scriptstyle a}\downarrow & & \downarrow{\scriptstyle p} \\ A & \xrightarrow{tg} & U. \end{array}$$

Now $a \in \mathbf{cell}(K)$ and $p \in \mathbf{inj}(K)$ so there is a lifting $A \xrightarrow{r} W$ with $ra = w$ and $pr = tg$. This gives a map $Y \xrightarrow{rsb} W$ such that $rsbf = ra = w$ and $prsb = tgsb = tb = u$, which is the required right lifting property showing that p is a fibration. □

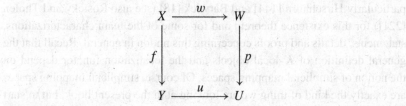

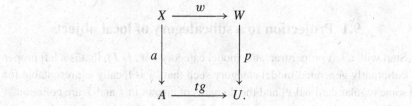

9

Direct left Bousfield localization

Suppose $\mathcal{M}$ is a model category and K a subset of morphisms. A *left Bousfield localization* is a left Quillen functor $\mathcal{M} \to \mathcal{N}$ sending elements of K to weak equivalences, universal for this property, and furthermore which induces an isomorphism of underlying categories and an isomorphism of classes of cofibrations. If it exists, it is unique up to isomorphism.

It is pretty well known that the left Bousfield localization exists whenever $\mathcal{M}$ is a left proper combinatorial model category. We refer to the references and particularly Hirschhorn [144] and Barwick [18] (see also Rosický and Tholen [221]) for this existence theorem and for some of the main characterizations, statements, details and proofs concerning this notion in general. Recall that the general definition of K-local objects and the localization functor depend on the notion of simplicial mapping spaces. Of course, simplicial mapping spaces are exactly the kind of thing we are looking at in the present book, but to start with these as basic building blocks would stretch the notion of "bootstrapping" pretty far. Therefore, in the present chapter, we consider a special case of left Bousfield localization in which everything is much more explicit.

9.1 Projection to a subcategory of local objects

Start with a left proper tractable model category $(\mathcal{M}, I, J)$, that is a left proper cofibrantly generated model category such that $\mathcal{M}$ is locally κ-presentable for some regular cardinal κ, and the domains of arrows in I and J are cofibrant.

Suppose we are given a subclass of objects considered as a full subcategory $\mathcal{R} \subset \mathcal{M}$, and a subset of morphisms $K \subset \mathrm{Mor}(\mathcal{M})$. We assume that:

(A1)–K is a small subset;
(A2)–$J \subset K$;

(A3)–$K \subset \mathbf{cof}(I)$ and the domains of arrows in K are cofibrant;
(A4)–if $X \in \mathcal{R}$ and $X \cong Y$ in $\text{ho}(\mathcal{M})$ then $Y \in \mathcal{R}$; and
(A5)–$\mathbf{inj}(K) \subset \mathcal{R}$.

Say that $(\mathcal{R}, K)$ is *directly localizing* if in addition to the above conditions:

(A6)–for all $X \in \mathcal{R}$ such that X is fibrant (i.e. J-injective), and for any $X \to Y$ which is a pushout by an element of K, there exists $Y \to Z$ in $\mathbf{cell}(K)$ such that $X \to Z$ is a weak equivalence.

Lemma 9.1.1 *Under the above hypotheses, we can remove the requirement that X be fibrant in the direct localizing condition: if $X \in \mathcal{R}$ and $X \to Y$ is a pushout by an element of K then there exists $Y \to Z$ in* $\mathbf{cell}(K)$ *such that $X \to Z$ is a weak equivalence.*

Proof Choose a map $X \to X'$ in $\mathbf{cell}(J)$, in particular a trivial cofibration, such that X' is fibrant. By invariance of $\mathcal{R}$ under weak equivalences (A3), $X' \in \mathcal{R}$. Let $Y' := Y \cup^X X'$, then $X' \to Y'$ is again a pushout by the same element of K. The above condition now applies: there is a map $Y' \to Z$ in $\mathbf{cell}(K)$ such that $X' \to Z$ is a weak equivalence. Note that $X \to Z$ is then also a weak equivalence. On the other hand, $Y \to Y'$ is in $\mathbf{cell}(J)$ hence also in $\mathbf{cell}(K)$ because $J \subset K$ by (A2). Thus the composition $Y \to Z$ is in $\mathbf{cell}(K)$. We get the desired properties. □

In our main examples, $\mathcal{R}$ will be the subcategory of precategories which satisfy the Segal conditions. Pellissier [211] called this the subcategory of "regal objects." In the terminology of localization, $\mathcal{R}$ is the category of K-local objects–this is why the present discussion is significantly simpler than the general left Bousfield localization: we assume already knowning something about the local objects and the problem is just to describe the localized model structure. The main step is to use K to construct a "monadic projection" (see Section 9.2) from $\mathcal{M}$ to $\mathcal{R}$ up to homotopy.

Let $G : \mathcal{M} \to \mathcal{M}$ with $\eta_X : X \to G(X)$ denote a K-injective replacement functor, such that $\eta_X \in \mathbf{cell}(K)$ for all X. This exists by the small object argument for the locally presentable category $\mathcal{M}$, Theorem 8.2.1.

For any $X \in \text{Ob}(\mathcal{M})$, since $G(X) \in \mathbf{inj}(K)$, condition (A5) implies that $G(X) \in \mathcal{R}$.

Our first step will be to augment the direct localizing property (A6), from morphisms in K to morphisms in $\mathbf{cell}(K)$:

Proposition 9.1.2 *Under the above assumptions, if $X \in \mathcal{R}$ and $X \to Y$ is in* $\mathbf{cell}(K)$ *then there exists $Y \to Z$ also in* $\mathbf{cell}(K)$ *such that $X \to Z$ is a weak equivalence.*

Proof Write $Y = \mathrm{colim}_i X_i$ with $X_0 = X$, and the colimit ranges over an ordinal β. Suppose this is a standard presentation of a cell complex, that is $X_i \to X_{i+1}$ is a pushout by an element of K, and for a limit ordinal i we have $X_i = \mathrm{colim}_{j<i} X_j$. At the end $Y = X_\beta$.

We construct a system of morphisms $v_i : X_i \to Z_i$ (for $i \in [\beta]$) together with compatible maps $g_{ij} : Z_i \to Z_j$ for $i < j \leq \beta$ forming a transitive system, and starting with $Z_0 = X_0$ (and v_0 is the identity), such that:

- each v_i is an element of $\mathbf{cell}(K)$;
- for a limit ordinal j we have $Z_j = \mathrm{colim}_{i<j} Z_i$;
- each $Z_i \in \mathcal{R}$;
- each $Z_i \to Z_j$ is a weak equivalence, and an element of $\mathbf{cell}(K)$;
- for $i < j \leq \beta$ the map $X_j \cup^{X_i} Z_i \to Z_j$ is in $\mathbf{cell}(K)$.

For any ordinal $\alpha \leq \beta$ let $\mathcal{V}(\alpha)$ denote the set of such collections of (Z_i, v_i, g_{ij}) for $i \leq \alpha$. We have restriction maps $\mathcal{V}(\alpha') \to \mathcal{V}(\alpha)$ for $\alpha \leq \alpha'$. We will show that for any $\alpha \leq \beta$, the map $\mathcal{V}(\alpha) \to \lim_{\eta<\alpha} \mathcal{V}(\eta)$ is surjective. At successor ordinals this says that $\mathcal{V}(\eta+1) \to \mathcal{V}(\eta)$ is surjective.

Assuming this, it follows by transfinite induction on $\alpha \leq \beta$ that for any $\eta \leq \alpha$ the map $\mathcal{V}(\alpha) \to \mathcal{V}(\eta)$ is surjective. Since $\mathcal{V}(0)$ is nonempty, this implies that $\mathcal{V}(\beta)$ is nonempty, i.e. there exists a system of the required kind, which implies the statement of the proposition.

Turn now to showing the claim: that $\mathcal{V}(\alpha) \to \lim_{\eta<\alpha} \mathcal{V}(\eta)$ is surjective. We are given data v_i for all $i < \alpha$ and need to construct $v_\alpha : X_\alpha \to Z_\alpha$.

If α is a limit ordinal, put

$$Z_\alpha := \mathrm{colim}_{i<\alpha} Z_i.$$

Let v_α be the natural map from $X_\alpha \cong \mathrm{colim}_{i<\alpha} X_i$ to Z_α. We claim that this is again in $\mathbf{cell}(K)$. Consider the system of maps

$$X_\alpha \to X_\alpha \cup^{X_i} Z_i \to X_\alpha \cup^{X_j} Z_j$$

for $i < j < \alpha$. By the fifth condition on our system (taking the pushout with $X_j \to X_\alpha$), these are in $\mathbf{cell}(K)$. However, the map $X_\alpha \to Z_\alpha$ is the colimit of this system, so by Lemma 8.3.2 it is also in $\mathbf{cell}(K)$. All the maps $X_\alpha \cup^{X_j} Z_j \to Z_\alpha$ are again in $\mathbf{cell}(K)$ for the same reason, giving back the fifth condition at α.

The second condition holds automatically by construction. For the fourth condition, note that the maps $Z_i \to Z_\alpha$ are in $\mathbf{cell}(K)$ because Z_α is by construction a transfinite composition of the previous maps which were in $\mathbf{cell}(K)$, see Lemma 8.3.2. Furthermore, the maps in the system $\{Z_i\}_{i<\alpha}$ are trivial cofibrations, and a directed limit of trivial cofibrations satisfies the required lifting property against fibrations, so it is again a trivial cofibration. Thus $Z_i \to Z_\alpha$ are trivial cofibrations, in particular they are weak equivalences. This finishes the fourth condition, and it also gives the third condition: since $Z_i \in \mathcal{R}$ we have $Z_\alpha \in \mathcal{R}$ by invariance of $\mathcal{R}$ under weak equivalences.

We now treat the other case of the claim: where α is a successor ordinal, say $\alpha = \eta + 1$. In this case $\lim_{\zeta < \alpha} \mathcal{V}(\zeta) = \mathcal{V}(\eta)$ so the problem is to show that $\mathcal{V}(\eta+1) \to \mathcal{V}(\eta)$ is surjective. In other words, to construct $v_{\eta+1} : X_{\eta+1} \to Z_{\eta+1}$ starting from v_η. Note that $X_\eta \to X_{\eta+1}$ is a pushout along an element of K. Let W be the pushout in the square

$$\begin{array}{ccc} X_\eta & \longrightarrow & X_{\eta+1} \\ \downarrow & & \downarrow \\ Z_\eta & \longrightarrow & W. \end{array}$$

The map along the bottom is again a pushout by the same element of K as the map along the top. The map $X_{\eta+1} \to W$ is in $\mathbf{cell}(K)$, since it is the pushout of the left vertical arrow by the map on top. By our inductive hypothesis, $Z_\eta \in \mathcal{R}$. By the assumed properties (improved in Lemma 9.1.1) there is a new $Z_{\eta+1}$ and a map $W \to Z_{\eta+1}$ in $\mathbf{cell}(K)$ such that $Z_\eta \to Z_{\eta+1}$ is a weak equivalence. Let $v_{\eta+1}$ be the composed map $X_{\eta+1} \to Z_{\eta+1}$. This map is the composition of $X_{\eta+1} \to W$ and $W \to Z_{\eta+1}$ which are both K-cell complexes, so $v_{\eta+1} \in \mathbf{cell}(K)$. If $j \leq \eta$ then $X_{\eta+1} \cup^{X_j} Z_j \to W$ is in $\mathbf{cell}(K)$ by the fifth condition up to η, so by construction the map $X_{\eta+1} \cup^{X_j} Z_j \to Z_{\eta+1}$ is in $\mathbf{cell}(K)$, the fifth condition at $\eta+1$. The second required property doesn't say anything because $\eta+1$ is not a limit ordinal; the fourth property comes from the above construction, and as usual the third property follows from the condition $Z_\eta \in \mathcal{R}$ and stability of $\mathcal{R}$ under weak equivalences.

This finishes the proof of the proposition. □

Corollary 9.1.3 *Suppose $f : X \to Y$ is a morphism in* $\mathbf{cell}(K)$*, such that $X, Y \in \mathcal{R}$. Then f is a weak equivalence.*

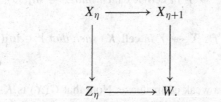

Proof By the proposition, there exists a morphism $\varphi : Y \to Z$ in $\mathbf{cell}(K)$ such that $\varphi \circ f : X \to Z$ is a weak equivalence. Applying the proposition again, there exists a morphism $\xi : Z \to W$ in $\mathbf{cell}(K)$ such that $\xi \circ \varphi : Y \to W$ is a weak equivalence. Looking at the images of everything in the homotopy category ho($\mathcal{M}$) we find that the image of φ is a map with both left and right inverses. It follows that φ goes to an isomorphism in the homotopy category, hence φ is a weak equivalence. By 3 for 2 we get that f is a weak equivalence. □

Corollary 9.1.4 *Under the above assumptions, for an object $X \in \mathcal{M}$ the following are equivalent:*

(1) $X \in \mathcal{R}$;
(2) $\eta_X : X \to G(X)$ is a weak equivalence;
(3) for any map $f : X \to Y$ in **cell**(K) *such that $Y \in$* **inj**(K), *f is a weak equivalence;*
(4) there exists a map $f : X \to Y$ in **cell**(K) *such that $Y \in$* **inj**(K) *and f is a weak equivalence.*

Proof Suppose η_X is a weak equivalence. Note that $G(X)$ is K-injective, so by our hypothesis on $\mathcal{R}$ we have $G(X) \in \mathcal{R}$. Then since $\mathcal{R}$ is supposed to be stable under weak equivalences, we get $X \in \mathcal{R}$.

Suppose $X \in \mathcal{R}$. Apply the previous Corollary 9.1.3 to the map $\eta_X : X \to G(X)$, which is in **cell**(K), between elements of $\mathcal{R}$. It says that η_X is a weak equivalence.

We have now shown (1) ⇔ (2). It is clear that (3) → (2) → (4). Suppose (4) with $f : X \to Y$ a weak equivalence and $Y \in$ **inj**(K). Then $Y \in \mathcal{R}$ by (A5) and $X \in \mathcal{R}$ by (A4) which is (1). □

This corollary says that $\mathcal{R}$ is determined by K. Thus, we can say that K is *directly localizing* if there exists a class of objects $\mathcal{R}$ satisfying properties (A1)–(A6). The class $\mathcal{R}$ can be assumed to be defined by conditions (2), (3) or (4) of the corollary. Notice that condition (A6) concerns only pushouts of objects which are already in $\mathcal{R}$, so it doesn't give any kind of splitting of the localizing system K itself.

Lemma 9.1.5 *Suppose given a diagram*

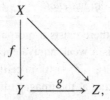

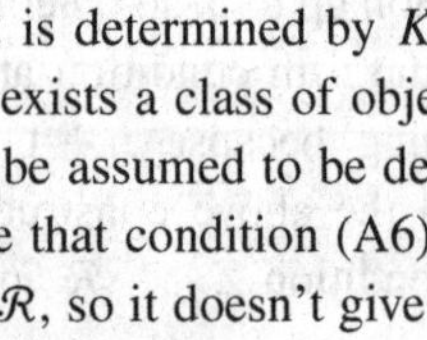

with $Y, Z \in \mathcal{R}$ such that $f : X \to Y$ and $gf : X \to Z$ are in $\mathbf{cell}(K)$*. Then g is a weak equivalence.*

Proof Let $U := Y \cup^X G(X)$, denote the two morphisms by $a : Y \to U$ and $b : G(X) \to U$ and compose with the map $\eta_U : U \to G(U)$ to get a diagram

$$\begin{array}{ccc} X & \xrightarrow{f} & Y \\ \downarrow{\scriptstyle \eta_X} & & \downarrow{\scriptstyle \eta_U a} \\ G(X) & \xrightarrow{\eta_U b} & G(U). \end{array}$$

Note here that $\eta_U b$ is not necessarily equal to $G(b\eta_X)$ (this kind of problem is the difficulty of the present proof). By hypothesis $f \in \mathbf{cell}(K)$, so all of the maps in this square are in $\mathbf{cell}(K)$.

Next, put $V := Z \cup^Y G(U)$ and let $c : Z \to V$ and $d : G(U) \to V$ denote the morphisms. Compose again with η_V to get the diagram

$$\begin{array}{ccc} Y & \xrightarrow{g} & Z \\ \downarrow{\scriptstyle \eta_U a} & & \downarrow{\scriptstyle \eta_V c} \\ G(U) & \xrightarrow{\eta_V d} & G(V). \end{array}$$

The map c comes by pushout from the map $\eta_U a : Y \to G(U)$, which is in $\mathbf{cell}(K)$, so $c \in \mathbf{cell}(K)$ and $\eta_V c \in \mathbf{cell}(K)$. However, we don't know that g, d or $\eta_V d$ are in $\mathbf{cell}(K)$.

Put these together into a big diagram of the form

$$\begin{array}{ccccc} X & \xrightarrow{f} & Y & \xrightarrow{g} & Z \\ \downarrow{\scriptstyle \eta_X} & & \downarrow{\scriptstyle \eta_U a} & & \downarrow{\scriptstyle \eta_V c} \\ G(X) & \xrightarrow{\eta_U b} & G(U) & \xrightarrow{\eta_V d} & G(V). \end{array} \qquad (9.1.1)$$

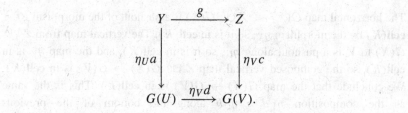

All of the vertical maps are in $\mathbf{cell}(K)$. The horizontal maps in the square on the left are in $\mathbf{cell}(K)$. The composition gf along the top is in $\mathbf{cell}(K)$; we would like to show the same for the composition along the bottom. For this, note that we have a diagram

$$
\begin{array}{ccccc}
X & \xrightarrow{f} & Y & \xrightarrow{g} & Z \\
{\scriptstyle \eta_X}\downarrow & & \downarrow{\scriptstyle a} & & \downarrow \\
G(X) & \xrightarrow{b} & U & \longrightarrow & Z \cup^Y U = Z \cup^X G(X) \\
& & {\scriptstyle \eta_U}\downarrow & & \downarrow{\scriptstyle Z \cup^Y \eta_U} \\
& & G(U) & \xrightarrow{d} & G(U) \cup^U (Z \cup^Y U) = V \\
& & & & \downarrow{\scriptstyle \eta_V} \\
& & & & G(V).
\end{array}
$$

The horizontal map $G(X) \to Z \cup^X G(X)$ is a pushout of the morphism $gf \in$ $\mathbf{cell}(K)$ by the morphism η_X, so it is in $\mathbf{cell}(K)$. The vertical map from $Z \cup^X G(X)$ to V is a pushout along η_U, so it is in $\mathbf{cell}(K)$, and the map η_V is in $\mathbf{cell}(K)$, so the composed vertical map $Z \cup^X G(X) \to G(V)$ is in $\mathbf{cell}(K)$. We conclude that the map $G(X) \to G(V)$ is in $\mathbf{cell}(K)$. This is the same as the composition $\eta_V d \circ \eta_U b$ along the bottom of the previous diagram (9.1.1).

This map is in $\mathbf{cell}(K)$ and goes between elements of $\mathcal{R}$, so it is a weak equivalence by Corollary 9.1.3. Furthermore, the map $\eta_U b$ on the bottom left of (9.1.1) is in $\mathbf{cell}(K)$ and goes between elements of $\mathcal{R}$, so it is a weak equivalence. We conclude by 3 for 2 that the map $G(U) \xrightarrow{\eta_V d} G(V)$ is a weak equivalence.

In our previous diagram (9.1.1) the vertical maps are in $\mathbf{cell}(K)$, and in the Y and Z columns these maps go between elements of $\mathcal{R}$, so the center and right vertical maps are weak equivalences. In the previous paragraph we have seen that the bottom of this rightward square is a weak equivalence, so by 3 for 2 we conclude that $g : Y \to Z$ is a weak equivalence. This proves the lemma. □

Corollary 9.1.6 *Suppose $f : X \to Y$ is a morphism in* $\mathbf{cell}(K)$. *Then $G(f)$: $G(X) \to G(Y)$ is a weak equivalence.*

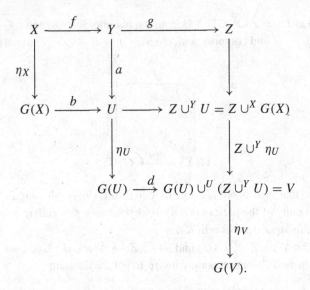

Proof Apply the previous lemma to the triangle

$$\begin{array}{ccc} X & & \\ {\scriptstyle \eta_X}\downarrow & \searrow & \\ G(X) & \xrightarrow{G(f)} & G(Y). \end{array}$$

Naturality for the transformation η says that the composition $G(f) \circ \eta_X$ is the same as $\eta_Y \circ f$. We know that $\eta_Y \in \mathbf{cell}(K)$, and $f \in \mathbf{cell}(K)$ by hypothesis, so $G(f) \circ \eta_X \in \mathbf{cell}(K)$. On the other hand, both $G(X)$ and $G(Y)$ are in $\mathcal{R}$, so Lemma 9.1.5 applies to show that $G(f)$ is a weak equivalence. □

Corollary 9.1.7 *For any object $X \in \mathcal{M}$, both maps $G(\eta_X)$ and $\eta_{G(X)}$ from $G(X)$ to $G(G(X))$ are weak equivalences.*

Proof The map η_X is in $\mathbf{cell}(K)$ so the previous corollary shows that $G(\eta_X)$ is a weak equivalence. The map $\eta_{G(X)}$ is in $\mathbf{cell}(K)$ and goes between two objects in $\mathcal{R}$, so it is a weak equivalence by Corollary 9.1.3. □

9.2 Weak monadic projection

It is useful to look at the subcategory $\mathcal{R}$ and the functor G in general terms. One can axiomatize their properties. Consider first the corresponding notion in plain 1-categories.

9.2.1 Monadic projection

Suppose $\mathcal{C}$ is a category, and $\mathcal{R} \subset \mathcal{C}$ a full subcategory. We assume that $\mathcal{R}$ is stable under isomorphisms.

A *monadic projection* from $\mathcal{C}$ to $\mathcal{R}$ is a functor $F : \mathcal{C} \to \mathcal{C}$ together with a natural transformation $\eta_X : X \to F(X)$, such that:

(Pr1)–$F(X) \in \mathcal{R}$ for all $X \in \mathcal{C}$;

(Pr2)–for any $X \in \mathcal{R}$, η_X is an isomorphism; and

(Pr3)–for any $X \in \mathcal{C}$, the map $F(\eta_X) : F(X) \to F(F(X))$ is an isomorphism.

Lemma 9.2.1 *Suppose (F, η) is a monadic projection. Then the two isomorphisms $F(\eta_X)$ and $\eta_{F(X)}$ from $F(X)$ to $F(F(X))$ are equal.*

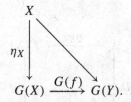

Proof Naturality of η with respect to the morphism η_X gives the commutative diagram

$$\begin{array}{ccc} X & \xrightarrow{\eta_X} & F(X) \\ \downarrow{\scriptstyle \eta_X} & & \downarrow{\scriptstyle \eta_{F(X)}} \\ F(X) & \xrightarrow{F(\eta_X)} & F(F(X)). \end{array}$$

For any $X \in \mathcal{R}$, η_X is an isomorphism so composing with its inverse we get $F(\eta_X) = \eta_{F(X)}$.

On the other hand, we can also apply F to the above diagram. By (Pr3), for any $X \in \mathcal{C}$ we have that $F(\eta_X)$ is an isomorphism, so again composing with its inverse we conclude that $F(F(\eta_X)) = F(\eta_{F(X)})$ for any $X \in \mathcal{C}$.

For arbitrary $X \in \mathcal{C}$, consider the diagram

$$\begin{array}{ccc} F(X) & \xrightarrow{F(\eta_X)} & F(F(X)) \\ \downarrow{\scriptstyle \eta_{F(X)}} & & \downarrow{\scriptstyle \eta_{F(F(X))}} \\ F(F(X)) & \xrightarrow{F(F(\eta_X))} & F(F(F(X))). \end{array}$$

It commutes by naturality of η with respect to the morphism $F(\eta_X)$. On the other hand, by the first statement we proved above, and noting that $F(X) \in \mathcal{R}$ by (Pr1), we have $\eta_{F(F(X))} = F(\eta_{F(X)})$. Then by the second statement we proved above, $F(F(\eta_X)) = F(\eta_{F(X)})$. We have now shown that both second maps in the two equal compositions of this diagram are the same isomorphism. It follows that the two first maps along the top and the left-hand side are the same. This proves the lemma. □

The following proposition shows that the monadic projection is essentially uniquely determined by the subcategory $\mathcal{R}$:

Proposition 9.2.2 *Suppose $\mathcal{R} \subset \mathcal{C}$ are as above, and (F, η) and (G, φ) are two monadic projections from $\mathcal{C}$ to $\mathcal{R}$. Then, for any $X \in \mathcal{M}$ the maps $F(\varphi_X) : F(X) \to F(G(X))$ and $G(\eta_X) : G(X) \to G(F(X))$ are isomorphisms. The diagram of isomorphisms*

$$\begin{array}{ccc} F(X) & \xrightarrow{F(\varphi_X)} & F(G(X)) \\ {\scriptstyle \varphi_{F(X)}}\big\downarrow & & \big\uparrow{\scriptstyle \eta_{G(X)}} \\ G(F(X)) & \xleftarrow{G(\eta_X)} & G(X) \end{array}$$

commutes.

Proof Define the functor $H(X) := G(F(X))$, with a natural transformation $\psi_X : X \to H(X)$ defined as the composition

$$X \xrightarrow{\varphi_X} G(X) \xrightarrow{G(\eta_X)} G(F(X)).$$

The naturality of the transformation φ with respect to the morphism η_X gives a commutative diagram

$$\begin{array}{ccc} X & \xrightarrow{\varphi_X} & G(X) \\ {\scriptstyle \eta_X}\big\downarrow & & \big\downarrow{\scriptstyle G(\eta_X)} \\ F(X) & \xrightarrow{\varphi_{F(X)}} & G(F(X)) \end{array}$$

giving the alternative expression for ψ,

$$\psi_X = G(\eta_X)\varphi_X = \varphi_{F(X)}\eta_X. \tag{9.2.1}$$

We claim that (H, ψ) is again a monadic projection from $\mathcal{C}$ to $\mathcal{R}$. Condition (Pr1) is a direct consequence of the same conditions for F and G.

For the second condition, suppose $X \in \mathcal{R}$. Then η_X is an isomorphism by (Pr2) for (F, η), and $\varphi_{F(X)}$ is an isomorphism by (Pr2) for (G, φ) plus (Pr1) for F. By the expression (9.2.1) we get that $\psi_X = \varphi_{F(X)}\eta_X$ is an isomorphism, which is (Pr2) for (H, ψ). One could instead use the expression $\psi_X = G(\eta_X)\varphi_X$ and the fact that a functor G preserves isomorphisms.

For the third condition, note from (9.2.1) again that $H(\psi_X)$ is obtained by applying G to the composed map

$$F(X) \xrightarrow{F(\eta_X)} F(F(X)) \xrightarrow{F(\varphi_{F(X)})} F(G(F(X))). \tag{9.2.2}$$

The first arrow $F(\eta_X)$ is an isomorphism by (Pr3) for (F, η). Consider the diagram

$$\begin{array}{ccc} F(X) & \xrightarrow{\eta_{F(X)}} & F(F(X)) \\ \Big\downarrow{\scriptstyle \varphi_{F(X)}} & & \Big\downarrow{\scriptstyle F(\varphi_{F(X)})} \\ G(F(X)) & \xrightarrow{\eta_{G(F(X))}} & F(G(F(X))). \end{array}$$

It commutes by naturality of η with respect to the morphism $\varphi_{F(X)}$. The right vertical arrow is the the second arrow in the composition (9.2.2). The top arrow is an isomorphism by (Pr1) and (Pr2) for (F, η). The left arrow is an isomorphism by (Pr1) for F and (Pr2) for (G, φ). The bottom arrow is an isomorphism by (Pr1) for G and (Pr2) for (F, η). It follows that the right vertical arrow $F(\varphi_{F(X)})$ is an isomorphism.

Now apply the above conclusions, together with the fact that G preserves isomorphisms, in the expression

$$H(\psi_X) = G\left(F(\varphi_{F(X)}) \circ F(\eta_X)\right),$$

to conclude that $H(\psi_X)$ is an isomorphism. This completes the proof of (Pr3) to show that (H, ψ) is a monadic projection.

Continue now the proof of the proposition. We have a morphism

$$G(X) \xrightarrow{G(\eta_X)} G(F(X)) = H(X).$$

Consider the diagram

$$\begin{array}{ccccc} G(X) & \xrightarrow{G(\varphi_X)} & G(G(X)) & & \\ \Big\downarrow{\scriptstyle G(\eta_X)} & & \Big\downarrow{\scriptstyle G(\eta_{G(X)})} & & \\ H(X) & \xrightarrow{H(\varphi_X)} & H(G(X)) & \xrightarrow{H(G(\eta_X))} & H(H(X)). \end{array}$$

The square commutes because it is obtained by applying G to the naturality diagram for η with respect to the morphism φ_X. The top arrow is an isomorphism by (Pr3) for (G, φ). The middle vertical arrow is an isomorphism by applying G to (Pr2) for (F, η) and using (Pr1) for G. This shows that the composition in the square is an isomorphism, so we deduce that the composition of the left bottom arrow with the left vertical arrow is an isomorphism.

The composition along the bottom is equal to $H(\psi_X)$, indeed $\psi_X = G(\eta_X)\varphi_X$ by definition. Hence, by (Pr3) for (H, ψ), which was proven above, the composition along the bottom is an isomorphism.

The morphism $H(\varphi_X)$ at the bottom of the square now has maps which compose on the left and the right to isomorphisms. It follows that this map is an isomorphism, and in turn that the left vertical arrow $G(\eta_X)$ is an isomorphism. This is one of the statements to be proven in the proposition.

The other statement, that $F(\varphi_X)$ is an isomorphism, is obtained by symmetry with the roles of F and G reversed.

To finish the proof, we have to show that the square diagram of isomorphisms commutes (which means that it commutes as a usually shaped diagram when the inverses of the isomorphisms are included).

Apply FG to the diagram in question, and add on another square to get the diagram of isomorphisms

$$\begin{array}{ccc}
FGF(X) & \xrightarrow{FGF(\varphi_X)} & FGFG(X) \\
{\scriptstyle FG(\varphi_{F(X)})}\downarrow & & \uparrow{\scriptstyle FG(\eta_{G(X)})} \\
FGGF(X) & \xleftarrow{FGG(\eta_X)} & FGG(X) \\
{\scriptstyle FG(\varphi_{F(X)})}\uparrow & & \uparrow{\scriptstyle FG(\varphi_X)} \\
FGF(X)) & \xleftarrow{FG(\eta_X)} & FG(X).
\end{array}$$

The bottom square commutes by FG applied to the naturality square for φ with respect to η_X. On the left-hand side, we have the same isomorphism going in both directions. Consider the outer square

$$\begin{array}{ccc}
FGF(X) & \xrightarrow{FGF(\varphi_X)} & FGFG(X)) \\
\Vert & & \uparrow{\scriptstyle FG(\eta_{G(X)}\varphi_X)} \\
FGF(X)) & \xleftarrow{FG(\eta_X)} & FG(X).
\end{array}$$

It is obtained by applying FG to the square

$$\begin{array}{ccc}
X & \xrightarrow{\eta_X} & F(X) \\
{\scriptstyle \varphi_X}\downarrow & & \downarrow{\scriptstyle F(\varphi_X)} \\
G(X) & \xrightarrow{\eta_{G(X)}} & FG(X)
\end{array}$$

which commutes by naturality of η with respect to φ_X. Since the outer square, and the bottom square of the above diagram of isomorphisms commute, it follows that the upper square commutes. The upper square was obtained by applying FG to the square of isomorphisms in $\mathcal{R}$ in question, and FG is a functor which is naturally isomorphic to the identity functor on $\mathcal{R}$. Therefore, the diagram of the proposition, which is a diagram of isomorphisms in $\mathcal{R}$, commutes. □

9.2.2 The weak version

Suppose $\mathcal{M}$ is a model category, and $\mathcal{R} \subset \mathcal{M}$ a full subcategory. We assume that $\mathcal{R}$ is stable under weak equivalences, which means that it comes from a subset of isomorphism classes in $\text{ho}(\mathcal{M})$. The aim is to apply the general discussion of this section to the subcategory $\mathcal{R}$ of the previous section; however, formally $\mathcal{R}$, and the functor G which sometimes shows up below, are not necessarily those of the previous section.

A *weak monadic projection* from $\mathcal{M}$ to $\mathcal{R}$ is a functor $F : \mathcal{M} \to \mathcal{M}$ together with a natural transformation $\eta_X : X \to F(X)$, such that:

(WPr1)–$F(X) \in \mathcal{R}$ for all $X \in \mathcal{M}$;
(WPr2)–for any $X \in \mathcal{R}$, η_X is a weak equivalence;
(WPr3)–for any $X \in \mathcal{M}$, the map $F(\eta_X) : F(X) \to F(F(X))$ is a weak equivalence;
(WPr4)–if $f : X \to Y$ is a weak equivalence between cofibrant objects then $F(f) : F(X) \to F(Y)$ is a weak equivalence; and
(WPr5)–$F(X)$ is cofibrant for any cofibrant $X \in \mathcal{M}$.

Note that the map in (WPr3) is different from the map $\eta_{F(X)} : F(X) \to F(F(X))$ which itself is a weak equivalence by (WPr1) and (WPr2). This differentiates the weak situation from Lemma 9.2.1 for the case of monadic projection considered in Chapter 8. Another notable detail is that in condition (WPr4) the objects are required to be cofibrant; this is done in order to make the proof of Corollary 9.3.1 work below.

Suppose (F, η) is a weak monadic projection from $\mathcal{M}$ to $\mathcal{R}$. Let $\text{ho}(\mathcal{R})$ denote the image of $\mathcal{R}$ in $\text{ho}(\mathcal{M})$. Then we can construct a monadic projection $(\text{ho}(F), \text{ho}(\eta))$ from $\text{ho}(\mathcal{M})$ to $\text{ho}(\mathcal{R})$. Because of this restriction to cofibrant objects in (WPr4), we need to compose with a cofibrant replacement in order to define $\text{ho}(F)$. This process, and hence the proof of Lemma 9.2.3 below, can be simplified if $\mathcal{M}$ is an injective model category where all objects are cofibrant–and (WPr5) would be superfluous as well.

Let $P : \mathcal{M} \to \mathcal{M}$ be a functor with a natural transformation $\xi_X : P(X) \to X$ such that $P(X)$ is cofibrant and ξ_X is a trivial fibration for all $X \in \mathcal{M}$. Then FP is invariant under weak equivalences: if $f : X \to Y$ is a weak equivalence, we have a commutative diagram

$$\begin{array}{ccc} P(X) & \longrightarrow & P(Y) \\ \big\downarrow & & \big\downarrow \\ X & \longrightarrow & Y \end{array}$$

such that three sides are weak equivalences. Therefore, $P(X) \to P(Y)$ is a weak equivalence between cofibrant objects, and, by (WPr4), the map $FP(X) \to FP(Y)$ is again a weak equivalence. It follows that FP descends to a functor which we denote by $\mathrm{ho}(F) : \mathrm{ho}(\mathcal{M}) \to \mathrm{ho}(\mathcal{M})$. For any $X \in \mathcal{M}$, consider the diagram

$$X \xleftarrow{\xi_X} P(X) \xrightarrow{\eta_{P(X)}} FP(X).$$

It projects in $\mathrm{ho}(\mathcal{M})$ to a diagram where the first arrow is invertible; we can define

$$\mathrm{ho}(\eta)_X := \mathrm{ho}(\eta_{P(X)}) \circ \mathrm{ho}(\xi_X)^{-1} : \mathrm{ho}(X) \to \mathrm{ho}(FP(X)) =: \mathrm{ho}(F)(X).$$

Here we denote also by ho the functor from $\mathcal{M}$ to $\mathrm{ho}(\mathcal{M})$.

Lemma 9.2.3 *Given a weak monadic projection (F, η) and choosing a cofibrant replacement functor (P, ξ), the collection* $\mathrm{ho}(\eta)$ *defined above is a natural transformation, and the pair* $(\mathrm{ho}(F), \mathrm{ho}(\eta))$ *is a monadic projection from* $\mathrm{ho}(\mathcal{M})$ *to* $\mathrm{ho}(\mathcal{R})$.

Proof For the naturality of $\mathrm{ho}(\eta)$, suppose $f : X \to Y$ is a morphism in $\mathcal{M}$. Then we have a diagram whose vertical arrows come from f,

$$\begin{array}{ccccc} X & \xleftarrow{\xi_X} & P(X) & \xrightarrow{\eta_{P(X)}} & FP(X) \\ \big\downarrow & & \big\downarrow & & \big\downarrow \\ Y & \xleftarrow{\xi_Y} & P(Y) & \xrightarrow{\eta_{P(Y)}} & FP(Y) \end{array}$$

which commutes by the naturality of ξ and η. The image of this is a commutative diagram in $\mathrm{ho}(\mathcal{M})$, where the first horizontal arrows are isomorphisms;

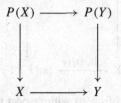

replacing them by their inverses and taking the outer square we get a commutative diagram

$$\begin{array}{ccc} X & \xrightarrow{\mathrm{ho}(\eta)_X} & \mathrm{ho}(F)(X) \\ \downarrow & & \downarrow \\ Y & \xrightarrow{\mathrm{ho}(\eta)_Y} & \mathrm{ho}(F)(Y) \end{array}$$

which shows the naturality of $\mathrm{ho}(\eta)$ with respect to morphisms coming from $\mathcal{M}$. The same diagram implies naturality with respect to the inverse of such a morphism, when f is a weak equivalence. These two classes of morphisms generate the morphisms of $\mathrm{ho}(\mathcal{M})$ so $\mathrm{ho}(\eta)$ is a natural transformation.

It remains to check the conditions (Pr1) – (Pr3) of a monadic projection. Condition (WPr1) implies that $\mathrm{ho}(F)(X) = \mathrm{ho}(FP(X))$ is in $\mathrm{ho}(\mathcal{R})$ for any X, giving (Pr1).

For property (Pr2), suppose $X \in \mathcal{R}$ (note that $\mathcal{R}$ and $\mathrm{ho}(\mathcal{R})$ have the same objects). Then $P(X) \to X$ is a weak equivalence, so by the invariance of $\mathcal{R}$ with respect to weak equivalences $P(X) \in \mathcal{R}$, and condition (WPr2) says that $\eta_{P(X)}$ is a weak equivalence. Therefore $\mathrm{ho}(\eta)_X = \mathrm{ho}(\eta_{P(X)}) \circ \mathrm{ho}(\xi_X)^{-1}$ is an isomorphism.

For property (Pr3), suppose $X \in \mathcal{M}$. Recall that $\mathrm{ho}(F)$ is obtained by descending the functor FP to the homotopy category. In order to apply this to a composition such as $\mathrm{ho}(\eta)_X = \mathrm{ho}(\eta_{P(X)}) \circ \mathrm{ho}(\xi_X)^{-1}$, we use the fact that FP takes weak equivalences to weak equivalences to say that $FP(\xi_X)$ is a weak equivalence, and then

$$\mathrm{ho}(F)(\mathrm{ho}(\eta)_X) = \mathrm{ho}(F)\left(\mathrm{ho}(\eta_{P(X)}) \circ \mathrm{ho}(\xi_X)^{-1}\right),$$

where the right-hand side is defined to be $\mathrm{ho}(FP(\eta_{P(X)})) \circ \mathrm{ho}(FP(\xi_X))^{-1}$. Consider the commutative diagram obtained by applying F to the naturality diagram for ξ with respect to $\eta_{P(X)}$:

$$\begin{array}{ccc} FP(P(X)) & \xrightarrow{FP(\eta_{P(X)})} & FP(FP(X)) \\ {\scriptstyle F(\xi_{P(X)})}\downarrow & & \downarrow{\scriptstyle F(\xi_{FP(X)})} \\ FP(X) & \xrightarrow{F(\eta_{P(X)})} & F(FP(X)). \end{array}$$

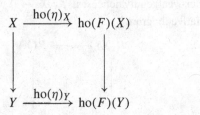

By condition (WPr3), the bottom arrow $F(\eta_{P(X)})$ is a weak equivalence. On the other hand, $P(X)$ is cofibrant, so $\xi_{P(X)}$ is a weak equivalence between cofibrant objects; by (WPr4), $F(\xi_{P(X)})$ is a weak equivalence. Similarly, (WPr5) says that $FP(X)$ is cofibrant, so $F(\xi_{FP(X)})$ is a weak equivalence. Three of the four sides of the square are weak equivalences, so the fourth side $FP(\eta_{P(X)})$ is a weak equivalence. Hence $\mathrm{ho}(FP(\eta_{P(X)}))$ is an isomorphism, therefore $\mathrm{ho}(F)(\mathrm{ho}(\eta)_X)$ is an isomorphism. This proves (Pr3). □

Proposition 9.2.4 *Suppose $\mathcal{R} \subset \mathcal{M}$ are as above, and (F, η) and (G, φ) are two weak monadic projections from $\mathcal{M}$ to $\mathcal{R}$. Then, for any cofibrant object $X \in \mathcal{M}$ the maps $F(\varphi_X) : F(X) \to F(G(X))$ and $G(\eta_X) : G(X) \to G(F(X))$ are weak equivalences, and the diagram of weak equivalences*

$$\begin{array}{ccc} F(X) & \xrightarrow{F(\varphi_X)} & F(G(X)) \\ {\scriptstyle \varphi_{F(X)}} \downarrow & & \uparrow {\scriptstyle \eta_{G(X)}} \\ G(F(X)) & \xleftarrow{G(\eta_X)} & G(X) \end{array}$$

becomes a commuting diagram of isomorphisms in the homotopy category $\mathrm{ho}(\mathcal{M})$.

Proof Use Proposition 9.2.2, which is the same as the present statement but for for monadic projections. In the case where all objects of $\mathcal{M}$ are cofibrant, we wouldn't need a cofibrant replacement in the construction for Lemma 9.2.3 above and the conclusion of the proposition follows directly. Otherwise, we need to unwind the occurrences of the cofibrant replacement functor in the conclusion.

According to Lemma 9.2.3, $(\mathrm{ho}(F), \mathrm{ho}(\eta))$ and $(\mathrm{ho}(G), \mathrm{ho}(\varphi))$ are monadic projections from $\mathrm{ho}(\mathcal{M})$ to $\mathrm{ho}(\mathcal{R})$. Use the same cofibrant replacement (P, ξ) to define both of these. Recall that

$$\mathrm{ho}(\varphi)_X := \mathrm{ho}(\varphi_{P(X)}) \circ \mathrm{ho}(\xi_X)^{-1}$$

and

$$\mathrm{ho}(F)(\mathrm{ho}(\varphi)_X) := \mathrm{ho}(FP(\varphi_{P(X)})) \circ \mathrm{ho}(FP(\xi_X))^{-1}.$$

Assume that X is cofibrant. Apply Proposition 9.2.2 to conclude that $\mathrm{ho}(F)$ $(\mathrm{ho}(\varphi)_X)$ is an isomorphism in $\mathrm{ho}(\mathcal{M})$. It follows that the same holds for $\mathrm{ho}(FP(\varphi_{P(X)}))$, so $FP(\varphi_{P(X)}$ is a weak equivalence. Look at F applied to the diagram of naturality for ξ with respect to $\varphi_{P(X)}$

$$\begin{array}{ccc} FP(P(X)) & \xrightarrow{FP(\varphi_{P(X)})} & FP(GP(X)) \\ {\scriptstyle F(\xi_{P(X)})}\downarrow & & \downarrow{\scriptstyle F(\xi_{GP(X)})} \\ FP(X) & \xrightarrow{F(\varphi_{P(X)})} & F(GP(X)). \end{array}$$

The vertical arrows are obtained by applying F to weak equivalences between cofibrant objects: use condition (WPr5) for $GP(X)$. The top is a weak equivalence as stated above, so we conclude that the bottom arrow $F(\varphi_{P(X)})$ is a weak equivalence. Now look at the diagram

$$\begin{array}{ccc} FP(X) & \xrightarrow{F(\varphi_{P(X)})} & F(GP(X)) \\ {\scriptstyle F(\xi_X)}\downarrow & & \downarrow{\scriptstyle FG(\xi_X)} \\ F(X) & \xrightarrow{F(\varphi_X)} & F(G(X)). \end{array}$$

By the assumption that X is cofibrant, and by (WPr5) for both F and G, we get that the vertical arrows are weak equivalences. The top is a weak equivalence by the previous square diagram, so we conclude that the bottom arrow $F(\varphi_X)$ is a weak equivalence.

The other statement that $G(\eta_X)$ is a weak equivalence, and is obtained by symmetry.

The commutativity of the square diagram follows from the corresponding part of Proposition 9.2.2, using the same technique as above for removing occurences of P. □

9.3 New weak equivalences

We now go back to the situation of the first section, with a combinatorial model category $\mathcal{M}$ with subcategory $\mathcal{R} \subset \mathcal{M}$ and set of morphisms K satisfying axioms (A1) – (A6), and K-injective replacement functor (G, η).

Lemma 9.3.1 *The pair (G, η) is a weak monadic projection from $\mathcal{M}$ to $\mathcal{R}$.*

Proof For (WPr1), if $X \in \mathcal{M}$ then by definition of G we have $G(X) \in$ $\mathbf{inj}(K)$, but $\mathbf{inj}(K) \subset \mathcal{R}$ by condition (A5).

For (WPr2), suppose $X \in \mathcal{R}$. Then η_X is a weak equivalence by Corollary 9.1.4.

For (WPr3), for any $X \in \mathcal{M}$ we have $G(\eta_X)$ a weak equivalence by the previous corollary.

For (WPr4), we first consider the case when f is a trivial cofibration between cofibrant objects. Choose a factorization

$$X \xrightarrow{h} Z \xrightarrow{g} Y,$$

such that $h \in \mathbf{cell}(J)$ and $g \in \mathbf{inj}(J)$. Then, since $f \in \mathbf{cof}(J)$ there is a lifting $s : Y \to Z$ such that $sf = h$ and $gs = 1_Y$. Thus, f is a retract of h in the category of objects under X. Applying the functor G we get that $G(f)$ is a retract of $G(h)$ in the category of objects under $G(X)$. On the other hand, $J \subseteq K$ by hypothesis (A2), so $\mathbf{cell}(J) \subset \mathbf{cell}(K)$ and $h \in \mathbf{cell}(K)$. By Corollary 9.1.6, $G(h)$ is a weak equivalence. But weak equivalences are closed under retracts, so $G(f)$ is a weak equivalence. This treats the case of a trivial cofibration.

Now for a general weak equivalence $f : X \to Y$ between cofibrant objects, consider the map $f \sqcup 1_Y : X \cup^{\emptyset} Y \to Y$. Choose a factorization

$$X \cup^{\emptyset} Y \xrightarrow{h \sqcup s} Z \xrightarrow{g} Y,$$

such that $h \sqcup s$ is an I-cofibration and g is a trivial fibration. Using the fact that X and Y are cofibrant objects, and also that f is a weak equivalence, we get that both maps $h : X \to Z$ and $s : Y \to Z$ are trivial cofibrations between cofibrant objects. Hence $G(h)$ and $G(s)$ are weak equivalences. But $G(g)$, being a left inverse to $G(s)$, is therefore a weak equivalence, so $G(f) = G(g)G(h)$ is a weak equivalence as desired to show (WPr4).

To show (WPr5), note simply that $\eta_X : X \to G(X)$ is in $\mathbf{cell}(K)$, so if X is cofibrant then $G(X)$ is cofibrant too. □

Fix a cofibrant replacement functor and natural transformation (P, ξ). We say that a morphism $f : X \to Y$ is a *new weak equivalence* if $GP(f) : GP(X) \to GP(Y)$ is a weak equivalence. Using the theory of the previous section, this notion depends only on $\mathcal{R}$ and not on K. In passing we also show that it doesn't depend on P.

Lemma 9.3.2 *Suppose* (H, ψ) *is a monadic projection from* $\mathrm{ho}(\mathcal{M})$ *to* $\mathrm{ho}(\mathcal{R})$. *Then* f *is a new weak equivalence if and only if* $H(\mathrm{ho}(f))$ *is an isomorphism.*

Proof By Lemma 9.2.3, $(\mathrm{ho}(G), \mathrm{ho}(\eta))$ defined using the original cofibrant replacement (P, ξ) is also a monadic projection from $\mathrm{ho}(\mathcal{M})$ to $\mathrm{ho}(\mathcal{R})$. Apply Proposition 9.2.2 to compare these, giving that $H(\mathrm{ho}(f))$ is an isomorphism if and only if $\mathrm{ho}(G)(f)$ is an isomorphism. In turn, $\mathrm{ho}(G)(f) = \mathrm{ho}(GP(f))$ is an isomorphism if and only if $GP(f)$ is a weak equivalence. □

We can choose $(H, \psi) = (\text{ho}(G), \text{ho}(\eta))$, so we can say that f is a new weak equivalence if and only if $\text{ho}(G)(\text{ho}(f))$ is an isomorphism.

If (F, φ) is any other weak monadic projection and (P', ξ') is a cofibrant replacement functor not necessarily the same as (P, ξ) then $H = \text{ho}'(F)$ and $\psi = \text{ho}'(\varphi)$ defined as above using (P', ξ') form a monadic projection from $\text{ho}(\mathcal{M})$ to $\text{ho}(\mathcal{R})$. So we can also say that $f : X \to Y$ is a new weak equivalance if and only if $\text{ho}'(F)(\text{ho}(f))$ is an isomorphism, which is true if and only if $FP'(f)$ is a weak equivalence.

Corollary 9.3.3 *The notion of new weak equivalence depends only on $\mathcal{R} \subset \mathcal{M}$.*

Proof The condition of the lemma clearly depends only on $\mathcal{R}$. □

Corollary 9.3.4 *If f is a weak equivalence in the original model structure for $\mathcal{M}$, then it is a new weak equivalence.*

Proof In this case $\text{ho}(f)$ is an isomorphism, so after application of a functor $\text{ho}(G)(\text{ho}(f))$ remains a weak equivalence. □

Corollary 9.3.5 *Suppose $f : X \to Y$ is a morphism between cofibrant objects. Then it is a new weak equivalence, if and only if $G(f) : G(X) \to G(Y)$ is an old weak equivalence. If (F, φ) is a different weak monadic projection then f is a new weak equivalence if and only if $F(f)$ is a weak equivalence.*

Proof If X and Y are cofibrant then $\text{ho}(F)(f)$ is isomorphic to the projection to $\text{ho}(\mathcal{M})$ of the morphism $F(f)$. This applies in particular to $F = G$. Conclude by using Lemma 9.3.2. □

The same can be said without specifying a full functor, but just looking at X and Y:

Corollary 9.3.6 *Given a morphism $f : X \to Y$ between cofibrant objects and a diagram*

$$\begin{array}{ccc} X & \xrightarrow{a} & A \\ {\scriptstyle f}\downarrow & & \downarrow{\scriptstyle p} \\ Y & \xrightarrow{b} & B \end{array}$$

*with $a, b \in$ **cell**(K) and $A, B \in \mathcal{R}$, the map f is a new weak equivalence if and only if p is a weak equivalence for the original model structure.*

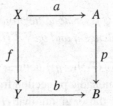

Proof Apply G to the above diagram. Corollary 9.1.6 says that the resulting horizontal arrows are old weak equivalences. Thus, by Corollary 9.3.5, f is a new weak equivalence if and only if $G(A) \xrightarrow{G(p)} G(B)$ is an old weak equivalence. However, in the naturality square for η with respect to p

$$\begin{array}{ccc} A & \xrightarrow{\eta_A} & G(A) \\ \downarrow & & \downarrow \\ B & \xrightarrow{\eta_B} & G(B) \end{array}$$

the horizontal arrows are old weak equivalences, since A and B are cofibrant objects in $\mathcal{R}$ (see Corollary 9.1.4). Putting these statements together using 3 for 2 gives the proof that f is a new weak equivalence if and only if p is an old weak equivalence. □

Lemma 9.3.7 *The notion of new weak equivalence is stable under retracts, and satisfies 3 for 2.*

Proof It is the pullback of the notion of isomorphism, by the monadic projection $\mathrm{ho}(G)$. □

9.4 Invariance properties

The transfinite consequence of the left properness condition, Proposition 7.3.3, allows us to prove invariance of the new trivial cofibrations with respect to transfinite composition. Say that a morphism is a *new trivial cofibration* if it is a cofibration and a new weak equivalence.

Lemma 9.4.1 *Suppose $\{X_n\}_{n\leq\beta}$ is a continuous transfinite sequence indexed by an ordinal β, such that for any limit ordinal n we have $X_m = \mathrm{colim}_{n<m} X_n$, and for any n with $n+1\leq\beta$ we have that $X_n \to X_{n+1}$ is a new trivial cofibration. Then $X_0 \to X_\beta$ is a new trivial cofibration.*

Proof Assume first that X_0 and hence all the X_n are cofibrant. Choose a sequence $\{Z_n\}_{n\leq\beta}$ and a morphism of sequences $X_n \to Z_n$ in the following way: let $Z_0 := G(X_0)$; if $n+1\leq\beta$ and the morphisms up to $X_n \to Z_n$ are chosen, let $Z_{n+1} := G\left(Z_n \cup^{X_n} X_{n+1}\right)$; and if m is a limit ordinal and the morphisms are chosen for all $n<m$, put $Z_m := \mathrm{colim}_{n<m} Z_n$. This defines the full sequence by transfinite induction. Furthermore, by induction we see that the morphisms $X_n \to Z_n$ are in $\mathbf{cell}(K)$. This is clear at $n=0$ and at $n+1$ if

we know it at n, by the definition. In the case where m is a limit ordinal, note that $Z_m = \mathrm{colim}_{n<m}(X_m \cup^{X_n} Z_n)$ using the continuity property of $X.$. The transition maps in this colimit are of the form

$$\begin{array}{c} X_m \cup^{X_n} Z_n = X_m \cup^{X_{n+1}} \left(Z_n \cup^{X_n} X_{n+1}\right) \\ \Big\downarrow \\ X_m \cup^{X_{n+1}} Z_{n+1} = X_m \cup^{X_{n+1}} G\left(Z_n \cup^{X_n} X_{n+1}\right), \end{array}$$

which is in $\mathbf{cell}(K)$. By definition $\mathbf{cell}(K)$ is closed under transfinite composition, so the map $X_m \to Z_m$ is in $\mathbf{cell}(K)$ as claimed. It follows (from Corollaries 9.1.6 and 9.3.5 together using the fact that X_m are cofibrant) that the maps $X_n \to Z_n$ are new weak equivalences.

We now argue by induction that the Z_n are in $\mathcal{R}$ and the maps $Z_0 \to Z_n$ are old trivial cofibrations. If it is known for Z_n then the map $Z_n \to Z_{n+1}$ is a new weak equivalence between fibrant objects in $\mathcal{R}$, so by Corollary 9.4.5 it is an old weak equivalence, thus $Z_0 \to Z_n$ is an old weak equivalence. If m is a limit ordinal and we know the statement for all smaller $n < m$, then the transition maps in the system $\{Z_n\}_{n<m}$ are old trivial cofibrations; so the map $Z_0 \to Z_m = \mathrm{colim}_{n<m} Z_n$ is an old trivial cofibration (trivial cofibrations are closed under transfinite composition as can be seen from their characterization by the lifting property with respect to fibrations). Since $Z_0 \in \mathcal{R}$ we get that $Z_n \in \mathcal{R}$ by the invariance property (A4). This completes the inductive proof for the statement given at the start of the paragraph.

Applying it to $n = \beta$ we conclude that $Z_0 \to Z_\beta$ is a new trivial cofibration, proving the lemma under the beginning assumption that X_0 was cofibrant.

Recall that $\mathcal{M}$ is left proper, which implies a left properness statement for transfinite compositions also, Proposition 7.3.3.

In the arbitrary case, choose a sequence $\{Y_n\}_{n\leq\beta}$ with a morphism of sequences consisting of old weak equivalences $p_n : Y_n \to X_n$, such that the Y_n are cofibrant and the transition morphisms $Y_n \to Y_{n+1}$ are cofibrations. This is done by using the factorization into $\mathbf{inj}(I) \circ \mathbf{cell}(I)$ at each stage. At a limit ordinal m given the choice for all $n < m$, we set $Y_m := \mathrm{colim}_{n<m} Y_n$. By the transfinite left properness property $Y_m \to X_m$ is an old weak equivalence. This allows us to make the inductive choice of the sequence $Y.$. Then, the $Y_n \to Y_{n+1}$ are new trivial cofibrations by Lemma 9.3.7 and the objects Y_n are cofibrant, so the first part of the proof applies: $Y_0 \to Y_\beta$ is a new weak equivalence. By 9.3.7 again, $X_0 \to X_\beta$ is a new weak equivalence. □

Using the left properness hypothesis on $\mathcal{M}$ we can show that morphisms in **cell**(K) are new weak equivalences, and also improve the earlier criteria by removing the conditions that the morphism goes between cofibrant objects:

Proposition 9.4.2 *A morphism $f : X \to Y$ in* **cell**(K) *is a new weak equivalence.*

Proof Corollaries 9.1.6 and 9.3.5 together immediately imply this when X is cofibrant. However, this is not sufficient in general, because new weak equivalences are defined using a cofibrant replacement. Of course, in case $\mathcal{M}$ satisfies the additional hypothesis that all objects are cofibrant, then this part of the argument (like many others) is considerably simplified.

To prove that **cell**(K) is contained in the new weak equivalences, in view of the closure of new trivial cofibrations under transfinite composition (Lemma 9.4.1 above), it suffices to treat morphisms which are pushouts along a single arrow in K. Suppose $X \leftarrow U \to V$ is a diagram with $U \to V$ in K, then we need to show that $X \to Y := X \cup^U V$ is a new weak equivalence. The source U is assumed to be cofibrant by Condition (A3). Choose a cofibrant replacement via a trivial fibration $P \to X$. The map from U lifts to a map $U \to P$. Put $Z := P \cup^U V$, then we have a cocartesian square

$$\begin{array}{ccc} P & \longrightarrow & Z \\ \downarrow & & \downarrow \\ X & \longrightarrow & Y \end{array}$$

where the left vertical arrow is an old weak equivalence. By left properness of the original model structure, the right vertical arrow is also an old weak equivalence. The top map is in **cell**(K) and goes between cofibrant objects, so as mentioned above it is a new weak equivalence. Now 3 for 2 given by Lemma 9.3.7 implies that the bottom map is a new weak equivalence. □

We can now obtain a criterion for a morphism to be a weak equivalence. Notice that this condition coincides with the definition of weak equivalences (PG) used in the construction of a model category by pseudo-generating sets in Section 8.7.

Corollary 9.4.3 *A morphism $f : X \to Y$ is a new weak equivalence if and only if there exists a diagram*

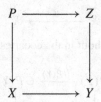

$$\begin{array}{ccc} X & \xrightarrow{a} & A \\ {\scriptstyle f}\downarrow & & \downarrow{\scriptstyle g} \\ Y & \xrightarrow{b} & B \end{array}$$

such that a and b are in **cell**(K) *and* $g \in$ **inj**(I).

Proof If such a diagram exists, then g is an old weak equivalence hence a new one; also a and b are new weak equivalences by Corollary 9.4.2, so by 3 for 2 for the new weak equivalences, Lemma 9.3.7, we get that f is a new weak equivalence.

Suppose f is a new weak equivalence. Fix a cofibrant replacement functor and natural transformation (P, ξ). Then $GP(f) : GP(X) \to GP(Y)$ is an old weak equivalence. The map $P(X) \to GP(X)$ is in **cell**(K). Let X' be the pushout in the cocartesian diagram

$$\begin{array}{ccc} P(X) & \xrightarrow{\eta_{P(X)}} & GP(X) \\ {\scriptstyle \xi_X}\downarrow & & \downarrow{\scriptstyle r} \\ X & \xrightarrow{u} & X' \end{array}$$

and similarly let Y' be the pushout in the cocartesian diagram

$$\begin{array}{ccc} P(Y) & \xrightarrow{\eta_{P(Y)}} & GP(Y) \\ {\scriptstyle \xi_Y}\downarrow & & \downarrow{\scriptstyle s} \\ Y & \xrightarrow{v} & Y'. \end{array}$$

The pushout maps u and v are in **cell**(K). The maps ξ_X and ξ_Y are old weak equivalences, and $\eta_{P(X)}$ and $\eta_{P(Y)}$ are old cofibrations, so the left properness hypothesis on the old model structure implies that the maps $r : GP(X) \to X'$ and $s : GP(Y) \to Y'$ are weak equivalences. The pushouts fit into a commutative cube, and the condition that $GP(f) : GP(X) \to GP(Y)$ is an old weak equivalence implies, by 3 for 2 in the old model structure, that the map $X' \to Y'$ is an old weak equivalence. We can factor this map as

$$X' \xrightarrow{h} X'' \xrightarrow{g} Y',$$

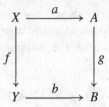

such that $h \in \mathbf{cell}(J)$ and g is an old fibration; but it is an old weak equivalence too so $g \in \mathbf{inj}(I)$. We have obtained the desired diagram

$$
\begin{array}{ccc}
X & \xrightarrow{hu} & X'' \\
{\scriptstyle f}\downarrow & & \downarrow{\scriptstyle g} \\
Y & \xrightarrow{v} & Y'
\end{array}
$$

with $hu \in \mathbf{cell}(K)$ because it is the composition of $u \in \mathbf{cell}(K)$ with $h \in \mathbf{cell}(J) \subset \mathbf{cell}(K)$, with $v \in \mathbf{cell}(K)$, and with $g \in \mathbf{inj}(I)$. □

We have the following criterion for new trivial cofibrations, which is the same as in Proposition 8.7.4. We repeat the proof here in order to verify that it works in our current situation:

Corollary 9.4.4 *A cofibration $X \stackrel{f}{\to} Y$ is a new trivial cofibration if and only if it fits into a diagram*

$$
\begin{array}{ccc}
X & \xrightarrow{a} & A \\
{\scriptstyle f}\downarrow & & {\scriptstyle s}\uparrow\ \downarrow{\scriptstyle g} \\
Y & \xrightarrow{b} & B
\end{array}
$$

commutative using either arrow on the right, with $gs = 1_B$, such that a and b are in $\mathbf{cell}(K)$.

Proof The same as the first part of Proposition 8.7.4, using some things that we already know: new weak equivalences are closed under retracts and satisfy 3 for 2 by Lemma 9.3.7, and they contain $\mathbf{cell}(K)$ by Proposition 9.4.2. □

Corollary 9.4.5 *A morphism $f : X \to Y$ such that $X, Y \in \mathcal{R}$, is a new weak equivalence if and only if it is a weak equivalence in the original model structure.*

Proof The "if" direction is given by Corollary 9.3.4. Suppose given a new weak equivalence f such that $X, Y \in \mathcal{R}$. Then $P(f) : P(X) \to P(Y)$ is a new weak equivalence between cofibrant objects which are again in $\mathcal{R}$ (since $\mathcal{R}$ is invariant under weak equivalences). In the diagram

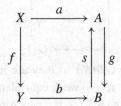

$$\begin{array}{ccc} P(X) & \longrightarrow & P(Y) \\ \downarrow & & \downarrow \\ GP(X) & \longrightarrow & GP(Y) \end{array}$$

the bottom arrow is an old weak equivalence by Corollary 9.3.5. The vertical arrows are old weak equivalences by Corollary 9.1.4, (1) $\rightarrow$ (2). By 3 for 2 in the original model structure, the top arrow $P(f)$ is an old weak equivalence, hence f is also. □

We can remove the hypothesis of cofibrancy from the statement of Corollary 9.3.5.

Corollary 9.4.6 *A morphism $f : X \rightarrow Y$ is a new weak equivalence, if and only if $G(f) : G(X) \rightarrow G(Y)$ is an old weak equivalence.*

Proof Look at the diagram

$$\begin{array}{ccc} X & \longrightarrow & G(X) \\ \downarrow & & \downarrow \\ Y & \longrightarrow & G(Y). \end{array}$$

The horizontal maps are in $\mathbf{cell}(K)$ so they are new weak equivalences by Proposition 9.4.2. If $G(f)$ is an old weak equivalence, then it is a new one by Corollary 9.3.4 and by Lemma 9.3.7 f is a new weak equivalence.

Suppose f is a new weak equivalence. As above this implies that $G(f)$ is a new weak equivalence, but noting that $G(X)$ and $G(Y)$ are in $\mathbf{inj}(K)$ hence in $\mathcal{R}$, Corollary 9.4.5 implies that $G(f)$ is an old weak equivalence. □

9.5 New fibrations

Recall that a new trivial cofibration is a cofibration (which means the same thing in the original and new structures) and also a new weak equivalence. Say that a morphism is a *new fibration* if it satisfies the right lifting property with respect to new trivial cofibrations.

Lemma 9.5.1 *The class of new trivial cofibrations is closed under composition and retracts, and contains* $\mathbf{cof}(K)$. *In particular, trivial fibrations from the original model structure, which are exactly* $\mathbf{cof}(J)$, *are also new trivial fibrations.*

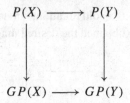

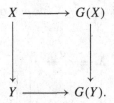

Proof By Lemma 9.3.7, the new weak equivalences are closed under retracts and compositions; the same holds for the cofibrations, so the intersection of these classes is closed under retracts and compositions. Elements of $\mathbf{cell}(K)$ are cofibrations by (A3) and new weak equivalences by 9.1.6 so they are new trivial cofibrations, and $\mathbf{cof}(K)$ is the closure of $\mathbf{cell}(K)$ under right retracts. Finally, since $J \subset K$ by assumption (A2) we get that $\mathbf{cof}(J) \subset \mathbf{cof}(K)$. □

The reader might hope that $\mathbf{cof}(K)$ is equal to the class of new trivial cofibrations; however, in general it will be smaller and the next section below will be needed to remedy this problem.

Corollary 9.5.2 *A new fibration is also a fibration in the old model structure.*

Proof A new fibration satisfies right lifting with respect to the class of old trivial cofibrations, since that is contained in the class of new ones. □

Lemma 9.5.3 *A morphism which is a new fibration and a new weak equivalence is a trivial fibration in the original model structure, in particular it satisfies the right lifting property with respect to cofibrations.*

Proof Suppose $f : X \to Y$ is a new fibration and a new weak equivalence. Choose a factorization $f = gh$ in the original model category structure $X \xrightarrow{h} Z \xrightarrow{g} Y$ with h a cofibration and g a trivial fibration. Then g is an old weak equivalence, hence it is a new weak equivalence by Corollary 9.3.4. By 3 for 2 for the new weak equivalences Lemma 9.3.7 we conclude that h is a new weak equivalence, hence by definition it is a new trivial cofibration. The condition that f be a new fibration implies that it satisfies lifting with respect to h, so there is a morphism $u : Z \to X$ such that $uh = 1_X$ and $fu = g$. In this way, f becomes a retract of g in the category of objects over Y. By closure of the original weak equivalences under retracts, f is an original weak equivalence. Since it is also a fibration in the original structure by Corollary 9.5.2, we conclude that it is a trivial fibration in the original model structure. □

Corollary 9.5.4 *The class of new trivial fibrations, defined as the intersection of the new fibrations and the new weak equivalences, is equal to the original class of trivial fibrations.*

Proof One inclusion is given by the preceding lemma. In the other direction, suppose $f : X \to Y$ is an original trivial fibration. Then it satisfies lifting with respect to cofibrations, in particular with respect to new trivial cofibrations. So it is a new fibration. It is an old weak equivalence, hence a new one by Corollary 9.3.4. Thus it is a new trivial fibration. □

We can sum up the preceding discussion as follows.

Scholium 9.5.5 *The three classes of morphisms: new weak equivalences, new fibrations and cofibrations (which are the same as the old cofibrations), generate the notions of new trivial fibration and new trivial cofibration by intersection of new fibrations and cofibrations, with new weak equivalences. All of these classes are closed under composition and retracts, and new weak equivalences satisfy 3 for 2. If*

$$\begin{array}{ccc} X & \xrightarrow{a} & U \\ {\scriptstyle i}\downarrow & & \downarrow{\scriptstyle p} \\ Y & \xrightarrow{b} & V \end{array}$$

is a diagram, and if either i is a new trivial cofibration and p a new fibration, or i a cofibration and p a new trivial fibration, then there exists a lifting $f : Y \to U$ with $fi = a$ and $pf = b$.

We can add that the original model structure provides one of the two required factorizations: if $f : X \to Y$ is any map, then it can be factored as

$$f = gh : X \xrightarrow{h} Z \xrightarrow{g} Y,$$

where h is a cofibration and g is an old or new trivial fibration (these being the same as Corollary 9.5.4).

9.6 Pushouts by new trivial cofibrations

The class of new trivial cofibrations, defined to be the intersection of the new weak equivalences with the cofibrations, is closed under pushout:

Lemma 9.6.1 *If*

$$\begin{array}{ccc} X & \xrightarrow{f} & Y \\ {\scriptstyle a}\downarrow & & \downarrow{\scriptstyle b} \\ Z & \xrightarrow{g} & W \end{array}$$

is a pushout square with f a new trivial cofibration, then g is a new trivial cofibration.

Proof First reduce to the case when X and hence Y are cofibrant objects. Choose a cofibrant replacement $p : X' \to X$ with $p \in \mathbf{inj}(I)$, and a factorization of the map from X' to Y to give a commutative square

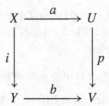

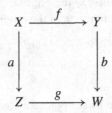

$$\begin{array}{ccc} X' & \xrightarrow{f'} & Y' \\ {\scriptstyle p}\downarrow & & \downarrow{\scriptstyle q} \\ X & \xrightarrow{f} & Y \end{array}$$

such that f' is a cofibration and $q \in \mathbf{inj}(I)$. Note that f' is a new trivial cofibration, by 3 for 2 for new weak equivalences (Lemma 9.3.7).

By the hypothesis that $\mathcal{M}$ is left proper, the map $Y' \to X \cup^{X'} Y'$ is an old weak equivalence, so by 3 for 2 the map $X \cup^{X'} Y' \to Y$ is an old weak equivalence. We get a map

$$Z \cup^{X'} Y' = Z \cup^{X} (X \cup^{X'} Y') \to Z \cup^{X} Y = W.$$

The maps from X to $X \cup^{X'} Y'$ and Y are cofibrations, so Lemma 7.3.1 applies: the map $Z \cup^{X'} Y' \to Z \cup^{X} Y = W$ is an old weak equivalence. Assume we know the statement of the lemma for cofibrant objects; then $Z \to Z \cup^{X'} Y'$ is a new weak equivalence. The composition of these two maps is the same as the map $g : Z \to W$ so by 3 for 2 for new weak equivalences, Lemma 9.3.7, we get that g is a new weak equivalence.

This reduces the statement of the lemma to the case where X and Y are cofibrant, which we now suppose. Let $\eta_X : X \to G(X)$ be the map in $\mathbf{cell}(K)$ to $G(X) \in \mathbf{inj}(K)$. Let $V := G(Y \cup^X G(X))$. The map $Y \to V$ is in $\mathbf{cell}(K)$ and $V \in \mathbf{inj}(K)$. Taking the pushout of the commutative square

$$\begin{array}{ccc} X & \xrightarrow{f} & Y \\ {\scriptstyle \eta_X}\downarrow & & \downarrow{\scriptstyle s} \\ G(X) & \xrightarrow{t} & V \end{array}$$

along the map $a : X \to Z$ gives a square

$$\begin{array}{ccc} Z & \xrightarrow{g} & Z \cup^X Y \\ {\scriptstyle Z \cup^X \eta_X}\downarrow & & \downarrow{\scriptstyle Z \cup^X s} \\ Z \cup^X G(X) & \xrightarrow{Z \cup^X t} & Z \cup^X V \end{array}$$

and this fits into a cube with the previous diagram. Since f is a new weak equivalence between cofibrant objects, the vertical maps in the first square

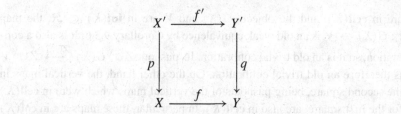

are in **cell**(K), and the objects $G(X)$ and V are in **inj**$(K) \subset \mathcal{R}$, the map $t : G(X) \to V$ is an old weak equivalence by Corollary 9.3.6. It is also a cofibration, so it is an old trivial cofibration. Its pushout $Z \cup^X G(X) \xrightarrow{Z \cup^X t} Z \cup^X V$ is therefore an old trivial cofibration. On the other hand, the vertical maps in the second square, being pushouts of the vertical maps which were in **cell**(K) for the first square, are also in **cell**(K). In particular, these maps are in **cof**(K) hence new weak equivalences by Lemma 9.5.1. The bottom map is an old, hence a new weak equivalence. By 3 for 2, Lemma 9.3.7, the map $g : Z \to Z \cup^X Y = W$ is a new weak equivalence. This proves the lemma since g is clearly a cofibration. □

9.7 The model category structure

We can now put together everything above to obtain the new model category structure on $\mathcal{M}$, as an application of Theorem 8.7.3 in the previous chapter, which was our version of Smith's recognition theorem as exposed by Barwick [18].

Theorem 9.7.1 *Suppose $\mathcal{M}$ is a left proper tractable model category, and $(\mathcal{R}, K)$ is a directly localizing system according to axioms (A1) – (A6) at the start of the chapter. The classes of original cofibrations, new weak equivalences and new fibrations constructed above provide $\mathcal{M}$ with a structure of model category, cofibrantly generated, and indeed combinatorial and even tractable. It is left proper. Furthermore, this structure is the left Bousfield localization of $\mathcal{M}$ by the original set of maps K. The fibrant objects are the K-injective objects, and a morphism $W \to U$ to a fibrant object is a fibration if and only if it is in* **inj**(K).

Proof We will be applying the discussion of Section 8.7 in the previous chapter, about pseudo-generating sets. We are given subsets I and K; the cofibrations are **cof**(I), and by Corollary 9.4.3 the new weak equivalences are exactly the class defined by the condition (PG). The trivial cofibrations are as defined in Section 8.7, hence the fibrations too. We verify the axioms for pseudo-generating sets.

(PGM1)–The hypothesis that $\mathcal{M}$ is locally presentable, is contained in the tractability hypothesis of the present statement; I is a small set as it is one of the generating sets for the old structure, and K is a small set by (A1).

(PGM2)–The domains of arrows in I are cofibrant, by the assumption that the old structure is tractable (and I is part of a tractable pair of generating sets); the domains of arrows in K are cofibrant, and $K \subset \mathbf{cof}(I)$, by (A3).

(PGM3)–The class of weak equivalences is closed under retracts by Lemma 9.3.7.

(PGM4)–The class of weak equivalences satisfies 3 for 2 by Lemma 9.3.7.

(PGM5)–The class of trivial cofibrations is closed under pushouts by Lemma 9.6.1.

(PGM6)–The class of trivial cofibrations is closed under transfinite composition by Lemma 9.4.1.

By Theorem 8.7.3, $\mathcal{M}$ with these classes of morphisms is a tractable model category.

To prove left properness, suppose given a cocartesian diagram

$$\begin{array}{ccc} X & \xrightarrow{u} & U \\ {\scriptstyle f}\downarrow & & \downarrow{\scriptstyle g} \\ Y & \xrightarrow{v} & V \end{array}$$

where f is a new weak equivalence and u is a cofibration. Consider a fibrant replacement $h : Y \to Y'$ and put $V' := Y' \cup^Y V = Y' \cup^X U$. Since h is a new trivial cofibration, so is the map $V \to V'$. Factor the composed map hf as $X \xrightarrow{i} X' \xrightarrow{p} Y'$, where i is a new trivial cofibration and p is a trivial fibration. In particular p is a weak equivalence in the old model structure. The map $U \to X' \cup^X U$ is a trivial cofibration by Lemma 9.6.1, and left properness of the original model structure implies that

$$X' \cup^X U \to Y' \cup^{X'} (X' \cup^X U) = V'$$

is an old weak equivalence (hence a new one). Composing these we get that $U \to V'$ is a new weak equivalence, and by 3 for 2 we conclude that g is a new weak equivalence.

The resulting model category is the left Bousfield localization of $\mathcal{M}$ along the subset K. The morphisms of K go to weak equivalences in the new structure. On the other hand, by the criterion of Corollary 9.4.3, given a model structure whose class of weak equivalences contains the old ones plus K, that class must also contain the new weak equivalences. So the class of new weak equivalences is the smallest one which can create a model structure along with the old cofibrations. This says that the new structure is the left Bousfield localization, see Hirschhorn [144] and Barwick [18]. Alternatively, the argument given at the end of the proof of Theorem 9.8.1 below gives explicitly the left Bousfield property.

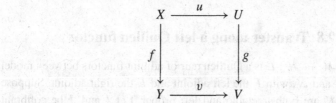

The characterization of fibrant objects and of fibrations to fibrant objects in the last paragraph is given by Proposition 8.7.4. □

The left Bousfield localization depends only on the class $\mathcal{R}$ and not on the choice of subset K:

Proposition 9.7.2 *Suppose $(\mathcal{R}, K)$ and $(\mathcal{R}, K')$ are two direct localizing systems for the same class of objects $\mathcal{R}$ in a left proper tractable model category $\mathcal{M}$. Then the two new model structures given by the previous theorem are the same.*

Proof By Corollary 9.3.3 the classes of new weak equivalences are the same, and by definition the classes of cofibrations are the same, so the classes of fibrations are the same. □

9.8 Transfer along a left Quillen functor

Suppose $F : \mathcal{M} \leftrightarrow \mathcal{N} : E$ is a Quillen pair of adjoint functors between model categories $\mathcal{M}$ and $\mathcal{N}$, with F the left adjoint and E the right adjoint. Suppose that $\mathcal{M}$ and $\mathcal{N}$ are both tractable and left proper. Let I and J be cofibrant generating sets for $\mathcal{M}$, and I' and J' cofibrant generating sets for $\mathcal{N}$.

Suppose that $\mathcal{R} \subset \mathcal{M}$ is a full subcategory, and $K \subset \mathrm{ARR}(\mathcal{M})$ is a small subset of arrows. Let $K' := F(K) \cup J' \subset \mathrm{ARR}(\mathcal{M})$ be the set consisting of J' plus all the $F(f)$ for $f \in K$. Let $\mathcal{R}' := RE^{-1}(\mathcal{R}) \subset \mathcal{N}$ be the full subcategory consisting of those objects $Y \in \mathcal{N}$, such that for a fibrant replacement $Y \to Y_1$ we have $E(Y_1) \in \mathcal{R}$. By condition (A4) for $\mathcal{R}$, and the fact that E preserves equivalences between fibrant objets, membership of Y in $\mathcal{R}$ is independent of the choice of fibrant replacement Y_1.

Theorem 9.8.1 *In the above situation, suppose that $(\mathcal{R}, K)$ is a direct localizing system in $\mathcal{M}$. Define $(\mathcal{R}', K')$ as above. Then $(\mathcal{R}', K')$ is a direct localizing system in $\mathcal{N}$. The functor F is a left Quillen functor from the left Bousfield localization of $\mathcal{M}$ along K, to the left Bousfield localization of $\mathcal{N}$ along K'.*

Proof We verify the conditions (A1)–(A6):

(A1)–$K' = F(K) \cup J$ is clearly a small subset.

(A2)–$J' \subset K'$ by definition.

(A3)–Since F is a left Quillen functor, it preserves cofibrations. But by hypothesis $K \subset \mathbf{cof}(I)$, so $F(K) \subset \mathbf{cof}(I')$. Also $J' \subset \mathbf{cof}(I')$ so $K' \subset \mathbf{cof}(I')$. The domains of arrows in $F(K)$ are cofibrant again because F preserves cofibrant objects.

(A4)–If $X \in \mathcal{R}'$ and $X \cong Y$ in $\text{ho}(\mathcal{M})$, choose a cofibrant (resp. fibrant) replacement by a weak equivalence $X \leftarrow X_1$ (resp. $Y \rightarrow Y_1$) in $\mathcal{N}$. The isomorphism in the homotopy category lifts to a diagram of the form $X_1 \leftarrow Z \rightarrow Y_1$, where both maps are weak equivalences and the first is a fibration. Hence Z is fibrant, and $E(X_1) \leftarrow E(Z) \rightarrow E(Y_1)$ is a diagram of weak equivalences in $\mathcal{M}$. The condition $X \in \mathcal{R}'$ means $E(X_1) \in \mathcal{R}$. Condition (A4) for $\mathcal{R}$ says that $E(Z)$, $E(Y_1) \in \mathcal{R}$ hence $Y \in \mathcal{R}'$.

(A5)–Suppose $Y \in \mathbf{inj}(K')$. In particular $Y \in \mathbf{inj}(J')$ so Y is fibrant. By the adjunction between F and E, the lifting property for Y with respect to $F(K)$ is equivalent to the lifting property for $E(Y)$ with respect to K. Hence $E(Y) \in \mathbf{inj}(K)$ so $E(Y) \in \mathcal{R}$. Here Y is its own fibrant replacement so this shows that $Y \in \mathcal{R}'$.

We now get to the important part of the argument, which is the verification of (A6). Suppose $X \in \mathcal{R}'$ and X is fibrant (i.e. J'-injective), and that $X \rightarrow Y$ is a pushout by an element of K'. If it is a pushout by an element of J' then it is already a weak equivalence so we can take $Y = Z$ (the identity is by definition in $\mathbf{cell}(K')$).

Suppose it is a pushout by an element $F(g) : F(A) \rightarrow F(B)$, where $g \in K$. That is to say we have a map $i' : F(A) \rightarrow X$ and $Y = X \cup^{F(A)} F(B)$. By adjunction this gives a map $i : A \rightarrow E(X)$. Since X is its own fibrant replacement, by definition of $\mathcal{R}'$ we have $E(X) \in \mathcal{R}$. Also E is a right Quillen functor so it preserves fibrant objects: $E(X)$ is fibrant. Apply (A6) to the pushout $E(X) \rightarrow E(X) \cup^A B$ in $\mathcal{M}$. That gives a map $h : E(X) \cup^A B \rightarrow Z$ in $\mathbf{cell}(K)$ such that $E(X) \rightarrow Z$ is a weak equivalence. Note that both g and h are cofibrations, so $F(g)$ and $F(h)$ are cofibrations, and $F(h) : F(E(X) \cup^A B) \rightarrow F(Z)$ is in $\mathbf{cell}(K)$. The pushout $E(X) \rightarrow E(X) \cup^A B$ is a cofibration so the composed map $E(X) \rightarrow Z$ is a cofibration, hence a trivial cofibration. These are preserved by F, so $F(E(X)) \rightarrow F(Z)$ is a trivial cofibration.

The adjunction map $F(E(X)) \rightarrow X$ induces a map $F(E(X)) \cup^A B) \rightarrow X \cup^{F(A)} F(B)$ and the pushout of $F(h)$ along here gives a map

$$Y = X \cup^{F(A)} F(B) \rightarrow (X \cup^{F(A)} F(B)) \cup^{F(E(X) \cup^A B)} F(Z) =: Z'$$

in $\mathbf{cell}(K)$. On the other hand,

$$Z' = (X \cup^{F(A)} F(B)) \cup^{F(E(X) \cup^A B)} F(Z) = X \cup^{F(E(X))} F(Z),$$

but that is a pushout along the trivial cofibration $F(E(X)) \rightarrow F(Z)$, so the map $X \rightarrow Z'$ is a trivial cofibration. This completes the verification of (A6).

To show that F is a left Quillen functor, note that the cofibrations are the same in both cases so we just have to show that F preserves new trivial cofibrations. Suppose $X \xrightarrow{f} Y$ is a new trivial cofibration. It fits into a diagram as in Corollary 9.4.4. Applying the functor F takes **cell**(K) to **cell**(K') so $F(f)$, which is a cofibration by the original left Quillen property of F, again fits into a diagram of the same form. Thus by Corollary 9.4.4, $F(f)$ is a new trivial cofibration. □

Applying the last paragraph of Theorem 9.7.1, which in turn comes from Proposition 8.7.4, gives the following remark. It is the abstract version appropriate to the present stage of our construction, of Bergner's result [42] characterizing fibrant Segal categories.

Remark 9.8.2 In the situation of Theorem 9.8.1, an object $U \in \mathcal{N}$ is fibrant for the new model structure corresponding to $(\mathcal{R}', K')$, if and only if it is fibrant in the original structure of $\mathcal{N}$ and $E(U)$ is K-injective. If U is a new fibrant object of $\mathcal{N}$ then a morphism $W \xrightarrow{p} U$ is a new fibration if and only if it is an old fibration and $E(p)$ is in **inj**(K).

Suppose now that we have a set Q and a collection of tractable left proper model categories $\mathcal{M}_q$ for $q \in Q$, with a collection of Quillen functors $F_q : \mathcal{M}_q \leftrightarrow \mathcal{N} : E_q$ to a fixed tractable left proper $\mathcal{N}$. Suppose $(\mathcal{R}_q, K_q)$ are direct localizing systems for the $\mathcal{M}_q$. Let (I', J') be generators for $\mathcal{N}$, and define $(\mathcal{R}', K')$ as follows. First, $\mathcal{R}' \subset \mathcal{N}$ is the full subcategory of objects $Y \in \mathcal{N}$ such that for a fibrant replacement $Y \to Y'$, we have $EE =_q (Y') \in \mathcal{R}_q$ for all $q \in Q$. Then let

$$K' := J' \cup \bigcup_{q \in Q} F_q(K_q).$$

Theorem 9.8.3 *In this situation, $(\mathcal{R}', K')$ is a direct localizing system for $\mathcal{N}$. An object $U \in \mathcal{N}$ (resp. a morphism $W \xrightarrow{p} U$ to a fibrant object) is fibrant for the resulting new model structure (resp. a new fibration), if and only if it is fibrant in the old structure and if each $E_q(U)$ (resp. $E_q(p)$) is fibrant in the direct left Bousfield localization corresponding to $(\mathcal{R}_q, K_q)$.*

Proof The same as above. For (A6) note that we have to treat pushout by a single map in K', which is either in J' or of the form $F_q(g)$ for $g \in K_q$. Use the argument of the previous proof for the functors F_q and E_q.

The characterization of fibrant objects and fibrations to fibrant objects, comes from the corresponding part of Theorem 9.7.1 going back to Proposition 8.7.4. □

PART III

Generators and relations

10

Precategories

This chapter introduces the main object of study, the notion of $\mathcal{M}$-*precategory*. The terminology "precategory" has been used in several different ways, notably by Janelidze [152]. The idea of the word is to invoke a structure coming prior to the full structure of a category. For us, a categorical structure means a category weakly enriched over $\mathcal{M}$ following Segal's method. Then a "precategory" will be a kind of simplicial object without imposing the Segal conditions. The passage from a precategory to a weakly enriched category consists of enforcing the Segal conditions using the small object argument. The basic philosophy behind this construction is that the precategory contains the necessary information for defining the category, by generators and relations. The small object argument then corresponds to the calculus whereby the generators and relations determine a category, this operation being generically denoted **Seg**. Splitting up things this way is motivated by the fact that simplicial objects satisfying the Segal condition are not in any obvious way closed under colimits. When we take colimits we get to arbitrary simplicial objects or precategories, which then have to generate a category by the **Seg** operation.

The calculus of generators and relations is the main subject of several upcoming chapters. In the present chapter, intended as a reference, we introduce the definition of precategory appropriate to our situation, and indicate the construction of some important examples which will be used later. For the purposes of the present chapter we don't need to be too specific about the hypotheses on $\mathcal{M}$; it will generally be supposed to be a tractable left proper cartesian model category, but this will be discussed in detail in Chapter 12.

10.1 Enriched precategories with a fixed set of objects

Suppose X is a set. Following Lurie [192], define the category Δ_X to have objects which are sequences of $x_i \in X$ denoted by $(x_0, \ldots, x_n)$ for any $n \in \Delta$, and morphisms

$$(x_0, \ldots, x_n) \xrightarrow{\phi} (y_0, \ldots, y_m),$$

for any $\phi : n \to m$ in Δ, whenever $y_{\varphi(i)} = x_i$ for $i = 0, \ldots, n$. Write Δ^o_X for the opposite category.

If $f : X \to Y$ is a map of sets, we obtain a functor $\Delta_f : \Delta_X \to \Delta_Y$ defined by $\Delta_f(x_0, \ldots, x_n) := (f(x_0), \ldots, f(x_n))$ and the corresponding functor on opposite categories denoted $\Delta^o_f : \Delta^o_X \to \Delta^o_Y$.

The basic objects of study will be functors $\mathcal{F} : \Delta^o_X \to \mathcal{M}$. Such a functor specifies for each sequence of elements $x_0, \ldots, x_n \in X$, an object $\mathcal{F}(x_0, \ldots, x_n)$ and for each arrow $\phi : n \to m$ and sequence $y_0, \ldots, y_m$ a morphism

$$\mathcal{F}(y_0, \ldots, y_m) \to \mathcal{F}(y_{\phi(0)}, \ldots, y_{\phi(n)})$$

compatible with compositions and identities on the level of ϕ. Recall that the category of such functors is denoted $\text{FUNC}\left(\Delta^o_X, \mathcal{M}\right)$.

The interpretation of precategories as functors on Δ^o_X seems to have been known as folklore for some time, for example Joachim Kock reports having heard of it in a talk by Bertrand Toën as early as 2001. In the literature it appears in the papers of Bergner [45] and Lurie [192].

For reasons which will become clear with the counterexample of Section 17.3.1 having to do with products, we want to impose a *unitality condition*, saying that $\mathcal{F}(y_0) = *$ for single-object sequences. The unitality condition corresponds, in a certain sense, to requiring strict units even though the composition of morphisms is not yet specified.

Definition 10.1.1 An *$\mathcal{M}$-precategory over object-set X* is a functor $\mathcal{F} : \Delta^o_X \to \mathcal{M}$ such that $\mathcal{F}(x_0) \cong *$ is the coinitial object of $\mathcal{M}$, for any $x_0 \in X$. Let $\mathcal{PC}(X, \mathcal{M})$ denote the category of $\mathcal{M}$-precategories over X with morphisms which are natural transformations of diagrams.

We occasionally use various terminologies such as *weakly $\mathcal{M}$-enriched precategory with object set X*, or just *$\mathcal{M}$-precategory over X*, for elements of $\mathcal{PC}(X, \mathcal{M})$.

In Section 11.4 of the next chapter we shall consider more generally a category of *diagrams with unitality condition* denoted $\text{FUNC}(\Phi/\Phi_0, \mathcal{M})$ whenever Φ is a small category and Φ_0 a subset of objects or equivalently a full subcategory. In the present case $\Phi = \Delta^o_X$ and the subset Φ_0 consists of all sequences of length zero (x_0). Note that Φ_0 is isomorphic to the discrete category corresponding to the set X, so in the notation of Section 11.4,

$$\mathcal{PC}(X, \mathcal{M}) := \text{FUNC}\left(\Delta^o_X / X, \mathcal{M}\right).$$

This is related to the category of all (not necessarily unital) diagrams $\text{FUNC}\left(\Delta_X^o, \mathcal{M}\right)$ by the *unitalization adjunction*

$$U_{X,!} : \text{FUNC}\left(\Delta_X^o, \mathcal{M}\right) \leftrightarrow \mathcal{PC}(X, \mathcal{M}) : U_X^*,$$

in which the right adjoint U_X^* just forgets the unitality property of a diagram, and the left adjoint takes a non-unital diagram and enforces the unitality condition. This will be discussed further in Sections 11.4 and 12.2.

It would of course be interesting to investigate further what would happen if we considered all objects in $\text{FUNC}\left(\Delta_X^o, \mathcal{M}\right)$, with a weak unitality condition imposed by the Segal condition for $n = 0$, saying that $\mathcal{F}(x_0)$ should be contractible. For the present purposes this doesn't lead directly to a cartesian model category, indeed the degeneracies of the single points $\mathcal{F}(x_0) = *$ play an important role in assuring the compatibility of weak equivalences with products. It is likely that this problem could be worked around, and that the resulting theory would be closely related to Kock's weak unit condition [172] as well as to Paoli's special Cat^n-groups [210].

For the present exposition it will be convenient to proceed as directly as possible without considering these other possibilities. When necessary, we refer to the objects of $\text{FUNC}\left(\Delta_X^o, \mathcal{M}\right)$ as "non-unital precategories."

Being a category which consists of some diagrams, $\mathcal{PC}(X, \mathcal{M})$ inherits the notions of *levelwise weak equivalences*, *levelwise cofibrations* and *levelwise fibrations*. See Section 11.4 for a discussion of the resulting injective and projective model structures.

10.2 The Segal conditions

Given an $\mathcal{M}$-precategory $\mathcal{A}$ over a set X, we can look at the *Segal maps*. If $(x_0, \ldots, x_n)$ is a sequence in X, the principal edges of the n-simplex are maps in Δ_X

$$(x_i, x_{i+1}) \to (x_0, \ldots, x_n),$$

and these give maps

$$\mathcal{A}(x_0, \ldots, x_n) \to \mathcal{A}(x_i, x_{i+1}).$$

Put these together to get the *Segal map at* $(x_0, \ldots, x_n)$

$$\mathcal{A}(x_0, \ldots, x_n) \to \mathcal{A}(x_0, x_1) \times \cdots \times \mathcal{A}(x_{n-1}, x_n).$$

Note that for $n = 0$ the Segal map at (x_0) is

$$\mathcal{A}(X_0) \to *.$$

Thus the unitality condition says that the $n = 0$ Segal maps are isomorphisms.

Say that an $\mathcal{M}$-precategory *satisfies the Segal conditions* if the Segal maps are weak equivalences, in other words they are contained in the subcategory of weak equivalences $\mathcal{W} \subset \mathcal{M}$. A precategory satisfying this condition will be called an *weakly $\mathcal{M}$-enriched category* or just *weak $\mathcal{M}$-category*. An $\mathcal{M}$-precategory is said to be a *strict $\mathcal{M}$-category* if the Segal maps are isomorphisms.

As was amply pointed out in Part I, the Segal condition at $n = 2$ serves to define a weak composition operation in the following sense. For any three objects $x_0, x_1, x_2 \in X$ we have a diagram

$$\begin{array}{ccc} \mathcal{A}(x_0, x_1, x_2) & \rightarrow & \mathcal{A}(x_0, x_1) \times \mathcal{A}(x_1, x_2) \\ \downarrow & & \\ \mathcal{A}(x_0, x_2) & & \end{array}$$

where the horizontal arrow is the Segal map; if it is a weak equivalence then the vertical arrow projects to a map

$$\mathcal{A}(x_0, x_1) \times \mathcal{A}(x_1, x_2) \rightarrow \mathcal{A}(x_0, x_2)$$

in $\text{ho}(\mathcal{M})$. The Segal conditions for higher n serve to fix the higher homotopy coherences necessary starting with associativity at $n = 3$. It is necessary to include all of the higher coherences in order to obtain a theory compatible with products, see Chapter 17 below.

10.3 Varying the set of objects

Up until now we have discussed the category of $\mathcal{M}$-precategories on a fixed set of objects. This way of splitting off a first part of the argument was originally suggested by Barwick [16], and that idea will be continued in Section 12.1, where we consider model structures on $\mathcal{PC}(X, \mathcal{M})$.

The next step is to investigate what happens under maps between sets of objects, and to define a category $\mathcal{PC}(\mathcal{M})$ of $\mathcal{M}$-precategories without specified set of objects.

If $f : X \rightarrow Y$ is a map of sets, we can *pull back* a structure of $\mathcal{M}$-precategory on Y, to a structure of $\mathcal{M}$-precategory on X, just by composing a diagram $\mathcal{A} : \Delta^o_Y \rightarrow \mathcal{M}$ with the functor Δ^o_f:

$$\left(\Delta^o_f\right)^* : \text{FUNC}\left(\Delta^o_Y, \mathcal{M}\right) \rightarrow \text{FUNC}\left(\Delta^o_X, \mathcal{M}\right),$$

$$\left(\Delta_f^o\right)^*(\mathcal{A})(x_0, \ldots, x_n) = \mathcal{A}(f(x_0), \ldots, f(x_n)).$$

This clearly preserves the unitality condition, so it restricts to a functor which, if no confusion arises, will be denoted just by

$$f^* : \mathcal{PC}(Y, \mathcal{M}) \to \mathcal{PC}(X, \mathcal{M}).$$

We also get a left adjoint to f^* denoted $f_! : \mathcal{PC}(X, \mathcal{M}) \to \mathcal{PC}(Y, \mathcal{M})$. The general formula is $f_! = U_{Y,!} \circ \left(\Delta_f^o\right)_!$, where $U_{Y,!}$ is the unitalization operation, see Sections 11.4 and 12.2.

If $f : X \hookrightarrow Y$ is an inclusion, the expression can be made more explicit. Writing $Y = f(X) \sqcup Z$, where Z is the complement of the image, we have $f_!(\mathcal{A})(y_0, \ldots, y_n) = \mathcal{A}(x_0, \ldots, x_n)$ if all $y_i \in f(X)$ and x_i is the preimage of y_i. If $y_0 = \ldots = y_n = z \in Z$ then $f_!(\mathcal{A})(y_0, \ldots, y_n) = *$, and $f_!(\mathcal{A})(y_0, \ldots, y_n) = \emptyset$ in all other cases. Thus, $f_!(\mathcal{A})$ is the precategory $\mathcal{A}$, extended by adding on the discrete set Z considered as a discrete $\mathcal{M}$-enriched category.

Lemma 10.3.1 *Suppose $f : X \to Y$ is a map of sets. If $\mathcal{A}$ is an $\mathcal{M}$-precategory on object set Y, which satisfies the Segal condition, then $f^*(\mathcal{A})$ satisfies the Segal condition as an $\mathcal{M}$-precategory on X.*

Assume at least condition (DCL) which is part of the cartesian condition 7.7.1 on $\mathcal{M}$. If f is injective and if $\mathcal{B}$ is an $\mathcal{M}$-precategory on X satisfying the Segal condition, then $f_!(\mathcal{B})$ satisfies the Segal condition as an $\mathcal{M}$-precategory on Y.

Proof The Segal maps for $f^*(\mathcal{A})$ are some among the Segal maps for $\mathcal{A}$, so the Segal conditions for $\mathcal{A}$ imply the same for $f^*(\mathcal{A})$. Suppose f is injective and $\mathcal{B} \in \mathcal{PC}(X, \mathcal{M})$ satisfies the Segal conditions. Use the notations $Y = f(X) \sqcup Z$ of the paragraph before the lemma. Given a sequence $(y_0, \ldots, y_n)$ of objects $y_i \in Y$, if $y_i = f(x_i)$ for all i then the Segal map for $f_!(\mathcal{B})$ at $(y_0, \ldots, y_n)$ is the same as that for $\mathcal{A}$ at $(x_0, \ldots, x_n)$. If $y_0 = \ldots = y_n = z \in Z$ then the Segal map for $f_!(\mathcal{B})$ at $(y_0, \ldots, y_n)$ is the identity of $*$. If $(y_0, \ldots, y_n)$ is a sequence which is not constant and which contains at least one element of Z, then one of the adjacent pairs (y_{i-1}, y_i) has to be nonconstant and contain an element of Z. Using condition (DCL) via Lemma 7.7.2, the cartesian product of anything with $\emptyset$ is again $\emptyset$, so in these cases the Segal maps are the identity of $\emptyset$. In all cases, the Segal maps remain weak equivalences. □

Lemma 10.3.2 *Suppose $f : X \to Y$ is a map of sets. The pullback f^* preserves levelwise weak equivalences, that is it takes levelwise weak equivalences in $\mathcal{PC}(Y, \mathcal{M})$ to levelwise weak equivalences in $\mathcal{PC}(X, \mathcal{M})$. Similarly, f^* preserves levelwise cofibrations and trivial cofibrations.*

Proof The pullback f^* of diagrams will preserve any levelwise properties. □

10.4 The category of precategories

We would like to define a notion of $\mathcal{M}$-precategory on an unspecified or variable set of objects, just as for usual categories. An *$\mathcal{M}$-precategory* $\mathcal{A}$ consists of a set denoted $\mathrm{Ob}(\mathcal{A})$, and an $\mathcal{M}$-precategory over $\mathrm{Ob}(\mathcal{A})$ denoted

$$(x_0, \ldots, x_n) \mapsto \mathcal{A}(x_0, \ldots, x_n) \in \mathcal{M},$$

for $x_i \in \mathrm{Ob}(\mathcal{A})$.

A *morphism* $f : \mathcal{A} \to \mathcal{B}$ between two $\mathcal{M}$-precategories consists of a map of sets $\mathrm{Ob}(f) : \mathrm{Ob}(\mathcal{A}) \to \mathrm{Ob}(\mathcal{B})$, and for any $x_0, \ldots, x_n \in \mathrm{Ob}(\mathcal{A})$ a morphism

$$f(x_0, \ldots, x_n) : \mathcal{A}(x_0, \ldots, x_n) \to \mathcal{B}(\mathrm{Ob}(f)(x_0), \ldots, \mathrm{Ob}(f)(x_n));$$

these are required to satisfy naturality with respect to maps $\phi : n \to m$ in Δ. Henceforth, if no confusion is possible, we denote by $f(x_i) := \mathrm{Ob}(f)(x_i)$ the action on objects, and use f also to denote $f(x_0, \ldots, x_n)$.

In terms of the notation given in Section 10.3, we can think of f as a morphism of $\mathcal{M}$-precategories over object set $\mathrm{Ob}(\mathcal{A})$, denoted

$$\mathrm{Mor}_f : \mathcal{A} \to \mathrm{Ob}(f)^*(\mathcal{B}).$$

Composition of morphisms is defined in the obvious way.

Let $\mathcal{PC}(\mathcal{M})$ denote the category of $\mathcal{M}$-precategories thusly defined. Taking the "underlying set of objects" is a functor

$$\mathrm{Ob} : \mathcal{PC}(\mathcal{M}) \to \mathrm{SET}.$$

This is a fibered category as we now explain. A map of sets $g : X \to Y$ induces a functor $\Delta_g : \Delta_X \to \Delta_Y$ hence an adjoint pair

$$\left(\Delta_g^o\right)_! : \mathrm{FUNC}\left(\Delta_X^o, \mathcal{M}\right) \leftrightarrow \mathrm{FUNC}\left(\Delta_Y^o, \mathcal{M}\right) : \left(\Delta_g^o\right)^*.$$

We have $\left(\Delta_g^o\right)^* (\mathcal{F})(x_0, \ldots, x_n) = \mathcal{F}(g(x_0), \ldots, g(x_n))$. The adjoint pair on the categories of $\mathcal{M}$-precategories is

$$g_! : \mathcal{PC}(X, \mathcal{M}) \leftrightarrow \mathcal{PC}(Y, \mathcal{M}) : g^*,$$

with $g^* = \left(\Delta_g^o\right)^*$ being the same and $g_! = U_{Y,!} \circ \Delta_{g,!,u}$ obtained by composing with the unitalization operation. These were discussed previously.

Consider the fibered category $\mathcal{F} \to \text{SET}$ whose fiber over a given set X is $\mathcal{PC}(X, \mathcal{M})$ and whose pullback maps are the Δ_g^*. An object of $\mathcal{F}$ is by definition a pair $(X, \mathcal{A})$, where $X \in \text{SET}$ and $\mathcal{A} \in \mathcal{PC}(X, \mathcal{M})$. A morphism from $(X, \mathcal{A})$ to $(Y, \mathcal{B})$ is a pair (g, h), where $g : X \to Y$ is a morphism in SET and $h : \mathcal{A} \to g^*(\mathcal{B})$ is a morphism in $\mathcal{PC}(Y, \mathcal{M})$. By inspection, this is the same structure as defined above, that is $\mathcal{F} = \mathcal{PC}(\mathcal{M})$, with functor Ob being the projection to SET. A morphism $f = (\text{Ob}(f), f)$ is cartesian if and only if Mor_f induces isomorphisms

$$f(x_0, \ldots, x_n) : \mathcal{A}(x_0, \ldots, x_n) \xrightarrow{\cong} \mathcal{B}(\text{Ob}(f)(x_0), \ldots, \text{Ob}(f)(x_n)).$$

10.5 Basic examples

In this section we indicate many of the basic examples of precategories which will be important at various places in the subsequent chapters. Rather than give a full definition and discussion here, we refer to the appropriate places when necessary.

Suppose X is a set. The *discrete precategory* $\textbf{disc}(X)$ is defined to have $\text{Ob}(\textbf{disc}(X)) := X$ and

$$\textbf{disc}(X)(x_0, \ldots, x_n) := \begin{cases} * & \text{if } x_0 = \cdots = x_n, \\ \emptyset & \text{otherwise.} \end{cases}$$

Here $\emptyset$ and $*$ denote the initial and coinitial objects of $\mathcal{M}$ respectively. The functor $X \mapsto \textbf{disc}(X)$ provides a functor $\text{SET} \to \mathcal{PC}(\mathcal{M})$, fully faithful and left adjoint to $\text{Ob} : \mathcal{PC}(\mathcal{M}) \to \text{SET}$.

The *codiscrete precategory* $\textbf{codsc}(X)$ is defined to have

$$\text{Ob}(\textbf{codsc}(X)) := X$$

and

$$\textbf{codsc}(X)(x_0, \ldots, x_n) := *$$

for all sequences $x_.$. The functor $\textbf{codsc} : \text{SET} \to \mathcal{PC}(\mathcal{M})$ is fully faithful and right adjoint to Ob.

The Segal maps are isomorphisms for both $\textbf{disc}(X)$ and $\textbf{codsc}(X)$, so they are strictly $\mathcal{M}$-enriched categories.

Suppose $k \in \mathbb{N}$. Denote by

$$[k] := \{v_0, \ldots, v_k\}, \quad v_0 < v_1 < \cdots < v_k$$

the standard linearly ordered set with $k + 1$ elements. The $[k]$ are the objects of Δ. It will be important to consider various different precategories whose underlying set of objects is $[k]$. These examples will all be *ordered precategories* in the sense that if $(y_0, \ldots, y_p)$ is any sequence of elements of the set $[k]$ with $y_j = \upsilon_{i_j}$, and if the sequence $i_1, \ldots, i_p$ is strictly decreasing at any place (i.e. if it is not an increasing sequence), then

$$\mathcal{A}(y_0, \ldots, y_p) = \emptyset.$$

For $B \in \mathcal{M}$ we have the precategory $h([k], B)$ which has the following concrete description (see Lemma 13.2.1): suppose $(y_0, \ldots, y_p)$ is any sequence of elements of the set $[k]$ with $y_j = \upsilon_{i_j}$. Then:

- if $(y_0, \ldots, y_p)$ is increasing but not constant, i.e. $i_{j-1} \leq i_j$ but $i_0 < i_p$, then

$$h([k]; B)(y_0, \ldots, y_p) = B;$$

- if $(y_0, \ldots, y_p)$ is constant, i.e. $i_0 = i_1 = \ldots = i_p$, then

$$h([k]; B)(y_0, \ldots, y_p) = *;$$

- and otherwise, that is if there exists $1 \leq j \leq p$ such that $i_{j-1} > i_j$, then

$$h([k]; B)(y_0, \ldots, y_p) = \emptyset.$$

For any precategory $\mathcal{A} \in \mathcal{PC}(\mathcal{M})$, a map $h([k]; B) \to \mathcal{A}$ is the same thing as a collection of elements $x_0, \ldots, x_k \in \mathrm{Ob}(\mathcal{A})$ together with a map $B \to \mathcal{A}(x_0, \ldots, x_k)$ in $\mathcal{M}$, so we can think of $h([k], B)$ as being a "representable" object in a certain sense.

It has as "boundary" the precategory $h(\partial[k], B)$, defined using the skeleton construction in Chapter 13, with the following description (see Lemma 13.2.3):

- if $(y_0, \ldots, y_p)$ is increasing but not constant, i.e. $i_{j-1} \leq i_j$ but $i_0 < i_p$, and if there is any $0 \leq m \leq k$ such that $i_j \neq m$ for all $0 \leq j \leq k$, then

$$h(\partial[k]; B)(y_0, \ldots, y_p) = B;$$

- if $(y_0, \ldots, y_p)$ is constant, i.e. $i_0 = i_1 = \ldots = i_p$, then

$$h(\partial[k]; B)(y_0, \ldots, y_p) = *;$$

- and otherwise, that is if either there exists $1 \leq j \leq p$ such that $i_{j-1} > i_j$ or else if the map $j \mapsto y_j$ is a surjection from $\{0, \ldots, p\}$ to $[k]$, then

$$h(\partial[k]; B)(y_0, \ldots, y_p) = \emptyset.$$

More generally, the pushouts

$$h([k], \partial[k]; A \xrightarrow{f} B) := h([k]; A) \cup^{h(\partial[k]; A)} h(\partial[k]; B)$$

are also useful, being the domains of generators of the Reedy cofibrations in $\mathcal{PC}(\mathcal{M})$.

In Section 14.1 and later in Chapter 16 we consider precategories $\Upsilon(B_1, \ldots, B_k)$ with the same set of objects $[k]$, depending on a sequence $B_1, \ldots, B_k \in \mathcal{M}$. As a matter of notation, in later chapters this will sometimes be denoted with a subscript indicating the number of arguments: $\Upsilon_k(B_1, \ldots, B_k)$.

The basic idea is to put B_i in as space of morphisms from υ_{i-1} to υ_i. Thus, the main part of the structure of precategory is given by

$$\Upsilon(B_1, \ldots, B_k)(\upsilon_{i-1}, \upsilon_i) := B_i.$$

This is extended whenever there is a constant string of points on either side:

$$\Upsilon(B_1, \ldots, B_k)(\upsilon_{i-1}, \ldots, \upsilon_{i-1}, \upsilon_i, \ldots, \upsilon_i) := B_i.$$

The unitality condition on the diagram $\Delta_{\{\upsilon_0, \ldots, \upsilon_k\}} \to \mathcal{M}$ means that for $0 \leq i \leq k$ we have

$$\Upsilon(B_1, \ldots, B_k)(\upsilon_i, \ldots, \upsilon_i) := *.$$

In all other cases,

$$\Upsilon(B_1, \ldots, B_k)(x_0, \ldots, x_n) := \emptyset.$$

If $\mathcal{A} \in \mathcal{PC}(\mathcal{M})$, a map $\Upsilon(B_1, \ldots, B_k) \to \mathcal{A}$ is the same thing as a collection of objects $x_0, \ldots, x_k \in \mathrm{Ob}(\mathcal{A})$ together with maps $B_i \to \mathcal{A}(x_{i-1}, x_i)$ in $\mathcal{M}$.

The $\mathcal{M}$-category associated to $\Upsilon(B_1, \ldots, B_k)$ can be described explicitly; it is denoted by $\widetilde{\Upsilon}_k(B_1, \ldots, B_k)$ in Section 16.1. For any sequence $\upsilon_{i_0}, \ldots, \upsilon_{i_n}$ with $i_0 \leq \cdots \leq i_n$, we put

$$\widetilde{\Upsilon}_k(B_1, \ldots, B_k)(\upsilon_{i_0}, \ldots, \upsilon_{i_n}) := B_{i_0+1} \times B_{i_0+2} \times \cdots \times B_{i_n-1} \times B_{i_n}. \quad (10.5.1)$$

For any other sequence, that is to say any sequence which is not increasing, the value is $\emptyset$. The value on a constant sequence is $*$.

As a rough approximation, the calculus of generators and relations can be understood as being the small object argument applied using the inclusions $\Upsilon(B_1, \ldots, B_k) \hookrightarrow \widetilde{\Upsilon}_k(B_1, \ldots, B_k)$. Starting with a precategory $\mathcal{A}$, for every map $\Upsilon(B_1, \ldots, B_k) \to \mathcal{A}$, take the pushout with $\widetilde{\Upsilon}_k(B_1, \ldots, B_k)$. Keep doing this and eventually we get to a precategory which satisfies the Segal conditions.

10.6 Limits, colimits and local presentability

In the diagram category $\text{FUNC}\left(\Delta_X^o, \mathcal{M}\right)$ limits and colimits are computed levelwise: if $\{\mathcal{A}_i\}_{i\in\alpha}$ is a diagram of objects $\mathcal{A}_i \in \text{FUNC}\left(\Delta_X^o, \mathcal{M}\right)$ then

$$(\text{colim}_{i\in\alpha}\mathcal{A}_i)(x_0,\ldots,x_n) = \text{colim}_{i\in\alpha}(\mathcal{A}_i(x_0,\ldots,x_n)),$$

and

$$(\lim_{i\in\alpha}\mathcal{A}_i)(x_0,\ldots,x_n) = \lim_{i\in\alpha}(\mathcal{A}_i(x_0,\ldots,x_n)).$$

The right adjoint functor U^* being the identity means that the same holds true for all limits in $\mathcal{PC}(X,\mathcal{M})$: if $\{\mathcal{A}_i\}_{i\in\alpha}$ is a diagram of objects $\mathcal{A}_i \in \mathcal{PC}(X,\mathcal{M})$ then

$$(\lim_{i\in\alpha}\mathcal{A}_i)(x_0,\ldots,x_n) = \lim_{i\in\alpha}(\mathcal{A}_i(x_0,\ldots,x_n)).$$

On the other hand, for colimits in $\mathcal{PC}(X,\mathcal{M})$ we have to reapply the functor $U_!$. Write this with superscripts $\mathcal{PC}$ or FUNC to indicate in which category the colimit is taken:

$$\text{colim}^{\mathcal{PC}}_{i\in\alpha}\mathcal{A}_i = U_!\left(\text{colim}^{\text{FUNC}}_{i\in\alpha}\mathcal{A}_i\right).$$

In the special case where $\text{colim}^{\text{FUNC}}_{i\in\alpha}\mathcal{A}_i$ is already in $\mathcal{PC}(X,\mathcal{M})$ then the functor $U_!$ acts as the identity. The condition that $\text{colim}^{\text{FUNC}}_{i\in\alpha}\mathcal{A}_i$ already be in $\mathcal{PC}(X,\mathcal{M})$ says that the value on sequences of length zero (x_0) is $*$. As the $\text{colim}^{\text{FUNC}}$ is calculated levelwise, it is sufficient to require that $\text{colim}_\alpha * = *$, which holds when α is a connected category by Lemma 8.1.9. This condition is necessary and sufficient unless X is the empty set.

We expand this discussion by giving more explicit descriptions of limits and colimits in $\mathcal{PC}(\mathcal{M})$. We can then show local presentability.

Suppose $\{\mathcal{A}_i\}_{i\in\alpha}$ is a diagram of objects $\mathcal{A}_i \in \mathcal{PC}(\mathcal{M})$, that is a functor $\alpha \to \mathcal{PC}(\mathcal{M})$. Let $X_i := \text{Ob}(\mathcal{A}_i)$ denote the object sets so we can consider $\mathcal{A}_i \in \mathcal{PC}(X_i,\mathcal{M})$. For any $f : i \to j$ in α denote by $\phi_f : X_i \to X_j$ the transition map on object sets, and $\rho_f : \mathcal{A}_i \to \phi_f^*\mathcal{A}_j$ the transition maps on the level of precategories.

Start by constructing the limit in $\mathcal{PC}(\mathcal{M})$. The object set of the limit will be

$$X := \lim_{i\in\alpha} X_i.$$

We have maps $p_i : X \to X_i$, hence $p_i^*(\mathcal{A}_i) \in \mathcal{PC}(X,\mathcal{M})$. These are provided with transition maps, indeed for $f : i \to j$ in α, $\phi_f p_i = p_j$ so

$p_i^*\left(\phi_f^*\mathcal{A}_j\right) = p_j^*\mathcal{A}_j$ and $p_i^*(\rho_f) : p_i^*(\mathcal{A}_i) \to p_j^*\mathcal{A}_j$ provide the transition maps for the system of $p_i^*(\mathcal{A}_i)$ considered as a diagram $\alpha \to \mathcal{PC}(X, \mathcal{M})$. Put

$$\mathcal{A} := \lim_{i\in\alpha}^{\mathcal{PC}(X,\mathcal{M})} p_i^*(\mathcal{A}_i) \in \mathcal{PC}(X, \mathcal{M}).$$

We claim that this is the limit of the diagram $\{\mathcal{A}_i\}_{i\in\alpha}$ in $\mathcal{PC}(\mathcal{M})$.

Suppose $\mathcal{B} \in \mathcal{PC}(\mathcal{M})$ with $\mathrm{Ob}(\mathcal{B}) = Y$ and suppose given a system of maps $\mathcal{B} \to \mathcal{A}_i$. These correspond to maps $r_i : Y \to X_i$ and $\varphi_i : \mathcal{B} \to r_i^*(\mathcal{A}_i)$. The collection of r_i gives a uniquely determined map $r : Y \to X$, and $r^* p_i^*(\mathcal{A}_i) = r_i^*(\mathcal{A}_i)$ so the φ_i correspond to a collection of maps $\mathcal{B} \to r^*\left(p_i^*(\mathcal{A}_i)\right)$. This gives a uniquely determined map $\varphi : \mathcal{B} \to r^*(\mathcal{A})$ whose composition with the projections of the limit expression for $\mathcal{A}$, are the φ_i. The pair (r, φ) is a map $\mathcal{B} \to \mathcal{A}$ in $\mathcal{PC}(\mathcal{M})$ uniquely solving the universal problem to show that $(X, \mathcal{A})$ is the limit of the $\mathcal{A}_i$.

The limit $\mathcal{A}$ can be described explicitly as an element of $\mathcal{PC}(X, \mathcal{M})$: for any $x_0, \ldots, x_n \in X$,

$$\mathcal{A}(x_0, \ldots, x_n) = \lim_{i\in\alpha}\mathcal{A}_i(p_i x_0, \ldots, p_i x_n).$$

Turn now to the construction of the colimit. The object set will be

$$Z := \mathrm{colim}_{i\in\alpha} X_i,$$

with maps $q_i : X_i \to Z$. These give $q_{i,!}(\mathcal{A}_i) \in \mathcal{PC}(X, \mathcal{M})$, with transition maps defined as follows. If $f : i \to j$ is an arrow in α then $q_j\phi_f = q_i$ so $q_{j,!}(\phi_{f,!}(\mathcal{A}_i)) = q_{i,!}(\mathcal{A}_i)$. The adjunction between $\phi_{f,!}$ and ϕ_f^* means that the transition map ρ_f for the system of $\mathcal{A}_i$ may be viewed as a map denoted $\tilde{\rho}_f : \phi_{f,!}(\mathcal{A}_i) \to \mathcal{A}_j$, in particular we get

$$q_{j,!}(\tilde{\rho}_f) : q_{i,!}(\mathcal{A}_i) = q_{j,!}(\phi_{f,!}(\mathcal{A}_i)) \to q_{j,!}(\mathcal{A}_j).$$

These provide transition maps for the diagram $\{q_{i,!}(\mathcal{A}_i)\}_{i\in\alpha}$ with values in $\mathcal{PC}(Z, \mathcal{M})$ and set

$$\mathcal{C} := \mathrm{colim}_{i\in\alpha}^{\mathcal{PC}(Z,\mathcal{M})} q_{i,!}(\mathcal{A}_i) \in \mathcal{PC}(Z, \mathcal{M}).$$

The object $(Z, \mathcal{C})$ is the colimit of the diagram of $\mathcal{A}_i$, for the same formal reason as in the previous discussion of the limit.

The colimit to define $\mathcal{C}$ is taken in the category $\mathcal{PC}(Z, \mathcal{M})$, which presupposes in general applying the unitalization operation $U_{Z,!}$ if α is disconnected. However, when the definition is unwound explicitly this phenomenon disappears, being absorbed in the calculation of the value of $\mathcal{C}$ on a sequence of points $(z_0, \ldots, z_n)$ via the introduction of a new category $\alpha/(z_0, \ldots, z_n)$.

Suppose $(z_0, \ldots, z_n)$ is a sequence of elements of Z. Let $\alpha/(z_0, \ldots, z_n)$ denote the category of pairs $(i, (x_0, \ldots, x_n))$ where $i \in \alpha$ and $(x_0, \ldots, x_n)$ is a sequence of points in X_i such that $q_i(x_k) = z_k$ for $k = 0, \ldots, n$. The association $(i, (x_0, \ldots, x_n)) \mapsto \mathcal{A}_i(x_0, \ldots, x_n)$ is a diagram from $\alpha/(z_0, \ldots, z_n)$ to $\mathcal{M}$.

Lemma 10.6.1 *In the above situation, the value of the colimiting object $\mathcal{C}$ on the sequence $(z_0, \ldots, z_n)$ is calculated as a colimit of the diagram $\alpha/(z_0, \ldots, z_n) \to \mathcal{M}$:*

$$\mathcal{C}(z_0, \ldots, z_n) = \mathrm{colim}_{(i,(x_0,\ldots,x_n))\in\alpha/(z_0,\ldots,z_n)} \mathcal{A}_i(x_0, \ldots, x_n).$$

Proof Let $q_{i,!}^{wu} : \mathrm{FUNC}(\Delta^o_{X_i}, \mathcal{M}) \to \mathrm{FUNC}(\Delta^o_Z, \mathcal{M})$ denote the pushforward in the world of non-unital precategories. It is the pushforward for diagrams valued in $\mathcal{M}$ along the functor

$$\Delta^o_{X_i} \to \Delta^o_Z.$$

If $(z_0, \ldots, z_n)$ is a sequence of points in Z, then $q_{i,!}^{wu}(\mathcal{A}_i)(z_0, \ldots, z_n)$ is the colimit of the $\mathcal{A}_i(x_0, \ldots, x_k)$ over the category of pairs $(u, (x_0, \ldots, x_k))$ where $u : (q_i x_0, \ldots, q_i x_k) \to (z_0, \ldots, z_n)$ is a map in Δ^o_Z or equivalently $(z_0, \ldots, z_n) \to (q_i x_0, \ldots, q_i x_k)$ is a map in Δ_Z. Such a map factors through a unique map $(z_0, \ldots, z_n) \to (q_i x_{i_0}, \ldots, q_i x_{i_n})$ where $(i_0, \ldots, i_n)$ is a multi-index representing a map $[n] \to [k]$ in Δ. Hence the category of pairs in question is a disjoint union of categories having coinitial objects. This yields the expression of $q_{i,!}^{wu}(\mathcal{A}_i)(z_0, \ldots, z_n)$ as the disjoint sum of $\mathcal{A}_i(x_0, \ldots, x_n)$ over all sequences $(x_0, \ldots, x_n)$ of objects in X_i such that $q_i(x_j) = z_j$. Putting these together over all $i \in \alpha$ gives

$$\mathrm{colim}_{i\in\alpha}\left(q_{i,!}^{wu}(\mathcal{A}_i)(z_0, \ldots, z_n)\right) = \mathrm{colim}_{(i,x.)\in\alpha/z.} \mathcal{A}_i(x_0, \ldots, x_n).$$

Now $U_{Z,!}$ is a left adjoint so it preserves colimits, in particular

$$\begin{aligned}\mathcal{C} = \mathrm{colim}^{\mathcal{PC}(Z,\mathcal{M})}_{i\in\alpha} q_{i,!}(\mathcal{A}_i) &= \mathrm{colim}_{i\in\alpha} U_{Z,!}\left(q_{i,!}^{wu}(\mathcal{A}_i)\right)\\ &= U_{Z,!}\left(\mathrm{colim}_{i\in\alpha} q_{i,!}^{wu}(\mathcal{A}_i)\right) = U_{Z,!}(\mathcal{C}'),\end{aligned}$$

where $\mathcal{C}' \in \mathrm{FUNC}(\Delta^o_Z, \mathcal{M})$ is the object defined by

$$\mathcal{C}'(z_0, \ldots, z_n) := \mathrm{colim}_{(i,(x_0,\ldots,x_n))\in\alpha/(z_0,\ldots,z_n)} \mathcal{A}_i(x_0, \ldots, x_n).$$

To finish the proof, note that $\mathcal{C}'$ is already in $\mathcal{PC}(Z, \mathcal{M})$. Indeed for a single element $z_0 \in Z$ the category $\alpha/(z_0)$ of pairs (i, x_0) with $q_i(x_0) = z_0$ is connected. This is a factoid about colimits of sets. Since $\mathcal{A}_i$ is unital we have $\mathcal{A}_i(x_0) = *$, thus by Lemma 8.1.9

$$\mathcal{C}'(z_0) = \mathrm{colim}_{(i,x_0)\in\alpha/(z_0)} * = *,$$

in other words $\mathcal{C}'$ is unital. Therefore $\mathcal{C} = U_{Z,!}(\mathcal{C}') = \mathcal{C}'$, which is the statement of the lemma. An alternative proof would be to note that $\mathcal{C}'$ satisfies the required universal property for defining a colimit. □

Corollary 10.6.2 *If $\mathcal{A}_{\cdot} : \alpha \to \mathcal{PC}(X, \mathcal{M})$ is a diagram of $\mathcal{M}$-precategories with a common object set X, indexed by a connected category α, then the natural map*

$$\mathrm{colim}^{\mathcal{PC}(\mathcal{M})}_{i\in\alpha} \mathcal{A}_i \to \mathrm{colim}^{\mathcal{PC}(X,\mathcal{M})}_{i\in\alpha} \mathcal{A}_i$$

is an isomorphism. In other words the colimit in $\mathcal{PC}(X, \mathcal{M})$ may be calculated levelwise:

$$\left(\mathrm{colim}^{\mathcal{PC}}_{i\in\alpha} \mathcal{A}_i\right)(x_0, \ldots, x_n) = \mathrm{colim}_{i\in\alpha}(\mathcal{A}_i(x_0, \ldots, x_n)).$$

Proof The explicit descriptions of both sides are the same. □

It is worthwhile to discuss explicitly some special cases. For example, $\mathrm{Ob}(\mathcal{A} \times \mathcal{B}) = \mathrm{Ob}(\mathcal{A}) \times \mathrm{Ob}(\mathcal{B})$ and for any sequence $((x_0, y_0), \ldots, (x_n, y_n))$ of elements of $\mathrm{Ob}(\mathcal{A}) \times \mathrm{Ob}(\mathcal{B})$ we have

$$\mathcal{A} \times \mathcal{B}((x_0, y_0), \ldots, (x_n, y_n)) = \mathcal{A}(x_0, \ldots, x_n) \times \mathcal{B}(y_0, \ldots, y_n).$$

This extends to fiber products: if $\mathcal{A} \to \mathcal{C}$ and $\mathcal{B} \to \mathcal{C}$ are maps, then

$$\mathrm{Ob}(\mathcal{A} \times_{\mathcal{C}} \mathcal{B}) = \mathrm{Ob}(\mathcal{A}) \times_{\mathrm{Ob}(\mathcal{C})} \mathrm{Ob}(\mathcal{B}),$$

and again for any sequence $((x_0, y_0), \ldots, (x_n, y_n))$ of elements in the fiber product of object sets we have

$$\mathcal{A} \times_{\mathcal{C}} \mathcal{B}((x_0, y_0), \ldots, (x_n, y_n)) = \mathcal{A}(x_0, \ldots, x_n) \times_{\mathcal{C}(z_0,\ldots,z_n)} \mathcal{B}(y_0, \ldots, y_n),$$

where the z_i are the common images of x_i and y_i in $\mathrm{Ob}(\mathcal{C})$.

For colimits, the disjoint sum $\mathcal{A} \sqcup \mathcal{B}$ has object set $\mathrm{Ob}(\mathcal{A}) \sqcup \mathrm{Ob}(\mathcal{B})$, and for a sequence of elements $(z_0, \ldots, z_n)$ in here we have

$$(\mathcal{A} \sqcup \mathcal{B})(z_0, \ldots, z_n) = \begin{cases} \mathcal{A}(z_0, \ldots, z_n) & \text{if all } z_i \in \mathrm{Ob}(\mathcal{A}), \\ \mathcal{B}(z_0, \ldots, z_n) & \text{if all } z_i \in \mathrm{Ob}(\mathcal{B}), \\ \emptyset & \text{otherwise.} \end{cases}$$

Look at the amalgamated sum or pushout of two maps $u : \mathcal{C} \to \mathcal{A}$ and $v : \mathcal{C} \to \mathcal{B}$, supposing that one of the maps say v is injective on the set of objects. This will usually be the case in our applications because we usually look at pushouts along cofibrations. The amalgamated sum $\mathcal{A} \cup^{\mathcal{C}} \mathcal{B}$ has object set

$$\mathrm{Ob}(\mathcal{A}) \cup^{\mathrm{Ob}(\mathcal{C})} \mathrm{Ob}(\mathcal{B}) = \mathrm{Ob}(\mathcal{A}) \sqcup (\mathrm{Ob}(\mathcal{B}) - \mathrm{Ob}(\mathcal{C})).$$

The category α indexing the colimit has three objects denoted a, b, c with arrows $a \leftarrow c \to b$. Given a sequence $(z_0, \ldots, z_n)$ of elements in here, the category $\alpha/(z_0, \ldots, z_n)$ has objects of three kinds denoted a, b and c, and objects of a given kind correspond to sequences in $\mathrm{Ob}(\mathcal{A})$, $\mathrm{Ob}(\mathcal{B})$ or $\mathrm{Ob}(\mathcal{C})$ respectively mapping to the given $z_\cdot$. Our general formula of Lemma 10.6.1 expresses $(\mathcal{A} \cup^{\mathcal{C}} \mathcal{B})(z_0, \ldots, z_n)$ as the pushout of disjoint sums

$$\coprod_{x_\cdot \mapsto z_\cdot} \mathcal{A}(x_0, \ldots, x_n) \cup^{\coprod_{w_\cdot \mapsto z_\cdot} \mathcal{C}(w_0, \ldots, w_n)} \coprod_{y_\cdot \mapsto z_\cdot} \mathcal{B}(y_0, \ldots, y_n).$$

Another important case is that of filtered colimits. Suppose α is a filtered (resp. κ-filtered) category and $\{\mathcal{A}_i\}_{i \in \alpha}$ is a diagram in $\mathcal{PC}(\mathcal{M})$. Put $X_i := \mathrm{Ob}(\mathcal{A}_i)$ and $Z := \mathrm{colim}_{i \in \alpha} X_i$. Then for $(z_0, \ldots, z_n)$ any sequence of elements of Z, the category $\alpha/(z_0, \ldots, z_n)$ is again filtered (resp. κ-filtered), so

$$\mathcal{A}(z_0, \ldots, z_n) = \mathrm{colim}_{(i,(x_0 \ldots, x_n)) \in \alpha/(z_0, \ldots, z_n)} \mathcal{A}(x_0 \ldots, x_n)$$

is a filtered (resp. κ-filtered) colimit in $\mathcal{M}$.

Lemma 10.6.3 *Suppose $\mathcal{M}$ is a locally κ-presentable category whose coinitial object $*$ a κ-presentable, and X is a set with $< \kappa$ elements. An object $\mathcal{A} \in \mathcal{PC}(X, \mathcal{M})$ is κ-presentable if and only if each $\mathcal{A}(x_0, \ldots, x_p)$ is a κ-presentable object of $\mathcal{M}$. The category $\mathcal{PC}(X, \mathcal{M})$ is locally κ-presentable.*

Proof We refer to the argument phrased in more general terms in Theorem 11.4.1 of the next section. □

Lemma 10.6.4 *Suppose $\mathcal{M}$ is locally presentable, whose coinitial object $*$ is κ-presentable. An object $\mathcal{A} \in \mathcal{PC}(\mathcal{M})$ is κ-presentable if and only if $X := \mathrm{Ob}(\mathcal{A})$ is a set of cardinality $< \kappa$, and each $\mathcal{A}(x_0, \ldots, x_p)$ is a κ-presentable object of $\mathcal{M}$.*

Proof Suppose $X := \mathrm{Ob}(\mathcal{A})$ is a set of cardinality $< \kappa$, and each $\mathcal{A}(x_0, \ldots, x_p)$ is a κ-presentable object of $\mathcal{M}$. Suppose $\{\mathcal{B}_i\}_{i \in \beta}$ is a diagram in $\mathcal{PC}(\mathcal{M})$ indexed by a κ-filtered category β, and suppose given a map

$$\mathcal{A} \to \mathcal{B} := \mathrm{colim}_{i \in \beta} \mathcal{B}_i.$$

In particular, we get a map $\mathrm{Ob}(\mathcal{A}) \to \mathrm{Ob}(\mathcal{B}) = \mathrm{colim}_{i \in \beta} \mathrm{Ob}(\mathcal{B}_i)$ and the condition $|\mathrm{Ob}(\mathcal{A})| < \kappa$ implies that this map factors through a map $\mathrm{Ob}(\mathcal{A}) \to \mathrm{Ob}(\mathcal{B}_j)$ for some $j \in \beta$. Given a sequence of objects $(x_0, \ldots, x_n) \in \mathrm{Ob}(\mathcal{A})$, let $(y_0, \ldots, y_n)$ denote the corresponding sequence of objects in $\mathrm{Ob}(\mathcal{B}_j)$ and $(z_0, \ldots, z_n)$ the sequence in $\mathrm{Ob}(\mathcal{B})$. Let $j\backslash\alpha$ denote the category of objects under j in α. Given $j \to i$ the image of $(y_0, \ldots, y_n)$ is a sequence denoted $(y_0^i, \ldots, y_n^i)$ in $\mathrm{Ob}(\mathcal{B}_i)$ mapping to $(z_0, \ldots, z_n)$. This gives a functor

$$j\backslash\alpha \to \alpha/(z_0, \ldots, z_n),$$

and the κ-filtered property of α implies that this functor is cofinal. Hence by Lemma 10.6.1 and the invariance of colimits under cofinal functors,

$$\mathcal{B}(z_0, \ldots, z_n) = \mathrm{colim}_{(j \to i) \in j\backslash\alpha} \mathcal{B}_i \left(y_0^i, \ldots, y_n^i \right).$$

The category $j\backslash\alpha$ is κ-filtered, so the map

$$\mathcal{A}(x_0, \ldots, x_n) \to \mathcal{B}(z_0, \ldots, z_n)$$

factors through one of the $\mathcal{B}_i \left(y_0^i, \ldots, y_n^i \right)$ for $j \to i$ in α. The cardinality of the set of possible sequences $(x_0, \ldots, x_n)$ is $< \kappa$, so the κ-filtered property says that we can choose a single i. Then the standard kind of argument shows that, by going further along, the maps $\mathcal{A}(x_0, \ldots, x_n) \to \mathcal{B}_i \left(y_0^i, \ldots, y_n^i \right)$ can be assumed to all fit together into a natural transformation in terms of $(x_0, \ldots, x_n) \in \Delta_X^o$. Thus we get a factorization of our map $\mathcal{A} \to \mathcal{B}$ through some $\mathcal{A} \to \mathcal{B}_i$. This shows that $\mathcal{A}$ is κ-presentable.

Suppose on the other hand that $\mathcal{A}$ is κ-presentable. We note first of all that $|X| < \kappa$. Indeed, if $|X| \geq \kappa$ then we could consider a κ-filtered system of subsets $Z_i \subset X$ with $|Z_i| < \kappa$, but $\mathrm{colim} Z_i = X$. Let $\mathbf{codsc}(Z_i)$ and $\mathbf{codsc}(X)$ denote the codiscrete precategories on these object sets, that is the precategories whose value is $*$ on any sequence of objects. Then $\mathrm{colim}_i \mathbf{codsc}(Z_i) = \mathbf{codsc}(X)$ in $\mathcal{PC}(\mathcal{M})$ (see Lemma 8.1.9), but the identity map on underlying object sets gives a map $\mathcal{A} \to \mathbf{codsc}(X)$ not factoring through any $\mathbf{codsc}(Z_i)$, contradicting the assumed κ-presentability of $\mathcal{A}$. Hence we may assume that $|X| < \kappa$.

We claim that $\mathcal{A}$ is κ-presentable when considered as an object in $\mathcal{PC}(X, \mathcal{M})$. Indeed, if $\{\mathcal{B}_i\}_{i \in \beta}$ is a κ-filtered diagram in $\mathcal{PC}(X, \mathcal{M})$ then since β is connected, any map

$$\mathcal{A} \to \mathrm{colim}_{i \in \beta}^{\mathcal{PC}(X,\mathcal{M})} \mathcal{B}_i$$

is also a map $\mathcal{A} \to \mathrm{colim}_{i \in \beta}^{\mathcal{PC}(\mathcal{M})} \mathcal{B}_i$ by Corollary 10.6.2. By the assumed κ-presentability of $\mathcal{A}$ this would have to factor through one of the $\mathcal{B}_i$, necessarily as a map inducing the identity on underlying object sets X. This shows that $\mathcal{A}$ is κ-presentable when considered as an object in $\mathcal{PC}(X, \mathcal{M})$. Now Corollary 10.6.3 tells us that the $\mathcal{A}(x_0, \ldots, x_p)$ are κ-presentable objects of $\mathcal{M}$. □

Proposition 10.6.5 *Suppose $\mathcal{M}$ is locally presentable. Then the category of $\mathcal{M}$-precategories $\mathcal{PC}(\mathcal{M})$ is also locally presentable.*

Proof Let κ be a regular cardinal such that $\mathcal{M}$ is locally κ-presentable and $*$ is a κ-presentable object. Note that, for any set X the category $\mathcal{PC}(X, \mathcal{M})$

is locally κ-presentable by its adjunction with the diagram category FUNC $(\Delta_X^o, \mathcal{M})$, which in turn is locally κ-presentable by Lemma 8.1.3.

Existence of arbitrary colimits in $\mathcal{PC}(\mathcal{M})$ was shown at the start of the section. It is clear from the description of κ-presentable objects in Lemma 10.6.4 that the isomorphism classes of κ-presentable objects of $\mathcal{PC}(\mathcal{M})$ form a set. Furthermore, any object $(Z, \mathcal{A})$ is κ-accessible. Indeed, consider the category of triples $(X, \mathcal{B}, u)$, where $X \subset Z$ is a subset of cardinality $|X| < \kappa$, $\mathcal{B} \in \mathcal{PC}(X, \mathcal{M})$ is a κ-presentable object and $u : \mathcal{B} \to \mathcal{A}|_X$ is a morphism to the pullback of $\mathcal{A}$ along the inclusion $X \subset Z$. Using the local κ-presentability of each $\mathcal{PC}(X, \mathcal{M})$ and the expression of Z as a κ-filtered union of the subsets X, the category of triples is κ-filtered and the colimit of the tautological functor is $(Z, \mathcal{A})$. □

10.7 Interpretations as presheaf categories

With some additional hypotheses on $\mathcal{M}$, the trick of introducing the categories Δ_X becomes unnecessary. This corresponds more closely with the notation of Tamsamani [250] and some of my preprints [234, 235], and will be useful in establishing notation for iterated n-precatgories. Starting from this first discussion we show later on that if $\mathcal{M}$ itself is a presheaf category then $\mathcal{PC}(\mathcal{M})$ is a presheaf category.

This consideration will not usually enter into our argument although it does provide a convenient change of notation for Chapter 15. Nevertheless, many arguments become considerably simpler in the case of a presheaf category–as we have already seen for cell complexes in Chapter 8. The passage from $\mathcal{M}$ to $\mathcal{PC}(\mathcal{M})$ preserves the condition of being a presheaf category, and many important initial cases such as the model category of simplicial sets $\mathcal{K}$ satisfy this condition. So it should be helpful throughout the book to be able to think of the case of presheaf categories.

The first part of our discussion follows Pellissier [211].

Suppose $\mathcal{M}$ is a locally presentable category. Define a functor **disc** : SET $\to \mathcal{M}$ by

$$\mathbf{disc}(U) := \coprod_{u \in U} *.$$

If necessary we shall denote by $*_u$ the term corresponding to $u \in U$ in the coproduct. The discrete object functor has a right adjoint $\mathbf{disc}^* : \mathcal{M} \to$ SET defined by $\mathbf{disc}^*(X) := \mathrm{HOM}_{\mathcal{M}}(*, X)$, with

$$\mathrm{HOM}_{\mathcal{M}}(\mathbf{disc}(U), X) = \mathrm{HOM}_{\mathrm{SET}}(U, \mathbf{disc}^*(X)).$$

In particular, **disc** commutes with colimits.

Given an expression $X = \coprod_{u\in U} X_u$ the universal property of the coproduct applied to the maps $X_u \to *_u$ yields a map $X \to \mathbf{disc}(U)$. On the other hand, given a map $f : X \to \mathbf{disc}(U)$ we can put

$$f^{-1}(u) := X \times_{\mathbf{disc}(U)} *_u.$$

We have a natural map $\coprod_{u\in U} f^{-1}(u) \to X$ compatible with the maps to $\mathbf{disc}(U)$.

Within the category $\mathcal{PC}(\mathcal{M})$, this notation $\mathbf{disc}(U)$ coincides with that of Section 10.5.

Consider the following hypothesis on $\mathcal{M}$, saying that disjoint unions behave well. As a matter of terminology, a colimit over a discrete category is a "coproduct"; the hypotheses are intended to make sure that it is reasonable to think of this as a direct union. This kind of hypothesis was introduced for the same purpose by Pellissier [211, definition 1.1.4].

Condition 10.7.1 (DISJ)

(a) If $f : X \to \mathbf{disc}(U)$ is a map, then we get an isomorphism $\coprod_{u\in U} f^{-1}(u) \cong X$ from the natural map. If $X = \coprod_{u\in U} X_u$ is a coproduct expression then $X_u = f^{-1}(u)$ for the corresponding map $f : X \to \mathbf{disc}(U)$.

(b) The coinitial object $*$ is nonempty and indecomposable, that is to say it cannot be written as a coproduct of two nonempty objects.

(c) The category $\mathcal{M}$ has more than just a single object up to isomorphism.

To see why condition (b) is necessary, note for example that if $\mathcal{M} = \mathrm{PRESH}(\Phi)$ is the category of presheaves on a category Φ which has more than one connected component (for example the discrete category with two objects) then $*$ is decomposable.

Condition (c) is required to rule out the trivial category $\mathcal{M} = *$ which satisfies all of our other hypotheses.

Lemma 10.7.2 *Assume that $\mathcal{M}$ is locally presentable, and satisfies Condition (DCL) of the cartesian condition 7.7.1. Assume that $\mathcal{M}$ satisfies Condition 10.7.1 (DISJ) above. Then we have the following further properties:*

(1) For any object $X \in \mathcal{M}$, giving an expression $X \cong \coprod_{u\in U} X_u$ is equivalent to giving a map $X \to \mathbf{disc}(U)$.

(2) If $Y \to \coprod_{u\in U} X_u$ is a map from a single object to a coproduct, then setting $Y_u := Y \times_{\coprod_{u\in U} X_u} X_u$ the natural map $\coprod_{u\in U} Y_u \to Y$ is an isomorphism (this basically says that coproducts are universal).

(3) The map $\emptyset \to *$ is not an isomorphism.

(4) *For any set U the adjunction map* $U \to \mathbf{disc}^*\mathbf{disc}(U)$ *is an isomorphism.*

(5) *The functor* **disc** *is fully faithful, and* $\mathbf{disc}^*$ *gives an inverse on its essential image.*

(6) *If* $X \cong \coprod_{u\in U} X_u$ *and* $Y \cong \coprod_{u\in U} Y_u$ *are decompositions corresponding to maps* $X, Y \to \mathbf{disc}(U)$ *then*

$$X \times_{\mathbf{disc}(U)} Y = \coprod_{u\in U} X_u \times Y_u.$$

(7) *Coproducts are disjoint: if* $\{X_u\}_{u\in U}$ *is a collection of objects and* $u \neq v$ *then*

$$X_u \times_{\coprod_{u\in U} X_u} X_v = \emptyset.$$

(8) *The functor* **disc** *preserves finite limits.*

Proof Condition (1) is just a restatement of the first part of (DISJ).

For (2), suppose $Y \to X = \coprod_{u\in U} X_u$ is a map. Compose with the map $X \to \mathbf{disc}(U)$ given by (1), to get $Y \to \mathbf{disc}(U)$. In turn this corresponds to a decomposition $Y = \coprod_{u\in U} Y_u$ with

$$Y_u = Y \times_{\mathbf{disc}(U)} *_u,$$

but $X_u = X \times_{\mathbf{disc}(U)} *_u$ so $Y_u = Y \times_X X_u$, which is the desired statement.

For (3), if $\emptyset \to *$ were an isomorphism, then we would have $\emptyset \times X = X$ for all objects X, but in view of Lemma 7.7.2 following from (DCL) this would imply that all objects are $\emptyset$ contradicting the nontriviality hypothesis (DISJ) (c) on $\mathcal{M}$.

For (4), suppose first we are given a map $f : * \to \mathbf{disc}(U)$. By (1) this corresponds to a decomposition $* = \coprod_{u\in U} * \times_{\mathbf{disc}(U)} *_u$. By Condion (DISJ)(b), all but one of the summands must be $\emptyset$. From (3) it follows that one of the summands is different, so there is unique one of the summands which is $*$. We get that our map factors through a unique $*_u$. This shows that the adjunction map $U \to \mathbf{disc}^*\mathbf{disc}(U)$ is an isomorphism.

For (5), suppose given a map $f : \mathbf{disc}(U) \to \mathbf{disc}(V)$. By property (4) there is a unique map $U \xrightarrow{g} V$ compatible with $\mathbf{disc}^*(f)$ by the adjunction isomorphisms. Uniqueness shows that **disc** is faithful. Applying **disc** to this comptibility diagram and composing with the naturality square for the other adjunction map gives a square

$$\begin{array}{ccc} \mathbf{disc}(U) & \xrightarrow{\mathbf{disc}(g)} & \mathbf{disc}(V) \\ \downarrow & & \downarrow \\ \mathbf{disc}(U) & \xrightarrow{f} & \mathbf{disc}(V) \end{array}$$

where the vertical maps are the adjunction compositions

$$\mathbf{disc}(U) \to \mathbf{disc}(\mathbf{disc}^*(\mathbf{disc}(U))) \to \mathbf{disc}(U),$$

and the same for V. These are the identities so $f = \mathbf{disc}(g)$. This shows that **disc** is fully faithful, and the $\mathbf{disc}^*$ gives an essential inverse by (4).

In the situation of (6), given $Z \stackrel{h}{\to} \mathbf{disc}(U)$ corresponding to $Z = \coprod_{u \in U} Z_u$, a map $Z \to X$ compatible with h is the same thing as a collection of maps $Z_u \to X_u$ by (2). Similarly for a map to Y. It follows that $\coprod_{u \in U} X_u \times Y_u$ satisfies the universal property for the fiber product, giving (6).

For (7), it suffices to show that $*_u \times_{\mathbf{disc}(U)} *_v = \emptyset$ for $v \neq u$, in view of the consequence of Lemma 7.7.2 of Condition (DCL) saying that no nonempty object can map to $\emptyset$. But $*_u = \coprod_{v \in U} (*_u) \times_{\mathbf{disc}(U)} *_v$ and, as was seen in the proof of (4), condition (DISJ) (b) implies that all but one of these terms must be $\emptyset$. Since there is a diagonal map from $*$ to the term $v = u$, the terms for $v \neq u$ must be $\emptyset$.

For (8), the functor **disc** preserves finite cartesian products: given maps $Z \to \mathbf{disc}(U)$ and $Z \to \mathbf{disc}(V)$ they correspond to decompositions as in (1). The maps $Z_u \to \mathbf{disc}(V)$ correspond to decompositions

$$Z_u = \coprod_{v \in V} Z_{u,v}, \quad Z_{u,v} = Z_u \times_{\mathbf{disc}(V)} *_v.$$

Putting these together over all $u \in U$ we get a decomposition of Z corresponding to a unique map $Z \to \mathbf{disc}(U \times V)$. This shows that $\mathbf{disc}(U \times V)$ satisfies the universal property to be the product $\mathbf{disc}(U) \times \mathbf{disc}(V)$. A similar argument shows that **disc** preserves equalizers. □

Suppose now that $\mathcal{M}$ satisfies condition (DISJ) in addition to being tractable left proper and cartesian, so the properties of the preceding lemma apply. Given $\mathcal{A} \in \mathcal{PC}(\mathcal{M})$ define a functor $\Delta^o \to \mathcal{M}$ denoted $[n] \mapsto \mathcal{A}_{n/}$ by

$$\mathcal{A}_{n/} := \coprod_{(x_0,\ldots,x_n) \in \mathrm{Ob}(\mathcal{A})^{n+1}} \mathcal{A}(x_0, \ldots, x_n).$$

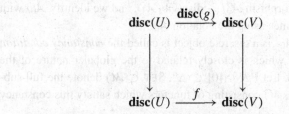

The functoriality maps are defined using those of $\mathcal{A}$. This has the property that $\mathcal{A}_{0/}$ is a discrete object, indeed the unitality conditions say that $\mathcal{A}(x_0) = *$ so there is a natural isomorphism $\mathcal{A}_{0/} \cong \mathbf{disc}(\mathrm{Ob}(\mathcal{A}))$, and we identify $\mathcal{A}_{0/}$ with the set $\mathrm{Ob}(\mathcal{A})$, sometimes using the notation $\mathcal{A}_0$.

The property that $\mathcal{A}_{0/}$ is a discrete object is called the *constancy condition* by Tamsamani [250], which is closely related to the globular nature of the theory of n-categories. Let $\mathrm{FUNC}([0] \subset \Delta^o, \mathrm{SET} \subset \mathcal{M})$ denote the full subcategory of $\mathrm{FUNC}(\Delta^o, \mathcal{M})$ consisting of functors which satisfy this constancy condition.

Suppose, on the other hand, that $n \mapsto \mathcal{A}_{n/}$ is a functor $\Delta^o \to \mathcal{M}$ which satisfies the constancy condition. Set $X := \mathrm{Ob}(\mathcal{A}) := \mathbf{disc}^*(\mathcal{A}_{0/})$, then $\mathcal{A}_{0/} = \mathbf{disc}(\mathrm{Ob}(\mathcal{A}))$. For any sequence $x_0, \ldots, x_n \in X = \mathrm{Ob}(\mathcal{A})$, we get a map

$$(x_0, \ldots, x_n) : * \to \mathcal{A}_{0/} \times \cdots \times \mathcal{A}_{0/},$$

and define

$$\mathcal{A}(x_0, \ldots, x_n) := \mathcal{A}_{n/} \times_{\mathcal{A}_{0/} \times \cdots \times \mathcal{A}_{0/}} *,$$

where the map $\mathcal{A}_{n/} \to \mathcal{A}_{0/} \times \cdots \times \mathcal{A}_{0/}$ is obtained using the $n+1$ vertices of the simplex $[n]$. Notice that

$$\mathcal{A}_{0/} \times \cdots \times \mathcal{A}_{0/} = \mathbf{disc}(\mathrm{Ob}(\mathcal{A}) \times \cdots \times \mathrm{Ob}(\mathcal{A})),$$

since **disc** preserves finite products (Lemma 10.7.2 (8)). The simplicial maps for $\mathcal{A}_{\cdot/}$ provide transition maps to make $\mathcal{A}(\cdots)$ into a functor $\Delta_X^p \to \mathcal{M}$.

Theorem 10.7.3 *Suppose $\mathcal{M}$ satisfies Condition 10.7.1 (DISJ). The above constructions provide essentially inverse functors*

$$\mathcal{PC}(\mathcal{M}) \leftrightarrow \mathrm{FUNC}([0] \subset \Delta^o, \mathrm{SET} \subset \mathcal{M}).$$

The full subcategory $\mathrm{FUNC}([0] \subset \Delta^o, \mathrm{SET} \subset \mathcal{M}) \subset \mathrm{FUNC}(\Delta^o, \mathcal{M})$ *is closed under limits and colimits, and the above essentially inverse functors preserve limits and colimits.*

Proof This follows from Lemma 10.7.2: suppose fixed the set of objects X, then giving $\mathcal{A}_{n/}$ with map to the cartesian product

$$\mathcal{A}_{n/} \to \mathbf{disc}(X \times \cdots X) = \mathbf{disc}(X) \times \cdots \times \mathbf{disc}(X)$$

is the same as giving the pieces of the decomposition

$$\mathcal{A}_{n/} = \coprod_{(x_0, \ldots, x_n)} \mathcal{A}(x_0, \ldots, x_n).$$

□

For iteration of our basic construction, the following lemmas show that the above notational reinterpretation is very reasonable:

Lemma 10.7.4 *Suppose $\mathcal{M}$ satisfies the hypothesis (DCL) of Definition 7.7.1. Then the category $\mathcal{PC}(\mathcal{M})$ satisfies condition (DISJ).*

Proof The discrete precategories are constructed as follows. The coinitial object $* \in \mathcal{PC}(\mathcal{M})$ has a single element $\mathrm{Ob}(*) = * = \{x\}$ and $*(x, \ldots, x) = * \in \mathcal{M}$. Thus, if U is any set then $\mathbf{disc}(U)$ calculated in $\mathcal{PC}(\mathcal{M})$ has object set $\mathbf{disc}(U) \cong U$, and $\mathbf{disc}(U)(x_0, \ldots, x_n) = *$ if $x_0 = \cdots = x_n$, otherwise $\mathbf{disc}(U)(x_0, \ldots, x_n) = \emptyset$ if the sequence is not constant.

For part (a), suppose $\mathcal{A} \in \mathcal{PC}(\mathcal{M})$ and $\mathcal{A} \to \mathbf{disc}(U)$ is a map. Put $X := \mathrm{Ob}(\mathcal{A})$, with $X \to U$ which corresponds to a decomposition $X = \coprod_{u \in U} X_u$. Let $\mathcal{A}_u \subset \mathcal{A}$ be the pullback of $\mathcal{A}$ along $X_u \to X$, that is it is the "full subprecategory" with object set $X_u \subset X$. This is indeed $\mathcal{A} \times_{\mathbf{disc}(U)} *_u$. The map

$$\coprod_{u \in U} \mathcal{A}_u \to \mathcal{A}$$

is an isomorphism. This follows from the fact that $\mathbf{disc}(U)(x_0, \ldots, x_n) = \emptyset$ for any non-constant sequence, plus the consequence Lemma 7.7.2 of (DCL). On the other hand, given a decomposition in coproduct $\mathcal{A} = \coprod_{u \in U} \mathcal{A}_u$ we get a map $\mathcal{A} \to \mathbf{disc}(U)$ for which the components $\mathcal{A}_u$ are the fibers.

Conditions (b) and (c) are easy. □

Lemma 10.7.5 *Suppose Ψ is a connected category (i.e. its nerve is a connected space, or equivalently any two objects are joined by a zig-zag of arrows). Then the category* PRESH(Ψ) = FUNC(Ψ^o, SET) *satisfies condition (DISJ), with the discrete objects being those equivalent to constant functors. In particular, the model category* SET *where all morphisms are fibrations and cofibrations, and the weak equivalences are isomorphisms, satisfies (DISJ). And the model category $\mathcal{K}$ of simplicial sets satisfies (DISJ).*

Proof If U is a set, the discrete $\mathbf{disc}(U)$ is the constant presheaf $x \mapsto U$ for all $x \in \Psi$. For (a), a map $\mathcal{A} \to \mathbf{disc}(U)$ is the same thing as a decomposition $\mathcal{A} = \coprod_{u \in U} \mathcal{A}_u$, as can be seen levelwise. Condition (b) follows from the connectedness of Ψ (indeed it is equivalent), and Condition (c) is easy. □

We now turn to the further situation where $\mathcal{M}$ is a presheaf category, say $\mathcal{M} = \text{PRESH}(\Psi) = \text{FUNC}(\Psi^o, \text{SET})$. In this case

$$\text{FUNC}\left(\Delta_X^o, \mathcal{M}\right) = \text{FUNC}\left(\Delta_X^o \times \Psi^o, \text{SET}\right) = \text{PRESH}(\Delta_X \times \Psi)$$

is again a presheaf category.

We construct a new category denoted $\mathbf{C}(\Psi)$, which looks somewhat like a "cone" on Φ. It is defined by contracting $\{[0]\} \times \Psi \subset \Delta \times \Psi$ to a single object denoted 0. Thus, the objects of $\mathbf{C}(\Psi)$ are of the form $([n], \psi)$ for $n \geq 1$ and $\psi \in \Psi$, or the object 0. The morphisms are as follows:

- the identity is the only endomorphism of 0;
- for $n \geq 1$ and $\psi \in \Psi$ there is a unique morphisms from any $([n], \psi)$ to 0;
- the morphisms from 0 to $([n], \psi)$ are the same as the morphisms $[0] \to [n]$ in Δ (i.e. there are $n + 1$ of them); and
- the morphisms from $([n], \psi)$ to $([n'], \psi')$ are of two kinds: either (a, f) where $a : [n] \to [n']$ is a morphism in Δ such that a doesn't factor through $[0]$ and $f : \psi \to \psi'$ is a morphism in Ψ; or else (a) where $a : [n] \to [n']$ is a morphism in Δ which factors as $[n] \to [0] \to [n']$.

The composition of morphisms is defined in an obvious way. Notice that the composition of anything with a morphism which factors through $[0]$ will again factor through $[0]$, which allows us to define compositions of the form $(a)\circ(a', f')$ or $(a, f)\circ(a')$ in the last case. The division of the morphisms from $([n], \psi)$ to $([n'], \psi')$ into two cases is necessary in order to define composition.

Proposition 10.7.6 *Suppose $\mathcal{M} = \text{PRESH}(\Psi)$ is a presheaf category. There is a natural isomorphism between $\text{PRESH}(\mathbf{C}(\Psi))$ and the category of unital $\mathcal{M}$-precategories $\mathcal{PC}(\mathcal{M})$. Thus, $\mathcal{PC}(\mathcal{M})$ is again a presheaf category.*

Proof Suppose $\mathcal{F} : \mathbf{C}(\Psi)^o \to \text{SET}$ is a presheaf. Let $X := \mathcal{F}(0)$. For $n \geq 1$ and any $(x_0, \ldots, x_n) \in \Delta_X$, let $\mathcal{A}(x_0, \ldots, x_n)$ be the presheaf on Ψ which assigns to $\psi \in \Psi$ the subset of elements of $\mathcal{F}([n], \psi)$ which project to $(x_0, \ldots, x_n)$ under the $n+1$ projection maps $\mathcal{F}([n], \psi) \to \mathcal{F}(\iota) = X$ corresponding to the $n+1$ maps $0 \to ([n], \psi)$. At $n = 0$, set $\mathcal{A}(x_0) := \{x_0\}$. The pair $(X, \mathcal{A})$ is an element of $\mathcal{PC}(\mathcal{M})$. Conversely, given any $(X, \mathcal{A}) \in \mathcal{PC}(\mathcal{M})$, define a presheaf $\mathcal{F} : \mathbf{C}(\Psi)^o \to \text{SET}$ by setting $\mathcal{F}(0) := X$ and

$$\mathcal{F}([n], \psi) := \coprod_{(x_0,\ldots,x_n)\in X^{n+1}} \mathcal{A}(x_0, \ldots, x_n)(\psi).$$

These constructions are inverses. □

Let $\mathbf{c}_\Psi : \Delta \times \Psi \to \mathbf{C}(\Psi)$ denote the projection. If Ψ is connected, then the category of presheaves $\text{PRESH}(\mathbf{C}(\Psi))$ may be identified, via $\mathbf{c}^*_\Psi$, with the full subcategory of $\text{PRESH}(\Delta \times \Psi)$ consisting of presheaves $\mathcal{A}$ which satisfy *Tamsamani's constancy condition* that $\mathcal{A}(0, \psi)$ is a constant set independent of $\psi \in \Psi$.

The construction $\Psi \mapsto \mathbf{C}(\Psi)$ was the iterative step in the construction of the sequence of categories denoted Θ^n in my preprint [234]. In that notation, $\Theta^0 = *$ and $\Theta^{n+1} = \mathbf{C}(\Theta^n)$. However, the notation Θ^n was subsequently used by Joyal [158] to denote a related but different sequence of categories; in order to avoid confusion we will use the cone notation $\mathbf{C}$.

For the theory of non-unital $\mathcal{M}$-precategories, one can also construct a category denoted $\mathbf{C}^+(\Psi)$ with the property that $\mathrm{PRESH}(\mathbf{C}^+(\Psi))$ is the category of non-unital $\mathcal{M}$-precategories. This is in fact even more straightforward than the construction of $\mathbf{C}$ for the unital theory. The following discussion is optional, but serves to put the previous discussion of $\mathbf{C}$ into a better perspective.

Recall that $\mathcal{PC}(\mathcal{M})$ was defined as a fibered category over SET whose fiber over a set X was $\mathcal{PC}(X, \mathcal{M})$. In the same way, the *category of non-unital $\mathcal{M}$-precategories* is the fibered category over SET whose fiber over X is $\mathrm{FUNC}\left(\Delta_X^o, \mathcal{M}\right)$.

The construction of $\mathbf{C}^+(\Psi)$ is to formally add an object denoted ι to $\Delta \times \Psi$. The objects of $\mathbf{C}^+(\Psi)$ are of the form either ι or $([n], \psi)$ for $n \in \Delta$ and $\psi \in \Psi$. The morphisms of $\mathbf{C}^+(\Psi)$ are defined as follows:

- there is a single identity morphism between ι and itself;
- there are no morphisms from $([n], \psi)$ to ι;
- the morphisms from $([n], \psi)$ to $([n'], \psi')$ are the same as the morphisms in $\Delta \times \Psi$; and
- the morphisms from ι to $([n], \psi)$ are the same as the morphisms $[0] \to [n]$ in Δ (i.e. there are $n+1$ of them).

Composition of morphisms is defined so that the single automorphism of ι is the identity, so that composition of morphisms within $\Delta \times \Psi$ is the same as from that category, and a composition of the form

$$\iota \to ([n], \psi) \to ([n'], \psi')$$

is given by the corresponding composition $[0] \to [n] \to [n']$.

Proposition 10.7.7 *Suppose $\mathcal{M} = \mathrm{PRESH}(\Psi)$ is a presheaf category. There is a natural isomorphism between $\mathrm{PRESH}(\mathbf{C}^+(\Psi))$ and the category of non-unital $\mathcal{M}$-precategories. In particular, the latter is a presheaf category.*

Proof Suppose $\mathcal{F} : \mathbf{C}^+(\Psi)^o \to \mathrm{SET}$ is a presheaf. Let $X := \mathcal{F}(\iota)$. For any $(x_0, \ldots, x_n) \in \Delta_X$, let $\mathcal{A}(x_0, \ldots, x_n)$ be the presheaf on Ψ which assigns to $\psi \in \Psi$ the subset of elements of $\mathcal{F}([n], \psi)$ which project to $(x_0, \ldots, x_n)$ under the $n+1$ projection maps $\mathcal{F}([n], \psi) \to \mathcal{F}(\iota) = X$ corresponding to the $n+1$ maps $\iota \to ([n], \psi)$. Then pair $\mathcal{A}$ is an element of $\mathrm{FUNC}\left(\Delta_X^o, \mathcal{M}\right)$. Conversely, given any $\mathcal{A} \in \text{FUNC}\left(\Delta_X^o, \mathcal{M}\right)$, define a presheaf $\mathcal{F} : \mathbf{C}^+(\Psi)^o \to \text{SET}$ by setting $\mathcal{F}(\iota) := X$ and

$$\mathcal{F}([n], \psi) := \coprod_{(x_0,\ldots,x_n)\in X^{n+1}} \mathcal{A}(x_0, \ldots, x_n)(\psi).$$

These constructions are inverses. □

The inclusion functor U^* from $\mathcal{PC}(\mathcal{M})$ to the category of non-unital precategories, can be viewed as a pullback. Indeed, there is a projection functor

$$\pi : \mathbf{C}^+(\Psi) \to \mathbf{C}(\Psi)$$

defined by $\pi([n], \psi) = ([n], \psi)$ if $n \geq 1$, $\pi([0], \psi) = 0$ and $\pi(\iota) = 0$. The pullback U^* is just the pullback functor

$$\pi^* : \mathcal{PC}(\mathcal{M}) \cong \text{PRESH}(\mathbf{C}(\Psi)) \to \text{PRESH}(\mathbf{C}^+(\Psi)).$$

The left adjoint of the pull back or inclusion, denoted pushforward $U_!$ above, is also the pushforward $\pi_!$ for the functor π.

Using the operations $\mathbf{C}^+(\Psi)$ and $\mathbf{C}(\Psi)$ we may stay essentially entirely within the realm of presheaf categories. The only place where we go outside of there is when we speak of the category of unital $\mathcal{M}$-precategories over a fixed set of objects X; even if $\mathcal{M}$ is a presheaf category, $\mathcal{PC}(X, \mathcal{M})$ will not generally be a presheaf category. On the other hand, the non-unital version $\text{FUNC}\left(\Delta_X^o, \mathcal{M}\right) = \text{PRESH}(\Delta_X \times \Psi)$ remains a presheaf category.

11
Algebraic theories in model categories

In this chapter we consider algebraic diagram theories consisting of a collection of finite product conditions imposed on diagrams $\Phi \to \mathcal{M}$. This is motivated by the situation considered in the previous Chapter 10. There we defined the notion of precategory on a fixed set of objects, which is a diagram $\mathcal{A} : \Delta^o_X \to \mathcal{M}$. The Segal conditions require that certain maps be weak equivalences. Imposing these conditions amounts to a homotopical analogue of the "finite product theories" often considered in category theory by Lawvere [181, 182] for example, see also further historical remarks by Adámek and Rosický [2, pp 171–172]. The homotopical analogue, whose origins go back to the various theories of H-spaces and loop spaces, was considered by Badzioch [7], and his treatment was used by Bergner for Segal categories [40, 45]. She also [43] generalized this to models in simplicial categories. Rosicky [222] carried these ideas further, and he points out more references too. Cranch [84] has investigated algebraic theories in $(\infty, 1)$-categories.

Let $\epsilon(n)$ denote the category

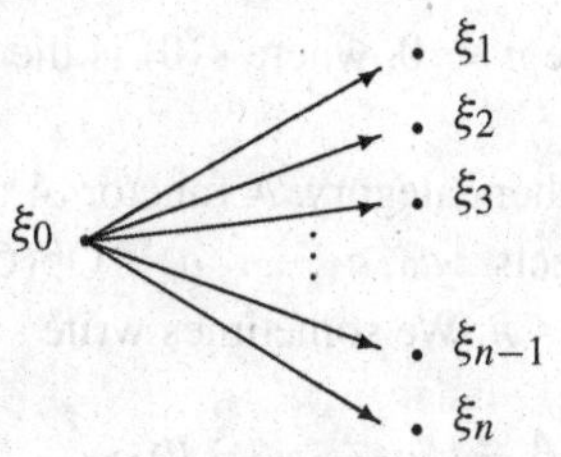

The Segal maps are obtained by pulling back $\mathcal{A}$ along various functors $\epsilon(n) \to \Delta^0_X$. This sets up a localization problem which can be phrased in more general terms. We treat the general situation in the present chapter. Even ignoring any possible other applications, this simplifies notations for the general aspects of the problem of enforcing the given collection of finite product conditions.

By an *algebraic theory* we mean a category Φ provided with a collection of "cartesian product diagrams," that is diagrams with the shape of a cartesian product, which are functors $\epsilon(n) \xrightarrow{P} \Phi$. A *realization* of the theory in a classical 1-category $\mathcal{C}$ is a functor $\Phi \to \mathcal{C}$ which sends these diagrams to cartesian products in $\mathcal{C}$ (the perhaps more standard terminology "model" would introduce some confusion in the present context). Many of the easiest kinds of structures can be written this way, although it is well understood that to get more complicated structures one needs to go to the notion of *sketch*, which is a category provided with more generally shaped limit diagrams. For our purposes, it will suffice to consider cartesian product diagrams.

Suppose $\mathcal{M}$ is an appropriate kind of model category. Then a homotopy realization of the theory in $\mathcal{M}$ is a functor $\Phi \to \mathcal{M}$ which sends the cartesian product diagrams, to homotopy cartesian products in $\mathcal{M}$.

These notions lead to a "calculus of generators and relations" where we start with an arbitrary functor $\Phi \to \mathcal{C}$ (resp. $\Phi \to \mathcal{M}$) and try to enforce the cartesian product (resp. homotopy cartesian product) condition. The main work of this chapter will be to do this for the case of homotopy realizations in a model category $\mathcal{M}$.

11.1 Diagrams over the categories $\epsilon(n)$

The first task is to take a close look at the categories indexing cartesian products. Let $\epsilon(n)$ denote the category with objects $\xi_0, \xi_1, \ldots, \xi_n$; and whose only morphisms, apart from the identities, are single morphisms $\rho_i : \xi_0 \to \xi_i$. We also let ξ_i denote the functors $\xi_i : \{*\} \to \epsilon(n)$ sending the point $*$ to the object $\xi_i(*) = \xi_i$.

This includes the case $n = 0$, where $\epsilon(0)$ is the discrete category with one object ξ_0.

Suppose $\mathcal{C}$ is some other category. A functor $\mathcal{A} : \epsilon(n) \to \mathcal{C}$ corresponds to a collection of objects $a_0, a_1, \ldots, a_n \in \mathrm{Ob}(\mathcal{C})$ together with maps $p_i : a_0 \to a_i$ for $1 \le i \le n$. We sometimes write

$$\mathcal{A} = (a_0, \ldots, a_n; p_1, \ldots, p_n).$$

Say that $\mathcal{A}$ is a *cartesian product diagram* if the collection of maps p_i expresses a_0 as a cartesian product of the $a_1, \ldots, a_n$ in $\mathcal{C}$.

Suppose $\mathcal{M}$ is a model category, in particular cartesian products exist. We say that a diagram $\mathcal{A} = (a_0, \ldots, a_n; p_1, \ldots, p_n)$ from $\epsilon(n)$ to $\mathcal{M}$ is a *homotopy cartesian product diagram* if the morphism

$$(p_1, \dots, p_n) : a_0 \to a_1 \times \cdots \times a_n$$

is a weak equivalence.

Consider now the diagram category FUNC($\epsilon(n)$, $\mathcal{M}$). We describe explicitly its injective and projective model structures, which are well-known to exist (see Hirschhorn [144], Barwick [18]).

A morphism of diagrams

$$\mathcal{A} = (a_0, \dots, a_n; p_1, \dots, p_n) \to \mathcal{B} = (b_0, \dots, b_n; q_1, \dots, q_n)$$

is a collection of maps $g = (g_0, \dots, g_n)$ with $g_i : a_i \to b_i$ such that $q_i g_0 = g_i p_i$. We can also think of a morphism g as consisting of the maps $(g_1, \dots, g_n)$ plus a map

$$g_P : a_0 \to b_0 \times_{b_1 \times \cdots \times b_n} (a_1 \times \cdots \times a_n),$$

such that the second projection is the structural map for $\mathcal{A}$. We then write $g = \langle g_1, \dots, g_n; g_P \rangle$.

A morphism g is a fibration in the projective structure, if and only if each $g_0, \dots, g_n$ is a fibration in $\mathcal{M}$. Similarly, g is a cofibration in the injective structure, if and only if each $g_0, \dots, g_n$ is a cofibration in $\mathcal{M}$.

To study fibrations in the injective structure, suppose

$$\mathcal{C} = (c_0, \dots, c_n; r_1, \dots, r_n), \quad \mathcal{D} = (d_0, \dots, d_n; s_1, \dots, s_n)$$

are two diagrams, with morphisms forming a square

$$\begin{array}{ccc} \mathcal{A} & \xrightarrow{u} & \mathcal{C} \\ \downarrow g & & \downarrow h \\ \mathcal{B} & \xrightarrow{v} & \mathcal{D}. \end{array}$$

These give diagrams

$$\begin{array}{ccc} a_i & \xrightarrow{u_i} & c_i \\ \downarrow g_i & & \downarrow h_i \\ b_i & \xrightarrow{v_i} & d_i. \end{array}$$

We look for a lifting $f : \mathcal{B} \to \mathcal{C}$ such that $fg = u$ and $hf = v$. This amounts to asking for liftings $f_i : b_i \to c_i$ such that $f_i g_i = u_i$ and $h_i f_i = v_i$, and also such that $s_i f_0 = f_i r_i$.

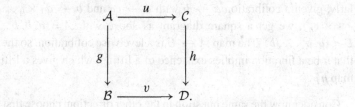

Suppose given already the liftings $f_1, \ldots, f_n$. Then to specify a full lifting f as above we need to find $f_0 : b_0 \to c_0$ such that the diagram

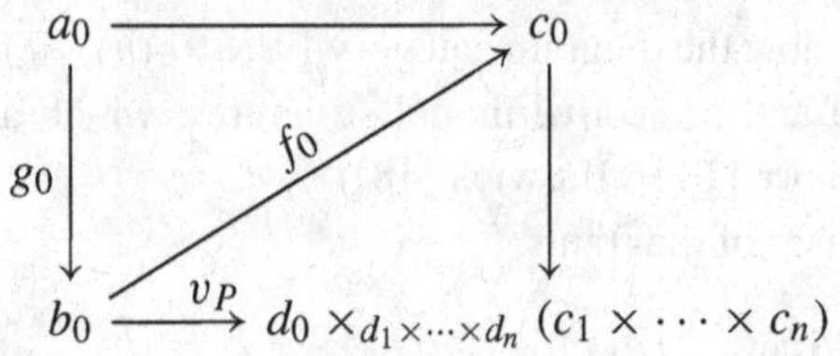

commutes.

Lemma 11.1.1 *A morphism $h : \mathcal{C} \to \mathcal{D}$, with notations as above, is a fibration in the injective model structure on* $\text{FUNC}(\epsilon(n), \mathcal{M})$ *if and only if each $h_1, \ldots, h_n$ is a fibration in $\mathcal{M}$, and the map*

$$h_P : c_0 \to d_0 \times_{d_1 \times \cdots \times d_n} (c_1 \times \cdots \times c_n)$$

is a fibration in $\mathcal{M}$.

Proof If h satisfies the conditions stated in the lemma, and f is a levelwise cofibration, i.e. each f_i is a cofibration, then we can first choose the liftings f_i for $i = 1, \ldots, n$, then choose f_0 lifting h_P. Therefore h is a fibration. Suppose, on the other hand, that h is a fibration. For any cofibration $a \to b$ and any $i = 1, \ldots, n$ we have an injective cofibration between objects with a (resp. b) placed at i and the remaining places filled in with $\emptyset$. Since h satisfies lifting along any such cofibration, it follows that h_i is a fibration in $\mathcal{M}$. Similarly, given a cofibration $a \to b$ with $a \to c_0$ and $b \to d_0 \times_{d_1 \times \cdots \times d_n} (c_1 \times \cdots \times c_n)$, we get a square diagram as above with $\mathcal{A} = (a, b, b \ldots, b)$ and $\mathcal{B} = (b, b, \ldots, b)$. The map $\mathcal{A} \to \mathcal{B}$ is a levelwise cofibration, so the condition that h be a fibration implies existence of a lifting, which gives a lifting for the map h_P. □

Consider now the same question in the other direction: choose first the lifting f_0. For any $i \geq 1$ we get a map $b_0 \sqcup^{a_0} a_i \to c_i$, and the choices of f_i for $i = 1, \ldots, n$ correspond to choices of lifting in the diagrams

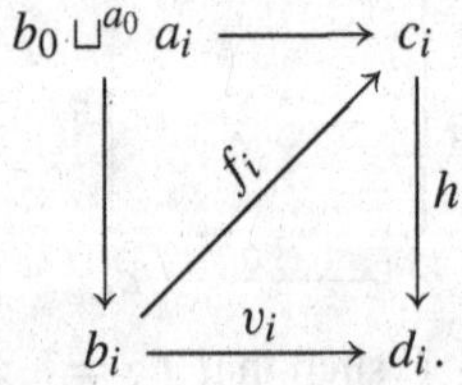

Lemma 11.1.2 *A morphism $g : \mathcal{A} \to \mathcal{B}$, with notations as above, is a cofibration in the projective model structure on* $\text{FUNC}(\epsilon(n), \mathcal{M})$ *if and only if g_0*

is a cofibration in $\mathcal{M}$, and for each $i = 1, \ldots, n$ the map $b_0 \sqcup^{a_0} a_i \to b_i$ is a cofibration in $\mathcal{M}$. In particular an object $\mathcal{B}$ is cofibrant if and only if b_0 is a cofibration and each $b_0 \to b_i$ is a cofibration.

Proof Similar to the previous proof. □

One can describe explicit generating sets for the cofibrations and trivial cofibrations in both the injective structure $\mathrm{FUNC}_{\mathrm{inj}}(\epsilon(n), \mathcal{M})$ and the projective structure $\mathrm{FUNC}_{\mathrm{proj}}(\epsilon(n), \mathcal{M})$.

Recall the standard adjunctions for the functors $\xi_i : * \to \mathcal{M}$. If $X \in \mathcal{M}$ is an object, we obtain a diagram $\xi_{i,!}(X) : \epsilon(n) \to \mathcal{M}$. Explicitly, if $i = 0$ then $\xi_{0,!}(X)$ is the constant diagram $(X, X, \ldots, X, 1_X, \ldots, 1_X)$ with values X.

If $i \geq 1$ then $\xi_{i,!}(X)$ is the diagram $(\emptyset, \ldots, \emptyset, X, \emptyset \ldots; \iota, \ldots, \iota)$, where ι denote the unique maps from $\emptyset$ to anything else, and here X is at the i-th place. The adjunction says that, for any diagram $\mathcal{A} = (a_0, \ldots, a_n; p_1, \ldots, p_n)$, a morphism $\xi_{i,!}(X) \to \mathcal{A}$ is the same as a morphism $X \to \xi_i^*(\mathcal{A}) = a_i$.

Suppose given generating sets I for the cofibrations of $\mathcal{M}$ and J for the trivial cofibrations.

For $f : X \to Y$ in I, we obtain a cofibration $\xi_{0,!}(f) : \xi_{0,!}(X) \to \xi_{0,!}(Y)$ in the projective model structure. To see this, use Lemma 11.1.2 on $g = \xi_{0,!}(f)$ and note that g_0 is just f so it is a cofibration; and $b_0 \sqcup^{a_0} a_i = Y \sqcup^X X = Y$ maps to $b_i = Y$ by a cofibration. If $f \in J$ then $\xi_{0,!}(f)$ is a trivial cofibration: over each object of $\epsilon(n)$ we just get back the map f so it is an levelwise weak equivalence.

For $f \in I$ as above, at any $i \geq 1$, $\xi_{i,!}(f)$ is a cofibration, again using Lemma 11.1.2 on $g = \xi_{0,!}(f)$. The map g_0 is the identity of $\emptyset$, and the maps $b_0 \sqcup^{a_0} a_j \to b_j$ are either the identity of $\emptyset$ for $j \neq i$, or f when $j = i$, so these are cofibrations. Again, if $f \in J$ then $\xi_{i,!}(f)$ is a trivial cofibration: over each object of $\epsilon(n)$ it gives either the identity of $\emptyset$ which is automatically a weak equivalence, or else f.

Define $I_{\epsilon(n)}$ (resp. $J_{\epsilon(n)}$) to be the set consisting of diagrams of the form $\xi_{i,!}(f)$ for $0 \leq i \leq n$ and $f \in I$ (resp. $f \in J$).

Proposition 11.1.3 *The sets $I_{\epsilon(n)}$ and $J_{\epsilon(n)}$ are generators for the projective model category structure $\mathrm{FUNC}_{\mathrm{proj}}(\epsilon(n), \mathcal{M})$.*

Proof By the adjunction, a morphism g in $\mathrm{FUNC}(\epsilon(n), \mathcal{M})$ satisfies the right lifting property with respect to $I_{\epsilon(n)}$ (resp. $J_{\epsilon(n)}$) if and only if $\xi_i^*(g)$ satisfies the right lifting property with respect to I (resp. J) for all $0 \leq i \leq n$. Since I (resp. J) is a set of generators for the cofibrations (resp. trivial cofibrations) of $\mathcal{M}$, this lifting property is equivalent to saying that each $\xi_i^*(g)$ is a trivial fibration (resp. a fibration). By the definition of the projective model structure, this is equivalent to saying that g is a trivial fibration (resp. a fibration). Hence $\mathbf{inj}(I_{\epsilon(n)})$ is the class of trivial fibrations and $\mathbf{inj}(J_{\epsilon(n)})$ is the class of fibrations, so $\mathbf{cof}(I_{\epsilon(n)})$ is the class of cofibrations and $\mathbf{cof}(J_{\epsilon(n)})$ is the class of trivial cofibrations. □

To get generators for the injective model structure, we need to add a new kind of injective cofibration. If $f : X \to Y$ is a cofibration, consider the diagrams

$$\varpi(f) := \xi_{0,!}(X) \cup^{\coprod_i \xi_{i,!}(X)} \coprod_i \xi_{i,!}(Y) = (X, Y, \ldots, Y; f, \ldots, f) \quad (11.1.1)$$

and $\xi_{0,!}(Y)$. We have a map $\varpi(f) \to \xi_{0,!}(Y)$ which is f at the object $\xi_0(*)$ and 1_Y at the other objects. Denote this map by $\rho(f) = (f, 1, \ldots, 1)$.

Proposition 11.1.4 *Let $I^+_{\epsilon(n)}$ denote the union of $I_{\epsilon(n)}$ with the set of maps of the form $\rho(f) = (f, 1, \ldots, 1)$ for $f \in I$. Let $J^+_{\epsilon(n)}$ denote the union of $J_{\epsilon(n)}$ with the set of maps of the form $\rho(f)$ for $f \in J$. Then $I^+_{\epsilon(n)}$ and $J^+_{\epsilon(n)}$ are generating sets for the injective model category structure* $\mathrm{FUNC}_{\mathrm{inj}}(\epsilon(n), \mathcal{M})$.

Proof If $f : X \to Y$ is a cofibration, then $(f, 1, \ldots, 1) : \mathcal{U} \to \mathcal{V}$ is a cofibration with the notations as above. Suppose that $g : \mathcal{A} \to \mathcal{B}$ satisfies right lifting with respect to f, where $g = (g_0, \ldots, g_n)$ goes from $\mathcal{A} = (a_0, \ldots, a_n; p_1, \ldots, p_n)$ to $\mathcal{B} = (b_0, \ldots, b_n; q_1, \ldots, q_n)$. The lifting property says that for any map $u_0 : X \to a_0$ and maps $u_i : Y \to a_i$ for $i \geq 1$, and maps $v_i : Y \to b_i$ for $i \geq 0$ such that $p_i u_0 = u_i f$, $v_i = g_i u_i$ for $i \geq 1$, and $v_0 f = g_0 u_0$, then there should exist a map $u'_0 : Y \to a_0$ such that $u'_0 f = u_0$, $g_0 u'_0 = v_0$, and $p_i u'_0 = u_i$. This is the same as the right lifting property for the square

$$\begin{array}{ccc} X & \longrightarrow & a_0 \\ \downarrow & & \downarrow \\ Y & \longrightarrow & b_0 \times_{b_1 \times \cdots \times b_n} (a_1 \times \cdots \times a_n) \end{array}$$

so the condition that g satisfy right lifting for any $(f, 1, \ldots, 1)$ for $f \in J$ is equivalent to the condition that $a_0 \to b_0 \times_{b_1 \times \cdots \times b_n} (a_1 \times \cdots \times a_n)$ is a fibration. Thus, $\mathbf{inj}(J^+_{\epsilon(n)})$ consists of maps which are levelwise fibrations (because of lifting with respect to $J_{\epsilon(n)}$) and such that the map of Lemma 11.1.1 is a fibration. By Lemma 11.1.1, $\mathbf{inj}(J^+_{\epsilon(n)})$ is the class of fibrations.

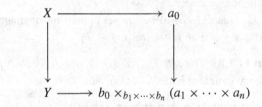

Similarly, the fact that g satisfies right lifting with respect to any $(f, 1, \ldots, 1)$ for $f \in I$ is equivalent to the conditoin that

$$a_0 \to b_0 \times_{b_1 \times \cdots \times b_n} (a_1 \times \cdots \times a_n)$$

be a trivial fibration. Thus, $\mathbf{inj}(I^+_{\epsilon(n)})$ consists of maps which are levelwise trivial fibrations (because of lifting with respect to $I_{\epsilon(n)}$) and such that the map of Lemma 11.1.1 is a trivial fibration.

We claim that this is equal to the class of fibrations. If $g \in \mathbf{inj}(I^+_{\epsilon(n)})$, then it is a fibration for the injective structure since $J^+_{\epsilon(n)} \subset I^+_{\epsilon(n)}$, and also levelwise a weak equivalence, so it is a trivial fibration. If g is a trivial fibration, then it is levelwise a trivial fibration, and the map $a_0 \to b_0 \times_{b_1 \times \cdots \times b_n} (a_1 \times \cdots \times a_n)$ is a fibration. However, the maps $a_i \to b_i$ are trivial fibrations, so $a_1 \times \cdots \times a_n \to b_1 \times \cdots \times b_n$ is a trivial fibration. Thus, the map $b_0 \times_{b_1 \times \cdots \times b_n} (a_1 \times \cdots \times a_n) \to b_0$ is a trivial fibration, and by 3 for 2 we conclude that $a_0 \to b_0 \times_{b_1 \times \cdots \times b_n} (a_1 \times \cdots \times a_n)$ is a weak equivalence. Hence it is a trivial fibration as required for the claim.

This identifies $\mathbf{inj}(I^+_{\epsilon(n)})$ and $\mathbf{inj}(J^+_{\epsilon(n)})$ with the classes of trivial fibrations and fibrations respectively, so $\mathbf{cof}(I^+_{\epsilon(n)})$ and $\mathbf{cof}(J^+_{\epsilon(n)})$ are the classes of cofibrations and trivial cofibrations respectively. □

Scholium 11.1.5 *If $(\mathcal{M}, I, J)$ is a cofibrantly generated (resp. combinatorial, tractable, left proper) model category, then* $\mathrm{FUNC}_{\mathrm{inj}}(\epsilon(n), \mathcal{M})$ *and* $\mathrm{FUNC}_{\mathrm{proj}}(\epsilon(n), \mathcal{M})$ *are cofibrantly generated (resp. combinatorial, tractable, left proper) model categories.*

See Theorem 7.6.3. Tractability is seen by inspection of the explicit generators.

11.2 Imposing the product condition

Assume that $\mathcal{M}$ is a tractable left proper cartesian model category. Recall that the cartesian condition (7.7.1) implies that for any $X \in \mathcal{M}$ and weak equivalence $f : Y \to Z$ the induced map $X \times Y \to X \times Z$ is a weak equivalence.

We are going to apply the direct left Bousfield localization theory of Chapter 9. Say that an object $\mathcal{A} \in \mathrm{FUNC}(\epsilon(n), \mathcal{M})$ is *product-compatible* if the map $\mathcal{A}(\xi_0) \to \mathcal{A}(\xi_1) \times \ldots \times \mathcal{A}(\xi_n)$ is a weak equivalence. Let $\mathcal{R}_{\epsilon(n)} \subset \mathrm{FUNC}(\epsilon(n), \mathcal{M})$ denote the full subcategory of product-compatible objects.

Lemma 11.2.1 *Suppose $\mathcal{M}$ is cartesian. The subcategory $\mathcal{R}_{\epsilon(n)}$ is invariant under weak equivalence: if $\mathcal{A}$ is product-compatible and its image in*

$\text{hoFUNC}(\epsilon(n), \mathcal{M})$ *is isomorphic to the image of another object* $\mathcal{B}$*, then* $\mathcal{B}$ *is also product-compatible.*

Proof If $\mathcal{A} \to \mathcal{B}$ is a levelwise weak equivalence, then the horizontal arrows in the diagram

$$\begin{array}{ccc} \mathcal{A}(\xi_0) & \longrightarrow & \mathcal{B}(\xi_0) \\ \downarrow & & \downarrow \\ \mathcal{A}(\xi_1) \times \cdots \times \mathcal{A}(\xi_n) & \to & \mathcal{B}(\xi_1) \times \cdots \times \mathcal{B}(\xi_n) \end{array}$$

are weak equivalences, using the cartesian condition for the bottom arrow. Hence, one of the vertical arrows is a weak equivalence if and only if the other one is. □

11.2.1 Direct localization of the projective structure

We now define the direct localizing system which goes with the full subcategory $\mathcal{R}_{\epsilon(n)}$ of product-compatible objects in the projective model category structure. For each generating cofibration $f : X \to Y$ in I, recall the morphism $\rho(f) = (f, 1, \ldots, 1)$ defined above (see (11.1.1)):

$$\varpi(f) = (X, Y, \ldots, Y; f, \ldots, f) \xrightarrow{\rho(f)} \xi_{0,!}(Y) = (Y, \ldots, Y; 1, \ldots, 1).$$

It is obviously an injective cofibration, and the domain $\varpi(f)$ is projectively cofibrant (apply Lemma 11.1.2). However, $\rho(f)$ will not in general be a projective cofibration. Choose a factorization

$$Y \cup^X Y \xrightarrow{a(f)} Z \xrightarrow{b(f)} Y,$$

such that $a(f)$ is a cofibration and $b(f)$ is a trivial fibration. Denoting by i_2 the first inclusion $Y \to Y \cup^X Y$ we get a trivial cofibration $a(f)i_2 : Y \to Z$. Let

$$\varpi(f) = (X, Y \cdots Y; f \cdots f) \xrightarrow{\zeta(f)} \psi(f) := (Y, Z \cdots Z; a(f)i_2 \cdots a(f)i_2)$$

be the map given by f over ξ_0 and $a(f)i_1$ over $\xi_1, \ldots, \xi_n$. Then the map occuring in Lemma 11.1.2 is exactly $a(f)$, so $\zeta(f)$ is a projective cofibration.

Recall the explicit generating set $J_{\epsilon(n)}$ for the trivial cofibrations in the projective model structure. Put

$$K^{\text{proj}}_{\epsilon(n)} := J_{\epsilon(n)} \cup \{\zeta(f)\}_{f \in I}.$$

Theorem 11.2.2 *The pair* $(\mathcal{R}_{\epsilon(n)}, K^{\text{proj}}_{\epsilon(n)})$ *is a direct localizing system for the model category* $\text{FUNC}_{\text{proj}}(\epsilon(n), \mathcal{M})$ *of* $\epsilon(n)$*-diagrams in* $\mathcal{M}$ *with the projective model structure. Let* $\text{FUNC}_{\text{proj},\Pi}(\epsilon(n), \mathcal{M})$ *denote the left Bousfield localized model structure constructed in Chapter 9. A morphism* $\mathcal{A} \to \mathcal{B}$ *is a weak equivalence if and only if it induces weak equivalences levelwise over the objects* $\xi_1, \ldots, \xi_n$. *An object* $\mathcal{A}$ *is fibrant in the localized structure if and only if it is product-compatible and each* $\mathcal{A}(\xi_i)$ *is fibrant.*

Proof We verify the properties (A1) – (A6). Properties (A1) and (A2) are immediate. For (A3) we have chosen $\zeta(f)$ so as to be a projectively cofibration, and its domain is projectively cofibrant. Condition (A4) is given by Lemma 11.2.1.

Suppose a diagram $\mathcal{A}$ is in $\mathbf{inj}(K^{\text{proj}}_{\epsilon(n)})$. In particular it is $J_{\epsilon(n)}$-injective, which is to say fibrant in the projective model structure. This means that each $\mathcal{A}(\xi_i)$ is a fibrant object in $\mathcal{M}$. The lifting property along the $\zeta(f)$ implies the following homotopy lifting property of the map

$$p : \mathcal{A}(\xi_0) \to \mathcal{A}(\xi_1) \times \cdots \times \mathcal{A}(\xi_n),$$

along a generating cofibration $f : X \to Y$ in I. Recall that $Y \cup^X Y \xrightarrow{a} Z \xrightarrow{b} Y$ was chosen above. If we are given a diagram

$$\begin{array}{ccc} X & \xrightarrow{u} & \mathcal{A}(\xi_0) \\ {\scriptstyle f}\downarrow & & \downarrow{\scriptstyle p} \\ Y & \xrightarrow{v} & \mathcal{A}(\xi_1) \times \cdots \times \mathcal{A}(\xi_n) \end{array}$$

there is a map $Y \xrightarrow{w} \mathcal{A}(\xi_0)$ such that $wf = u$, and a map $Z \xrightarrow{h} \mathcal{A}(\xi_1) \times \cdots \times \mathcal{A}(\xi_n)$ such that $hai_2 = v$ but $hai_1 = pw$. This homotopy lifting property implies that p is a weak equivalence, see Lemma 7.5.1. Thus, $\mathcal{A} \in \mathcal{R}_{\epsilon(n)}$, which gives (A5).

For condition (A6), suppose $\mathcal{A}$ is in $\mathcal{R}_{\epsilon(n)}$ and $\mathcal{A} \to \mathcal{B}$ is a pushout along an element of $K^{\text{proj}}_{\epsilon(n)}$. Applying the small object argument we can find a map $\mathcal{B} \to \mathcal{C}$ in $\mathbf{cell}(K^{\text{proj}}_{\epsilon(n)})$ such that $\mathcal{C} \in \mathbf{inj}(K^{\text{proj}}_{\epsilon(n)})$. Then $\mathcal{A} \to \mathcal{C}$ is also in $\mathbf{cell}(K^{\text{proj}}_{\epsilon(n)})$. Notice, however, that the elements of $K^{\text{proj}}_{\epsilon(n)}$ are levelwise trivial cofibrations over the objects $\xi_1, \ldots, \xi_n \in \epsilon(n)$ (but not over ξ_0). Therefore, the map $\mathcal{A} \to \mathcal{C}$ induces weak equivalences over each $\xi_1, \ldots, \xi_n$. Using the cartesian condition on $\mathcal{M}$, this implies that the right vertical map in the square

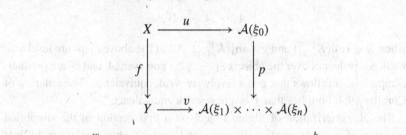

$$\begin{array}{ccc}\mathcal{A}(\xi_0) & \longrightarrow & \mathcal{A}(\xi_1) \times \cdots \times \mathcal{A}(\xi_n) \\ \downarrow & & \downarrow \\ \mathcal{C}(\xi_0) & \longrightarrow & \mathcal{C}(\xi_1) \times \cdots \times \mathcal{C}(\xi_n)\end{array}$$

is a weak equivalence. The hypothesis that $\mathcal{A} \in \mathcal{R}_{\epsilon(n)}$ says that the top map is a weak equivalence, and the fact that $\mathcal{C} \in \mathbf{inj}(K^{\text{proj}}_{\epsilon(n)})$ and part (A5) proved above say that $\mathcal{C} \in \mathcal{R}_{\epsilon(n)}$, so the bottom map is a weak equivalence. By 3 for 2 the left vertical arrow is a weak equivalence, showing that $\mathcal{A} \to \mathcal{C}$ is a levelwise weak equivalence of diagrams. This shows (A6).

Our direct localizing system leads to a left Bousfield localization by Theorem 9.7.1.

We now look at the characterizations of new weak equivalences. As seen above, the elements of $\mathbf{cell}(K^{\text{proj}}_{\epsilon(n)})$ are weak equivalences levelwise over the $\xi_1, \ldots, \xi_n$. Using the characterizatin of Corollary 9.4.3 we see that all new weak equivalences are levelwise weak equivalences over the $\xi_1, \ldots, \xi_n$. Suppose $\mathcal{A} \xrightarrow{f} \mathcal{B}$ is a morphism inducing a weak equivalence over each $\xi_1, \ldots, \xi_n$. Choose $\mathcal{B} \xrightarrow{b} \mathcal{B}'$ in $\mathbf{cell}(K^{\text{proj}}_{\epsilon(n)})$ such that $\mathcal{B}'$ is in $\mathbf{inj}(K^{\text{proj}}_{\epsilon(n)})$, in particular it is product-compatible. Factor the composed map as

$$\mathcal{A} \xrightarrow{a} \mathcal{A}' \xrightarrow{g} \mathcal{B}',$$

where $a \in \mathbf{cell}(K^{\text{proj}}_{\epsilon(n)})$ and $g \in \mathbf{inj}(K^{\text{proj}}_{\epsilon(n)})$. All of the above maps are levelwise weak equivalences over the objects $\xi_1, \ldots, \xi_n$; however, $\mathcal{A}'$ and $\mathcal{B}'$ are product-compatible. It follows that g is a levelwise weak equivalence. The criterion of Corollary 9.4.3 implies that f is a new weak equivalence.

The characterization of fibrant objects is a first version in the simplified situation of a single product diagram, of Bergner's characterization [42] of fibrant Segal categories. The new fibrant objects are in $\mathbf{inj}(K^{\text{proj}}_{\epsilon(n)})$ so they are in $\mathcal{R}_{\epsilon(n)}$, i.e. product-compatible, and levelwise fibrant. Suppose $\mathcal{A}$ is product-compatible and levelwise fibrant. Suppose $\mathcal{U} \xrightarrow{f} \mathcal{V}$ is a new trivial cofibration, and we are given a map $\mathcal{U} \xrightarrow{u} \mathcal{A}$. By the previous paragraph it induces a levelwise trivial cofibration over the objects $\xi_1, \ldots, \xi_n$. Hence the components $u_1, \ldots, u_n$ extend to maps $\mathcal{V}(\xi_i) \xrightarrow{v_i'} \mathcal{A}(\xi_i)$. Putting these together, the composition

$$\mathcal{V}(\xi_0) \to \mathcal{V}(\xi_1) \times \cdots \times \mathcal{V}(\xi_n) \to \mathcal{A}(\xi_1) \times \cdots \times \mathcal{A}(\xi_n)$$

gives the bottom arrow of the diagram

$$\begin{array}{ccc}\mathcal{U}(\xi_0) & \xrightarrow{u_0} & \mathcal{A}(\xi_0) \\ f_0 \downarrow & & \downarrow \\ \mathcal{V}(\xi_0) & \longrightarrow & \mathcal{A}(\xi_1) \times \cdots \times \mathcal{A}(\xi_n).\end{array}$$

The right vertical arrow is a weak equivalence between fibrant objects, so by Lemma 7.5.1 there is a homotopy lifting relative $\mathcal{U}(\xi_0)$, in other words a map $\mathcal{V}(\xi_0) \stackrel{v_0}{\to} \mathcal{A}(\xi_0)$ such that $rf_0 = u_0$, and the other triangle commutes up to a homotopy relative $\mathcal{U}(\xi_0)$. Lemma 7.5.2 says we can change the maps v_i' to maps v_i, still restricting to u_i on $\mathcal{U}(\xi_i)$, but compatible with v_0. We have now constructed the required extension $\mathcal{V} \to \mathcal{A}$, showing that $\mathcal{A}$ is a new fibant object. □

11.2.2 Direct localization of the injective structure

When possible, it is more convenient to use the injective model structure on $\text{FUNC}(\epsilon(n), \mathcal{M})$. Consider the explicit generating set $J^+_{\epsilon(n)}$ for the trivial cofibrations in the injective model structure, given by Proposition 11.1.4. Note that these have cofibrant domains, so in this case the injective model structure is always tractable. We can define two different sets of cofibrations, the first extending $K^{\text{proj}}_{\epsilon(n)}$:

$$K^{\text{inj}+}_{\epsilon(n)} := K^{\text{proj}}_{\epsilon(n)} \cup J^+_{\epsilon(n)} = J^+_{\epsilon(n)} \cup \{\zeta(f)\}_{f\in I};$$

and the second defined using the simpler maps $\rho(f)$ which were already injective cofibrations:

$$K^{\text{inj}}_{\epsilon(n)} := J^+_{\epsilon(n)} \cup \{\rho(f)\}_{f\in I}.$$

Theorem 11.2.3 *The pairs $(\mathcal{R}_{\epsilon(n)}, K^{\text{inj}+}_{\epsilon(n)})$ and $(\mathcal{R}_{\epsilon(n)}, K^{\text{inj}}_{\epsilon(n)})$ are both direct localizing systems for the category $\text{FUNC}_{\text{inj}}(\epsilon(n), \mathcal{M})$ of diagrams $\epsilon(n) \to \mathcal{M}$ with the injective model structure. Let $\text{FUNC}_{\text{inj},\Pi}(\epsilon(n), \mathcal{M})$ denote the left Bousfield localized model structure, which is the same in both cases. The weak equivalences are the same as for the projective structure. An object $\mathcal{A}$ is fibrant in the localized structure if and only if it is product-compatible and satisfies the fibrancy criterion of Lemma 11.1.1 for the injective model structure. The identity functor is a left Quillen functor*

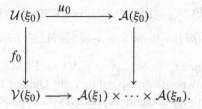

$$\text{FUNC}_{\text{proj},\Pi}(\epsilon(n), \mathcal{M}) \to \text{FUNC}_{\text{inj},\Pi}(\epsilon(n), \mathcal{M})$$

from the new projective to the new injective model structure.

Proof The functor $\text{FUNC}_{\text{proj}}(\epsilon(n), \mathcal{M}) \to \text{FUNC}_{\text{inj}}(\epsilon(n), \mathcal{M})$ is a left Quillen functor, whose corresponding right Quillen functor (both being the identity on underlying categories) preserves the class $\mathcal{R}_{\epsilon(n)}$ of product-compatible diagrams. In other words the transfered class is the same. The subset $K^{\text{inj}+}_{\epsilon(n)}$ is the transfered subset given in Theorem 9.8.1, so by that theorem $(\mathcal{R}_{\epsilon(n)}, K^{\text{inj}+}_{\epsilon(n)})$ is a direct localizing system and the identity functor is a left Quillen functor from the previous new projective model structure to the resulting left Bousfield localization of the injective structure

$$\text{FUNC}_{\text{proj},\Pi}(\epsilon(n), \mathcal{M}) \to \text{FUNC}_{\text{inj},\Pi}(\epsilon(n), \mathcal{M}).$$

The proof that $(\mathcal{R}_{\epsilon(n)}, K^{\text{inj}}_{\epsilon(n)})$ is a direct localizing system is the same as in the proof of the previous Theorem 11.2.2, but in fact easier since an object which satisfies lifting with respect to the $\rho(f)$ has the stronger property that

$$\mathcal{A}(\xi_0) \to \mathcal{A}(\xi_1) \times \cdots \times \mathcal{A}(\xi_n)$$

is in $\mathbf{inj}(I)$, that is it is a trivial fibration. So in this case we don't need to rely on the notion of homotopy lifting property as was done in the previous proof. We get conditions (A1) – (A6) and also the same description of weak equivalences, and the corresponding description of fibrant objects.

The two model structures given by Theorem 9.7.1 applied to the direct localizing systems $\left(\mathcal{R}_{\epsilon(n)}, K^{\text{inj}}_{\epsilon(n)}\right)$ and $\left(\mathcal{R}_{\epsilon(n)}, K^{\text{inj}+}_{\epsilon(n)}\right)$ are the same, by Proposition 9.7.2. □

Suppose $\mathcal{A} \in \text{FUNC}(\epsilon(n), \mathcal{M})$. Suppose given a factorization

$$\mathcal{A}(\xi_0) \stackrel{e_0}{\to} E_0 \stackrel{p}{\to} \mathcal{A}(\xi_1) \times \cdots \times \mathcal{A}(\xi_n)$$

in $\mathcal{M}$. Let $E_i := \mathcal{A}(\xi_i)$ for $i = 1, \ldots, n$. The structural map p gives a structure of $\epsilon(n)$-diagram to the collection $(E_0, \ldots, E_n)$, call it $\mathcal{E}$. The map e gives a map $e : \mathcal{A} \to \mathcal{E}$. If e_0 is a cofibration in $\mathcal{M}$ then e is a cofibration in $\text{FUNC}_{\text{inj}}(\epsilon(n), \mathcal{M})$.

Lemma 11.2.4 *In the above situation, if e_0 is a cofibration and p is a weak equivalence in $\mathcal{M}$ then $e : \mathcal{A} \to \mathcal{E}$ is a trivial cofibration in* $\text{FUNC}_{\text{inj},\Pi}(\epsilon(n), \mathcal{M})$.

Proof The map e is levelwise a cofibration by construction. It is a weak equivalence since it induces a weak equivalence levelwise over the objects $\xi_1, \ldots, \xi_n$. □

11.2.3 Transfering these structures

Putting together the above analysis of diagrams over $\epsilon(n)$ with the transfer along a Quillen functor gives the following general picture. Suppose we are given a set Q, integers $n(q) \geq 0$ for $q \in Q$, a family of tractable left proper cartesian model categories $\mathcal{M}_q$ for $q \in Q$, a tractable left proper model category $\mathcal{N}$, and a family of Quillen functors $F_q : \mathrm{FUNC}_{\mathrm{proj}}(\epsilon(n(q)), \mathcal{M}_q) \leftrightarrow \mathcal{N} : G_q$.

Let $\mathcal{R}' \subset \mathcal{N}$ be the full subcategory of objects Y such that, for a fibrant replacement $Y \to Y'$, the diagrams $G_q(Y') : \epsilon(n(q)) \to \mathcal{M}_q$ are product-compatible. Let (I^q, J^q) be generating sets for $\mathcal{M}_q$, and (I', J') generators for $\mathcal{N}$.

Corollary 11.2.5 *Let K' be the union of J', of the set of morphisms of the form $F_q(g)$ for $g \in J^q_{\epsilon(n(q))}$, and of the set of morphisms of the form $F_q(\zeta(f))$ for $f \in I^q$. Then $(\mathcal{R}', K')$ is a direct localizing system for $\mathcal{N}$.*

If F_q are left Quillen functors from $\mathrm{FUNC}_{\mathrm{proj}}(\epsilon(n(q)), \mathcal{M}_q)$ to $\mathcal{N}$, then we can consider K^{inj}, the union of J' with the set of morphisms of the form $F_q(\rho(f))$ for $f \in I^q$, and $(\mathcal{R}', K^{\mathrm{inj}})$ is a direct localizing system for $\mathcal{N}$ giving the same model structure as $(\mathcal{R}', K')$.

Proof Let K'' be the union of K' with the set of morphisms of the form $F_q(g)$ for $g \in J^q_{\epsilon(n(q))}$. Then $(\mathcal{R}', K'')$ is a direct localizing system for $\mathcal{N}$, by Theorem 9.8.3 applied to the direct localizing systems of Theorem 11.2.2. However, the $F_q(g)$ for $g \in J^q_{\epsilon(n(q))}$ are trivial cofibrations between cofibrant objects in the original model structure of $\mathcal{N}$, so they could be included in a bigger generating set J'' for the original trivial cofibrations of $\mathcal{N}$. But one can note that in the construction of a direct localizing system by adding on some new morphisms to the original generating set, the properties are independent of the choice of original generating set. So K' works as well as K''.

Suppose now that F_q remain left Quillen functors when we use the injective model structures on their sources. Then, with a similar discussion for leaving out the images of the morphisms in $J^{q,+}_{\epsilon(n(q))}$, Theorem 9.8.3 applies to the direct localizing systems of Theorem 11.2.3 to conclude that $(\mathcal{R}', K^{\mathrm{inj}})$ is a direct localizing system. As pointed out in Proposition 9.7.2, the resulting model structure is the same as for $(\mathcal{R}', K')$. □

11.3 Algebraic diagram theories

Classically, an "algebraic theory" is given by a small category Φ and a collection of product diagrams $P_q : \epsilon(n(q)) \to \Phi$. The objects of the theory are the functors $\mathcal{A} : \Phi \to \text{SET}$ with the property that $P_q^*(\mathcal{A})$ is a cartesian product, that is *product-compatible* in the above terminology. Of course this theory has since been much generalized, to include the notion of "finite limit sketches" among other things. However, for our purposes it will be sufficient to consider just the basic version of the theory, and to give it a weak-enriched counterpart using the notion of direct left Bousfield localization we have developed so far.

So, suppose Φ is a small category, Q is a small set, we have integers $n(q) \geq 0$ for $q \in Q$, and suppose given functors $P_q : \epsilon(n(q)) \to \Phi$ for $q \in Q$.

For the coefficients, fix a tractable left proper model category $\mathcal{M}$ satisfying condition (PROD) of Definition 7.7.1. Let (I, J) be a set of generators for $\mathcal{M}$. Playing the role of the model category $\mathcal{N}$ will be the category of Φ-diagrams in $\mathcal{M}$ with its projective or injective model structure, $\mathcal{N} = \text{FUNC}_{\text{proj}}(\Phi, \mathcal{M})$ (resp. $\mathcal{N} = \text{FUNC}_{\text{inj}}(\Phi, \mathcal{M})$). Let $(I_{\Phi,\text{proj}}, J_{\Phi,\text{proj}})$ and $(I_{\Phi,\text{inj}}, J_{\Phi,\text{inj}})$ be the sets of generators for the projective and injective model structures respectively, see the discussion of references for Theorem 7.6.3. The generators for the injective structure might not in general have cofibrant domains, which, however, is not a problem when all objects of $\mathcal{M}$ are cofibrant.

Recall that the identity functor is a Quillen adjunction between the projective and injective diagram categories

$$\mathbf{1} : \text{FUNC}_{\text{proj}}(\Phi, \mathcal{M}) \leftrightarrow \text{FUNC}_{\text{inj}}(\Phi, \mathcal{M}) : \mathbf{1}.$$

Lemma 11.3.1 *With the above notations, for each $q \in Q$ we get a Quillen adjunction*

$$P_{q,!} : \text{FUNC}_{\text{proj}}(\epsilon(n(q)), \mathcal{M}) \leftrightarrow \text{FUNC}_{\text{proj}}(\Phi, \mathcal{M}) : P_q^*.$$

This composes with the identity functor to give a Quillen adjunction

$$\mathbf{1}P_{q,!} : \text{FUNC}_{\text{proj}}(\epsilon(n(q)), \mathcal{M}) \leftrightarrow \text{FUNC}_{\text{inj}}(\Phi, \mathcal{M}) : P_q^*\mathbf{1}.$$

Proof This is the standard Quillen adjunction between Bousfield projective model structures coming from the functor $P_q : \epsilon(n(q)) \to \Phi$. □

In the above situation, let $\mathcal{R}(\Phi, P_\cdot, \mathcal{M})$ denote the full subcategory of $\text{FUNC}(\Phi, \mathcal{M})$ consisting of diagrams $\mathcal{A}$ such that for a fibrant replacement $\mathcal{A} \to \mathcal{A}'$ in the injective model structure (which is also a fibrant replacement in the projective model structure), for all $q \in Q$, $P_q^*(\mathcal{A}') \in \text{FUNC}(\epsilon(n(q)), \mathcal{M})$ is product-compatible. Let $K_{\text{proj/inj}}(\Phi, P_\cdot, \mathcal{M}, I, J)$ be the sets given by Corollary 11.2.5 for the projective/injective structure, consisting of the elements of $J_{\Phi,\text{proj/inj}}$, of the $P_{q,!}(J_{\epsilon(n(q))})$, and of the $P_{q,!}(\rho_{\epsilon(n(q))}(f))$ for $f \in I$.

Here a choice of subscript proj or inj is proposed when necessary. Make the convention that when speaking of the injective structure we assume that all objects of $\mathcal{M}$ are cofibrant so the injective diagram structures are tractable by Corollary 8.6.5.

Theorem 11.3.2 *Use the above notations; for the injective structure suppose that $\mathcal{M}$ satisfies additionally the condition that all objects be cofibrant. The pair $(\mathcal{R}(\Phi, P., \mathcal{M}), K_{\text{proj/inj}}(\Phi, P., \mathcal{M}, I, J))$ is a direct localizing system. Define the* model category of weak $(\Phi, P.)$-algebras *in $\mathcal{M}$ denoted by* $\text{ALG}_{\text{proj/inj}}(\Phi, P.; \mathcal{M})$, *to be the direct left Bousfield localization of the projective or injective diagram model category* $\text{FUNC}_{\text{proj/inj}}(\Phi, \mathcal{M})$ *with respect to the full subcategory $\mathcal{R}(\Phi, P., \mathcal{M})$ and sets $K_{\text{proj/inj}}(\Phi, P., \mathcal{M}, I, J)$.*

The cofibrations are levelwise cofibrations in the injective structure, and projective diagram cofibrations in the projective structure. The fibrant objects of $\text{ALG}_{\text{proj/inj}}(\Phi, P.; \mathcal{M})$ *are the diagrams $\mathcal{A} : \Phi \to \mathcal{M}$ such that $\mathcal{A}$ is levelwise fibrant (for the projective structure) or fibrant in the injective diagram structure, and for each $q \in Q$ the pullback $P_q^*(\mathcal{A}) : \epsilon(n(q)) \to \mathcal{M}$ is product-compatible. Given a map $f : \mathcal{A} \to \mathcal{B}$ of Φ-diagrams in $\mathcal{M}$, the following conditions are equivalent:*

- *f is a weak equivalence in* $\text{ALG}_{\text{proj/inj}}(\Phi, P.; \mathcal{M})$;
- *for any square diagram*

$$\begin{array}{ccccc} \mathcal{A} & \longleftarrow & \mathcal{A}' & \longrightarrow & \mathcal{A}'' \\ \downarrow & & \downarrow & & \downarrow \\ \mathcal{B} & \longleftarrow & \mathcal{B}' & \longrightarrow & \mathcal{B}'' \end{array}$$

  such that the left horizontal arrows are projective/levelwise cofibrant replacements, and $\mathcal{A}'', \mathcal{B}'' \in \mathcal{R}(\Phi, P., \mathcal{M})$, then the morphism $\mathcal{A}'' \to \mathcal{B}''$ is an levelwise weak equivalence;
- *there exists a square diagram as above with $\mathcal{A}'' \to \mathcal{B}''$ an levelwise weak equivalence;*
- *there exists a square diagram as above with $\mathcal{A}'' \to \mathcal{B}''$ an levelwise weak equivalence, but without the requirement $\mathcal{A}'', \mathcal{B}'' \in \mathcal{R}(\Phi, P., \mathcal{M})$.*

These projective and injective model categories of weak algebras are tractable and left proper.

Proof Apply the construction of the direct left Bousfield localization given in the previous chapter, starting with either $\text{FUNC}_{\text{proj}}(\Phi, \mathcal{M})$ or $\text{FUNC}_{\text{inj}}(\Phi, \mathcal{M})$ of Theorem 7.6.3, and transfering the direct localizing systems as in Corollary 11.2.5. For the injective structure, the hypothesis that all objects of $\mathcal{M}$ are cofibrant means that the generating morphisms have cofibrant domains.

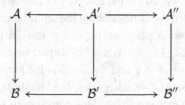

For the characterization of fibrant objects, see the successive statements of Proposition 8.7.4, Theorem 9.7.1, Remark 9.8.2 and Theorem 9.8.3. Apply these together with the characterizations of fibrant objects in the product-compatible $\epsilon(n)$-diagram model structures. □

11.4 Unitality

Suppose given a full subcategory $\Phi_0 \subset \Phi$. Typically, these will be the $P_q(\xi_0)$ for $q \in Q$ such that $n(q) = 0$. We would like to consider diagrams $\mathcal{A} : \Phi \to \mathcal{M}$ such that $\mathcal{A}(x) = *$ for $x \in \Phi_0$. Call such a diagram *unital along* Φ_0.

Let $\text{FUNC}(\Phi/\Phi_0, \mathcal{M}) \subset \text{FUNC}(\Phi, \mathcal{M})$ denote the full subcategory of diagrams which are unital along Φ_0. Denote by

$$U^*_{\Phi_0} : \text{FUNC}(\Phi/\Phi_0, \mathcal{M}) \to \text{FUNC}(\Phi, \mathcal{M})$$

the identity inclusion functor.

The idea for this notation is that Φ/Φ_0 represents the contraction of Φ_0 to a point, the result being a pointed category, i.e. a category with distinguished object; and $\text{FUNC}(\Phi/\Phi_0, \mathcal{M})$ is the category of pointed functors from here to the category $\mathcal{M}$ pointed by distinguishing the coinitial object $*$.

The present discussion plays an important role in the theory of weakly enriched precategories. The motivating example, first introduced in Section 10.1 above, is when $\Phi = \Delta_X$ and $\Phi_0 = X$ is the subcategory of sequences of length 0. In the case of n-precategories, the unitality condition corresponds to Tamsamani's constancy condition. This corresponds to the idea of having a globular theory in which the objects form a discrete set. Even though the composition operation is only weakly defined, the unitality condition provides, via the degeneracies of Δ, diagrams representing composition with a strict unit–whence the terminology.

For any $x \in \Phi$ denote by Φ_0/x the category of arrows $z \to x$ with $z \in \Phi_0$.

Theorem 11.4.1 *If $\mathcal{M}$ is a locally presentable category, then the category* $\text{FUNC}(\Phi/\Phi_0, \mathcal{M})$ *is locally presentable and $U^*_{\Phi_0}$ has a left adjoint $U_{\Phi_0,!}$. The left adjoint is given as follows: if $\mathcal{A} \in \text{FUNC}(\Phi, \mathcal{M})$ then $U_{\Phi_0,!}\mathcal{A}$ is the diagram which sends an object $x \in \Phi$ to the amalgamated sum of $\mathcal{A}(x)$ and* $\text{colim}_{\Phi_0/x} *$ *over* $\text{colim}_{z \in \Phi_0/x} \mathcal{A}(z)$. *The adjunction map $U_{\Phi_0,!}U^*_{\Phi_0}\mathcal{B} \to \mathcal{B}$ is the identity for any $\mathcal{B} \in \text{FUNC}(\Phi/\Phi_0, \mathcal{M})$, so $U_{\Phi_0,!}$ is a monadic projection in the terminology of Section 9.2.1.*

The full subcategory $\text{FUNC}(\Phi/\Phi_0, \mathcal{M}) \subset \text{FUNC}(\Phi, \mathcal{M})$ *is closed under small limits and over colimits with small nonempty connected index categories, in particular it is closed under amalgamated sums, filtered colimits, and transfinite composition.*

For any regular cardinal κ such that $\mathcal{M}$ is locally κ-presentable, $*$ is a κ-presentable object of $\mathcal{M}$, and $|\Phi| < \kappa$, an object $\mathcal{A} \in \mathrm{FUNC}(\Phi/\Phi_0, \mathcal{M})$ is κ-presentable if and only if each of the $\mathcal{A}(x)$ are κ-presentable in $\mathcal{M}$.

Proof Put

$$U_{\Phi_0,!}(\mathcal{A})(x) := \mathcal{A}(x) \cup^{\mathrm{colim}_{z\in\Phi_0/x}\mathcal{A}(z)} \mathrm{colim}_{\Phi_0/x} * .$$

Given a morphism $x \to y$, we obtain morphisms

$$\mathrm{colim}_{\Phi_0/x} * \to \mathrm{colim}_{\Phi_0/y} *$$

and

$$\mathrm{colim}_{z\in\Phi_0/x}\mathcal{A}(z) \to \mathrm{colim}_{z\in\Phi_0/y}\mathcal{A}(z).$$

These are compatible with the maps in the above amalgamated sum so they give $U_{\Phi_0,!}(\mathcal{A})$ a structure of diagram (i.e. functor). If $x \in \Phi_0$ then x is the coinitial object of Φ_0/x, and we get $U_{\Phi_0,!}(\mathcal{A})(x) = \mathcal{A}(x) \cup^{\mathcal{A}(x)} * = *$. Thus $U_{\Phi_0,!}(\mathcal{A}) \in \mathrm{FUNC}(\Phi/\Phi_0, \mathcal{M})$.

To show adjunction, suppose $\mathcal{B} \in \mathrm{FUNC}(\Phi/\Phi_0, \mathcal{M})$. Given a map $\mathcal{A} \to U^*_{\Phi_0}\mathcal{B}$ then for any $z \in \Phi_0$ the map $\mathcal{A}(z) \to \mathcal{B}(z)$ factors through $*$ (since indeed $\mathcal{B}(z) = *$). For any x this gives a factorization

$$\begin{array}{ccc} \mathrm{colim}_{z\in\Phi_0/x}\mathcal{A}(z) & \to & \mathrm{colim}_{z\in\Phi_0/x} * \\ \downarrow & & \downarrow \\ \mathcal{A}(x) & \longrightarrow & \mathcal{B}(x) \end{array}$$

so our map of diagrams factors through a unique map $U_{\Phi_0,!}(\mathcal{A}) \to \mathcal{B}$.

If $\mathcal{A} = U^*_{\Phi_0}\mathcal{B}$ is a diagram with $\mathcal{A}(z) = *$ for $z \in \Phi_0$ already, then the second map in the amalgamated sum defining $U_{\Phi_0,!}(\mathcal{A})$ is the identity, so $U_{\Phi_0,!}(\mathcal{A}) = \mathcal{A}$, i.e. the adjunction is a monadic projection.

Closure under arbitrary small limits is automatic since $U^*_{\Phi_0}$ is a right adjoint. For closure under connected colimits, suppose α is an index category with connected nerve. Then $\mathrm{colim}_\alpha * = *$ where $*$ is the coinitial object of $\mathcal{M}$ and the colimit is taken over the constant functor $\alpha \to \mathcal{M}$ (Lemma 8.1.9). As colimits in $\mathrm{FUNC}(\Phi, \mathcal{M})$ are calculated levelwise, it follows that colimits over α preserve the condition for inclusion in $\mathrm{FUNC}(\Phi/\Phi_0, \mathcal{M})$, which is that the diagram take values $*$ levelwise over Φ_0. Note that this property says that connected colimits in $\mathrm{FUNC}(\Phi/\Phi_0, \mathcal{M})$ are calculated levelwise. That wouldn't be true, however, for disconnected colimits such as disjoint sums (i.e. coproducts).

We now identify the κ-presentable objects of $\text{FUNC}(\Phi/\Phi_0, \mathcal{M})$. If each $\mathcal{A}(x)$ is a κ-presentable object of $\mathcal{M}$, then by Lemma 8.1.3 $\mathcal{A}$ is κ-presentable in $\text{FUNC}(\Phi, \mathcal{M})$. If we are given a κ-filtered system $\{\mathcal{B}_i\}_{i\in\beta}$ in FUNC $(\Phi/\Phi_0, \mathcal{M})$, any map

$$\mathcal{A} \to \text{colim}_{i\in\beta}^{\text{FUNC}(\Phi/\Phi_0\mathcal{M})} \mathcal{B}_i$$

is also a map to $\text{colim}_{i\in\beta}^{\text{FUNC}(\Phi,\mathcal{M})} \mathcal{B}_i$ by the closure under connected colimits; hence it factors through one of the $\mathcal{B}_i$. This factorization is a morphism in the full subcategory $\text{FUNC}(\Phi/\Phi_0, \mathcal{M})$, which shows that $\mathcal{A}$ is κ-presentable in $\text{FUNC}(\Phi/\Phi_0, \mathcal{M})$.

Suppose on the other hand $\mathcal{A}$ is κ-presentable in $\text{FUNC}(\Phi/\Phi_0, \mathcal{M})$. Given our assumption that $\mathcal{M}$ is locally κ-presentable and $|\Phi| < \kappa$, the category $\text{FUNC}(\Phi, \mathcal{M})$ is locally κ-presentable, and its κ-presentable objects are exactly the diagrams $\mathcal{B}$ such that $\mathcal{B}(x)$ is κ-presentable in $\mathcal{M}$. This was stated above as Lemma 8.1.3 with reference to Adámek and Rosický [2]. In particular, we can express $\mathcal{A}$ as a colimit in the category $\text{FUNC}(\Phi, \mathcal{M})$

$$\mathcal{A} = \text{colim}_{i\in\beta}\mathcal{B}_i,$$

with $\mathcal{B}_i(x)$ being κ-presentable, and indexed by a κ-filtered category β. The unitalization functor $U_{\Phi_0,!}$ being a left adjoint, we get

$$\mathcal{A} = \text{colim}_{i\in\beta} U_{\Phi_0,!}(\mathcal{B}_i) \text{ in } \text{FUNC}(\Phi/\Phi_0\mathcal{M}). \tag{11.4.1}$$

The hypothesis that $\mathcal{A}$ is κ-presentable in $\text{FUNC}(\Phi/\Phi_0, \mathcal{M})$, applied to the identity map of $\mathcal{A}$, says that the identity factors through a map $\mathcal{A} \to U_{\Phi_0,!}(\mathcal{B}_i)$. On the other hand, using the hypothesis that $*$ is κ-presentable, the explicit description of $U_{\Phi_0,!}$ shows that each $U_{\Phi_0,!}(\mathcal{B}_i)(x)$ is κ-presentable. We have a retraction

$$\mathcal{A}(x) \to U_{\Phi_0,!}(\mathcal{B}_i)(x) \to \mathcal{A}(x),$$

the composition being the identity of $\mathcal{A}(x)$. It easily follows that $\mathcal{A}(x)$ is κ-presentable in $\mathcal{M}$. This completes the proof of the identification of κ-presentable objects of $\text{FUNC}(\Phi/\Phi_0\mathcal{M})$.

It is clear from this description that the κ-presentable objects form a small set. The above argument shows that any $\mathcal{A} \in \text{FUNC}(\Phi/\Phi_0\mathcal{M})$ is a κ-filtered colimit of κ-presentable objects, indeed we obtained the expression (11.4.1). □

We can construct the projective model category structure:

Proposition 11.4.2 *If $\Phi_0 \subset \Phi$ and $\mathcal{M}$ is a tractable left proper model category, we get a projective model structure* $\text{FUNC}_{\text{proj}}(\Phi/\Phi_0, \mathcal{M})$ *which is a tractable left proper model category. Furthermore the unitalization construction is a left Quillen functor*

$$\text{FUNC}_{\text{proj}}(\Phi, \mathcal{M}) \xrightarrow{U_{\Phi_0,!}} \text{FUNC}_{\text{proj}}(\Phi/\Phi_0, \mathcal{M}).$$

Proof The weak equivalences in $\text{FUNC}_{\text{proj}}(\Phi/\Phi_0, \mathcal{M})$ are defined to be the levelwise weak equivalences. These satisfy 3 for 2 and are closed under retracts. The fibrations are defined to be the levelwise fibrations. The trivial fibrations are the intersection of these classes. The cofibrations are determined by the left lifting property with respect to trivial fibrations. We construct explicitly a generating set, by a small variant of Bousfield's original construction.

Choose a generating set I for the cofibrations of $\mathcal{M}$. It leads to the set I_Φ of generators for cofibrations in the projective model structure $\text{FUNC}_{\text{proj}}(\Phi, \mathcal{M})$ discussed in Theorem 7.6.3. Recall that I_Φ consists of all morphisms of the form $i_{x,!}(f)$ where $f : A \to B$ is in I, where $i_x : \{x\} \to \Phi$ is the inclusion of a discrete single object and where

$$i_{x,!} : \mathcal{M} = \text{FUNC}(\{x\}, \mathcal{M}) \to \text{FUNC}(\Phi, \mathcal{M})$$

is the corresponding left adjoint functor. This is just Bousfield's classic generating set for projective diagram cofibrations. Set

$$I_{\Phi/\Phi_0} := U_{\Phi_0,!}(I_\Phi).$$

Notice that

$$U_{\Phi_0,!}i_{x,!} : \mathcal{M} = \text{FUNC}(\{x\}, \mathcal{M}) \to \text{FUNC}(\Phi/\Phi_0, \mathcal{M})$$

is the left adjoint functor for inducing unital diagrams from objects of $\mathcal{M}$ placed over $x \in \text{Ob}(\Phi)$. The set I_{Φ/Φ_0} consists of all $U_{\Phi_0,!}i_{x,!}(f)$ where f runs through the set I of generating cofibrations for $\mathcal{M}$ and x runs through $\text{Ob}(\Phi)$.

For x and $f : A \to B$ fixed,

$$U_{\Phi_0,!}i_{x,!}(f) : U_{\Phi_0,!}i_{x,!}(A) \to U_{\Phi_0,!}i_{x,!}(B)$$

has the following explicit description. For any object $y \in \Phi$, let $\Phi_{\text{nf}}(x, y)$ denote the set of arrows from x to y which don't factor through an objects of Φ_0, and let $\Phi_{\text{f}}(x, y)$ denote the set of arrows which factor through an element of Φ_0. Thus $\Phi(x, y) = \Phi_{\text{nf}}(x, y) \sqcup \Phi_{\text{f}}(x, y)$. Then,

$$U_{\Phi_0,!}i_{x,!}(A)(y) = \coprod_{\Phi_{\text{nf}}(x,y)} A, \quad \text{if } \Phi_{\text{f}}(x, y) = \emptyset,$$

$$U_{\Phi_0,!}i_{x,!}(A)(y) = * \sqcup \coprod_{\Phi_{\rm nf}(x,y)} A, \quad \text{if } \Phi_{\rm f}(x,y) \neq \emptyset.$$

For an arrow $y \to z$, composition induces $\Phi_{\rm f}(x, y) \to \Phi(x, z)$ but only $\Phi_{\rm nf}(x, y) \to \Phi_{\rm nf}(x, y) \sqcup \Phi_{\rm f}(x, y)$. The morphisms of functoriality for the diagram $U_{\Phi_0,!}i_{x,!}(A)$ are either the identity on A or on $*$, or else the projection $A \to *$ in the case of an arrow in $\Phi_{\rm nf}(x, y)$ which composes with $y \to z$ to give an arrow in $\Phi_{\rm f}(x, z)$.

Note that if $u \in \Phi_0$ then $\Phi_{\rm nf}(x, u) = \emptyset$ and $U_{\Phi_0,!}i_{x,!}(A)(u) = *$ so the above formula defines a unital diagram. One can check by hand that the explicit construction described above is adjoint to the functor $\text{FUNC}(\Phi/\Phi_0, \mathcal{M}) \to \mathcal{M}$ of evaluation at x, which serves to show that the explicit construction is indeed $U_{\Phi_0,!}i_{x,!}$.

The same description holds for $U_{\Phi_0,!}i_{x,!}(B)$ and the map $U_{\Phi_0,!}i_{x,!}(f)$ is obtained by applying either f or 1_* on the various factors.

A couple of things are immediate from this description:

(1) if f is any cofibration in $\mathcal{M}$ then $U_{\Phi_0,!}i_{x,!}(f)$ is an injective (i.e. levelwise) cofibration in $\text{FUNC}(\Phi/\Phi_0; \mathcal{M})$, indeed the $U_{\Phi_0,!}i_{x,!}(f)(y)$ are disjoint unions of copies of f and of the isomorphism 1_*; and
(2) if f is a trivial cofibration in $\mathcal{M}$ then $U_{\Phi_0,!}i_{x,!}(f)$ is an injective (i.e. levelwise) trivial cofibration in $\text{FUNC}(\Phi/\Phi_0, \mathcal{M})$, for the same reason.

On the other hand, the adjunction formula says that $U_{\Phi_0,!}i_{x,!}$ is left adjoint to the restriction

$$i_x^* : \text{FUNC}(\Phi/\Phi_0, \mathcal{M}) \to \mathcal{M},$$

i.e. the evaluation at $x \in \Phi$. Hence, a morphism g of diagrams in FUNC $(\Phi/\Phi_0, \mathcal{M})$ satisfies right lifting with respect to $U_{\Phi_0,!}i_{x,!}(f)$, if and only if $i_x^*(g) = g(x)$ satisfies the right lifting property with respect to f.

The previous paragraph implies that $\mathbf{inj}(I_{\Phi/\Phi_0})$ is equal to the class of levelwise trivial fibrations, hence $\mathbf{cof}(I_{\Phi/\Phi_0})$ is the class of cofibrations. Thus I_{Φ/Φ_0} is a set of generators for the cofibrations.

If we had started with a set J generating the trivial cofibrations of $\mathcal{M}$, then defining J_{Φ/Φ_0} to be the set of all $U_{\Phi_0,!}i_{x,!}(f)$ for $f \in J$, gives by the same argument $\mathbf{inj}(J_{\Phi/\Phi_0})$ equal to the class of levelwise fibrations. We claim that $\mathbf{cof}(J_{\Phi/\Phi_0})$ is then equal to the class of trivial cofibrations. By property (2) above, J_{Φ/Φ_0} and hence $\mathbf{cof}(J_{\Phi/\Phi_0})$ consist of levelwise weak equivalences, so they are contained in the class of trivial cofibrations. Suppose $g : R \to S$ is a trivial cofibration. By the small object argument it can be factored as $g = ph$ where $p \in \mathbf{inj}(J_{\Phi/\Phi_0})$ and $h \in \mathbf{cell}(J_{\Phi/\Phi_0})$. In particular, p is an levelwise fibration, but it is also an levelwise weak equivalence by 3 for 2, so it is a trivial fibration hence satisfies lifting with respect to cofibrations. As g is assumed to be a cofibration, there is a lifting which shows g to be a retract of h. Thus $g \in \mathbf{cof}(J_{\Phi/\Phi_0})$. We have now shown conditions (CG1) – (CG3b) for I and J with respect to the given three classes of morphisms, and we know that weak equivalences are closed under retracts and satisfy 3 for 2. These give a cofibrantly generated model structure (see Proposition 8.4.1). It is tractable since the elements of the generating sets have cofibrant domains. Left properness is checked levelwise.

To see that unitalization is a left Quillen functor, note that it is a left adjoint. Furthermore, it preserves cofibrations and trivial cofibrations, indeed by our construction the generating sets I_Φ and J_Φ for cofibrations and trivial cofibrations in $\text{FUNC}_{\text{proj}}(\Phi, \mathcal{M})$ are mapped by $U_{\Phi_0,!}$ to the generating sets for cofibrations and trivial cofibrations in $\text{FUNC}_{\text{proj}}(\Phi/\Phi_0, \mathcal{M})$. □

And the injective structure:

Proposition 11.4.3 *Suppose $\mathcal{M}$ is combinatorial and all objects are cofibrant, hence it is tractable. There exists a tractable injective model category structure denoted $\text{FUNC}_{\text{inj}}(\Phi/\Phi_0, \mathcal{M})$, where the weak equivalences are levelwise weak equivalences, and cofibrations are levelwise cofibrations. If $\mathcal{M}$ is left proper then so is $\text{FUNC}_{\text{inj}}(\Phi/\Phi_0, \mathcal{M})$. The identity on the underlying category constitutes a left Quillen functor*

$$\text{FUNC}_{\text{proj}}(\Phi/\Phi_0, \mathcal{M}) \to \text{FUNC}_{\text{inj}}(\Phi/\Phi_0, \mathcal{M}),$$

comparing this with the injective structure.

Proof Weak equivalences are clearly closed under retracts and satisfy 3 for 2. The class of trivial cofibrations, that is the intersection of the classes of cofibrations and weak equivalences, is closed under pushout and transfinite composition since these colimits are calculated levelwise.

The sets of injective cofibrations and injective trivial cofibrations have generating sets, as was shown in Theorem 8.6.3 using Lurie's technique of Theorems 8.6.1 and 8.6.2. This gives the necessary accessibility argument which allows to apply Smith's recognition lemma to obtain the model structure, such as described by Barwick [18]. If $\mathcal{M}$ is left proper, the colimits involved in this condition are connected so they are computed levelwise, hence the same condition holds for $\text{FUNC}_{\text{inj}}(\Phi/\Phi_0, \mathcal{M})$.

The identity is adjoint to itself, and by (1) and (2) in the proof of the previous proposition, the generating cofibrations of $\text{FUNC}_{\text{proj}}(\Phi/\Phi_0, \mathcal{M})$ are also injective cofibrations; so the identity may be seen as a left Quillen functor from the projective to the injective structure. □

Remark 11.4.4 Unfortunately $U_!$ is not necessarily a left Quillen functor between the injective structures.

In the next chapter when our discussion is applied to the special case Δ^o_X/X we can impose an additional condition 12.2.2 on $\mathcal{M}$ so that it works, or alternatively use the Reedy structure on the category of diagrams.

11.5 Unital algebraic diagram theories

Combine the previous discussions: suppose Φ is a small category, $\Phi_0 \subset \Phi$ is a full subcategory, Q is a small set, we have integers $n(q) \geq 0$ for $q \in Q$, and suppose given functors $P_q : \epsilon(n(q)) \to \Phi$ for $q \in Q$. Suppose that $\mathcal{M}$ is a tractable left proper cartesian model category with generating sets I and J. For considerations of the injective model structure, suppose furthermore that all objects of $\mathcal{M}$ are cofibrant. We obtain left Quillen functors (the leftmost varying in a family indexed by $q \in Q$):

$$\begin{array}{ccc} \mathrm{FUNC}_{\mathrm{proj}}(\epsilon(n(q)), \mathcal{M}) \xrightarrow{P_{q,!}} \mathrm{FUNC}_{\mathrm{proj}}(\Phi, \mathcal{M}) & & \\ \downarrow U_{\Phi_0,!} & & \\ \mathrm{FUNC}_{\mathrm{proj}}(\Phi/\Phi_0, \mathcal{M}) \xrightarrow{1} \mathrm{FUNC}_{\mathrm{inj}}(\Phi/\Phi_0, \mathcal{M}). & & \end{array}$$

By the property of transfer of families of direct localizing systems along Quillen functors (Theorem 9.8.3), we obtain left Bousfield localizations of $\mathrm{FUNC}_{\mathrm{proj}}$ $(\Phi/\Phi_0, \mathcal{M})$ and $\mathrm{FUNC}_{\mathrm{inj}}(\Phi/\Phi_0, \mathcal{M})$ along the images of the $\zeta_{\epsilon(n(q))}(f)$ for $f \in I$. Denote these respectively by $\mathrm{ALG}_{\mathrm{proj}}(\Phi/\Phi_0, P.; \mathcal{M})$ and $\mathrm{ALG}_{\mathrm{inj}}$ $(\Phi/\Phi_0, P.; \mathcal{M})$. They are tractable left proper model categories whose underlying categories are the unital diagram category $\mathrm{FUNC}(\Phi/\Phi_0, \mathcal{M})$. In the projective structure, the cofibrations are generated by $U_{\Phi_0,!}(I_\Phi)$, whereas in the injective structure the cofibrations are the levelwise cofibrations. The fibrant objects in the projective structure are the levelwise fibrant objects whose pullback to each $\epsilon(n(q))$ satisfies the product condition.

In the projective case this compares with the non-unital algebraic diagram theories by a Quillen adjunction

$$U_{\Phi_0,!} : \mathrm{ALG}_{\mathrm{proj}}(\Phi, P.; \mathcal{M}) \leftrightarrow \mathrm{ALG}_{\mathrm{proj}}(\Phi/\Phi_0, P.; \mathcal{M}) : U^*_{\Phi_0}.$$

Indeed, the direct left Bousfield localization of the unital theory can be seen as coming from the localization of the non-unital theory given in Theorem 11.3.2, by transfer along the left Quillen functor $U_{\Phi_0,!}$, and in the situation of Theorem 9.8.1 we still get a left Quillen functor.

Remark 11.5.1 If furthermore we know that $U_{\Phi_0,!}$ gives a left Quillen functor on the injective diagram structures, then this completes to a Quillen adjunction

$$U_{\Phi_0,!} : \mathrm{ALG}_{\mathrm{inj}}(\Phi, P.; \mathcal{M}) \leftrightarrow \mathrm{ALG}_{\mathrm{inj}}(\Phi/\Phi_0, P.; \mathcal{M}) : U^*_{\Phi_0}.$$

Lemma 11.5.2 *Suppose $r : \mathcal{A} \to \mathcal{A}'$ is a trivial cofibration towards a fibrant object, in either of the projective or injective model structures on* $\mathrm{ALG}(\Phi/\Phi_0, P.; \mathcal{M})$. *If $\mathcal{A}$ satisfies the product condition, then r is levelwise a weak equivalence, that is $r(x) : \mathcal{A}(x) \to \mathcal{A}'(x)$ is a weak equivalence in $\mathcal{M}$ for any $x \in \Phi$.*

Proof This follows from Corollary 9.4.5, noting that the model structures on $\mathrm{ALG}(\Phi/\Phi_0, P.; \mathcal{M})$ are obtained from direct left Bousfield localizing systems with $\mathcal{R}$ being the class of objects satisfying the product condition. □

11.6 The generation operation

Suppose Φ is a small category, $\Phi_0 \subset \Phi$ is a full subcategory, Q is a small set, we have integers $n(q) \geq 0$ for $q \in Q$, and suppose given functors $P_q : \epsilon(n(q)) \to \Phi$ for $q \in Q$. Suppose that $\mathcal{M}$ is a tractable left proper model category with generating sets I and J.

Make the following assumption:

(INJ)–the functors $U_{\Phi_0,!}P_{q,!}$ send cofibrations (resp. trivial cofibrations) in $\mathrm{FUNC}_{\mathrm{inj}}(\epsilon(n_q), \mathcal{M})$ to levelwise cofibrations (resp. levelwise trivial cofibrations).

In other words we have left Quillen functors

$$\mathrm{FUNC}_{\mathrm{inj}}(\epsilon(n_q), \mathcal{M}) \xrightarrow{U_{\Phi_0,!}P_{q,!}} \mathrm{FUNC}_{\mathrm{inj}}(\Phi/\Phi_0, \mathcal{M}).$$

In this case we can use the generating set for the new model structure on $\mathrm{FUNC}_{\mathrm{inj}}(\epsilon(n_q), \mathcal{M})$ made from the simpler cofibrations $\rho(f)$.

Suppose $\mathcal{A} \in \mathrm{FUNC}(\Phi/\Phi_0, \mathcal{M})$ and $q \in Q$. Let $n := n_q$. Define a trivial cofibration $\mathcal{A} \to \mathbf{Gen}(\mathcal{A}; q)$ as follows: choose a factorization

$$P_q^*(\mathcal{A})(\xi_0) \stackrel{e_0}{\to} E_0 \stackrel{p}{\to} P_q^*(\mathcal{A})(\xi_1) \times \cdots \times P_q^*(\mathcal{A})(\xi_n),$$

with e_0 a cofibration and p a weak equivalence, in $\mathcal{M}$. This gives a trivial cofibration $P_q^*(\mathcal{A}) \xrightarrow{e} \mathcal{E}$ in $\text{FUNC}_{\text{inj},\Pi}(\epsilon(n), \mathcal{M})$. Using condition (INJ) we obtain a cofibration

$$\mathcal{A} \to \mathbf{Gen}(\mathcal{A}; q) := \mathcal{A} \cup^{U_{\Phi_0,!}P_{q,!}(\mathcal{A})} U_{\Phi_0,!}P_{q,!}(E) \tag{11.6.1}$$

in the injective model structure $\text{FUNC}_{\text{inj}}(\Phi/\Phi_0, \mathcal{M})$.

Note that $\mathbf{Gen}(\mathcal{A}, q)$ doesn't depend, up to equivalence, on the choice of factorization E. If necessary we can include the factorization in the notation $\mathbf{Gen}(\mathcal{A}, q; e_0, p)$.

The weak monadic projection from $\text{FUNC}(\Phi/\Phi_0, \mathcal{M})$ to the class of objects satisfying the product condition may be thought of as a transfinite iteration of the operation $\mathcal{A} \mapsto \mathbf{Gen}(\mathcal{A}, q)$ over all $q \in Q$.

11.7 Reedy structures

In the main situation where the theory of this chapter will be applied, the underlying category Φ is a *Reedy category*, and we can give the category of diagrams $\Phi \to \mathcal{M}$ the Reedy model category structure denoted by $\text{FUNC}_{\text{Reedy}}(\Phi, \mathcal{M})$. This allows us to remove the condition that all objects of $\mathcal{M}$ be cofibrant, which was needed in order to get tractability of the injective model structures.

Assume that Φ_0 consists of objects of the bottom degree in the Reedy structure. Then there is a corresponding model structure again called "Reedy" and denoted $\text{FUNC}_{\text{Reedy}}(\Phi/\Phi_0, \mathcal{M})$ on the category of unital diagrams, such that $(U_!, U^*)$ remains a Quillen adjunction. The Reedy structures $\text{FUNC}_{\text{Reedy}}(\Phi, \mathcal{M})$ and $\text{FUNC}_{\text{Reedy}}(\Phi/\Phi_0, \mathcal{M})$ can again by localized by direct left Bousfield localization, to give model categories denoted $\text{ALG}_{\text{Reedy}}(\Phi, P.; \mathcal{M})$ and $\text{ALG}_{\text{Reedy}}(\Phi/\Phi_0, P.; \mathcal{M})$. These fit in between the projective and the injective structures above.

The Reedy structures will be discussed more in the case we need, $\Phi = \Delta_X^o$ and Φ_0 is the subcategory X of sequences of length 0, in Proposition 12.3.1. The reader is refered to there for the proof, and the general case Φ/Φ_0 may be filled in as an exercise. In many cases of interest, the Reedy and injective structures coincide, see Proposition 13.7.2.

12
Weak equivalences

This chapter continues the study of weakly enriched categories using Segal's method. We use the model category for algebraic theories, developed in the previous chapter, to get model structures for Segal precategories on a fixed set of objects. This structure will be studied in detail later, to deal with the passage from a Segal precategory to the Segal category it generates.

Then we consider the full category of Segal precategories, with movable sets of objects, giving various definitions and notations. Constructing a model structure in that case is the main subject of the subsequent chapters.

The reader will note that this division of the global argument into two pieces, was present already in Dwyer–Kan's treatment of the model category for simplicial categories. They discussed the model category for simplicial categories on a fixed set of objects in a series of papers (Dwyer and Kan [101, 102, 103]); but it wasn't until some time later in their unpublished manuscript with Hirschhorn [99], which subsequently became Dwyer *et al.* [100], and then Bergner's paper [39] that the global case was treated.

For the theory of weak enrichment following Segal's method, the corresponding division and introduction of the notion of left Bousfield localization for the first part was suggested in Barwick's thesis [16].

Assume throughout that $\mathcal{M}$ is a tractable left proper cartesian model category. See Section 7.7 for an explanation and first consequences of the cartesian condition.

12.1 Local weak equivalences

The Segal conditions (Section 10.2) for $\mathcal{M}$-precategories can be expressed in terms of the algebraic diagram theory of the previous chapter, which was the motivation for introducing that notion.

Let $\Phi := \Delta^o_X$, and let $\Phi_0 = \mathbf{disc}(X)$ be the discrete subcategory on object set X, considered as a subcategory by letting $x \in X$ correspond to the sequence (x). An $\mathcal{M}$-precategory $\mathcal{A} \in \mathcal{PC}(X, \mathcal{M})$ is by definition the same thing as a functor $\Phi \to \mathcal{M}$ sending the objects of Φ_0 to $*$, which is to say

$$\mathcal{PC}(X, \mathcal{M}) = \text{FUNC}\left(\Delta^o_X / X, \mathcal{M}\right).$$

The Segal conditions are a collection of finite product conditions as was considered in the previous chapter. The set of product conditions Q consists of the full set of objects of Δ^o_X. For $q = (x_0, \ldots, x_n)$ the integer $n(q)$ is equal to n, and we define a functor

$$P_{(x_0,\ldots,x_n)} : \epsilon(n) \to \Delta^o_X$$

by

$$P_{(x_0,\ldots,x_n)}(\xi_0) := (x_0, \ldots, x_n),$$

$$P_{(x_0,\ldots,x_n)}(\xi_i) := (x_{i-1}, x_i) \text{ for } 1 \leq i \leq n.$$

The images of the projection maps in $\epsilon(n)$ are the opposites of the inclusion maps $(x_{i-1}, x_i) \hookrightarrow (x_0, \ldots, x_n)$ in Δ_X. An $\mathcal{M}$-precategory is an $\mathcal{M}$-enriched Segal category, if and only if it satisfies the product condition with respect to the collection of functors $P_{\cdot}$, indeed the two conditions are identical. Unitality gives the product condition whenever $n = 0$, whereas the product condition is automatically true whenever $n = 1$ because the Segal maps are the identity in this case. Thus, this condition needs only to be imposed for $n \geq 2$.

We obtain adjoint functors

$$P_{(x_0,\ldots,x_n),!} : \text{FUNC}(\epsilon(n), \mathcal{M}) \leftrightarrow \text{FUNC}\left(\Delta^o_X, \mathcal{M}\right) : P^*_{(x_0,\ldots,x_n)},$$

and the unital versions

$$U_! P_{(x_0,\ldots,x_n),!} : \text{FUNC}(\epsilon(n), \mathcal{M}) \leftrightarrow \mathcal{PC}(X, \mathcal{M}) : U_! P^*_{(x_0,\ldots,x_n)},$$

where the right adjoint is just the pullback of a diagram $\mathcal{A} : \Delta^o_X \to \mathcal{M}$ to the category $\epsilon(n)$. An $\mathcal{M}$-precategory $\mathcal{A}$ satisfies the Segal conditions, if and only if $U_! P^*_{(x_0,\ldots,x_n)}(\mathcal{A})$ is a product-compatible diagram $\epsilon(n) \to \mathcal{M}$, for each sequence $(x_0, \ldots, x_n)$.

A direct application of the construction of Chapter 11 gives two model structures (projective and injective) on $\mathcal{PC}(X, \mathcal{M})$ such that the fibrant objects satisfy the Segal condition. We add a third *Reedy* structure since Δ^o_X is a Reedy category.

These model structures are intermediate with respect to our main goal of constructing global model structures on $\mathcal{PC}(\mathcal{M})$: the maps in $\mathcal{PC}(X, \mathcal{M})$ are ones which induce the identity on the set of objects X. Nonetheless, these easier model structures on $\mathcal{PC}(X, \mathcal{M})$ will be very useful in numerous arguments later.

The idea of introducing the intermediate model category $\mathcal{PC}(X, \mathcal{M})$, and of expressing it as a left Bousfield localization, is due to Barwick [16].

Start by discussing the case of non-unital diagrams. If we fix generating sets (I, J) for $\mathcal{M}$, then we obtain generating sets for the projective model structure $\text{FUNC}_{\text{proj}}\left(\Delta_X^o, \mathcal{M}\right)$ and injective model structure $\text{FUNC}_{\text{inj}}\left(\Delta_X^o, \mathcal{M}\right)$, recalled in Theorem 7.6.3. Notice that if $\mathcal{M}$ is tractable and I and J consist of arrows with cofibrant domains, the explicit generators for the projective model structure also have cofibrant domains. Thus $\text{FUNC}_{\text{proj}}\left(\Delta_X^o, \mathcal{M}\right)$ will again be tractable. For the injective model structure the construction of generating sets of Barwick [18] and Lurie [190] (as discussed in Theorem 8.6.2 above) was complicated, and probably doesn't allow to choose generators with cofibrant domains. This problem can be bypassed, either by simply assuming that all objects of $\mathcal{M}$ are cofibrant, a hypothesis preserved under the iteration of $\mathcal{M} \mapsto \mathcal{PC}(\mathcal{M})$ and verified in the examples of interest; or later with the introduction of Reedy model structures where again the generators become explicit.

Within the projective or injective model categories of non-unital diagrams, we obtain a direct left Bousfield localizing system $(\mathcal{R}^{\text{nu}}, K^{\text{nu}})$ where $\mathcal{R}$ is the class of non-unital diagrams satisfying the Segal conditions, and K^{nu} is given by the generating trivial cofibrations plus the maps of the form $P_{(x_0,\ldots,x_n),!}(\zeta_n(f))$ for f in the generating set I of cofibrations of $\mathcal{M}$. Here $\zeta_n(f)$ is the diagram $\epsilon(n) \to \mathcal{M}$ considered in Section 11.2.1. This yields the direct left Bousfield localized model structures which were designated by the notation $\text{ALG}(\ldots)$ in the previous chapter. Denote these model categories of weakly unital precategories now by

$$\text{ALG}_{\text{proj}}\left(\Delta_X^o, P_{\cdot}; \mathcal{M}\right), \quad \text{ALG}_{\text{inj}}\left(\Delta_X^o, P_{\cdot}; \mathcal{M}\right).$$

The underlying categories are both the same $\text{FUNC}\left(\Delta_X^o, \mathcal{M}\right)$. The fibrant objects are diagrams which are fibrant in the projective or injective model structures for diagrams, and which are Segal categories.

The same will work for for $\mathcal{M}$-precategories where the unitality condition is imposed. The model structure on $\mathcal{M}$-precategories for a fixed set of objects is given by the following theorem. Note the introduction of the notation $\mathcal{A} \to \mathbf{Seg}(\mathcal{A})$ for a choice of fibrant replacement in either of the model categories. This notation will be used later, but can mean that a choice is made at each usage, rather than fixing a global choice once and for all. Most constructions will be independent of the choice, up to equivalence.

Theorem 12.1.1 *Supose $\mathcal{M}$ is a tractable left proper cartesian model category, then there are left proper combinatorial model category structures on the unital $\mathcal{M}$-precategories*

$$\mathcal{PC}_{\text{proj}}(X, \mathcal{M}) := \text{ALG}_{\text{proj}}\left(\Delta^o_X/X, P.; \mathcal{M}\right),$$

and assuming that all objects of $\mathcal{M}$ are cofibrant,

$$\mathcal{PC}_{\text{inj}}(X, \mathcal{M}) := \text{ALG}_{\text{inj}}\left(\Delta^o_X/X, P.; \mathcal{M}\right).$$

The fibrant objects are unital fibrant diagrams which satisfy the Segal condition. The cofibrations are the projective or injective cofibrations in the unital diagram category FUNC $\left(\Delta^o_X/X, \mathcal{M}\right)$. *The weak equivalences are the same in both structures.*

Let $\mathcal{A} \to$ **Seg**$(\mathcal{A})$ *denote a trivial cofibration towards a fibrant replacement of $\mathcal{A}$ in the projective model structure $\mathcal{PC}_{\text{proj}}(X, \mathcal{M})$; this can be chosen functorially. A map $\mathcal{A} \to \mathcal{B}$ is a weak equivalence if and only if* **Seg**$(\mathcal{A}) \to$ **Seg**$(\mathcal{B})$ *is a levelwise weak equivalence when considered as a map of diagrams $\Delta^o_X \to \mathcal{M}$.*

Proof Propositions 11.4.2 and 11.4.3 give projective and injective diagram model structures $\text{FUNC}_{\text{proj}}\left(\Delta^o_X/X, \mathcal{M}\right)$ and $\text{FUNC}_{\text{inj}}\left(\Delta^o_X/X, \mathcal{M}\right)$ on the category of unital diagrams, which is the same underlying category as $\mathcal{PC}(X, \mathcal{M})$.

We get direct left Bousfield localizing systems $(\mathcal{R}, K_{\text{proj}})$ and $(\mathcal{R}, K_{\text{inj}})$ for these model structures, by transfering the direct localizing systems for $\epsilon(n)$-diagrams of Theorem 11.2.2, as was discussed in Section 11.5 using Theorem 9.8.3.

In both cases $\mathcal{R}$ is the class of $\mathcal{M}$-precategories which satisfy the Segal conditions; then K_{proj} (resp. K_{inj}) is the union of the set of generators for trivial cofibrations in the projective (resp. injective) diagram structure, plus the morphisms of the form $U_! P_{(x_0,\ldots,x_n),!}(\zeta_n(f))$ for f in the generating set I of cofibrations of $\mathcal{M}$. Note that the images $PU_{(x_0,\ldots,x_n),!}(g)$ of generating trivial cofibrations for the diagram categories on $\epsilon(n(q))$ are already trivial cofibrations in FUNC $\left(\Delta^o_X/X, \mathcal{M}\right)$, so we don't need to include them again.

Now Theorem 9.7.1 applies to give the required model structures. The characterization of weak equivalences comes from Lemma 9.4.5. □

Definition 12.1.2 A morphism in $\mathcal{PC}(X, \mathcal{M})$ is a *local weak equivalence* if it is a weak equivalence in one of the model structures of the previous theorem.

A morphism $f : \mathcal{A} \to \mathcal{B}$ in $\mathcal{PC}(\mathcal{M})$ is said to be a local weak equivalence, if $\mathrm{Ob}(f) : \mathrm{Ob}(\mathcal{A}) \xrightarrow{\cong} \mathrm{Ob}(\mathcal{B})$ and if the corresponding map $\mathcal{A} \to \mathrm{Ob}(f)^*\mathcal{B}$ in $\mathcal{PC}(\mathrm{Ob}(\mathcal{A}), \mathcal{M})$ is a local weak equivalence.

The operation **Seg** can be chosen to be functorial over morphisms in $\mathcal{PC}(\mathcal{M})$, since it is given by applying the small object argument. A morphism of $\mathcal{M}$-precategories $\mathcal{A} \xrightarrow{f} \mathcal{B}$ is a local weak equivalence, if and only if $\mathrm{Ob}(f)$ induces an isomorphism on sets of objects, and $\mathbf{Seg}(f) : \mathbf{Seg}(\mathcal{A}) \to \mathbf{Seg}(\mathcal{B})$ induces a levelwise weak equivalence of $\Delta^o_{\mathrm{Ob}(\mathcal{A})}$-diagrams.

In case it is necessary, we use the terminology *levelwise weak equivalence* to speak of weak equivalences in the unital diagram category $\mathrm{FUNC}\left(\Delta^o_X/X, \mathcal{M}\right)$.

Lemma 12.1.3 *If $\mathcal{A}, \mathcal{B} \in \mathcal{PC}(X, \mathcal{M})$ satisfy the Segal conditions, then a map $f : \mathcal{A} \to \mathcal{B}$ in $\mathcal{PC}(X, \mathcal{M})$ is a local weak equivalence if and only if it is a levelwise weak equivalence.*

If $\mathcal{B}$ satisfies the Segal conditions and $\mathcal{A} \to \mathcal{B}$ is a trivial fibration in either the projective or injective model structures on $\mathcal{PC}(X, \mathcal{M})$, then $\mathcal{A}$ also satisfies the Segal conditions and f is a levelwise weak equivalence.

Proof Consider the square

$$\begin{array}{ccc} \mathcal{A} & \longrightarrow & \mathbf{Seg}(\mathcal{A}) \\ {\scriptstyle f}\downarrow & & \downarrow{\scriptstyle \mathbf{Seg}(f)} \\ \mathcal{B} & \longrightarrow & \mathbf{Seg}(\mathcal{B}). \end{array}$$

If $\mathcal{A}$ and $\mathcal{B}$ both satisfy the Segal conditions, then the horizontal arrows are levelwise weak equivalences. By the definition of the model structure of Theorem 12.1.1 the map f is a local weak equivalence if and only if $\mathbf{Seg}(f)$ is a levelwise weak equivalence, and by 3 for 2 this is equivalent to requiring that f be a levelwise weak equivalence.

Suppose $\mathcal{B}$ satisfies the Segal conditions, and $f : \mathcal{A} \to \mathcal{B}$ is a trivial fibration in the projective model structure on $\mathcal{PC}(X, \mathcal{M})$. This means that f satisfies right lifting with respect to cofibrations, but the cofibrations are the same as those of the unital diagram category. Thus, f is a trivial fibration in $\mathrm{FUNC}\left(\Delta^o_X/X, \mathcal{M}\right)$, in particular it is a levelwise weak equivalence. Under our hypotheses on $\mathcal{M}$, Lemma 7.7.3 says that cartesian product is invariant under weak equivalences in $\mathcal{M}$, from which it follows that $\mathcal{A}$ satisfies the Segal conditions too. □

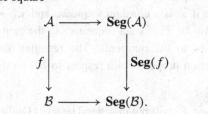

12.2 Unitalization adjunctions

The projective model structure of Theorem 12.1.1 is related to the nonunital version by a Quillen adjunction of unitalization

$$U_! : \mathrm{ALG}_{\mathrm{proj}}\left(\Delta^o_X, P.; \mathcal{M}\right) \leftrightarrow \mathcal{PC}_{\mathrm{proj}}(X, \mathcal{M}) : U^*.$$

This follows from the application of Theorem 9.7.1.

It is useful to describe the unitalization operation for Δ^o_X/X:

Lemma 12.2.1 *Suppose $\mathcal{A} : \Delta^o_X \to \mathcal{M}$ is a functor. Then $U_!\mathcal{A}$ has the following explicit description:*

- *if $x_0 = \ldots = x_n = x$ is a constant sequence then the full degeneracy gives a map $\mathcal{A}(x) \to \mathcal{A}(x, \ldots, x)$ and*

$$(U_!\mathcal{A})(x, \ldots, x) = \mathcal{A}(x, \ldots, x) \cup^{\mathcal{A}(x)} *;$$

- *otherwise, if $x_0, \ldots, x_n$ is not a constant sequence then*

$$(U_!\mathcal{A})(x_0, \ldots, x_n) = \mathcal{A}(x_0, \ldots, x_n).$$

Proof The object explicitly constructed in this way is again a functor $\Delta^o_X \to \mathcal{M}$ because if $x.$ is a constant sequence and $y. \to x.$ is any map in Δ_X then $y.$ must also be a constant sequence, so the map $\mathcal{A}(x_0, \ldots, x_n) \to \mathcal{A}(y_0, \ldots, y_k)$ passes to the quotients. The resulting functor satisfies the required left adjunction property with respect to U^* on the right, so it must be $U_!\mathcal{A}$. □

In the injective case, $U_!$ will not in general be a left Quillen functor; we need to impose an additional hypothesis:

Condition 12.2.2 *Suppose*

$$\begin{array}{ccccc} A & \longrightarrow & X & \longrightarrow & A \\ \downarrow & & \downarrow & & \downarrow \\ B & \longrightarrow & Y & \longrightarrow & B \end{array}$$

is a diagram in $\mathcal{M}$ such that the vertical arrows are cofibrations (resp. trivial cofibrations) and the horizontal compositions are the identity. Then $X \cup^A * \to Y \cup^B *$ is a cofibration (resp. trivial cofibration). Include also the hypothesis that all objects of $\mathcal{M}$ are cofibrant.

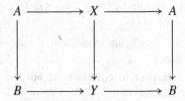

The following observations allow us to use the injective model categories in many cases. In fact, in these cases the injective structure also coincides with the Reedy structure; however, it seems comforting to be able to use the injective structure, which is conceptually simpler, instead.

Lemma 12.2.3 *Suppose $\mathcal{M}$ is a presheaf category, is left proper, and the class of cofibrations is the class of monomorphisms of presheaves. Then Condition 12.2.2 holds.*

Proof Given a diagram of sets as in Condition 12.2.2 where the vertical maps are injections, then the map $X \cup^A * \to Y \cup^B *$ is an injection of sets. Indeed if $x \in X$ maps to an element of B the image of x by the projection to A, maps to the same element of B. By injectivity of the map $X \to Y$ it follows that $x \in A$, which shows injectivity of the map on quotients.

Now if $\mathcal{M}$ is a presheaf category and the cofibrations are the monomorphisms, applying the previous paragraph levelwise we obtain the desired result for cofibrations. Suppose given a diagram whose vertical arrows are trivial cofibrations. The split injections $A \to X$ and $B \to Y$ are monomorphisms, hence cofibrations, so Corollary 7.3.2 applies to conclude that the pushout map is a weak equivalence, hence it is a trivial cofibration. □

Lemma 12.2.4 *If $\mathcal{M}$ satisfies Condition 12.2.2 then the unitalization functors give a Quillen adjunction between injective diagram structures*

$$U_! : \mathrm{ALG}_{\mathrm{inj}}\left(\Delta^o_X, P_\cdot; \mathcal{M}\right) \leftrightarrow \mathcal{PC}_{\mathrm{inj}}(X, \mathcal{M}) : U^*,$$

where U^* is just the identity inclusion of unital precategories in all precategories.

Proof We first show that unitalization is a Quillen adjunction between levelwise injective diagram categories

$$U_! : \mathrm{FUNC}_{\mathrm{inj}}\left(\Delta^o_X, \mathcal{M}\right) \leftrightarrow \mathrm{FUNC}_{\mathrm{inj}}\left(\Delta^o_X/X, \mathcal{M}\right) : U^*.$$

Suppose $\mathcal{A} \to \mathcal{B}$ is a levelwise cofibration (resp. trivial cofibration) of diagrams, the claim is that $U_!\mathcal{A} \to U_!\mathcal{B}$ is a levelwise cofibration (resp. trivial cofibration). In view of the description of Lemma 12.2.1 it suffices to look at the values over a constant sequence $(x, \ldots, x)$. We have a diagram

$$\begin{array}{ccccc} \mathcal{A}(x) & \longrightarrow & \mathcal{A}(x,\ldots,x) & \longrightarrow & \mathcal{A}(x) \\ \downarrow & & \downarrow & & \downarrow \\ \mathcal{B}(x) & \longrightarrow & \mathcal{B}(x,\ldots,x) & \longrightarrow & \mathcal{B}(x) \end{array}$$

where the vertical arrows are cofibrations (resp. trivial cofibrations), and the second horizontal arrows are, say, the projections corresponding to the first object of the sequence. Condition 12.2.2 now says exactly that

$$\begin{array}{c} U_!\mathcal{A}(x,\ldots,x) = \mathcal{A}(x,\ldots,x) \cup^{\mathcal{A}(x)} * \\ \downarrow \\ U_!\mathcal{B}(x,\ldots,x) = \mathcal{B}(x,\ldots,x) \cup^{\mathcal{B}(x)} * \end{array}$$

is a cofibration (resp. trivial cofibration). This shows the Quillen adjunction for the diagram categories.

Now, the categories in question for the lemma are obtained by direct left Bousfield localization using the sets of generators plus the morphisms of the form $P_{q,!}(\zeta_{n(q)}(f))$ for $q \in Q$ and f in a generating set for cofibrations of $\mathcal{M}$. The left Quillen functor passes to a left Quillen functor between localizations by Theorem 9.8.1. □

12.3 The Reedy structure

The category Δ^o_X is a Reedy category, using the subcategories of injective and surjective maps of finite ordered sets, as direct and inverse subcategories, and the length function $(x_0, \ldots, x_n) \mapsto n$. This leads to a *Reedy model structure* on the category of diagrams, using levelwise weak equivalences, denoted $\text{FUNC}_{\text{Reedy}}(\Delta^o_X, \mathcal{M})$ (Proposition 7.6.5).

The notion of Reedy structure on diagram categories is a slightly more technical area of the theory of model categories–it is impossible to give an exhaustive list of references, but see Reedy [217], Hirschhorn [144], Hovey [149], Barwick [19], Berger and Moerdijk [38] and Johnson [155] for example. Nonethelesss, these structures are very natural and turn out to be the best ones for our theory of precategories. In many useful examples the Reedy structure coincides with the injective structure. This is the case for example if $\mathcal{M} = \text{PRESH}(\Psi)$ is a presheaf category and the cofibrations of $\mathcal{M}$ are the monomorphisms of presheaves, see Proposition 13.7.2.

There is a unital version of the Reedy model structure:

Proposition 12.3.1 *For any tractable left proper model category $\mathcal{M}$, the unital diagram category has a tractable left proper model structure denoted* $\text{FUNC}_{\text{Reedy}}(\Delta^o_X/X, \mathcal{M})$, *related to the non-unital Reedy structure by a Quillen adjunction*

$$U_! : \mathrm{FUNC}_{\mathrm{Reedy}}\left(\Delta^o_X, \mathcal{M}\right) \leftrightarrow \mathrm{FUNC}_{\mathrm{Reedy}}\left(\Delta^o_X/X, \mathcal{M}\right) : U^*$$

using the levelwise weak equivalences. The fibrations (resp. cofibrations) of $\mathrm{FUNC}_{\mathrm{Reedy}}\left(\Delta^o_X/X, \mathcal{M}\right)$ *are exactly the maps f such that $U^*(f)$ is a Reedy fibration (resp. cofibration) in* $\mathrm{FUNC}_{\mathrm{Reedy}}\left(\Delta^o_X, \mathcal{M}\right)$*, in particular they have the same description in terms of latching and matching objects. The generating sets for the unital Reedy structure are obtained by applying $U_!$ to the generating sets for the usual Reedy diagram structure.*

The Reedy structure lies in between the projective and injective structures with a diagram of left Quillen functors

$$\begin{array}{ccccc}
\mathrm{FUNC}_{\mathrm{proj}}\left(\Delta^o_X, \mathcal{M}\right) & \longrightarrow & \mathrm{FUNC}_{\mathrm{Reedy}}\left(\Delta^o_X, \mathcal{M}\right) & \longrightarrow & \mathrm{FUNC}_{\mathrm{inj}}\left(\Delta^o_X, \mathcal{M}\right) \\
\downarrow U_! & & \downarrow U_! & & \\
\mathrm{FUNC}_{\mathrm{proj}}\left(\Delta^o_X/X, \mathcal{M}\right) & \to & \mathrm{FUNC}_{\mathrm{Reedy}}\left(\Delta^o_X/X, \mathcal{M}\right) & \to & \mathrm{FUNC}_{\mathrm{inj}}\left(\Delta^o_X/X, \mathcal{M}\right)
\end{array}$$

where the horizontal rows are identity functors. If $\mathcal{M}$ satisfies Condition 12.2.2 then this can be completed by putting in the rightmost vertical arrow.

Proof The first thing to show is that if $\mathcal{A} \xrightarrow{f} \mathcal{B}$ is a Reedy cofibration of diagrams on Δ^o_X, then $U^*U_!\mathcal{A} \xrightarrow{U^*U_!f} U^*U_!\mathcal{B}$ is again a Reedy cofibration. The latching objects over a sequence $(x_0, \ldots, x_n) \in \Delta^o_X$ involve maps which correspond to surjections $(x_0, \ldots, x_n) \xrightarrow{\sigma} (y_0, \ldots, y_k)$ of sequences, with $\sigma : [n] \twoheadrightarrow [k]$ with $y_{\sigma(i)} = x_i$. In particular, any $(y_0, \ldots, y_k)$ involved in the latching map at a non-constant sequence $(x_0, \ldots, x_n)$, is also not constant. Over these sequences the relative latching map for $U^*U_!f$ is the same as for f so the Reedy condition is preserved. We may therefore concentrate on the case of a constant sequence. Let $x^n := (x_0, \ldots, x_n)$ with $x_i = x$. Let $\mathcal{A}' := U^*U_!\mathcal{A}$, $\mathcal{B}' := U^*U_!\mathcal{B}$, and $f' := U^*U_!f$. Thus

$$\mathcal{A}'(x^n) = \mathcal{A}(x^n) \cup^{\mathcal{A}(x)} *, \quad \mathcal{B}'(x^n) = \mathcal{B}(x^n) \cup^{\mathcal{B}(x)} *.$$

The latching object for $\mathcal{A}$ is expressed as the pushout

$$\begin{array}{ccc}
\coprod_{0<i<j\leq n} \mathcal{A}(x^{n-2}) & \to & \coprod_{0<i\leq n} \mathcal{A}(x^{n-1}) \\
\downarrow & & \downarrow \\
\coprod_{0<i\leq n} \mathcal{A}(x^{n-1}) & \longrightarrow & \mathrm{latch}(\mathcal{A}, x^n)
\end{array}$$

and similarly for $\mathrm{latch}(\mathcal{B}, x^n)$. The coproducts may be considered as being taken over contractions of adjacent pairs in the sequence $(x_0, \ldots, x_n)$ with all $x_i = x$.

We get a map from $\mathcal{A}(x)$ into all elements of the above cocartesian diagram, and then the same diagram for $\mathcal{A}'$ gives

$$\mathrm{latch}(\mathcal{A}', x^n) = \mathrm{latch}(\mathcal{A}, x^n) \cup^{\mathcal{A}(x)} *.$$

Similarly

$$\mathrm{latch}(\mathcal{B}', x^n) = \mathrm{latch}(\mathcal{B}, x^n) \cup^{\mathcal{B}(x)} *.$$

Recall that

$$\mathrm{latch}(f, x^n) := \mathrm{latch}(\mathcal{B}, x^n) \cup^{\mathrm{latch}(\mathcal{A}, x^n)} \mathcal{A}(x^n) \to \mathcal{B}(x^n).$$

Thus $\mathrm{latch}(f', x^n)$ is obtained by contracting out $\mathcal{B}(x)$ in the domain and range, in particular

$$\mathrm{latch}(\mathcal{B}', x^n) \cup^{\mathrm{latch}(\mathcal{A}', x^n)} \mathcal{A}'(x^n) = * \cup^{\mathcal{B}(x)} \mathrm{latch}(\mathcal{B}, x^n) \cup^{\mathrm{latch}(\mathcal{A}, x^n)} \mathcal{A}(x^n),$$

as can be seen by using the universal property of amalgamated sums. In the diagram

$$\begin{array}{ccccc}
\mathcal{B}(x) & \longrightarrow & \mathrm{latch}(\mathcal{B}, x^n) \cup^{\mathrm{latch}(\mathcal{A}, x^n)} \mathcal{A}(x^n) & \longrightarrow & \mathcal{B}(x^n) \\
\downarrow & & \downarrow & & \downarrow \\
* & \longrightarrow & \mathrm{latch}(\mathcal{B}', x^n) \cup^{\mathrm{latch}(\mathcal{A}', x^n)} \mathcal{A}'(x^n) & \longrightarrow & \mathcal{B}'(x^n)
\end{array}$$

the left square is cocartesian, and the outer square is cocartesian, therefore the right square is cocartesian. The upper right map is assumed to be a cofibration, so the bottom right map is also a cofibration.

This shows that $U^*U_!f$ is a Reedy cofibration whenever f is. The same argument shows that if f is a Reedy trivial cofibration then $U^*U_!f$ is a Reedy trivial cofibration.

To construct the model structure on the unital diagram category, define the weak equivalences (resp. cofibrations, fibrations) to be those arrows f in $\mathrm{FUNC}_{\mathrm{Reedy}}(\Delta^o_X/X, \mathcal{M})$ such that U^*f is a levelwise weak equivalence (resp. Reedy cofibration, Reedy fibration). In view of this definition, the same holds for the intersection classes of trivial fibrations and trivial cofibrations. Since U^* takes compositions to compositions and retracts to retracts, the classes are all closed under retracts, and levelwise weak equivalences satisfy 3 for 2.

Suppose I_R and J_R are generating sets for the Reedy structure on non-unital diagrams $\text{FUNC}_{\text{Reedy}}(\Delta_X^o, \mathcal{M})$, and let $I'_R := U_! I_R$ and $J'_R := U_! J_R$. Apply Proposition 8.4.1. Note that elements of I'_R are cofibrations (CG2a) and elements of J'_R are trivial cofibrations (CG3a), by the above arguments. All objects are small (CG1).

To complete, we need to show that a map $g : U \to V$ is a fibration (resp. trivial fibration) if and only if it satisfies lifting with respect to I'_R (resp. J'_R). But this is an immediate consequence of the definitions and the adjunction property. For example, g is a fibration $\Leftrightarrow U^* g$ is a fibration $\Leftrightarrow U^* g$ satisfies right lifting with respect to $J_R \Leftrightarrow g$ satisfies right lifting with respect to $U_! J_R = J'_R$. The proof for trivial fibrations is the same.

Applying Proposition 8.4.1 we get a cofibrantly generated model category, from which it also follows that $\mathbf{cof}(I'_R)$ is the class of cofibrations, and $\mathbf{cof}(J'_R)$ is the class of trivial cofibrations.

The Reedy diagram structure is tractable, so we may assume that the domains of elements of I_R and J_R are cofibrant, from which it follows that the same is true of I'_R and J'_R. Reedy cofibrations are in jective cofibrations. But Reedy-unital cofibrations (i.e. cofibrations in $\text{FUNC}_{\text{Reedy}}(\Delta_X^o/X, \mathcal{M})$) are maps whose U^* are Reedy diagram cofibrations, hence they are levelwise cofibrations. Also weak equivalences are defined levelwise; so left properness may be verified levelwise.

For the second paragraph of the theorem, the upper sequence of left Quillen functors

$$\text{FUNC}_{\text{proj}}(\Delta_X^o, \mathcal{M}) \to \text{FUNC}_{\text{Reedy}}(\Delta_X^o, \mathcal{M}) \to \text{FUNC}_{\text{inj}}(\Delta_X^o, \mathcal{M})$$

is classical. From this the lower sequence follows too. To go from the projective to the Reedy structure, the generators of the projective unital diagram structure are obtained by applying $U_!$ to generators of the projective non-unital structure, but these go to Reedy cofibrations (resp. trivial cofibrations) in the non-unital Reedy structure, which in turn get sent to the same in the unital Reedy structure. To go from the Reedy to the injective structure, note that the unital Reedy cofibrations are also non-unital Reedy cofibrations by definition, and these are levelwise cofibrations by the upper sequence. The same is true for trivial cofibrations because the weak equivalences are the same in all cases.

If $\mathcal{M}$ satisfies Condition 12.2.2 then $U_!$ is a left Quillen functor by Lemma 12.2.4 and it clearly makes the diagram commute. This completes the proof. □

Theorem 12.3.2 *Suppose $\mathcal{M}$ is a tractable left proper cartesian model category. The model structure of* $\text{FUNC}_{\text{Reedy}}(\Delta_X^o/X, \mathcal{M})$ *admits a direct left*

Bousfield localization denoted $\mathcal{PC}_{\mathrm{Reedy}}(X, \mathcal{M})$ *by a direct localizing system* $(\mathcal{R}, K_{\mathrm{Reedy}})$*, where* $\mathcal{R}$ *is the class* $\mathcal{M}$*-precategories satisfying the Segal conditions, and* K_{Reedy} *is the generating set for trivial Reedy cofibrations, plus the set of cofibrations of the form* $U_! P_{(x_0,\ldots,x_n),!}(\rho_n(f))$ *for* f *a generating cofibration of* $\mathcal{M}$*. The cofibrations of* $\mathcal{PC}_{\mathrm{Reedy}}(X, \mathcal{M})$ *are the Reedy cofibrations in the diagram category* $\mathrm{FUNC}_{\mathrm{Reedy}}\left(\Delta^o_X, \mathcal{M}\right)$*. The fibrant objects of* $\mathcal{PC}_{\mathrm{Reedy}}(X, \mathcal{M})$ *are the Reedy fibrant diagrams which satisfy the Segal conditions. The Reedy structure lies between the projective and injective structures, with the identity functors being left Quillen*

$$\mathcal{PC}_{\mathrm{proj}}(X, \mathcal{M}) \to \mathcal{PC}_{\mathrm{Reedy}}(X, \mathcal{M}) \to \mathcal{PC}_{\mathrm{inj}}(X, \mathcal{M}).$$

The weak equivalences are the same (the local weak equivalences) for the three structures.

Proof This is analogous to the proof of Theorem 12.1.1 using Theorem 9.7.1 and Theorem 9.8.3. The only difference worth pointing out here is that we can use the morphisms $\rho_n(f)$ defined just before Proposition 11.1.4, in place of the $\zeta_n(f)$ from Section 11.2.1.

We claim that

$$U_! P_{(x_0,\ldots,x_n),!}(\varpi_n(f)) \xrightarrow{U_! P_{(x_0,\ldots,x_n),!}(\rho_n(f))} U_! P_{(x_0,\ldots,x_n),!}(\xi_{0,!}(V))$$

is a Reedy cofibration if $U \xrightarrow{f} V$ is a cofibration in $\mathcal{M}$. Recall from Equation (11.1.1) that

$$\varpi_n(f) := \xi_{0,!}(U) \cup^{\coprod_i \xi_{i,!}(U)} \coprod_i \xi_{i,!}(V) = (U, V, \ldots, V; f, \ldots, f).$$

It will be more convenient to consider an arbitrary $\mathcal{A} \in \mathcal{PC}(X, \mathcal{M})$ with a map $U_! P_{(x_0,\ldots,x_n),!}(\varpi_n(f)) \xrightarrow{a} \mathcal{A}$ and let

$$\mathcal{B} := \mathcal{A} \cup^{U_! P_{(x_0,\ldots,x_n),!}(\varpi_n(f))} U_! P_{(x_0,\ldots,x_n),!}(\xi_{0,!}(V))$$

be the pushout along $U_! P_{(x_0,\ldots,x_n),!}(\rho_n(f))$. We would like to show that $\mathcal{A} \to \mathcal{B}$ is a cofibration.

The map a corresponds to a collection of maps $a_i : V \to \mathcal{A}(x_{i-1}, x_i)$ and $a_0 : U \to \mathcal{A}(x_0, \ldots, x_n)$ such that the diagrams

$$\begin{array}{ccc} U & \longrightarrow & \mathcal{A}(x_0, \ldots, x_n) \\ \downarrow & & \downarrow \\ V & \longrightarrow & \mathcal{A}(x_{i-1}, x_i) \end{array}$$

commute. We can describe $\mathcal{B}$ explicitly. For any sequence $(y_0, \ldots, y_k)$,

$$\mathcal{B}(y_0, \ldots, y_k) = \mathcal{A}(y_0, \ldots, y_k) \cup^{\coprod_{y.\to x.} U} \left(\coprod_{y.\to x.} V \right),$$

where the coproduct is taken over all arrows $y. \to x.$ which don't factor through one of the adjacent pairs $(x_{i-1}, x_i) \to (x_0, \ldots, x_n)$. The left map in the amalgamated sum uses a_0, whereas the functoriality maps to define the functor $\mathcal{B} : \Delta^o_X \to \mathcal{M}$ are made using the a_i for $i = 1, \ldots, n$. Suppose $(z_0, \ldots, z_l) \to (y_0, \ldots, y_k)$ is a map. For any $h : y. \to x.$ which doesn't factor through an adjacent pair, if the composition $z. \to x.$ does factor through an adjacent pair (x_{i-1}, x_i) then we use the map $a_i : V \to \mathcal{A}(x_{i-1}, x_i) \to \mathcal{A}(z_0, \ldots, x_l)$ on the component V of $\mathcal{B}(y_0, \ldots, y_k)$ corresponding to h. The other maps of functoriality are straightforward.

From this description we already see easily that $\mathcal{A} \to \mathcal{B}$ is an injective (i.e. levelwise) cofibration. To show that it is a Reedy cofibration, consider $(y_0, \ldots, y_k)$. The latching objects are colimits over the category of surjections $(y_0, \ldots, y_k) \stackrel{\sigma}{\twoheadrightarrow} (z_0, \ldots, z_l)$ with $l < k$. For any σ, the arrows $z. \to x.$ which don't factor through an adjacent pair form a subset of the similar arrows $y. \to x.$. We can write

$$\mathrm{latch}(\mathcal{B}, y.) = \mathrm{latch}(\mathcal{A}, y.) \cup^{\coprod_{y.\to x.,\mathrm{fact}} U} \left(\coprod_{y.\to x.,\mathrm{fact}} V \right),$$

where the coproduct is now taken over $y. \to x.$ which factor through some surjection $y. \to z.$ in the latching category. The relative latching object for the map $\mathcal{A} \to \mathcal{B}$ is then

$$\mathcal{A}(y.) \cup^{\coprod_{y.\to x.,\mathrm{fact}} U} \left(\coprod_{y.\to x.,\mathrm{fact}} V \right) \to \mathcal{A}(y.) \cup^{\coprod_{y.\to x.} U} \left(\coprod_{y.\to x.} V \right).$$

Thus, the relative latching map consists just of adding on some additional pushouts along $U \to V$. It is a cofibration, which shows that $\mathcal{A} \to \mathcal{B}$ is a Reedy cofibration.

Now the same proof as for Theorem 12.1.1 shows that we can get a direct localizing system for the class $\mathcal{R}$ of $\mathcal{M}$-enriched Segal categories by starting with the generating set for trivial cofibrations in $\mathrm{FUNC}_{\mathrm{Reedy}}(\Delta^o_X/X, \mathcal{M})$ and adding the $U_! P_{(x_0,\ldots,x_n),!}(\rho_n(f))$ to get the associated K_{Reedy}.

We could also have added the $U_! P_{(x_0,\ldots,x_n),!}(\zeta_n(f))$, or both. These all give the same left Bousfield localization (Proposition 9.7.2). Using the

$U_!P_{(x_0,\ldots,x_n),!}(\zeta_n(f))$, Theorem 9.8.1 gives the left Quillen functors from the projective to the Reedy and then to the injective structure.

We can remark that the above proof also shows that we could have used the $U_!P_{(x_0,\ldots,x_n),!}(\rho_n(f))$ to form a direct localizing system for the injective structure. □

For the proof in Chapter 19 it will be useful to consider the relationship between the direct localizing systems for different object sets X. Let $K^X_{\text{Reedy}} \subset \text{ARR}(\mathcal{PC}(X,\mathcal{M}))$ denote the direct localizing system of Theorem 12.3.2. If $f : Y \to X$ is a map of sets, let $f_!(K^Y_{\text{Reedy}})$ denote the collection of pushforwards. This will specially be used with $Y = [k] = \{\upsilon_0,\ldots,\upsilon_k\}$.

Proposition 12.3.3 *For any $\mathcal{A} \in \mathcal{PC}(X,\mathcal{M})$ the following conditions are equivalent:*

(1) $\mathcal{A}$ is fibrant in $\mathcal{PC}_{\text{Reedy}}(X,\mathcal{M})$;
(2) $\mathcal{A}$ is a Reedy fibrant diagram in $\text{FUNC}_{\text{Reedy}}(\Delta^o_X/X,\mathcal{M})$ and satisfies the Segal conditions;
(3) $\mathcal{A}$ is K^X_{Reedy}-injective;
(4) $\mathcal{A}$ is $(z_!(K^{[k]}_{\text{Reedy}}))$-injective for any sequence of objects $z : [k] \to X$;
(5) $\mathcal{A} \to *$ satisfies the right lifting property in $\mathcal{PC}(\mathcal{M})$ with respect to the morphisms of $K^{[k]}_{\text{Reedy}}$ for all k.

Suppose $\mathcal{A} \xrightarrow{f} \mathcal{B}$ is a morphism in $\mathcal{PC}_{\text{Reedy}}(X,\mathcal{M})$ such that $\mathcal{B}$ is fibrant. Then the following conditions are equivalent:

(a) f is a fibration;
(b) f is a fibration in $\text{FUNC}_{\text{Reedy}}(\Delta^o_X/X,\mathcal{M})$ and for any sequence $x_0,\ldots,x_n \in X$ the pullback $P^*_{(x_0,\ldots,x_n)}(f)$ is a fibration in $\text{FUNC}_{\text{inj},\sqcap}(\epsilon(n),\mathcal{M})$;
(c) f is K^X_{Reedy}-injective;
(d) f is $(z_!(K^{[k]}_{\text{Reedy}}))$-injective for any sequence of objects $z : [k] \to X$;
(e) f satisfies the right lifting property in $\mathcal{PC}(\mathcal{M})$ with respect to the morphisms of $K^{[k]}_{\text{Reedy}}$ for all k.

Proof The equivalence between (1) and (2) is part of the statement of the preceding theorem. The equivalence with (3) comes from Proposition 8.7.4, which is applicable because the construction via direct left Bousfield localization used the pseudo-generating recognition theorem 8.7.3. To show the equivalence between (2) and (4), note that $\mathcal{A}$ is Reedy fibrant as a diagram $\Delta^o_X \to \mathcal{M}$ (unital along X), if and only if the pullback $z^*(\mathcal{A})$ is Reedy fibrant in $\text{FUNC}_{\text{Reedy}}(\Delta^o_{[k]}/[k],\mathcal{M})$ for any k and any sequence of points $z : [k] \to X$.

Indeed the Reedy fibration condition for diagrams defined using the matching object construction, is a pointwise condition to be verified at each sequence, and the matching object for $\mathcal{A}$ at a sequence $z : [k] \to X$ is the same as the matching object for $z^*(\mathcal{A})$ at the tautological sequence in $[k]$. Similarly, the Segal condition for $\mathcal{A}$ depends only on the restriction to the various sequences. Thus, condition (2) for $\mathcal{A}$ is equivalent to the conjunction of conditions (2) for all $z^*(\mathcal{A})$, and this in turn is equivalent to (4) by the equivalences (2)$\Leftrightarrow$(3) for all the $z^*(\mathcal{A})$. Condition (5) is identical to (4).

The proof of the second part is analogous, taking note of Theorem 9.8.3 for the equivalence with condition (b). □

Remark 12.3.4 The same result holds for the projective model structure of Theorem 12.1.1, with analogous proof.

12.4 Global weak equivalences

We turn now from consideration of the various model categories for $\mathcal{M}$-precategories with a fixed set of objects, to the global category $\mathcal{PC}(\mathcal{M})$ of $\mathcal{M}$-precategories with arbitrary variable set of objects. We'll get back to a closer analysis of the generation operation **Seg** on $\mathcal{PC}(X, \mathcal{M})$ in Chapter 14 on the calculus of generators and relations, but first we consider weak equivalences in the global case, and in the next Chapter 13, various types of cofibrations in the global case.

In this section we formalize Tamsamani's induction step for defining equivalences of n-categories, as applied in the $\mathcal{M}$-enriched case. For classical Segal categories which is the $\mathcal{K}$-enriched case, this notion of equivalence goes back at least to Dwyer and Kan and is known as "Dwyer–Kan equivalence" in the work of Rezk [218] and Bergner [44]. The general case requires a functor $\tau_{\leq 0} : \mathcal{M} \to \text{SET}$ in order to be able to talk about the essential surjectivity condition; this issue was explored by Pellissier, but we have modified his situation a little bit by introducing sets X of objects external to $\mathcal{M}$. Our discussion here specializes the more general discussion in Section 3.3.

Define the *zeroth truncation functor* $\tau_{\leq 0} : \mathcal{M} \to \text{SET}$ by

$$\tau_{\leq 0}(\mathcal{A}) := \text{HOM}_{\text{ho}(\mathcal{M})}(*, \mathcal{A}).$$

Then define a *truncation functor* $\tau_{\leq 1} : \mathcal{PC}(\mathcal{M}) \to \text{CAT}$ as follows: for $\mathcal{A} \in \mathcal{M}$ with $\text{Ob}(\mathcal{A}) = X$, choose a local weak equivalence in $\mathcal{PC}(X, \mathcal{M})$ from $\mathcal{A}$ to an $\mathcal{M}$-enriched Segal category $\textbf{Seg}(\mathcal{A})$ (i.e. a precategory which satisfies the Segal conditions). Consider the functor from Δ_X to SET defined by

$$\tau_{\leq 1}(\mathcal{A})(x_0, \ldots, x_n) := \tau_{\leq 0}(\mathbf{Seg}(\mathcal{A})(x_0, \ldots, x_n))$$
$$= \mathrm{HOM}_{\mathrm{ho}(\mathcal{M})}(*, \mathbf{Seg}(\mathcal{A})(x_0, \ldots, x_n)).$$

The Segal maps for this functor are isomorphisms (proven in the next lemma), so it defines a 1-category whose object set is X and such that the set of morphisms from x_0 to x_1 is $\tau_{\leq 1}(\mathcal{A})(x_0, x_1) = \tau_{\leq 0}(\mathbf{Seg}(\mathcal{A})(x_0, x_1)$.

One good choice for $\mathbf{Seg}(\mathcal{A})$ would be to take a fibrant replacement, however other smaller choices can be useful too (see Chapter 14 below).

Lemma 12.4.1 *The truncation $\tau_{\leq 0} : \mathcal{M} \to$ SET sends weak equivalences to isomorphisms and products to products. The SET-precategory $\tau_{\leq 1}(\mathcal{A})$ defined above satisfies the Segal conditions so it corresponds to a 1-category, and this category is independent of the choice of $\mathbf{Seg}(\mathcal{A})$. It gives a functor $\tau_{\leq 1} : \mathcal{PC}(\mathcal{M}) \to$ CAT, and for a fixed object set X the functor takes local weak equivalences in $\mathcal{PC}(X, \mathcal{M})$ to isomorphisms.*

Proof The $\tau_{\leq 0}$ factors through the projection to the homotopy category so it clearly preserves weak equivalences. It preserves products since $\mathcal{M}$ is assumed to be cartesian. In particular the Segal condition for $\mathbf{Seg}(\mathcal{A})$ yields the Segal condition for the Δ^o_X-set $\tau_{\leq 1}(\mathcal{A})$ to be the nerve of a category, independent of the choice of $\mathbf{Seg}(\mathcal{A})$ because $\tau_{\leq 0}$ sends weak equivalences to isomorphisms. We get a functor $\tau_{\leq 1}$. If $\mathcal{A} \to \mathcal{B}$ is a local weak equivalence in $\mathcal{PC}(X, \mathcal{M})$ then $\mathbf{Seg}(\mathcal{A}) \to \mathbf{Seg}(\mathcal{B})$ is a levelwise weak equivalence (see Theorem 12.1.1) so the resulting $\tau_{\leq 1}(\mathcal{A}) \to \tau_{\leq 1}(\mathcal{B})$ is fully faithful, but in the present case it is also the identity on the set X of objects so it is an isomorphism. □

Lemma 12.4.2 *If $\mathcal{A}$ and $\mathcal{B}$ are $\mathcal{M}$-precategories satisfying the Segal condition then there is a natural isomorphism*

$$\tau_{\leq 1}(\mathcal{A} \times \mathcal{B}) \cong \tau_{\leq 1}(\mathcal{A}) \times \tau_{\leq 1}(\mathcal{B}).$$

Proof As $\tau_{\leq 0}$ is compatible with products this follows from the definition of $\tau_{\leq 1}$. □

The truncation operation is crucial to Tamsamani's construction [250] of n-categories, since it allows us to define equivalences between enriched category objects. In the case of n-categories, $\mathcal{M}$ corresponds to the model category for $(n-1)$-categories and $\mathrm{ho}(\mathcal{M})$ can be identified with the homotopy category of Segal $(n-1)$-categories. For Segal n-categories, the truncation as we have defined it here coincides with Tamsamani's truncation operation which was constructed by induction on n. The general iteration procedure requires a general definition of truncation starting from a model category $\mathcal{M}$. This more general case was considered by Pellissier [211, definition 1.4.1].

We now proceed with the definition of global weak equivalence, combining two conditions: "fully faithful" is defined by asking for weak equivalences in $\mathcal{M}$ between morphism objects, and "essentially surjective" is defined using the above truncation operation.

A map $f : (X, \mathcal{A}) \to (Y, \mathcal{B})$ consists of a map $\mathrm{Ob}(f) : X \to Y$, which we sometimes denote by f for short, together with maps $f : \mathcal{A}(x_0, \ldots, x_p) \to \mathcal{B}(f(x_0), \ldots, f(x_p))$ in $\mathcal{M}$ for each $(x_0, \ldots, x_p) \in \Delta_X$, compatible with restrictions along maps in Δ_X.

Suppose that $(X, \mathcal{A})$ and $(Y, \mathcal{B})$ are $\mathcal{M}$-precategories satisfying the Segal condition. A map $f : (X, \mathcal{A}) \to (Y, \mathcal{B})$ is *fully faithful* if, for every sequence $(x_0, \ldots, x_n)$ of objects in X, the map

$$f(x_0, \ldots, x_n) : \mathcal{A}(x_0, \ldots, x_n) \to \mathcal{B}(f(x_0), \ldots, f(x_n))$$

is a weak equivalence in $\mathcal{M}$.

If $(X, \mathcal{A})$ and $(Y, \mathcal{B})$ are only $\mathcal{M}$-precategories, let $\mathcal{A} \to \mathbf{Seg}(\mathcal{A})$ and $\mathcal{B} \to \mathbf{Seg}(\mathcal{B})$ denote injective trivial cofibrations towards objects satisfying the Segal conditions–for example, fibrant replacements in the Reedy model structures on $\mathcal{PC}_{\text{Reedy}}(X, \mathcal{M})$ and $\mathcal{PC}_{\text{Reedy}}(Y, \mathcal{M})$. These were constructed in Theorem 12.3.2, although one could also use the projective structure of Theorem 12.1.1 or the injective one when it exists. The technique goes back to Section 11.5, which in turn applies the direct left Bousfield localisation of Chapter 9.

Recall that $\mathrm{Ob}(f)^*(\mathcal{B})$ is an $\mathcal{M}$-precategory on $X = \mathrm{Ob}(\mathcal{A})$, as is $\mathrm{Ob}(f)^*(\mathbf{Seg}(\mathcal{B}))$. Furthermore, the map $\mathrm{Ob}(f)^*(\mathcal{B}) \to \mathrm{Ob}(f)^*(\mathbf{Seg}(\mathcal{B}))$ is again a local weak equivalence in the model structure $\mathcal{PC}_{\text{Reedy}}(X, \mathcal{M})$ (see Lemma 10.3.2). It is also again a levelwise cofibration, so it is again a trivial cofibration.

By making an appropriate choice of construction of the operation $\mathbf{Seg}(\cdot)$ we can obtain that f induces a map $\mathbf{Seg}(\mathcal{A}) \to \mathrm{Ob}(f)^*(\mathbf{Seg}(\mathcal{B}))$; however, in any case we could modify the choice of $\mathbf{Seg}(\mathcal{A})$ in order to obtain such a factorization. So we shall assume that this factorization is given. This results in a map $\mathbf{Seg}(\mathcal{A}) \to \mathbf{Seg}(\mathcal{B})$.

We say that $f : \mathcal{A} \to \mathcal{B}$ is *fully faithful* if the map $\mathbf{Seg}(\mathcal{A}) \to \mathbf{Seg}(\mathcal{B})$ is fully faithful in the previous sense, i.e. it induces a levelwise equivalence

$$\mathbf{Seg}(\mathcal{A})(x_0, \ldots, x_p) \xrightarrow{\sim} \mathbf{Seg}(\mathcal{B})(f(x_0), \ldots, f(x_p)),$$

for each sequence of objects $(x_0, \ldots, x_p) \in \Delta_{\mathrm{Ob}(\mathcal{A})}$. Since $\mathbf{Seg}(\mathcal{A})$ and $\mathbf{Seg}(\mathcal{B})$ both satisfy the Segal condition, it is easy to see that it is sufficient to check this on sequences of length one, that is it suffices to require that

$$\mathbf{Seg}(\mathcal{A})(x, y) \xrightarrow{\sim} \mathbf{Seg}(\mathcal{B})(f(x), f(y)),$$

for every pair of objects $x, y \in \mathrm{Ob}(\mathcal{A})$.

We say that a map $f : (X, \mathcal{A}) \to (Y, \mathcal{B})$ is *essentially surjective* if the functor of categories obtained by the truncation operation $\tau_{\leq 1}(f) : \tau_{\leq 1}(X, \mathcal{A}) \to \tau_{\leq 1}(Y, \mathcal{B})$ is an essentially surjective map of categories, in other words if the set of isomorphism classes $\mathbf{Iso}\tau_{\leq 1}(\mathcal{A})$ of $\tau_{\leq 1}(X, \mathcal{A})$ surjects to the set of isomorphism classes $\mathbf{Iso}\tau_{\leq 1}(\mathcal{B})$ of $\tau_{\leq 1}(Y, \mathcal{B})$.

Definition 12.4.3 We say that $f : (X, \mathcal{A}) \to (Y, \mathcal{B})$ is a *global weak equivalence* if it is fully faithful and essentially surjective.

Lemma 12.4.4 *A morphism $f : \mathcal{A} \to \mathcal{B}$ which induces an isomorphism on sets of objects $\mathrm{Ob}(f) : \mathrm{Ob}(\mathcal{A}) \cong \mathrm{Ob}(\mathcal{B})$ is a global weak equivalence if and only if the corresponding map $f_\sharp : \mathrm{Ob}(f)_!\mathcal{A} \to \mathcal{B}$ in $\mathcal{PC}(\mathrm{Ob}(\mathcal{B}), \mathcal{M})$ is a local weak equivalence in the model structures of Theorems 12.1.1 and 12.3.2.*

In particular, if $\mathrm{Ob}(f)$ is an isomorphism and $f_\sharp$ is a levelwise weak equivalence of $\Delta^o_{\mathrm{Ob}(\mathcal{B})}$-diagrams in $\mathcal{M}$, then f is a global weak equivalence.

Proof This is immediate from the definitions, since the essential surjectivity is automatically guaranteed by the condition that $\mathrm{Ob}(f)$ be an isomorphism. For the last paragraph note that Corollary 9.3.4 applies in the direct left Bousfield localizations of Theorems 12.1.1 and 12.3.2. □

12.5 Categories enriched over ho($\mathcal{M}$)

As in the previous section, the small object argument for the pseudo-generating sets for trivial cofibrations in $\mathcal{PC}_{\mathrm{proj}}(X, \mathcal{M})$, leads to a functorial construction which associates to any $\mathcal{A} \in \mathcal{PC}(\mathcal{M})$ another precategory $\mathbf{Seg}(\mathcal{A}) \in \mathcal{PC}(\mathrm{Ob}(\mathcal{A}), \mathcal{M})$ together with a natural transformation $\mathcal{A} \to \mathbf{Seg}(\mathcal{A})$, which is a local weak equivalence in $\mathcal{PC}(\mathrm{Ob}(\mathcal{A}), \mathcal{M})$, such that $\mathbf{Seg}(\mathcal{A})$ satisfies the Segal condition. For the present section, fix one such construction, although the discussion is left invariant if we make a different choice.

The category $\mathrm{ho}(\mathcal{M})$ admits finite cartesian products, and these are calculated by cartesian products in $\mathcal{M}$ using conditions (DCL) and (PROD) (Lemma 7.7.3). Therefore, it makes sense to look at the category $\mathrm{C}\textsc{at}(\mathrm{ho}(\mathcal{M}))$ of $\mathrm{ho}(\mathcal{M})$-enriched categories, see Section 7.9. Recall that a $\mathrm{ho}(\mathcal{M})$-enriched category C consists of a set $\mathrm{Ob}(C)$ and for any $x, y \in \mathrm{Ob}(C)$, a morphism object $C(x, y) \in \mathrm{ho}(\mathcal{M})$, plus composition maps satisfying strict associativity.

Lemma 12.5.1 *Suppose $\mathcal{A} \in \mathcal{PC}(\mathcal{M})$. Then we obtain a* ho($\mathcal{M}$)*-enriched category denoted* **eh**($\mathcal{A}$) *with the same set of objects as $\mathcal{A}$, by putting* **eh**($\mathcal{A}$)(x, y) *equal to the class of* **Seg**($\mathcal{A}$)(x, y) *in* ho($\mathcal{M}$). *The remainder of the simplicial diagram structure of* **Seg**($\mathcal{A}$) *(up to* **Seg**($\mathcal{A}$)(x_0, x_1, x_2, x_3)*) provides this with composition maps which are unital and strictly associative. This gives a functor from $\mathcal{PC}(\mathcal{M})$ to* $\mathrm{CAT}(\mathrm{ho}(\mathcal{M}))$.

Proof Composing $\Delta^o_X \xrightarrow{\mathbf{Seg}(\mathcal{A})} \mathcal{M}$ with the projection $\mathcal{M} \to \mathrm{ho}(\mathcal{M})$ we get a functor $\mathbf{eh}(\mathcal{A}) : \Delta^o_X \to \mathrm{ho}(\mathcal{M})$. The homotopy Segal conditions for **Seg**($\mathcal{A}$) imply that the Segal maps for **eh**($\mathcal{A}$) are isomorphisms. By the interpretation of Theorem 7.9.4 (the reader is encouraged by now to have looked at the proof of this theorem, which was left as an exercise), this says that **eh**($\mathcal{A}$) corresponds to a ho($\mathcal{M}$)-enriched category. □

Recall that the truncation $\tau_{\leq 0}$ on $\mathcal{M}$ factors through a functor denoted the same way $\tau_{\leq 0} : \mathrm{ho}(\mathcal{M}) \to \mathrm{SET}$ compatible with cartesian products, so and applying that to the enrichment category gives a functor

$$\tau^h_{\leq 1} : \mathrm{CAT}(\mathrm{ho}(\mathcal{M})) \to \mathrm{CAT}.$$

The image of **eh**($\mathcal{A}$) under this functor is exactly $\tau_{\leq 1}(\mathcal{A})$ according to its construction, that is to say

$$\tau^h_{\leq 1}(\mathbf{eh}(\mathcal{A})) = \tau_{\leq 1}(\mathcal{A}). \tag{12.5.1}$$

Recall from Section 7.9 that a functor $g : C \to C'$ in $\mathrm{CAT}(\mathrm{ho}(\mathcal{M}))$ is an *equivalence of categories* if g is essentially surjective, i.e.

$$\mathbf{Iso}\tau^h_{\leq 1}(g) : \mathbf{Iso}\tau^h_{\leq 1}(C) \to \mathbf{Iso}\tau^h_{\leq 1}(C')$$

is surjective, and if g is fully faithful, i.e. for any $x, y \in \mathrm{Ob}(C)$ the map $C(x, y) \to C'(g(x), g(y))$ is an isomorphism in ho($\mathcal{M}$). As in Lemma 7.9.1, if g is an equivalence of categories then $\mathbf{Iso}\tau^h_{\leq 1}(g)$ is an isomorphism.

Proposition 12.5.2 *A morphism of $\mathcal{M}$-precategories $f : \mathcal{A} \to \mathcal{B}$ is a global weak equivalence in $\mathcal{PC}(\mathcal{M})$ if and only if the corresponding morphism* **eh**(f) : **eh**($\mathcal{A}$) $\to$ **eh**($\mathcal{B}$) *is an equivalence of* ho($\mathcal{M}$)*-enriched categories.*

Proof By the identification (12.5.1) between truncations, the essential surjectivity condition is the same. The fully faithful conditions are the same too, because a morphism in $\mathcal{M}$ is an equivalence if and only if its image in ho($\mathcal{M}$) is an isomorphism. □

This point of view allows us to apply the arguments of Section 7.9, which concerned a simpler 1-categorical notion of enrichment, to obtain some important first properties of the class of global weak equivalences.

Lemma 12.5.3 *Suppose f is a global weak equivalence. Then $\tau_{\leq 1}(f) : \tau_{\leq 1}(\mathcal{A}) \to \tau_{\leq 1}(\mathcal{B})$ is an equivalence of categories, in particular it induces an isomorphism of sets* $\mathbf{Iso}\tau_{\leq 1}(\mathcal{A}) \cong \mathbf{Iso}\tau_{\leq 1}(\mathcal{B})$.

Proof Use Lemma 7.9.1. □

Proposition 12.5.4 *The class of global weak equivalences is closed under retracts and satisfies 3 for 2. Furthermore, if $f : \mathcal{A} \to \mathcal{B}$ and $g : \mathcal{B} \to \mathcal{A}$ are morphisms such that fg and fg are global weak equivalences, then f and g are global weak equivalences.*

Proof Using Proposition 12.5.2, this is an immediate consequence of Theorem 7.9.3. □

12.6 Change of enrichment category

The following discussion is the key to resolving the issue that was raised by Pellissier [211]: he found a mistake in my preprint [234], in what will correspond to our construction of interval objects in Chapter 18 below. Pellissier solved the problem for $\mathcal{M} = \mathcal{K}$, but one can go pretty easily from there to any $\mathcal{M}$ by investigating what happens under change of enrichment category. This strategy will be used to construct interval objects in Chapter 18.

Suppose $F : \mathcal{M} \to \mathcal{M}'$ is a left Quillen functor between tractable left proper cartesian model categories; denote its right adjoint by F^*. Then F induces left Quillen functors between the levelwise diagram model categories

$$F^{\Delta_X^o} : \mathrm{FUNC}_{\mathrm{inj}}\left(\Delta_X^o, \mathcal{M}\right) \to \mathrm{FUNC}_{\mathrm{inj}}\left(\Delta_X^o, \mathcal{M}'\right)$$

and

$$F^{\Delta_X^o} : \mathrm{FUNC}_{\mathrm{proj}}\left(\Delta_X^o, \mathcal{M}\right) \to \mathrm{FUNC}_{\mathrm{proj}}\left(\Delta_X^o, \mathcal{M}'\right),$$

whose right adjoints are obtained by applying the right adjoint F^* levelwise. The Quillen property is verified levelwise on the left for the injective structure and levelwise on the right for the projective structure.

Define the functor

$$\mathcal{PC}(X, F) : \mathcal{PC}(X, \mathcal{M}) \to \mathcal{PC}(X, \mathcal{M}')$$

by putting

$$\mathcal{PC}(X, F)(\mathcal{A}) := U_!\left(F^{\Delta_X^o}(U^*\mathcal{A})\right).$$

The reader may note that if we assume furthermore that F is *unital*, meaning $F(*) = *$, then $\mathcal{PC}(X, F)$ is just the restriction of $F^{\Delta_X^o}$ to the unital diagrams, and the following discussion could be simplified. We don't make that assumption in general.

Lemma 12.6.1 *The functor $\mathcal{PC}(X, F)$ is left adjoint to the functor $\mathcal{PC}(X, F^*)$ obtained by restricting the levelwise adjoint to the unital diagram categories. It is compatible with $F^{\Delta_X^o}$ via the diagram*

$$\begin{array}{ccc} \mathrm{FUNC}\left(\Delta_X^o, \mathcal{M}\right) & \xrightarrow{F^{\Delta_X^o}} & \mathrm{FUNC}\left(\Delta_X^o, \mathcal{M}'\right) \\ \Big\downarrow U_! & & \Big\downarrow U_! \\ \mathcal{PC}(X, \mathcal{M}) & \xrightarrow{\mathcal{PC}(X, F)} & \mathcal{PC}(X, \mathcal{M}'). \end{array}$$

Proof Note that F^* commutes with limits so $F^*(*) = *$, in particular applying F^* levelwise preserves the unitality condition, defining a functor $\mathcal{PC}(X, \mathcal{M}') \xrightarrow{\mathcal{PC}(X,F^*)} \mathcal{PC}(X, \mathcal{M})$. We verify that $\mathcal{PC}(X, F)$ as defined above is its left adjoint. Suppose $\mathcal{A} \in \mathcal{PC}(X, \mathcal{M})$ and $\mathcal{B} \in \mathcal{PC}(X, \mathcal{M}')$. A map $\mathcal{PC}(X, F)(\mathcal{A}) \to \mathcal{B}$ in $\mathcal{PC}(X, \mathcal{M}')$ is the same as a map $F^{\Delta_X^o}(U^*\mathcal{A}) \to U^*\mathcal{B}$, which in turn is the same as a map $U^*\mathcal{A} \to F^{*,\Delta_X^o}U^*\mathcal{B}$, but this is the same as a map $\mathcal{A} \to \mathcal{PC}(X, F^*)\mathcal{B}$.

For the compatibility diagram note that

$$\mathcal{PC}(X, F)(U_!\mathcal{A}) = U_!\left(F^{\Delta_X^o}(U^*U_!\mathcal{A})\right) = U_!\left(F^{\Delta_X^o}(\mathcal{A})\right).$$

This can be seen by the explicit description of the unitalization operation $U^*U_!$ as a pushout at the constant sequences

$$\begin{aligned} U_!\left(F^{\Delta_X^o}(U^*U_!\mathcal{A})\right)(x^n) &= F(U^*U_!\mathcal{A}(x^n)) \cup^{F(U^*U_!\mathcal{A}(x))} * \\ &= F(\mathcal{A}(x^n) \cup^{\mathcal{A}(x)} *) \cup^{F(*)} * = F(\mathcal{A}(x^n)) \cup^{F(\mathcal{A}(x))} * \\ &= U_!\left(F^{\Delta_X^o}(\mathcal{A})\right)(x^n). \end{aligned}$$

Recall that the unitalization is trivial at non-constant sequences. □

Assume that $\mathcal{M}$ and $\mathcal{M}'$ are tractable left proper cartesian model categories. If $F : \mathcal{M} \to \mathcal{M}'$ is a left Quillen functor, then for any finite collection of objects $A_1, \ldots, A_m$ (including the empty collection with $m = 0$) we have

$$F(A_1 \times \cdots \times A_m) \to F(A_1) \times \cdots \times F(A_m).$$

We say that F is *weakly compatible with products* if these maps are weak equivalences for any sequence of objects. In particular, for the case $m = 0$ this says that $F(*)$ is contractible.

Although this natural condition does not seem to be needed for the following proposition, it will be used in Theorem 14.7.3 later.

Proposition 12.6.2 *Suppose that $\mathcal{M} \xrightarrow{F} \mathcal{M}'$ is a left Quillen functor between two tractable left proper cartesian model categories. Then the functor $\mathcal{PC}(X, F)$ is a left Quillen functor between the projective (resp. Reedy) levelwise model category structures on $\mathcal{PC}(X, \mathcal{M})$ and $\mathcal{PC}(X, \mathcal{M}')$. If $\mathcal{M}'$ also satisfies Condition 12.2.2 then $\mathcal{PC}(X, F)$ is also a left Quillen functor between the injective structures.*

Proof We first note that $\mathcal{PC}(X, F)$ is a left Quillen functor between unital diagram categories

$$\text{FUNC}\left(\Delta^o_X/X, \mathcal{M}\right) \to \text{FUNC}\left(\Delta^o_X/X, \mathcal{M}'\right)$$

in the projective and Reedy structures (and the injective structures if Condition 12.2.2 holds).

For the projective structure, the fibrations and trivial fibrations are defined levelwise, but the right adjoint $\mathcal{PC}(X, F^*)$ is also defined levelwise so it preserves levelwise (trivial) fibrations. For the Reedy structure, recall that the diagrammatic Reedy structure is defined using the inclusion U^* from unital to all diagrams: a morphism f of unital diagrams is a Reedy (trivial) cofibration if and only if $U^* f$ is. So $F^{\Delta^o_X}(U^*$ sends Reedy (trivial) cofibrations in $\text{FUNC}_{\text{Reedy}}\left(\Delta^o_X/X, \mathcal{M}\right)$ to Reedy (trivial) cofibrations in $\text{FUNC}_{\text{Reedy}}$ $\left(\Delta^o_X, \mathcal{M}\right)$. But now $U_!$ is left Quillen for the Reedy structure by Proposition 12.3.1, so $\mathcal{PC}(X, F) = U_! F^{\Delta^o_X}(U^*$ preserves Reedy (trivial) cofibrations.

If we assume furthermore Condition 12.2.2 then, by Lemma 12.2.4, $U_!$ is left Quillen for the injective structures so the same argument works. This completes the proof of what is claimed in the first paragraph.

Now $\mathcal{PC}(X, F)$ passes to a left Quillen functor between the direct left Bousfield localizations. We may assume that F sends the generating sets for $\mathcal{M}$ into ones for $\mathcal{M}'$. Then the functor $\mathcal{PC}(X, F)$ sends the pseudo-generators $P_{(x_0,\ldots,x_n),!}(\zeta_n(f))$ for the direct left Bousfield localized structure on $\mathcal{PC}(X, \mathcal{M})$, to pseudo-generators for $\mathcal{PC}(X, \mathcal{M}')$; in particular to trivial cofibrations. By the left Bousfield localization property, $\mathcal{PC}(X, F)$ are left Quillen functors between the direct left Bousfield localizations. □

13
Cofibrations

In this chapter, continuing the construction of model structures for the category $\mathcal{PC}(\mathcal{M})$ of $\mathcal{M}$-enriched precategories with variable set of objects, we define and discuss various classes of cofibrations. The corresponding classes of trivial cofibrations are then defined as the intersection of the cofibrations with the global weak equivalences defined in the preceding chapter. The fibrations are defined by the lifting property with respect to trivial cofibrations. It will have to be proven later that the class of trivial fibrations, defined as the intersection of the fibrations and the global weak equivalences, is also defined by the lifting property with respect to cofibrations.

In reality, we consider three model structures, called the *injective*, the *projective*, and the *Reedy* structures. These will be denoted by subscripts $\mathcal{PC}_{\text{inj}}$, $\mathcal{PC}_{\text{proj}}$ and $\mathcal{PC}_{\text{Reedy}}$ respectively when necessary. They share the same class of weak equivalences, but the cofibrations are different. It turns out that the Reedy structure is the best one for the purposes of iterating the construction. That is a somewhat subtle point because the Reedy structure coincides with the injective structure when $\mathcal{M}$ lies in a wide range of model categories where the monomorphisms are cofibrations, see Proposition 13.7.2. When they are different it is better to take the Reedy route.

13.1 Skeleta and coskeleta

The definition of Reedy cofibrations, as well as the study of the projective ones, is based on consideration of the skeleta and coskeleta of objects in $\mathcal{PC}(\mathcal{M})$. The Reedy structure will be defined in an explicit way without making use of the general definition of Reedy category. The category $\mathcal{PC}(\mathcal{M})$ consists of precategories with various different object sets X, so it isn't enough to just invoke the Reedy structure of each Δ_X. Considering instead the relevant structures explicitly has the added advantage of making the discussion accessible for readers who wish to avoid plunging into the full theory of Reedy categories. Those already familiar with those kinds of things can try to view the discussion in a more general language (see Barwick [19], Berger and Moerdijk [38] and others).

For a fixed set X, consider the subcategory $\Delta_{X,m} \subset \Delta_X$ consisting of all sequences $(x_0, \ldots, x_p)$ of length $p \leq m$. For example, $\Delta_{X,0}$ consists of sequences of length 0, i.e. (x_0) only. Let $\sigma(X, m) : \Delta_{X,m} \hookrightarrow \Delta_X$ denote the inclusion functor. Then we have a functor

$$\mathbf{sk}_m := \sigma(X, m)_! \circ \sigma(X, m)^* : \textsc{Func}(\Delta_X, \mathcal{M}) \to \textsc{Func}(\Delta_X, \mathcal{M}).$$

We call $\mathbf{sk}_m(\mathcal{A}) = \sigma(X, m)_!(\sigma(X, m)^*(\mathcal{A}))$ the *m-skeleton* of $\mathcal{A}$. There is a natural map $\mathbf{sk}_m(\mathcal{A}) \to \mathcal{A}$.

Given m and $\mathcal{A}$ we would like to calculate $\mathbf{sk}_m(\mathcal{A})$. Suppose $x. = (x_0, \ldots, x_k)$ is a sequence of elements of X of length k. Consider the category of all surjections $x. \twoheadrightarrow y.$, where $y. = (y_0, \ldots, y_p)$ is a sequence of length $p \leq m$. Morphisms are morphisms of objects $y.$ in Δ_X commuting with the maps from $x.$. Note that $y. \in \Delta_{X,m}$ so the pullback of $\mathcal{A}$ to $\Delta_{X,m}$ restricts, by the projection functor to the variable $y.$, to a contravariant functor from our category of surjections to $\mathcal{M}$. We claim that

$$\mathbf{sk}_m(\mathcal{A})(x.) = \operatorname{colim}_{x. \twoheadrightarrow y.} \mathcal{A}(y.). \tag{13.1.1}$$

Indeed, this is almost exactly the same as the expression of Lemma 7.6.1 for the pushforward $\sigma(X, m)_!$. In the general expression, the colimit is taken over all morphisms from $x.$ to some $y. \in \Delta_{X,m}$. However, any such morphism canonically factors as $x. \twoheadrightarrow y'_. \hookrightarrow y.$ with $y'_. \in \Delta_{X,m}$ too, so the category of surjections is cofinal and suffices for calculating the colimit.

From this expression we conclude that if $x. = (x_0, \ldots, x_k)$ is a sequence of length $k \leq m$ then the natural map

$$\mathbf{sk}_m(\mathcal{A})(x.) \to \mathcal{A}(x.)$$

is an isomorphism, in other words the restrictions of $\mathcal{A}$ and $\mathbf{sk}_m(\mathcal{A})$ to $\Delta_{X,m}$ are the same. It follows that $\mathbf{sk}_m$ is a monad, in other words $\mathbf{sk}_m(\mathbf{sk}_m(\mathcal{A})) = \mathbf{sk}_m(\mathcal{A})$ (or rather they are isomorphic in a natural way). We say that $\mathcal{A}$ is *m-skeletal* if the map $\mathbf{sk}_m(\mathcal{A}) \to \mathcal{A}$ is an isomorphism; and $\mathbf{sk}_m$ is a monadic projection to the full subcategory of m-skeletal objects.

If $n \geq m$ and $\mathcal{A}$ is m-skeletal then it is n-skeletal, as $\mathbf{sk}_n(\mathbf{sk}_m(\mathcal{A})) = \mathbf{sk}_m(\mathcal{A})$. This formula comes from the fact that the restriction to $\Delta_{X,n}$ of $\mathbf{sk}_m(\mathcal{A})$ is the same as the pushforward from $\Delta_{X,m}$ to $\Delta_{X,n}$ of the restriction

$\sigma(X,m)^*(\mathcal{A})$, as can be seen by the expressions of the pushforwards as colimits over surjective maps; then the composition of pushforwards from $\Delta_{X,m}$ to $\Delta_{X,n}$ and then to Δ_X, is the same as the pushforward $\sigma(X,m)_!$ to give the claimed formula.

Lemma 13.1.1 *The functor* $\mathbf{sk}_m$ *preserves unitality, that is if* $\mathcal{A} \in \mathcal{PC}(X,\mathcal{M})$ *then* $\mathbf{sk}_m(\mathcal{A}) \in \mathcal{PC}(X,\mathcal{M})$. *It has a right adjoint* $\mathbf{csk}_m : \mathcal{PC}(X,\mathcal{M}) \to \mathcal{PC}(X,\mathcal{M})$ *called the m-*coskeleton. *Both of these are compatible with changing X so they give functors* $\mathcal{PC}(\mathcal{M}) \to \mathcal{PC}(\mathcal{M})$. *The natural morphism induces an isomorphism* $\text{colim}_m\mathbf{sk}_m(\mathcal{A}) \cong \mathcal{A}$. *The morphisms* $\mathbf{sk}_{m-1}(\mathcal{A}) \to \mathbf{sk}_m(\mathcal{A}) \to \mathcal{A}$ *are monomorphisms if* $\mathcal{M}$ *is a presheaf category.*

Proof Suppose (x_0) is a sequence of length zero. Then $(x_0) \in \Delta_{X,m}$ for any $m \geq 0$ so the category of surjections $(x_0) \twoheadrightarrow y.$ considered above has as initial object (x_0) itself. The formula (13.1.1) then says

$$\mathbf{sk}_m(\mathcal{A})(x_0) = \mathcal{A}(x_0).$$

It follows that if $\mathcal{A}$ is unital, so is $\mathbf{sk}_m(\mathcal{A})$.

The functors $\sigma(X,m)_!$ and $\sigma(X,m)^*$ have right adjoints $\sigma(X,m)^*$ and $\sigma(X,m)_*$ respectively. So, considered as a functor between diagram categories, $\mathbf{sk}_m$ has as right adjoint:

$$\mathbf{csk}_m(\mathcal{B}) = \sigma(X,m)_*\sigma(X,m)^*(\mathcal{B}).$$

As before we can note that this functor has the expression

$$\mathbf{csk}_m(\mathcal{B})(x.) = \lim_{y.\hookrightarrow x.} \mathcal{B}(y.),$$

where the limit is taken over inclusions $y. \hookrightarrow x.$ in Δ_X such that $y.$ is in $\Delta_{X,m}$, i.e. has length $\leq m$. Then again note that if $x. = (x_0)$ is a sequence of length zero, the index category for the limit has an initial object (x_0) itself, so

$$\mathbf{csk}_m(\mathcal{B})(x_0) = \mathcal{B}(x_0).$$

Thus $\mathbf{csk}_m$ preserves the unitality condition and gives an endofunctor of $\mathcal{PC}(X,\mathcal{M})$, right adjoint to $\mathbf{sk}_m$.

Suppose $\psi : X \to Y$ is a morphism of sets. For any $\mathcal{A} \in \mathcal{PC}(X,\mathcal{M})$ and $\mathcal{B} \in \mathcal{PC}(Y,\mathcal{M})$ with a morphism $f : \mathcal{A} \to \psi^*\mathcal{B}$ in $\mathcal{PC}(X,\mathcal{M})$ we would like to get a map $\mathbf{sk}_m(\mathcal{A}) \to \psi^*\mathbf{sk}_m(\mathcal{B})$. In view of the definition of $\mathbf{sk}_m$ this is equivalent to giving a map of diagrams over $\Delta_{X,m}$

$$\sigma(X,m)^*(\mathcal{A}) \to \sigma(X,m)^*\psi^*\mathbf{sk}_m(\mathcal{B}). \tag{13.1.2}$$

Let $\psi_m : \Delta_{X,m} \to \Delta_{Y,m}$ denote the induced functor, then $\psi \circ \sigma(X,m) = \sigma(Y,m) \circ \psi_m$ so

$$\begin{aligned}\sigma(X,m)^*\psi^*\mathbf{sk}_m(\mathcal{B}) &= \psi_m^*\sigma(Y,m)^*\mathbf{sk}_m(\mathcal{B})\\ &= \psi_m^*\sigma(Y,m)^*(\mathcal{B}) = \sigma(X,m)^*\psi^*(\mathcal{B}),\end{aligned}$$

and functoriality for $\sigma(X,m)^*$ applied to f gives a map

$$\sigma(X,m)^*(f) : \sigma(X,m)^*(\mathcal{A}) \to \sigma(X,m)^*\psi^*(\mathcal{B}),$$

which yields the desired map (13.1.2). We also have a commutative square

$$\begin{array}{ccc}\mathbf{sk}_m(\mathcal{A}) & \longrightarrow & \psi^*\mathbf{sk}_m(\mathcal{B})\\ \Big\downarrow & & \Big\downarrow\\ \mathcal{A} & \longrightarrow & \psi^*\mathcal{B}.\end{array}$$

So, given a morphism $\mathcal{A} \to \mathcal{B}$ in $\mathcal{PC}(\mathcal{M})$ this construction has defined a map $\mathbf{sk}_m(\mathcal{A}) \to \mathbf{sk}_m(\mathcal{B})$. It is compatible with identities and compositions so $\mathbf{sk}_m$ is an endofunctor of $\mathcal{PC}(\mathcal{M})$, provided with a natural transformation $\mathbf{sk}_m(\mathcal{A}) \to \mathcal{A}$ giving a structure of monadic projection to the full subcategory of m-skeletal precategories.

Similarly for the coskeleton we look for a map

$$\psi_!\mathbf{csk}_m(\mathcal{A}) \to \sigma(Y,m)_*\sigma(Y,m)^*(\mathcal{B}). \tag{13.1.3}$$

This is equivalent to looking for

$$\sigma(Y,m)^*\psi_!\mathbf{csk}_m(\mathcal{A}) \to \sigma(Y,m)^*(\mathcal{B}).$$

But in general

$$\psi_!(\mathcal{C})(y.) = \mathrm{colim}_{y. \to \psi x.}\mathcal{C}(x.),$$

and the colimit can be taken over surjective maps $y. \twoheadrightarrow \psi x.$. Thus, if $y.$ has length m then it only depends on the restriction of $\mathcal{C}$ to $\Delta_{X,m}$, in particular

$$\sigma(Y,m)^*\psi_!\mathbf{csk}_m(\mathcal{A}) = \sigma(Y,m)^*\psi_!(\mathcal{A}),$$

and we can look for a map $\psi_!\mathcal{A} \to \mathbf{csk}_m(\mathcal{B})$, or equivalently by adjunction, $\mathcal{A} \to \psi^*\mathbf{csk}_m(\mathcal{B})$. But the natural map $\mathcal{B} \to \mathbf{csk}_m(\mathcal{B})$ gives by pullback under ψ and composition

$$\mathcal{A} \to \psi^*(\mathcal{B}) \to \psi^*\mathbf{csk}_m(\mathcal{B})$$

as required. We get the functoriality maps for defining

$$\mathbf{csk}_m : \mathcal{PC}(\mathcal{M}) \to \mathcal{PC}(\mathcal{M}),$$

again with a natural transformation $\mathcal{A} \to \mathbf{csk}_m(\mathcal{A})$.

Suppose $\mathcal{A} \in \mathcal{PC}(X, \mathcal{M})$. For any $x.$ a sequence of length p, then for $m \geq p$ we have $\mathbf{sk}_m(\mathcal{A})(x.) = \mathcal{A}(x.)$. Hence the morphism

$$\mathrm{colim}_m \mathbf{sk}_m(\mathcal{A}) \to \mathcal{A}$$

is an isomorphism on each object $x.$ of Δ_X. As colimits in diagram categories are computed levelwise and the unitalification operation $U_!$ preserves colimits, we have $\mathcal{A} \cong \mathrm{colim}_m \mathbf{sk}_m(\mathcal{A})$ in $\mathcal{PC}(X, \mathcal{M})$. Recall from Corollary 10.6.2 that connected colimits in $\mathcal{PC}(X, \mathcal{M})$ also are colimits in $\mathcal{PC}(\mathcal{M})$, so we get the same formula in $\mathcal{PC}(\mathcal{M})$.

One can remark that dually, and for the same reason,

$$\mathcal{A} \overset{\cong}{\to} \lim_m \mathbf{csk}_m(\mathcal{A}).$$

Finally, we note that if $\mathcal{M}$ is a presheaf category then we have a family of functors $h_u : \mathcal{M} \to \mathrm{SET}$, the evaluations on objects of the underlying category for the presheaves, such that h_u preserves colimits. It follows that $h_u \circ (\mathbf{sk}_m(\mathcal{A})) = \mathbf{sk}_m(h_u \circ \mathcal{A})$; however, $h_u \circ \mathcal{A}$ is a set-valued diagram over Δ_X. As is well known, the inclusion of the m-skeleton is an injection of simplicial sets, and this works equally well for Δ_X^o-sets. Thus, the map

$$h_u \circ (\mathbf{sk}_m(\mathcal{A})) = \mathbf{sk}_m(h_u \circ \mathcal{A}) \to h_u \circ \mathcal{A}$$

is a monomorphism. Therefore, the map $\mathbf{sk}_m(\mathcal{A}) \to \mathcal{A}$ induces a monomorphism upon application of all of the functors h_u; but as $\mathcal{M}$ is a presheaf category this implies that our map is a monomorphism as claimed in the last statement of the lemma. □

Notice that $\mathbf{sk}_0(\mathcal{A}) = \mathbf{disc}(\mathrm{Ob}(\mathcal{A}))$ is the set $\mathrm{Ob}(\mathcal{A})$ considered as a discrete precategory, that is one whose p-fold morphism objects are all $*$ for constant sequences or $\emptyset$ otherwise (see Section 10.5). Indeed if $x.$ is a sequence then there is a surjection to a unique (y_0) of length 0, if and only if $x.$ is constant, so $\mathbf{sk}_0(\mathcal{A})(x.) = \mathcal{A}(y_0) = *$ if $x.$ is constant of value y_0, and $\mathbf{sk}_0(\mathcal{A})(x.) = \emptyset$ if $x.$ is nonconstant.

If $\mathcal{A} \in \mathcal{PC}(\mathcal{M})$ and $(x_0, \ldots, x_m)$ is a sequence of objects with $m \geq 1$, define the *degenerate subobject*

$$\mathbf{d}(\mathcal{A}; x_0, \ldots, x_m) := \mathbf{sk}_{m-1}(\mathcal{A})(x_0, \ldots, x_m).$$

It has a map denoted

$$\delta(\mathcal{A}; x_0, \ldots, x_m) : \mathbf{d}(\mathcal{A}; x_0, \ldots, x_m) \to \mathcal{A}(x_0, \ldots, x_m),$$

and we can express D as a colimit:

Lemma 13.1.2 *With the above notations,*

$$\mathbf{d}(\mathcal{A}; x_0, \ldots, x_m) = \operatorname{colim}_{x. \to y.} \mathcal{A}(y_0, \ldots, y_p),$$

where the colimit is taken over surjective maps $x. \to y.$ in $\Delta_{\mathrm{Ob}(\mathcal{A})}$ such that $y.$ are sequences of length $p < m$. The map $\delta(\mathcal{A}; x_0, \ldots, x_m)$ is obtained from the restriction maps of $\mathcal{A}$ via the universal property of the colimit. The map $\delta(\mathcal{A}; x_0, \ldots, x_m)$ is a monomorphism if $\mathcal{M}$ is a presheaf category.

Proof This comes from the definition of D and the colimit expression for $\mathbf{sk}_{m-1}$. The monomorphism property comes from the previous lemma. □

13.2 Some natural precategories

Consider the ordered sets $[k] \in \mathrm{Ob}(\Delta)$ with the notation

$$[k] = \{\upsilon_0, \ldots, \upsilon_k\}, \quad \upsilon_0 < \cdots < \upsilon_k.$$

For any $B \in \mathcal{M}$ and any $[k] \in \mathrm{Ob}(\Delta)$ define the precategory $h([k]; B) \in \mathcal{PC}([k], \mathcal{M})$ as follows. Let $\mathbf{t}_k \in \mathrm{Ob}(\Delta_{[k]})$ denote the tautological object $\mathbf{t}_k := (\upsilon_0, \ldots, \upsilon_k)$ of length k, and let

$$\mathbf{i}\{\mathbf{t}_k\} : \{\mathbf{t}_k\} \hookrightarrow \Delta_{[k]}$$

denote the inclusion of the discrete category on a single point $\mathbf{t}_k \in \mathrm{Ob}(\Delta_{[k]})$. Let $B_{\mathbf{t}_k}$ denote the constant diagram with value B on the one-point category $\{\mathbf{t}_k\}$, and put

$$h([k]; B) := U_! \mathbf{i}\{\mathbf{t}_k\}_!(B_{\mathbf{t}_k}).$$

Recall that $U_!$ is the unitalization operation, necessary here because the Δ_X-diagram $\mathbf{i}\{\mathbf{t}_k\}_!(B_{\mathbf{t}_{[k]}})$ will not in general be unital. The following lemma gives a concrete description of $h([k]; B)$ and could be taken as its definition:

Lemma 13.2.1 *Suppose $(y_0, \ldots, y_p)$ is any sequence of elements of the set $[k]$ with $y_j = \upsilon_{i_j}$. Then:*

- *if $(y_0, \ldots, y_p)$ is increasing but not constant, i.e. $i_{j-1} \le i_j$ but $i_0 < i_p$, then*

$$h([k]; B)(y_0, \ldots, y_p) = B;$$

- *if $(y_0, \ldots, y_p)$ is constant, i.e. $i_0 = i_1 = \ldots = i_p$, then*

$$h([k]; B)(y_0, \ldots, y_p) = *;$$

- *and otherwise, that is if there exists $1 \le j \le p$ such that $i_{j-1} > i_j$ then*

$$h([k]; B)(y_0, \ldots, y_p) = \emptyset.$$

The pullback maps giving $h([k]; B)$ a structure of diagram are all either the unique maps of the form $\emptyset \to B$, $\emptyset \to *$, or $B \to *$, or the identity $B \to B$.

Proof The functor $B \mapsto h([k]; B)$ is by construction left adjoint to the functor $\mathbf{i}\{\mathbf{t}_k\}^*$ from $\mathcal{PC}([k], \mathcal{M})$ to $\mathcal{M}$ which sends $\mathcal{A}$ to $\mathcal{A}(\upsilon_0, \ldots, \upsilon_k)$. On the other hand we can check by hand (as in the proof of the next lemma below) that the functor sending B to the precategory defined explicitly in the statement of the lemma, is also adjoint to the same functor. □

Lemma 13.2.2 *If $B \in \mathcal{M}$ and $\mathcal{C} \in \mathcal{PC}(\mathcal{M})$ then a morphism $f : h([k], B) \to \mathcal{C}$ is the same thing as a sequence of objects $x_0, \ldots, x_k \in \mathrm{Ob}(\mathcal{C})$ together with a map $\varphi : B \to \mathcal{C}(x_0, \ldots, x_k)$ in $\mathcal{M}$.*

Proof Use the explicit description of the previous lemma. Given f we get $x_i := f(\upsilon_i)$ and the map φ is given by $f_{\upsilon_0,\ldots,\upsilon_k}$. On the other hand, given $x.$ and φ, the restrictions $\mathcal{C}(x_0, \ldots, x_k) \to \mathcal{C}(x_{i_0}, \ldots, x_{i_p})$ lead to maps

$$B \to \mathcal{C}(x_{i_0}, \ldots, x_{i_p})$$

for any sequence as in the first part of the previous lemma; if $i_0 = \cdots = i_p$ then this map factors through $*$ by the unitality condition for $\mathcal{C}$, treating the second part of the previous lemma; and nothing is needed for defining a map in the third case of the previous lemma. This defines the required map f. □

Note that the above construction is functorial in B, that is a map $A \to B$ induces $h([k]; A) \to h([k]; B)$. Define the "boundary" of $h([k]; B)$ by the skeleton operation:

$$h(\partial[k]; B) := \mathbf{sk}_{k-1} h([k]; B),$$

with the natural inclusion

$$h(\partial[k]; B) \to h([k]; B).$$

It is also functorial in B.

Lemma 13.2.3 *This boundary object has the following concrete description:*

- *if* $(y_0, \ldots, y_p)$ *is increasing but not constant, i.e.* $i_{j-1} \leq i_j$ *but* $i_0 < i_p$*, and if there is any* $0 \leq m \leq k$ *such that* $i_j \neq m$ *for all* $0 \leq j \leq k$*, then*

$$h(\partial[k]; B)(y_0, \ldots, y_p) = B;$$

- *if* $(y_0, \ldots, y_p)$ *is constant, i.e.* $i_0 = i_1 = \ldots = i_p$*, then*

$$h(\partial[k]; B)(y_0, \ldots, y_p) = *;$$

- *and otherwise, that is if either there exists* $1 \leq j \leq p$ *such that* $i_{j-1} > i_j$ *or else if the map* $j \mapsto y_j$ *is a surjection from* $\{0, \ldots, p\}$ *to* $[k]$*, then*

$$h(\partial[k]; B)(y_0, \ldots, y_p) = \emptyset.$$

Proof Use the description of Lemma 13.2.1 and the formula

$$h(\partial[k]; B)(y.) = \mathrm{colim}_{y. \twoheadrightarrow z.} h(B)(z.). \qquad \square$$

If $f : A \to B$ is a cofibration in $\mathcal{M}$, put

$$h([k], \partial[k]; A \xrightarrow{f} B) := h([k]; A) \cup^{h(\partial[k]; A)} h(\partial[k]; B).$$

We therefore obtain two natural maps coming from f, the first is

$$P([k]; f) : h([k]; A) \to h([k]; B),$$

and the second is

$$R([k]; f) : h([k], \partial[k]; A \xrightarrow{f} B) \to h([k]; B).$$

The maps $P([k]; f)$ will form the generators for the projective cofibrations, while the $R([k]; f)$ form the generators for the Reedy cofibrations. Unfortunately we don't have an easy way to describe generators for the injective cofibrations.

13.3 Projective cofibrations

The different model structures are characterized and differentiated by their notions of cofibrations. In the projective structure, the generating set is easiest to describe but on the other hand there is no easy criterion for being a cofibration.

A map $f : \mathcal{A} \to \mathcal{B}$ is a *projective cofibration* if on the set of objects it is an injective map of sets $\mathrm{Ob}(f) : \mathrm{Ob}(\mathcal{A}) \to \mathrm{Ob}(\mathcal{B})$, and if the map $\mathrm{Ob}(f)_!(\mathcal{A}) \to \mathcal{B}$ is a cofibration in the projective model category structure

$\mathcal{PC}_{\text{proj}}(\text{Ob}(\mathcal{B}), \mathcal{M})$. Recall that $\text{Ob}(f)_!(\mathcal{A})$ is the precategory $\mathcal{A}$ transported to the subset $f(\text{Ob}(\mathcal{A})) \subset \text{Ob}(\mathcal{B})$, then extended by adding on the discrete or initial precategory over the complementary subset $\text{Ob}(\mathcal{B}) - \text{Ob}(f)(\text{Ob}(\mathcal{A}))$.

Say that f is a *projective trivial cofibration* if it is a projective cofibration, and a global weak equivalence.

For now, we say that a morphism $u : U \to V$ in $\mathcal{PC}(\mathcal{M})$ is a *projective fibration* if it satisfies the right lifting property with respect to all projective trivial cofibrations; we say that u is a *projective trivial fibration* if it is a projective fibration and a global weak equivalence.

For the time being we also need a separate notation: say that $u : U \to V$ in $\mathcal{PC}(\mathcal{M})$ is an *apparent projective trivial fibration* if it satisfies the right lifting property with respect to all projective cofibrations. One of our tasks in future chapters will be to identify the class of projective trivial fibrations with the apparent ones. For now we can characterize and use the apparent ones:

Lemma 13.3.1 *A morphism $u : U \to V$ in $\mathcal{PC}(\mathcal{M})$ is an apparent projective trivial fibration if and only if $\text{Ob}(u)$ is surjective and $U \to \text{Ob}(u)^*(V)$ is a levelwise trivial fibration. A morphism $f : \mathcal{A} \to \mathcal{B}$ is a projective cofibration if and only if it satisfies the left lifting property with respect to the apparent projective trivial fibrations.*

Proof Consider the class $\mathscr{A}$ of morphisms u such that $\text{Ob}(u)$ is surjective and $U \to \text{Ob}(u)^*(V)$ is a levelwise trivial fibration. If f is a projective cofibration then it breaks down as a composition $f = f' \circ d$, where $f' : \text{Ob}(f)_!(\mathcal{A}) \to \mathcal{B}$ is a morphism in $\mathcal{PC}(\text{Ob}(\mathcal{B}), \mathcal{M})$ and d is the extension by adding on the discrete precategory $\text{Ob}(\mathcal{B}) - \text{Ob}(f)(\text{Ob}(\mathcal{A}))$. Similarly, if $u \in \mathscr{A}$ then $u = p \circ u'$, where $u' : U \to \text{Ob}(u)^*(V)$ is a levelwise trivial fibration in $\mathcal{PC}(\text{Ob}(U), \mathcal{M})$ and p is the tautological map $\text{Ob}(u)^*(V) \to V$. Notice that p satisfies the right lifting property with respect to any morphism which induces an isomorphism on sets of objects, and also with respect to extensions by adding discrete sets; and dually f' satisfies the left lifting property with respect to tautological maps such as p, while d satisfies the left lifting property with respect to any map surjective on objects. All told, the lifting property with f on the left and u on the right, is equivalent to the lifting property with f' on the left and u' on the right. This holds whenever u' is a levelwise trivial fibration and f' is a projective cofibration in $\mathcal{PC}(\text{Ob}(\mathcal{B}), \mathcal{M})$.

These considerations show that $\mathscr{A}$ is contained in the class of apparent projective trivial fibrations. Furthermore, if u satisfies right lifting for any projective cofibration f then in particular $\text{Ob}(u)$ must be surjective (using for f any extension by a nonempty discrete set); and u' must satisfy right lifting with respect to projective trivial cofibrations inducing isomorphisms on sets of objects, so u' is a levelwise trivial fibration (by the projective model structure on the $\mathcal{PC}(X, \mathcal{M})$). This shows that the apparent projective trivial cofibrations are contained in $\mathcal{A}$ so the two classes coincide.

Then, similarly, if f satisfies left lifting with respect to $\mathcal{A}$ then first of all it must be injective on sets of objects, as seen by considering u which are surjective maps of codiscrete objects (i.e. objects U with $U(y.) = *$ for all sequences $y.$). And then decomposing $f = f' \circ d$, where d is the extension by the complementary subset, we see that f' should be a projective cofibration in $\mathcal{PC}(\mathrm{Ob}(\mathcal{B}), \mathcal{M})$. Thus f is a projective cofibration by definition. But all projective cofibrations satisfy lifting with respect to the apparent projective trivial fibrations, by the definition of this latter class, and equality with the class $\mathcal{A}$ shows that the class of projective cofibrations is exactly that which satisfies left lifting with respect to $\mathcal{A}$. □

We have stability under pushouts and retracts:

Corollary 13.3.2 *Assume that $\mathcal{M}$ is a tractable left proper cartesian model category. Suppose $f : \mathcal{A} \to \mathcal{B}$ and $g : \mathcal{A} \to \mathcal{C}$ are morphisms in $\mathcal{PC}(\mathcal{M})$ such that f is a projective cofibration. Then the map*

$$u : \mathcal{C} \to \mathcal{B} \cup^{\mathcal{A}} \mathcal{C}$$

is a projective cofibration. Furthermore, the projective cofibrations are stable under retracts and transfinite composition. The class of projective trivial cofibrations is closed under transfinite composition and retracts.

Proof Stability under pushouts, retracts and transfinite composition come from the characterization of the projective cofibrations as the maps which satisfy left lifting with respect to the class of apparent projective trivial fibrations (Lemma 13.3.1). The last sentence now follows immediately from Proposition 12.5.4. □

The advantage of the notion of projective cofibration is that the generating set is very easy to describe.

Proposition 13.3.3 *Fix a generating set I for the cofibrations of $\mathcal{M}$. Then for any k and any $f : A \to B$ in I, consider the map*

$$P([k]; f) : h([k]; A) \to h([k]; B).$$

The collection of these for integers k and all $f \in I$, together with the map $\emptyset \to *$, forms a generating set for the class of projective cofibrations in $\mathcal{PC}(\mathcal{M})$.

Proof Any projective cofibration $f : \mathcal{A} \to \mathcal{B}$ factors as $f = f'd$, where

$$\mathcal{A} \xrightarrow{d} \mathcal{A}' \sqcup \mathbf{disc}(Z) \xrightarrow{f'} \mathcal{B},$$

where $Z = \mathrm{Ob}(\mathcal{B}) = \mathrm{Ob}(f)(\mathrm{Ob}(\mathcal{A}))$, where $\mathcal{A}' = \mathrm{Ob}(f)_!(\mathcal{A})$ is the precategory on the set $\mathrm{Ob}(f)(\mathrm{Ob}(\mathcal{A}))$ obtained by transport of structure, $\mathbf{disc}(Z)$ is the discrete precategory on the set Z, and d is the extension morphism. In this situation furthermore, f' is a projective cofibration in $\mathcal{PC}(X, \mathcal{M})$ where $X = \mathrm{Ob}(\mathcal{B}) = \mathrm{Ob}(\mathcal{A}' \sqcup \mathbf{disc}(Z))$.

Now d is obtained by successive pushout along $\emptyset \to *$, while the $w_! P([k]; f)$ for various maps $w : [k] \to X$ form the set of generators for the projective cofibrations in $\mathcal{PC}(X, \mathcal{M})$. Thus, f' is a retract of a transfinite pushout along the $w_! P([k]; f)$. Putting these together gives the expression of f as a retract of a transfinite pushout of morphisms in our generating set. □

13.4 Injective cofibrations

It is easier to see whether a given map is an injective cofibration, since this condition is defined levelwise, but the only way to get a generating set in general is by an accessibility argument.

A morphism $f : \mathcal{A} \to \mathcal{B}$ is an *injective cofibration* if $\mathrm{Ob}(f)$ is an injective map of sets, and if the map $\mathrm{Ob}(f)_!(\mathcal{A}) \to \mathcal{B}$ is a cofibration in the injective model category structure $\mathcal{PC}_{\mathrm{inj}}(\mathrm{Ob}(\mathcal{B}), \mathcal{M})$. Since injective cofibrations are defined levelwise, we can be more explicit about this condition: it means that for any sequence $(x_0, \ldots, x_p)$ in $\mathrm{Ob}(\mathcal{A})$, the map $\mathcal{A}(x_0, \ldots, x_p) \to \mathcal{B}(f(x_0), \ldots, f(x_p))$ is a cofibration in $\mathcal{M}$, for any constant sequence $(y, \ldots, y)$ at a point $y \in Y - f(X)$ the map $* \to \mathcal{B}(y, \ldots, y)$ is a cofibration in $\mathcal{M}$, and for any nonconstant sequence $(y_0, \ldots, y_p)$ such that at least one of the y_i is not in $f(X)$, the map $\emptyset \to \mathcal{B}(y_0, \ldots, y_p)$ is a cofibration, i.e. $\mathcal{B}(y_0, \ldots, y_p)$ is a cofibrant object.

Say that f is an *injective trivial cofibration* if it is an injective cofibration and a global weak equivalence.

We again have the stability proposition:

Proposition 13.4.1 *Assume that $\mathcal{M}$ is a tractable left proper cartesian model category. Suppose $f : \mathcal{A} \to \mathcal{B}$ and $g : \mathcal{A} \to \mathcal{C}$ are morphisms in $\mathcal{PC}(\mathcal{M})$ such that f is an injective cofibration. Then the map*

$$\mathcal{C} \to \mathcal{B} \cup^{\mathcal{A}} \mathcal{C}$$

is an injective cofibration. Furthermore, the injective cofibrations are stable under retracts and transfinite composition. The class of injective trivial cofibrations is closed under transfinite composition and retracts.

Proof The explicit conditions given in the paragraph defining the injective cofibrations are defined levelwise over Δ_Z, where Z is the set of objects of the target precategory. The conditions are preserved by pushouts, retracts and transfinite composition. Again, the last sentence now follows immediately from Proposition 12.5.4. □

Lemma 13.4.2 *Suppose $\mathcal{M}$ is a tractable model category. Then the class of injective cofibrations in $\mathcal{PC}(\mathcal{M})$ admits a small set of generators.*

Proof Follow the argument given in Barwick [18] for the proof using a general accessibility argument. □

Unfortunately, we don't get very much information on the set of generators. In Section 13.7 below, if $\mathcal{M}$ satisfies some further hypotheses which hold for presheaf categories, then the injective cofibrations are the same as the Reedy cofibrations and the generators will be described explicitly.

A similar problem with the injective structure is that if $\mathcal{M}$ is tractable, we don't know whether the generating cofibrations for $\mathcal{PC}(\mathcal{M})$ have cofibrant domains. For diagram categories, this isn't true, as was pointed out at the end of Section 8.6; the question for $\mathcal{PC}(\mathcal{M})$ remains open.

For our purposes this question is one further reason for introducing the Reedy model structure which has an explicit set of generators; in many cases (such as when $\mathcal{M}$ is a presheaf category, Proposition 13.7.2) the Reedy and injective structures will coincide.

13.5 A pushout expression for the skeleta

The main observation crucial for understanding the Reedy cofibrations, is an expression for the successive skeleta as pushouts along maps of the form $R([k]; f)$. If $x_0, \ldots, x_m$ is a sequence of objects, recall the notation

$$\mathbf{d}(\mathcal{A}; x_0, \ldots, x_m) \xrightarrow{\delta(\mathcal{A}, x.)} \mathcal{A}(x_0, \ldots, x_m)$$

from Section 13.2. The identity map

$$\mathbf{d}(\mathcal{A}; x.) = \mathbf{sk}_{m-1}(\mathcal{A})(x_0, \ldots, x_m)$$

corresponds by the universal property of $h([m]; \cdot)$ to a map

$$h([m]; \mathbf{d}(\mathcal{A}; x.)) \to \mathbf{sk}_{m-1}(\mathcal{A})$$

in $\mathcal{PC}(\mathcal{M})$. On the other hand, by the definition of $h(\partial[m];\cdot)$ and functoriality of the skeleton operation, the map

$$h([m];\mathcal{A}(x.)) \rightarrow \mathcal{A}$$

yields a map

$$h(\partial[m];\mathcal{A}(x.)) \rightarrow \mathbf{sk}_{m-1}(\mathcal{A}).$$

These two maps agree on $h(\partial[m];\mathbf{d}(\mathcal{A};x.))$ so they give a map defined on the amalgamated sum

$$\begin{aligned} h([m],\partial[m];\mathbf{d}(\mathcal{A};x.) &\stackrel{\delta(\mathcal{A};x.)}{\longrightarrow} \mathcal{A}(x.)) \\ &= h([m];\mathbf{d}(\mathcal{A};x.)) \cup^{h(\partial[m];\mathbf{d}(\mathcal{A};x.))} h(\partial[m];\mathcal{A}(x.)), \end{aligned}$$

giving the top map in the commutative square

$$\begin{array}{ccc} h([m],\partial[m];\mathbf{d}(\mathcal{A};x.) \rightarrow \mathcal{A}(x.)) & \rightarrow & \mathbf{sk}_{m-1}(\mathcal{A}) \\ {\scriptstyle R([m];\delta(\mathcal{A};x.))}\downarrow & & \downarrow \\ h([m];\mathcal{A}(x.)) & \longrightarrow & \mathbf{sk}_m(\mathcal{A}). \end{array}$$

The map on the bottom is given by adjunction from the equality

$$\mathcal{A}(x_0,\ldots,x_m) = \mathbf{sk}_m(\mathcal{A})(x_0,\ldots,x_m).$$

Putting these together over all sequences $x_0,\ldots,x_m$ of length m we get an expression for $\mathbf{sk}_m(\mathcal{A})$.

Proposition 13.5.1 *For any m we have an expression of $\mathbf{sk}_m(\mathcal{A})$ as a pushout of $\mathbf{sk}_{m-1}(\mathcal{A})$ by copies of the standard maps $R([m];\cdot)$ indexed by sequences $x. = (x_0,\ldots,x_m)$:*

$$\mathbf{sk}_m(\mathcal{A}) = \mathbf{sk}_{m-1}(\mathcal{A}) \cup^{\coprod_{x.} h([k],\partial[k];\delta(\mathcal{A};x.))} \coprod_{x.} h([m],\mathcal{A}(x.)).$$

This is a pushout, and uses amalgamated sums, in the category $\mathcal{PC}(\mathcal{M})$.

Proof This is a classical fact about simplicial objects. The pushout is m-skeletal; its $(m-1)$-skeleton agrees with $\mathbf{sk}_{m-1}(\mathcal{A})$ since the maps $R([m];\cdot)$ induce isomorphisms on $(m-1)$-skeleta; and on sequences of length m the pushout expression is designed exactly to give the same values as $\mathcal{A}$. These imply that the pushout is equal to $\mathbf{sk}_m(\mathcal{A})$. $\square$

13.6 Reedy cofibrations

Consider a map $f : \mathcal{A} \to \mathcal{B}$ in $\mathcal{PC}(\mathcal{M})$, giving for each k a map on skeleta $\mathbf{sk}_k(\mathcal{A}) \to \mathbf{sk}_k(\mathcal{B})$. Define the *relative skeleton of* f

$$\mathcal{A} \cup^{\mathbf{sk}_k(\mathcal{A})} \mathbf{sk}_k(\mathcal{B}) \xrightarrow{\mathbf{sk}^{\mathrm{rel}}_k(f)} \mathcal{B}.$$

We say that f is a *Reedy cofibration* if $\mathrm{Ob}(f)$ is injective and the relative skeleton maps are injective cofibrations for every k.

Lemma 13.6.1 *The class of Reedy cofibrations is closed under pushout, transfinite composition and retracts.*

Proof The skeleton operation is given by a pushforward which is a kind of colimit, so it commutes with colimits over connected categories which are computed levelwise. The relative skeleton map of a retract is again a retract so closure under retracts comes from the same property in $\mathcal{M}$ levelwise. □

Theorem 13.6.2 *Suppose $f : \mathcal{A} \to \mathcal{B}$ is map such that $\mathrm{Ob}(f)$ is injective and view $\mathrm{Ob}(\mathcal{A})$ as a subset of $\mathrm{Ob}(\mathcal{B})$. The following conditions are equivalent:*

(a) f is a Reedy cofibration;

(b) for any $m \geq 1$ the map

$$\mathbf{sk}_m(\mathcal{A}) \cup^{\mathbf{sk}_{m-1}(\mathcal{A})} \mathbf{sk}_{m-1}(\mathcal{B}) \to \mathbf{sk}_m(\mathcal{B}) \tag{13.6.1}$$

is an injective cofibration;

(c) for any sequence $(x_0, \ldots, x_p)$ of objects in $\mathrm{Ob}(\mathcal{A})$, the map

$$\mathcal{A}(x_0, \ldots, x_p) \cup^{\mathbf{d}(\mathcal{A}; x_0, \ldots, x_p)} \mathbf{d}(\mathcal{B}; x_0, \ldots, x_p) \to \mathcal{B}(x_0, \ldots, x_p) \tag{13.6.2}$$

is a cofibration in $\mathcal{M}$, and for any sequence $(y_0, \ldots, y_p)$ of objects in $\mathrm{Ob}(\mathcal{B})$ not all in $\mathrm{Ob}(\mathcal{A})$, the map $\mathbf{d}(\mathcal{B}; y_0, \ldots, y_p) \to \mathcal{B}(y_0, \ldots, y_p)$ is a cofibration in $\mathcal{M}$;

(d) letting $X := \mathrm{Ob}(\mathcal{B})$, the map $\mathrm{Ob}(f)_!(\mathcal{A}) \to \mathcal{B}$ is a Reedy cofibration in the model structure $\mathcal{PC}_{\mathrm{Reedy}}(X, \mathcal{M})$ of Theorem 12.3.2.

Proof First note that (a) implies (c), indeed (c) is the statement of (a) for $k = p - 1$ at the sequence of objects $(x_0, \ldots, x_p)$ or $(y_0, \ldots, y_p)$.

Next we show that (b) implies the following more general statement: for any $0 \leq n \leq m$ the map

$$\mathbf{sk}_m(\mathcal{A}) \cup^{\mathbf{sk}_n(\mathcal{A})} \mathbf{sk}_n(\mathcal{B}) \to \mathbf{sk}_m(\mathcal{B}) \tag{13.6.3}$$

is an injective cofibration. This is tautological for $m = n$. Let n be fixed, and suppose we know this statement for some $m \geq n$. Then

$$\mathbf{sk}_{m+1}(\mathcal{A}) \cup^{\mathbf{sk}_n(\mathcal{A})} \mathbf{sk}_n(\mathcal{B}) = \mathbf{sk}_{m+1}(\mathcal{A}) \cup^{\mathbf{sk}_m(\mathcal{A})} (\mathbf{sk}_m(\mathcal{A}) \cup^{\mathbf{sk}_n(\mathcal{A})} \mathbf{sk}_n(\mathcal{B})).$$

Injective cofibrations are stable under pushout (Lemma 13.4.2), and our inductive hypothesis says that

$$\mathbf{sk}_m(\mathcal{A}) \cup^{\mathbf{sk}_n(\mathcal{A})} \mathbf{sk}_n(\mathcal{B}) \to \mathbf{sk}_m(\mathcal{B})$$

is an injective cofibration. Take the pushout of this by $\mathbf{sk}_{m+1}(\mathcal{A})$ over $\mathbf{sk}_m(\mathcal{A})$ and use the previous identification, to get that

$$\mathbf{sk}_{m+1}(\mathcal{A}) \cup^{\mathbf{sk}_n(\mathcal{A})} \mathbf{sk}_n(\mathcal{B}) \to \mathbf{sk}_{m+1}(\mathcal{A}) \cup^{\mathbf{sk}_m(\mathcal{A})} \mathbf{sk}_m(\mathcal{B})$$

is an injective cofibration. On the other hand condition (b) says that

$$\mathbf{sk}_{m+1}(\mathcal{A}) \cup^{\mathbf{sk}_m(\mathcal{A})} \mathbf{sk}_m(\mathcal{B}) \to \mathbf{sk}_{m+1}(\mathcal{B})$$

is an injective cofibration, so composing these gives that

$$\mathbf{sk}_{m+1}(\mathcal{A}) \cup^{\mathbf{sk}_n(\mathcal{A})} \mathbf{sk}_n(\mathcal{B}) \to \mathbf{sk}_{m+1}(\mathcal{B})$$

is an injective cofibration. This proves that the maps (13.6.3) are injective cofibrations by induction.

This statement now implies condition (a), indeed if $(x_0, \ldots, x_p)$ is any sequence of objects in $\mathrm{Ob}(\mathcal{B})$ then for any $m \geq p$, $m \geq k$ we have

$$\mathcal{B}(x_0, \ldots, x_p) = \mathbf{sk}_m(\mathcal{B})(x_0, \ldots, x_p).$$

The same is true for $\mathcal{A}$ if the x_i are all in $\mathrm{Ob}(\mathcal{A})$. Hence

$$\mathcal{A} \cup^{\mathbf{sk}_k(\mathcal{A})} \mathbf{sk}_k(\mathcal{B})(x_0, \ldots, x_p) = \mathbf{sk}_m(\mathcal{A}) \cup^{\mathbf{sk}_k(\mathcal{A})} \mathbf{sk}_k(\mathcal{B})(x_0, \ldots, x_p),$$

so the map

$$\mathcal{A} \cup^{\mathbf{sk}_k(\mathcal{A})} \mathbf{sk}_k(\mathcal{B})(x_0, \ldots, x_p) \to \mathcal{B}(x_0, \ldots, x_p)$$

is the same as the map

$$\mathbf{sk}_m(\mathcal{A}) \cup^{\mathbf{sk}_k(\mathcal{A})} \mathbf{sk}_k(\mathcal{B})(x_0, \ldots, x_p) \to \mathcal{B}(x_0, \ldots, x_p).$$

This latter is just the value of (13.6.3) on the sequence $(x_0, \ldots, x_p)$ so it is a cofibration in $\mathcal{M}$, as needed to show the Reedy condition (a). This shows that (b) implies (a).

To complete the proof we need to show that (c) implies (b). For this, use the expression of Proposition 13.5.1 for $\mathbf{sk}_m(\mathcal{B})$ as a pushout of $\mathbf{sk}_{m-1}(\mathcal{B})$ and the standard inclusions $R([m], \delta(\mathcal{B}; y_0, \ldots, y_p))$ and similarly for $\mathcal{A}$. We have a diagram

$$\begin{array}{ccc} \mathbf{sk}_{m-1}(\mathcal{A}) & \longrightarrow & \mathbf{sk}_m(\mathcal{A}) \\ \downarrow & & \downarrow \\ \mathbf{sk}_{m-1}(\mathcal{B}) & \longrightarrow & \mathbf{sk}_m(\mathcal{B}) \end{array}$$

where the top arrow is pushout along the $R([m], \delta(\mathcal{A}; x_0, \ldots, x_p))$ for sequences of objects $(x_0, \ldots, x_p)$ of $\mathcal{A}$, and the bottom arrow is pushout along the $R([m], \delta(\mathcal{B}; x_0, \ldots, x_p))$ for sequences of objects $(x_0, \ldots, x_p)$ of $\mathcal{B}$. Taking the pushout of the upper left corner of the diagram gives the expression

$$\begin{aligned} &\mathbf{sk}_m(\mathcal{A}) \cup^{\mathbf{sk}_{m-1}(\mathcal{A})} \mathbf{sk}_{m-1}(\mathcal{B}) \\ &\quad = \mathbf{sk}_{m-1}(\mathcal{B}) \cup^{\coprod_{x.} h([m], \partial[m]; \delta(\mathcal{A}; x.))} \coprod_{x.} h([m], \mathcal{A}(x.)). \end{aligned}$$

The coproducts are over sequences $x. = (x_0, \ldots, x_m)$ of length m of objects of $\mathcal{A}$; however, it can be extended to a coproduct over sequences of objects of $\mathcal{B}$ by setting $\mathcal{A}(x_0, \ldots, x_m) := \emptyset$ as well as $\mathbf{d}(\mathcal{A}; x_0, \ldots, x_m) := \emptyset$ if any of the x_i are not in $\mathrm{Ob}(\mathcal{A})$. On the other hand,

$$\mathbf{sk}_m(\mathcal{B}) = \mathbf{sk}_{m-1}(\mathcal{B}) \cup^{\coprod_{x.} \partial h([m], \partial[m]; \delta(\mathcal{B}; x.))} \coprod_{x.} h([m], \mathcal{B}(x.)).$$

Putting these two together, we conclude that

$$\mathbf{sk}_m(\mathcal{B}) = \left(\mathbf{sk}_m(\mathcal{A}) \cup^{\mathbf{sk}_{m-1}(\mathcal{A})} \mathbf{sk}_{m-1}(\mathcal{B})\right) \cup^{\coprod_{x.} \mathcal{C}(x.)} \coprod_{x.} h([m], \mathcal{B}(x.)),$$

where

$$\mathcal{C}(x.) := h([m], \mathcal{A}(x.)) \cup^{h([m], \partial[m]; \delta(\mathcal{A}; x.))} h([m], \partial[m]; \delta(\mathcal{B}; x.)).$$

Hence, to prove that the map $\mathbf{sk}_m(\mathcal{A}) \cup^{\mathbf{sk}_{m-1}(\mathcal{A})} \mathbf{sk}_{m-1}(\mathcal{B}) \to \mathbf{sk}_m(\mathcal{B})$ is an injective cofibration, using stability of injective cofibrations under pushouts, it suffices to show that each of the maps

$$\mathcal{C}(x_0, \ldots, x_m) \to h([m], \mathcal{B}(x_0, \ldots, x_m)) \tag{13.6.4}$$

is an injective cofibration. For both sides, the set of objects is now our standard set $[m] = \{v_0, \ldots, v_m\}$. Consider the value on a sequence of objects $(y_0, \ldots, y_p)$ in $[m]$. There are four possible cases:

(i) If it is a constant sequence, then

$$\begin{aligned} h([m], \mathcal{A}(x.))(y.) &= h([m], \partial[m]; \delta(\mathcal{A}; x.))(y.) \\ &= h([m], \partial[m]; \delta(\mathcal{B}; x.))(y.) \\ &= h([m], \mathcal{B}(x.))(y.) = * \end{aligned}$$

and the map (13.6.4) is the identity.

(ii) If the sequence is somewhere decreasing, i.e. there is some j with $y_j < y_{j-1}$, then

$$\begin{aligned} h([m], \mathcal{A}(x.))(y.) &= h([m], \partial[m]; \delta(\mathcal{A}; x.))(y.) \\ &= h([m], \partial[m]; \delta(\mathcal{B}; x.))(y.) \\ &= h([m], \mathcal{B}(x.))(y.) = \emptyset \end{aligned}$$

and again the map (13.6.4) is the identity.

(iii) If the sequence is nondecreasing, but there is some v_i not contained in $y.$, then

$$h([m], \mathcal{A}(x.))(y.) = h([m], \partial[m]; \delta(\mathcal{A}; x.))(y.) = \mathcal{A}(x.),$$

and

$$h([m], \mathcal{B}(x.))(y.) = h([m], \partial[m]; \delta(\mathcal{B}; x.))(y.) = \mathcal{B}(x.),$$

so in this case

$$\mathcal{C}(x.)(y.) = \mathcal{A}(x.) \cup^{\mathcal{A}(x.)} \mathcal{B}(x.) = \mathcal{B}(x.),$$

and once again (13.6.4) is the identity.

(iv) If the sequence is nondecreasing and surjects onto the full set of objects, then

$$h([m], \mathcal{A}(x.))(y.) = \mathcal{A}(x.), \quad h([m], \partial[m]; \delta(\mathcal{A}; x.))(y.) = \mathbf{d}(\mathcal{A}; x.),$$

and

$$h([m], \mathcal{B}(x.))(y.) = \mathcal{B}(x.), \quad h([m], \partial[m]; \delta(\mathcal{B}; x.))(y.) = \mathbf{d}(\mathcal{B}; x.),$$

so

$$\mathcal{C}(x.)(y.) = \mathcal{A}(x.) \cup^{\mathbf{d}(\mathcal{A}; x.)} \mathbf{d}(\mathcal{B}; x.),$$

so the map

$$\mathcal{C}(x.)(y.) \to h([m], \mathcal{B}(x.))(y.)$$

is exactly the map (13.6.2)

$$\mathcal{A}(x.) \cup^{\mathbf{d}(\mathcal{A}; x.)} \mathbf{d}(\mathcal{B}; x.) \to \mathcal{B}(x.),$$

which is known to be a cofibration in $\mathcal{M}$ because we are assuming condition (c) of the theorem. This completes the proof that the map (13.6.1) is an injective cofibration, showing (c)⇒(b). This completes the proof of the equivalence of (a), (b) and (c).

For the equivalence with (d), we have $\mathbf{sk}_m(\mathrm{Ob}(f)_!\mathcal{A}) = \mathrm{Ob}(f)_!\mathbf{sk}_m(\mathcal{A})$ by commutation of pushforwards. Now

$$\mathcal{A} \cup^{\mathbf{sk}_m(\mathcal{A})} \mathcal{B} = \mathrm{Ob}(f)_!(\mathcal{A}) \cup^{\mathrm{Ob}(f)_!\mathbf{sk}_m(\mathcal{A})} \mathcal{B}$$

from the definition of colimits in $\mathcal{PC}(\mathcal{M})$ (Section 10.6), so $\mathcal{A} \to \mathcal{B}$ is a Reedy cofibration if and only if $\mathrm{Ob}(f)_!\mathcal{A} \to \mathcal{B}$ is. However, this latter induces an isomorphism on sets of objects, and for such maps the criterion (c) above is the same as the Reedy condition that the relative latching maps be cofibrations in $\mathcal{M}$. This shows that (d) is equivalent to (a), (b), (c). □

The map (13.6.4) occuring above is of the form $R([m], g)$ as shown in the following lemma:

Lemma 13.6.3 *Suppose*

$$\begin{array}{ccc} E & \xrightarrow{a} & F \\ {\scriptstyle u}\downarrow & & \downarrow{\scriptstyle v} \\ U & \xrightarrow{b} & V \end{array}$$

is a diagram in $\mathcal{M}$. Consider the induced map $g : U \cup^E F \to V$. Then the two maps

$$h([m], U) \cup^{h([m],\partial[m]; E \stackrel{u}{\to} U)} h([m], \partial[m]; F \stackrel{u}{\to} V) \to h([m], V)$$

and

$$R([m], g) : h([m], \partial[m]; g) \to h([m], V)$$

are the same.

Proof Look at the values on any sequence $y. = (y_0, \ldots, y_p)$ of objects in $[m] = \{\upsilon_0, \ldots, \upsilon_m\}$. There are several possibilities:

- if the sequence is constant then both maps are $* \to *$;
- if the sequence is anywhere decreasing then both maps are $\emptyset \to \emptyset$;

- if the sequence is nondecreasing but misses some element v_j, then we are in the boundary $\partial[m]$ and the first map is
$$U \cup^U V \to V$$
and the second map is $U \to V$, these are the same;
- if the sequence is nondecreasing and surjects onto $[m]$ then both maps are
$$U \cup^E F \to V.$$

We need to point out that these identifications of the maps, and in particular of their sources, are functorial in the restriction maps for diagrams over $\Delta^o_{[m]}$, so they give an identification

$$h([m], U) \cup^{h([m],\partial[m]; E \stackrel{u}{\to} U)} h([m], \partial[m]; F \stackrel{u}{\to} V) \cong h([m], \partial[m]; g)$$

and both maps from here to $h([m], V)$ are the same. □

Corollary 13.6.4 *For an object $\mathcal{A} \in \mathcal{PC}(\mathcal{M})$, the following are equivalent:*

(a) $\mathcal{A}$ is Reedy cofibrant;
(b) for any $m \geq 1$ the map $\mathbf{sk}_{m-1}(\mathcal{A}) \to \mathbf{sk}_m(\mathcal{A})$ is an injective cofibration;
(c) for any sequence of objects $(x_0, \ldots, x_p)$ the map
$$\mathbf{sk}_{p-1}(\mathcal{A})(x_0, \ldots, x_p) \to \mathcal{A}(x_0, \ldots, x_p)$$
is a cofibration in $\mathcal{M}$;
(d) $\mathcal{A}$ is a Reedy cofibrant object in $\mathcal{PC}_{\mathrm{Reedy}}(X, \mathcal{M})$ for $X := \mathrm{Ob}(\mathcal{A})$;
(e) for any subset $S \subset \mathrm{Ob}(A)$, the natural map $f : \mathbf{disc}(S) \to \mathcal{A}$ is a Reedy cofibration.

Proof Apply Theorem 13.6.2 to the map $\emptyset \to \mathcal{A}$. Note that $\mathrm{Ob}(\emptyset) = \emptyset$ and $\mathbf{sk}_m(\emptyset) = \emptyset$. Condition (c) here is the part of condition (c) of the theorem, concerning sequences of objects not all coming from the source. The theorem gives that (a)–(d) are equivalent.

For the equivalence with (e), put $X := \mathrm{Ob}(\mathcal{A})$. The map $\mathrm{Ob}(f) : \mathrm{Ob}(\mathbf{disc}(S)) \to \mathrm{Ob}(\mathcal{A})$ is just the inclusion $S \subset X$. From the explicit description of $\mathrm{Ob}(f)_!$ we get $\mathrm{Ob}(f)_!(\mathbf{disc}(S)) = \mathbf{disc}(X)$. This holds also for $\emptyset \subset X$, and in fact $\mathbf{disc}(X) \in \mathcal{PC}(X, \mathcal{M})$ is the initial object. Now f is a Reedy cofibration if and only if $\mathrm{Ob}(f)_!(\mathbf{disc}(S)) \to \mathcal{A}$ is a Reedy cofibration in $\mathcal{PC}(X, \mathcal{M})$ (part (d) of Theorem 13.6.2), which holds if and only if $\mathcal{A}$ is a Reedy cofibrant object of $\mathcal{PC}_{\mathrm{Reedy}}(X, \mathcal{M})$ which is (d) here. □

Corollary 13.6.5 *If $f : A \to B$ is a cofibration in $\mathcal{M}$, then $R([k]; f)$ (page 304) is a Reedy cofibration.*

Proof Recall that

$$h([k], \partial[k], f) = h([k], A) \cup^{\mathbf{sk}_{k-1}h([k],A)} \mathbf{sk}_{k-1}h([k], B),$$

and $R([k]; f)$ is the map from here to $h([k], B)$. If $m \leq k-1$ then $\mathbf{sk}_m h([k], \partial[k], f) = \mathbf{sk}_m([k], B)$, so the map occuring in condition (b) of Theorem 13.6.2 is the identity in this case. For $m \geq k$,

$$\begin{aligned}\mathbf{sk}_m h([k], \partial[k], f) &\cup^{\mathbf{sk}_{m-1}h([k],\partial[k],f)} \mathbf{sk}_{m-1}h([k], B)\\ &= \mathbf{sk}_m h([k], A) \cup^{\mathbf{sk}_{m-1}h([k],A)} \mathbf{sk}_{m-1}h([k], B),\end{aligned}$$

so the map occuring in condition (b) is

$$\mathbf{sk}_m h([k], A) \cup^{\mathbf{sk}_{m-1}h([k],A)} \mathbf{sk}_{m-1}h([k], B) \to \mathbf{sk}_m(h[k], B).$$

As for the equivalence with condition (c), it suffices to note that this induces a cofibration over sequences $x_0, \ldots, x_m$ of length m in $\{\upsilon_0, \ldots, \upsilon_k\}$ (similar to the proof of Proposition 13.7.1 below). □

Corollary 13.6.6 *Many maps between the $h([k], B)$ are Reedy cofibrations due to the previous corollary. For example, if $f : A \to B$ is a cofibration in $\mathcal{M}$ then the map*

$$h([k-1], A) \to h([k], B),$$

induced by applying f at one of the faces of the k-simplex $[k-1] \subset [k]$, is a Reedy cofibration.

Proof Either calculate directly the skeleta, or use the previous corollary inductively. □

The following statement is somewhat similar to giving a set of generators for the Reedy cofibrations. On the one hand, it refers to the full class of morphisms of the form $R([m], g)$ for cofibrations g, while on the other it gives a stronger expression without refering to retracts:

Proposition 13.6.7 *A morphism $f : \mathcal{A} \to \mathcal{B}$ is a Reedy cofibration if and only if it is a transfinite composition of a disjoint union with a discrete set, and then pushouts along morphisms of the form $R([m], g)$ for cofibrations g in $\mathcal{M}$.*

Proof A Reedy cofibration $f : \mathcal{A} \to \mathcal{B}$ can be expressed as the countable composition of the morphisms $\mathcal{A} \cup^{\mathbf{sk}_{m-1}(\mathcal{A})} \mathbf{sk}_{m-1}(\mathcal{B}) \to \mathcal{A} \cup^{\mathbf{sk}_m(\mathcal{A})} \mathbf{sk}_m(\mathcal{B})$, which are themselves Reedy cofibrations. At the start, $\mathcal{A} \cup^{\mathbf{sk}_0(\mathcal{A})} \mathbf{sk}_0(\mathcal{B})$ is the disjoint union of $\mathcal{A}$ with the discrete set $\mathrm{Ob}(\mathcal{B}) - \mathrm{Ob}(\mathcal{A})$. Then, as we have seen in the proof of Theorem 13.6.2, at each stage the morphism in question is obtained by simultaneous pushout along morphisms (13.6.4) of the form

$$h([m], U) \cup^{h([m],\partial[m]; E \xrightarrow{u} U)} h([m], \partial[m]; F \xrightarrow{v} V) \to h([m], V), \quad (13.6.5)$$

where from the notations of (13.6.4) we put $U := \mathcal{A}(x_0, \ldots, x_m)$, $V := \mathcal{B}(x_0, \ldots, x_m)$, $E := \mathbf{d}(\mathcal{A}; x_0, \ldots, x_m)$ and $F := \mathbf{d}(\mathcal{B}; x_0, \ldots, x_m)$, with

$$u := \delta(\mathcal{A}; x_0, \ldots, x_m) : E \to U$$

and

$$v := \delta(\mathcal{B}; x_0, \ldots, x_m) : F \to V.$$

The maps u and v, as well as the maps $U \to V$ and $E \to F$, all fit into a commutative square. Condition (c) of the theorem says that the map $g : U \cup^E F \to V$ is a cofibration. By Lemma 13.6.3, the map (13.6.5) is the same as $R([m]; g)$. □

The the notion of Reedy cofibration is similar to that of projective cofibration, in that we can give explicitly the set of generators:

Lemma 13.6.8 *Suppose I is a set of maps in $\mathcal{M}$. Let*

$$R(I) \subset \mathrm{ARR}(\mathcal{PC}(\mathcal{M}))$$

denote the set of all arrows of the form $R([k]; g)$ for $k \in \mathbb{N}$ and $g \in I$. If $f \in \mathbf{cell}(I)$ and $m \in \mathbb{N}$ then $R([m]; f) \in \mathbf{cell}(R(I))$.

Proof Look at the behavior of $R([k]; f)$ under pushouts and transfinite composition. Suppose

$$\begin{array}{ccc} A & \xrightarrow{f} & B \\ {\scriptstyle g}\downarrow & & \downarrow{\scriptstyle u} \\ C & \xrightarrow{v} & P \end{array}$$

is a pushout diagram in $\mathcal{M}$, that is $P = B \cup^A C$. This induces a diagram

$$\begin{array}{ccc} h([k], \partial[k]; g) & \to & h([k], \partial[k]; u) \\ {\scriptstyle R([k]; g)}\downarrow & & \downarrow{\scriptstyle R([k]; u)} \\ h([k]; C) & \longrightarrow & h([k]; P). \end{array}$$

We claim that this second diagram is then also a pushout in $\mathcal{PC}(\mathcal{M})$. In fact it is a diagram in $\mathcal{PC}([k], \mathcal{M})$, and connected colimits of diagrams in $\mathcal{PC}([k], \mathcal{M})$ are the same as the corresponding colimits in $\mathcal{PC}(\mathcal{M})$, also in turn they are the same as the corresponding colimits in $\text{FUNC}(\Delta^o_{[k]}, \mathcal{M})$ (Section 10.6), so the pushout of the second diagram can be computed levelwise. Then it is easy to see that it is a pushout, using the explicit description of the values of $h(([k], \partial[k]); \cdot)$ and $h([k]; \cdot)$. The conclusion from this discussion, is that any pushout along $R([k]; u)$ will also be a pushout along $R([k]; g)$.

Consider now a transfinite composition: suppose we have a series

$$\cdots \to A_i \xrightarrow{f_{i,i+1}} A_{i+1} \to \cdots$$

in $\mathcal{M}$ indexed by $i \in \beta$ for some ordinal β. To treat limit ordinals we need also to consider the transition maps $f_{i,j} : A_i \to A_j$ for any $i < j$. Assume that if j is a limit ordinal then $A_j = \text{colim}_{i<j} A_i$, and let $A_\beta := \text{colim}_{i\in\beta} A_i$. Consider the map

$$f : A_0 \to A_\beta.$$

We can express $R([k]; f)$ as a transfinite composition of pushouts along $R([k]; \cdot)$. Consider the series

$$G_i := h([k], \partial[k]; A_i \to A_\beta),$$
$$\cdots \to G_i \xrightarrow{g_{i,i+1}} G_{i+1} \to \cdots$$

with the more general transition maps $g_{i,j} : G_i \to G_j$. This is still a transfinite series: if j is a limit ordinal then $G_j = \text{colim}_{i<j} G_i$, and $G_\beta := \text{colim}_{i\in\beta} G_i$ is equal to $h([k], \partial[k]; 1_{A_\beta}) = h([k]; A_\beta)$. These can be seen by calculating the colimits levelwise over $\Delta_{[k]}$. The map

$$G_0 \to G_\beta$$

is equal to $R([k]; f)$. Furthermore, G_{i+1} is the pushout of G_i along the map $R([k]; f_{i,i+1})$. Thus, $R([k]; A_0 \to A_\beta)$ is a transfinite composition of pushouts along the $R([k]; f_{i,i+1})$. In turn, if $f_{i,i+1}$ is a pushout along an element of I then by the discussion of pushouts above, $R([k]; f_{i,i+1})$ is a pushout along an element of $R(I)$. Thus, if our series gives an expression for f as an element of **cell**(I), then $R([k]; f)$ is seen to be in **cell**$(R(I))$. □

A similar statement is needed for retracts. Suppose

$$\begin{array}{ccc} & & B \\ & \overset{g}{\nearrow} & {\scriptstyle p}\downarrow\uparrow{\scriptstyle s} \\ A & \xrightarrow{f} & C \end{array}$$

is a retract diagram in $\mathcal{M}$, that is $f = pg$, $g = sf$, and $ps = 1$. This gives a diagram

$$\begin{array}{ccc} h([k], \partial[k]; g) & \xrightarrow{R([k]; g)} & h([k]; B) \\ \downarrow\uparrow & & \downarrow\uparrow \\ h([k], \partial[k]; f) & \xrightarrow{R([k]; f)} & h([k]; C) \end{array}$$

where the vertical arrows are induced by p and s.

Lemma 13.6.9 *In the above situation, suppose we are given a pushout diagram*

$$\begin{array}{ccc} h([k], \partial[k]; f) & \longrightarrow & \mathcal{U} \\ {\scriptstyle R([k]; f)}\downarrow & & \downarrow \\ h([k]; C) & \longrightarrow & \mathcal{V}. \end{array}$$

Then $\mathcal{U} \to \mathcal{V}$ is a retract of the pushout of $\mathcal{U}$ along $R([k]; g)$, in the category of objects under $\mathcal{U}$.

Proof Using the left downward map in the previous diagram, we get a map $h([k], \partial[k]; g) \to \mathcal{U}$ so we can form the pushout $\mathcal{V}'$ of $\mathcal{U}$ along the map $R([k]; g)$. The vertical maps on the right of the previous diagram induce maps $\mathcal{V} \to \mathcal{V}'$ and $\mathcal{V}' \to \mathcal{V}$, compatible with the maps from $\mathcal{U}$, and the composition $\mathcal{V} \to \mathcal{V}' \to \mathcal{V}$ is the identity as desired. To see all of these things, note that in

$$\begin{array}{ccc} & & h([k], \partial[k]; g) \\ & \swarrow & \downarrow\uparrow \\ \mathcal{U} & \longleftarrow & h([k], \partial[k]; f) \end{array}$$

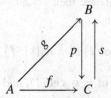

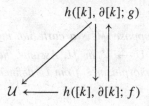

both triangles, obtained by using the upward or the downward arrows, commute. The required statements are then obtained by functoriality of pushout diagrams using the identity on $\mathcal{U}$. □

Corollary 13.6.10 *If $f \in \mathbf{cof}(I)$ then $R([k]; f) \in \mathbf{cof}(R(I))$.*

Proof Write f as a retract of $g \in \mathbf{cell}(I)$, and apply the previous Lemma 13.6.9 to $R([k]; f)$ seen as a pushout of itself and the identity; thus $R([k]; f)$ is a retract of a pushout along $R([k]; g)$. On the other hand, Lemma 13.6.8 shows that $R([k]; g) \in \mathbf{cell}(R(I))$. Thus, $R([k]; f)$ is a retract of a map in $\mathbf{cell}(R(I))$, so it is in $\mathbf{cof}(R(I))$. □

Proposition 13.6.11 *Fix a generating set I for the cofibrations of $\mathcal{M}$. Then for any k and any $f : A \to B$ in I, consider the map*

$$R([k]; f) : h(([k], \partial[k]); A \xrightarrow{f} B) \to h([k]; B).$$

The collection of these for integers k and all $f \in I$, forms a generating set $R(I)$ for the class of Reedy cofibrations in $\mathcal{PC}(\mathcal{M})$. If the elements of I have cofibrant domains, then the elements of $R(I)$ have Reedy cofibrant domains.

Proof We have seen in Proposition 13.6.7 that any Reedy cofibration can be written as a successive pushout by maps of the form $R([k]; f)$ for various k and various cofibrations f in $\mathcal{M}$. As $\mathbf{cof}(R(I))$ is closed under pushout and transfinite composition, it suffices to show that, for any cofibration f, the map $R([k]; f)$ is in $\mathbf{cof}(R(I))$. This is exactly the statement of the previous Corollary 13.6.10. Furthermore, if f has cofibrant domain, then $R([k]; f)$ will have a Reedy cofibrant domain. □

If $\mathcal{M}$ is cartesian, then the Reedy cofibrations also satisfy the cartesian property. This is one of the main reasons for introducing the Reedy objects. This is closely related to the corresponding result for diagrams over a Reedy category, see for example Barwick [19] and Berger and Moerdijk [38]. I would like to thank several people, including Clemens Berger, Clark Barwick, Ieke Moerdijk and Mark Johnson, for replying to a query about this on the topology mailing list.

Proposition 13.6.12 *Suppose $\mathcal{M}$ is a cartesian model category. Consider morphisms $A \xrightarrow{f} B$ and $U \xrightarrow{g} V$ in $\mathcal{M}$. Assume that they are cofibrations. Then, for any k, m the morphism ξ from $\mathcal{U}$ defined as the pushout in the cocartesian diagram*

$$\begin{array}{ccc} h([k],\partial[k];f)\times h([m],\partial[m];g) & \to & h([k],\partial[k];f)\times h([m];V) \\ \downarrow & & \downarrow \\ h([k];B)\times h([m],\partial[m];g) & \longrightarrow & \mathcal{U}, \end{array}$$

to $\mathcal{F} := h([k];B)\times h([m];V)$, is a Reedy cofibration.

Proof Use the criterion (c) of Theorem 13.6.2. Consider any sequence of objects $z. = ((x_0, y_0), \ldots, (x_p, y_p))$, where $x_i \in [k]$ and $y_j \in [m]$. We first need to calculate $\mathbf{sk}_{p-1}(\mathcal{U})(z.)$ and $\mathbf{sk}_{p-1}(\mathcal{F})(z.)$.

If either one of the sequences $x.$ or $y.$ is decreasing at any index, then (using that the product of anything with $\emptyset$ is again $\emptyset$)

$$\mathbf{sk}_{p-1}(\mathcal{U})(z.) = \mathbf{sk}_{p-1}(\mathcal{F})(z.) = \mathcal{U}(z.) = \mathcal{F}(z.) = \emptyset,$$

and the map (13.6.2) for ξ is an isomorphism hence a cofibration. Similarly, if $z.$ is constant then

$$\mathbf{sk}_{p-1}(\mathcal{U})(z.) = \mathbf{sk}_{p-1}(\mathcal{F})(z.) = \mathcal{U}(z.) = \mathcal{F}(z.) = *,$$

so again the map (13.6.2) is an isomorphism hence a cofibration.

Thus, we may assume that both sequences $x.$ and $y.$ are nondecreasing and at least one of them is nonconstant.

However, if one or the other of the sequences is constant, then the morphism in question becomes the same as the map (13.6.2) for the other side, and we know that $R([k];f)$ or $R([m];g)$ are Reedy cofibrations by 13.6.4.

So, we may now assume that both sequences are nonconstant.

If $z.$ is strictly increasing, then it has no quotient $z. \to w.$ of length $\leq p-1$, so

$$\mathbf{sk}_{p-1}(\mathcal{F})(z.) = \mathbf{sk}_{p-1}(\mathcal{U})(z.) = \emptyset.$$

So in this case the map (13.6.2) is just the map

$$\mathcal{U}(z.) \to \mathcal{F}(z.).$$

First note that $\mathcal{F}(z.) = B \times V$. The calculation of $\mathcal{U}(z.)$ breaks into several cases. If $x.$ lies in $\partial[k]$ (i.e. it misses at least one object of $[k]$) and $y.$ lies in $\partial[m]$ then

$$\mathcal{U}(z.) = B \times V \cup^{B\times V} B \times V = B \times V,$$

so the map from here to $\mathcal{F}(z.)$ is an isomorphism hence a cofibration. If $x.$ lies in $\partial[k]$ but $y.$ surjects onto $[m]$ then

$$\mathcal{U}(z.) = B \times V \cup^{B\times U} B \times U = B \times V,$$

so again the map to $\mathcal{F}(z.)$ is an isomorphism. Similarly, if $x.$ surjects onto $[k]$ but $y.$ lies in $\partial[m]$, then

$$\mathcal{U}(z.) = A \times V \cup^{A\times V} B \times V = B \times V,$$

and the map to $\mathcal{F}(z.)$ is an isomorphism. Finally, suppose that $x.$ surjects onto $[k]$ and $y.$ surjects onto $[m]$. Then

$$\mathcal{U}(z.) = A \times V \cup^{A\times U} B \times U,$$

so the map from here to $B \times V$ is a cofibration by the cartesian axiom for $\mathcal{M}$ and the assumption that f and g were cofibrations of $\mathcal{M}$. This completes the proof that the map (13.6.2) for ξ is a cofibration in the case where the sequence $z.$ is strictly increasing.

Assume therefore that $z.$ is not strictly increasing, i.e. it has at least one adjacent pair of objects that are equal. We have

$$\mathbf{sk}_{p-1}(\mathcal{F})(z.) = \mathbf{d}(\mathcal{F}; z.) = \mathrm{colim}_{z. \rightarrow w.}\mathcal{F}(w.),$$

$$\mathbf{sk}_{p-1}(\mathcal{U})(z.) = \mathbf{d}(\mathcal{U}; z.) = \mathrm{colim}_{z. \rightarrow w.}\mathcal{U}(w.),$$

where the colimits are taken over surjective maps $z. \rightarrow w.$ such that $w.$ has length less than or equal to $p-1$, see Lemma 13.1.2. The category of quotients $z. \rightarrow w.$ of length $q \leq p-1$ is nonempty, because we are assuming that $z.$ is not strictly increasing. The opposite category of this category of quotients has an initial object, corresponding to the quotient of minimal length obtained by identifying all adjacent equal objects.

The diagram which to $z. \rightarrow w.$ associates $\mathcal{F}(w.)$ is constant, taking values $B \times V$. This uses the definitions of $h([k]; B)$ and $h([m]; V)$, and the fact that both sequences $x.$ and $y.$ are nondecreasing and nonconstant. Hence,

$$\mathbf{sk}_{p-1}(\mathcal{F})(z.) = \mathrm{colim}_{z. \rightarrow w.}B \times V = B \times V = \mathcal{F}(z.),$$

since the colimit of a constant diagram over a category with initial object is equal to the constant value of the diagram.

Next, look at

$$\mathbf{sk}_{p-1}(\mathcal{U})(z.) = \mathbf{d}(\mathcal{U}; z.) = \mathrm{colim}_{z. \rightarrow w.}\mathcal{U}(w.).$$

Suppose given a quotient $z. \rightarrow w.$ of length $q \leq p-1$. Write $w. = ((r_0, s_0), \ldots, (r_q, s_q))$. Note that $r.$ is a quotient sequence of $x.$ and $s.$ is a quotient sequence of $y.$. The question of whether the sequence of first elements $r.$ lies in $\partial[k]$ or $[k]$, or whether the sequence of second elements $s.$ lies in $\partial[m]$ or $[m]$, is independent of the choice of quotients and depends only on $x.$ or $y.$. Hence, the values $h([k], \partial[k]; f)(r.)$ and $h([m], \partial[m]; g)(s.)$ are independent of the choice of $w.$, and the colimit defining $\mathbf{sk}_{p-1}(\mathcal{U})(z.)$ is equal to its constant value on any of the objects $w.$. This breaks into exactly the same cases as considered previously, and by the same reasoning we see that

$$\mathbf{sk}_{p-1}(\mathcal{U})(z.) = \mathcal{U}(z.).$$

Now the map (13.6.2) for ξ is written as

$$\mathcal{U}(z.) \cup^{\mathbf{sk}_{p-1}(\mathcal{U})(z.)} \mathbf{sk}_{p-1}(\mathcal{F})(z.) \to \mathcal{F}(z.),$$

but in view of the identifications given above this map is just the identity of $\mathcal{F}(z.) = B \times V$ so it is a cofibration. This completes the proof of the proposition. □

Corollary 13.6.13 *Suppose $f : \mathcal{A} \to \mathcal{B}$ and $g : \mathcal{U} \to \mathcal{V}$ are Reedy cofibrations in $\mathcal{PC}(\mathcal{M})$. Then*

$$\mathcal{A} \times \mathcal{V} \cup^{\mathcal{A} \times \mathcal{U}} \mathcal{B} \times \mathcal{U} \to \mathcal{B} \times \mathcal{V}$$

is a Reedy cofibration in $\mathcal{PC}(\mathcal{M})$. In particular, the product of two Reedy cofibrant objects is again Reedy cofibrant.

Proof Both f and g may be expressed as transfinite compositions of pushouts along elementary Reedy cofibrations of the form $R([k], h)$. The previous proposition gives the cartesian property for these. Since Reedy cofibrations are closed under pushout and transfinite composition, we get the cartesian property for any f and g. □

One of the main steps in our proof will be to give the same property for trivial Reedy cofibrations, in Chapter 17 below.

13.7 Relationship between the classes of cofibrations

Proposition 13.7.1 *A projective cofibration is a Reedy cofibration, and a Reedy cofibration is an injective cofibration.*

Proof Starting from generators for the projective cofibrations, if $f : A \to B$ is a cofibration in $\mathcal{M}$ then $h([k], f)$ is a Reedy cofibration. Indeed, the set of objects is $\{v_0, \ldots, v_k\}$. Suppose given a sequence of the form $x. = (v_{i_0}, \ldots, v_{i_m})$. If the sequence is increasing and nonconstant, then the same is true of any surjective image, and we get

$$h([k], A)(x.) = A, \quad \mathbf{sk}_m h([k], A)(x.) = A \text{ or } \emptyset,$$

with $\emptyset$ occuring if there are no surjections to sequences of length $\leq m$. Similarly for B. The relative skeleton map for $h([k], f)$ at $x.$ is either f or the identity of B in this case. If the sequence is constant, then the same is true of any surjective image and the relative skeleton map is the identity of $*$. If the sequence is anywhere strictly decreasing, again the same is true of any surjective image and the relative skeleton map is the identity of $\emptyset$. Thus, $h([k], f)$ is a Reedy cofibration.

Using $m = 0$ in the definition of Reedy cofibrations, we get that they are levelwise cofibrations. □

Proposition 13.7.2 *Suppose $\mathcal{M}$ is the category of presheaves over a connected category, and the cofibrations are the monomorphisms of $\mathcal{M}$. Then the Reedy and injective cofibrations of $\mathcal{PC}(\mathcal{M})$ coincide, and if I is a generating set of cofibrations for $\mathcal{M}$ then the set $R(I)$ consisting of the $R([k]; f)$ for $f \in I$ is a generating set of cofibrations for the injective cofibrations.*

Proof If $\mathcal{M} = \text{PRESH}(\Phi)$ we can verify the property that the relative skeleton maps are cofibrations, levelwise over Φ. It then reduces to the classical statement that the skeleton of a simplicial set is a simplicial subset. □

Theorem 13.7.3 *Suppose a map $f : \mathcal{A} \to \mathcal{B}$ in $\mathcal{PC}(\mathcal{M})$ satisfies the right lifting property with respect to the class of projective (resp. injective, Reedy) cofibrations. Then f is a global weak equivalence.*

Proof It suffices to treat the case of projective cofibrations, since the other ones contain this class.

Recall that any set $X \in \text{SET}$ corresponds to a discrete precategory $\mathbf{disc}(X) \in \mathcal{PC}(\mathcal{M})$ whose object set is X itself, and whose morphism objects are defined by

$$\mathbf{disc}(X)(x_0, \ldots, x_n) = \begin{cases} * & \text{if } x_0 = \cdots = x_n, \\ \emptyset & \text{otherwise.} \end{cases}$$

Included among the projective cofibrations is $\emptyset \to \mathbf{disc}(\{x\})$. The morphism $f : \mathcal{A} \to \mathcal{B}$ in $\mathcal{PC}(\mathcal{M})$ satisfies the right lifting property with respect to $\emptyset \to \{x\}$, if and only if $\text{Ob}(\mathcal{A}) \to \text{Ob}(\mathcal{B})$ is surjective. In particular f is essentially surjective.

Next, suppose g is a generating cofibration of $\mathcal{M}$. If f satisfies the right lifting property with respect to a given $h([k], g)$, Lemma 13.2.2 implies that for any $[k]$-sequence of objects $x_0, \ldots, x_k \in \text{Ob}(\mathcal{A})$, the map

$$\mathcal{A}(x_0, \ldots, x_k) \to \mathcal{B}(f(x_0), \ldots, f(x_k))$$

satisfies the right lifting property with respect to g. If f satisfies the right lifting property with respect to the generators of projective cofibrations of $\mathcal{PC}(\mathcal{M})$, which is to say all the $h([k], g)$ as g runs over a generating set of cofibrations of $\mathcal{M}$ and k is any positive integer, it follows that for any $x_0, \ldots, x_k \in \mathrm{Ob}(\mathcal{A})$, the map

$$\mathcal{A}(x_0, \ldots, x_k) \to \mathcal{B}(f(x_0), \ldots, f(x_k))$$

is a trivial fibration in $\mathcal{M}$–in particular it is a weak equivalence. This shows that f is fully faithful. □

The cofibrations in the local categories $\mathcal{PC}(X, \mathcal{M})$ are related to those in the global $\mathcal{PC}(\mathcal{M})$, leading to a relationship between the apparent trivial fibrations.

Lemma 13.7.4 *Suppose $\mathcal{A} \xrightarrow{f} \mathcal{B}$ is a map in $\mathcal{PC}(X, \mathcal{M})$ which satisfies the right lifting property with respect to all Reedy (resp. projective) cofibrations in $\mathcal{PC}(X, \mathcal{M})$. Then f also satisfies the right lifting property with respect to all Reedy (resp. projective) cofibrations in $\mathcal{PC}(\mathcal{M})$.*

Proof Suppose given a square

$$\begin{array}{ccc} \mathcal{U} & \longrightarrow & \mathcal{A} \\ {\scriptstyle g}\downarrow & & \downarrow{\scriptstyle f} \\ \mathcal{V} & \longrightarrow & \mathcal{B} \end{array}$$

where the map g on the left is a cofibration in $\mathcal{PC}(\mathcal{M})$. Let $\mathcal{U}' = \mathcal{U} \sqcup \mathbf{disc}(\mathrm{Ob}(\mathcal{V}) - \mathrm{Ob}(\mathcal{U}))$ so we have a factorization

$$\mathcal{U} \to \mathcal{U}' \xrightarrow{g'} \mathcal{V}$$

and g' induces an isomorphism on objects. Furthermore, g' is still a cofibration of the required type. Since f is the identity of $X = \mathrm{Ob}(\mathcal{A}) = \mathrm{Ob}(\mathcal{B})$, the top map extends in a unique way to $\mathcal{U}' \to \mathcal{A}$. Then take the pushouts with $\mathcal{A}$ to get a diagram

$$\begin{array}{ccc} \mathcal{A} & \longrightarrow & \mathcal{A} \\ {\scriptstyle g}\downarrow & & \downarrow{\scriptstyle f} \\ \mathcal{A} \cup^{\mathcal{U}'} \mathcal{V} & \longrightarrow & \mathcal{B} \end{array}$$

in $\mathcal{PC}(X, \mathcal{M})$. The left map $\mathcal{A} \to \mathcal{A} \cup^{\mathcal{U}'} \mathcal{V}$ is a Reedy (resp. projective) cofibration, but inducing an isomorphism on the set of objects which may be naturally identified as X. Hence, it is a cofibration in $\mathcal{PC}(X, \mathcal{M})$ and by hypothesis a lifting exists. This gives a lifting for the original problem. □

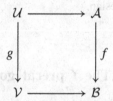

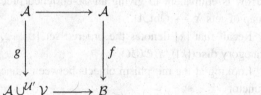

14

Calculus of generators and relations

In this chapter we look more closely at the specific calculus of generators and relations corresponding to the direct left Bousfield localization of $\mathcal{PC}(X, \mathcal{M})$ discussed in Chapter 12. Throughout, the model category $\mathcal{M}$ is assumed to be tractable left proper and cartesian.

14.1 The Υ precategories

Recall that $\emptyset$ denotes the initial object and $*$ the coinitial object of $\mathcal{M}$. We introduce some $\mathcal{M}$-enriched precategories $\Upsilon_k(B_1, \ldots, B_k)$ via an adjunction. The subscript k may be dropped when the context allows, particularly for a more compact notation in the case $k = 1$.

If X is a set, it may be considered as a discrete precategory $\mathbf{disc}(X)$ with object set itself, and morphism objects $\mathbf{disc}(X)(x_0, \ldots, x_n) := \emptyset$ when some $x_i \neq x_j$ but $\mathbf{disc}(X)(x_0, \ldots, x_n) := *$ when $x_0 = \cdots = x_n$. We can consider the category $\mathbf{disc}(X)/\mathcal{PC}(\mathcal{M})$ of arrows $\mathbf{disc}(X) \to \mathcal{A}$ in $\mathcal{PC}(\mathcal{M})$. Such an arrow is equivalent to giving an $\mathcal{M}$-enriched precategory $\mathcal{A}$ together with a map of sets $X \to \mathrm{Ob}(\mathcal{A})$.

Recall that $[k]$ denotes the ordered set $\{\upsilon_0, \ldots, \upsilon_k\}$, so we can form the category $\mathbf{disc}([k])/\mathcal{PC}(\mathcal{M})$.

Looking at the morphism objects between adjacent elements of $[k]$ gives a functor

$$(\mathcal{E}_1, \ldots, \mathcal{E}_k) : \mathbf{disc}([k])/\mathcal{PC}(\mathcal{M}) \to \mathcal{M} \times \cdots \times \mathcal{M},$$

which is defined by

$$(f : [k] \to \mathrm{Ob}(\mathcal{A})) \mapsto (\mathcal{A}(f(0), f(1)), \ldots, \mathcal{A}(f(k-1), f(k))) .$$

It has a left adjoint, consisting of a functor

$$\Upsilon_k : \mathcal{M}^k \to \mathcal{PC}(\mathcal{M}),$$

together with a natural transformation υ : $\mathbf{disc}([k]) \to \Upsilon_k(B_1, \ldots, B_k)$ so that the resulting functor

$$\mathcal{M}^k \to \mathbf{disc}([k])/\mathcal{PC}(\mathcal{M}),$$

$$(B_1, \ldots, B_k) \mapsto \left(\mathbf{disc}([k]) \stackrel{\upsilon}{\to} \Upsilon_k(B_1, \ldots, B_k)\right)$$

is left adjoint to $(\mathcal{E}_1, \ldots, \mathcal{E}_k)$.

After this abstract introduction, we can describe $\Upsilon(B_1, \ldots, B_k)$ more explicitly, and the reader could well skip the above discussion at first reading and consider just the following construction. The main part of the structure of precategory is given by

$$\Upsilon(B_1, \ldots, B_k)(\upsilon_{i-1}, \upsilon_i) = B_i.$$

This is extended whenever there is a constant string of points on either side:

$$\Upsilon(B_1, \ldots, B_k)(\upsilon_{i-1}, \ldots, \upsilon_{i-1}, \upsilon_i, \ldots, \upsilon_i) = B_i.$$

The unitality condition on the diagram $\Delta^o_{\{\upsilon_0,\ldots,\upsilon_k\}} \to \mathcal{M}$ implies, by minimality of Υ_k, that for $0 \leq i \leq k$ we have

$$\Upsilon(B_1, \ldots, B_k)(\upsilon_i, \ldots, \upsilon_i) = *.$$

In all other cases, and

$$\Upsilon(B_1, \ldots, B_k)(x_0, \ldots, x_n) = \emptyset.$$

The reader is invited to check that the obvious maps turn $\Upsilon(B_1, \ldots, B_k)$ as defined above, into a functor from $\Delta^o_{\{\upsilon_0,\ldots,\upsilon_k\}}$ to $\mathcal{M}$, which is unital, in other words it is an element of $\mathcal{PC}(\mathcal{M})$. The tautological map

$$\upsilon : [k] \to \mathrm{Ob}(\Upsilon(B_1, \ldots, B_k)) = \{\upsilon_0, \ldots, \upsilon_k\}, \quad i \mapsto \upsilon_i$$

provides $\Upsilon(B_1, \ldots, B_k)$ with a structure of element of the over-category $\mathbf{disc}([k])/\mathcal{PC}(\mathcal{M})$.

The construction is clearly functorial in $(B_1, \ldots, B_k)$, and υ is a natural transformation. We get a functor $\Upsilon_k : \mathcal{M}^k \to \mathbf{disc}([k])/\mathcal{PC}(\mathcal{M})$.

Lemma 14.1.1 *The explicitly constructed Υ_k is left adjoint to the functor $(\mathcal{E}_1, \ldots, \mathcal{E}_k)$. Furthermore, it satisfies a more global adjunction property: if $\mathcal{R} \in \mathcal{PC}(\mathcal{M})$ then a morphism $\Upsilon_k(B_1, \ldots, B_k) \to \mathcal{R}$ is the same thing as a string of objects $x_0, \ldots, x_k \in \mathrm{Ob}(\mathcal{R})$, and maps $B_i \to \mathcal{R}(x_{i-1}, x_i)$.*

Proof Given a string of objects $x_0, \ldots, x_k \in \mathrm{Ob}(\mathcal{R})$, and a sequence of maps $B_i \to \mathcal{R}(x_{i-1}, x_i)$, the functoriality maps for the diagram $\mathcal{R} : \Delta^o_{\mathrm{Ob}(\mathcal{R})} \to \mathcal{M}$ provide the required maps to define $\Upsilon_k(B_1, \ldots, B_k) \to \mathcal{R}$ and this is inverse to the obvious construction in the other direction. □

Corollary 14.1.2 *There is also an expression as a pushout of the standard representables:*

$$\Upsilon_k(B_1, \ldots, B_k) = h([1], B_1) \cup^{v_1} h([1], B_2) \cup^{v_2} \cdots \cup^{v_{k-1}} h([1], B_k),$$

where the maps in the amalgamated sums go alternately from the single point precategories to the second or first objects of $h([1], -)$.

Proof The pushout expression satisfies the same universal property as given for Υ in the previous lemma. □

The construction Υ sends sequences of cofibrations in $\mathcal{M}$ to Reedy cofibrations:

Lemma 14.1.3 *If* $f_i : A_i \to B_i$ *are cofibrations in* $\mathcal{M}$*, then the induced map*

$$\Upsilon(A_1, \ldots, A_k) \to \Upsilon(B_1, \ldots, B_k)$$

is a Reedy cofibration.

Proof Put together in a string the cofibrations $R([1], f_i)$ of Corollary 13.6.5. □

One could form a more complicated Reedy cofibration out of the f_i, for example when $k = 2$,

$$\Upsilon(B_1, A_2) \cup^{\Upsilon(A_1, A_2)} \Upsilon(A_1, B_2) \to \Upsilon(B_1, B_2)$$

is a Reedy cofibration. We leave it to the reader to elucidate notation for the most general possibility.

As a particular case of the Υ construction, which will enter into our discussion in the next section, note that for $(B_1, \ldots, B_k) = (B, \ldots, B)$ there is a tautological map $\Upsilon_k(B, \ldots, B) \to h([k], B)$. In terms of universal properties, if $\mathcal{R} \in \mathcal{PC}(\mathcal{M})$ and we are given a sequence of objects $x_0, \ldots, x_n \in \mathrm{Ob}(\mathcal{R})$ and $B \to \mathcal{R}(x_0, \ldots, x_n)$, which corresponds to $h([k], B) \to \mathcal{R}$, then composing with the principal edge maps (i.e. those which make up the Segal map) we get $B \to \mathcal{R}(x_{i-1}, x_i)$ and this collection corresponds to the map $\Upsilon_k(B, \ldots, B) \to \mathcal{R}$.

14.2 Some trivial cofibrations

There is an important link between the pseudo-generating set used to define global weak equivalences in Chapters 11 and 12, and the objects Υ defined above.

A morphism of sets $g : X \to Y$ induces a functor $\Delta^o_g : \Delta^o_X \to \Delta^o_Y$. Consider a sequence of objects $x_0, \ldots, x_n \in X$ and the resulting Segal functor $P_{(x_0,\ldots,x_n)} : \epsilon(n) \to \Delta^o_X$. Then the composition with Δ^o_g is equal to the functor corresponding to the sequence $g(x_0), \ldots, g(x_n) \in Y$:

$$\Delta^o_g \circ P_{(x_0,\ldots,x_n)} = P_{(g(x_0),\ldots,g(x_n))} : \epsilon(n) \to \Delta^o_Y .$$

We obtain a diagram

$$\begin{array}{ccccc} \mathrm{FUNC}(\epsilon(n), \mathcal{M}) & \to & \mathrm{FUNC}\left(\Delta^o_X, \mathcal{M}\right) & \stackrel{g_!}{\to} & \mathrm{FUNC}\left(\Delta^o_Y, \mathcal{M}\right) \\ & \searrow & \downarrow & & \downarrow \\ & & \mathcal{PC}(X, \mathcal{M}) & \xrightarrow{g_!} & \mathcal{PC}(Y, \mathcal{M}) \end{array}$$

where the vertical arrows are the unitalization operators $U_!$.

If $f : A \to B$ is a generating cofibration for $\mathcal{M}$, look at what happens to the map

$$\varpi_n(f) = (A, B, \ldots, B; f, \ldots, f) \xrightarrow{\rho_n(f)} \xi_{0,!}(B) = (B, \ldots, B; 1, \ldots, 1)$$

in $\mathrm{FUNC}(\epsilon(n), \mathcal{M})$, which was considered in Chapter 11. The image in $\mathcal{PC}(X, \mathcal{M})$ is the pseudo-generator

$$U_! P_{(x_0,\ldots,x_n),!}(\rho_n(f)) : U_! P_{(x_0,\ldots,x_n),!}(\varpi_n(f)) \to U_! P_{(x_0,\ldots,x_n),!}(\xi_{0,!}(B))$$

for the Reedy model structure considered in Theorem 12.3.2. When we project to $\mathcal{PC}(Y, \mathcal{M})$ using $g_!$ this gives the corresponding pseudo-generator in $\mathcal{PC}(Y, \mathcal{M})$ for the sequence $g(x_0), \ldots, g(x_n)$:

$$g_! \left(U_! P_{(x_0,\ldots,x_n),!}(\rho_n(f))\right) = U_! P_{(g(x_0),\ldots,g(x_n)),!}(\rho_n(f)).$$

This remark is particularly useful when $X = [n]$ is the universal set having a sequence of objects $\upsilon_0, \ldots, \upsilon_n \in [n]$. For any sequence $y_0, \ldots, y_n \in Y$ there is a unique map $g_{(y_0,\ldots,y_n)} : [n] \to Y$ sending υ_i to y_i, and the standard generator for the cofibration f at the sequence $(y_0, \ldots, y_n)$ is then expressed as

$$U_! P_{(y_0,\ldots,y_n),!}(\rho_n(f)) = g_{(y_0,\ldots,y_n),!} \left(U_! P_{(\upsilon_0,\ldots,\upsilon_n),!}(\rho_n(f))\right). \qquad (14.2.1)$$

So, in order to understand these generators it suffices to look at the corresponding ones in $\mathcal{PC}([n], \mathcal{M})$.

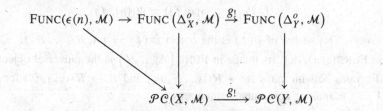

Using $\zeta_n(f)$ instead of $\rho_n(f)$ gives the pseudo-generators for the projective model structure (Theorem 12.1.1). However, there will not be such a canonical description of the pushouts by $U_! P_{(\upsilon_0,\ldots,\upsilon_n),!}(\zeta_n(f))$ because of the additional choice necessary to define $\zeta_n(f)$. The reader is invited to modify the following discussion accordingly in that case.

Lemma 14.2.1 *Fix n and suppose $f : A \to B$ is a cofibration in $\mathcal{M}$. Define the arrow*

$$\Psi([n], f) := \left(h([n], A) \cup^{\Upsilon_n(A,\ldots,A)} \Upsilon_n(B, \ldots, B) \to h([n], B)\right)$$

in $\mathcal{PC}([n], \mathcal{M})$. Then

$$U_! P_{(\upsilon_0,\ldots,\upsilon_n),!}(\rho(f)) = \Psi([n], f).$$

Proof The source of $\rho(f)$ is the object $\varpi(f) = (A, B, \ldots, B; f, \ldots, f)$ in $\text{FUNC}(\epsilon(n), \mathcal{M})$. Its image in $\text{FUNC}\left(\Delta^o_{[n]}, \mathcal{M}\right)$ is the universal object for diagrams $\mathcal{R}$ with maps $A \to \mathcal{R}(\upsilon_0, \ldots, \upsilon_n)$ and $B \to \mathcal{R}(\upsilon_{i-1}, \upsilon_i)$ for $i = 1, \ldots, n$, making commutative diagrams

$$\begin{array}{ccc} A & \longrightarrow & B \\ \downarrow & & \downarrow \\ \mathcal{R}(\upsilon_0, \ldots, \upsilon_n) & \to & \mathcal{R}(\upsilon_{i-1}, \upsilon_i). \end{array}$$

After applying $U_!$ the image in $\mathcal{PC}([n], \mathcal{M})$ is again the universal object for unital diagrams $\mathcal{R}$ as above.

The collection of maps $B \to \mathcal{R}(\upsilon_{i-1}, \upsilon_i)$ corresponds by adjunction to a map $\Upsilon_n(B, \ldots, B) \to \mathcal{R}$, the map $A \to \mathcal{R}(\upsilon_0, \ldots, \upsilon_n)$ corresponds to a map $h([n], A) \to \mathcal{R}$, and the commutative diagrams amount to requiring that they lead to the same map $\Upsilon_n(A, \ldots, A) \to \mathcal{R}$. Hence, our universal diagram is the pushout

$$U_! P_{(\upsilon_0,\ldots,\upsilon_n),!}(\varphi(f)) = h([n], A) \cup^{\Upsilon_n(A,\ldots,A)} \Upsilon_n(B, \ldots, B).$$

Similarly, the target of $\rho(f)$ is the universal object for unital diagrams S with maps $B \to S(\upsilon_0, \ldots, \upsilon_n)$, thus

$$U_! P_{(\upsilon_0,\ldots,\upsilon_n),!}(\xi_{0,!}(B)) = h([n], B).$$

The map between them is the one induced by $h([n], A) \to h([n], B)$ and $\Upsilon_n(B, \ldots, B) \to h([n], B)$. □

Corollary 14.2.2 *If $y_0, \ldots, y_n$ is any sequence of objects in Y and f is a cofibration in $\mathcal{M}$ then $U_! P_{(y_0,\ldots,y_n),!}(\rho(f))$ is obtained by applying $g_{(y_0,\ldots,y_n),!}$ to the map of the lemma.*

Proof Apply (14.2.1) to the previous lemma. □

Lemma 14.2.3 *Suppose $f : A \to B$ is a cofibration in $\mathcal{M}$ and $n \geq 0$. Then $\Psi([n], f)$ is a Reedy trivial cofibration in $\mathcal{PC}([n], \mathcal{M})$ and a globally trivial Reedy cofibration in $\mathcal{PC}(\mathcal{M})$.*

Proof This follows from Theorem 12.3.2. □

For brevity, the source of $\Psi([n], f)$ will be denoted

$$\mathrm{src}\Psi([n], f) := h([n], A) \cup^{\Upsilon_n(A,\ldots,A)} \Upsilon_n(B, \ldots, B).$$

The target is $\mathrm{targ}\Psi([n], f) = h([n], B)$.

Proposition 14.2.4 *Suppose $f : A \to B$ is a cofibration in $\mathcal{M}$, and suppose $\mathcal{R} \in \mathcal{PC}(\mathcal{M})$. To give a map from the source of $\Psi([n], f)$ to $\mathcal{R}$ is the same as to give a sequence of objects $x_0, \ldots, x_n \in \mathrm{Ob}(\mathcal{R})$ together with a commutative diagram*

$$\begin{array}{ccc} A & \longrightarrow & B \\ \downarrow & & \downarrow \\ \mathcal{R}(x_0, \ldots, x_n) & \to & \mathcal{R}(x_0, x_1) \times \cdots \times \mathcal{R}(x_{n-1}, x_n) \end{array}$$

with f on the top and the Segal map on the bottom. The pushout

$$\mathcal{R} \cup^{\mathrm{src}\Psi([n],f)} h([n], B)$$

computed in $\mathcal{PC}(\mathcal{M})$ has a set of objects isomorphic to $\mathrm{Ob}(\mathcal{R})$ and transporting by this identification (which will usually be made tacitly) this pushout is the same as the pushout of $\mathcal{R}$ along $U_! P_{(y_0,\ldots,y_n),!}(\rho(f))$ in the category $\mathcal{PC}(\mathrm{Ob}(\mathcal{R}), \mathcal{M})$.

Proof The description of maps $\mathrm{src}\Psi([n], f) \to \mathcal{R}$ was done in the proof of Lemma 14.2.1. Since $\Psi([n], f)$ is an isomorphism on sets of objects, pushout along it preserves the set of objects up to canonical isomorphism.

In general if $g : X \to Y$ is a map of sets, if $\mathcal{T} \to \mathcal{T}'$ is a morphism in $\mathcal{PC}(X, \mathcal{M})$ and $\mathcal{R} \in \mathcal{PC}(Y, \mathcal{M})$ with a map $g_!(\mathcal{T}) \to \mathcal{R}$, then the pushout $\mathcal{R} \cup^{\mathcal{T}} \mathcal{T}'$ in $\mathcal{PC}(\mathcal{M})$ corresponds (under the canonical isomorphism between

Y and the pushout of Y along 1_X) to the pushout $\mathcal{R} \cup^{g_!(\mathcal{T})} g_!(\mathcal{T}')$ in $\mathcal{PC}(Y, \mathcal{M})$. Indeed they both satisfy the same universal property in $\mathcal{PC}(Y, \mathcal{M})$ because for any $\mathcal{R}' \in \mathcal{PC}(Y, \mathcal{M})$ a map $g_!(\mathcal{T}) \to \mathcal{R}'$ is the same thing as a map $\mathcal{T} \to \mathcal{R}'$ in $\mathcal{PC}(\mathcal{M})$ inducing g on sets of objects; and the same for $\mathcal{T}'$.

Apply this general fact to pushouts along the map $\Psi([n], f)$ to get the last statement of the proposition. □

Corollary 14.2.5 *If $\mathcal{A} \in \mathcal{PC}(\mathcal{M})$ then there is a global trivial Reedy cofibration $\mathcal{A} \to$ **Seg**$(\mathcal{A})$ which is obtained as a transfinite composition of pushouts along morphisms either of the form $\Psi([n], f)$, or levelwise trivial Reedy cofibrations, such that* **Seg**$(\mathcal{A})$ *satisfies the Segal condition.*

Proof If $X = \mathrm{Ob}(\mathcal{A})$ then there is a transfinite composition of pushouts along elements of the standard generating set for the direct left Bousfield localization considered in Chapters 11 and 12, see Theorem 12.3.2. The elements of the standard generating set are either generating trivial cofibrations for the unital diagram theory without the product condition, i.e. generating levelwise trivial projective cofibrations; or else maps of the form $U_! P_{(x_0,\ldots,x_n),!}(\rho(f))$. We have seen above that pushout in $\mathcal{PC}(X, \mathcal{M})$ along $U_! P_{(x_0,\ldots,x_n),!}(\rho(f))$ is the same as pushout along $\Psi([n], f)$ in the global category $\mathcal{PC}(\mathcal{M})$. □

Notice from our discussion that a transfinite composition of pushouts along maps as used in the corollary, is a global trivial cofibration, indeed it is a trivial cofibration in the direct Bousfield localized projective model structure constructed in Chapters 11 and 12. In particular, if $\mathcal{A} \to \mathcal{A}''$ is some such transfinite composition, then applying the corollary to $\mathcal{A}''$ we obtain another such transfinite composition $\mathcal{A}'' \to \mathbf{Seg}(\mathcal{A}'')$ such that $\mathbf{Seg}(\mathcal{A}'')$ satisfies the Segal condition. In this case all the maps

$$\mathcal{A} \to \mathcal{A}'' \to \mathbf{Seg}(\mathcal{A}'')$$

are global trivial cofibrations.

14.3 Pushout by isotrivial cofibrations

One of the main problems is to prove that pushout by a global weak equivalence is again a global weak equivalence. An important first case is pushout by a global weak equivalence which is an isomorphism on objects. In the following discussion we use the generic term "cofibration" for either a projective,

Reedy or injective[1] cofibration: the statements come in three versions one for each of the model structures of Theorem 12.1.1 or Theorem 12.3.2.

An *isotrivial cofibration* is a cofibration $\mathcal{A} \stackrel{f}{\to} \mathcal{B}$ (in whichever of the projective, Reedy or injective structures we are using), such that $\mathrm{Ob}(f)$ induces an isomorphism $\mathrm{Ob}(\mathcal{A}) \cong \mathrm{Ob}(\mathcal{B})$, and f is a global weak equivalence.

Lemma 14.3.1 *A morphism $\mathcal{A} \stackrel{f}{\to} \mathcal{B}$ is an isotrivial cofibration, if and only if $\mathrm{Ob}(f)$ is an isomorphism and $\mathrm{Ob}(f)_!\mathcal{A} \stackrel{f_\natural}{\to} \mathcal{B}$ is a trivial cofibration in the appropriate model structure of Theorem 12.1.1 or Theorem 12.3.2 on $\mathcal{PC}(X, \mathcal{M})$, where $X = \mathrm{Ob}(\mathcal{B})$.*

Proof The condition that $\mathrm{Ob}(f)$ is an isomorphism is tautologically necessary so we assume it. By transport of structure using $\mathrm{Ob}(f)_!$ we may assume that f is a morphism in $\mathcal{PC}(X, \mathcal{M})$. Considered as a morphism of $\mathcal{M}$-precategories f is then automatically essentially surjective. We get the diagram

$$\begin{array}{ccc} \mathcal{A} & \longrightarrow & \mathcal{B} \\ \downarrow & & \downarrow \\ \mathbf{Seg}(\mathcal{A}) & \longrightarrow & \mathbf{Seg}(\mathcal{B}) \end{array}$$

in $\mathcal{PC}(X, \mathcal{M})$, where the vertical arrows are local weak equivalences in the model structures of 12.1.1 or 12.3.2. The fully faithful condition for f is by definition the condition that the bottom map be a levelwise weak equivalence; but this is equivalent to being a local weak equivalence since the objects satisfy the Segal conditions. By 3 for 2 this condition is equivalent to the top map being a local weak equivalence in $\mathcal{PC}(X, \mathcal{M})$. □

Lemma 14.3.2 *Suppose $\mathcal{A} \in \mathcal{PC}(\mathcal{M})$ and $f : \mathcal{B} \to \mathcal{C}$ is an isotrivial cofibration. Suppose furthermore that f is a levelwise weak equivalence of diagrams, which means that for any sequence of objects $(x_0, \ldots, x_p) \in \mathrm{Ob}(\mathcal{B})$ the map*

$$\mathcal{B}(x_0, \ldots, x_p) \to \mathcal{C}(f(x_0), \ldots, f(x_p))$$

is a weak equivalence in $\mathcal{M}$. Suppose given a map $g : \mathcal{B} \to \mathcal{A}$. Then

$$\mathcal{A} \to \mathcal{A} \cup^{\mathcal{B}} \mathcal{C}$$

[1] For the injective model structure to exist we need tractability of the appropriate injective diagram categories, but in any case one can speak of injective cofibrations, which mean levelwise ones.

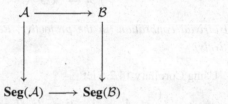

also induces an isomorphism on sets of objects, and is a levelwise weak equivalence of diagrams. In particular, it is an isotrivial cofibration.

Proof By transport of structure we may assume that $\mathrm{Ob}(f)$ is the identity of $X = \mathrm{Ob}(\mathcal{B}) = \mathrm{Ob}(\mathcal{C})$ and think of $\mathcal{B}, \mathcal{C} \in \mathcal{PC}(X, \mathcal{M})$. Let $Y = \mathrm{Ob}(\mathcal{A})$ and denote also by $g : X \to Y$ the map induced by g on sets of objects. Then

$$\mathcal{A} \cup^{\mathcal{B}} \mathcal{C} = \mathcal{A} \cup^{g_!(\mathcal{B})} g_!(\mathcal{C}) \text{ in } \mathcal{PC}(Y, \mathcal{M})$$

but $g_!(\mathcal{B}) \to g_!(\mathcal{C})$ is a levelwise trivial cofibration, so the pushout is a levelwise trivial cofibration, hence in particular a trivial cofibration in the model structure of Theorem 12.1.1 or Theorem 12.3.2. □

Theorem 14.3.3 *Suppose $\mathcal{A} \in \mathcal{PC}(\mathcal{M})$ and $f : \mathcal{B} \to \mathcal{C}$ is an isotrivial cofibration (in the projective, Reedy or injective structures). Suppose given a map $g : \mathcal{B} \to \mathcal{A}$. Then*

$$\mathcal{A} \to \mathcal{A} \cup^{\mathcal{B}} \mathcal{C}$$

is an isotrivial cofibration (in the projective, Reedy or injective structures respectively).

Proof Using Corollary 14.2.5 let

$$\mathcal{A} \to \mathcal{A}', \quad \mathcal{B} \to \mathcal{B}', \mathcal{C} \to \mathcal{C}'$$

be global trivial cofibrations towards objects which satisfy the Segal property, obtained by series of pushouts along maps which are either levelwise trivial cofibrations, or of the form $\Psi([n], u)$, where u are generating cofibrations of $\mathcal{M}$. Make the choice for $\mathcal{B}$ first, and let $\mathcal{A}'' = \mathcal{A} \cup^{\mathcal{B}} \mathcal{B}'$ and $\mathcal{C}'' = \mathcal{C} \cup^{\mathcal{B}} \mathcal{B}'$. Then, as in the remark after Corollary 14.2.5, we can continue with $\mathcal{A}'' \to \mathcal{A}'$ and $\mathcal{C}'' \to \mathcal{C}'$. That way, there are maps

$$\mathcal{A}' \leftarrow \mathcal{B}' \to \mathcal{C}',$$

the second one still being a global trivial cofibration (in any of the projective, Reedy or injective structures), and

$$\mathcal{A} \cup^{\mathcal{B}} \mathcal{C} \to \mathcal{A}' \cup^{\mathcal{B}'} \mathcal{C}'$$

is itself obtained by pushout along maps of the same form. In particular, this latter map is also a global trivial cofibration, so by the 3 for 2 property for global weak equivalences, it suffices to show that the map

$$\mathcal{A}' \to \mathcal{A}' \cup^{\mathcal{B}'} \mathcal{C}'$$

is a global weak equivalence. But now, the map $\mathcal{B}' \to \mathcal{C}'$ satisfies the hypothesis of Lemma 14.3.2, exactly by the fully faithful condition. So, by that lemma, the map $\mathcal{A}' \to \mathcal{A}' \cup^{\mathcal{B}'} \mathcal{C}'$ is a global trivial cofibration as required. □

Cofibrant pushouts are invariant under global weak equivalences inducing isomorphisms on sets of objects:

Lemma 14.3.4 *Suppose given a diagram*

$$\begin{array}{ccccc} \mathcal{A} & \leftarrow & \mathcal{B} & \rightarrow & \mathcal{C} \\ \downarrow & & \downarrow & & \downarrow \\ \mathcal{A}' & \leftarrow & \mathcal{B}' & \rightarrow & \mathcal{C}' \end{array}$$

such that the left horizontal arrows are Reedy cofibrations, and the vertical arrows are local weak equivalences, i.e. global weak equivalences inducing isomorphisms on sets of objects. Then the induced map

$$\mathcal{A} \cup^{\mathcal{B}} \mathcal{C} \to \mathcal{A}' \cup^{\mathcal{B}'} \mathcal{C}'$$

is a local weak equivalence.

Proof Choose a diagram

$$\begin{array}{ccccc} \mathcal{A} & \leftarrow & \mathcal{B} & \rightarrow & \mathcal{C} \\ \downarrow & & \downarrow & & \downarrow \\ \mathcal{A}'' & \leftarrow & \mathcal{B}'' & \rightarrow & \mathcal{C}'' \end{array}$$

such that the vertical maps are Reedy cofibrations and local weak equivalences, and the precategories along the bottom row satisfy the Segal condition. We can do this by first choosing $\mathcal{B}'' := \mathbf{Seg}(\mathcal{B})$, which is a transfinite composition of pushouts along standard morphisms in the direct localizing system K_{Reedy} of Theorem 12.3.2, which in the global context of $\mathcal{PC}(\mathcal{M})$ may be interpreted as the standard maps of Corollary 14.2.5. Let

$$\mathcal{A}'' := \mathbf{Seg}(\mathcal{A} \cup^{\mathcal{B}} \mathcal{B}''),$$

and similarly for $\mathcal{C}''$. That way the map $\mathcal{B}'' \to \mathcal{A}''$ is again a Reedy cofibration. Now put

$$\mathcal{B}^3 := \mathbf{Seg}(\mathcal{B}' \cup^{\mathcal{B}} \mathcal{B}''),$$

then

$$\mathcal{A}^3 := \mathbf{Seg}\left(\mathcal{B}^3 \cup^{\mathcal{B}' \cup^{\mathcal{B}} \mathcal{B}''} \mathcal{A}' \cup^{\mathcal{A}} \mathcal{A}''\right)$$

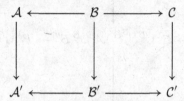

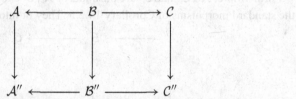

and

$$\mathcal{C}^3 := \mathbf{Seg}\left(\mathcal{B}^3 \cup^{\mathcal{B}'\cup^{\mathcal{B}}\mathcal{B}''} \mathcal{C}' \cup^{\mathcal{C}} \mathcal{C}''\right).$$

We have the diagram

$$\begin{array}{ccccc}
\mathcal{B}' \cup^{\mathcal{B}} \mathcal{B}'' & \longrightarrow & \mathcal{A}' \cup^{\mathcal{A}} (\mathcal{A} \cup^{\mathcal{B}} \mathcal{B}'') & \longrightarrow & \mathcal{A}' \cup^{\mathcal{A}} \mathcal{A}'' \\
\downarrow & & \downarrow & & \downarrow \\
\mathcal{B}^3 & \longrightarrow & \mathcal{B}^3 \cup^{\mathcal{B}'\cup^{\mathcal{B}}\mathcal{B}''} \mathcal{A}' \cup^{\mathcal{B}} \mathcal{B}'' & \rightarrow & \mathcal{B}^3 \cup^{\mathcal{B}'\cup^{\mathcal{B}}\mathcal{B}''} \mathcal{A}' \cup^{\mathcal{A}} \mathcal{A}'' \\
\| & & \downarrow & & \downarrow \\
\mathcal{B}^3 & \longrightarrow & \mathbf{Seg}(\ldots) & \longrightarrow & \mathcal{A}^3
\end{array}$$

in which the two upper squares are cartesian, the vertical maps are obtained by pushouts along standard maps of Corollary 14.2.5, and the top vertical arrows are Reedy cofibrations. It follows that $\mathcal{B}^3 \to \mathcal{A}^3$ is again a Reedy cofibration. There is a similar diagram for $\mathcal{C}^3$ (but the horizontal arrows are not necessarily cofibrations).

Furthermore, $\mathcal{A}^3$, $\mathcal{B}^3$ and $\mathcal{C}^3$ are pushouts of $\mathcal{A}'$, $\mathcal{B}'$ and $\mathcal{C}'$ respectively along the standard morphisms of Corollary 14.2.5. They fit into a diagram

$$\begin{array}{ccccc}
\mathcal{A}'' & \longleftarrow & \mathcal{B}'' & \longrightarrow & \mathcal{C}'' \\
\downarrow & & \downarrow & & \downarrow \\
\mathcal{A}^3 & \longleftarrow & \mathcal{B}^3 & \longrightarrow & \mathcal{C}^3
\end{array}$$

consisting of objects satisfying the Segal condition. The vertical arrows are global weak equivalences inducing isomorphisms on objects. Indeed on the left, for example, pushout along the isotrivial cofibration $\mathcal{A} \to \mathcal{A}''$ is again an isotrivial cofibration (Lemma 14.3.3), from which it follows that the map $\mathcal{A}' \to \mathcal{A}^3$ is a global weak equivalence. By 3 for 2, the map $\mathcal{A}'' \to \mathcal{A}^3$ is a global weak equivalence.

We get that the vertical arrows in the previous diagram are levelwise weak equivalences of diagrams. These are preserved by pushout when one of the maps in the pushout is a levelwise cofibration, as is verified levelwise (recall that $\mathcal{M}$ is assumed to be left proper). Therefore, the map

$$\mathcal{A}'' \cup^{\mathcal{B}''} \mathcal{C}'' \to \mathcal{A}^3 \cup^{\mathcal{B}^3} \mathcal{C}^3$$

is a levelwise weak equivalence of diagrams, so it is a global weak equivalence.

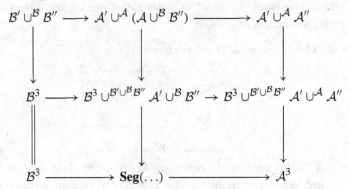

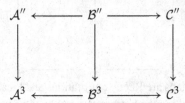

Now, the map

$$\mathcal{A} \cup^{\mathcal{B}} \mathcal{C} \to \mathcal{A}'' \cup^{\mathcal{B}''} \mathcal{C}''$$

is obtained by a transfinite composition of pushouts along the standard maps of Corollary 14.2.5, so it is a global weak equivalence. On the other hand, the map

$$\mathcal{A}' \cup^{\mathcal{B}'} \mathcal{C}' \to \mathcal{A}^3 \cup^{\mathcal{B}^3} \mathcal{C}^3$$

is also obtained by a transfinite composition of pushouts along the standard maps, so it is a global weak equivalence. By 3 for 2 we conclude that the map

$$\mathcal{A} \cup^{\mathcal{B}} \mathcal{C} \to \mathcal{A}' \cup^{\mathcal{B}'} \mathcal{C}'$$

is a global weak equivalence. It induces an isomorphism on sets of objects, so it is a local weak equivalence. □

A similar argument gives closure under transfinite composition. A cofibrancy condition can be avoided in $\mathcal{PC}(\mathcal{M})$ under the condition that it could be avoided in $\mathcal{M}$ already:

Lemma 14.3.5 *The notion of global weak equivalence in $\mathcal{PC}(\mathcal{M})$ is closed under transfinite compositions such that the transition maps are any kind of cofibrations. If weak equivalences of $\mathcal{M}$ are closed under transfinite composition, then the notion of global weak equivalence in $\mathcal{PC}(\mathcal{M})$ is also closed under transfinite composition.*

Proof Suppose we are given a transfinite sequence $\{\mathcal{A}_i\}_{i \in \alpha}$ indexed by an ordinal α, with continuity at limit ordinals $< \alpha$. Put $\mathcal{A}_\alpha := \text{colim}_{i<\alpha} \mathcal{A}_i$. Suppose that the transition maps $\mathcal{A}_i \to \mathcal{A}_{i+1}$ are global weak equivalences; and suppose either:

(i) that the transition maps are injective cofibrations, or else
(ii) that weak equivalences in $\mathcal{M}$ are closed under transfinite composition.

We want to prove that $\mathcal{A}_0 \to \mathcal{A}_\alpha$ is a global weak equivalence. By induction on α we may assume that this statement is known for all sequences indexed by strictly smaller ordinals, in particular we get that the transition maps $\mathcal{A}_i \to \mathcal{A}_j$ are global weak equivalences for all $i < j < \alpha$.

Set $X_i := \text{Ob}(\mathcal{A}_i)$ and $X := \text{Ob}(\mathcal{A}_\alpha)$. Then X is the filtered colimit of the X_i. Operating by induction on i, we will choose a sequence of morphisms $\mathcal{A}_i \to \mathcal{A}'_i$ where $\mathcal{A}'_i \in \mathcal{PC}(X_i, \mathcal{M})$ is a weak equivalent replacement of $\mathcal{A}_i$ in the injective model structure $\mathcal{PC}_{\text{Reedy}}(X_i, \mathcal{M})$ for each i, such that $\mathcal{A}'_i$ satisfies the Segal conditions and such that these are compatible in the sense that for $i < j < \alpha$ we have a commutative diagram

$$\begin{array}{ccc} \mathcal{A}_i & \longrightarrow & \mathcal{A}_j \\ \downarrow & & \downarrow \\ \mathcal{A}'_i & \longrightarrow & \mathcal{A}'_j. \end{array}$$

To make this choice, suppose it is done for all $i < j$.

If j is a limit ordinal, set $\mathcal{A}'_j := \mathrm{colim}_{i<j}\mathcal{A}'_i$. By an argument which we will use below for the colimit at α (which we don't repeat here because the notations will be more comfortable later), the map $\mathcal{A}_j \rightarrow \mathcal{A}'_j$ is a weak equivalence in $\mathcal{PC}(X_j, \mathcal{M})$ and $\mathcal{A}'_j$ satisfies the Segal conditions. This treats the case of a limit ordinal.

Suppose $j = i + 1$ is a successor ordinal. Consider the diagram

$$\begin{array}{ccc} \mathcal{A}_i & \longrightarrow & \mathcal{A}_j \\ \downarrow & & \downarrow \\ \mathcal{A}'_i & \longrightarrow & P_j \end{array}$$

where $\mathbf{P}_j := \mathcal{A}'_i \cup^{\mathcal{A}_i} \mathcal{A}_j$ is defined as the pushout. Now, the left vertical map is a weak equivalence, by the inductive hypothesis on the statement we are trying to prove, and it is also a Reedy cofibration in $\mathcal{PC}(X_i, \mathcal{M})$. By Lemma 14.3.3, pushout along a trivial cofibration which induces an isomorphism on sets of objects, is again a trivial cofibration. Thus $\mathcal{A}_j \rightarrow \mathbf{P}_j$ is a trivial cofibration, and we can let $\mathbf{P}_j \rightarrow \mathcal{A}'_j$ be a K-injective replacement of $\mathbf{P}_j$, which will also be a K-injective replacement of $\mathcal{A}_j$ accepting compatible maps from all the $\mathcal{A}'_i$ for $i < j$. Here K denotes the pseudo-generating set for $\mathcal{PC}(X_j, \mathcal{M})$ considered in Chapter 12.

This completes the choice of the sequence $\mathcal{A}'_i$, which by construction is continuous at limit ordinals. Put $\mathcal{A}'_\alpha := \mathrm{colim}_{i<\alpha}\mathcal{A}'_i$. As promised above, we show that $\mathcal{A}'_\alpha$ satisfies the Segal conditions and $\mathcal{A}_\alpha \rightarrow \mathcal{A}'_\alpha$ is a weak equivalence in $\mathcal{PC}(X, \mathcal{M})$.

Suppose given a sequence of objects $x_0, \ldots, x_n \in X$. It comes from a sequence $x_0^k, \ldots, x_n^k \in X_k$ for some $k < \alpha$ (which depends on the sequence), and denote by $x_0^i, \ldots, x_n^i \in X_i$ the image sequence for any $i \geq k$. The colimit morphism objects for $\mathcal{A}'_\alpha$ are

$$\mathcal{A}'_\alpha(x_0, \ldots, x_n) = \mathrm{colim}_{k \leq i < \alpha}\mathcal{A}'_i\left(x_0^i, \ldots, x_n^i\right).$$

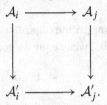

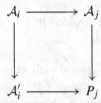

Commutation of cartesian products with colimits, condition (DCL) on $\mathcal{M}$, now tells us that

$$\mathcal{A}'_\alpha(x_0, x_1) \times \cdots \times \mathcal{A}'_\alpha(x_{n-1}, x_n)$$
$$= \mathrm{colim}_{k\le i<\alpha}\mathcal{A}'_i\left(x_0^i, x_1^i\right) \times \cdots \times \mathcal{A}'_i\left(x_{n-1}^i, x_n^i\right).$$

Transfinite composition of a sequence of weak equivalences is again a weak equivalence, so the Segal maps for $\mathcal{A}'_\alpha$ are weak equivalences. Thus, $\mathcal{A}'_\alpha$ satisfies the Segal conditions.

By the construction above, the map $\mathcal{A}_\alpha \to \mathcal{A}'_\alpha$ is obtained by a transfinite composition of pushouts along pseudo-generating trivial cofibrations for the various $\mathcal{PC}(X_i, \mathcal{M})$, thus it is a trivial cofibration.

We can now show that $\mathcal{A}_0 \to \mathcal{A}_\alpha$ is a global weak equivalence. Use $\mathcal{A}_\alpha \to \mathcal{A}'_\alpha$ as replacement satisfying the Segal conditions, so the problem is to show that $\mathcal{A}'_0 \to \mathcal{A}'_\alpha$ is a weak equivalence.

It is essentially surjective, indeed given any $x \in X = \mathrm{Ob}\left(\mathcal{A}'_\alpha\right)$ there is some $i < \alpha$ such that x comes from $x^i \in X_i$. Hence the isomorphism class of x in the truncation, is in the image of

$$\mathbf{Iso}\tau_{\le 1}\left(\mathcal{A}'_i\right) \to \mathbf{Iso}\tau_{\le 1}\left(\mathcal{A}'_\alpha\right).$$

However, since $\mathcal{A}'_0 \to \mathcal{A}'_i$ is a global weak equivalence, it is essentially surjective–in other words

$$\mathbf{Iso}\tau_{\le 1}\left(\mathcal{A}'_0\right) \to \mathbf{Iso}\tau_{\le 1}\left(\mathcal{A}'_i\right)$$

is surjective. It follows that the isomorphism class of x is in the image of

$$\mathbf{Iso}\tau_{\le 1}\left(\mathcal{A}'_0\right) \to \mathbf{Iso}\tau_{\le 1}\left(\mathcal{A}'_\alpha\right).$$

This shows essential surjectivity.

To show that the map is fully faithful, let $x_0, \ldots, x_n$ be a sequence of objects in $X_0 = \mathrm{Ob}\left(\mathcal{A}'_0\right)$. Let x_j^i denote their images in X_i including $i = \alpha$. Now

$$\mathcal{A}'_\alpha\left(x_0^\alpha, \ldots, x_n^\alpha\right) = \mathrm{colim}_{i<\alpha}\mathcal{A}'_i\left(x_0^i, \ldots, x_n^i\right).$$

To show full faithfulness we need to show that

$$\mathcal{A}'_0(x_0, \ldots, x_n) \to \mathrm{colim}_{i<\alpha}\mathcal{A}'_i\left(x_0^i, \ldots, x_n^i\right) \tag{14.3.1}$$

is a weak equivalence.

Recall that we are assuming either (i) or (ii) above. In case (i), the transition maps $\mathcal{A}_i \to \mathcal{A}_j$ are injective (i.e. levelwise) cofibrations, and by construction of $\mathcal{A}'_i$ the same is true of the transition maps $\mathcal{A}'_i \to \mathcal{A}'_j$. Hence the colimit expression above is a sequential colimit whose transition maps are trivial cofibrations, thus the map (14.3.1) is a trivial cofibration, which shows that $\mathcal{A}'_0 \to \mathcal{A}'_\alpha$ is fully faithful.

In case (ii), we just know that the transition maps in the colimit are weak equivalences, but the hypothesis (ii) says again that the map (14.3.1) is a weak equivalence.

In either case, $\mathcal{A}'_0 \to \mathcal{A}'_\alpha$ is fully faithful. □

14.4 An elementary generation step Gen

Corollary 14.2.5 looks at the replacement $\mathcal{A} \to \mathbf{Seg}(\mathcal{A})$ from the point of view of direct left Bousfield localizing systems of Chapters 11 and 12. It will also be useful to have an approach which is more homotopically canonical, in other words some kind of process which looks canonical when viewed in the homotopy category ho $\left(\text{FUNC}\left(\Delta^o_X/X, \mathcal{M}\right)\right)$ of unital diagrams up to levelwise weak equivalences. The full process will be broken up into elementary steps denoted $\mathcal{A} \to \mathbf{Gen}(\mathcal{A}, q)$. These should be thought of as "calculating generators and relations at the Segal map corresponding to q."

Fix a set of objects X. Recall that the underlying category used is $\Phi = \Delta^o_X$; the subcategory on which the unitality condition will be imposed is $\Phi_0 = \{(x_0)\}_{x_0 \in X}$, and the set Q used to determine the algebraic theory of Segal categories over X, is just the object set of Δ_X that is the set of sequences $(x_0, \ldots, x_n)$ of $x_i \in X$. For $q = (x_0, \ldots, x_n)$, we have $n_q := n$; the functor $P_q : \epsilon(n) \to \Delta^o_X$ sends ξ_0 to $q = (x_0, \ldots, x_n)$ and for $1 \leq j \leq n$ it sends ξ_j to the adjacent pair (x_{j-1}, x_j). The structural maps for P_q come from inclusions of each adjacent pair in the full sequence; they are the maps which together make up the Segal maps.

An $\mathcal{A} \in \mathcal{PC}(X, \mathcal{M})$ is a functor $\mathcal{A} : \Delta^o_X \to \mathcal{M}$ such that $\mathcal{A}(x) = *$ for any single element sequence (x).

For any $q = (x_0, \ldots, x_n) \in Q = \text{Ob}\left(\Delta^o_X\right)$, the functor

$$U_! P_{q,!} : \text{FUNC}(\epsilon(n), \mathcal{M}) \to \mathcal{PC}(X, \mathcal{M})$$

sends injective cofibrations to Reedy cofibrations. Recall that it sends the standard $\rho_n(f)$ to the cofibrations $\Psi([n], f)$, see Corollary 14.2.5. Abbreviate the right adjoint $P^*_q U^*$ by just P^*_q.

For $\mathcal{A} \in \mathcal{PC}(X, \mathcal{M})$ and $q = (x_0, \ldots, x_n)$, we have considered in Section 11.6 the cofibration $\mathcal{A} \to \mathbf{Gen}(\mathcal{A}, q)$, which depends on a choice of factorization:

$$P^*_q(\mathcal{A})_0 \to E \to P^*_q(\mathcal{A})_1 \times \cdots \times P^*_q(\mathcal{A})_n.$$

Note that $P_q^*(\mathcal{A})_0 = \mathcal{A}(x_0, \ldots, x_n)$, while $P_q^*(\mathcal{A})_j = \mathcal{A}(x_{j-1}, x_j)$ for $1 \leq j \leq n$. Thus, the factorization we need to choose may be written as

$$\mathcal{A}(x_0, \ldots, x_n) \xrightarrow{e} E \xrightarrow{(p_1,\ldots,p_n)} \mathcal{A}(x_0, x_1) \times \cdots \times \mathcal{A}(x_{n-1}, x_n).$$

Given such a factorization, which will generally be chosen so that e is a cofibration and $(p_1, \ldots, p_n)$ is a weak equivalence, we get a map

$$\mathbf{gen}(\mathcal{A}, q) : \mathcal{A} \to \mathbf{Gen}(\mathcal{A}, q)[E, e, p_1, \ldots, p_n].$$

If no confusion arises, we abbreviate the right-hand side to $\mathbf{Gen}(\mathcal{A}, q)$. In the notation of the previous section,

$$\mathbf{Gen}(\mathcal{A}, q)[E, e, p_1, \ldots, p_n] = \mathcal{A} \cup^{\mathrm{src}\Psi([n],e)} h([n], E). \tag{14.4.1}$$

Lemma 14.4.1 *Let $\mathcal{A} \in \mathcal{PC}(X, \mathcal{M})$ and $q = (x_0, \ldots, x_n)$, and suppose given a choice of factorization $E, e, p_1, \ldots, p_n$ as above. Suppose e is a cofibration and $(p_1, \ldots, p_n) : E \to \mathcal{A}(x_0, x_1) \times \cdots \times \mathcal{A}(x_{n-1}, x_n)$ is a weak equivalence in $\mathcal{M}$. Then $\mathbf{gen}(\mathcal{A}, q)$ is a trivial cofibration from $\mathcal{A}$ to $\mathbf{Gen}(\mathcal{A}, q) = \mathbf{Gen}(\mathcal{A}, q)[E, e, p_1, \ldots, p_n]$ in the model structure $\mathcal{PC}_{\mathrm{Reedy}}(X, \mathcal{M})$ constructed in Theorem 12.3.2.*

Proof Indeed $\mathbf{gen}(\mathcal{A}, q)$ is a pushout along $\Psi([n], e)$ by (14.4.1). □

We can describe explicitly the structure of $\mathbf{Gen}(\mathcal{A}; x_0, \ldots, x_n)$ depending on the choice of $E, e, p_1, \ldots, p_n$. For any sequence $(z_0, \ldots, z_m)$, let

$$\Delta_X^{NS}((z_0, \ldots, z_m), (x_0, \ldots, x_n)) \subset \Delta_X((z_0, \ldots, z_m), (x_0, \ldots, x_n))$$

be the subset of maps $\phi : (z_0, \ldots, z_m) \to (x_0, \ldots, x_m)$ in Δ_X which don't factor through any of the adjacent pairs (x_{j-1}, x_j). Abbreviate this by $\Delta_X^{NS}(z., x.)$ when convenient. Then $\mathbf{Gen}(\mathcal{A}; x_0, \ldots, x_n)(z_0, \ldots, z_m)$ is a pushout in the diagram

$$\begin{array}{ccc}
\coprod_{\phi \in \Delta_X^{NS}(z.,x.)} \mathcal{A}(x_0, \ldots, x_n) & \longrightarrow & \coprod_{\phi \in \Delta_X^{NS}(z.,x.)} E \\
\downarrow & & \downarrow \\
\mathcal{A}(z_0, \ldots, z_m) & \longrightarrow & \mathbf{Gen}(\mathcal{A}; x_0, \ldots, x_n)(z_0, \ldots, z_m).
\end{array}$$

In order to define the functoriality it is convenient to rewrite this as follows: for any object $B \in \mathcal{M}$ and any $\phi \in \Delta_X((z_0, \ldots, z_m), (x_0, \ldots, x_n))$ denote by $\mathbf{alt}(\phi, B)$ either: B if $\phi \in \Delta_X^{NS}(z., x.)$; or $*$ if ϕ factors through a singleton

$$(z_0, \ldots, z_m) \to (x_j) \to (x_0, \ldots, x_n);$$

or $\mathcal{A}(x_{j-1}, x_j)$ if ϕ factors through a map

$$(z_0, \ldots, z_m) \to (x_{j-1}, x_j) \to (x_0, \ldots, x_n),$$

but not through a singleton (in which case the choice of j is unique). Now, for either $B = \mathcal{A}(x_0, \ldots, x_n)$ or $B = E$, we have structural maps $B \to \mathcal{A}(x_{j-1}, x_j)$. Using these, given a map

$$\psi : (y_0, \ldots, y_p) \to (z_0, \ldots, z_m),$$

then for any $\phi \in \Delta_X((z_0, \ldots, z_m), (x_0, \ldots, x_n))$ we get a map

$$\mathbf{alt}(\phi, B) \to \mathbf{alt}(\phi\psi, B),$$

and these are compatible with composition in ψ. Indeed, if ϕ factors through an adjacent pair or a singleton, then so does $\phi\psi$. If $\phi\psi$ factors through a singleton then the map from $\mathbf{alt}(\phi, B)$ is the unique map to $\mathbf{alt}(\phi\psi, B) = *$. If $\phi\psi$ factors through an adjacent pair but ϕ didn't factor through an adjacent pair, then the map

$$\mathbf{alt}(\phi, B) = \mathcal{A}(x_0, \ldots, x_n) \text{ or } E \to \mathbf{alt}(\phi\psi, B) = \mathcal{A}(x_{j-1}, x_j)$$

is the given structural map.

Now with these notations, $\mathbf{Gen}(\mathcal{A}; x_0, \ldots, x_n)(z_0, \ldots, z_m)$ can also be expressed as a pushout of the form

$$\begin{array}{ccc}
\coprod_{\phi \in \Delta_X(z.,x.)} \mathbf{alt}(\phi, \mathcal{A}(x_0, \ldots, x_n)) & \longrightarrow & \coprod_{\phi \in \Delta_X(z.,x.)} \mathbf{alt}(\phi, E) \\
\downarrow & & \downarrow \\
\mathcal{A}(z_0, \ldots, z_m) & \longrightarrow & \mathbf{Gen}(\mathcal{A}; x_0, \ldots, x_n)(z_0, \ldots, z_m).
\end{array}$$

Note that for a map $\phi : z. \to x.$ factoring through (x_{j-1}, x_j) (resp. (x_j)) we get a map $(z_0, \ldots, z_m) \to (x_{j-1}, x_j)$ (resp. to (x_j)) and hence a map $\mathcal{A}(x_{j-1}, x_j) \to \mathcal{A}(z_0, \ldots, z_m)$ (resp. a map $* = \mathcal{A}(x_j) \to \mathcal{A}(z_0, \ldots, z_m)$). These combine to give the left vertical map.

Now, in case of a map $\psi : (y_0, \ldots, y_p) \to (z_0, \ldots, z_m)$ we get maps

$$\coprod_{\phi \in \Delta_X(z.,x.)} \mathbf{alt}(\phi, \mathcal{A}(x_0, \ldots, x_n)) \to \coprod_{\phi \in \Delta_X(z.,x.)} \mathbf{alt}(\phi\psi, \mathcal{A}(x_0, \ldots, x_n))$$

$$\to \coprod_{\zeta \in \Delta_X(y.,x.)} \mathbf{alt}(\zeta, \mathcal{A}(x_0, \ldots, x_n))$$

and

$$\coprod_{\phi \in \Delta_X(z.,x.)} \mathbf{alt}(\phi, E) \to \coprod_{\phi \in \Delta_X(z.,x.)} \mathbf{alt}(\phi\psi, E) \to \coprod_{\zeta \in \Delta_X(y.,x.)} \mathbf{alt}(\zeta, E),$$

which induce the map of functoriality

$$\mathbf{Gen}(\mathcal{A}; x_0, \ldots, x_n)(z_0, \ldots, z_m) \to \mathbf{Gen}(\mathcal{A}; x_0, \ldots, x_n)(y_0, \ldots, y_p).$$

This whole discussion can be simplified considerably in the case when the $x_0, \ldots, x_n$ are all distinct in other words $\forall 0 \le i \ne j \le n,\ x_i \ne x_j$. Then, for any sequence of objects $(z_0, \ldots, z_m) \in \Delta_X$ there is at most one map $\phi : (z_0, \ldots, z_m) \to (x_0, \ldots, x_n)$. It factors through an adjacent pair if and only if $z.$ is of the form $(x_{j-1}, \ldots, x_{j-1}, x_j, \ldots, x_j)$ and it factors through a singleton if and only if $z. = (x_j, \ldots, x_j)$. There is a map, but not factoring through an adjacent pair or a singleton, if and only if

$$z. = (x_{i_0}, \ldots, x_{i_m}),$$

where $0 \le i_0 \le i_1 \le \cdots \le i_m \le n$ and $i_0 + 1 < i_m$. We state the result in this case as a lemma:

Lemma 14.4.2 *Suppose $\mathcal{A} \in \mathcal{PC}(X, \mathcal{M})$ and $(x_0, \ldots, x_n)$ is a sequence of pairwise disjoint objects. Suppose given a choice of factorization $E, e, p_1, \ldots, p_n$ as above. Then, for any increasing sequence of indices $0 \le i_0 \le i_1 \le \cdots \le i_m \le n$ with $i_0 + 1 < i_m$,*

$$\mathbf{Gen}(\mathcal{A}; x_0, \ldots, x_n)(x_{i_0}, \ldots, x_{i_m}) = \mathcal{A}(x_{i_0}, \ldots, x_{i_m}) \cup^{\mathcal{A}(x_0, \ldots, x_n)} E.$$

For any other sequence of objects $(z_0, \ldots, z_m)$ we have

$$\mathbf{Gen}(\mathcal{A}; x_0, \ldots, x_n)(z_0, \ldots, z_m) = \mathcal{A}(z_0, \ldots, z_m).$$

These expressions are compatible with the maps from $\mathcal{A}(z_0, \ldots, z_m)$, and the maps of functoriality are given by the structural maps for E in the case of a map $\psi : (z_0, \ldots, z_m) \to (x_{i_0}, \ldots, x_{i_m})$ which factors through a sequence projecting to an adjacent pair in $(x_0, \ldots, x_n)$. The maps of functoriality are given by the degeneracy maps of $\mathcal{A}$ in case ψ factors through a repeated singleton.

Proof Left to the reader. □

14.5 Fixing the fibrant condition locally

Applying the operation **Gen** may destroy the fibrancy condition in the plain diagram Reedy structure. Let $\mathbf{fix} : \mathcal{A} \to \mathbf{Fix}(\mathcal{A})$ be a replacement by a fibrant object in the Reedy model structure on the plain diagram category $\mathrm{FUNC}_{\mathrm{Reedy}}(\Delta^o_X/X, \mathcal{M})$, where $X = \mathrm{Ob}(\mathcal{A})$. The map $\mathbf{fix}(\mathcal{A})$ is a pushout along elements of K_{Reedy}, namely the generators for the trivial cofibrations of $\mathrm{FUNC}_{\mathrm{Reedy}}(\Delta^o_X/X, \mathcal{M})$. In particular, it is an isotrivial cofibration and levelwise a trivial cofibration. This doesn't change the homotopy type in $\mathrm{ho}(\mathrm{FUNC}(\Delta^o_X/X, \mathcal{M}))$.

14.6 Combining generation steps

We can put together the elementary generation steps defined above, to obtain a model for the replacement to a Segal object:

Theorem 14.6.1 *Suppose $\mathcal{A} \in \mathcal{PC}(X, \mathcal{M})$. There is a transfinite composition sequence $\{\mathcal{A}_i\}_{i\in[\alpha]}$ indexed by an ordinal $[\alpha] = \alpha + 1$, with $\mathcal{A}_0 = \mathcal{A}$, such that $\mathcal{A}_\cdot$ is continuous at limit ordinals, such that $\mathcal{A}_\alpha$ satisfies the Segal conditions, and such that each $\mathcal{A}_i \to \mathcal{A}_{i+1}$ has the form $\mathbf{fixgen}(\mathcal{A}_i, q_i)$ for some $q_i = (x_{i,0}, \ldots, x_{i,n_i})$. In other words, $\mathcal{A}_{i+1} = \mathbf{FixGen}(\mathcal{A}_i, q_i)[E_i, e_i, p_{i,1}, \ldots, p_{i,n_i}]$ as considered above. Furthermore, the transition maps are all trivial cofibrations in $\mathcal{PC}_{\mathrm{Reedy}}(X, \mathcal{M})$; in particular $\mathcal{A} \to \mathcal{A}_\alpha$ is a trivial cofibration. Denote the end result by $\mathbf{Seg}^{FG}(\mathcal{A})$; it is a fibrant object in the Reedy model structure $\mathcal{PC}(X, \mathcal{M})$ and equivalent to $\mathbf{Seg}(\mathcal{A})$ in the Reedy model structure on the plain diagram category $\mathrm{FUNC}_{\mathrm{Reedy}}(\Delta^o_X/X, \mathcal{M})$, indeed it is a possible choice for the fibrant replacement $\mathbf{Seg}(\mathcal{A})$.*

Proof The $\mathbf{gen}(\mathcal{A}_i, q_i)$ are trivial cofibrations which may be chosen to contain a pushout along any particular cofibration of the form $\Psi([n], f)$. Similarly the maps **fix** may be chosen to contain any of the pushouts along generating trivial cofibrations for the level structure. Thus, the maps envisioned above may contain all pushouts along elements of the pseudo-generating set K_{Reedy}. Choose the sequence using the small object argument so that the end result $\mathcal{A}_\alpha$ is in $\mathbf{inj}(K_{\mathrm{Reedy}})$. In particular it satisfies the Segal conditions. Note that K_{Reedy}-injective objects are also fibrant, see Theorem 12.3.2 where that statement really comes from Theorem 9.7.1. Thus, $\mathcal{A} \to \mathbf{Seg}^{FG}(\mathcal{A})$ is a trivial cofibration towards a fibrant object in $\mathcal{PC}_{\mathrm{Reedy}}(X, \mathcal{M})$. □

An important corollary of this procedure is the following statement, tautological for maps between Segal categories:

Corollary 14.6.2 *Suppose a map $f : \mathcal{A} \to \mathcal{B}$ in $\mathcal{PC}(\mathcal{M})$ induces a weak equivalence*

$$\mathcal{A}(x_0, \ldots, x_p) \xrightarrow{\sim} \mathcal{B}(f(x_0), \ldots, f(x_p)),$$

for any sequence $(x_0, \ldots, x_p) \in \Delta_{\mathrm{Ob}(\mathcal{A})}$. Then f is fully faithful.

Proof In the process of Theorem 14.6.1 this condition is preserved at each step. It follows that $\mathbf{Seg}^{FG}(\mathcal{A}) \to \mathbf{Seg}^{FG}(\mathcal{B})$ is fully faithful, which is by definition the full faithfulness criterion for f. □

14.7 Functoriality of the generation process

Suppose $\mathcal{M}$ and $\mathcal{N}$ are tractable left proper cartesian model categories. Suppose $F : \mathcal{M} \to \mathcal{N}$ is a left Quillen functor. We say that F is *weakly compatible with products* if, for any $A, B \in \mathcal{M}$, the natural map $F(A \times B) \to F(A) \times F(B)$ is a weak equivalence in $\mathcal{N}$. Say that F is *unital* if $F(*) = *$.

The functor F induces functors

$$\mathcal{PC}(X, F) : \mathcal{PC}(X, \mathcal{M}) \to \mathcal{PC}(X, \mathcal{N}),$$

and hence, as X varies, it induces a functor

$$\mathcal{PC}(F) : \mathcal{PC}(\mathcal{M}) \to \mathcal{PC}(\mathcal{N}).$$

In general this should be defined by applying F levelwise over Δ_X^o, then following up with the unitalization operation $U_!$ as in Lemma 12.6.1. If F is unital, however, then for $\mathcal{A} \in \mathcal{PC}(X, \mathcal{M})$ and a sequence of objects $x_0, \ldots, x_n$ in X, the simpler expression

$$F(\mathcal{A})(x_0, \ldots, x_n) := F(\mathcal{A}(x_0, \ldots, x_n))$$

holds.

Lemma 14.7.1 *In the above situation with F unital, suppose $\mathcal{A} \in \mathcal{PC}(X, \mathcal{M})$, choose $q = (x_0, \ldots, x_n)$, and suppose given a choice of factorization $E, e, p_1, \ldots, p_n$ as above. Suppose e is a cofibration and $(p_1, \ldots, p_n) : E \to \mathcal{A}(x_0, x_1) \times \cdots \times \mathcal{A}(x_{n-1}, x_n)$ is a weak equivalence in $\mathcal{M}$. Then $F(e)$ is a cofibration and*

$$(Fp_1, \ldots, Fp_n) : F(E) \to F(\mathcal{A})(x_0, x_1) \times \cdots \times F(\mathcal{A})(x_{n-1}, x_n)$$

is a weak equivalence in $\mathcal{N}$, so $(F(E), F(e), F(p_1), \ldots, F(p_n))$ constitute data for defining

$$\mathbf{Gen}(F(\mathcal{A}), q) = \mathbf{Gen}(F(\mathcal{A}), q)[F(E), F(e), F(p_1), \ldots, F(p_n)]$$

with its map

$$F(\mathcal{A}) \xrightarrow{\mathbf{gen}(F(\mathcal{A}), q)} \mathbf{Gen}(F(\mathcal{A}), q)$$

in $\mathcal{PC}(X, \mathcal{N})$. *In these terms we have* $\mathbf{gen}(F(\mathcal{A}), q) = F(\mathbf{gen}(\mathcal{A}, q))$ *and*

$$\begin{aligned}&\mathbf{Gen}(F(\mathcal{A}), q)[F(E), F(e), F(p_1), \ldots, F(p_n)]\\&\quad = F(\mathbf{Gen}(\mathcal{A}, q)[E, e, p_1, \ldots, p_n]),\end{aligned}$$

which can be written more succinctly as $\mathbf{Gen}(F(\mathcal{A}), q) = F(\mathbf{Gen}(\mathcal{A}, q))$.

Proof By inspection. □

Corollary 14.7.2 *In our above situation of a left Quillen functor* $F : \mathcal{M} \to \mathcal{N}$ *unital and weakly compatible with products, between two tractable left proper cartesian model categories, suppose* $\mathcal{A} \in \mathcal{PC}(X, \mathcal{M})$ *and let* $\{\mathcal{A}_i\}_{i \in [\alpha]}$ *be a transfinite composition of elementary generation steps in* $\mathcal{PC}(X, \mathcal{M})$ *with* $\mathcal{A} = \mathcal{A}_0 \to \mathcal{A}_\alpha$ *a trivial cofibration in* $\mathcal{PC}_{\text{Reedy}}(X, \mathcal{M})$ *and* $\mathcal{A}_\alpha$ *satisfying the Segal conditions. Then* $\{F(\mathcal{A}_i)\}_{i \in [\alpha]}$ *constitutes a transfinite composition of elementary generation steps in* $\mathcal{PC}(X, \mathcal{N})$, *the map* $F(\mathcal{A}) \to F(\mathcal{A}_\alpha)$ *is a trivial cofibration in the Reedy structure* $\mathcal{PC}_{\text{Reedy}}(X, \mathcal{N})$, *and* $F(\mathcal{A}_\alpha)$ *satisfies the Segal conditions as an* $\mathcal{N}$*-enriched precategory.*

Proof Just apply F to the sequence of Theorem 14.6.1. □

We state the next corollary as a theorem, since it is the main result we need from this section. Recall from Proposition 12.6.2 that, given a left Quillen functor between tractable left proper cartesian model categories, it induces a left Quillen functor on the Reedy model categories of precategories. One didn't even need to suppose that F preserves products. If that is the case, the statement can be slightly improved to say that $\mathcal{PC}(X, F)$ preserves local weak equivalences in the precategory model structures:

Theorem 14.7.3 *Suppose* $F : \mathcal{M} \to \mathcal{N}$ *is a left Quillen functor, preserving weak equivalences, unital and weakly compatible with products, between two tractable left proper cartesian model categories. Then* $\mathcal{PC}(X, F) : \mathcal{PC}_{\text{Reedy}}(X, \mathcal{M}) \to \mathcal{PC}_{\text{Reedy}}(X, \mathcal{N})$ *is a left Quillen functor which takes local weak equivalences to local weak equivalences.*

Proof Suppose $\mathcal{A} \to \mathcal{B}$ is a local weak equivalence in $\mathcal{PC}_{\text{Reedy}}(X, \mathcal{M})$. This means that $\mathbf{Seg}^{FG}(\mathcal{A}) \to \mathbf{Seg}^{FG}(\mathcal{B})$ is a levelwise weak equivalence of diagrams, so applying $\mathcal{PC}(X, F)$ gives a new levelwise weak equivalence of diagrams towards $\mathcal{N}$. By the previous corollary, when we apply $\mathcal{PC}(X, F)$ we get $\mathbf{Seg}^{FG}(\mathcal{PC}(X, F)\mathcal{A}) \to \mathbf{Seg}^{FG}(\mathcal{PC}(X, F)\mathcal{B})$, it follows that $\mathcal{PC}(X, F)\mathcal{A}) \to \mathcal{PC}(X, F)\mathcal{B}$ is a local weak equivalence in $\mathcal{PC}_{\text{Reedy}}(X, \mathcal{N})$. □

14.8 Example: generators and relations for 1-categories

It is interesting and instructive to consider the case where $\mathcal{M} = \text{SET}$ is the model category of sets. Here, the weak equivalences are isomorphisms, and the fibrations and cofibrations are arbitrary maps. It is easily seen to be tractable, left proper and cartesian. The category of SET-enriched precategories $\mathcal{PC}(\text{SET})$ may be identified with the category of simplicial sets. For $\mathcal{A} \in \mathcal{PC}(\text{SET})$, if we make this identification then the 0-simplices correspond to the objects; the 1-simplices give generators for the morphisms, and the 2-simplices give relations among the morphisms. The Segal conditions for $\mathcal{A} \in \mathcal{PC}(\text{SET})$ are exactly the classical conditions which are equivalent to stating that $\mathcal{A}$ is the nerve of a category. The calculus of generators and relations constructed above, reduces in this case to the classical calculus of generators and relations for 1-categories.

Fix a set X. For this section, the main part of the data of $\mathcal{A} \in \mathcal{PC}(X, \text{SET})$ consists of sets $\mathcal{A}(x, y)$ for each pair $x, y \in X$, and $\mathcal{A}(x, y, z)$ for each triple $x, y, z \in Z$. Recall that $\mathcal{A}(x)$ is a singleton, so the degeneracy maps yield elements denoted $1_x \in \mathcal{A}(x, x)$. Similarly, for any $f \in \mathcal{A}(x, y)$ the degeneracies yield elements denoted $[1_y f] \in \mathcal{A}(x, y, y)$ and $[f 1_x] \in \mathcal{A}(x, x, y)$. In the case $f = 1_x$ there is no confusion as both elements denoted $[1_x 1_x]$ correspond to the same element of $\mathcal{A}(x, x, x)$ obtained from the singleton $\mathcal{A}(x)$ by degeneracy. We have projections

$$i_{01}^* : \mathcal{A}(x, y, z) \to \mathcal{A}(x, y), \quad i_{12}^* : \mathcal{A}(x, y, z) \to \mathcal{A}(y, z),$$
$$I_{02}^* : \mathcal{A}(x, y, z) \to \mathcal{A}(x, z).$$

These are compatible with the degeneracies in an obvious way, for example $i_{02}^*[1_y f] = i_{02}^*[f 1_x] = f$.

The elements of $\mathcal{A}(x, y)$ are the "generating arrows" between x and y. An element $r \in \mathcal{A}(x, y, z)$ corresponds to an "elementary relation" of the form $f = gh$, where

$$f = i_{02}^*(r) \in \mathcal{A}(x, z), \quad g = i_{12}^*(r) \in \mathcal{A}(y, z), \quad h = i_{01}^*(r) \in \mathcal{A}(x, y).$$

This can be seen by looking more closely at what happens under the generation step $\mathbf{Gen}(\mathcal{A}, q)$ for a triple $q = (x, y, z)$. Such a step involves a choice of factorization

$$\mathcal{A}(x, y, z) \xrightarrow{e} E \xrightarrow{(p_1, p_2)} \mathcal{A}(x, y) \times \mathcal{A}(y, z).$$

All maps are cofibrations in SET so there is no restriction on e; on the other hand the weak equivalences are the isomorphisms, so (p_1, p_2) has to be an isomorphism and we may in effect suppose $E = \mathcal{A}(x, y) \times \mathcal{A}(y, z)$.

We can describe explicitly the resulting precategory $\mathcal{A}' := \mathbf{Gen}(\mathcal{A}, q)$. Suppose (u, v) is a pair of elements of X. We need to consider the maps $\phi : (u, v) \to (x, y, z)$ in Δ_X. There are six possibilities:

$$(u, v) = (x, y) \xrightarrow{i_{01}} (x, y, z),$$

$$(u, v) = (y, z) \xrightarrow{i_{12}} (x, y, z),$$

$$(u, v) = (x, z) \xrightarrow{i_{02}} (x, y, z),$$

$$(u, v) = (x, x) \xrightarrow{i_{00}} (x, y, z),$$

$$(u, v) = (y, y) \xrightarrow{i_{11}} (x, y, z),$$

$$(u, v) = (z, z) \xrightarrow{i_{22}} (x, y, z).$$

In case of coincidences among the x, y, z these can overlap in the sense that the same (u, v) could have several maps to (x, y, z). However, in the notation of Section 14.4, the only one of these maps which is in $\Delta_X^{NS}((u, v), (x, y, z))$ is i_{02}. Thus, using $E = \mathcal{A}(x, y) \times \mathcal{A}(y, z)$, we have a pushout diagram

$$\begin{array}{ccc} \mathcal{A}(x, y, z) & \to & \mathcal{A}(x, y) \times \mathcal{A}(y, z) \\ \downarrow & & \downarrow \\ \mathcal{A}(x, z) & \longrightarrow & \mathcal{A}'(x, z). \end{array}$$

In other words, $\mathcal{A}'(x, z)$ is obtained by taking the old $\mathcal{A}(x, z)$ and adding the symbols gh for $g \in \mathcal{A}(y, z)$ and $h \in \mathcal{A}(x, y)$, subject to the relations that $f = gh$ whenever there is an element of $\mathcal{A}(x, y, z)$ mapping to $f \in \mathcal{A}(x, z)$, $h \in \mathcal{A}(x, y)$ and $g \in \mathcal{A}(y, z)$.

Similarly, if (u, v, w) is a triple of elements, the only maps

$$(u, v, w) \to (x, y, z)$$

which are in $\Delta_X^{NS}((u, v), (x, y, z))$ are the identity $(x, y, z) \to (x, y, z)$, and the degeneracies of the previous map i_{02}

$$(x, z, z) \to (x, y, z), \quad (x, x, z) \to (x, y, z).$$

The pushout expression along the identity says that for any such symbol gh which is added to $\mathcal{A}'(x, z)$ following the previous discussion, there will also be a corresponding element of $\mathcal{A}'(x, y, z)$ saying that it is the composition of g and h, in other words the formal composition gh will be recorded as a composition of g and h. The degeneracies of i_{02} take care of keeping track of left and right identities for the new morphisms gh which are added.

We leave as an exercise to verify that the generation steps $\mathbf{Gen}(\mathcal{A}, q)$ at quadruples $q = (w, x, y, z)$ correspond to enforcing the associativity axiom.

Putting this all together, we see that after repeating the generation steps infinitely many times for each uplet of objects, the resulting precategory $\mathbf{Seg}^{FG}(\mathcal{A})$ is the nerve of a category, and it is the category generated by the original arrows $\mathcal{A}(x, y)$ subject to the original relations $\mathcal{A}(x, y, z)$.

15

Generators and relations for Segal categories

In this chapter, we consider one of the main examples of the theory: when $\mathcal{M} = \mathcal{K}$ is the Kan–Quillen model category of simplicial sets. The notion of weakly $\mathcal{M}$-enriched category then becomes the notion of *Segal category*, one version of the $(\infty, 1)$-categories which are ubiquitous in applications of higher category theory. The first section is devoted to a review of the basic definitions and notations in this important case. Then, we take the opportunity to illustrate the general calculus of generators and relations as it applies to the problem of creating the loop space of a space X, together with its Segal delooping structure. This collection of structure is the *Poincaré–Segal groupoid* of X, a Segal category analogous to the classical Poincaré groupoid. The word "groupoid" in the terminology indicates that even the 1-morphisms are invertible up to equivalence. Keeping the higher categorical structure allows this object to reflect the full weak homotopy type of X, indeed that is what Segal's basic theorem says, as will be reviewed in the second section below. As will be pointed out in the third section, the Poincaré–Segal groupoid has a fairly simple abstract description generalizing the space of loops on X. However, explicit calculation of the infinite-dimensional space of loops is not easy, and the subsequent sections are devoted to showing how the calculus of generators and relations, developed in preceding chapters, allows us to view the loop space as something deduced algebraically from X. This will be happening in the world of Segal groupoids, just because the direction of a path is reversible. A first example, the calculation of $\pi_3(S^2)$, is worked out in detail.

15.1 Segal categories

Theorem 15.1.1 *The Kan–Quillen model category $\mathcal{K}$ of simplicial sets, is a tractable left proper cartesian model category.*

Proof Cofibrations are just monomorphisms of simplicial sets, so the cartesian and tractability conditions are easy to check. The Eilenberg–Zilber theorem on products of simplicial sets gives Condition (PROD) of Definition 7.7.1 for trivial cofibrations. □

Corollary 15.1.2 *For a fixed set of objects X, we get model categories of $\mathcal{K}$-enriched precategories over X with either the projective structure $\mathcal{PC}_{\text{proj}}(X, \mathcal{K})$, or the Reedy structure $\mathcal{PC}_{\text{Reedy}}(X, \mathcal{K})$, which is the same as the injective structure $\mathcal{PC}_{\text{inj}}(X, \mathcal{K})$.*

Proof Apply Theorems 12.1.1 and 12.3.2. Note that $\mathcal{K}$ is a presheaf category and the cofibrations are monomorphisms, so by Proposition 13.7.2 the Reedy and injective cofibrations coincide. The local weak equivalences are the same in all structures so the Reedy and injective structures are the same. □

A $\mathcal{K}$-precategory is called a *Segal precategory*; those satisfying the Segal conditions are called *Segal categories*. We have defined the classes of global weak equivalences, and two flavors of cofibrations (hence fibrations), on the way to constructing the projective and Reedy or equivalently injective model structures on the category $\mathcal{PC}(\mathcal{K})$ of Segal precategories in the next part.

The theory of Segal categories goes back a long way. They were first considered by Dwyer *et al.* [104] and Schwänzl and Vogt [224], but of course the definition goes essentially back to Segal [228], who restricted it to the case when there is only one object. Other references are Adams [3], Thomason [253] and Cordier and Porter [83]. Bergner [40] and Pellissier [211] have already given complete constructions of the global model category structure. The model categories $\mathcal{PC}_{\text{proj}}(\mathcal{K})$ and $\mathcal{PC}_{\text{Reedy}}(\mathcal{K})$, which we are in the process of constructing, are the same as those of Bergner [40].

The notations of Section 10.7 apply, so a Segal precategory may be considered as a bisimplicial set–denoted either $m \mapsto \mathcal{A}_{m/} \in \mathcal{K}$ or $m, n \mapsto \mathcal{A}_{m,n} \in \text{SET}$, as will be described in more detail next–whose simplicial set $\mathcal{A}_{0/}$ in degree 0 is the discrete set $\text{Ob}(\mathcal{A})$.

The category $\mathcal{K}$ satisfies the additional Condition 10.7.1 (DISJ) on disjoint unions. Therefore, $\mathcal{PC}(\mathcal{K})$ may be viewed as a presheaf category. More precisely, as $\mathcal{K}$ is the category of presheaves on Δ, by Theorem 10.7.3 we have a natural identification

$$\mathcal{PC}(\mathcal{K}) \cong \text{PRESH}(\mathbf{C}(\Delta)) \subset \text{PRESH}(\Delta \times \Delta). \tag{15.1.1}$$

A Segal precategory for us is a pair $(X, \mathcal{A})$, where $X \in \text{SET}$ and $\mathcal{A}$ is a collection of simplicial sets $\mathcal{A}(x_0, \ldots, x_n)$ for sequences of $x_i \in X$. Recall that

$\mathbf{C}(\Delta)$ is a quotient of $\Delta \times \Delta$, thus the subset relation in (15.1.1). Under the identification (15.1.1) this corresponds to a bisimplicial set given by the formula

$$\widetilde{\mathcal{A}} : (n, m) \in \mathbf{C}(\Delta) \mapsto \coprod_{(x_0,\ldots,x_n)\in X^{n+1}} \mathcal{A}(x_0, \ldots, x_n)_m.$$

If $n = 0$ then it is constant in $m \in \Delta$, which is *Tamsamani's constancy condition* in this case. The constancy condition characterizes the bisimplicial sets which are presheaves on the quotient $\mathbf{C}(\Delta)$.

In what follows, we use the following notation of Section 10.7 for a Segal precategory $\mathcal{A}$:

- $\mathcal{A}_0 = \mathrm{Ob}(\mathcal{A}) = \widetilde{\mathcal{A}}(0, m)$ for any m;
- $\mathcal{A}_{n/}$ is the simplicial set $m \mapsto \widetilde{\mathcal{A}}(n, m)$.

The assignment $n \mapsto \mathcal{A}_{n/}$ is a functor $\Delta^o \to \mathcal{K}$, which is the same as the bisimplicial set $\widetilde{\mathcal{A}}$. We denote the first component by $\mathcal{A}_0$ rather than $\mathcal{A}_{0/}$ to emphasize the constancy condition, that it is a set considered as a constant simplicial set. This notation is convenient because colimits (resp. limits) in $\mathcal{PC}(\mathcal{K})$ correspond to levelwise colimits (resp. limits) of functors $\Delta^o \to \mathcal{K}$.

For a pictorial point of view, it is often more intuitive to replace the model category $\mathcal{K}$ by the category TOP of topological spaces, and to think of Segal (pre)categories as functors $\Delta^o_X \to$ TOP. As usual, for technical details it is more convenient to use simplicial sets, and we don't treat the thorny questions surrounding model category structures on TOP. Below we will sometimes replace $\mathcal{K}$ by TOP and leave to the reader to insert the appropriate realization and singular complex functors between these two.

A Segal precategory may therefore be thought of as a functor $\mathcal{A} : \Delta^o \to$ TOP denoted $n \mapsto \mathcal{A}_{n/}$, which is to say a simplicial space, satisfying the "constancy" or "globular" condition that $\mathcal{A}_0 = \mathrm{Ob}(\mathcal{A})$ is a discrete set.

15.2 The Poincaré–Segal groupoid

Recall that if $\mathcal{A}$ is a Segal category, then the *truncation* $\tau_{\leq 1}(\mathcal{A})$ was defined in Section 12.4. It is the usual 1-category whose nerve is the simplicial set $m \mapsto \pi_0(\mathcal{A}_{m/})$. A *Segal groupoid* is a Segal category such that $\tau_{\leq 1}(\mathcal{A})$ is a groupoid.

Given a Kan simplicial set X, in other words a fibrant object in $\mathcal{K}$, we can define its *Poincaré–Segal groupoid* $\Pi_S(X)$, which is a Segal groupoid. It is constructed as the right adjoint of the diagonal realization functor, which we consider first.

The *diagonal* $\mathbf{d} : \Delta \to \Delta \times \Delta$ provides a pullback functor

$$\mathbf{d}^* : \text{PRESH}(\Delta \times \Delta) \to \text{PRESH}(\Delta) = \mathcal{K};$$

so composing with (15.1.1) we obtain the *realization functor*

$$|\ | : \mathcal{PC}(\mathcal{K}) \to \mathcal{K},$$

$$|(X, \mathcal{A})| := \mathbf{d}^* \widetilde{\mathcal{A}} = \left(m \mapsto \widetilde{\mathcal{A}}_{m,m}\right).$$

Recall that $\mathbf{c}_\Delta : \Delta \times \Delta \to \mathbf{C}(\Delta)$ denotes the projection map. Composition with the diagonal yields a functor

$$\mathbf{c}_\Delta \circ \mathbf{d} : \Delta \to \mathbf{C}(\Delta),$$

and the realization functor is just the pullback

$$|\mathcal{A}| = (\mathbf{c}_\Delta \circ \mathbf{d})^*(\mathcal{A}) \tag{15.2.1}$$

from $\mathcal{PC}(\mathcal{K}) \cong \text{PRESH}(\mathbf{C}(\Delta))$ to $\mathcal{K} = \text{PRESH}(\Delta)$. It follows that $|\ |$ preserves limits and colimits, indeed it has both left and right adjoints.

Taking the right adjoint of the expression (15.2.1), define

$$\Pi_S := (\mathbf{c}_\Delta \circ \mathbf{d})_* : \mathcal{K} \to \mathcal{PC}(\mathcal{K}).$$

Note that Π_S commutes with limits. For a fibrant simplicial set X, call $\Pi_S(X)$ the *Poincaré–Segal groupoid* of X. A more concrete description is available when X is a topological space, to be discussed in the next section.

We now state the general theorem relating the homotopy theory in $\mathcal{PC}(\mathcal{K})$ of Segal groupoids, with the classical homotopy theory of simplicial sets. It is essentially due to Segal, although Segal's arguments were mostly stated for the situation of categories with a single object. These were reviewed in the more general categorical setting by Tamsamani [250]. Tamsamani then furthermore iterated the result to obtain an equivalence between the theory of n-truncated homotopy types, and their Poincaré n-groupoids. This will be discussed further in Section 20.4.

The main goal in Part IV is to show that the given classes of morphisms provide a model structure for $\mathcal{PC}(\mathcal{K})$. Of course that statement is also contained in the references, see Bergner [40] and Pellissier [211]. In that context, the following theorem essentially says that the above functors form a Quillen adjunction, although the model structure isn't necessary for stating the present theorem.

Theorem 15.2.1 *The realization functor $|\ \ |$ sends injective cofibrations (resp. injective global trivial cofibrations) to cofibrations (resp. trivial cofibrations) of simplicial sets. The Poincaré–Segal groupoid functor sends fibrations in $\mathcal{K}$ to new fibrations in $\mathcal{PC}(\mathcal{K})$, and takes weak equivalences between fibrant simplicial sets, to global weak equivalences in $\mathcal{PC}(\mathcal{K})$.*

If X is a fibrant simplicial set then $\Pi_S(X)$ is fibrant, so it is a Segal category. It is, in fact, a Segal groupoid, and furthermore the counit of the adjunction

$$|\Pi_S(X)| \to X$$

is a weak equivalence.

Conversely, if $\mathcal{A} \in \mathcal{PC}(\mathcal{K})$ is a Segal groupoid and $|\mathcal{A}| \to Y$ is a fibrant replacement in $\mathcal{K}$, then the map obtained by composing $\Pi_S(g)$ with the unit of the adjunction

$$\mathcal{A} \to \Pi_S(Y)$$

is a global equivalence of Segal categories.

Proof We refer to Tamsamani [250] for the interpretation of Segal's original results [228] in the context of many-object groupoids. See also Bergner [40], Dwyer *et al.* [104], Berger [36], Duskin [98], Baković and Jurčo [14], Cabello and Garzon [71] and others. Further details on the looping and delooping operations which enter in here, will be given below and in the next section, but to discuss the full proof would go beyond the scope of the present book. □

Tamsamani [250] defines the homotopy groups of an n-nerve. This will be explained further in Section 20.4.1. Taken at the first level, we can similarly define the homotopy groups of a Segal groupoid $\mathcal{A}$ at any $x \in \mathrm{Ob}(\mathcal{A})$ by putting

$$\pi_0(\mathcal{A}) := \mathrm{Iso}\tau_{\leq 1}\mathcal{A},$$

and for $i \geq 1$,

$$\pi_i(\mathcal{A}, x) := \pi_{i-1}(\mathcal{A}(x, x), 1_x)$$

where $1_x : * \to \mathcal{A}(x, x)$ is the degeneracy map using $\mathcal{A}(x) = *$.

From the Segal groupoid condition it follows that the relation of homotopy on $\mathrm{Ob}(\mathcal{A})$ may be characterized as saying that $x \sim y$ if and only if $\mathcal{A}(x, y) \neq \emptyset$, and then $\pi_0(\mathcal{A}) = \mathrm{Ob}(\mathcal{A})/\sim$.

Lemma 15.2.2 *A morphism $f : \mathcal{A} \to \mathcal{B}$ between Segal groupoids is a global weak equivalence, if and only if $\pi_0(\mathcal{A}) \to \pi_0(\mathcal{B})$ is surjective (resp. an isomorphism), and for each $i \geq 1$ and each object $x \in \mathrm{Ob}(\mathcal{A})$, the induced maps $\pi_i(\mathcal{A}, x) \to \pi_i(\mathcal{B}, f(x))$ are isomorphisms.*

Proof This follows directly from the definition using the criterion that a map between simplicial sets is a weak equivalence if and only if it induces an isomorphism on homotopy groups. □

Proposition 15.2.3 *If $\mathcal{A}$ is a Segal groupoid then $\pi_0(\mathcal{A}) = \pi_0(|\mathcal{A}|)$ and for any $i \geq 1$ and $x \in \mathrm{Ob}(\mathcal{A})$, $\pi_i(\mathcal{A}, x) = \pi_i(|\mathcal{A}|, |x|)$. Similarly, if Z is a simplicial set satisfying Kan's fibrancy condition then $\pi_0(\Pi_S(Z)) = \pi_0(Z)$ and for any vertex $z \in Z_0$, $\pi_i(\Pi_S(Z), z) = \pi_i(Z, z)$.*

A morphism $f : \mathcal{A} \to \mathcal{B}$ between Segal groupoids is a global weak equivalence if and only if the induced map on realizations is a weak equivalence $|\mathcal{A}| \sim |\mathcal{B}|$.

Proof The first part states some of the essential facts about Segal's construction [228], entering into the proof of Theorem 15.2.1, see *loc. cit.* [250]. The second paragraph follows immediately from the first part together with the previous lemma. □

15.3 Looping and delooping

After this first part, the remainder of the chapter is devoted to following out the calculus of generators and relations in the case of Segal precategories. In the spirit of Segal's "delooping machine," this process gives a good explanation of how the technical machinery introduced in the previous chapters works.

Before getting to the calculus of generators and relations, we explain how Segal's machine provides the structure required to recover a space X from its loop space ΩX.

Versions of the Poincaré–Segal groupoid when X is a topological space, have easy explicit descriptions. There are two possible points of view, depending on whether we would like to view the Segal category $\mathcal{A} = \Pi_S(X)$ as a simplicial topological space or as a bisimplicial set. In case of confusion the former will be denoted by $\Pi^t_S(X)$ and the latter by $\Pi^b_S(X)$.

In either case, the objects of $\Pi_S(X)$ are just the points of x:

$$\mathrm{Ob}(\Pi_S(X)) = X^{\mathrm{disc}},$$

the superscript emphasizing that this is just a plain set devoid of topology.

Given two points $x, y \in X = \mathrm{Ob}(\Pi_S(X))$, the space of maps between them is just the space of paths from x to y in X,

$$\Pi^t_S(X)(x, y) := \underline{\mathrm{HOM}}^{x,y}([0, 1], X),$$

where $\underline{\text{HOM}}$ means we take the topological mapping space, and the superscripts mean that the endpoints of the paths are constrained to be x and y respectively. For $x = y$ we recover exactly the space of loops based at x:

$$\Pi^t_S(X)(x, x) := \Omega_x(X).$$

The bisimplicial version is obtained by simply applying the singular simplicial set functor

$$\Pi^b_S(X)(x, y) := \mathbf{Sing}\left(\Pi^t_S(X)(x, y)\right)_\cdot,$$

but this can be given a more concrete description. Denote by $\mathbf{R}([k])$ the standard k-simplex. These organize into a cosimplicial object in the variable $[k] \in \Delta$, providing the typical example of a cosimplicial resolution $\Delta \to \text{TOP}$ as will be used later in a more general setting (Section 18.5). Note that $\mathbf{R}([1]) = [0, 1]$ is just the interval. The standard simplices define the singular simplicial set functor

$$\mathbf{Sing}(Y)_k = \text{HOM}(\mathbf{R}([k]), Y),$$

and applying this above we get

$$\begin{aligned}\Pi^b_S(X)(x, y)_k &= \text{HOM}\left(\mathbf{R}([k]), \Pi^t_S(X)(x, y)\right) \\ &= \text{HOM}(\mathbf{R}([k]), \underline{\text{HOM}}^{x,y}([0, 1], X)) \\ &= \text{HOM}^{x,y}(\mathbf{R}([k]) \times [0, 1], X).\end{aligned}$$

The superscript in the last line means that we look only at maps $f : \mathbf{R}([k]) \times [0, 1] \to X$ with $f(u, 0) = x$ and $f(u, 1) = y$, i.e. sending $\mathbf{R}([k]) \times \{0\}$ to x and $\mathbf{R}([k]) \times \{1\}$ to y.

Suppose now that $(x_0, \ldots, x_n)$ is a sequence of points of X. Define

$$\Pi^t_S(X)(x_0, \ldots, x_n) := \underline{\text{HOM}}^{x_0,\ldots,x_n}(\mathbf{R}([n]), X)$$

to be the space of maps $\mathbf{R}([n]) \to X$ such that the i-th vertex v_i of $\mathbf{R}([n])$ maps to x_i. These fit together into a functor

$$\Pi^t_S(X) : \Delta^o_X \to \text{TOP},$$

which is the topological version of the Poincaré–Segal groupoid. To get the bisimplicial set version, apply **Sing** as before:

$$\Pi^b_S(X)(x_0, \ldots, x_n) := \mathbf{Sing}\left(\Pi^t_S(X)(x_0, \ldots, x_n)\right)_\cdot.$$

By the same considerations as above, $\Pi^b_S(X)(x_0, \ldots, x_n)_k$ may be identified with the set of maps $\mathbf{R}([k]) \times \mathbf{R}([n]) \xrightarrow{f} X$ such that, for each vertex v_i of $\mathbf{R}([n])$, the restricted map

$$\mathbf{R}([k]) \times \{v_i\} \to X$$

is constant with value x_i. This condition is the start of Tamsamani's constancy condition [250].

To sum up, then, if X is a topological space we get a functor

$$\Pi^t_S(X) : \Delta^o_X \to \text{TOP},$$

and its corresponding composition with **Sing**

$$\Pi^b_S(X) : \Delta^o_X \to \mathcal{K}.$$

These may be transformed into a simplicial space and a bisimplicial set, respectively, by applying the construction discussed after equation (15.1.1). They are denoted $m \mapsto \Pi^t_S(X)_{m/}$ or $m \mapsto \Pi^b_S(X)_{m/}$. The bisimplicial case serves to explain the origin of the slash notation: it means that a further simplicial index is to be expected. In both cases, the zeroth object is just the discrete set X^{disc}, so the slash may be left out of $\Pi^t_S(X)_0 = \text{Ob}\left(\Pi^t_S(X)\right) = X$, etc.

An advantage of starting with a topological space X is that no fibrant replacement is needed before applying Π^t_S or Π^b_S.

Suppose X is path-connected. Choose a basepoint $x \in X$, and let

$$\Pi_S(X, x) \subset \Pi_S(X)$$

be the full sub-Segal category consisting of the single object x. Thus,

$$\text{Ob}(\Pi_S(X, x)) = \{x\}$$

and

$$\Pi_S(X, x)(x, \ldots, x) = \Pi_S(X)(x, \ldots, x).$$

These notations (which technically speaking concern the case when $X \in \mathcal{K}$ is a fibrant simplicial set) hold also for the topological versions $\Pi^t_S(X, x)$ and $\Pi^b_S(X, x)$.

This brings us back to the situation which Segal originally talked about. Notice that $\Delta_{\{x\}} \cong \Delta$ so the simplicial space or bisimplicial set versions are the same thing as the $\Delta^o_{\{x\}}$-diagrams:

$$\Pi^t_S(X, x)_{m/} = \Pi^t_S(X, x)(x, \ldots, x).$$

And, since there is only one object, $\Pi^t_S(X, x)_0 = *$. The idea that $\mathcal{A} = \Pi^t_S(X, x)$ represents a "delooping structure" on the loop space of X, comes from the fact that $\mathcal{A}_{1/} = \Omega_x(X)$ is the loop space itself. As explained at the beginning of the book, the remaining pieces of the simplicial space provide an up-to-homotopy multiplication law.

The Segal groupoid condition in this case says that the simplicial space is *grouplike*, in other words the monoid obtained by considering $\pi_0(\mathcal{A}_{1/})$ is a group, i.e. all elements are invertible.

The delooping operation is just the diagonal realization, and Segal's theorem may be restated as follows:

Corollary 15.3.1 *Suppose X is a connected CW complex. If* $\mathcal{A} = \Pi^t_S(X, x)$, *then* $\mathcal{A}_0 = *$, *the Segal maps for* $\mathcal{A}$ *are weak equivalences, the resulting monoid structure on* $\pi_0(\mathcal{A}_{1/})$ *is a group, and there is a natural weak homotopy equivalence*

$$X \sim |\mathcal{A}|$$

between X and the diagonal realization obtained by transforming $\mathcal{A}$ *to a bisimplicial set, restricting to the diagonal, and realizing this simplicial set back to a space.*

In the other direction, suppose $\mathcal{A}$ *is a simplicial space satisfying the first three conditions above. Then* $Z = |\mathcal{A}|$ *is a connected CW complex, and there is a natural equivalence of Segal groupoids* $\Pi^t_S(Z, z) \sim \mathcal{A}$, *where z is the basepoint of Z determined by the unique vertex of* $\mathcal{A}$.

Proof This was actually Segal's original statement, but can also be seen as a consequence of the multi-object statement, Theorem 15.2.1, because of the invariance of Proposition 15.2.3. □

Even though it sounds ferocious, the "delooping" operation is quite simple, being just a realization of a bisimplicial set, once the full amount of structure is given. There are other ways of doing the realization of a simplicial space, equivalent if the spaces involved aren't too pathological; the reader is invited to consult May [202], Goerss and Jardine [122], Segal's original papers [226, 227, 228], or any of a number of topology texts on this matter.

On the other hand, the looping operation $(X, x) \mapsto \Omega_x X$ is more difficult to calculate. In fact, if one could calculate the loop space then the homotopy groups would easily follow: $\pi_n(X, x)$ is just π_0 of the n-th iterated loop space of X. The calculus of generators and relations for a Segal precategory allows us to view the loop space as being the result of an algebraic process, and this will provide us with a good setting in which to illustrate the concrete topological content of the abstract theory which has been developed in previous chapters.

15.4 The calculus

A space X leads to a Segal precategory, for example if X comes about as a simplicial set then this simplicial set may be viewed as a simplicial space which is discrete in the second simplicial direction. To go from here to ΩX with its delooping structure, we should apply the calculus of generators and relations to enforce the Segal conditions.

This section will give a more detailed description of how this works to go from a Segal precategory to a Segal category, keeping track of the connectivity properties of the intervening spaces. In Section 15.5, we show how the process leads, in principle, to a calculation of the loop space of a space. In Section 15.6 below we will be able to follow along what happens as one of the first nontrivial homotopy groups $\pi_3(S^2)$ appears out of this process. The rest of the chapter constitutes a reworked version of my preprint [236].

In view of the topological motivation as described in the previous section, we concentrate on the case of Segal groupoids and more specifically the case when there is only one object and $\mathcal{A}_{1/}$ is connected. Note that if $\mathcal{A}_0 = *$ and $\mathcal{A}_{1/}$ is connected, then $\tau_{\leq 1}(\mathbf{Seg}(\mathcal{A}))$ is just the discrete category $\mathbf{disc}(*)$ with a single object, in particular it is a groupoid. In this case, a stronger finiteness property will hold: to arrange things up to a certain level of connectivity, it will suffice to do a finite number of elementary generation operations which we denote here by Arrg (or later Arrg2). Of course, if $\mathcal{A}_{1/}$ is not connected then we may be in the presence of a fundamental group and it requires, in principle, an infinite and even undecideable number of operations to compute the group. Some remarks on this aspect are given near the end of the chapter.

15.4.1 Arranging in degree *m*

Suppose $\mathcal{A}$ is a Segal precategory with $\mathcal{A}_0 = *$. We say that $\mathcal{A}$ is (m, k)-*arranged* if the Segal map

$$\mathcal{A}_{m/} \to \mathcal{A}_{1/} \times \cdots \times \mathcal{A}_{1/}$$

induces isomorphisms on π_i for $i < k$ and a surjection on π_k.

Note that for $l \geq k$, adding l-cells to $\mathcal{A}_{m/}$ or $(l+1)$-cells to $\mathcal{A}_{1/}$ doesn't affect this property.

We now define an operation where we try to "arrange" $\mathcal{A}$ in degree m. This operation is inspired by the operation **Gen** considered in Section 14.4. In the present version it will be called

$$\mathcal{A} \mapsto \mathrm{Arrg}(\mathcal{A}, m).$$

Fix m in what follows. Let C be the mapping cone of the Segal map

$$\mathcal{A}_{m/} \to \mathcal{A}_{1/} \times \cdots \times \mathcal{A}_{1/}.$$

To be precise, as a bisimplicial set

$$C = (I \times \mathcal{A}_{m/}) \cup^{\{1\}\times\mathcal{A}_{m/}} (\mathcal{A}_{1/} \times \cdots \times \mathcal{A}_{1/}),$$

where I is the standard simplicial interval, and the notation is the amalgamated sum of bisimplicial sets (note also that the globular condition is preserved, so it is an amalgamated sum of Segal precategories). Note that $\{1\} \times \mathcal{A}_{m/}$ denotes the second endpoint of the interval crossed with $\mathcal{A}_{m/}$. We have morphisms

$$\mathcal{A}_{m/} \overset{a}{\hookrightarrow} C \overset{b}{\to} \mathcal{A}_{1/} \times \cdots \times \mathcal{A}_{1/},$$

the morphism a being the inclusion of $\{0\} \times \mathcal{A}_{m/}$ into $I \times \mathcal{A}_{m/}$ (thus it is a cofibration, i.e. an injection, of simplicial sets) and the second morphism b coming from the projection $I \times \mathcal{A}_{m/} \to \mathcal{A}_{m/}$. The second morphism b is a weak equivalence.

We now define $\mathrm{Arrg}(\mathcal{A}, m)$: for any p, let

$$\mathrm{Arrg}(\mathcal{A}, m)_{p/} := \mathcal{A}_{p/} \cup^{(\bigcup \mathcal{A}_{m/})} \left(\bigcup_{p\to m} C \right)$$

be the combined amalgamated sum of $\mathcal{A}_{p/}$ with several copies of the morphism $a : \mathcal{A}_{m/} \to C$, one copy for each map $p \to m$ not factoring through a principal edge (see below for further discussion of this condition), these maps inducing $\mathcal{A}_{m/} \to \mathcal{A}_{p/}$.

We need to define $\mathrm{Arrg}(\mathcal{A}, m)$ as a Segal precategory, i.e. as a bisimplicial set. For this we need morphisms of functoriality

$$\mathrm{Arrg}(\mathcal{A}, m)_{p/} \to \mathrm{Arrg}(\mathcal{A}, m)_{q/},$$

for any $q \to p$. These are defined as follows. We consider a component of $\mathrm{Arrg}(\mathcal{A}, m)_{p/}$ which is a copy of C attached along a map $\mathcal{A}_{m/} \to \mathcal{A}_{p/}$ corresponding to $p \to m$ which doesn't factor through a principal edge. If the composed map $q \to p \to m$ doesn't factor through a principal edge then the component C maps to the corresponding component of $\mathrm{Arrg}(\mathcal{A}, m)_{1/}$. If the map does factor through a principal edge $q \to 1 \to m$ then we obtain a map $C \to \mathcal{A}_{1/}$ (the component of the map b corresponding to this principal edge). Compose with the map $\mathcal{A}_{1/} \to \mathcal{A}_{q/}$ to obtain a map $C \to \mathcal{A}_{q/}$. Note that if the map further factors

$$q \to 0 \to 1 \to m,$$

then the map $\mathcal{A}_{1/} \to \mathcal{A}_{q/}$ factors through the basepoint

$$\mathcal{A}_{1/} \to \mathcal{A}_0 \to \mathcal{A}_{q/},$$

and our map on C factors through the basepoint. This factorization doesn't depend on choice of principal edge containing the map $0 \to m$.

One verifies that this prescription defines a functor $p \mapsto \mathrm{Arrg}(\mathcal{A}, m)_{p/}$ from Δ to simplicial sets. This verification will be a consequence of the more conceptual description which follows.

Let $h(m)$ be the simplicial set representing the standard m-simplex; it is the contravariant functor on Δ represented by the object m. Let $\Sigma(m) \subset h(m)$ be the subcomplex which is the union of the principal edges.

Definition 15.4.1 If X is a simplicial set and B is another simplicial set, denote by $X \boxtimes B$ the bisimplicial set exterior product, defined by

$$(X \boxtimes B)_{p,q} := X_p \times B_q.$$

If B is any simplicial set then putting $h(m)$ or $\Sigma(m)$ in the first variable, we obtain an inclusion of bisimplicial sets, which we denote

$$\Sigma(m) \boxtimes B \hookrightarrow h(m) \boxtimes B.$$

Note that these bisimplicial sets are not Segal precategories because they don't satisfy the globular condition (they are not constant over 0 in the first variable). However, the morphism of simplicial sets

$$(\Sigma(m) \boxtimes B)_{0/} \hookrightarrow (h(m) \boxtimes B)_{0/}$$

is an isomorphism because Σ contains all of the vertices.

If $\mathcal{A}$ is a Segal precategory then a morphism $h(m) \boxtimes B \to \mathcal{A}$ is the same thing as a morphism $B \to \mathcal{A}_{m/}$. Similarly, a morphism

$$\Sigma(m) \boxtimes B \to \mathcal{A}$$

is the same thing as a morphism

$$B \to \mathcal{A}_{1/} \times_{\mathcal{A}_0} \cdots \times_{\mathcal{A}_0} \mathcal{A}_{1/}.$$

The morphism of realizations

$$|\Sigma(m) \boxtimes B| \to |h(m) \boxtimes B|$$

is a weak equivalence. To see this note that it is the product of $|B|$ and

$$|\Sigma(m)| \to |h(m)|,$$

and this last morphism is a weak equivalence (it is the inclusion from the "spine" of the m-simplex to the m-simplex; both are contractible).

Suppose $B' \subset B$ is an injection of simplicial sets. Put

$$\mathcal{U} := (\Sigma(m) \boxtimes B) \cup^{\Sigma(m)\boxtimes B'} \left(h(m) \boxtimes B'\right),$$

and

$$\mathcal{V} := h(m) \boxtimes B.$$

We have an injection $\mathcal{U} \hookrightarrow \mathcal{V}$. If $\mathcal{A}$ is a Segal precategory then a map $\mathcal{U} \to \mathcal{A}$ consists of a commutative diagram

$$\begin{array}{ccc} B' & \longrightarrow & B \\ \downarrow & & \downarrow \\ \mathcal{A}_{m/} & \longrightarrow & \mathcal{A}_{1/} \times_{\mathcal{A}_0} \cdots \times_{\mathcal{A}_0} \mathcal{A}_{1/}. \end{array}$$

The inclusion

$$\Sigma(m) \boxtimes B \to \mathcal{U}$$

induces a weak equivalence of realizations, because the inclusion $\Sigma(m) \boxtimes B' \to h(m) \boxtimes B'$ does. Therefore, the morphism $|\mathcal{U}| \to |\mathcal{V}|$ is a weak equivalence.

We can now interpret our operation $\text{Arrg}(\mathcal{A}, m)$ in these terms. Applying the previous paragraph to the inclusion $\mathcal{A}_{m/} \hookrightarrow C$, we obtain an inclusion of bisimplicial sets $\mathcal{U} \hookrightarrow \mathcal{V}$. We get a map $\mathcal{U} \to \mathcal{A}$ corresponding to the diagram

$$\begin{array}{ccc} \mathcal{A}_{m/} & \longrightarrow & C \\ \downarrow & & \downarrow \\ \mathcal{A}_{m/} & \longrightarrow & \mathcal{A}_{1/} \times_{\mathcal{A}_0} \cdots \times_{\mathcal{A}_0} \mathcal{A}_{1/}. \end{array}$$

The left vertical arrow is the identity map, the top arrow is a and the right vertical arrow is b. The bottom arrow is the Segal map.

It is easy to see that

$$\text{Arrg}(\mathcal{A}, m) = \mathcal{A} \cup^{\mathcal{U}} \mathcal{V}.$$

In passing, this proves associativity of the previous formulas for functoriality of $\text{Arrg}(\mathcal{A}, m)$.

We get

$$|\text{Arrg}(\mathcal{A}, m)| = |\mathcal{A}| \cup^{|\mathcal{U}|} |\mathcal{V}|.$$

Since $|\mathcal{U}| \to |\mathcal{V}|$ is a weak equivalence, this implies the following:

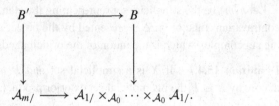

Lemma 15.4.2 *The morphism induced by the above inclusion on realizations,*

$$|\mathcal{A}| \hookrightarrow |\mathrm{Arrg}(\mathcal{A}, m)|$$

is a weak equivalence of spaces.

□

The key observation is the following proposition:

Proposition 15.4.3 *Suppose $\mathcal{A}$ is a Segal precategory with $\mathcal{A}_0 = *$ and $\mathcal{A}_{1/}$ connected. Suppose that $\mathcal{A}$ is $(m, k-1)$-arranged and (p, k)-arranged for some $p \neq m$. Then* $\mathrm{Arrg}(\mathcal{A}, m)$ *is (p, k)-arranged and (m, k)-arranged.*

Proof Keep the hypotheses of the proposition. Let C be the cone occuring in the construction $\mathrm{Arrg}(\mathcal{A}, m)$. Denote $\mathcal{B} := \mathrm{Arrg}(\mathcal{A}, m)$. Then the map

$$a : \mathcal{A}_{m/} \hookrightarrow C$$

is weakly equivalent to a map obtained by adding cells of dimension $\geq k$ to $\mathcal{A}_{m/}$. This is by the condition that $\mathcal{A}$ is $(m, k-1)$-arranged. Let $h_1, \ldots, h_u$ be the k-cells that are attached to $\mathcal{A}_{m/}$ to give C.

We first show that $\mathcal{B}$ is (p, k)-arranged. Note that $\mathcal{B}_{p/}$ is obtained from $\mathcal{A}_{p/}$ by attaching a certain number of k-cells, $h_i^{p\to m}$ for $i = 1, \ldots, u$ indexed by the maps $p \to m$ not factoring through the principal edges of m; plus some cells of dimension $\geq k+1$. The higher-dimensional cells don't have any effect on the question of whether $\mathcal{A}$ is (p, k)-arranged.

On the other hand, $\mathcal{B}_{1/}$ is obtained from $\mathcal{A}_{1/}$ by attaching cells $h_i^{1\to m}$ for $i = 1, \ldots, u$ and indexed by the maps $1 \to m$ not factoring through the principal edges. Note here that these maps cannot be degenerate, thus they are the nonprincipal edges.

Now $\mathcal{B}_{1/} \times \ldots \times \mathcal{B}_{1/}$ (product of p copies) is obtained from $\mathcal{A}_{1/} \times \cdots \times \mathcal{A}_{1/}$ by adding k-cells indexed as $\nu_{1\to p}\left(h_i^{1\to m}\right)$, where the indexing $1 \to p$ are principal edges and $1 \to m$ are nonprincipal edges. Then we add cells of dimension $\geq k+1$, which have no effect on the question. The notation $\nu_{1\to p}$ refers to the map

$$\mathcal{A}_{1/} \to \mathcal{A}_{1/} \times \cdots \times \mathcal{A}_{1/}$$

putting the base point (i.e. the degeneracy of the unique point in $\mathcal{A}_0$) in all of the factors except the one corresponding to the map $1 \to p$.

For every principal edge $1 \to p$ there is a unique degeneracy $p \to 1$ inducing an isomorphism $1 \to 1$ and this establishes a bijection between principal edges and degeneracies. Thus, we may rewrite our indexing of the k-cells attached to the product above as $\nu_{p\to 1}\left(h_i^{1\to m}\right)$.

Now for every pair $(p \to 1, 1 \to m)$ the composition $p \to m$ is a degenerate morphism, not factoring through a principal edge; and these degenerate morphisms are all different for different pairs $(p \to 1, 1 \to m)$. Thus $\mathcal{B}_{p/}$ contains a k-cell $h_i^{p\to m}$ for each $i = 1, \ldots, u$ and each of these maps $p \to m$ (plus possibly other cells for other maps $p \to m$ but we don't use these). Take such a cell $h_i^{p\to m}$, and look at its image in $\mathcal{B}_{1/} \times \cdots \times \mathcal{B}_{1/}$ by the Segal map. The projection to any factor $1 \to p$, other than the one which splits the degeneracy $p \to 1$, is totally degenerate coming from a factorization $1 \to 0 \to 1$, hence goes to the unique basepoint. The projection to the unique factor which splits the degeneracy is just the cell $h_i^{1\to m}$. Thus, the projection of our cell $h_i^{p\to m}$ to the product is exactly the cell $\nu_{p\to 1}\left(h_i^{1\to m}\right)$. This shows that all of the new k-cells which have been added to the product are lifted as new k-cells in $\mathcal{B}_{p/}$. Together with the fact that $\mathcal{A}_{p/} \to \mathcal{A}_{1/} \times \cdots \times \mathcal{A}_{1/}$ was an isomorphism on π_i for $i < k$ and a surjection for $i = k$, we obtain the same property for $\mathcal{B}_{p/} \to \mathcal{B}_{1/} \times \cdots \times \mathcal{B}_{1/}$. Note that the further k-cells which are attached to $\mathcal{B}_{p/}$ by morphisms $p \to m$, other than those we have considered above, don't affect this property. (In general, attaching k-cells to the domain of a map doesn't affect this property, but attaching cells to the range can affect it, which was why we had to look carefully at the cells attached to $\mathcal{B}_{1/}$.) This completes the proof that $\mathcal{B}$ remains (p, k)-arranged.

We now prove that $\mathcal{B}$ becomes (m, k)-arranged. Note that $\mathcal{B}_{m/}$ is obtained by first adding on C to $\mathcal{A}_{m/}$ via the identity map $m \to m$, then adding some other stuff which we treat in a minute. The Segal map for $\mathcal{B}$ maps this copy of C directly into $\mathcal{A}_{1/} \times \cdots \times \mathcal{A}_{1/}$. The fact that C is a mapping cone for the Segal map means that the map

$$C \to \mathcal{A}_{1/} \times \cdots \times \mathcal{A}_{1/}$$

is a homotopy equivalence. In particular, it is bijective on π_i for $i < k$ and surjective for $i = k$.

Now $\mathcal{B}_{m/}$ is obtained from C by adding various cells to C along degenerate maps $m \to m$. The new k-cells which are added to $\mathcal{B}_{1/} \times \cdots \times \mathcal{B}_{1/}$ (m-factors this time) are lifted to cells in $\mathcal{B}_{m/}$ added to C via the degeneracies $m \to m$ which factor through a principal edge. The argument is the same as above and we don't repeat it. We obtain that $\mathcal{B}$ is (m, k)-arranged.

This completes the proof of the proposition. □

15.4.2 Connectivity properties of (m,k)-arranged precategories

Suppose $\mathcal{A}$ is a Segal precategory with $\mathcal{A}_0 = *$ and $\mathcal{A}_{1/}$ connected, such that $\mathcal{A}$ is (m, k)-arranged for all $m + k \leq n$. We would like to obtain a map $\mathcal{A} \to \mathcal{A}'$

to a Segal category, which induces an equivalence in the domain covered by the (m, k)-arrangement conditions which are already known.

Let $\Lambda(m) \subset h(m)$ denote some "horn," i.e. a union of all faces but one. Use the notation of Definition 15.4.1. For any inclusion of simplicial sets $B' \subset B$ we can set

$$\mathcal{U} := (\Lambda(m) \boxtimes \mathcal{B}) \cup^{\Lambda(m)\boxtimes B'} (h(m) \boxtimes B')$$

and

$$\mathcal{V} := h(m) \boxtimes B,$$

we obtain an inclusion of bisimplicial sets $\mathcal{U} \subset \mathcal{V}$ such that $|\mathcal{U}| \to |\mathcal{V}|$ is a weak equivalence. If $\mathcal{A}$ is a Segal precategory and $\mathcal{U} \to \mathcal{A}$ a morphism, then set $\mathcal{A}' := \mathcal{A} \cup^{\mathcal{U}} \mathcal{V}$. The morphism $|\mathcal{A}| \to |\mathcal{A}'|$ is a weak equivalence. Note that $\mathcal{A}'$ is again a Segal precategory because the morphism $\Lambda(m) \to h(m)$ includes all of the vertices of $h(m)$. Finally, note that for $p \leq m - 2$ the morphism

$$\mathcal{A}_{p/} \to \mathcal{A}'_{p/}$$

is an isomorphism (because the same is true of $\Lambda(m) \to h(m)$). This last property allows us to conserve the homotopy type of the smaller $\mathcal{A}_{p/}$.

We would like to use operations of the above form, to (m, k)-arrange $\mathcal{A}$. In order to do this we analyze what a morphism from $\mathcal{U}$ to $\mathcal{A}$ means.

For any simplicial set X, we can form the simplicial set $[X, \mathcal{A}]$ with the property that a map $B \to [X, \mathcal{A}]$ is the same thing as a map $X \boxtimes B \to \mathcal{A}$. If $X \hookrightarrow Y$ is an inclusion obtained by adding on an m-simplex $h(m)$ over a map $Z \to X$ where $Z \subset h(m)$ is some subset of the boundary, then

$$[Y, \mathcal{A}] = [X, \mathcal{A}] \times_{[Z,\mathcal{A}]} \mathcal{A}_{m/}.$$

In this way, we can reduce $[X, \mathcal{A}]$ to a gigantic iterated fiber product of the various components $\mathcal{A}_{p/}$.

Lemma 15.4.4 *In the above situation, there is a cofibration $\mathcal{A} \to \mathcal{A}'$ such that $\mathcal{A}_{p/} \to \mathcal{A}'_{p/}$ is a weak equivalence for all p, and such that for any cofibration of simplicial sets $X \subset Y$ including all of the vertices of Y, the morphism*

$$[Y, \mathcal{A}'] \to [X, \mathcal{A}']$$

is a Kan fibration of simplicial sets.

Proof In order to construct $\mathcal{A}'$, we just "throw in" everything that is necessary. More precisely, suppose $B' \subset B$ is a trivial fibration of simplicial sets. A diagram

$$\begin{array}{ccc} B' & \longrightarrow & B \\ \downarrow & & \downarrow \\ [Y, \mathcal{A}'] & \longrightarrow & [X, \mathcal{A}'] \end{array}$$

corresponds to a diagram

$$\begin{array}{ccc} \mathcal{U} & \longrightarrow & X \boxtimes B \\ \downarrow & & \downarrow \\ \mathcal{A}' & \longrightarrow & \mathcal{A}' \end{array}$$

with $\mathcal{U} = (Y \boxtimes B') \cup^{X \boxtimes B'} (X \boxtimes B)$. The morphism of bisimplicial sets $\mathcal{U} \to X \boxtimes B$ is a weak equivalence on each vertical column $()_{p/}$. Therefore we can throw in to $\mathcal{A}'$ the pushout along this morphism, without changing the weak equivalence type of the $\mathcal{A}_{p/}$. Note that the new thing is again a Segal precategory because of the assumption that $X \subset Y$ contains all of the vertices. Keep doing this addition over all possible diagrams, an infinite number of times, until we get the required Kan fibration condition to prove the lemma. □

Theorem 15.4.5 *If $\mathcal{A}$ is a Segal precategory with $\mathcal{A}_0 = *$ and $\mathcal{A}_{1/}$ connected, such that $\mathcal{A}$ is (m, k)-arranged for all $m + k \le n$, then there exists a morphism $\mathcal{A} \to \mathcal{A}'$ such that:*

(1) the morphism $|\mathcal{A}| \to |\mathcal{A}'|$ is a weak equivalence;

(2) $\mathcal{A}'$ is a Segal groupoid; and

(3) the map of simplicial sets $\mathcal{A}_{m/} \to \mathcal{A}'_{m/}$ induces an isomorphism on π_i for $i + m < n - 1$.

Proof The answer $\mathcal{A}'$ is the result of a procedure which we now describe. At each step of the procedure, the construction of the previous lemma will be applied without necessarily saying so everywhere. Thus we may always assume that our Segal precategory satisfies the condition of Lemma 15.4.4.

Having described above an arranging operation $\mathcal{A} \mapsto \mathrm{Arrg}(\mathcal{A}, m)$, we now describe a second arranging operation, under the hypothesis that $\mathcal{A}$ satisfies the conclusion of Lemma 15.4.4. Fix m and fix a horn $\Lambda(m) \subset h(m)$ (complement of all but one of the faces, and the face that is left out should be neither the first nor the last face). Let C be the cone on the map

$$\mathcal{A}_{m/} = [h(m), \mathcal{A}] \to [\Lambda(m), \mathcal{A}].$$

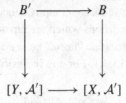

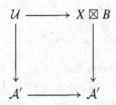

Thus we have a diagram

$$\mathcal{A}_{m/} \to C \to [\Lambda(m), \mathcal{A}].$$

Note that $\Lambda(m)$ is a gigantic iterated fiber product of various $\mathcal{A}_{p/}$ for $p < m$. This diagram corresponds to a map $\mathcal{U} \to \mathcal{A}$, where

$$\mathcal{U} := (\Lambda(m) \boxtimes C) \cup^{\Lambda(m)\boxtimes\mathcal{A}_{m/}} h(m) \boxtimes \mathcal{A}_{m/}.$$

Letting $\mathcal{V} := h(m) \boxtimes C$ we set

$$\mathrm{Arrg2}(\mathcal{A}, m) := \mathcal{A} \cup^{\mathcal{U}} \mathcal{V}.$$

Notice first of all that by the previous discussion, the map

$$|\mathcal{A}| \to |\mathrm{Arrg2}(\mathcal{A}, m)| \tag{15.4.1}$$

is a weak equivalence of spaces.

We have to try to figure out what effect $\mathrm{Arrg2}(\mathcal{A}, m)$ has. We do this under the following hypothesis on the utilisation of this operation: that $\mathcal{A}$ is $(m, k-1)$-arranged, and (p, k)-arranged for all $p < m$.

The first step is to notice that the fiber product in the expression of $[\Lambda(m), \mathcal{A}]$ is a homotopy fiber product, because of the condition of Lemma 15.4.4 which is imposed on $\mathcal{A}$. Furthermore, the elements in this fiber product all satisfy the Segal condition up to k (bijectivity on π_i for $i < k$ and surjectivity for π_k). Thus the morphism

$$[\Lambda(m), \mathcal{A}] \to \mathcal{A}_{1/} \times_{\mathcal{A}_0} \cdots \times_{\mathcal{A}_0} \mathcal{A}_{1/}$$

(the Segal fiber product for m, on the right) is an isomorphism on π_i for $i < k$ and a surjection for $i = k$. Thus, when we add the cone C to $\mathcal{A}_{m/}$, we obtain the condition of being (m, k)-arranged.

By hypothesis, $\mathcal{A}$ is $(m, k-1)$-arranged, so the map $\mathcal{A}_{m/} \to C$ is an isomorphism on π_i for $i < k-1$ and surjective for π_{k-1}. Thus C may be viewed as obtained from $\mathcal{A}_{m/}$ by adding on cells of dimension $\geq k$. Therefore, for all p the morphisms $\mathcal{U}_{p/} \to \mathcal{V}_{p/}$ are homotopically obtained by addition of cells of dimension $\geq k$.

From the previous paragraph, some extra cells of dimension $\geq k$ are added to various $\mathcal{A}_{p/}$ in the process. This doesn't spoil the condition of being (p, k)-arranged wherever it exists. However, the major advantage of this second operation is that the $\mathcal{A}_{p/}$ are left unchanged for $p \leq m-2$. This is because all p-faces of the m-simplex are then contained in the horn $\Lambda(m)$.

We review the above results. First, the hypotheses on $\mathcal{A}$ were:

(a) that $\mathcal{A}$ satisfies the lifting condition of Lemma 15.4.4;

(b) that $\mathcal{A}$ is (p, k)-arranged for $p < m$, and $(m, k - 1)$-arranged.

We then obtain a construction Arrg2($\mathcal{A}, m$) with the following properties:

(1) the map $\mathcal{A}_{p/} \to \mathrm{Arrg2}(\mathcal{A}, m)_{p/}$ is an isomorphism for $p \leq m - 2$;

(2) for any p the map $\mathcal{A}_{p/} \to \mathrm{Arrg2}(\mathcal{A}, m)_{p/}$ induces an isomorphism on π_i for $i < k - 1$;

(3) if $\mathcal{A}$ is (p, k)-arranged for any p then Arrg2($\mathcal{A}, m$) is also (p, k)-arranged; and

(4) Arrg2($\mathcal{A}, m$) is (m, k)-arranged.

Remark: at $m = 2$ the operations Arrg($\mathcal{A}$, 2) and Arrg2($\mathcal{A}$, 2) coincide.

With an infinite series of applications of the construction Arrg2($\mathcal{A}, m$) and the replacement operation of Lemma 15.4.4 we can prove Theorem 15.4.5. The reader may do this as an exercise or else read the explanation below.

For condition (1), note that the arranging operations Arrg and Arrg2 induce weak equivalences on realizations, by Lemma 15.4.2 and the similar fact that (15.4.1) is an equivalence.

Take an array of dots, one for each (p, k). Color the dots green if $\mathcal{A}$ is (p, k)-arranged, red otherwise (note that one red dot in a column implies red dots everywhere above). We do a sequence of operations of the form of Lemma 15.4.4 (which doesn't change the homotopy type levelwise) and then Arrg2($\mathcal{A}, m$). When we do this, change the colors of the dots appropriately.

Also mark an $\times$ at any dot (p, k) such that the $\pi_i(\mathcal{A}_{p/})$ change for any $i \leq k - 1$. (Keep any $\times$ which are marked, from one step to another.) If a dot (p, k) is never marked with a $\times$ it means that the $\pi_i(\mathcal{A}_{p/})$ remain unchanged for $i < k$.

We don't color the dots $(1, k)$ but we still might mark an $\times$.

Suppose the dot (m, k) is red, the dots (p, k) are green for $p < m$ and the dot $(m, k-1)$ is green. Then apply the replacement operation of Lemma 15.4.4 and the operation Arrg2($\mathcal{A}, m$). This has the following effects. Any green dot (p, j) for $p \leq m - 2$ (and arbitrary j) remains green. The dot $(m - 1, k)$ remains green. However, the dot $(m - 1, k + 1)$ becomes red. The dot (m, k) becomes green. The dots $(m-1, k)$ and (m, k), as well as all (p, k) for $p > m$, are marked with an $\times$. The dots above these are also marked with an $\times$ but no other dots are (newly) marked with an $\times$.

In the situation of Theorem 15.4.5, we start with green dots at (p, k) for $p+k \leq n$. We may as well assume that the rest of the dots are colored red. Start with $(m, k) = (n + 1, 0)$ and apply the procedure of the previous paragraph. The dot $(n + 1, 0)$ becomes green, the dot $(n, 0)$ stays green, and the dots $(n, 0), (n + 1, 0), \ldots$ are marked with an $\times$. Continue now at $(n, 1)$ and so on. At the end we have made all of the dots (p, k) with $p + k = n + 1$ green, and we will have marked with an $\times$ all of the dots (p, k) with $p+k = n$ (including the dot $(1, n - 1)$; and also all of the dots above this line).

We can now iterate the procedure. We successively get green dots on each of the lines $p + k = n + j$ for $j = 1, 2, 3, \ldots$. Furthermore, no new dots will be marked with a $\times$. After taking the union over all of these iterations, we obtain an $\mathcal{A}'$ which is (p, k)-arranged for all (p, k). Thus $\mathcal{A}'$ is a Segal category.

The $\mathcal{A}'_{p/}$ are connected, so $\tau_{\leq 1}(\mathcal{A}')_{p/} = *$, hence $\tau_{\leq 1}(\mathcal{A}')$ is the discrete category with a single object. It is a groupoid, so $\mathcal{A}'$ is a Segal groupoid, Condition (2) of the theorem.

Note that the morphism $|\mathcal{A}| \to |\mathcal{A}'|$ is a weak equivalence of spaces.

By looking at which dots are marked with an $\times$, we find that the morphisms

$$\mathcal{A}_{p/} \to \mathcal{A}'_{p/}$$

induce isomorphisms on π_i whenever $i < n - p - 1$. This gives Condition (3), so completes the proof of the theorem. □

15.4.3 Iteration

The following corollary to Theorem 15.4.5 says that in order to calculate the n-type of $\Omega|\mathcal{A}|$ we just have to change $\mathcal{A}$ by pushouts preserving the weak equivalence type of $|\mathcal{A}|$ in such a way that $\mathcal{A}$ is (m, k)-arranged for all $m + k \leq n + 2$.

Corollary 15.4.6 *Suppose $\mathcal{A}$ is a Segal precategory with $\mathcal{A}_0 = *$ and $\mathcal{A}_{1/}$ connected, such that $\mathcal{A}$ is (m, k)-arranged for all $m + k \leq n$. Then the natural morphism*

$$|\mathcal{A}_{1/}| \to \Omega|\mathcal{A}|$$

induces an isomorphism on π_i for $i < n - 1$.

Proof Use Theorem 15.4.5 to obtain a morphism $\mathcal{A} \to \mathcal{A}'$ with the properties stated there (which we refer to as (1)–(3)). We have a diagram

$$\begin{array}{ccc} |\mathcal{A}_{1/}| & \longrightarrow & \Omega|\mathcal{A}| \\ \downarrow & & \downarrow \\ |\mathcal{A}'_{1/}| & \longrightarrow & \Omega|\mathcal{A}'|. \end{array}$$

By property (1) the vertical morphism on the right is a weak equivalence. By property (2) and Theorem 5.3.1 the morphism on the bottom is a weak equivalence. By property (3) the vertical morphism on the right induces isomorphisms on π_i for $i < n-1$. This gives the required statement. □

Corollary 15.4.7 *Fix n, and suppose $\mathcal{A}$ is a Segal precategory with $\mathcal{A}_0 = *$ and $\mathcal{A}_{1/}$ connected. By applying the operations $\mathcal{A} \mapsto \mathrm{Arrg}(\mathcal{A}, m)$ for various m, a finite number of times (less than $(n+2)^2$) in a predetermined way, we can effectively get to a morphism of Segal precategories $\mathcal{A} \to \mathcal{B}$ such that*

$$|\mathcal{A}| \to |\mathcal{B}|$$

is a weak equivalence of spaces, and such that $\mathcal{B}$ is (m, k)-arranged for all $m + k \leq n + 2$. Furthermore, $\mathcal{B}_0 = *$ and $\mathcal{B}_{1/}$ is connected.

Proof By Corollary 15.4.2 any successive application of the operations $\mathcal{A} \mapsto \mathrm{Arrg}(\mathcal{A}, m)$ yields a morphism $|\mathcal{A}| \to |\mathcal{B}|$ which is a weak equivalence of spaces. By Proposition 15.4.3 it suffices, for example, to successively apply $\mathrm{Arrg}(\mathcal{A}, i)$ for $i = 2, 3, \ldots, n+2$ and to repeat this $n+2$ times. These operations preserve connectedness of the pieces in degree 1, so $\mathcal{B}_{1/}$ is connected. □

Corollary 15.4.8 *Fix n, and suppose $\mathcal{A}$ is a Segal precategory with $\mathcal{A}_0 = *$ and $\mathcal{A}_{1/}$ connected. Let $\mathcal{B}$ be the result of the operations of Corollary 15.4.7. Then the n-type of the simplicial set $\mathcal{B}_{1/}$ is equivalent to the n-type of $\Omega|\mathcal{A}|$.*

Proof Apply Corollaries 15.4.6 and 15.4.7. □

15.5 Computing the loop space

Suppose X is a simplicial set with $X_0 = X_1 = *$, and with finitely many nondegenerate simplices. Fix n. We will obtain, by iterating an operation closely related to the operation **gen** of Chapter 14, a finite complex representing the n-type of ΩX.

Let $\mathcal{A}$ be X considered as a Segal precategory constant in the second variable, in other words

$$\mathcal{A}_{p,k} := X_p.$$

Apply the arrangement process, iterated as in Corollary 15.4.7.

Corollary 15.5.1 *Fix n, and suppose X is a simplicial set with finitely many nondegenerate simplices, with $X_0 = X_1 = *$. Let $\mathcal{A}$ be X considered as a*

Segal precategory. Let $\mathcal{B}$ be the result of the operations of Corollary 15.4.7. Then the n-type of the simplicial set $\mathcal{B}_{1/}$ is equivalent to the n-type of ΩX.

Proof An immediate restatement of 15.4.8. □

Remark: Any finite region of the Segal precategory $\mathcal{B}$ is effectively computable. In fact it is just an iteration of operations pushout and mapping cone, arranged in a way which depends on combinatorics of simplicial sets. Thus the $(n+1)$-skeleton of the simplicial set $\mathcal{B}_{1/}$ is effectively calculable (in fact, one could bound the number of simplices in $\mathcal{B}_{1/}$).

Corollary 15.5.2 *Fix n, and suppose X is a simplicial set with finitely many nondegenerate simplices, with $X_0 = X_1 = *$. Then we can effectively calculate $H_i(\Omega|X|, \mathbb{Z})$ for $i \leq n$.*

Proof Immediate from above. □

In some sense this corollary is the "most effective" part of the present argument, since we can get at the calculation after a bounded number of easy steps of the form $\mathcal{A} \mapsto \mathrm{Arrg}(\mathcal{A}, m)$.

We describe how to use the above description of ΩX inductively to obtain the $\pi_i(X)$. This seems to be a new algorithm, different from those of E. Brown [61], Kan [163, 164] and Curtis [88, 89], and the technique of James [151].

There is an unboundedness to the resulting algorithm, coming essentially from a problem with π_1 at each stage. Even though we know in advance that the π_1 is abelian, we would need to know "why" it is abelian in a precise way in order to specify a strategy for making $\mathcal{A}_{1/}$ connected at the appropriate place in the loop. In the absence of a particular description of the proof we are forced to say "search over all proofs" at this stage. See Subsection 15.5.5 for further discussion.

15.5.1 Getting $\mathcal{A}_{1/}$ to be connected

In the general situation, we have to tackle the problem of computation of a fundamental group using generators and relations, known to be undecideable in general. Some sub-cases can still be treated effectively.

The first question is how to arrange $\mathcal{A}$ on the level of $\tau_{\leq 1}(\mathcal{A})$.

We define operations $\mathrm{Arrg}^{0\,\mathrm{only}}(\mathcal{A}, m)$ and $\mathrm{Arrg}^{1\,\mathrm{only}}(\mathcal{A}, m)$. These consist of doing the operation $\mathrm{Arrg}(\mathcal{A}, m)$ but instead of using the entire mapping cone C, only adding on 0-cells to $\mathcal{A}_{m/}$ to get a surjection on π_0; or only adding on 1-cells to get an injection on π_0. Note in the second case that we *don't* add extra 0-cells. This is an important point, because if we added further 0-cells every time we added some 1-cells, the process would never stop.

To define $\mathrm{Arrg}^{0\,\mathrm{only}}(\mathcal{A}, m)$, use the same construction as for $\mathrm{Arrg}(\mathcal{A}, M)$ but instead of setting C to be the mapping cone, we put

$$C' := \mathcal{A}_{m/} \cup sk_0(\mathcal{A}_{1/} \times \cdots \times \mathcal{A}_{1/}).$$

Here sk_0 denotes the 0-skeleton of the simplicial set, and im means the image under the Segal map. Let $C \subset C'$ be a subset where we choose only one point for each connected component of the product. With this C the same construction as previously gives $\mathrm{Arrg}^{0\,\mathrm{only}}(\mathcal{A}, m)$.

With the subset $C \subset C'$ chosen as above (note that this choice can effectively be made) the resulting simplicial set

$$p \mapsto \pi_0\left(\mathrm{Arrg}^{0\,\mathrm{only}}(\mathcal{A}, m)_{p/}\right)$$

may be described only in terms of the simplicial set

$$p \mapsto \pi_0(\mathcal{A}_{p/}).$$

That is to say, this operation $\mathrm{Arrg}^{0\,\mathrm{only}}(\mathcal{A}, m)$ commutes with the operation of componentwise applying π_0. We formalize this as

$$\tau_{\leq 1}\mathrm{Arrg}^{0\,\mathrm{only}}(\mathcal{A}, m) = \tau_{\leq 1}\mathrm{Arrg}^{0\,\mathrm{only}}(\tau_{\leq 1}\mathcal{A}, m).$$

To define $\mathrm{Arrg}^{1\,\mathrm{only}}(\mathcal{A}, m)$, let C be the cone of the map from $\mathcal{A}_{m/}$ to

$$im(\mathcal{A}_{m/}) \cup sk_1(\mathcal{A}_{1/} \times \cdots \times \mathcal{A}_{1/})^o,$$

where $sk_1(\mathcal{A}_{1/} \times \ldots \times \mathcal{A}_{1/})^o$ denotes the union of connected components of the 1-skeleton of the product, which touch $im(\mathcal{A}_{m/})$. In this case, note that the inclusion

$$\mathcal{A}_{m/} \hookrightarrow C$$

is 0-connected (all connected components of C contain elements of $\mathcal{A}_{m/}$). Using this C we obtain the operation $\mathrm{Arrg}^{1\,\mathrm{only}}(\mathcal{A}, m)$. It doesn't introduce any new connected components in the new simplicial sets $\mathcal{A}'_{p/}$, but may connect together some components which were disjoint in $\mathcal{A}_{p/}$.

Again, the operation $\mathrm{Arrg}^{1\,\mathrm{only}}(\mathcal{A}, m)$ commutes with truncation: we have

$$\tau_{\leq 1}\mathrm{Arrg}^{1\,\mathrm{only}}(\mathcal{A}, m) = \tau_{\leq 1}\mathrm{Arrg}^{1\,\mathrm{only}}(\tau_{\leq 1}\mathcal{A}, m).$$

Our goal in this section is to find a sequence of operations which makes $\tau_{\leq 1}(\mathcal{A})_1$ become trivial (equal to $*$). In view of this, and the above commutations, we may henceforth work with simplicial sets (which we denote $U = \tau_{\leq 1}\mathcal{A}$ for example) and use the above operations followed by the truncation $\tau_{\leq 1}$ as modifications of the simplicial set U. We try to obtain $U_1 = *$. This corresponds to making $\mathcal{A}_{1/}$ connected.

Our operations have the following interpretation. The operation

$$U \mapsto \tau_{\leq 1}\mathrm{Arrg}^{0\,\mathrm{only}}(U, 2)$$

has the effect of formally adding to $\mathcal{U}_1$ all binary products of pairs of elements in U_1. (We say that a binary product of $u, v \in U_1$ is defined if there is an element $c \in U_2$ with principal edges u and v in U_1; the product is then the image w of the third edge of c.)

The operation

$$U \mapsto \tau_{\leq 1}\mathrm{Arrg}^{1\,\mathrm{only}}(U, 2)$$

has the effect of identifying w and w' any time both w and w' are binary products of the same elements u, v.

The operation

$$U \mapsto \tau_{\leq 1}\mathrm{Arrg}^{0\,\mathrm{only}}(U, 3)$$

has the effect of introducing, for each triple (u, v, w), the various binary products one can make (keeping the same order) and giving a formula

$$(uv)w = u(vw)$$

for certain of the binary products thus introduced.

It is somewhat unclear whether blindly applying the composed operation

$$U \mapsto \tau_{\leq 1}\mathrm{Arrg}^{1\,\mathrm{only}}\left(\tau_{\leq 1}\mathrm{Arrg}^{0\,\mathrm{only}}(U, 3), 2\right)$$

many times must automatically lead to $U_1 = *$ in case the actual fundamental group is trivial. This is because in the process of adding the associativity, we also add in some new binary products; to which associativity might then have to be applied in order to get something trivial, and so on.

If the above doesn't work, then we may need a slightly revised version of the operation $\mathrm{Arrg}^{0\,\mathrm{only}}(U, 3)$ where we add in only certain triples u, v, w. This can be accomplished by choosing a subset of the original C at each time. Similarly for the $\mathrm{Arrg}^{0\,\mathrm{only}}(U, 2)$ for binary products. We now obtain a situation where we have operations which effect the appropriate changes on U corresponding to all of the various possible steps in an elementary proof that the associative unitary monoid generated by generators U_1 with relations U_2, is trivial. Thus, if we have an elementary proof that the associative unitary monoid generated by U_1 with relations U_2 is trivial, then we can read off from the steps in the proof, the necessary sequence of operations to apply to get $U_1 = *$. On the level of $\mathcal{A}$ these same steps will result in a new $\mathcal{A}$ with $\mathcal{A}_{1/}$ connected.

In our case we are interested in the *group completion* of the monoid: we want to obtain the condition of being a Segal groupoid not just a Segal category. It is possible that the simplicial set X we start with would yield a monoid which is not a group, when the above operations are applied. To fix this, we take note of another operation which can be applied to $\mathcal{A}$ which doesn't affect the weak type of the realization, and which guarantees that, when the monoid U is generated, it becomes a group.

Let I be the category with two objects and one morphism $0 \to 1$, and let $\overline{I}$ be the category with two objects and an isomorphism between them. Consider these as Segal categories (taking their nerve as bisimplicial sets constant in the second variable). Note that $|I|$ and $|\overline{I}|$ are both contractible, so the obvious inclusion $I \hookrightarrow \overline{I}$ induces an equivalence of realizations.

The bisimplicial set $\overline{I}$ is just that which is represented by $(1, 0) \in \Delta \times \Delta$. Thus for a Segal precategory $\mathcal{A}$, if $f \in \mathcal{A}_{1,0}$ is an object of $\mathcal{A}_{1/}$ (a "morphism" in $\mathcal{A}$) then it corresponds to a morphism $I \to \mathcal{A}$. Set

$$\mathcal{A}^f := \mathcal{A} \cup^I \overline{I}.$$

Now the morphism f is strictly invertible in the precategory $\tau_{\leq 1}(\mathcal{A}^f)$ and in particular, when we apply the operations described above, the image of f becomes invertible in the resulting category. If $\mathcal{A}_0 = *$ (whence $\mathcal{A}^f_0 = *$ too) then the image of f becomes invertible in the resulting monoid. Note finally that

$$|\mathcal{A}| \to |\mathcal{A}^f| = |\mathcal{A}| \cup^{|I|} |\overline{I}|$$

is a weak equivalence. In fact we want to invert all of the 1-morphisms. Let

$$\mathcal{A}' := \mathcal{A} \cup^{\bigcup_f I} \left(\bigcup_f \overline{I} \right),$$

where the union is taken over all $f \in \mathcal{A}_{1,0}$. Again $|\mathcal{A}| \to |\mathcal{A}'|$ is a weak equivalence. Now, when we apply the previous procedure to $\tau_{\leq 1}(\mathcal{A}')$ giving a category U (a monoid if $\mathcal{A}$ had only one object), all morphisms coming from $\mathcal{A}_{1,0}$ become invertible. Note that the morphisms in $\mathcal{A}'$, i.e. objects of $\mathcal{A}'_{1,0}$, are either morphisms in $\mathcal{A}$ or their newly added inverses. Thus all of the morphisms coming from $\mathcal{A}'_{1,0}$ become invertible in the category U. But it is clear from the operations described above that U is generated by the morphisms in $\mathcal{A}'_{1,0}$. Therefore U is a groupoid. In the case of only one object, U becomes a group.

By Segal's theorem we then have $U = \pi_1(|\mathcal{A}|)$. If we know for some reason that $|\mathcal{A}|$ is simply connected, then U is the trivial group. More precisely, search for a proof that $\pi_1 = 1$, and when such a proof is found, apply the corresponding series of operations to $\tau_{\leq 1}(\mathcal{A}')$ to obtain $U = *$. Applying the operations to $\mathcal{A}'$ upstairs, we obtain a new $\mathcal{A}''$ with $|\mathcal{A}''| \cong |\mathcal{A}'| \cong |\mathcal{A}|$ and $\mathcal{A}''_{1/}$ connected.

Another way of looking at this is to say that every time one needs to take the inverse of an element in the proof that the group is trivial, add on a copy of $\overline{I}$ over the corresponding copy of I.

15.5.2 The case of finite homotopy groups

We first present our algorithm for the case of finite homotopy groups. Suppose we want to calculate $\pi_n(X)$. We assume known that the $\pi_i(X)$ are finite for $i \leq n$.

Start: Fix n and start with a simplicial set X containing a finite number of nondegenerate simplices. Suppose we know that $\pi_1(X, x)$ is a given finite group; record this group, and set Y equal to the corresponding covering space of X. Thus Y is simply connected. Now contract out a maximal tree to obtain Z with $Z_0 = *$.

Step 1: Let $\mathcal{A}_{p,k} := Z_p$ be the corresponding Segal precategory. It has only one object.

Step 2: Let $\mathcal{A}'$ be the amalgamated sum of $\mathcal{A}$ with one copy of the nerve of the category $\overline{I}$ (containing two isomorphic objects), for each morphism $I \to \mathcal{A}$ (i.e. each point in $\mathcal{A}_{1,0}$).

Step 3: Apply the procedure of Subsection 15.5.1 to obtain a morphism $\mathcal{A}' \to \mathcal{A}''$ with $\mathcal{A}''_{1/}$ connected, and inducing a weak equivalence on realizations. (This step can only be bounded if we have a specific proof that $\pi_1(Y, y) = 1$.)

Step 4: Apply the procedure of Corollary 15.5.1 and Theorem 15.4.5 to obtain a morphism $\mathcal{A}'' \to \mathcal{B}$ inducing a weak equivalence on realizations, such that $\mathcal{B}$ is a Segal groupoid. Note that the $(n-1)$-type of $\mathcal{B}_{1/}$ is effectively calculable (the noneffective parts of the proof of Theorem 15.4.5 served only to prove the properties in question). By Segal's theorem,

$$|\mathcal{B}_{1/}| \sim \Omega|\mathcal{B}| \sim \Omega|Y|,$$

which in turn is the connected component of $\Omega|X|$. Thus

$$\pi_n(|X|) = \pi_{n-1}(|\mathcal{B}_{1/}|).$$

Step 5: Go back to the *Start* with the new n equal to the old $n-1$, and the new X equal to the simplicial set $\mathcal{B}_{1/}$ above. The new fundamental group is known to be abelian (since it is π_2 of the previous X). Thus we can calculate the new fundamental group as $H_1(X)$ and, under our hypothesis, it will be finite.

Keep repeating the procedure until we get down to $n = 1$ and have recorded the answer.

15.5.3 How to get rid of free abelian groups in π_2

In the case where the higher homotopy groups are infinite (i.e. they contain factors of the form $\mathbb{Z}^a$) we need to do something to get past these infinite groups. If we go down to the case where π_1 is infinite, then taking the universal covering no longer results in a finite complex. We prefer to avoid this by tackling the problem at the level of π_2, with a geometrical argument. Namely, if $H^2(X, \mathbb{Z})$ is nonzero then we can take a class there as giving a line bundle, and take the total space of the corresponding S^1-bundle. This amounts to taking the fiber of a map $X \to K(\mathbb{Z}, 2)$. This can be done explicitly and effectively, resulting again in a calculable finite complex. In the new complex we will have reduced the rank of $H_2(X, \mathbb{Z}) = \pi_2(X)$ (we are assuming that X is simply connected).

The original method of E. Brown [61] for effectively calculating the π_i was basically to do this at all i. The technical problems in *loc. cit.* [61] are caused by the fact that one doesn't have a finite complex representing $K(\mathbb{Z}, n)$. In the case $n = 2$ we don't have these technical problems because we can look at circle fibrations and the circle is a finite complex. For this section, then, we are in some sense reverting to an easy case of *loc. cit.* [61] and not using the Seifert–Van Kampen technique.

Suppose X is a simplicial set with finitely many nondegenerate simplices, and suppose $X_0 = X_1 = *$. We can calculate $H^2(X, \mathbb{Z})$ as the kernel of the differential

$$d : \mathbb{Z}^{X'_2} \to \mathbb{Z}^{X'_3}.$$

Here X'_i is the set of nondegenerate i-simplices. (Note that a basis of this kernel can effectively be computed using Gaussian elimination). Pick an element β of this basis, which is a collection of integers b_t for each 2-simplex (i.e. triangle) t. For each triangle t define an S^1-bundle L_t over t together with trivialization

$$L_t|_{\partial t} \cong \partial t \times S^1.$$

To do this, take $L_t = t \times S^1$ but change the trivialization along the boundary by a bundle automorphism

$$\partial t \times S^1 \to \partial t \times S^1$$

obtained from a map $\partial t \to S^1$ with winding number b_t. Let $L^{(2)}$ be the S^1-bundle over the 2-skeleton of X obtained by glueing together the L_t along the trivializations over their boundaries. We can do this effectively and obtain $L^{(2)}$ as a simplicial set with a finite number of nondegenerate simplices.

The fact that $d(\beta) = 0$ means that, for a 3-simplex e, the restriction of $L^{(2)}$ to ∂e (which is topologically an S^2) is a trivial S^1-bundle. Thus $L^{(2)}$ extends to an S^1-bundle $L^{(3)}$ on the 3-skeleton of X. Furthermore, it can be extended across any simplices of dimension ≥ 4 because all S^1-bundles on S^k for $k \geq 3$, are trivial ($H^2(S^k, \mathbb{Z}) = 0$). We obtain an S^1-bundle L on X. By subdividing things appropriately (including possibly subdividing X) we can assume that L is a simplicial set with a finite number of nondegenerate simplices. It depends on the choice of basis element β, so call it $L(\beta)$.

Let

$$T = L(\beta_1) \times_X \cdots \times_X L(\beta_r)$$

where $\beta_1, \ldots, \beta_r$ are our basis elements found above. It is a torus bundle with fiber $(S^1)^r$. The long exact homotopy sequence for the map $T \to X$ gives

$$\pi_i(T) = \pi_i(X), \quad i \geq 3;$$

and

$$\pi_2(T) = \ker(\pi_2(X) \to \mathbb{Z}^r).$$

Note that $\mathbb{Z}^r$ is the dual of $H^2(X, \mathbb{Z})$ so the kernel $\pi_2(T)$ is finite. Finally, $\pi_1(T) = 0$ since the map $\pi_2(X) \to \mathbb{Z}^r$ is surjective.

Note that we have a proof that $\pi_1(T) = 0$.

15.5.4 The general algorithm

Here is the general situation. Fix n. Suppose X is a simplicial set with finitely many nondegenerate simplices, with $X_0 = *$ and with a proof that $\pi_1(X) = \{1\}$. We will calculate $\pi_i(X)$ for $i \leq n$.

Step 1: Calculate $\pi_2(X) = H_2(X, \mathbb{Z})$ by Gaussian elimination.

Step 2: Apply the operation described in the previous subsection above, to obtain a new T with $\pi_i(T) = \pi_i(X)$ for $i \geq 3$, with $T_0 = *$, with $\pi_1(T) = 1$, and with $\pi_2(T)$ is finite.

Let $\mathcal{A}$ be the Segal precategory corresponding to T.

Step 3: Use the discussion of Section 15.5.1 to obtain a morphism $\mathcal{A} \to \mathcal{A}'$ inducing a weak equivalence of realizations, such that $\mathcal{A}'_{1/}$ is connected. For this step we need a proof that $\pi_1(T) = 1$. In the absence of a specific (finite) proof, search over all proofs.

Step 4: Use Corollary 15.4.7 to replace $\mathcal{A}'$ by a Segal precategory $\mathcal{B}$ with $|\mathcal{A}'| \to |\mathcal{B}|$ a weak equivalence, such that the $(n-1)$-type of $\mathcal{B}_{1/}$ is equivalent to $\Omega|\mathcal{B}|$, which in turn is equivalent to $\Omega|X|$. Let $Y = \mathcal{B}_{1/}$ as a simplicial set.

Note that Y is connected and $\pi_1(Y)$ is finite, being equal to $\pi_2(T)$. We have $\pi_i(X) = \pi_{i-1}(Y)$ for $3 \le i \le n$.

Step 5: Choose a universal cover of Y, and mod out by a maximal tree in the 1-skeleton to obtain a simplicial set Z, with finitely many nondegenerate simplices, with $Z_0 = *$, and with a proof that $\pi_1(Z) = 1$. We have $\pi_i(X) = \pi_{i-1}(Z)$ for $3 \le i \le n$.

Go back to the beginning of the algorithm and plug in $(n-1)$ and Z. Keep doing this until, at the step where we calculate π_2 of the new object, we end up having calculated $\pi_i(X)$ as desired.

15.5.5 Proofs of the Godement condition

We pose the following question: how could one obtain, in the process of applying the above algorithm, an explicit proof that at each stage the fundamental group (of the universal cover Z in step 5) is trivial? This could then be plugged into the machinery above to obtain an explicit strategy, thus we would avoid having to try all possible strategies. To do this we would need an explicit proof that $\pi_1(Y)$ is finite in step 4, and this in turn would be based on a proof that $\pi_1(Y) = \pi_2(T)$ as well as a proof of the Godement property that $\pi_2(T)$ is abelian.

15.6 Example: $\pi_3(S^2)$

The story behind my preprint [236] was that Ronnie Brown came by Toulouse for Jean Pradines' retirement party, and we were discussing Seifert–Van Kampen. He pointed out that the result of my previous preprint [234] didn't seem to lead to any actual calculations. After that, I tried to use that technique (in its simplified Segal-categoric version) to calculate $\pi_3(S^2)$. It was apparent from this calculation that the process was effective in general.

We describe here what happens for calculating $\pi_3(S^2)$. We take as simplicial model a simplicial set with the basepoint as unique 0-cell $*$ and with one nondegenerate simplex e in degree 2. Note that this leads to many degenerate simplices in degrees ≥ 2 (however, there is only one degenerate simplex which we denote $*$ in degree 1).

We follow out what happens in a language of cell-addition. Thus we don't feel required to take the whole cone C at each step of an operation $\mathrm{Arrg}(\mathcal{A}, m)$; we take any addition of cells to $\mathcal{A}_{m/}$ lifting cells in $\mathcal{A}_{1/} \times \cdots \times \mathcal{A}_{1/}$.

We keep the notation $\mathcal{A}$ for the result of each operation (since our discussion is linear, this shouldn't cause too much confusion).

The first step is to (2, 0)-arrange $\mathcal{A}$. We do this by adding a 1-cell joining the two 0-cells in $\mathcal{A}_{2/}$, in an operation of type $\mathrm{Arrg}(\mathcal{A}, 2)$. Note that both 0-cells map to the same point $\mathcal{A}_{1/} \times \mathcal{A}_{1/} = *$. The first result of this is to add on 1-cells in the $\mathcal{A}_{m/}$ connecting all of the various degeneracies of e, to the basepoint. Thus the $\mathcal{A}_{m/}$ become connected. Additionally we get a new 1-cell added onto to $\mathcal{A}_{1/}$ corresponding to the third face (02). Furthermore, we obtain all images of this cell by degeneracies $m \to 1$. Thus we get m circles attached to the pieces which became connected in the first part of this operation. Now each $\mathcal{A}_{m/}$ is a wedge of m circles.

In particular note that $\mathcal{A}$ is now $(m, 1)$-arranged for all m.

The next step is to (2, 2)-arrange $\mathcal{A}$. To do this, note that the Segal map is

$$S^1 \vee S^1 = \mathcal{A}_{2/} \to \mathcal{A}_{1/} \times \mathcal{A}_{1/} = S^1 \times S^1.$$

To arrange this map we have to add a 2-cell to $S^1 \vee S^1$ with attaching map the commutator relation. Again, this has the result of adding on 2-cells to all of the $\mathcal{A}_{m/}$ over the pairwise commutators of the loops. Furthermore, we obtain an extra 2-cell added onto $\mathcal{A}_{1/}$ via the edge (02). The attaching map here is the commutator of the generator with itself, so it is homotopically trivial and we have added on a 2-sphere. (Note in passing that this 2-sphere is what gives rise to the class of the Hopf map.) Again, we obtain the images of this S^2 by all of the degeneracy maps $m \to 1$. Now

$$\mathcal{A}_{1/} = S^1 \vee S^2,$$
$$\mathcal{A}_{2/} = (S^1 \times S^1) \vee S^2 \vee S^2,$$

and in general $\mathcal{A}$ is $(m, 2)$-arranged for all m. Looking forward to the next section, we see that adding 3-cells to $\mathcal{A}_{m/}$ for $m \geq 3$ in the appropriate way as described in the proof of 15.4.5, will end up resulting in the addition of 4-cells (or higher) to $\mathcal{A}_{1/}$ so this no longer affects the 2-type of $\mathcal{A}_{1/}$. Thus (for the purposes of getting $\pi_3(S^2)$) we may now ignore the $\mathcal{A}_{m/}$ for $m \geq 3$.

The remaining operation is to (2, 3)-arrange $\mathcal{A}$. For this, look at the Segal map

$$\mathcal{A}_{2/} = (S^1 \times S^1) \vee S^2 \vee S^2 \rightarrow$$

$$\mathcal{A}_{1/} \times \mathcal{A}_{1/} = (S^1 \vee S^2) \times (S^1 \vee S^2).$$

Let C be the mapping cone on this map. Then we end up attaching one copy of C to $\mathcal{A}_{1/}$ along the third edge map $\mathcal{A}_{2/} \rightarrow \mathcal{A}_{1/}$. This gives the answer for the 2-type of ΩS^2:

$$\tau_{\leq 2}(\Omega S^2) = \tau_{\leq 2}\left((S^1 \vee S^2) \cup^{(S^1 \times S^1) \vee S^2 \vee S^2} C\right).$$

To calculate $\pi_2(\Omega S^2)$ we revert to a homological formulation (because it isn't easy to "see" the cone C). In homology of degree ≤ 2, the above Segal map

$$(S^1 \times S^1) \vee S^2 \vee S^2 \rightarrow (S^1 \vee S^2) \times (S^1 \vee S^2)$$

is an isomorphism. Thus the map $\mathcal{A}_{2/} \rightarrow C$ is an isomorphism on homology in degrees ≤ 2, and adding in a copy of C along $\mathcal{A}_{2/}$ doesn't change the homology. Thus

$$H_2(\Omega S^2) = H_2(S^1 \vee S^2) = \mathbb{Z}.$$

Noting that (as we know from general principles) $\pi_1(\Omega S^2) = \mathbb{Z}$ acts trivially on $\pi_2(\Omega S^2)$ and π_1 itself has no homology in degree 2, we get that $\pi_2(\Omega S^2) = H_2(\Omega S^2) = \mathbb{Z}$.

Exercise: Calculate $\pi_4(S^3)$ using the above method.

Remark: our above recourse to homology calculations suggests that it might be interesting to do pushouts and the the arranging operations in the context of simplicial chain complexes.

15.6.1 Seeing Kan's simplicial free groups

Using the above procedure, we can actually see how Kan's simplicial free groups arise in the calculation for an arbitrary simplicial set X. They arise just from a first stage where we add on 1-cells. Namely, if in doing the procedure $\mathrm{Arrg}(\mathcal{A}, m)$ we replace C by a choice of 1-cell joining any two components of $\mathcal{A}_{m/}$ which go to the same component under the Segal map, then applying this operation for various m, we obtain a simplicial space whose components are connected and homotopic to wedges of circles. (We have to start with an X having $X_1 = *$.) The resulting simplicial space has the same realization as X. If X has only finitely many nondegenerate simplices then one can stop after a finite number of applications of this operation. Taking the fundamental groups of the component spaces (based at the degeneracy of the unique basepoint) gives a simplicial free group. Taking the classifying simplicial sets of these groups in each component we obtain a bisimplicial set whose realization is equivalent to X. This bisimplicial set actually satisfies $\mathcal{A}_{p,0} = \mathcal{A}_{0,k} = *$, in other words it satisfies the globular condition in both directions! We can therefore view it as a Segal precategory in two ways. The second way, interchanging the two variables, yields a Segal precategory where the Segal maps are *isomorphisms* (because at each stage it was the classifying simplicial set for a group). Thus, viewed in this way, it is a Segal groupoid and Segal's theorem implies that the simplicial set $p \mapsto \mathcal{A}_{p,1}$, which is the underlying set of a simplicial free group, has the homotopy type of ΩX.

PART IV

The model structure

16

Sequentially free precategories

To prove that the classes of Reedy cofibrations and global weak equivalences give a model structure on $\mathcal{PC}(\mathcal{M})$, some computational work is needed. It turns out that the basic objects to be studied are the categories with an ordered set of objects $x_0, \ldots, x_n$ and morphisms other than the identity from x_i to x_j only when $i < j$. More precisely, we consider the *free* categories of this type obtained by specifying an object $B_i \in \mathcal{M}$ of morphisms from x_{i-1} to x_i for $1 \leq i \leq n$; then the object of morphisms from x_i to x_j should be the product of the B_k for $k = i+1, \ldots, j$. One of the main tasks is to look at a notion of precategory which corresponds to this notion of category, and to follow through explicitly the calculus of generators and relations.

In the next chapter will be our main calculation, which is what happens when one takes the product of two such categories. These considerations lead to a general theorem on products–the cartesian condition for $\mathcal{PC}(\mathcal{M})$. It turns out that the cartesian condition then implies the preservation of global weak equivalences under pushout which is the crucial hypothesis of Smith's recognition theorem.

Throughout, the notion of local weak equivalence on $\mathcal{PC}(X, \mathcal{M})$ is the one given by the model structures of Theorem 12.1.1 and 12.3.2.

16.1 Imposing the Segal condition on Υ

Recall the $\mathcal{M}$-precategories $\Upsilon(B_1, \ldots, B_k)$ defined in Sections 10.5 and 14.1. These can be strictly categorified, which is to say that we can construct corresponding strict $\mathcal{M}$-categories which will be weakly equivalent (Theorem 16.1.2 below).

Define a precategory $\widetilde{\Upsilon}_k(B_1, \ldots, B_k)$ as follows: the set of objects is the same as for $\Upsilon_k(B_1, \ldots, B_k)$, that is $[k] = \{\upsilon_0, \ldots, \upsilon_k\}$. For any sequence $\upsilon_{i_0}, \ldots, \upsilon_{i_n}$ with $i_0 \leq \ldots \leq i_n$, we put

$$\widetilde{\Upsilon}_k(B_1, \ldots, B_k)(\upsilon_{i_0}, \ldots, \upsilon_{i_n}) := B_{i_0+1} \times B_{i_0+2} \times \cdots \times B_{i_n-1} \times B_{i_n}. \quad (16.1.1)$$

This includes the unitality condition $\widetilde{\Upsilon}_k(B_1, \ldots, B_k)(\upsilon_i, \ldots, \upsilon_i) = *$. For any other sequence, that is to say any sequence which is not increasing, the value is $\emptyset$. Note in particular that

$$\widetilde{\Upsilon}_k(B_1, \ldots, B_k)(\upsilon_{i-1}, \upsilon_i) = B_i = \Upsilon_k(B_1, \ldots, B_k)(\upsilon_{i-1}, \upsilon_i). \quad (16.1.2)$$

By the adjunction property of $\Upsilon_k(B_1, \ldots, B_k)$ there is a unique map

$$\Upsilon_k(B_1, \ldots, B_k) \to \widetilde{\Upsilon}_k(B_1, \ldots, B_k) \quad (16.1.3)$$

inducing the identity (16.1.2) on adjacent pairs of objects. In fact, we can consider $\Upsilon_k(B_1, \ldots, B_k)$ as a subobject of $\widetilde{\Upsilon}_k(B_1, \ldots, B_k)$ with this map as the inclusion.

Lemma 16.1.1 *The precategory $\widetilde{\Upsilon}_k(B_1, \ldots, B_k)$ is a strict $\mathcal{M}$-category, in particular it is a Segal $\mathcal{M}$-category.*

Proof If $\upsilon_{i_0}, \ldots, \upsilon_{i_n}$ is an increasing sequence, i.e. $i_0 \leq \cdots \leq i_n$, and if $1 \leq j \leq n-1$ then by the formula (16.1.1) the natural maps obtained by splitting the sequence of objects at υ_{i_j} give an isomorphism

$$\begin{array}{c}\widetilde{\Upsilon}_k(B_1, \ldots, B_k)(\upsilon_{i_0}, \ldots, \upsilon_{i_n}) \\ \downarrow \cong \\ \widetilde{\Upsilon}_k(B_1, \ldots, B_k)(\upsilon_{i_0}, \ldots, \upsilon_{i_j}) \times \widetilde{\Upsilon}_k(B_1, \ldots, B_k)(\upsilon_{i_j}, \ldots, \upsilon_{i_n}).\end{array}$$

By induction it follows that the Segal maps are isomorphisms. □

One should think of $\widetilde{\Upsilon}_k(B_1, \ldots, B_k)$ as being the *free $\mathcal{M}$-category generated by morphism objects B_i going from υ_{i-1} to υ_i*. Here the objects are linearly ordered, and the generating objects of $\mathcal{M}$ are placed between adjacent objects in the ordering.

We would like to make precise this intuition by showing that it is the Segal $\mathcal{M}$-category generated by the precategory $\Upsilon_k(B_1, \ldots, B_k)$ as stated in the following theorem:

Theorem 16.1.2 *For any k and any sequence of objects $B_1, \ldots, B_k$, the inclusion (16.1.3) is a local weak equivalence in $\mathcal{PC}(\{\upsilon_0, \ldots, \upsilon_k\}, \mathcal{M})$. If each B_i is cofibrant, then it is a trivial cofibration in the Reedy model structure.*

16.2 Sequentially free precategories in general

Before getting to the proof of the previous theorem, which will be done at the end of the chapter on page 395, it is useful to generalize the above situation by giving a criterion for when a precategory will lead to a free Segal category with linearly ordered object set and generators B_i between adjacent objects:

Definition 16.2.1 A *sequentially free $\mathcal{M}$-precategory* consists of a finite linearly ordered set $X = \{x_0, \ldots, x_k\}$ together with a structure of $\mathcal{M}$-precategory $\mathcal{A} \in \mathcal{PC}(X, \mathcal{M})$, satisfying the following properties:

(SF1)–if $x_{i_0}, \ldots, x_{i_n}$ is a sequence of objects which are not increasing, i.e. there is some $0 < j \leq n$ with $i_{j-1} > i_j$, then $\mathcal{A}(x_{i_0}, \ldots, x_{i_n}) = \emptyset$; and

(SF2)–if $x_{i_0}, \ldots, x_{i_n}$ is a sequence of objects in increasing order, i.e. $i_{j-1} \leq i_j$ for all $0 < j \leq n$, then the outer map for the n-simplex provides a weak equivalence

$$\mathcal{A}(x_{i_0}, \ldots, x_{i_n}) \xrightarrow{\sim} \mathcal{A}(x_{i_0}, x_{i_n}). \tag{16.2.1}$$

Remark 16.2.2 Note that condition (SF2) for a sequence (x_i) of length $n = 0$ says that the map $* = \mathcal{A}(x_i) \to \mathcal{A}(x_i, x_i)$ is a weak equivalence.

We often say that $\mathcal{A}$ is sequentially free *with respect to a given order on X* if $(X, \mathcal{A})$ is sequentially free for the ordering in question.

Lemma 16.2.3 *Both $\Upsilon_k(B_1, \ldots, B_k)$ and $\widetilde{\Upsilon}_k(B_1, \ldots, B_k)$ are sequentially free $\mathcal{M}$-precategories with respect to the ordering on $[k]$.*

Proof Both clearly satisfy (SF1). For $\widetilde{\Upsilon}_k(B_1, \ldots, B_k)$ the condition (SF2) holds by construction. For $\Upsilon_k(B_1, \ldots, B_k)$, suppose we have an increasing sequence of objects $\upsilon_{i_0} \leq \cdots \leq \upsilon_{i_n}$. If $i_n = i_0$ then the values on both sides of (16.2.1) are $*$. If $i_n = i_0 + 1$ then the values on both sides are B_{i_n}, whereas if $i_n > i_0 + 1$ the values on both sides are $\emptyset$. In all three cases the map (16.2.1) is an isomorphism, *a fortiori* a weak equivalence. □

Lemma 16.2.4 *Suppose X is a fixed linearly ordered finite set, β is an ordinal and $\{\mathcal{A}(b)\}_{b\in\beta}$ is a transfinite sequence of $\mathcal{M}$-precategories, that is to say a functor $\beta \to \mathcal{PC}(X, \mathcal{M})$. Suppose that each $(X, \mathcal{A}(b))$ is a sequentially free $\mathcal{M}$-precategory with respect to the given order on X, and that for any $b \leq b'$ the transition map $\mathcal{A}(b) \to \mathcal{A}(b')$ is a Reedy (hence injective) cofibration in $\mathcal{PC}(X, \mathcal{M})$. Then the colimit $(X, \mathrm{colim}_{b\in\beta}\mathcal{A}(b))$ is again a sequentially free $\mathcal{M}$-precategory with respect to the same ordering.*

Proof The colimit is calculated levelwise as a Δ^o_X-diagram in $\mathcal{M}$, since filtered colimits preserve the unitality condition. A colimit of objects $\emptyset$ is again $\emptyset$, so the colimit satisfies (SF1). The left properness hypotheses on $\mathcal{M}$ implies transfinite left properness (Proposition 7.3.3), so the weak equivalence (16.2.1) is preserved in the colimit whose transition maps are cofibrations, giving (SF2). □

We now come to one of the main steps where we gain some control over the process of generators and relations. Recall from Section 14.4 that the operation **Gen** consisting of applying one step of the calculus of generators and relations.

Lemma 16.2.5 *Suppose $(X, \mathcal{A})$ is a sequentially free $\mathcal{M}$-precategory with linearly ordered object set $X = \{x_0, \ldots, x_m\}$, and $x_a, x_{a+1}, \ldots, x_b$ is a strictly increasing sequence of adjacent objects with $0 \leq a < b \leq m$. Then the new $\mathcal{M}$-precategory* **Gen**$((X, \mathcal{A}); x_a, \ldots, x_b)$ *is also sequentially free with the same ordered set of objects X, and furthermore for any $a < j \leq b$ the map*

$$\mathcal{A}(x_{j-1}, x_j) \to \mathbf{Gen}((X, \mathcal{A}); x_a, \ldots, x_b)(x_{j-1}, x_j)$$

is an isomorphism (hence a weak equivalence) in $\mathcal{M}$.

Proof Fix a factorization $E, e, p_1, \ldots, p_{b-a}$ as used for the construction of **Gen**$((X, \mathcal{A}); x_a, \ldots, x_b)$. Note that

$$p : E \xrightarrow{\sim} \mathcal{A}(x_a, x_{a+1}) \times \cdots \times \mathcal{A}(x_{b-1}, x_b)$$

is a weak equivalence. The sequence of objects $x_a, \ldots, x_b$ is disjoint, so we can use the description of **Gen**$((X, \mathcal{A}); x_a, \ldots, x_b)$ given in Lemma 14.4.2. That says that for any sequence of the form $x_{i_0}, \ldots, x_{i_p}$ if $a \leq i_0 \leq \cdots \leq i_p \leq b$ with $i_0 + 2 \leq i_p$ then

$$\mathbf{Gen}(\mathcal{A}; x_a, \ldots, x_b)(x_{i_0}, \ldots, x_{i_p}) = \mathcal{A}(x_{i_0}, \ldots, x_{i_p}) \cup^{\mathcal{A}(x_a, \ldots, x_b)} E, \quad (16.2.2)$$

but for any other sequence,

$$\mathbf{Gen}(\mathcal{A}; x_a, \ldots, x_b)(x_{i_0}, \ldots, x_{i_p}) = \mathcal{A}(x_{i_0}, \ldots, x_{i_p}). \quad (16.2.3)$$

We can now check the conditions (SF1) and (SF2). If the sequence of objects $(x_{i_0}, \ldots, x_{i_p})$ is not increasing, then it falls into the second case (16.2.3), and $\mathcal{A}(x_{i_0}, \ldots, x_{i_p}) = \emptyset$ by (SF1) for $\mathcal{A}$, which gives (SF1) for **Gen** $(\mathcal{A}; x_a, \ldots, x_b)$. Suppose that $(x_{i_0}, \ldots, x_{i_p})$ is increasing. We need to check (SF2). If either $i_0 < a$ or $i_p > b$ then we again fall into case (16.2.3) for both the full sequence $(x_{i_0}, \ldots, x_{i_p})$ and also the pair of endpoints (x_{i_0}, x_{i_p}). Thus, by (SF2) for $\mathcal{A}$ we have a weak equivalence

$$\begin{array}{c} \mathbf{Gen}(\mathcal{A}; x_a, \ldots, x_b)(x_{i_0}, \ldots, x_{i_p}) = \mathcal{A}(x_{i_0}, \ldots, x_{i_p}) \\ \downarrow \sim \\ \mathcal{A}(x_{i_0}, x_{i_p}) = \mathbf{Gen}(\mathcal{A}; x_a, \ldots, x_b)(x_{i_0}, x_{i_p}) \end{array}$$

giving this case of (SF2) for $\mathbf{Gen}(\mathcal{A}; x_a, \ldots, x_b)$. Suppose, on the other hand, that $a \leq i_0 \leq i_p \leq b$. If $i_p \leq i_0 + 1$ then we again are in case (16.2.3) for both the full sequence and the pair of endpoints, so we get the condition (SF2) as above. Suppose therefore that $i_0 + 2 \leq i_p$. In the diagram

$$\begin{array}{ccccc} \mathcal{A}(x_{i_0}, \ldots, x_{i_p}) & \leftarrow & \mathcal{A}(x_a, \ldots, x_b) & \longrightarrow & E \\ \downarrow & & \| & & \| \\ \mathcal{A}(x_{i_0}, x_{i_p}) & \longleftarrow & \mathcal{A}(x_a, \ldots, x_b) & \longrightarrow & E \end{array}$$

the left vertical arrow is an equivalence by (SF2) for $\mathcal{A}$. By left properness of $\mathcal{M}$ via Corollary 7.3.2, the vertical maps therefore induce a weak equivalence from the pushout of the top row to the pushout of the bottom row. By equations (16.2.2), which apply both to the full sequence $(x_{i_0}, \ldots, x_{i_p})$ and to the pair of endpoints (x_{i_0}, x_{i_p}), these pushouts are $\mathbf{Gen}(\mathcal{A}; x_a, \ldots, x_b)(x_{i_0}, \ldots, x_{i_p})$ and $\mathbf{Gen}(\mathcal{A}; x_a, \ldots, x_b)(x_{i_0}, x_{i_p})$ respectively. It follows that the map from the one to the other is an equivalence, which gives condition (SF2) in this last case. This proves that $\mathbf{Gen}((X, \mathcal{A}); x_a, \ldots, x_b)$ is again a sequentially free $\mathcal{M}$-precategory.

For the last statement, note that the sequence (x_{j-1}, x_j) falls into case (16.2.3) because the space between the endpoints is only 1. Hence the formula (16.2.3) says that the map

$$\mathcal{A}(x_{j-1}, x_j) \to \mathbf{Gen}((X, \mathcal{A}); x_a, \ldots, x_b)(x_{j-1}, x_j)$$

is an isomorphism. □

Recall that the map $\mathcal{A} \to \mathbf{Gen}((X, \mathcal{A}); x_a, \ldots, x_b)$ is a local weak equivalence in $\mathcal{PC}(X, \mathcal{M})$.

Lemma 16.2.6 *Suppose $(X, \mathcal{A})$ is a sequentially free $\mathcal{M}$-precategory with ordered object set $X = \{x_0, \ldots, x_m\}$. By iterating a series of operations of the form $\mathcal{A} \mapsto \mathbf{Gen}((X, \mathcal{A}); x_a, \ldots, x_b)$ we can obtain a weak equivalent* *sequentially free $\mathcal{M}$-precategory $(X, \mathcal{A}) \to (X, \mathcal{A}')$ such that $\mathcal{A}'$ satisfies the Segal conditions.*

Proof We show how to obtain the Segal condition for strictly increasing sequences of adjacent objects. At the end of the proof we go from here to the Segal condition for general sequences.

Fix an integer n_0 and suppose $(X, \mathcal{A})$ satisfies the Segal condition for all adjacent sequences $(x_c, x_{c+1}, \ldots, x_d)$ with $d - c > n_0$, and a certain number of adjacent sequences $(x_{c_v}, x_{c_v+1}, \ldots, x_{d_v})$ for (c_v, d_v) indexed by $v \in V$ for some set v, with $d_v - c_v = n_0$. Suppose $0 \leq a < b \leq m$ with $b - a = n_0$. Then $\mathbf{Gen}((X, \mathcal{A}); x_a, \ldots, x_b)$ satisfies the Segal condition for all adjacent sequences $(x_c, x_{c+1}, \ldots, x_d)$ with $d-c > n_0$, for the given adjacent sequences $(x_{c_v}, x_{c_v+1}, \ldots, x_{d_v})$ with $v \in V$, and also for the sequence $(x_a, \ldots, x_b)$. Indeed, if $(x_c, x_{c+1}, \ldots, x_d)$ with $d - c > n_0$ then the terms entering into the Segal map for this sequence are all covered by the situation (16.2.3) in the explicit description of $\mathbf{Gen}((X, \mathcal{A}); x_a, \ldots, x_b)$ used in the previous proof (see Lemma 14.4.2). Note that the terms in the product on the right-hand side of the Segal map are of the form $\mathbf{Gen}((X, \mathcal{A}); x_a, \ldots, x_b)(x_{j-1}, x_j) = \mathcal{A}(x_{j-1}, x_j)$. Here we use the condition that we are only looking at adjacent sequences. By the recurrence hypothesis on $\mathcal{A}$, the Segal map for this sequence is an equivalence.

Similarly, if $(x_{c_v}, x_{c_v+1}, \ldots, x_{d_v})$ is one of our given sequences with $d_v - c_v = n_0$, and if it is different from the sequence $(x_a, \ldots, x_b)$, then either $c_v < a$ or $d_v > b$ so again everything entering into the Segal map for this sequence is the same as for $\mathcal{A}$, by (16.2.3). Thus, again the inductive hypothesis on $\mathcal{A}$ implies the Segal condition for $\mathbf{Gen}((X, \mathcal{A}); x_a, \ldots, x_b)(x_{c_v}, x_{c_v+1}, \ldots, x_{d_v})$.

At the sequence $(x_a, \ldots, x_b)$ equation (16.2.2) gives

$$\mathbf{Gen}(\mathcal{A}; x_a, \ldots, x_b)(x_a, \ldots, x_b) = \mathcal{A}(x_a, \ldots, x_b) \cup^{\mathcal{A}(x_a, \ldots, x_b)} E = E,$$

and the map

$$E \to \mathcal{A}(x_a, x_{a+1}) \times \cdots \times \mathcal{A}(x_{b-1}, x_b)$$

is an equivalence by hypothesis on the choice of E. Therefore, the Segal condition holds at the sequence $(x_a, \ldots, x_b)$ too and we can add this to our collection V of good sequences of lenght n_0.

So, the inductive procedure is to start with the maximal sequence $x_0, \ldots, x_m$ of length m, impose the Segal condition here by changing $\mathcal{A}$ to $\mathbf{Gen}(\mathcal{A}; x_0, \ldots, x_m)$; then successively impose the Segal condition on all sequences of length $m - 1$, then $m - 2$ and so on, using the above inductive observation. At each step $\mathcal{A}$ is replaced by $\mathbf{Gen}(\mathcal{A}; x_a, \ldots, x_b)$ and there is a map from the old to the new $\mathcal{A}$ which is a local weak equivalence in $\mathcal{PC}(X, \mathcal{M})$. By Lemma 16.2.5, the new $\mathcal{A}$ is always again a sequentially free $\mathcal{M}$-precategory. Combining these steps (there are a finite number) down to $n_0 = 2$ we arrive at a map $(X, \mathcal{A}) \to (X, \mathcal{A}')$ which is a local weak equivalence in $\mathcal{PC}(X, \mathcal{M})$, and such that $\mathcal{A}'$ is a sequentially free $\mathcal{M}$-precategory which satisfies the Segal condition for all sequences of the form $(x_c, \ldots, x_d)$.

We claim that this implies that $\mathcal{A}'$ satisfies the Segal condition for any sequence of the form $x_{i_0}, \ldots, x_{i_p}$. Note that if $p = 0$ then the Segal condition is automatic since $\mathcal{A}'(X) = *$. Suppose $p \geq 1$ and consider our sequence $(x_{i_0}, x_{i_1}, \ldots, x_{i_p})$. If any $i_{j-1} > i_j$ then $\mathcal{A}'(x_{i_0}, \ldots, x_{i_{p-1}}, x_{i_p}) = \emptyset$ and $\mathcal{A}'(x_{i_{j-1}}, x_{i_j}) = \emptyset$. The second statement, plus Lemma 7.7.2 on the cartesian product with $\emptyset$, imply that the right-hand side of the Segal map is $\emptyset$, which is the same as the left-hand side by the first statement. So in this case, the Segal condition is automatic. Hence we may assume that $i_{j-1} \leq i_j$ for all $1 \leq j \leq p$.

Let $a := i_0$ and $b := i_p$, and denote by $(x_a, \ldots, x_b)$ the full sequence of adjacent elements (each counted once) going from x_a to x_b. There is a unique map $\sigma : (x_{i_0}, \ldots, x_{i_{p-1}}, x_{i_p}) \to (x_a, \ldots, x_b)$ in Δ_X, sending each x_{i_j} to the same object at the unique place it occurs in $(x_a, \ldots, x_b)$. Note that we are using here the reduction of the previous paragraph that $i_{j-1} \leq i_j$ for all $1 \leq j \leq p$, guaranteeing that each x_{i_j} occurs in the list $(x_a, \ldots, x_b)$. This σ induces a map

$$\sigma^* : \mathcal{A}'(x_a, \ldots, x_b) \xrightarrow{\sim} \mathcal{A}'(x_{i_0}, \ldots, x_{i_{p-1}}, x_{i_p}). \tag{16.2.4}$$

That this is a weak equivalence can be seen by considering the diagram with maps to the spanning pair $(x_a, x_b) = (x_{i_0}, x_{i_p})$:

$$\begin{array}{ccc} \mathcal{A}'(x_a, \ldots, x_b) & \to & \mathcal{A}'(x_{i_0}, \ldots, x_{i_{p-1}}, x_{i_p}) \\ \downarrow & & \downarrow \\ \mathcal{A}'(x_a, x_b) & = & \mathcal{A}'(x_{i_0}, x_{i_p}). \end{array}$$

The vertical maps are weak equivalences by the sequentially free condition, so the top map is a weak equivalence by 3 for 2.

On the other hand, for each pair in the original sequence, consider the Segal map for it:

$$\sigma_{i_0,i_1} : \mathcal{A}'(x_{i_0}, \ldots, x_{i_1}) \to \mathcal{A}'(x_{i_0}, x_{i_0+1}) \times \cdots \times \mathcal{A}'(x_{i_1-1}, x_{i_1}),$$

$$\vdots$$

$$\sigma_{i_{p-1},i_p} : \mathcal{A}'(x_{i_{p-1}}, \ldots, x_{i_p}) \to \mathcal{A}'(x_{i_{p-1}}, x_{i_{p-1}+1}) \times \cdots \times \mathcal{A}'(x_{i_p-1}, x_{i_p}).$$

The mini-sequences appearing on the left-hand sides are the sequences of length $i_j - i_{j-1}$ going from i_{j-1} to i_j by intervals of step 1. Whenever $i_{j-1} = i_j$, we have a sequence of length zero and both sides of the map are equal to $*$. The case of the Segal condition, which we already know, says that the σ_{i_{j-1},i_j} are weak equivalences.

Putting these all together, we get a weak equivalence whose target is the full Segal product for the sequence $(x_a, \ldots, x_b)$ considered above:

$$\mathcal{A}'(x_{i_0} \ldots x_{i_1}) \times \cdots \times \mathcal{A}'(x_{i_{p-1}} \cdots x_{i_p}) \xrightarrow{\sim} \mathcal{A}'(x_a, x_{a+1}) \times \cdots \times \mathcal{A}'(x_{b-1}, x_b).$$

On the other hand, the map to the spanning interval gives a map for each mini-sequence

$$\mathcal{A}'(x_{i_{j-1}}, \ldots, x_{i_j}) \xrightarrow{\sim} \mathcal{A}'(x_{i_{j-1}}, x_{i_j}), \tag{16.2.5}$$

which is a weak equivalence, by the condition that $\mathcal{A}'$ is sequentially free. The mini-sequences map into the full adjacent sequence $(x_a, \ldots, x_b)$, giving maps

$$\mathcal{A}'(x_a, \ldots, x_b) \to \mathcal{A}'(x_{i_{j-1}}, \ldots, x_{i_j}).$$

The Segal map for $(x_a, \ldots, x_b)$ thus factors as

$$\begin{array}{ccc} \mathcal{A}'(x_a, \ldots, x_b) \xrightarrow{\sim} & \mathcal{A}'(x_{i_0}, \ldots, x_{i_1}) \times \cdots \times \mathcal{A}'(x_{i_{p-1}}, \ldots, x_{i_p}) \\ & \downarrow \sim \\ & \mathcal{A}'(x_a, x_{a+1}) \times \cdots \times \mathcal{A}'(x_{b-1}, x_b). \end{array} \tag{16.2.6}$$

The second arrow is a weak equivalence as pointed out previously, and the composition is a weak equivalence by the Segal condition which we already know for the sequence $(x_a, \ldots, x_b)$, so the first map is a weak equivalence by 3 for 2.

Now we have a commutative diagram

$$\begin{array}{ccc} \mathcal{A}'(x_a, \ldots, x_b) & \longrightarrow & \mathcal{A}'(x_{i_0}, \ldots, x_{i_1}) \times \cdots \times \mathcal{A}'(x_{i_{p-1}}, \ldots, x_{i_p}) \\ \downarrow & & \downarrow \\ \mathcal{A}'(x_{i_0}, \ldots, x_{i_{p-1}}, x_{i_p}) & \longrightarrow & \mathcal{A}'(x_{i_0}, x_{i_1}) \times \cdots \times \mathcal{A}'(x_{i_{p-1}}, x_{i_p}) \end{array}$$

where the left vertical arrow is a weak equivalence as seen above (16.2.4), the top arrow is a weak equivalence by (16.2.6), and the right vertical arrow is a weak equivalence by combining together the equivalences (16.2.5). By 3 for 2, it follows that the bottom map is a weak equivalence, which is the Segal condition for the sequence $(x_{i_0}, \ldots, x_{i_{p-1}}, x_{i_p})$. This completes the proof of the lemma. □

Corollary 16.2.7 *Suppose $(X, \mathcal{A})$ is a sequentially free ordered $\mathcal{M}$-precategory, and let $r : (X, \mathcal{A}) \to (X, \mathcal{A}')$ be a fibrant replacement in either the projective, Reedy or injective model structure on $\mathcal{PC}(X, \mathcal{M})$ constructed in Theorems 12.1.1 and 12.3.2. Then $(X, \mathcal{A}')$ is a sequentially free ordered $\mathcal{M}$-precategory for the same order on X, and for any $0 < j \leq n$ the map $\mathcal{A}(x_{j-1}, x_j) \to \mathcal{A}'(x_{j-1}, x_j)$ is a weak equivalence.*

The same conclusions holds if, instead of a fibrant replacement, r is a weak equivalence to an object $\mathcal{A}'$ which satisfies the Segal conditions.

Proof Use either the projective, Reedy, or the injective structure (if it exists) in what follows. Note that the sequentially free condition is preserved by levelwise weak equivalences of unital diagrams on Δ^o_X.

By Lemma 16.2.6, there is a map $\mathcal{A} \to \mathcal{A}''$, local weak equivalence in $\mathcal{PC}(X, \mathcal{M})$, such that $\mathcal{A}''$ satisfies the Segal conditions and is still sequentially free. Consider furthermore a map $s : \mathcal{A}'' \to \mathcal{A}^3$ which is a trivial cofibration to a fibrant object in $\mathcal{PC}(X, \mathcal{M})$. Then $\mathcal{A}^3$ also satisfies the Segal conditions, so Lemma 12.1.3 says that s is a levelwise weak equivalence. It follows that $\mathcal{A}^3$ is again sequentially free.

Our different fibrant replacement $r : \mathcal{A} \to \mathcal{A}'$ is a trivial cofibration, so there is a map $g : \mathcal{A}' \to \mathcal{A}^3$ compatible with the maps from $\mathcal{A}$. By 3 for 2 this map g is a local weak equivalence in $\mathcal{PC}(X, \mathcal{M})$, so again by Lemma 12.1.3 it is a levelwise weak equivalence, hence $\mathcal{A}'$ is sequentially free. Furthermore, $\mathcal{A}(x_{j-1}, x_j) \to \mathcal{A}''(x_{j-1}, x_j)$ is a weak equivalence in $\mathcal{M}$; it follows from the above levelwise weak equivalences compatible with the maps from $\mathcal{A}$, that the same is true for $\mathcal{A}^3$, then $\mathcal{A}'$.

For the last paragraph, suppose $r : \mathcal{A} \to \mathcal{A}'$ is a local weak equivalence in $\mathcal{PC}(X, \mathcal{M})$ and $\mathcal{A}'$ satisfies the Segal conditions. Then choosing first a fibrant replacement $\mathcal{A}^3$ of $\mathcal{A}'$, then a factorization, we can get a square

$$\begin{array}{ccc} \mathcal{A} & \longrightarrow & \mathcal{A}' \\ \downarrow & & \downarrow \\ \mathcal{A}'' & \longrightarrow & \mathcal{A}^3 \end{array}$$

such that the vertical arrows are fibrant replacements and all arrows are local weak equivalences. By Lemma 12.1.3, the right vertical and bottom maps are levelwise weak equivalences since $\mathcal{A}'$ and $\mathcal{A}''$ satisfy the Segal conditions. By the first part of the present lemma, $\mathcal{A}''$ is sequentially free, it follows that $\mathcal{A}^3$

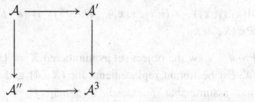

and then $\mathcal{A}'$ are sequentially free. On adjacent pairs of objects the left vertical map induces an equivalence, so by 3 for 2 the top map does too. This shows that the required statements hold for $\mathcal{A}'$. □

In the situation of the previous corollary, suppose $(x_{i_0}, \ldots, x_{i_n})$ is an increasing sequence of objects. Then we have weak equivalences

$$\mathcal{A}'(x_{i_0}, x_{i_n}) \xrightarrow{\sim} \mathcal{A}'(x_{i_0}, \ldots, x_{i_n}) \xrightarrow{\sim} \mathcal{A}'(x_{i_0}, x_{i_1}) \times \cdots \times \mathcal{A}'(x_{i_{n-1}}, x_{i_n}).$$

On the other hand, if the sequence is not increasing, the corresponding morphism space is $\emptyset$.

Suppose that our sequence of objects is a sequence of adjacent objects, that is to say look at a sequence of the form $(x_a, x_{a+1}, \ldots, x_b)$ for $a \leq b$. The condition of the corollary says furthermore that we have weak equivalences $\mathcal{A}(x_{j-1}, x_j) \xrightarrow{\sim} \mathcal{A}'(x_{j-1}, x_j)$ for $a < j \leq b$. These equivalences go together to give an equivalence on the level of the cartesian product, by the product conditions (PROD) and (DCL) for $\mathcal{M}$ as in Lemma 7.7.3. Thus, the pair of weak equivalences in the previous paragraph extends to a chain of equivalences

$$\begin{array}{ccc} \mathcal{A}'(x_a, \ldots, x_b) & \xrightarrow{\sim} & \mathcal{A}'(x_a, x_{a+1}) \times \cdots \times \mathcal{A}'(x_{b-1}, x_b) \\ \downarrow \wr & & \uparrow \wr \\ \mathcal{A}'(x_a, x_b) & & \mathcal{A}(x_a, x_{a+1}) \times \cdots \times \mathcal{A}(x_{b-1}, x_b). \end{array} \tag{16.2.7}$$

All in all, this says that $(X, \mathcal{A}')$ looks up to homotopy very much like the $\widetilde{\Upsilon}_k(B_1, \ldots, B_k)$ defined at the start, with $B_i = \mathcal{A}(x_{i-1}, x_i)$.

Corollary 16.2.8 *Suppose $f : (X, \mathcal{A}) \to (X, \mathcal{B})$ is a morphism in $\mathcal{PC}(X, \mathcal{M})$ between $\mathcal{M}$-precategories which are both sequentially free for the same ordering of the underlying set of objects X. Suppose that for any adjacent objects x_{j-1}, x_j in the ordering, f induces a weak equivalence $\mathcal{A}(x_{j-1}, x_j) \xrightarrow{\sim} \mathcal{B}(x_{j-1}, x_j)$. Then f is a local weak equivalence in $\mathcal{PC}(X, \mathcal{M})$.*

Proof View the object set as numbered $X = \{x_0, \ldots, x_k\}$. Let $(X, \mathcal{A}')$ and $(X, \mathcal{B}')$ be fibrant replacements for $(X, \mathcal{A})$ and $(X, \mathcal{B})$ respectively, and we may assume that f extends to a map $f' : (X, \mathcal{A}') \to (X, \mathcal{B}')$. For any $0 \leq a \leq b \leq k$, the maps f' and f induce a morphism between the chains of equivalences considered in (16.2.7), for $\mathcal{A}, \mathcal{A}'$ and $\mathcal{B}, \mathcal{B}'$. The hypothesis of the present corollary says that the induced map on the bottom right corner is a weak equivalence; thus the induced map $f' : \mathcal{A}'(x_a, x_b) \to \mathcal{B}'(x_a, x_b)$ is a weak equivalence. Note, on the other hand, that if $a > b$ then $\mathcal{A}'(x_a, x_b) = \emptyset$ and $\mathcal{B}'(x_a, x_b) = \emptyset$. Thus f' induces a weak equivalence on the morphism space for any pair of objects. By the Segal condition, it follows that f' is a levelwise weak equivalence in the category of Δ_X^o-diagrams in $\mathcal{M}$, so it is a local weak equivalence in the model structures. Since f' fits with f into a diagram of fibrant replacements for the source and target, this implies that f was a local weak equivalence. $\qquad\square$

Proof of Theorem 16.1.2 The map

$$f : \Upsilon_k(B_1, \ldots, B_k) \to \widetilde{\Upsilon}_k(B_1, \ldots, B_k)$$

satisfies the hyptheses of the previous corollary, so we conclude that it is a local weak equivalence. It is easy to see that f is a levelwise or injective cofibration, using the cartesian property of $\mathcal{M}$, the hypothesis that B_i are cofibrant, and Lemma 7.7.4 – so for the usual cases when Reedy and injective cofibrations coincide, this finishes the proof.

For the general case, it is also pretty straightforward to show that f is a Reedy cofibration. Introduce the notations

$$\mathcal{C} := \Upsilon_k(B_1, \ldots, B_k),$$

$$\widetilde{\mathcal{C}} := \widetilde{\Upsilon}_k(B_1, \ldots, B_k)$$

for brevity. Suppose $z. = (z_0, \ldots, z_p)$ is a sequence of objects with $z_j = \upsilon_{i_j}$. Recall from Lemma 13.1.2 that

$$\mathbf{d}(\mathcal{C}; z_0, \ldots, z_p) = \mathrm{colim}_{z. \to w.} \mathcal{C}(w_0, \ldots, w_q),$$

where the colimit is taken over surjective maps $z. \to w.$ to sequences of length $q < p$, and similarly for $\widetilde{\mathcal{C}}$. By Theorem 13.6.2 we have to show that the map (13.6.2)

$$\mathcal{C}(z_0, \ldots, z_p) \cup^{\mathbf{d}(\mathcal{C}; z_0, \ldots, z_p)} \mathbf{d}(\widetilde{\mathcal{C}}; z_0, \ldots, z_p) \to \widetilde{\mathcal{C}}(z_0, \ldots, z_p)$$

is a cofibration.

Notice that if $z.$ is anywhere decreasing, then all objects entering in here (including in the colimits defining the degenerate subobjects) are $\emptyset$ so the map in question is an isomorphism. Therefore we may assume that $z.$ is nondecreasing.

If $z.$ is a strictly increasing sequence, i.e. has no repetitions, then there are no surjections to shorter sequences $w.$ and the degenerate subobjects are trivial, both $\emptyset$. In this case, the map in question is just

$$\mathcal{C}(z_0, \ldots, z_p) \to \widetilde{\mathcal{C}}(z_0, \ldots, z_p).$$

If $p = 0$ we get the identity of $*$; if $p = 1$ and $i_0 = i_1 - 1$ then we get the identity of B_{i_0}; and if $i_0 < i_p - 1$ we get the inclusion

$$\emptyset \to B_{i_0+1} \times \cdots \times B_{i_p},$$

which is a cofibration, since the B_i are assumed cofibrant and $\mathcal{M}$ is cartesian (see Lemma 7.7.4).

The remaining case is when $z.$ is nondecreasing but has some repetitions. Let $y.$ be the sequence containing the same elements but with no repetitions. It is an initial object of the category of surjections used in the colimit defining the degenerate subobjects $\mathbf{d}(\mathcal{C}; z_0, \ldots, z_p)$ and $\mathbf{d}(\widetilde{\mathcal{C}}; z_0, \ldots, z_p)$. By the specific form of both $\mathcal{C}$ and $\widetilde{\mathcal{C}}$, the maps

$$\mathcal{C}(y.) \to \mathcal{C}(z.), \quad \widetilde{\mathcal{C}}(y.) \to \widetilde{\mathcal{C}}(z.)$$

are isomorphisms. Furthermore, for any other surjection which must factor $z. \to w. \to y.$, the induced maps

$$\mathcal{C}(y.) \to \mathcal{C}(w.), \quad \widetilde{\mathcal{C}}(y.) \to \widetilde{\mathcal{C}}(w.)$$

are again isomorphisms. This means that the colimits used to define the degenerate subobjects are colimits of constant diagrams indexed by a connected category. It follows from the cartesian condition on $\mathcal{M}$ that these colimits are equal to their constant values, indeed Lemma 8.1.9 says that for the constant diagram with values $*$, and the condition that the direct product commutes with colimits then gives it for any constant diagram. Thus, the maps

$$\mathbf{d}(\mathcal{C}; z_0, \ldots, z_p) \to \mathcal{C}(z_0, \ldots, z_p)$$

and

$$\mathbf{d}(\widetilde{\mathcal{C}}; z_0, \ldots, z_p) \to \widetilde{\mathcal{C}}(z_0, \ldots, z_p)$$

are isomorphisms. It follows that the map in question (13.6.2) is an isomorphism. This completes the verification that f is a Reedy cofibration using Theorem 13.6.2. □

17
Products

In this chapter, we consider the cartesian product of two $\mathcal{M}$-enriched precategories. On the one hand, we would like to maintain the cartesian or product condition (PROD) for the new model category we are constructing. On the other hand, compatibility with cartesian product provides the main technical tool we need in order to study pushouts by global weak equivalences in $\mathcal{PC}(\mathcal{M})$. For a fixed set of objects X, recall that we already have the projective (and sometimes injective) model structures of Theorem 12.1.1, as well as the Reedy structure $\mathcal{PC}_{\text{Reedy}}(X, \mathcal{M})$ of Theorem 12.3.2. We will mainly be using the Reedy structure.

The situation can be simpler to understand when $\mathcal{M}$ consists of presheaves over a connected category and the cofibrations are the monomorphisms; then Proposition 13.7.2 applies to say that the Reedy and injective structures are the same. That case would be sufficient for iterating our main construction $\mathcal{PC}$, a point of view which allows the reader to replace Reedy cofibrations by injective ones the first time through.

In this chapter is the place where we really use the full structure of the category Δ, as well as the unitality condition. At the end of the chapter, we'll discuss some counterexamples showing why these aspects are necessary.

17.1 Products of sequentially free precategories

Suppose $X = \{x_0, \ldots, x_m\}$ and $Y = \{y_0, \ldots, y_n\}$ are finite linearly ordered sets, and $(X, \mathcal{A})$ and $(Y, \mathcal{B})$ are sequentially free $\mathcal{M}$-precategories, with respect to the orderings. The product of these objects considered in the category $\mathcal{PC}(\mathcal{M})$, has the form $(X \times Y, \mathcal{A} \boxtimes \mathcal{B})$, where

$$(\mathcal{A} \boxtimes \mathcal{B})((x_{i_0}, y_{j_0}), \ldots, (x_{i_p}, y_{j_p}) := \mathcal{A}(x_{i_0}, \ldots, x_{i_p}) \times \mathcal{B}(y_{j_0}, \ldots, y_{j_p}).$$

Let $\mathcal{A} \to \mathcal{A}'$ and $\mathcal{B} \to \mathcal{B}'$ be local weak equivalences towards objects satisfying the Segal condition and which are therefore also sequentially free (Corollary 16.2.7). We obtain a map of $\mathcal{M}$-precategories $\mathcal{A} \boxtimes \mathcal{B} \to \mathcal{A}' \boxtimes \mathcal{B}'$ on the object set $X \times Y$. We would like to show that this is a local weak equivalence in $\mathcal{PC}(X \times Y, \mathcal{M})$.

For any subset $S \subset X \times Y$, let $\mathbf{i}(S) : S \hookrightarrow X \times Y$ denote the inclusion map, and

$$\mathbf{i}(S)^* : \mathcal{PC}(X \times Y, \mathcal{M}) \to \mathcal{PC}(S, \mathcal{M})$$

the pullback map on $\mathcal{M}$-precategory structures. This can be pushed back to a $\mathcal{M}$-precategory structure with object set $X \times Y$, and the resulting functor will be denoted as $\mathbf{i}(S)^*_!$ for brevity:

$$\mathbf{i}(S)^*_! := \mathbf{i}(S)_! \mathbf{i}(S)^* : \mathcal{PC}(X \times Y, \mathcal{M}) \to \mathcal{PC}(X \times Y, \mathcal{M}).$$

This may be described explicitly as follows. Suppose $\mathcal{C} \in \mathcal{PC}(X \times Y, \mathcal{M})$. Recall from the first paragraph of Section 10.3, that $\mathbf{i}(S)_! \mathbf{i}(S)^*(\mathcal{C})$ is the same as the precategory $\mathbf{i}(S)^*(\mathcal{C})$ over set of objects S, extended by adding on the discrete $\mathcal{M}$-enriched category on the complementary set $X \times Y - S$. Thus, for any sequence of objects $((x_{i_0}, y_{i_0}), \ldots, (x_{i_p}, y_{i_p}))$ we have

$$\mathbf{i}(S)^*_!(\mathcal{C})((x_{i_0}, y_{i_0}), \ldots, (x_{i_p}, y_{i_p})) = \mathcal{C}((x_{i_0}, y_{i_0}), \ldots, (x_{i_p}, y_{i_p}))$$

if all the (x_{i_j}, y_{i_j}) are in S; if any of the pairs is not in S and the sequence is not constant then

$$\mathbf{i}(S)^*_!(\mathcal{C})((x_{i_0}, y_{i_0}), \ldots, (x_{i_p}, y_{i_p})) = \emptyset,$$

while the unitality condition requires that

$$\mathbf{i}(S)^*_!(\mathcal{C})((x_{i_0}, y_{i_0}), \ldots, (x_{i_0}, y_{i_0})) = *,$$

for a constant sequence at any point $(x_{i_0}, y_{i_0}) \in X \times Y$.

Lemma 17.1.1 *If $\mathcal{C}$ satisfies the Segal condition, then so does $\mathbf{i}(S)^*(\mathcal{C})$ (on object set S) and $\mathbf{i}(S)^*_!(\mathcal{C})$ (on object set $X \times Y$).*

If $\mathcal{C} \xrightarrow{f} \mathcal{C}'$ is a Reedy cofibration in $\mathcal{PC}(X \times Y, \mathcal{M})$ then $\mathbf{i}(S)^*(f)$ is a Reedy cofibration in $\mathcal{PC}(S, \mathcal{M})$ and $\mathbf{i}(S)^*_!(f)$ is a Reedy cofibration back in $\mathcal{PC}(X \times Y, \mathcal{M})$. Furthermore, the functor $\mathbf{i}(S)^*_!$ preserves amalgamated sums and more generally colimits over connected index categories.

Proof The first part follows immediately from Lemma 10.3.1. The second part may be seen from the explicit descriptions, using the characterisations of Theorem 13.6.2 and Corollary 13.6.4. For the statement on colimits, use Lemma 8.1.9. □

The above discussion is usually applied to an object of the form $\mathcal{C} = \mathcal{A} \boxtimes \mathcal{B} \in \mathcal{PC}(X \times Y, \mathcal{M})$.

Here are the special types of subsets S we will consider. A *box* will be a subset of the form

$$\mathbf{B}_{a,b} := \{x_0, \ldots, x_a\} \times \{y_0, \ldots, y_b\} \subset X \times Y,$$

which can be pictured for example with $(a, b) = (5, 3)$ as

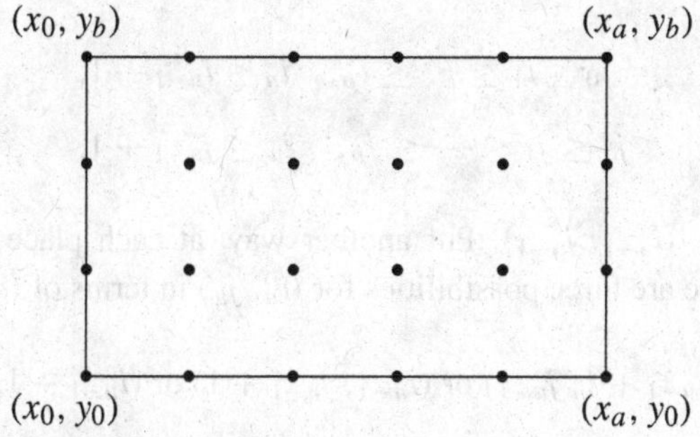

A *notched box* is a subset

$$\mathbf{B}_{a,b} \supset \mathbf{B}^{\nu}_{a,b} := \mathbf{B}_{a,b-1} \cup \{(x_\nu, y_b), \ldots, (x_a, y_b)\},$$

defined when $b \geq 1$. Thus, a notched box $\mathbf{B}^{\nu}_{a,b}$ is a box $\mathbf{B}_{a,b}$ minus a part of the top row $\{(x_0, y_b), \ldots, (x_{\nu-1}, y_b)\}$. For example, a notched box with $(a, b) = (5, 3)$ and $\nu = 3$ looks like

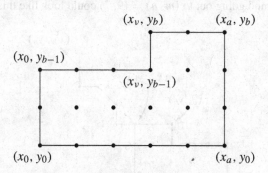

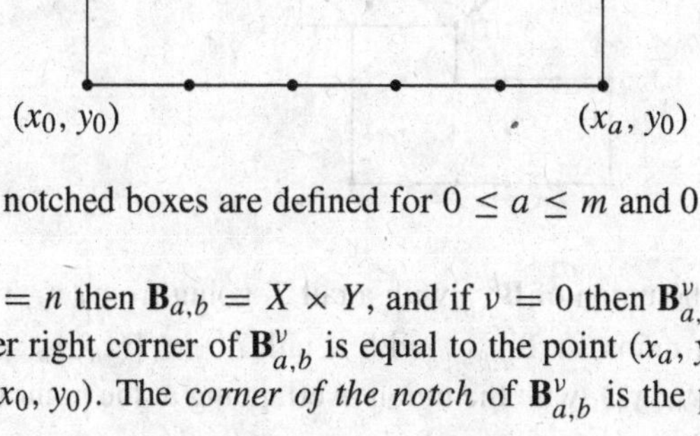

The boxes and notched boxes are defined for $0 \leq a \leq m$ and $0 \leq b \leq n$, and $0 \leq \nu \leq a$.

If $a = m$, $b = n$ then $\mathbf{B}_{a,b} = X \times Y$, and if $\nu = 0$ then $\mathbf{B}^{\nu}_{a,b} = \mathbf{B}_{a,b}$. For any ν, the upper right corner of $\mathbf{B}^{\nu}_{a,b}$ is equal to the point (x_a, y_b). The lower left corner is (x_0, y_0). The *corner of the notch* of $\mathbf{B}^{\nu}_{a,b}$ is the point (x_ν, y_b). Note that

$$\mathbf{B}^{\nu}_{a,b} - \{(x_\nu, y_b)\} = \mathbf{B}^{\nu+1}_{a,b},$$

whenever $0 \le \nu < a$. For $\nu = a$ we have $\mathbf{B}^{\nu}_{a,b} - \{(x_\nu, y_b)\} = \mathbf{B}_{a,b-1}$.

A *tail* is a subset of the form

$$T^{i_0,\ldots,i_p}_{j_0,\ldots,j_p} := \{(x_{i_0}, y_{j_0}), \ldots, (x_{i_p}, y_{j_p})\},$$

which will be considered whenever

$$i_0 \le i_1 \le \cdots \le i_p, \quad i_u \le i_{u-1} + 1,$$

$$j_0 \le j_1 \le \cdots \le j_p, \quad j_u \le j_{u-1} + 1,$$

and $(i_u, j_u) \neq (i_{u-1}, j_{u-1})$. Put another way, at each place in the pair of sequences, there are three possibilities for (i_u, j_u) in terms of (i_{u-1}, j_{u-1}):

$$(i_u, j_u) = (i_{u-1} + 1, j_{u-1}) \text{ or } (i_{u-1}, j_{u-1} + 1) \text{ or } (i_{u-1} + 1, j_{u-1} + 1).$$

Thus, the sequence moves in single horizontal, vertical or diagonal steps. We say that the tail *goes from* (x_{i_0}, y_{j_0}) *to* (x_{i_p}, y_{j_p}). A *full tail* is one which goes from (x_0, y_0) to (x_m, y_n).

A *dipper* is a subset S which is the union of a notched box $\mathbf{B}^{\nu}_{a,b}$, with a tail $T^{i_0,\ldots,i_p}_{j_0,\ldots,j_p}$ going from $(x_{i_0}, y_{j_0}) = (x_a, y_b)$ to $(x_{i_p}, y_{j_p}) = (x_m, y_n)$. These overlap at the point (x_a, y_b). For example, a dipper starting with the previous notched box, and going out to $(m, n) = (9, 7)$ could look like this:

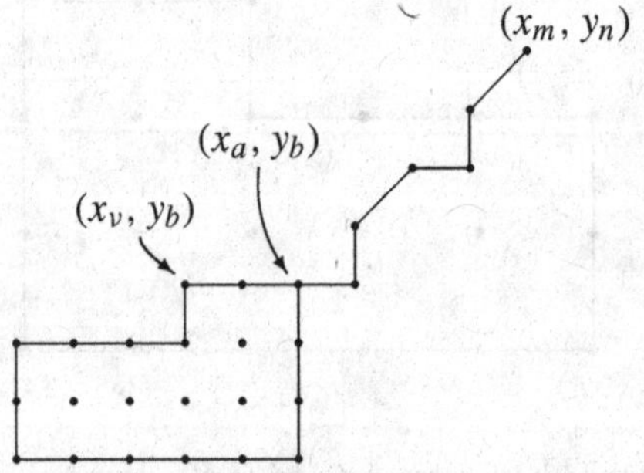

For $b \ge 1$, the union of $\mathbf{B}^{b}_{a,b}$ with a tail T going from (x_a, y_b) to (x_m, y_n), is equal to the union of $\mathbf{B}^{0}_{a,b-1}$ with a tail $T' = \{(x_a, x_{b-1})\}$ going from (x_a, x_{b-1}) to (x_m, y_n), with the union overlapping at the point (x_a, x_{b-1}). For example, in the previous picture if we set $\nu = b$ it becomes

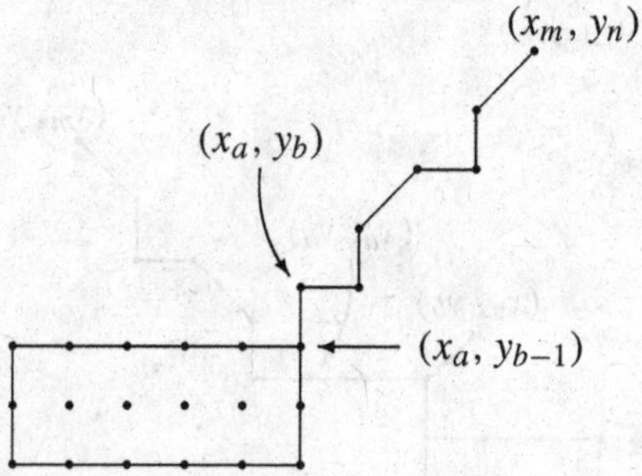

In this situation if furthermore $b - 1 = 0$, then the subset S becomes a full tail going from (x_0, y_0) to (x_m, y_n). Note also that a box with $a = 0$ plus a tail, is again equal to a full tail.

In view of this, we consider the variables a, b, ν for the notched box in a dipper S, only in the range $0 \leq \nu < a$ and $0 < b$. The corner of the notch (x_ν, y_b) is well defined, and $S - \{(x_\nu, y_b)\}$ is again either a dipper, or a full tail.

The product of the linear orders on X and Y is a partial order on $X \times Y$, where $(x_a, y_b) < (x_{a'}, y_{b'})$ whenever $a \leq a'$, $b \leq b'$ and $(a, b) \neq (a', b')$. Given a dipper S, let (x_ν, y_b) be the corner of the notch and define new subsets

$$S_{>(x_\nu, y_b)<} := \{(x', y') \in S, \text{ s.t. } (x', y') < (x_\nu, y_b) \text{ or } (x_\nu, y_b) < (x', y')\}$$

and

$$S_{\geq(x_\nu, y_b)\leq} := \{(x', y') \in S, \text{ s.t. } (x', y') \leq (x_\nu, y_b) \text{ or } (x_\nu, y_b) \geq (x', y')\}.$$

Note that

$$S_{\geq(x_\nu, y_b)\leq} = S_{>(x_\nu, y_b)<} \cup \{(x_\nu, y_b)\}.$$

For example, for the dipper pictured previously with $a = 5, b = 3, \nu = 3$, $m = 9, n = 7$, we have the picture for $S_{>(x_\nu, y_b)<}$

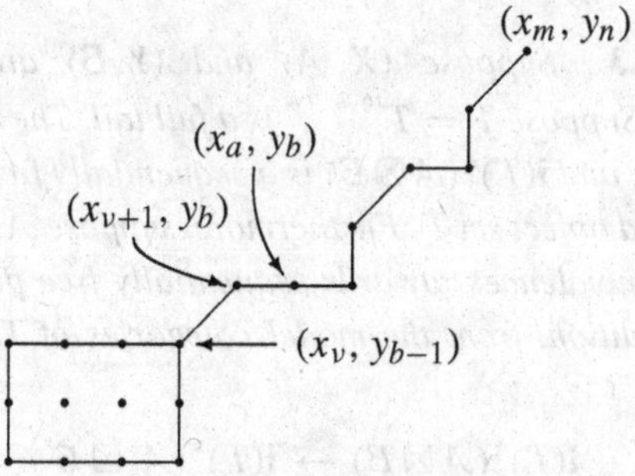

and the picture for $S_{\geq(x_\nu,y_b)\leq}$

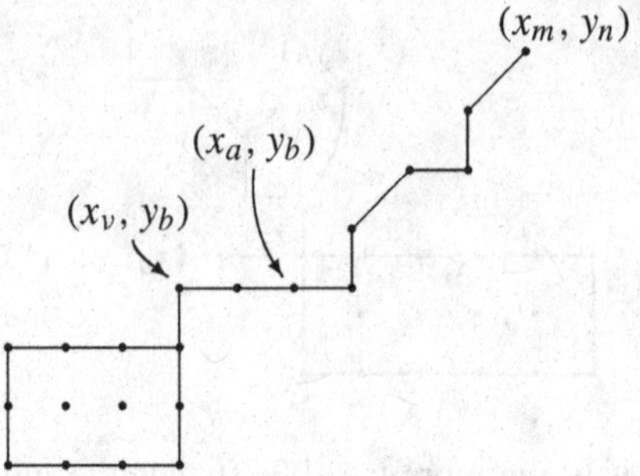

Lemma 17.1.2 *If S is a dipper with notched box $\mathbf{B}^{\nu}_{a,b}$ for $0 \leq \nu < a$ and $0 < b$, then either $\nu > 0$ and $b > 1$ in which case $S_{>(x_\nu,y_b)<}$ and $S_{\geq(x_\nu,y_b)\leq}$ are dippers with notched box $\mathbf{B}^{0}_{\nu,b-1}$ and tails going from (x_ν, y_{b-1}) to (x_m, y_n); or else either $\nu = 0$ or $b = 1$ in which case $S_{>(x_\nu,y_b)<}$ and $S_{\geq(x_\nu,y_b)\leq}$ are full tails.*

Proof Look at the above pictures. □

The idea of the proof of the product property is to consider subsets S which are dippers or full tails, and prove that $\mathbf{i}(S)^*(\mathcal{A} \boxtimes \mathcal{B}) \to \mathbf{i}(S)^*(\mathcal{A}' \boxtimes \mathcal{B}')$ is a local weak equivalence in $\mathcal{PC}(S, \mathcal{M})$.

We first discuss the case of a full tail. Then the case of dippers will be treated by induction on a, b, ν, eventually getting to the case $a = m, b = n$, which gives the desired theorem. The case of full tails, treated in the following proposition, encloses the case of all increasing single-step paths going from $(0, 0)$ to (m, n). This formalizes the intuition that to understand the product we should understand what happens on each path. This is similar to what is going on in the decomposition of a product of simplices: the product has a decomposition into simplices indexed by the same collection of paths–although the simplices of maximal dimension correspond to paths without diagonal steps.

Proposition 17.1.3 *Suppose $(X, \mathcal{A})$ and $(Y, \mathcal{B})$ are sequentially free $\mathcal{M}$-precategories. Suppose $T = T^{i_0,\ldots,i_p}_{j_0,\ldots,j_p}$ is a full tail. The induced ordering on T is a linear order, and $\mathbf{i}(T)^*(\mathcal{A} \boxtimes \mathcal{B})$ is a sequentially free $\mathcal{M}$-precategory on the linearly ordered object set T. Furthermore, suppose $\mathcal{A} \to \mathcal{A}'$ and $\mathcal{B} \to \mathcal{B}'$ are local weak equivalences towards sequentially free precategories satisfying the Segal condition, from the model categories of Theorems 12.1.1 and 12.3.2. Then*

$$\mathbf{i}(T)^*(\mathcal{A} \boxtimes \mathcal{B}) \to \mathbf{i}(T)^*(\mathcal{A}' \boxtimes \mathcal{B}')$$

is a local weak equivalence whose target satisfies the Segal condition. The same is also true of $\mathbf{i}(T)^*_!(\mathcal{A} \boxtimes \mathcal{B})$ *whose object set is* $X \times Y$.

Proof An increasing sequence of objects of T has the form $(z_0, \ldots, z_r)$ where $z_k = (x_{u(k)}, y_{v(k)})$ with $u(k) = i_{a(k)}$ and $v(k) = j_{a(k)}$ for $0 \leq a(0) \leq \ldots \leq a(r) \leq p$ an increasing sequence in the set of indices for the objects of T. The tail condition means that for any such increasing sequence, the sequences $x_{u(0)}, \ldots, x_{u(r)}$ and $y_{u(0)}, \ldots, y_{u(r)}$ are increasing sequences in X and Y respectively. Now

$$\mathbf{i}(T)^*(\mathcal{A} \boxtimes \mathcal{B})(z_0, \ldots, z_r) = \mathcal{A}(x_{u(0)}, \ldots, x_{u(r)}) \times \mathcal{B}(y_{u(0)}, \ldots, y_{u(r)}),$$

whereas

$$\mathbf{i}(T)^*(\mathcal{A} \boxtimes \mathcal{B})(z_0, z_r) = \mathcal{A}(x_{u(0)}, x_{u(r)}) \times \mathcal{B}(y_{u(0)}, y_{u(r)})$$

so the fact that $\mathcal{A}$ and $\mathcal{B}$ are sequentially free implies, via the cartesian condition for $\mathcal{M}$, that the map

$$\mathbf{i}(T)^*(\mathcal{A} \boxtimes \mathcal{B})(z_0, \ldots, z_r) \to \mathbf{i}(T)^*(\mathcal{A} \boxtimes \mathcal{B})(z_0, z_r)$$

is a weak equivalence.

Suppose $\mathcal{A} \to \mathcal{A}'$ and $\mathcal{B} \to \mathcal{B}'$ are local weak equivalences towards sequentially free precategories satisfying the Segal condition. It follows from Corollary 16.2.7 that, for any adjacent pair of objects $x_{i-1}, x_i \in X$, the map

$$\mathcal{A}(x_{i-1}, x_i) \to \mathcal{A}'(x_{i-1}, x_i)$$

is a weak equivalence. Sequential freeness implies that $\mathcal{A}(x_i, x_i)$ and $\mathcal{A}'(x_i, x_i)$ are contractible (see Remark 16.2.2). In particular, for any object x_i the map

$$\mathcal{A}(x_i, x_i) \to \mathcal{A}'(x_i, x_i)$$

is a weak equivalence. The same two statements hold for $\mathcal{B}$. But now the tail property of T says that an adjacent pair of objects in T is of the form $(x_i, y_j), (x_k, y_l)$, where x_i, x_k are either adjacent objects or the same object in X, and y_j, y_l are either adjacent objects or the same object in Y. It follows that the map

$$\mathcal{A}(x_i, x_i) \times \mathcal{B}(y_j, y_l) \to \mathcal{A}'(x_i, x_i) \times \mathcal{B}'(y_j, y_l)$$

is a weak equivalence. By Corollary 16.2.8, the map

$$\mathbf{i}(T)^*(\mathcal{A} \boxtimes \mathcal{B}) \to \mathbf{i}(T)^*(\mathcal{A}' \boxtimes \mathcal{B}')$$

is a local weak equivalence. The set of objects of T injects into $X \times Y$, so upon pushing forward to precategories on object set $X \times Y$, again

$$\mathbf{i}(T)_!^*(\mathcal{A} \boxtimes \mathcal{B}) \to \mathbf{i}(T)_!^*(\mathcal{A}' \boxtimes \mathcal{B}')$$

is a local weak equivalence. □

The next step is to note that when we remove the corner of the notch from a dipper S we have a pushout diagram. In order to give the statement with some generality, say that $\mathcal{C} \in \mathcal{PC}(X \times Y, \mathcal{M})$ is *ordered* if $\mathcal{C}((x_{i_0}, y_{j_0}), \ldots, (x_{i_p}, y_{j_p})) = \emptyset$ whenever we have a nonincreasing sequence $(x_{i_0}, y_{j_0}), \ldots, (x_{i_p}, y_{j_p})$ in the product order, that is to say whenever there is some k such that either $i_{k-1} > i_k$ or $j_{k-1} > j_k$. Let $\mathcal{PC}^{\mathrm{ord}}(X \times Y, \mathcal{M})$ denote the category of ordered $\mathcal{M}$-enriched precategories over $X \times Y$.

Proposition 17.1.4 *Suppose $S = \mathbf{B}^{v}_{a,b} \cup T$ is a dipper and (x_v, y_b) the corner of the notch. Assume $0 \le v < a$ and $0 < b$. Then the pushout expression for subsets of $X \times Y$*

$$S = (S - \{(x_v, y_b)\}) \cup^{S_{>(x_v,y_b)<}} S_{\geq(x_v,y_b)\leq}$$

extends to a pushout expression for any ordered $\mathcal{C} \in \mathcal{PC}^{\mathrm{ord}}(X \times Y, \mathcal{M})$: the square

$$\begin{array}{ccc} \mathbf{i}(S_{>(x_v,y_b)<})_!^*(\mathcal{C}) & \longrightarrow & \mathbf{i}(S_{\geq(x_v,y_b)\leq})_!^*(\mathcal{C}) \\ \downarrow & & \downarrow \\ \mathbf{i}(S - \{(x_v, y_b)\})_!^*(\mathcal{C}) & \longrightarrow & \mathbf{i}(S)_!^*(\mathcal{C}) \end{array}$$

is a pushout square in the category $\mathcal{PC}(X \times Y, \mathcal{M})$.

Proof It suffices to verify this levelwise on $\Delta_{X\times Y}$, in other words we have to verify it for every sequence of objects $(x_{i_0}, y_{i_0}), \ldots, (x_{i_p}, y_{i_p})$. If either of $x.$ or $y.$ is not increasing then it is trivially true, so we may assume that both are increasing. If none of the elements in the sequence are (x_v, y_b) then it is again trivially true. So we may assume that there is a j with $(x_{i_j}, y_{i_j}) = (x_v, y_b)$. Then the full sequence is contained in the region $S_{\geq(x_v,y_b)\leq}$. It follows that both vertical maps in the above diagram, at the level of the sequence $(x., y.)$, are isomorphisms. This implies that the diagram is a pushout. □

Looking at this before putting everything back onto the same object set $X \times Y$, one can also say that the square

$$\begin{array}{ccc} \mathbf{i}(S_{>(x_v,y_b)<})^*(\mathcal{C}) & \longrightarrow & \mathbf{i}(S_{\geq(x_v,y_b)\leq})^*(\mathcal{C}) \\ \downarrow & & \downarrow \\ \mathbf{i}(S - \{(x_v, y_b)\})^*(\mathcal{C}) & \longrightarrow & \mathbf{i}(S)^*(\mathcal{C}) \end{array}$$

is a pushout square in $\mathcal{PC}(\mathcal{M})$. This version of the statement perhaps explains better what is going on, but is less useful to us since we are currently working in the model category structure on $\mathcal{PC}(X \times Y, \mathcal{M})$ for a fixed set of objects $X \times Y$.

Lemma 17.1.5 *Suppose $S \subset S' \subset X \times Y$. If $\mathcal{C}$ is a cofibrant object in $\mathcal{PC}_{\text{Reedy}}(X \times Y, \mathcal{M})$ then the map*

$$g : \mathbf{i}(S)^*_!(\mathcal{C}) \to \mathbf{i}(S')^*_!(\mathcal{C})$$

is a cofibration in $\mathcal{PC}_{\text{Reedy}}(X \times Y, \mathcal{M})$.

Proof For simplicity look first at the argument in the situation where $\mathcal{M}$ is a category of presheaves over a connected category, with monomorphisms as cofibrations. Thus Proposition 13.7.2 says that the Reedy and injective cofibrations coincide. Notice also that all objects are cofibrant, and for any $z \in X \times Y$ the map $* \to \mathcal{C}(z, z, \ldots, z)$ is a cofibration. On any sequence $(z_0, \ldots, z_p)$ of points in $X \times Y$, the map g is either the identity, the inclusion from $\emptyset$ to $\mathcal{C}(z_0, \ldots, z_p)$, or the inclusion from $*$ to $\mathcal{C}(z, \ldots, z)$ in case of a constant sequence at a point $z \in S' - S$. It follows that g is an injective, hence Reedy cofibration.

Consider now the case where $\mathcal{M}$ is a general tractable left proper cartesian model category and we use the Reedy structure $\mathcal{PC}_{\text{Reedy}}(X \times Y, \mathcal{M})$. If $\mathcal{C}$ is a Reedy cofibrant object, then $\mathbf{disc}(X \times Y) \to \mathcal{C}$ is a Reedy cofibration (see part (e) of Corollary 13.6.4). In order to apply Theorem 13.6.2, look at the map (13.6.2) for

$$g : \mathbf{i}(S)^*_!(\mathcal{C}) \to \mathbf{i}(S')^*_!(\mathcal{C})$$

at some sequence of objects $(z_0, \ldots, z_p)$ in $X \times Y$. There are several cases: either the sequence is entirely in S, or entirely in S' but not entirely in S; or not entirely in S' either. It could also be constant, or nonconstant. This makes for six cases altogether. Depending on which case we are in, the map $g(z_0, \ldots, g_p)$ is either the identity of $\mathcal{C}(z_0, \ldots, z_p)$, the identity of $*$, the identity of $\emptyset$, the inclusion $\emptyset \to \mathcal{C}(z_0, \ldots, z_p)$, or (only in the case of a constant sequence at a point of $S' - S$), the inclusion $* \to \mathcal{C}(z, \ldots, z)$. In all cases, the map induced by g is the same as that induced either by the identity of $\mathcal{C}$, the identity of $\mathbf{disc}(X \times Y)$, or the inclusion $\mathbf{disc}(X \times Y) \to \mathcal{C}$. Furthermore, if $z. \to w.$ is any surjection of sequences, then $w.$ falls into the same case as $z.$. Apply the criterion of part (c) of Theorem 13.6.2. At a given sequence of points, the map (13.6.2) in question will be the same as the map corresponding either to the identity of $\mathcal{C}$, the identity of $\mathbf{disc}(X \times Y)$, or the inclusion $\mathbf{disc}(X \times Y) \to \mathcal{C}$. This is because the surjections $z. \to w.$ appearing in the index categories for the colimits defining the degenerate objects

$$\mathbf{d}(\mathbf{i}(S)^*_!(\mathcal{C}); z_0, \ldots, z_p), \quad \mathbf{d}(\mathbf{i}(S)^*_!(\mathcal{C}); z_0, \ldots, z_p)$$

fall into the same case as $z.$. Now the identity of $\mathcal{C}$, the identity of $\mathbf{disc}(X \times Y)$ and the inclusion $\mathbf{disc}(X \times Y) \to \mathcal{C}$ are all Reedy cofibrations, so in every case the map (13.6.2) in criterion (c) for g, is a cofibration. Therefore g is a Reedy cofibration. □

Corollary 17.1.6 *Suppose $g : \mathcal{C} \to \mathcal{C}'$ is a map of ordered $\mathcal{M}$-enriched precategories over object set $X \times Y$, such that both $\mathcal{C}$ and $\mathcal{C}'$ are Reedy cofibrant, that is cofibrant in $\mathcal{PC}_{\text{Reedy}}(X \times Y, \mathcal{M})$. Suppose that for any full tail T going from $(0, 0)$ to (m, n) the map*

$$\mathbf{i}(T)^*_!(\mathcal{C}) \to \mathbf{i}(T)^*_!(\mathcal{C}')$$

is a local weak equivalence. Then g is a local weak equivalence.

Proof We show by induction that for any dipper of the form $S = \mathbf{B}^{\nu}_{a,b} \cup T$, the map

$$\mathbf{i}(S)^*_!(\mathcal{C}) \to \mathbf{i}(S)^*_!(\mathcal{C}')$$

is a local weak equivalence. The induction is by $(b, a - \nu)$ in lexicographic order. In the initial case $b = 1$ and $\nu = a$, S is a tail so the inductive statement is one of the hypotheses. Suppose the statement is known for all dippers S' corresponding to (a', b', ν') with $b' < b$ or $b' = b$ and $a' - \nu' < a - \nu$. Use the expression of Proposition 17.1.4: $\mathbf{i}(S)^*_!(\mathcal{C})$ and $\mathbf{i}(S)^*_!(\mathcal{C}')$ are respectively the pushouts of the top and bottom rows in the diagram

$$\begin{array}{ccccc}
\mathbf{i}(S - \{(x_\nu, y_b)\})^*_!(\mathcal{C}) & \leftarrow & \mathbf{i}(S_{>(x_\nu, y_b)<})^*_!(\mathcal{C}) & \rightarrow & \mathbf{i}(S_{\geq(x_\nu, y_b)\leq})^*_!(\mathcal{C}) \\
\downarrow & & \downarrow & & \downarrow \\
\mathbf{i}(S - \{(x_\nu, y_b)\})^*_!(\mathcal{C}') & \leftarrow & \mathbf{i}(S_{>(x_\nu, y_b)<})^*_!(\mathcal{C}') & \rightarrow & \mathbf{i}(S_{\geq(x_\nu, y_b)\leq})^*_!(\mathcal{C}').
\end{array}$$

The vertical maps in the diagram induce, between the pushouts, the given map $\mathbf{i}(S)^*_!(\mathcal{C}) \to \mathbf{i}(S)^*_!(\mathcal{C}')$.

The horizontal maps are Reedy cofibrations by the preceding lemma.

The inductive hypothesis applies to each of the vertical maps. This is seen by noting that the invariants $(b', a' - v')$ for the dippers $S - \{(x_v, y_b)\}$, $S_{>(x_v, y_b)<}$ and $S_{\geq(x_v, y_b)\leq}$ are strictly smaller than $(b, a - v)$ in lexicographic order:

- for $S - \{(x_v, y_b)\}$ we have $a' = a$, $v' = v + 1$ and $b' = b$, unless $v = a - 1$ in which case $b' = b - 1$;
- for $S_{>(x_v, y_b)<}$ and $S_{\geq(x_v, y_b)\leq}$ we have $b' = b - 1$.

Hence, by the inductive hypothesis, each of the vertical maps is a local weak equivalence. Now $\mathcal{PC}_{\text{Reedy}}(X \times Y, \mathcal{M})$ is left proper because it is a left Bousfield localization cf. Theorem 12.3.2. This implies by Corollary 7.3.2 that a map of pushouts along cofibrations, which is induced by a triple of local weak equivalences, is a local weak equivalence (see alternatively Lemma 14.3.4). Therefore the map on pushouts $\mathbf{i}(S)^*_!(\mathcal{C}) \to \mathbf{i}(S)^*_!(\mathcal{C}')$ is a local weak equivalence. This completes the inductive proof.

At the last case $S = X \times Y$ we obtain the conclusion of the corollary, that $g : \mathcal{C} \to \mathcal{C}'$ is a local weak equivalence. □

This gives the first main result of this chapter:

Theorem 17.1.7 *Suppose $(X, \mathcal{A})$ and $(Y, \mathcal{B})$ are sequentially free $\mathcal{M}$-precategories, cofibrant in $\mathcal{PC}_{\text{Reedy}}(X \times Y, \mathcal{M})$. Suppose $\mathcal{A} \to \mathcal{A}'$ and $\mathcal{B} \to \mathcal{B}'$ are trivial cofibrations in $\mathcal{PC}_{\text{Reedy}}(X, \mathcal{M})$ and $\mathcal{PC}_{\text{Reedy}}(Y, \mathcal{M})$ respectively, towards sequentially free $\mathcal{M}$-precategories. Then the map*

$$(X \times Y, \mathcal{A} \boxtimes \mathcal{B}) \to (X \times Y, \mathcal{A}' \boxtimes \mathcal{B}')$$

is a trivial cofibration in $\mathcal{PC}_{\text{Reedy}}(X \times Y, \mathcal{M})$.

Proof Note first of all that the map is a Reedy cofibration, by Corollary 13.6.13. To show that it is a local weak equivalence, suppose first that $\mathcal{A}'$ and $\mathcal{B}'$ satisfy the Segal condition. In this case, Corollary 17.1.6 applies with $\mathcal{C} = \mathcal{A} \boxtimes \mathcal{B}$ and $\mathcal{C}' = \mathcal{A}' \boxtimes \mathcal{B}'$. The hypothesis on full tails used in Corollary 17.1.6 is provided by Proposition 17.1.3.

If $\mathcal{A}'$ and $\mathcal{B}'$ do not themselves satisfy the Segal condition, we can choose further morphisms $\mathcal{A}' \to \mathcal{A}''$ and $\mathcal{B}' \to \mathcal{B}''$ which are trivial cofibrations in $\mathcal{PC}_{\text{Reedy}}(X, \mathcal{M})$ and $\mathcal{PC}_{\text{Reedy}}(Y, \mathcal{M})$ respectively, such that $\mathcal{A}''$ and $\mathcal{B}''$ satisfy the Segal condition. The first case of this proof then applies to the maps $\mathcal{A} \to \mathcal{A}''$ and $\mathcal{B} \to \mathcal{B}''$, and also to the maps $\mathcal{A}' \to \mathcal{A}''$ and $\mathcal{B}' \to \mathcal{B}''$. These show that

$$(X \times Y, \mathcal{A} \boxtimes \mathcal{B}) \to (X \times Y, \mathcal{A}'' \boxtimes \mathcal{B}'')$$

and

$$(X \times Y, \mathcal{A}' \boxtimes \mathcal{B}') \to (X \times Y, \mathcal{A}'' \boxtimes \mathcal{B}'')$$

are local weak equivalences. By 3 for 2 it follows that

$$(X \times Y, \mathcal{A} \boxtimes \mathcal{B}) \to (X \times Y, \mathcal{A}' \boxtimes \mathcal{B}')$$

is a local weak equivalence, thus it is a trivial cofibration in $\mathcal{PC}_{\text{Reedy}}(X \times Y, \mathcal{M})$. □

17.2 Products of general precategories

The next step is to extend the result of Theorem 17.1.7 from the sequentially free case, to the product of arbitrary $\mathcal{M}$-enriched precategories.

Recall that we have defined morphisms

$$\Upsilon_k(B_1, \ldots, B_k) \to \widetilde{\Upsilon}_k(B_1, \ldots, B_k)$$

of sequentially free $\mathcal{M}$-precategories, the target satisfies the Segal conditions, and the morphism is a local weak equivalence in $\mathcal{PC}([k], \mathcal{M})$ by Theorem 16.1.2. Use the notation

$$\Sigma([k]; B) := \Upsilon_k(B, \ldots, B),$$

where the same object B occurs k times.

Lemma 17.2.1 *In the case where $B_1 = \ldots = B_k = B \in \mathcal{M}$, the map $\Upsilon \to \widetilde{\Upsilon}$ factors as*

$$\Sigma([k]; B) \to h([k]; B) \to \widetilde{\Upsilon}_k(B, \ldots, B),$$

and both maps are global weak equivalences between sequentially free $\mathcal{M}$-enriched precategories.

Proof All three are sequentially free, and Corollary 16.2.8 applies. For $\Sigma([k]; B)$ and $\widetilde{\Upsilon}_k(B, \ldots, B)$ this is exactly what we said in the proof of Theorem 16.1.2; for $h([k]; B)$ see the explicit description in Section 10.5. □

Proposition 17.2.2 *Suppose $\mathcal{A} \in \mathcal{PC}(\mathcal{M})$. Then we can obtain a global weak equivalence $\mathcal{A} \to \mathcal{A}'$ such that $\mathcal{A}'$ satisfies the Segal conditions, i.e. $\mathcal{A}' \in \mathcal{R}$, by taking a transfinite composition of pushouts along morphisms of the form*

$$\Sigma([k]; V) \cup^{\Sigma([k];U)} h([k]; U) \to h([k]; V) \tag{17.2.1}$$

for generating cofibrations $U \xrightarrow{f} V$ in $\mathcal{M}$.

Proof The maps (17.2.1) are the same as the $\Psi([k], f)$ considered in Corollary 14.2.5 and which make up the new pieces in K_{Reedy}. Note that this collection is missing the piece of K_{Reedy} consisting of the generators for levelwise trivial Reedy cofibrations. However, for the present statement that piece is not needed: if we apply the small object argument to the present collection of morphisms we can obtain a map $\mathcal{A} \to \mathcal{A}'$ which is a transfinite composition of pushouts along morphisms of the form (17.2.1), such that $\mathcal{A}'$ satisfies the left lifting property with respect to this collection. The pushouts in question preserve global weak equivalences, indeed the maps are a part of K_{Reedy} so that follows from the construction of the model structure of Theorem 12.3.2 by direct left Bousfield localization (in this point of view the pushouts can be seen as local weak equivalences); or else one could apply Theorem 14.3.3 and Lemma 14.3.5, which is really saying pretty much the same thing. Now, an object which satisfies the left lifting property with respect to the $\Psi([k], f)$, satisfies the Segal conditions because the product maps satisfy lifting along any generating cofibration $U \to V$ for $\mathcal{M}$, thus they are trivial fibrations in $\mathcal{M}$, which shows that $\mathcal{A}'$ satisfies the Segal condition. □

Notice that in the construction of the proposition, the resulting $\mathcal{A}'$ will not in general be even levelwise fibrant, one would have to include pushouts along morphisms of the form $h([k], f)$ for f a generating trivial cofibration of $\mathcal{M}$.

Theorem 17.2.3 *Suppose $\mathcal{A} \in \mathcal{PC}(\mathcal{M})$ is Reedy cofibrant, $k \in \mathbb{N}$ and $B \in \mathcal{M}$ is a cofibrant object. Then the map*

$$\mathcal{A} \times \Sigma([k]; B) \to \mathcal{A} \times h([k]; B)$$

is a global weak equivalence.

Proof The proof goes in several steps.

(i) Suppose $\mathcal{A}$ is a sequentially free $\mathcal{M}$-enriched precategory. Note that $\Sigma([k]; B) \to h([k]; B)$ is a trivial cofibration of sequentially free $\mathcal{M}$-enriched precategories, inducing an order-preserving isomorphism on objects. Apply Theorem 17.1.7 to this map and the identity of $\mathcal{A}$, to conclude that

$$\mathcal{A} \times \Sigma([k]; B) \to \mathcal{A} \times h([k]; B)$$

is a global weak equivalence. This completes the proof when $\mathcal{A}$ is sequentially free. This applies in particular to the $h([k]; B)$ which are sequentially free.

(ii) Recall from Theorem 14.3.3, that we already know that any pushout along a global trivial cofibration inducing an isomorphism on sets of objects, is again a global trivial cofibration.

(iii) Suppose we know the statement of the theorem for $\mathcal{A}$, $\mathcal{A}'$ and $\mathcal{A}''$ and suppose given a diagram

$$\mathcal{A}' \leftarrow \mathcal{A} \rightarrow \mathcal{A}'',$$

in which one of the arrows, say $\mathcal{A} \rightarrow \mathcal{A}''$, is a Reedy cofibration. Then we claim that the statement of the theorem is true for $\mathcal{Q} := \mathcal{A}' \cup^{\mathcal{A}} \mathcal{A}''$. Notice first that

$$\mathcal{Q} \times \Sigma([k]; B) = (\mathcal{A}' \times \Sigma([k]; B)) \cup^{\mathcal{A} \times \Sigma([k]; B)} (\mathcal{A}'' \times \Sigma([k]; B))$$

by commutation of colimits and cartesian products in $\mathcal{PC}(\mathcal{M})$ (which is part (DCL) of the cartesian condition 7.7.1). The same is true for the product with $h([k]; B)$. The map

$$\mathcal{Q} \times \Sigma([k]; B) \rightarrow \mathcal{Q} \times h([k]; B) \tag{17.2.2}$$

is therefore obtained by functoriality of the pushout of the columns in

$$\begin{array}{ccc}
\mathcal{A}' \times \Sigma([k]; B) & \longrightarrow & \mathcal{A}' \times h([k]; B) \\
\uparrow & & \uparrow \\
\mathcal{A} \times \Sigma([k]; B) & \longrightarrow & \mathcal{A} \times h([k]; B) \\
\downarrow & & \downarrow \\
\mathcal{A}'' \times \Sigma([k]; B) & \rightarrow & \mathcal{A}'' \times h([k]; B).
\end{array}$$

We are supposing that we know that each of the horizontal maps is a global weak equivalence, also they induce isomorphisms on objects. By the hypothesis of this part and the cartesian property for $\mathcal{M}$ applied levelwise, the bottom vertical maps, on both left and right, are Reedy cofibrations by Corollary 13.6.13.

By Lemma 14.3.4, the induced map on pushouts (17.2.2) is a global weak equivalence. This proves the claim for step (iii).

(iv) Suppose given a sequence $\mathcal{A}_i$ indexed by an ordinal β, with transition maps which are injective cofibrations. Suppose the statement of the theorem is true for each $\mathcal{A}_i$, then it is true for $\mathcal{Q} := \mathrm{colim}_{i \in \beta} \mathcal{A}_i$. Indeed, just as in the previous part $\mathcal{Q} \times h([k]; B)$ can be expressed as a transfinite composition of pushouts of $\mathcal{Q} \times \Sigma([k]; B)$ along maps which are by hypothesis global trivial cofibrations which induce isomorphisms on objects. By Theorem 14.3.3 and Lemma 14.3.5, the composition is a global weak equivalence.

(v) We show by induction on $m \in \mathbb{N}$ that if $\mathbf{sk}_m(\mathcal{A}) \cong \mathcal{A}$ then the statement of the theorem holds for $\mathcal{A}$. It is easy to see in case $m = 0$ because then $\mathcal{A}$ is just a discrete set. Suppose this is known for any $m \leq n$, and suppose $\mathcal{A} = \mathbf{sk}_n(\mathcal{A})$. By Proposition 13.5.1 we can express $\mathcal{A}$ as a transfinite composition of pushouts of $\mathbf{sk}_{n-1}(\mathcal{A})$ along Reedy cofibrations of the form $h([n], \partial[n]; U \rightarrow V) \rightarrow h([n]; V)$. On the other hand,

$$h([n], \partial[n]; U \rightarrow V) = h([n]; U) \cup^{h(\partial[n]; U)} \partial h(\partial[n]; V),$$

and $h(\partial[n]; U) = \mathbf{sk}_{n-1} h([n]; U)$. By the inductive hypothesis the statement of the theorem is known for $h(\partial[n]; U)$ and similarly for $h(\partial[n]; V)$. It is known for $h([n]; U)$ by (i). So by (iii) the statement of the theorem is known for $h([n], \partial[n]; U \rightarrow V)$. Furthermore, it is known for $\mathbf{sk}_{n-1}(\mathcal{A})$ by the inductive hypothesis. Again by (iii) and (iv) we conclude the statement for $\mathcal{A}$.

(vi) Any Reedy cofibrant $\mathcal{A}$ can be expressed as a transfinite composition of the maps $\mathbf{sk}_m(\mathcal{A}) \rightarrow \mathbf{sk}_{m-1}(\mathcal{A})$, so by (iv) and (v) we get the statement of the theorem for any Reedy cofibrant $\mathcal{A}$. This completes the proof. □

Recall from Corollary 14.2.5 and the remark at the beginning of the proof of Proposition 17.2.2 above, for any cofibration $f : U \rightarrow V$ we have the notation

$$\mathrm{src}\Psi([k], f) = \Sigma([k]; V) \cup^{\Sigma([k]; U)} h([k]; U),$$

and the map $\Psi([k], f)$ goes from here to $h([k]; V)$.

Corollary 17.2.4 *Suppose $\mathcal{A} \in \mathcal{PC}(\mathcal{M})$ is Reedy cofibrant, $k \in \mathbb{N}$, and $f : U \rightarrow V$ is a cofibration in $\mathcal{M}$. Then the map*

$$\mathcal{A} \times \mathrm{src}\Psi([k], f) \rightarrow \mathcal{A} \times h([k]; v)$$

is a global weak equivalence.

Proof In the cocartesian diagram

$$\begin{array}{ccc} \mathcal{A} \times \Sigma([k]; U) & \longrightarrow & \mathcal{A} \times h([k]; U) \\ \downarrow & & \downarrow \\ \mathcal{A} \times \Sigma([k]; V) & \rightarrow & \mathcal{A} \times \mathrm{src}\Psi([k], f) \end{array}$$

the upper arrow is a global weak equivalence inducing an isomorphism on the set of objects, by the previous theorem. Furthermore, it is a Reedy cofibration by Proposition 13.6.12, so it is a Reedy isotrivial cofibration. By Theorem 14.3.3, the bottom map is a global weak equivalence. The statement of the corollary now follows by again using the previous Theorem 17.2.3 as well as 3 for 2. □

Theorem 17.2.5 *Assume $\mathcal{M}$ is a tractable left proper cartesian model category. For any $\mathcal{A}, \mathcal{B} \in \mathcal{PC}(\mathcal{M})$, the map*

$$\mathcal{A} \times \mathcal{B} \to \mathbf{Seg}(\mathcal{A}) \times \mathbf{Seg}(\mathcal{B})$$

is a global weak equivalence.

Proof We suppose first that $\mathcal{A}$ and $\mathcal{B}$ are Reedy cofibrant. There is a map $\mathcal{A} \to \mathcal{A}'$ to an object satisfying the Segal conditions, which is a transfinite composition of pushouts along morphisms of the form $\mathrm{src}\Psi([k], f) \xrightarrow{\Psi([k],f)} h([k]; V)$. Each of these pushouts is a global trivial Reedy cofibration inducing an isomorphism on the set of objects, by Theorem 14.3.3. The map

$$\mathcal{A} \times \mathcal{B} \to \mathcal{A}' \times \mathcal{B}$$

is the corresponding transfinite composition of pushouts along morphisms of the form

$$\mathrm{src}\Psi([k], f) \times \mathcal{B} \to h([k]; U) \times \mathcal{B}.$$

This is because part of the cartesian hypothesis for $\mathcal{M}$ is (DCL) commutation of cartesian products and colimits. By Corollary 17.2.4, the morphisms $\mathrm{src}\Psi([k], f) \times \mathcal{B} \to h([k]; U) \times \mathcal{B}$ are global weak equivalences; they are also Reedy cofibrations by Proposition 13.6.12 because we assumed that $\mathcal{B}$ is Reedy cofibrant. These maps are again isomorphisms on objects, so we can apply Theorem 14.3.3, which says that global trivial cofibrations which induce isomorphisms on the set of objects are preserved under pushout (and see Lemma 14.3.5 for the transfinite composition). Therefore $\mathcal{A} \times \mathcal{B} \to \mathcal{A}' \times \mathcal{B}$ is a global weak equivalence.

Arguing in the same way for the product of a map $\mathcal{B} \to \mathcal{B}'$ with $\mathcal{A}'$, then composing the two equivalences we conclude that the map

$$\mathcal{A} \times \mathcal{B} \to \mathcal{A}' \times \mathcal{B}'$$

is a global weak equivalence. On the other hand, $\mathcal{A}' \to \mathbf{Seg}(\mathcal{A}')$ and $\mathcal{B}' \to \mathbf{Seg}(\mathcal{B}')$ are levelwise weak equivalences, so

$$\mathcal{A}' \times \mathcal{B}' \to \mathbf{Seg}(\mathcal{A}') \times \mathbf{Seg}(\mathcal{B}')$$

is a levelwise weak equivalence. Similarly, the fact that $\mathcal{A} \to \mathcal{A}'$ is a global weak equivalence inducing an isomorphism on objects, implies that

$$\mathbf{Seg}(\mathcal{A}) \to \mathbf{Seg}(\mathcal{A}')$$

is a levelwise weak equivalence, and by the same remark for $\mathcal{B}$ then taking the product, we get that

$$\mathbf{Seg}(\mathcal{A}) \times \mathbf{Seg}(\mathcal{B}) \to \mathbf{Seg}(\mathcal{A}') \times \mathbf{Seg}(\mathcal{B}')$$

is a levelwise weak equivalence. Thus we obtain a diagram

$$\begin{array}{ccc} \mathcal{A} \times \mathcal{B} & \longrightarrow & \mathcal{A}' \times \mathcal{B}' \\ \downarrow & & \downarrow \\ \mathbf{Seg}(\mathcal{A}) \times \mathbf{Seg}(\mathcal{B}) & \to & \mathbf{Seg}(\mathcal{A}') \times \mathbf{Seg}(\mathcal{B}') \end{array}$$

where the top map is a global weak equivalence, and the right vertical and bottom maps are levelwise hence global weak equivalences (Lemma 12.4.4). By 3 for 2 it follows that the left vertical map is a global weak equivalence as required for the theorem. This completes the proof for the case of Reedy cofibrant objects.

Now suppose $\mathcal{A}$ and $\mathcal{B}$ are general objects of $\mathcal{PC}(\mathcal{M})$. Consider Reedy cofibrant replacements $\mathcal{A}' \to \mathcal{A}$ and $\mathcal{B}' \to \mathcal{B}$; these may be chosen as levelwise equivalences of diagrams, which are then global weak equivalences by Lemma 12.4.4. In particular, $\mathbf{Seg}(\mathcal{A}') \to \mathbf{Seg}(\mathcal{A})$ is a levelwise weak equivalence and the same for $\mathcal{B}'$ so

$$\mathbf{Seg}(\mathcal{A}') \times \mathbf{Seg}(\mathcal{B}') \to \mathbf{Seg}(\mathcal{A}) \times \mathbf{Seg}(\mathcal{B})$$

is a levelwise weak equivalence. Note also that

$$\mathcal{A}' \times \mathcal{B}' \to \mathcal{A} \times \mathcal{B}$$

is a levelwise weak equivalence of diagrams over $\Delta^o_{\mathrm{Ob}(\mathcal{A})\times\mathrm{Ob}(\mathcal{B})}$, so it is a global weak equivalence by Lemma 12.4.4. The first part of the proof treating the Reedy cofibrant case shows that

$$\mathcal{A}' \times \mathcal{B}' \to \mathbf{Seg}(\mathcal{A}') \times \mathbf{Seg}(\mathcal{B}')$$

is a global weak equivalence. In the square diagram

$$\begin{array}{ccc} \mathcal{A}' \times \mathcal{B}' & \longrightarrow & \mathcal{A} \times \mathcal{B} \\ \downarrow & & \downarrow \\ \mathbf{Seg}(\mathcal{A}') \times \mathbf{Seg}(\mathcal{B}') & \rightarrow & \mathbf{Seg}(\mathcal{A}) \times \mathbf{Seg}(\mathcal{B}) \end{array}$$

the top, bottom and left vertical arrows are global weak equivalences, so by 3 for 2 the right vertical arrow is a global weak equivalence. This completes the proof. □

Corollary 17.2.6 *Suppose $\mathcal{A} \rightarrow \mathcal{B}$ and $\mathcal{C} \rightarrow \mathcal{D}$ are global weak equivalences. Then the map*

$$\mathcal{A} \times \mathcal{C} \rightarrow \mathcal{B} \times \mathcal{D}$$

is a global weak equivalence.

Proof Suppose first of all that $\mathcal{A}$, $\mathcal{B}$, $\mathcal{C}$ and $\mathcal{D}$ are objects satisfying the Segal conditions. Then the products also satisfy the Segal conditions. Truncation of these is compatible with cartesian products, by Lemma 12.4.2, so the map in question is essentially surjective. By looking at the morphism objects we see that it is fully faithful, so it is a global weak equivalence by following the definition.

Next suppose that $\mathcal{A}$, $\mathcal{B}$, $\mathcal{C}$ and $\mathcal{D}$ are any objects, and look at the diagram

$$\begin{array}{ccc} \mathcal{A} \times \mathcal{C} & \longrightarrow & \mathcal{B} \times \mathcal{D} \\ \downarrow & & \downarrow \\ \mathbf{Seg}(\mathcal{A}) \times \mathbf{Seg}(\mathcal{C}) & \rightarrow & \mathbf{Seg}(\mathcal{B}) \times \mathbf{Seg}(\mathcal{D}). \end{array}$$

The vertical maps are global weak equivalences by Theorem 17.2.5, while the bottom map is a global weak equivalence by the first paragraph of the proof. By 3 for 2, the top map is a global weak equivalence. □

Recall that $\mathbf{I}$ denotes the $\mathcal{M}$-category with two objects υ_0, υ_1 and a single isomorphism between them, in other terms $\mathbf{I} = \mathbf{codsc}(\{\upsilon_0, \upsilon_1\})$.

Corollary 17.2.7 *For any $\mathcal{M}$-precategory $\mathcal{A}$, the morphisms*

$$\mathcal{A} = \mathcal{A} \times \{\upsilon_0\} \rightarrow \mathcal{A} \times \mathbf{I} \rightarrow \mathcal{A}$$

are global weak equivalences.

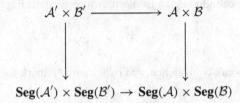

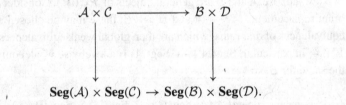

Proof The inclusion $\{\upsilon_0\} \hookrightarrow \mathbf{I}$ is a global weak equivalence; apply the previous corollary. □

This may be extended as follows. If $\mathcal{A}$ is an $\mathcal{M}$-precategory and $F : Z \to \mathrm{Ob}(\mathcal{A})$ is a map of sets, recall that we obtain an $\mathcal{M}$-precategory denoted $F^*(\mathcal{A})$ whose set of objects is Z, with

$$F^*(\mathcal{A})(z_0, \ldots, z_n) := \mathcal{A}(F(z_0), \ldots, F(z_n)).$$

These identifications mean that F corresponds to a natural morphism f of precategories appearing in the following statement:

Corollary 17.2.8 *Suppose $\mathcal{A}$ is an $\mathcal{M}$-precategory and $F : Z \twoheadrightarrow \mathrm{Ob}(\mathcal{A})$ is a surjection. Then the natural map*

$$F^*(\mathcal{A}) \xrightarrow{f} \mathcal{A}$$

is a global weak equivalence.

Proof The fact that f is a global weak equivalence follows immediately from the definition if $\mathcal{A}$ satisfies the Segal conditions. The theory of products developed in the present chapter is needed to treat the case when $\mathcal{A}$ is just a precategory. Choose a section $S : \mathrm{Ob}(\mathcal{A}) \to Z$ with $F \circ S = 1_{\mathrm{Ob}(\mathcal{A})}$. This gives a morphism $\mathcal{A} \xrightarrow{s} F^*(\mathcal{A})$ with $f \circ s = 1_{\mathcal{A}}$. More generally any map of sets lying over $\mathrm{Ob}(\mathcal{A})$ induces a map between corresponding pulled-back precategories, equal to the identity on the various $\mathcal{A}(x_0, \ldots, x_n)$, in the same way as for f. This will be used several times below.

We would like to show that $s \circ f$ is homotopic to the identity in an appropriate sense (although we don't yet have the full model structure). This is based on the observation that

$$\mathcal{A} \times \mathbf{I} = (P_1)^*(\mathcal{A}),$$

for the projection $P_1 : \mathrm{Ob}(\mathcal{A}) \times \{\upsilon_0, \upsilon_1\} \to \mathrm{Ob}(\mathcal{A})$. That should be applied upstairs on $F^*(\mathcal{A})$. Letting

$$Q_1 : Z \times \{\upsilon_0, \upsilon_1\} \to Z$$

denote the projection there (which induces a morphism on precategories denoted q_1), we have

$$F^*(\mathcal{A}) \times \mathbf{I} = (Q_1)^*(F^*(\mathcal{A})) = (F Q_1)^*(\mathcal{A}).$$

Consider the function

$$H : Z \times \{\upsilon_0, \upsilon_1\} \to Z \times \{\upsilon_0, \upsilon_1\}$$

defined by $H(z, \upsilon_0) := z$, and $H(z, \upsilon_1) := SF(z)$. This function induces a morphism of $\mathcal{M}$-precategories

$$F^*(\mathcal{A}) \times \mathbf{I} \xrightarrow{h} F^*(\mathcal{A}) \times \mathbf{I}$$

restricting to H on the set of objects. The composition of h with the global weak equivalence of the previous corollary

$$F^*(\mathcal{A}) \times \{\upsilon_0\} \hookrightarrow F^*(\mathcal{A}) \times \mathbf{I}$$

is the the same global weak equivalence; this implies that h is a global weak equivalence by 3 for 2 (Proposition 12.5.4). Therefore the composition

$$F^*(\mathcal{A}) \times \{\upsilon_1\} \hookrightarrow F^*(\mathcal{A}) \times \mathbf{I} \xrightarrow{h} F^*(\mathcal{A}) \times \mathbf{I} \xrightarrow{q_1} F^*(\mathcal{A})$$

is a global weak equivlance. This composition is $s \circ f$. It now follows by Proposition 12.5.4 that f is a global weak equivalence. □

A similar argument will be used in the next chapter along the way to showing that pushout by trivial cofibrations is a trivial cofibration.

17.3 The role of unitality, degeneracies and higher coherences

In this section, we point out why we need to impose the unitality condition $A(x_0) = *$, to include the degeneracy maps in Δ (which also correspond to some sort of unit condition), and why we can't truncate Δ by, say, dropping the objects $[n]$ for $n \geq 4$. These all have to do with the arguments of this chapter about products. In some sense it goes back to the Eilenberg–Zilber theorem; our product condition can be viewed as a generalization to the present context where the information of direction of arrows is retained.

17.3.1 The unitality condition

Suppose we tried to use non-unital precategories. These would be pairs $(X, \mathcal{A})$ where $\mathcal{A} : \Delta^o_X \to \mathcal{M}$ is an arbitrary functor. The Segal condition would include, for sequences of length $n = 0$, the fact that $\mathcal{A}(x_0) \to *$ should be a weak equivalence, in other words $\mathcal{A}(x_0)$ is weakly contractible. So, this would constitute a weak version of the unitality condition. We would proceed much as above, imposing the Segal condition by the small object argument in an operation denoted $\mathcal{A} \mapsto \mathbf{Seg}^n(\mathcal{A})$. It seems likely that this would lead to a model category, conjecturally Quillen equivalent to the model categories on unital precategories which we are constructing here.

However, even if the model structure existed, it could not be cartesian. The reason for this occurs at some very degenerate objects: consider the non-unital precategory with object set $\mathrm{Ob}(\mathcal{B}) = \{y\}$ a singleton, but with functor the constant functor with values the initial object:

$$\mathcal{B}(y, \ldots, y) := \emptyset.$$

This includes the case of sequences of length 0: $\mathcal{B}(y) = \emptyset$, so $\mathcal{B}$ doesn't satisfy the unitality condition. Now $\mathbf{Seg}^n(\mathcal{B})$ would be some kind of $\mathcal{M}$-enriched category with a single object; it seems clear that it would be the coinitial $*$ but in any case has to contain $*$ as a retract.

Suppose $\mathcal{A}$ is another non-unital precategory (which might in fact be unital). Consider $\mathcal{A}\times\mathcal{B}$. The object set is $\mathrm{Ob}(\mathcal{A})\times\{y\} \cong \mathrm{Ob}(\mathcal{A})$. But for any sequence $((x_0, y), \ldots, (x_n, y))$ we have

$$\begin{aligned}(\mathcal{A}\times\mathcal{B})((x_0, y), \ldots, (x_n, y)) &= \mathcal{A}(x_0, \ldots, x_n) \times \mathcal{B}(y, \ldots, y)\\ &= \mathcal{A}(x_0, \ldots, x_n) \times \emptyset = \emptyset.\end{aligned}$$

In particular, the structure of $\mathcal{A}\times\mathcal{B}$ depends only on $\mathrm{Ob}(\mathcal{A})$ and not on $\mathcal{A}$ itself. This would be incompatible with the cartesian condition (for any reasonable choice of $\mathcal{M}$), because

$$\mathbf{Seg}^n(\mathcal{A}) \times * \to \mathbf{Seg}^n(\mathcal{A}) \times \mathbf{Seg}^n(\mathcal{B})$$

is contained as a retract, but $\mathbf{Seg}^n(\mathcal{A}\times\mathcal{B})$ is essentially trivial.

To make the last step of the above argument precise we would need to investigate $\mathbf{Seg}^n$ explicitly. In the case $\mathcal{M} = \textsc{Set}$, the same discussion as in Section 14.8 applies, expressing $\mathbf{Seg}^n(\mathcal{A})$ as the category generated by $\mathcal{A}$ considered as a system of generators and relations; the first step would be to impose the Segal condition at $n = 0$ which, for $\mathcal{M} = \textsc{Set}$, is exactly the unitality condition; from there the rest is the same. In this case we see that $\mathbf{Seg}^n(\mathcal{B})$ is really just $*$, $\mathbf{Seg}^n(\mathcal{A}\times\mathcal{B})$ is the discrete category on object set $\mathrm{Ob}(\mathcal{A}) \times \{y\}$, and it cannot contain $\mathbf{Seg}^n(\mathcal{A})$ as a retract in general.

17.3.2 Degeneracies

Let $\Phi \subset \Delta$ denote the category consisting only of face maps, in other words the objects of Φ are the nonempty finite linearly ordered sets $[k]$ for $k \in \mathbb{N}$, but the maps are the injective order-preserving maps. We could try to create a theory of weak categories based on Φ rather than Δ. It would be appropriate to call these "weak semicategories," because the degeneracy maps in Δ

correspond to inserting identity morphisms into a composable sequence. This theory is certainly interesting and important, and has not been fully worked out as far as I know.

This would undoubtedly be related to the work of J. Kock [172, 173] on weakly unital higher categories, as one could start by considering weak semicategories, then impose a weak unitality condition.

Unfortunately the theory of products again doesn't work if Δ is replaced by Φ. Indeed, a slight modification of the example of the previous subsection again provides a counterexample. Let $\mathcal{B}$ be the precategory with a single object y, with $\mathcal{B}(y) = *$ so it is unital, but with $\mathcal{B}(y, \ldots, y) = \emptyset$ for any sequence of length $n \geq 1$ (that is to say, with $n+1$ elements). This will still be a valid functor from $\Phi_{\{y\}}$ to $\mathcal{M}$; however, taking the product $\mathcal{A} \times \mathcal{B}$ will destroy the structure of $\mathcal{A}$.

As in the previous subsection, this can be made precise in the case $\mathcal{M} = \text{SET}$. The non-unital precategories may then be considered as systems of generators and relations for a category, but the system doesn't contain the degeneracies.

It is interesting to look more closely at how this works in the case of systems of generators for a monoid, that is to say for a category with a single object. The 1-cells of a precategory $\mathcal{A}$ correspond to generators of the monoid, and the 2-cells correspond to relations of the form $f = gh$ among the generators. In this case, the system of generators and relations is just given by two sets $\mathcal{A}(x, x)$ and $\mathcal{A}(x, x, x)$ with three maps

$$\mathcal{A}(x, x, x) \substack{\longrightarrow \\ \longrightarrow \\ \longrightarrow} \mathcal{A}(x, x).$$

The process of generators and relations would have to include the addition of identities.

In this case, for example, if $\mathcal{A}$ has a single generator and no relations, then its product with itself $\mathcal{A} \times \mathcal{A}$ will again be a system with a single generator and no relations; but $\mathcal{A}$ generates the monoid $\mathbb{N}$ and $\mathbb{N} \times \mathbb{N}$ is different from $\mathbb{N}$, so the product of systems of generators and relations doesn't generate the product of the corresponding monoids.

One can see in this simple example how the degeneracies come to the rescue. A system of generators and relations with unitality and degeneracies corresponds to a diagram of the form

$$\mathcal{A}(x, x, x) \substack{\longrightarrow \\ \longleftarrow \\ \longrightarrow \\ \longleftarrow \\ \longrightarrow} \mathcal{A}(x, x) \substack{\longrightarrow \\ \longleftarrow \\ \longrightarrow} \mathcal{A}(x) = *.$$

This means that there is an explicit element 1 among the generators, with the relations $1 \cdot f = f$ and $f \cdot 1 = f$ for any other generator f.

Now let's look again at $\mathbb{N}$ generated by a system $\mathcal{A}$ consisting of a single generator. We have $\mathcal{A}(x, x) = \{f, 1\}$ with relations corresponding to the left and right identities for both f and 1 itself. The product now has generators

$$\mathcal{A} \times \mathcal{A}(x, x) = \{f \times f, f \times 1, 1 \times f, 1 \times 1\},$$

where we have noted the single object of $\mathcal{A} \times \mathcal{A}$ as x again rather than (x, x). The unit generator is 1×1. The relations include the left and right identities with the unit generator 1×1, plus two new relations of the form

$$(f \times 1) \cdot (1 \times f) = f \times f$$

and

$$(1 \times f) \cdot (f \times 1) = f \times f.$$

The first of these two relations serves to eliminate the generator $f \times f$ so we get to a monoid with two generators $f \times 1$ and $1 \times f$, then the second relation gives the commutativity $(f \times 1) \cdot (1 \times f) = (1 \times f) \cdot (f \times 1)$. So, the monoid generated by the system $\mathcal{A} \times \mathcal{A}$ is indeed $\mathbb{N} \times \mathbb{N}$.

Working out this example demonstrates how the degeneracies of Δ enter into the cartesian condition in an important way.

17.4 Why we can't truncate Δ

The above examples could all be done in the case of $\mathcal{M} = \text{SET}$, where the passage from precategories to categories is the process of generating a category by generators and relations. For that, we didn't need to consider the part of Δ involving $[n]$ for $n \geq 4$ (the case $n = 3$ being needed for the associativity condition).

On the other hand, for weak enrichment in a general model category $\mathcal{M}$, we can't replace Δ by any finite truncation, that is by a subcategory $\Delta^{\leq m}$ of finite ordered sets of size $\leq m$. Our argument for products is one place where this can be seen: for the product of sequentially free precategories of lengths m and n, we end up having to look at tails which are sequentially free precategories of length up to $m + n$.

Another reason is the requirement that there should be a higher Poincaré groupoid construction. In the case when $\mathcal{M} = \mathcal{K}$ is the model category of simplicial sets, the $\mathcal{K}$-enriched precategories should be realizable into arbitrary homotopy types; and in particular Segal groupoids should be eqivalent to homotopy types by a pair of functors including Poincaré–Segal category and realization. These should be compatible with homotopy groups in a way similar to that described in Chapters 2 and 3.

If we impose these conditions, it becomes easy to see that the Segal groupoids defined using only $\Delta^{\leq m}$ (say for $m \geq 3$) don't model all homotopy types. This means that, for the program we are pursuing here, the category $\Delta^{\leq m}$ cannot be sufficient.

It should also be possible to show that $\Delta^{\leq m}$ can't be used to model homotopy n-types for $n > m$, in any way at all; however, it doesn't seem completely clear how to formulate a good statement of this kind.

If for 1-categories it suffices to look at $\Delta^{\leq 3}$, we could expect more generally that in order to consider n-categories it would suffice to look at $\Delta^{\leq n+2}$, indeed this showed up in the explicit example of Chapter 15 and will show up again in our discussion of stabilization in Chapter 23.

18
Intervals

Given our tractable, left proper and cartesian model category $\mathcal{M}$, the main remaining problem in order to construct the global model structure on $\mathcal{PC}(\mathcal{M})$ is to consider the notion of *interval*, which should be an $\mathcal{M}$-precategory (to be called $\Xi(N|N')$ in our notations below), weak equivalent to the usual category $\mathbf{I}$ with two isomorphic objects $\upsilon_0, \upsilon_1 \in \mathbf{I}$, and with a single morphism between any pair of objects.

If $\mathcal{A} \in \mathcal{PC}(\mathcal{M})$ is a weakly $\mathcal{M}$-enriched category, an *internal equivalence* between $x_0, x_1 \in \mathrm{Ob}(\mathcal{A})$ is a "morphism from x_0 to x_1" (see (18.2.1) below), which projects to an isomorphism in the truncated category $\tau_{\leq 1}(\mathcal{A})$. This terminology was introduced by Tamsamani [250]. It plays a vital role in the study of global weak equivalences. Essential surjectivity of a morphism $f : \mathcal{A} \to \mathcal{B}$ means (assuming that $\mathcal{B}$ is levelwise fibrant) that, for any object $y \in \mathrm{Ob}(\mathcal{B})$, there is an object $x \in \mathrm{Ob}(\mathcal{A})$ and an internal equivalence between $f(x)$ and y.

Unfortunately, an internal equivalence between x_0 and x_1 in $\mathcal{A}$ doesn't necessarily translate into the existence of a morphism $\mathbf{I} \to \mathcal{A}$. This will work after we have established the model structure on $\mathcal{PC}(\mathcal{M})$ if we assume that $\mathcal{A}$ is a fibrant object. However, in order to finish the construction of the model structure, we should start with the weaker hypothesis[1] that $\mathcal{A}$ satisfies the Segal conditions and is levelwise fibrant. The "interval object" $\Xi(N|N')$ should be contractible, and have the versality property that whenever x_0 and x_1 are internally equivalent, there is a morphism $\Xi(N|N') \to \mathcal{A}$ relating them.

The construction of such a versal interval was the subject of an error in my preprint [234], found and corrected by Pellissier [211]. This was somewhat similar to a mistake in Dwyer–Hirschhorn–Kan's original construction [99] of

[1] As observed by Bergner [42] this hypothesis will be equivalent to fibrancy in the global projective model structure, once we know that it exists.

the model category structure for simplicial categories, pointed out by Toën and subsequently fixed for simplicial categories by Bergner [39]. Pellissier fixed this problem for the model category of Segal categories by constructing an explicit interval object and verifying its topological properties using the comparison between Segal 1-groupoids and spaces. Drinfeld [93] has constructed an interval object for differential graded categories.

Pellissier's correction as written covered only the case of $\mathcal{K}$-enriched weak categories, and one of our purposes here is to point out that his argument serves to construct the required intervals in general, by functoriality with respect to a left Quillen functor $\mathcal{K} \to \mathcal{M}$. For the main result, which is contractibility of the $\Xi(N|N')$, we proceed therefore in two steps: first considering the problem for the case of Segal categories, i.e. $\mathcal{K}$-enriched weak categories, as was done by Pellissier [211]; then going to the case of $\mathcal{M}$-enriched weak categories by transfer along $\mathcal{K} \to \mathcal{M}$. Sections 9.8, 12.6 and 14.7 about transfer along a left Quillen functor were motivated by this movement. The possibility of doing that is one of the advantages of the fully iterative point of view originally suggested by André Hirschowitz in Pellissier's thesis topic, in which $\mathcal{M}$ is a general input into the construction. It should also be possible to adapt Pellissier's correction directly to the original n-nerves considered by Tamsamani [250] and used in my preprint [234], by using Tamsamani's theorems on the topological realization of weak n-groupoids, which in turn applied Segal's original results in a partially iterative way. That would be more geometrically motivated, but for the present treatment the fully iterative approach is both more general and more direct.

I would like to thank Regis Pellissier for finding and correcting this problem.

18.1 Contractible objects and intervals in $\mathcal{M}$

An object $A \in \mathcal{M}$ is *contractible* if the unique morphism $A \to *$ is a weak equivalence. An *interval object* is a triple (B, i_0, i_1) where $B \in \mathcal{M}$ and $i_0, i_1 : * \to B$ such that B is contractible and $i_0 \sqcup i_1 : * \cup^{\emptyset} * \to B$ is a cofibration.

Assumption (AST) in the cartesian condition 7.7.1 says that $*$ is a cofibrant object, so an interval object is itself cofibrant.

A *morphism between intervals* from (B, i_0, i_1) to $\left(B', i'_0, i'_1\right)$ means a morphism $f : B \to B'$ such that $f \circ i_0 = i'_0$ and $f \circ i_1 = i'_1$. Since B and B' are contractible, a morphism f is automatically a weak equivalence.

Lemma 18.1.1 *Suppose (B, i_0, i_1) and $\left(B', i'_0, i'_1\right)$ are two interval objects. Then there is a third one $\left(B'', i''_0, i''_1\right)$ and morphisms of intervals $f : B \to B''$ and $f' : B' \to B''$. These may be assumed to be trivial cofibrations.*

Proof Put $A := B \cup^{* \cup^{\emptyset} *} B'$ and choose a factorization

$$A \xrightarrow{f} B'' \to *,$$

where the first morphism is a cofibration and the second morphism is a weak equivalence. Now i_0 and i'_0 are the same when considered as maps $* \to A$ because of the amalgamated sum in the definition of A. Thus $f \circ i_0 = f \circ i'_0$ gives a map $i''_0 : * \to B''$. Similarly, $i''_1 := f \circ i_1 = f \circ i'_1$. The map $* \cup^{\emptyset} * \to A$ is a cofibration, and since f is a cofibration, the composition into B'' is a cofibration. Note that the maps $B \to A$ and $B' \to A$ are cofibrations, so the same is true of the maps to B'', and since the intervals are weakly equivalent to $*$ these maps are trivial cofibrations. □

Recall that we defined in Chapter 12 a functor $\tau_{\leq 0} : \mathcal{M} \to \text{SET}$ by $\tau_{\leq 0}(A) := \text{HOM}_{\text{ho}(\mathcal{M})}(*, A')$, where $A \to A'$ is a fibrant replacement.

Lemma 18.1.2 *Suppose A is a fibrant object and $a, b : * \to A$. The following conditions are equivalent:*

(a) the classes of a and b in $\tau_{\leq 0}(A)$ coincide;
(b) for any interval object (B, i_0, i_1) there exists a map $B \to A$ sending i_0 to a and i_1 to b; and
(c) there exists an interval object (B, i_0, i_1) and a map $B \to A$ sending i_0 to a and i_1 to b.

Proof This is an exercise in Quillen's theory of the homotopy category of a model category, which we do for the reader's convenience.

Note that (b) ⇒ (c) ⇒ (a) easily. Assume that A is also cofibrant. To prove that (a) ⇒ (c), suppose that the classes of a and b coincide in $\tau_{\leq 0}(A)$. This is equivalent to saying that the two maps $a, b : * \to A$ project to the same map in $\text{ho}(\mathcal{M})$. Recall from Quillen [215] that $\text{ho}(\mathcal{M})$ is also the category of fibrant and cofibrant objects of $\mathcal{M}$, with homotopy classes of maps. Since $*$ is automatically fibrant, and cofibrant by hypothesis, and since we are assuming that A is cofibrant and fibrant, condition (a) says that the two maps a and b are homotopic in the sense of Quillen [215], which says exactly condition (c). For the implication (a) ⇒ (c), but with A assumed only to be fibrant, choose a trivial fibration from a cofibrant object $A' \to A$. Lift to maps $a', b' : * \to A'$. Since $A' \to A$ projects to an isomorphism in $\text{ho}(\mathcal{M})$, the maps a' and b' are equivalent in $\tau_{\leq 0}(A')$ so by condition (c) proven for A' previously, there exists an interval object (B, i_0, i_1) and an extension of $a' \sqcup b'$ to $B \to A'$. Composing gives the required map $B \to A$.

To finish the proof it suffices to show (c) ⇒ (b). Suppose (B, i_0, i_1) and (B', i'_0, i'_1) are two interval objects. Applying Lemma 18.1.1, there is an interval object (B'', i''_0, i''_1) with trivial cofibrations from both B and B'. If $a \sqcup b : * \cup^* * \to A$ extends to a map $B \to A$, and if A is fibrant, then the lifting property for A gives the extension to $B'' \to A$, which then restricts to a map $B' \to A$ as required to show (c) $\Rightarrow$ (b). ☐

Using the assumption that $\mathcal{M}$ is cartesian, we can make a similar statement explaining the relation of homotopy between morphisms using an interval, if the target is a fibrant object of $\mathcal{M}$:

Lemma 18.1.3 *Suppose $\mathcal{M}$ is a cartesian model category. Suppose A is a cofibrant object and C is a fibrant object. Then, for two morphisms $f, g : A \to B$ the following statements are equivalent:*

(a) f and g are homotopic in Quillen's sense, meaning that the classes of f and g in $\mathrm{HOM}_{\mathrm{ho}(\mathcal{M})}(A, C)$ coincide;
(b) for any interval object (B, i_0, i_1) there exists a map $h : A \times B \to C$ such that $h \circ (1_A \times i_0) = f$ and $h \circ (1_A \times i_1) = g$; and
(c) there exists an interval object (B, i_0, i_1) and a map $h : A \times B \to C$ such that $h \circ (1_A \times i_0) = f$ and $h \circ (1_A \times i_1) = g$.

Proof The cartesian property of $\mathcal{M}$ implies that for any interval object B, the diagram

$$A \times (* \cup^{\emptyset} *) = A \cup^{\emptyset} A \to A \times B \to A$$

is an $A \times I$-object in Quillen's sense, and so can be used to measure homotopy between our two maps. ☐

18.2 Intervals for $\mathcal{M}$-enriched precategories

Let $\mathbb{I} := \Upsilon(*)$ denote the category with two objects υ_0, υ_1 and a single morphism between them. Thus, $\mathbb{I}(\upsilon_0, \ldots, \upsilon_0) = *$, $\mathbb{I}(\upsilon_1, \ldots, \upsilon_1) = *$, $\mathbb{I}(\upsilon_0, \ldots, \upsilon_0, \upsilon_1, \ldots, \upsilon_1) = *$ and the remaining values are $\emptyset$. This is the image of the usual category $[0 \to 1]$ under $\mathcal{PC}(\mathrm{SET}) \to \mathcal{PC}(\mathcal{M})$.

An alternative description of $\mathbb{I}$ in terms of the representable object notation of Section 10.5 is $\mathbb{I} = h([1], *)$. If $\mathcal{A}$ is any $\mathcal{M}$-enriched precategory, a map $\mathbb{I} \to \mathcal{A}$ is the same thing as a triple (x_0, x_1, a) where $x_0, x_1 \in \mathrm{Ob}(\mathcal{A})$ and $a : * \to \mathcal{A}(x_0, x_1)$ is an element of the "set of morphisms from x_0 to x_1." This "set of morphisms" may be denoted by

$$\mathrm{Mor}^1_{\mathcal{A}}(x_0, x_1) := \mathrm{HOM}_{\mathcal{M}}(*, \mathcal{A}(x_0, x_1)) = \mathrm{HOM}^{x_0, x_1}_{\mathcal{PC}(\mathcal{M})}(\mathbb{I}, \mathcal{A}), \qquad (18.2.1)$$

where the superscript on the right designates the subset of maps $\mathbb{I} \to \mathcal{A}$ sending υ_0 to x_0 and υ_1 to x_1.

Let $\mathbf{I}$ denote the image of the category with two isomorphic objects under the map $\mathcal{PC}(\text{SET}) \to \mathcal{PC}(\mathcal{M})$. We think of $\mathbf{I}$ as containing $\mathbb{I}$ as a subcategory. Thus $\mathbf{I}$ again has objects υ_0, υ_1, but $\mathbb{I}(x_0, \ldots, x_p) = *$ for any sequence of objects. One can also view it as the codiscrete precategory with two objects, $\mathbf{I} = \mathbf{codsc}([1]) = \mathbf{codsc}(\{\upsilon_0, \upsilon_1\})$ in the notation of Section 10.5.

If $\mathcal{A} \in \mathcal{PC}(\mathcal{M})$, a map $\mathbf{I} \to \mathcal{A}$ is sure to correspond to an internal equivalence between the images of the two endpoints υ_0, υ_1. Say that a map $\mathbb{I} \to \mathcal{A}$ which extends to $\mathbf{I} \to \mathcal{A}$ is *strongly invertible*. An important little observation is that the identity morphisms (i.e. those given by the images of the degeneracies $* = \mathcal{A}(x_0) \to \mathcal{A}(x_0, x_0)$) are strongly invertible.

Unfortunately, given a general morphism from x_0 to x_1 in $\mathcal{A}$, the corresponding map $\mathbb{I} \to \mathcal{A}$ will not in general extend to $\mathbf{I} \to \mathcal{A}$. That is to say, not all internal equivalences will be strongly invertible. This is why we need to do some further work to construct versal interval objects.

Assuming that $\mathcal{A}$ satisfies the Segal condition and is levelwise fibrant, suppose $x_0, x_1 \in \text{Ob}(\mathcal{A})$ and suppose $a : * \to \mathcal{A}(x_0, x_1)$ is a morphism from x_0 to x_1. The condition of a being an internal equivalence means that there should be morphisms b and c from x_1 to x_0, such that ba is homotopic to the identity of x_0 and ac is homotopic to the identity of x_1. In turn, these homotopies can be represented by maps from interval objects in $\mathcal{M}$ which we shall denote by N and N' respectively. We will build up a big amalgamated sum representing this collection of data.

It turns out to be convenient to relax slightly the conditions that the homotopies go between ab and the identity (resp. ca and the identity). Instead, we say that the homotopies go between ab or ca and strongly invertible morphisms. In particular, the source of c could be an object x_1' different from x_1 and the target of b could be an object x_0' different from x_0.

This situation can be represented diagramatically by

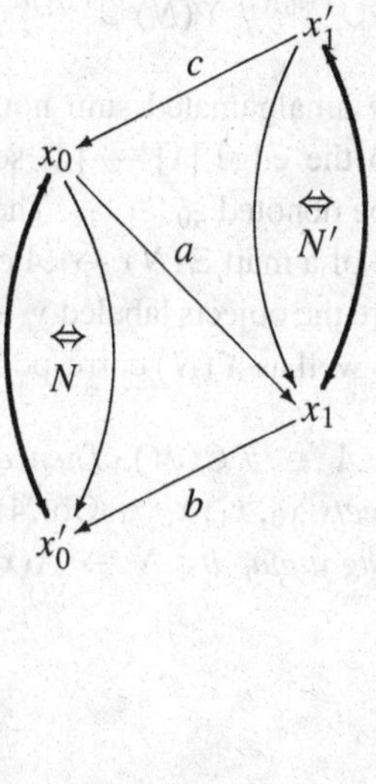

The goal in this section is to construct a precategory $\Xi(N|N')$ such that a map $\Xi(N|N') \to \mathcal{A}$ is the same thing as such a diagram. Notice that the diagram may be divided into two triangles, which are independent except for the fact that they share the same edge labeled a. Our $\Xi(N|N')$ will be an amalgamated sum of two precategories $\Xi(N)$ and $\Xi(N')$ along $\mathbb{I}$, where each of the pieces represents a triangular diagram.

So to start, suppose given an interval object (N, i_0, i_1) with $N \in \mathcal{M}$ and $i_0, i_1 : * \to N$. We will construct a precategory $\Xi(N)$ such that a map from $\Xi(N)$ to $\mathcal{A}$ is the same thing as a diagram of the form

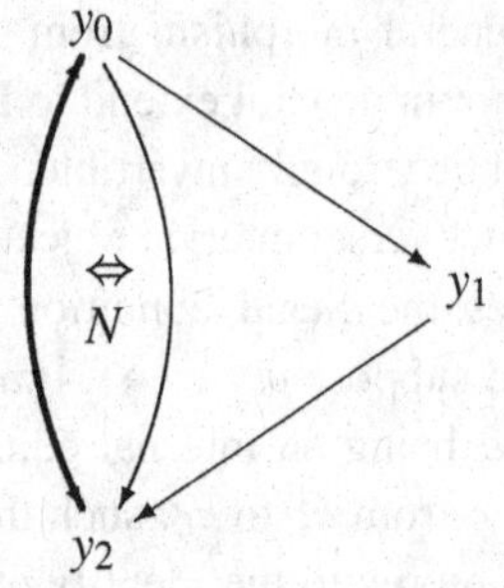

in $\mathcal{A}$. There are three pieces. The part involving N is a map to $\mathcal{A}$ from an $\mathcal{M}$-enriched precategory of the form $\Upsilon(N)$ (see Section 14.1 and Chapter 16), which comes with two maps $\Upsilon(i_0)$ and $\Upsilon(i_1)$ from $\mathbb{I}$ to $\Upsilon(N)$. The commutative triangle corresponds to a map from a representable precategory of the form $h([2], *)$ to $\mathcal{A}$. The strongly invertible morphism on the left corresponds to an extension of one of the $\mathbb{I} \to \Upsilon(N) \to \mathcal{A}$ to a map $\mathbf{I} \to \mathcal{A}$.

Motivated by this picture, define the $\mathcal{M}$-enriched precategory $\Xi(N)$ to be the amalgamated sum of three terms corresponding to these three pieces:

$$\Xi(N) := \mathbf{I} \cup^{\Upsilon(i_0)(\mathbb{I})} \Upsilon(N) \cup^{\Upsilon(i_1)(\mathbb{I})} h([2], *).$$

The map at the end of the amalgamated sum notation is $\mathbb{I} = h([1], *) \to h([2], *)$, corresponding to the edge $[1] \to [2]$ sending 0 to 0 and 1 to 2. The objects of $\Xi(N)$ will be denoted ξ_0, ξ_1, ξ_2. These correspond to the three objects of $h([2], *)$. In case of a map $\Xi(N) \to \mathcal{A}$ corresponding to a diagram as above, the images of ξ_i are the objects labeled y_i above. Thus the two objects υ_0, υ_1 of both copies of $\mathbb{I}$ as well as $\Upsilon(N)$ correspond to ξ_0 and ξ_2 respectively.

Lemma 18.2.1 *Suppose $\mathcal{A} \in \mathcal{PC}(\mathcal{M})$. Then a map $\Xi(N) \to \mathcal{A}$ corresponds to giving three objects $x_0, x_1, x_2 \in \mathrm{Ob}(\mathcal{A})$, to giving an element t : $* \to \mathcal{A}(x_0, x_1, x_2)$, to giving a map $b : N \to A(x_0, x_2)$ and to giving a map*

$g : \mathbf{I} \to \mathcal{A}$ such that $b \circ \Upsilon(i_1) = \partial_{02}(t)$, and $b \circ \Upsilon(i_0) = g(e_{01})$, where $e_{01} : * \to \mathbf{I}(\upsilon_0, \upsilon_1)$ is the unique map.

Proof This comes from the amalgamated sum description for $\Xi(N)$. □

We think of $t : * \to \mathcal{A}(x_0, x_1, x_2)$ as corresponding to a commutative triangle with maps $\partial_{01}(t)$ and $\partial_{12}(t)$ whose "composition" is

$$\partial_{12}(t) \circ \partial_{01}(t) = \partial_{02}(t).$$

Then N can be a homotopy from $\partial_{02}(t)$ to the map $g(e_{01})$ (in our application N will be contractible). Then the extension of this map to g defined on $\mathbf{I}$ says that $g(e_{01})$ is strictly invertible. So, roughly speaking, when we look at a map $\Xi(N) \to \mathcal{A}$ we are looking at two morphisms whose composition $\partial_{12}(t) \circ \partial_{01}(t)$ is equivalent to an invertible map.

The two different maps in question correspond to maps $\zeta_{01}, \zeta_{12} : \mathbb{I} \to \Xi(N)$ with ζ_{01} corresponding to $\partial_{01}(t)$ and ζ_{12} to $\partial_{12}(t)$.

The construction Ξ also works for the other half of $\Xi(N|N')$. We distinguish the two interval objects which are used here, for clarity of notation. Obviously one could choose the same on both sides.

Given two interval objects N and N', we can form

$$\Xi(N|N') := \Xi(N) \cup^{\mathbb{I}} \Xi(N'),$$

where the map $\mathbb{I} \to \Xi(N)$ is ζ_{01} and the map $\mathbb{I} \to \Xi(N')$ is ζ_{12}. These become the same map denoted $\eta : \mathbb{I} \to \Xi(N|N')$. Denote the four objects of $\Xi(N|N')$ by $\xi_{|0}$, $\xi_{0|1}$, $\xi_{1|2}$ and $\xi_{2|}$, these correspond with the objects of $\Xi(N)$ or $\Xi(N')$ by saying that $\xi_{i|j}$ corresponds to ξ_i in the left piece $\Xi(N)$ and to ξ_j in the right piece $\Xi(N')$ to give the following picture of $\Xi(N|N')$:

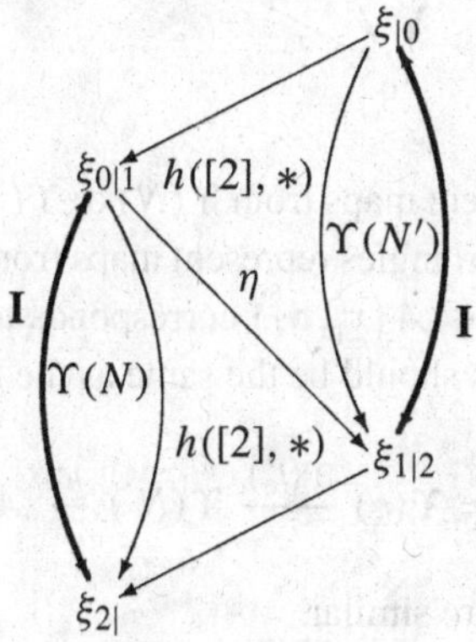

displaying η as a morphism from $\xi_{0|1}$ to $\xi_{1|2}$, i.e. an element of the set $\mathrm{Mor}^1_{\Xi(N|N')}(\xi_{0|1}, \xi_{1|2})$ defined in (18.2.1).

Lemma 18.2.2 *Suppose (N, i_0, i_1) and $\left(N', i'_0, i'_1\right)$ are interval objects of $\mathcal{M}$. If $\mathcal{A}$ is an $\mathcal{M}$-enriched precategory, then a map $\Xi(N|N') \to \mathcal{A}$ corresponds to the data of a morphism (x_0, x_1, a) in $\mathcal{A}$, of two other objects x'_0 and x'_1, together with commutative triangles*

$$s : * \to \mathcal{A}\left(x_0, x_1, x'_0\right), \quad t : * \to \mathcal{A}\left(x'_1, x_0, x_1\right),$$

with maps $h : \Upsilon(N) \to \mathcal{A}$ and $h' : \Upsilon(N') \to \mathcal{A}$ and two maps $u, v : \mathbf{I} \to \mathcal{A}$ such that various maps $\mathbb{I} \to \mathcal{A}$ induced by these data coincide (see the diagram in the proof below).

Proof This comes from the amalgamated sum description of $\Xi(N|N')$ and the corresponding properties for $\Xi(N)$ and $\Xi(N')$. The objects x_0 and x_1 are the images of $\xi_{0|1}$ and $\xi_{1|2}$ while x'_0 is the image of $\xi_{2|}$ and x'_1 is the image of $\xi_{|0}$.

The maps, which are supposed to coincide, may be read off from the diagram

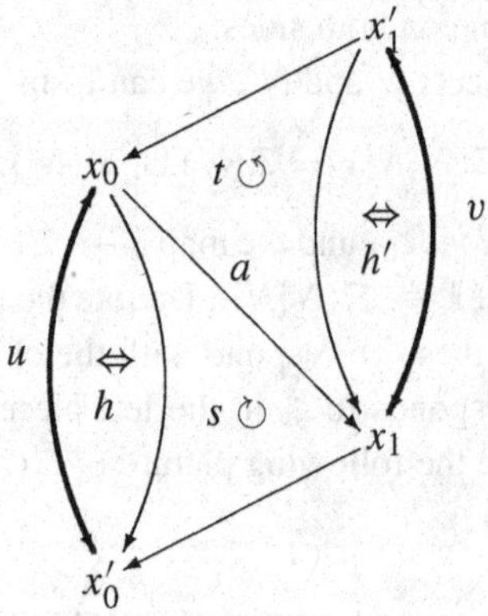

in which the 2-cells represent maps from $\Upsilon(N)$ or $\Upsilon(N')$, the thick lines represent maps from $\mathbf{I}$, and the triangles represent maps from $h([2], *)$. For example, the boundary $\partial_{02} \circ t : * \to \mathcal{A}\left(x'_1, x_1\right)$ corresponds to a map $\mathbb{I} \to \mathcal{A}$ sending υ_0 to x'_1 and υ_1 to x_1. This should be the same as the map

$$\mathbb{I} = \Upsilon(*) \xrightarrow{\Upsilon(i'_0)} \Upsilon(N') \xrightarrow{h'} \mathcal{A}.$$

The other identifications are similar. □

Lemma 18.2.3 *For any two interval objects N, N', the $\mathcal{M}$-precategory $\Xi(N|N')$ is Reedy cofibrant, and indeed the inclusion*

$$\mathbf{disc}\{\xi_{0|1}, \xi_{1|2}\} \to \Xi(N|N')$$

is a Reedy cofibration, hence an injective one.

Proof Use Corollaries 13.6.5 and 13.6.6, and Lemma 14.1.3. □

However, $\Xi(N|N')$ is not projectively cofibrant, because the inclusions of edges $\mathbb{I} \to h([2], N)$ are Reedy but not projective cofibrations. This issue will be addressed further in the comments after Remark 18.3.2 below.

Record here what happens when we change the intervals used in the construction:

Lemma 18.2.4 *Suppose $f : N \to P$ is a morphism between interval objects (N, i_0, i_1) and (P, j_0, j_1), that is $f \circ i_0 = j_0$, $f \circ i_1 = j_1$. Suppose similarly $f' : N' \to P'$ is a morphism between interval objects (N', i'_0, i'_1) and (P', j'_0, j'_1). Then these induce a global weak equivalence*

$$\Xi(N|N') \to \Xi(P|P').$$

Proof It is a levelwise weak equivalnce, being a pushout of maps which are levelwise a weak equivalences. □

18.3 The versality property

From the universal property of Lemma 18.2.2, we obtain the versality property of $\Xi(N|N')$:

Theorem 18.3.1 *Suppose $\mathcal{A} \in \mathcal{PC}(X, \mathcal{M})$ satisfies the Segal condition and is a fibrant object in the Reedy unital diagram model structure $\text{FUNC}_{\text{Reedy}}(\Delta^o_X/X, \mathcal{M})$. Suppose that $x, y \in X = \text{Ob}(\mathcal{A})$ and $a : * \to \mathcal{A}(x, y)$ is an element of $\text{Mor}^1_{\mathcal{A}}(x, y)$. Suppose that a is an inner equivalence, in other words the image of a in the truncated category $\tau_{\leq 1}(\mathcal{A}) \in \text{CAT}$, is invertible. Then for any interval objects N and N' in $\mathcal{M}$ there exists a morphism $\Xi(N|N') \to \mathcal{A}$ sending $\xi_{0|1}$ and $\xi_{2|}$ to x and $\xi_{1|2}$ and $\xi_{|0}$ to y, sending the tautological morphism η to a, and sending the two copies of $\mathbf{I}$ to the identities of x and y respectively.*

Proof Since $\mathcal{A}$ satisfies the Segal condition, the truncation $\tau_{\leq 1}(\mathcal{A})$ may be defined using $\mathcal{A}$ itself, that is to say that the truncation is the 1-category with $\text{Ob}(\mathcal{A})$ as set of objects, and whose nerve relative to this set is the functor

$$\Delta^o_x \to \text{SET}, \quad (x_0, \ldots, x_n) \mapsto \tau_{\leq 0}\mathcal{A}(x_0, \ldots, x_n).$$

The Reedy fibrant condition for $\mathcal{A}$ implies that it is levelwise fibrant, thus for any sequence $(x_0, \ldots, x_n) \in \Delta^o_X$ the image $\mathcal{A}(x_0, \ldots, x_n)$ is a fibrant object of $\mathcal{M}$, hence

$$\tau_{\leq 0}\mathcal{A}(x_0, \ldots, x_n) = \mathrm{HOM}_{\mathcal{M}}(*, \mathcal{A}(x_0, \ldots, x_n))/\sim,$$

where $\sim$ is the relation of homotopy occuring in Lemma 18.1.2.

The fact that a maps to an isomorphism in $\tau_{\leq 1}\mathcal{A}$ therefore means that there is an inverse $b \in \tau_{\leq 0}\mathcal{A}(y, x)$; and by the levelwise fibrant condition it can be represented by $b : * \to \mathcal{A}(y, x)$.

By the Segal condition the morphism

$$\mathcal{A}(x, y, x) \to \mathcal{A}(x, y) \times \mathcal{A}(y, x)$$

is a weak equivalence. On the other hand, the Reedy fibration condition in the diagram category means that the matching map at (x, y, z) is a fibration in $\mathcal{M}$, which in turn implies that the Segal map above is a fibration. Hence it is a trivial fibration, in particular the element $(a, b) : * \to \mathcal{A}(x, y) \times \mathcal{A}(y, x)$ lifts to a map $* \to \mathcal{A}(x, y, x)$. This gives a diagram

$$s : h([2], *) \to \mathcal{A}$$

representing "the composition $b \circ a$," fitting into the lower left triangle in the picture on page 428. The image of s in the nerve of $\tau_{\leq 1}\mathcal{A}$ is the commutative triangle for the composition of the images of b and a. We chose b as representing an inverse to a in the truncated category, so the 02 edge of s is homotopic to the identity of x; by Lemma 18.1.2 there exists a map $N \to \mathcal{A}(x, x)$ representing this homotopy, or by adjunction

$$h : \Upsilon(N) \to \mathcal{A}.$$

This gives the lower or $\Xi(N)$ part of the required diagram. A similar discussion using the fact that $a \circ b$ is homotopic to the identity of y, gives the upper or $\Xi(N')$ part, and these glue together to give the required map $\Xi(N|N') \to \mathcal{A}$. □

Remark 18.3.2 Let $\Xi^{\mathrm{proj}}(N|N')$ denote a cofibrant replacement for $\Xi(N|N')$ in the projective unital diagram category $\mathrm{FUNC}_{\mathrm{proj}}\left(\Delta^o_X/X, \mathcal{M}\right)$. Then it has the same versality property with respect to any $\mathcal{A}$ which is levelwise fibrant and satisfies the Segal condition.

The projectively cofibrant version $\Xi^{\mathrm{proj}}(N|N')$ could be constructed explicitly by inserting objects of the form $\Upsilon(L)$ and $\Upsilon(L')$ in between ζ and the two triangles, for intervals L and L', according to the picture

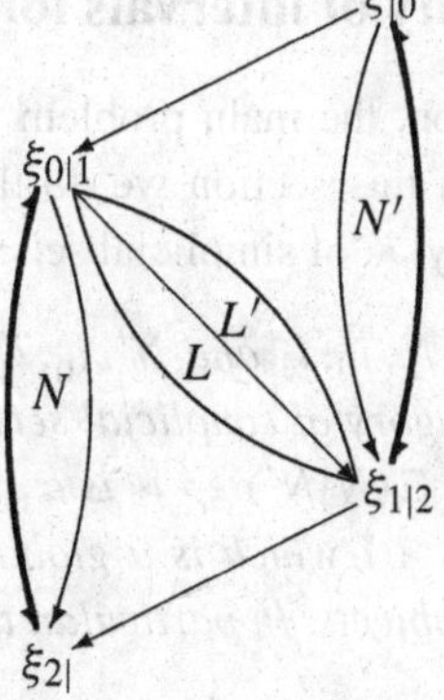

This corresponds to the step in the proof of the theorem where we used the Reedy fibrant property to lift (a, b) to an element $s : * \to \mathcal{A}(x, y, x)$. If $\mathcal{A}$ is assumed only fibrant in the projective structure (i.e. levelwise fibrant) then (a, b) only lifts up to a homotopy given by $L \to \mathcal{A}(x, y) \times \mathcal{A}(y, x)$. The second term may be neglected since we don't care about the bottom arrow of the big diagram, and the piece $L \to \mathcal{A}(x, y)$ corresponds to a map $\Upsilon(L) \to \mathcal{A}$. We presented the Reedy version in our main discussion above because the diagrams are easier to picture.

Let $\widetilde{\Xi}(N|N') \subset \mathbf{Seg}(\Xi(N|N'))$ be the full subcategory containing only the two objects $\xi_{|0}$ and $\xi_{2|}$. In the situation of the theorem, we get by restriction plus functoriality of **Seg** a map

$$\widetilde{\Xi}(N|N') \to \mathbf{Seg}(\mathcal{A}).$$

Similarly, if $\widetilde{\Xi}^{\mathrm{proj}}(N|N') \subset \mathbf{Seg}(\Xi^{\mathrm{proj}}(N|N'))$ is the full subcategory containing only $\xi_{|0}$ and $\xi_{2|}$, then in the situation of the remark we get a map as stated in the following corollary:

Corollary 18.3.3 *Suppose $\mathcal{A}$ is levelwise fibrant and satisfies the Segal conditions, and suppose $x, y \in \mathrm{Ob}(\mathcal{A})$ are two internally equivalent objects. Then there is a map*

$$\widetilde{\Xi}^{\mathrm{proj}}(N|N') \to \mathbf{Seg}(\mathcal{A})$$

sending the two objects of $\widetilde{\Xi}^{\mathrm{proj}}(N|N')$ to x and y respectively.

Proof As in the above remark we get a map $\Xi^{\mathrm{proj}}(N|N') \to \mathcal{A}$, hence by functoriality of **Seg** and composition with the inclusion,

$$\widetilde{\Xi}^{\mathrm{proj}}(N|N') \subset \mathbf{Seg}(\Xi^{\mathrm{proj}}(N|N')) \to \mathbf{Seg}(\mathcal{A})$$

as required. □

It remains to be seen that $\Xi(N|N')$) and thus $\widetilde{\Xi}(N|N')$ are contractible.

18.4 Contractibility of intervals for $\mathcal{K}$-precategories

Given the above construction, the main problem is to prove that the interval $\Xi(N|N')$ is contractible. In this section we do that for enrichment over the Kan–Quillen model category $\mathcal{K}$ of simplicial sets:

Theorem 18.4.1 *Suppose N, i_0, i_1 and N', i'_0, i'_1 are two interval objects in the Kan–Quillen model category of simplicial sets $\mathcal{K}$. Then $\Xi(N|N')$ is contractible in $\mathcal{PC}(\mathcal{K})$, that is $\Xi(N|N') \to *$ is a global weak equivalence. We have a map $\Xi(N|N') \to \mathbf{I} \times \mathbf{I}$ which is a global weak equivalence and an isomorphism on the sets of objects. In particular, the map*

$$\mathbf{Seg}(\Xi(N|N')) \to \mathbf{I} \times \mathbf{I}$$

induces a levelwise weak equivalence, which is to say that

$$\mathbf{Seg}(\Xi(N|N'))(x_0, \ldots, x_p) \text{ is contractible in } \mathcal{M}$$

for any sequence of objects $x_0, \ldots, x_p \in \{\xi_{|0}, \xi_{0|1}, \xi_{1|2}, \xi_{2|}\}$.

Proof This was treated in the last chapter of Pellissier's thesis [211] and our present version is only slightly different in that we have expanded somewhat Ξ as something with four objects. Our present picture is perhaps closer to Drinfeld's intervals for DG-categories [93].

Elements of $\mathcal{PC}(\mathcal{K})$ may be considered as certain kinds of bisimplicial sets (see Section 10.7 and Chapter 15), and this commutes with amalgamated sums. Similarly the diagonal realization from bisimplicial sets to simplicial sets commutes with amalgamated sums and takes Reedy or injective cofibrations[2] in $\mathcal{PC}(\mathcal{K})$ to cofibrations in $\mathcal{K}$ (which are just the monomorphisms). Call the composition of these two operations $|\cdot| : \mathcal{PC}(\mathcal{K}) \to \mathcal{K}$. Note that $|\mathbb{I}|$, $|\mathbf{I}|$, and $|h([2], *)|$ are contractible simplicial sets, and if N is an interval object in $\mathcal{K}$ then $|\Upsilon(N)|$ is contractible. Thus, $|\Xi(N)|$ is a successive pushout along cofibrations between of contractible objects, so it is contractible. Similarly, the amalgamated sum of two of these over the contractible $|\mathbb{I}|$ (mapping into both sides by cofibrations) is contractible, so $|\Xi(N|N')|$ is contractible. In general, a map $A \to \mathbf{Seg}(A)$ induces a weak equivalence of simplicial sets $|A| \xrightarrow{\sim} |\mathbf{Seg}(A)|$. Thus in our case, $|\mathbf{Seg}(\Xi(N|N')|$ is contractible. On the other hand, all of the 1-morphisms in $\Xi(N|N')$ go to invertible morphisms in $\mathbf{Seg}(\Xi(N|N')$, in effect the main middle morphism η has by construction a left and a right inverse up to equivalence; so it goes to an equivalence, and its inverses go to equivalences too. Thus $\mathbf{Seg}(\Xi(N|N')$ is a Segal groupoid. Now,

[2] The Reedy and injective cofibrations are the same in $\mathcal{PC}(\mathcal{K})$ by Proposition 13.7.2 as was pointed out in Corollary 15.1.2.

a Segal groupoid whose realization is contractible, is contractible (see Proposition 15.2.3). Thus $\mathbf{Seg}(\Xi(N|N')$ is contractible, which proves the theorem in the case of $\mathcal{K}$. □

18.5 Construction of a left Quillen functor $\mathcal{K} \to \mathcal{M}$

In order to transfer the above contractibility result for $\Xi(N|N')$ in the $\mathcal{K}$-enriched case, to the general case, we explain in this section the essentially well-known construction of a left Quillen functor $\mathcal{K} \to \mathcal{M}$. Hovey [149] explained the intuition that every monoidal model category is a module over $\mathcal{K}$, and even without the monoidal structure there is a left Quillen functor from $\mathcal{K}$ into $\mathcal{M}$. The construction is based on a choice of contractible cosimplicial object in $\mathcal{M}$, or more precisely a *cosimplicial resolution* in the sense of Hirschhorn [144], see also Bousfield [57]. That means a functor $\mathbf{R} : \Delta \to \mathcal{M}$ which is cofibrant in the Reedy model structure $\text{FUNC}_{\text{Reedy}}(\Delta, \mathcal{M})$.

Recall that an object $A \in \mathcal{M}$ is *contractible* if the unique morphism $A \to *$ is a weak equivalence. We say that a cosimplicial object $\mathbf{R} : \Delta \to \mathcal{M}$ is *levelwise contractible* if $\mathbf{R}([n])$ is contractible for each object $[n] \in \text{Ob}(\Delta)$.

Lemma 18.5.1 *There exists a choice of Reedy cofibrant levelwise contractible cosimplicial object* $\mathbf{R} : \Delta \to \mathcal{M}$*, such that* $\mathbf{R}([0]) = *$.

Proof See Hirschhorn [144]. Note that the construction proceeds inductively on n, and we can start out with $\mathbf{R}([0]) = *$. Assuming $\mathbf{R}([k])$ is defined for $k < n$ then the latching and matching objects at $[n]$ are defined, and one chooses a factorization

$$\text{latch}(\mathbf{R}, [n]) \to \mathbf{R}([n]) \to \text{match}(\mathbf{R}, [n])$$

as a trivial fibration composed with a cofibration. The $\text{match}(\mathbf{R}, [n])$ are contractible at each stage as may be seen inductively. This factorization defines the structural morphisms to extend $\mathbf{R}$ up to $[n]$. □

Remark 18.5.2 In the inductive situation where $\mathcal{M} = \mathcal{PC}(\mathcal{M}')$ is a model category of the kind we are presently constructing, another more concrete choice is to let $\mathbf{R}([n]) := \mathbf{codsc}([n])$ be the codiscrete 1-category on the set $[n]$ consisting of $n + 1$ objects, considered as an $\mathcal{M}'$-precategory via $\text{SET} \to \mathcal{M}'$. This works notably for m-precategories with $\mathcal{M} = \mathcal{PC}^m(\text{SET})$.

Fix one choice of cosimplicial resolution $\mathbf{R}$ as in Lemma 18.5.1, from now on. The objects $\mathbf{R}([n])$ may be thought of as the "standard n-simplices" in $\mathcal{M}$. If $A \in \mathcal{M}$, define $\mathbf{R}^*(A) : \Delta^o \to \text{SET}$ to be the functor

$$\mathbf{R}^*(A) : [n] \mapsto \text{HOM}_{\mathcal{M}}(\mathbf{R}([n]), A).$$

Theorem 18.5.3 *Suppose* $\mathbf{R}$ *is a Reedy cofibrant levelwise contractible cosimplicial object, then* $\mathbf{R}^*$ *is a right Quillen functor from* $\mathcal{M}$ *to* $\mathcal{K} = \text{FUNC}(\Delta^o, \text{SET})$. *Its left adjoint*

$$\mathbf{R}_! : \mathcal{K} \to \mathcal{M}$$

is a left Quillen functor given by the usual formula for the topological realization of a simplicial set, but with the standard n-simplex replaced by $\mathbf{R}([n]) \in \mathcal{M}$. *Assuming that* $\mathbf{R}([0]) = *$, *the functor* $\mathbf{R}_!$ *is unital* $\mathbf{R}_!(*) = *$, *and weakly compatible with products.*

Proof See Hirschhorn [144]. If $h([n])$ denotes the representable simplicial set corresponding to the object $[n]$, i.e. the standard n-simplex, then $\mathbf{R}_!(h([n])) = \mathbf{R}([n])$. Assuming $\mathbf{R}([0]) = *$, this gives the unitality condition for $\mathbf{R}_!$. For weak compatibility with products, notice that $h([n]) \times h([m])$ is a contractible simplicial set. If v is any one of its vertices, then $\{v\} \to h([n]) \times h([m])$ is a trivial cofibration, and $\mathbf{R}_!$ preserves trivial cofibrations. Hence

$$* = \mathbf{R}_!(\{v\}) \to \mathbf{R}_!(h([n]) \times h([m]))$$

is a trivial cofibration, showing that $\mathbf{R}_!(h([n]) \times h([m]))$ is contractible. In particular, the map

$$\mathbf{R}_!(h([n]) \times h([m])) \to \mathbf{R}_!(h([n])) \times \mathbf{R}_!(h([m]))$$

is a weak equivalence. A standard inductive argument on dimensions of cells and using the pushout-product diagrams, similar in outline to the discussion in Chapter 17, shows that $\mathbf{R}_!$ is weakly compatible with products. □

Corollary 18.5.4 *The realization functor induces, for every set X, a functor*

$$\mathcal{PC}(X, \mathbf{R}_!) : \mathcal{PC}_c(X, \mathcal{K}) \to \mathcal{PC}_c(X, \mathcal{M}).$$

This is a left Quillen functor for c denoting either the projective, injective or Reedy model structures on $\mathcal{K}$ *and* $\mathcal{M}$*-enriched precategories over X. It is compatible with change of set X, and gives a functor*

$$\mathcal{PC}(\mathbf{R}_!) : \mathcal{PC}(\mathcal{K}) \to \mathcal{PC}(\mathcal{M})$$

from the Segal precategories to the $\mathcal{M}$*-enriched precategories.*

Proof Combine Theorem 18.5.3 with the discussion of Proposition 12.6.2 and Theorem 14.7.3. If $\mathcal{A} \in \mathcal{PC}(X, \mathcal{K})$ then for any sequence $(x_0, \ldots, x_n)$ of elements of X,

$$(\mathcal{PC}(\mathbf{R}_!)\mathcal{A})(x_0, \ldots, x_n) = \mathbf{R}_!(\mathcal{A}(x_0, \ldots, x_n)).$$ □

The corresponding right Quillen functor

$$\mathcal{PC}(\mathbf{R}^*) : \mathcal{PC}(\mathcal{M}) \to \mathcal{PC}(\mathcal{K})$$

should be applied to $\mathcal{A} \in \mathcal{PC}(\mathcal{M})$ only after taking a fibrant replacement $\mathcal{A} \to \mathcal{A}'$. Define $\mathbf{Int}_{\mathbf{R}}(\mathcal{A}) := \mathcal{PC}(\mathbf{R}^*)(\mathcal{A}')$. We call this the **R**-*interior* of $\mathcal{A}$, since it is obtained by looking at maps from the standard simplices $\mathbf{R}([n])$ into $\mathcal{A}(x_0, \ldots, x_n)$ so it measures $\mathcal{A}$ "from the inside." This construction is compatible with truncation:

Lemma 18.5.5 *For $\mathcal{A} \in \mathcal{PC}(\mathcal{M})$ we have an isomorphism of categories $\tau_{\leq 1}(\mathcal{A}) \cong \tau_{\leq 1}(\mathbf{Int}_{\mathbf{R}}(\mathcal{A}))$.*

Proof This follows from the definition of $\tau_{\leq 1}$ in Section 12.4. □

By its construction as a colimit, $\mathcal{PC}(\mathbf{R}_!)$ preserves amalgamated sums, preserves constructions Υ and h, preserves the various notions of cofibrancy. Since $\mathcal{M}$ is cartesian, Theorem 14.7.3 says that $\mathcal{PC}(\mathbf{R}_!)$ takes weak equivalences in $\mathcal{PC}(X, \mathcal{K})$ to weak equivalences in $\mathcal{PC}(X, \mathcal{M})$. Furthermore, since it preserves truncations, $\mathcal{PC}(\mathbf{R}_!)$ preserves global weak equivalences, and preserves the notion of interval objects.

The choice of $\mathbf{R}_!$ doesn't give a simplicial model category structure on $\mathcal{M}$, because it isn't necessarily strictly compatible with products. One level up on $\mathcal{PC}(\mathcal{M})$, Rezk [219] points out that the model category we are constructing will not in general be simplicial, a property enjoyed by his structure. In terms of the choice discussed in Remark 18.5.2, the problem is that the product $\mathbf{R}([m]) \times \mathbf{R}([n])$ is the codiscrete category $\mathbf{codsc}([m] \times [n])$, which isn't exactly equal to the Eilenberg–Zilber amalgamated sum of the many simplices $\mathbf{codsc}[m+n]$. The Reedy structure on $\mathcal{PC}(\mathcal{M})$, which we are in the process of constructing, should, however, be an *approximate simplicial model category* in the sense of Bauer and Datuashvili [31].

18.6 Contractibility in general

We can now use the functor $\mathcal{PC}(X, \mathbf{R}_!)$ to transfer the the contractibility result for $\mathcal{K}$-enriched precategories, to $\mathcal{PC}(\mathcal{M})$. This yields the first main theorem of the present chapter saying that $\Xi(N|N')$ is a good interval object in $\mathcal{PC}(\mathcal{M})$. This was the last step missing in Pellissier's [211] correction of my preprint [234], but which is actually straightforward from a Quillen-functorial point of view.

The contractibility statement is made before we have completely finished the construction of the model structure, although it is the penultimate step. Some care is therefore still necessary in using only the parts of the model structure which are already known.

Theorem 18.6.1 *Suppose N, i_0, i_1 and N', i'_0, i'_1 are two interval objects. Then $\Xi(N|N')$ is contractible in $\mathcal{PC}(\mathcal{M})$, that is $\Xi(N|N') \rightarrow *$ is a global weak equivalence. We have a map $\Xi(N|N') \rightarrow \mathbf{I} \times \mathbf{I}$ which is a global weak equivalence and an isomorphism on the sets of objects. In particular, the map*

$$\mathbf{Seg}(\Xi(N|N')) \rightarrow \mathbf{I} \times \mathbf{I}$$

induces a levelwise weak equivalence, which is to say that

$$\mathbf{Seg}(\Xi(N|N'))(x_0, \ldots, x_p) \text{ is contractible in } \mathcal{M}$$

for any sequence of objects $x_0, \ldots, x_p \in \{\xi_{|0}, \xi_{0|1}, \xi_{1|2}, \xi_{2|}\}$.

Proof First notice that the statement of the theorem is independent of the choice of interval object: if $N \rightarrow P$ and $N' \rightarrow P'$ are maps of interval objects then the statement of the theorem for (N, N') is equivalent to the statement for (P, P'). See Lemma 18.2.4. In particular it suffices to prove the statement for one pair of intervals.

Theorem 18.4.1 gives the same statement for precategories enriched over the Kan–Quillen model category $\mathcal{K}$ of simplicial sets. Then choose a left Quillen functor $\mathbf{R}_! : \mathcal{K} \rightarrow \mathcal{M}$ as in the previous Section 18.5. This gives a functor $\mathcal{PC}(\mathbf{R}_!) : \mathcal{PC}(\mathcal{K}) \rightarrow \mathcal{PC}(\mathcal{M})$ which preserves amalgamated sums. Suppose (B, i_0, i_1) is an interval object in $\mathcal{K}$. Then $\mathbf{R}_!(B)$ is an interval object in $\mathcal{M}$ and

$$\mathcal{PC}(\mathbf{R}_!)(\Xi(B|B)) = \Xi(\mathbf{R}_!(B)|\mathbf{R}_!(B)).$$

Now since $\mathcal{PC}(\mathbf{R}_!)$ preserves global weak equivalences, we obtain the statement of the theorem for the pair of interval objects $\mathbf{R}_!(B)|\mathbf{R}_!(B)$. By the invariance discussed in the first paragraph of the proof, this implies the statement of the theorem for all N, N'. □

Recall that $\widetilde{\Xi}(N|N') \subset \mathbf{Seg}(\Xi(N|N'))$ is the full subcategory containing only the two objects $\xi_{|0}$ and $\xi_{2|}$. Since all objects of $\mathbf{Seg}(\Xi(N|N'))$ are equivalent, the inclusion

$$\widetilde{\Xi}(N|N') \hookrightarrow \mathbf{Seg}(\Xi(N|N'))$$

is a global weak equivalence (it is by definition fully faithful and both sides satisfy the Segal conditions). It follows from Theorem 18.6.1 and the 3 for 2 property of global weak equivalences, that the functor

$$p_{N,N'} : \widetilde{\Xi}(N|N') \to \mathbf{I}$$

is a global weak equivalence; furthermore this induces isomorphisms on the sets of objects (there are exactly two objects on each side), and both sides satisfy the Segal conditions, so $p_{N,N'}$ is a levelwise weak equivalence of diagrams.

18.7 Pushout of trivial cofibrations

These interval objects allow us to analyze pushouts along trivial cofibrations which are not isomorphisms on objects. In this discussion, we use Reedy cofibrations, but they could be replaced by injective ones when the injective structure exists.

We start by considering the pushout along the standard interval **I**:

Lemma 18.7.1 *Suppose $\mathcal{A} \in \mathcal{PC}(\mathcal{M})$, and $y \in \mathrm{Ob}(\mathcal{A})$. Then the pushout morphism*

$$a : \mathcal{A} \to \mathcal{A} \cup^{\{y\}} \mathbf{I},$$

obtained by identifying υ_0 and y, is a global weak equivalence.

Proof By Corollary 17.2.6 applied to the identity of $\mathcal{A}$ and the map $p : \mathbf{I} \to *$, the map

$$1_{\mathcal{A}} \times p : \mathcal{A} \times \mathbf{I} \to \mathcal{A}$$

is a global weak equivalence. Let $i_0, i_1 : * \to \mathbf{I}$ be the two inclusions of objects υ_0 and υ_1. The two maps

$$1_{\mathcal{A}} \times i_0, 1_{\mathcal{A}} \times i_1 : \mathcal{A} \to \mathcal{A} \times \mathbf{I}$$

are global weak equivalences, as can be seen by composing with $1_{\mathcal{A}} \times p$ and using 3 for 2.

Now, consider the morphism $g : \mathbf{I} \times \mathbf{I} \to \mathbf{I} \times \mathbf{I}$ equal to the identity on $\mathbf{I} \times \{\upsilon_0\}$ and sending $\mathbf{I} \times \{\upsilon_1\}$ to the single object (υ_0, υ_1). Set

$$\mathcal{B} := \mathcal{A} \cup^{\{y\}} \mathbf{I}.$$

Let $q : \mathcal{B} \to \mathcal{A}$ denote the projection obtained by sending all of **I** to the single object $y \in \mathrm{Ob}(\mathcal{A})$. Then

$$\mathcal{B} \times \mathbf{I} = (\mathcal{A} \times \mathbf{I}) \cup^{\{\upsilon_0\} \times \mathbf{I}} (\mathbf{I} \times \mathbf{I}).$$

Apply the map g to the second factor of this pushout, to obtain a map

$$f : \mathcal{B} \times \mathbf{I} \to \mathcal{B} \times \mathbf{I}$$

such that f restricts to the identity on $\mathcal{B} \times \{\upsilon_0\}$, while $f|_{\mathcal{B} \times \{\upsilon_1\}}$ is the projection $q : \mathcal{B} \to \mathcal{A}$. By the first paragraph of the proof, the maps induced by f,

$$\mathcal{B} \times \{\upsilon_0\} \to \mathcal{B} \times \mathbf{I}$$

and

$$\mathcal{B} \times \{\upsilon_1\} \to \mathcal{B} \times \mathbf{I}$$

are global weak equivalences. The composition of $1_{\mathcal{B}} \times p : \mathcal{B} \times \mathbf{I} \to \mathcal{B}$ with the morphism f considered above, is a morphism

$$(1_{\mathcal{B}} \times p) \circ f : \mathcal{B} \times \mathbf{I} \to \mathcal{B}$$

such that the composition $(1_{\mathcal{B}} \times p) \circ f \circ (1_{\mathcal{B}} \times i_0)$ is the identity of $\mathcal{B}$, and the composition $(1_{\mathcal{B}} \times p) \circ f \circ (1_{\mathcal{B}} \times i_0)$ is the composition

$$\mathcal{B} \xrightarrow{q} \mathcal{A} \xrightarrow{a} \mathcal{B}.$$

The facts that $(1_{\mathcal{B}} \times p) \circ f \circ (1_{\mathcal{B}} \times i_0)$ is the identity of $\mathcal{B}$, and that $(1_{\mathcal{B}} \times i_0)$ is a global weak equivalence, imply by 3 for 2 that $(1_{\mathcal{B}} \times p) \circ f$ is a global weak equivalence. But then, composing with the global weak equivalence $(1_{\mathcal{B}} \times i_1)$ we see that $(1_{\mathcal{B}} \times p) \circ f \circ (1_{\mathcal{B}} \times i_0)$ is a global weak equivalence, in other words that the composition aq is a global weak equivalence. In the other direction, the composition

$$\mathcal{A} \xrightarrow{a} \mathcal{B} \xrightarrow{q} \mathcal{A}$$

is the identity of $\mathcal{A}$. Thus, we conclude from the last sentence of Theorem 12.5.4 that both q and the inclusion $\mathcal{A} \to \mathcal{B}$ are global weak equivalences. This last statement is what we are supposed to prove. □

Corollary 18.7.2 *Suppose $\mathcal{B}$ is an $\mathcal{M}$-enriched precategory with two objects b_0, b_1. Suppose $\mathcal{B}$ satisfies the Segal conditions and is contractible, that is the map $\mathcal{B} \to *$ is a global weak equivalence. Then for any $\mathcal{A} \in \mathcal{PC}(\mathcal{M})$ and any object $y \in \mathrm{Ob}(\mathcal{A})$ the map*

$$\mathcal{A} \to \mathcal{A} \cup^{\{b_0\}} \mathcal{B}$$

obtained by identifying b_0 to y, is a global weak equivalence.

Proof There is a unique map $f : \mathcal{B} \to \mathbf{I}$ sending b_0 to υ_0 and b_1 to υ_1. This map is a global weak equivalence, as seen by applying 3 for 2 to the composition

$$\mathcal{B} \to \mathbf{I} \to *.$$

But since it induces an isomorphism on objects, and both sides satisfy the Segal conditions, it is a levelwise weak equivalence of diagrams. Applying f to the second piece of the given pushout, we get a map

$$g : \mathcal{A} \cup^{\{b_0\}} \mathcal{B} \to \mathcal{A} \cup^{\{y\}} \mathbf{I}$$

to the pushout considered in the previous corollary. However, g is a levelwise weak equivalence of diagrams, so it is a global weak equivalence. Note that g commutes with the maps from $\mathcal{A}$, so by the previous corollary and 3 for 2 we conclude that the map of the present statement is a global weak equivalence. □

Suppose $f : \mathcal{A} \to \mathcal{B}$ is a trivial Reedy cofibration, and suppose $\mathcal{B}$ is levelwise fibrant and satisfies the Segal conditions. Let $Z := \mathrm{Ob}(\mathcal{B}) - f(\mathrm{Ob}(\mathcal{A}))$. For each $z \in Z$ choose $e(z) \in \mathrm{Ob}(A)$ and

$$a(z) \in \mathcal{B}(f(e(z)), z),$$

such that the image of $a(z)$ is invertible in the truncated category. This is possible by the definition of essential surjectivity of $\mathcal{A} \to \mathcal{B}$.

Applying Theorem 18.3.1 there exist collections of interval objects N_z, N'_z indexed by $z \in Z$, and functors $t_i : \Xi\left(N_z|N'_z\right) \to \mathcal{B}$ sending $\xi_{|0}$ to $e(z)$, $\xi_{2|}$ to z, and sending the tautological morphism η to $a(z)$. By functoriality of the construction **Seg** we get

$$\mathbf{Seg}\left(\Xi\left(N_z|N'_z\right)\right) \to \mathbf{Seg}(\mathcal{B}),$$

and restricting this gives $\tilde{t}_i : \widetilde{\Xi}\left(N_z|N'_z\right) \to \mathbf{Seg}(\mathcal{B})$. Now, $\tilde{t}_i$ sends the first object to $e(z) \in \mathcal{A}$ and the second object to z. Putting these all together we get a morphism in $\mathcal{PC}(\mathcal{M})$,

$$\mathcal{A} \cup^{\coprod_{z\in Z}\{\tilde{t}_i\xi(|0)\}} \coprod_{z\in Z} \widetilde{\Xi}\left(N_z|N'_z\right) \xrightarrow{T} \mathbf{Seg}(\mathcal{B}),$$

and now T induces an isomorphism on sets of objects. It is no longer necessarily a cofibration. We would like to show that T is a global weak equivalence.

Corollary 18.7.3 *Suppose $\mathcal{A} \in \mathcal{PC}(\mathcal{M})$, and $y \in \mathrm{Ob}(\mathcal{A})$. Suppose N, N' are interval objects in $\mathcal{M}$. Then the pushout morphism*

$$\mathcal{A} \to \mathcal{A} \cup^{\{y\}} \widetilde{\Xi}\left(N_z|N'_z\right)$$

obtained by identifying $\xi(|0)$ and y, is a global weak equivalence.

Proof Apply Corollary 18.7.2 with $\mathcal{B} = \mathbf{Seg}\left(\Xi\left(N_z|N'_z\right)\right)$. □

Corollary 18.7.4 *In the situation described above the preceding corollary, the morphism*

$$\mathcal{A} \to \mathcal{A} \cup^{\coprod_{z\in Z}\{\tilde{r}_i\xi(|0)\}} \coprod_{z\in Z} \widetilde{\Xi}\left(N_z|N'_z\right)$$

is a global weak equivalence. Given that $\mathcal{A} \to \mathcal{B}$ was a global weak equivalence, the functor

$$T : \mathcal{A} \cup^{\coprod_{z\in Z}\{\tilde{r}_i\xi(|0)\}} \coprod_{z\in Z} \widetilde{\Xi}\left(N_z|N'_z\right) \to \mathbf{Seg}(\mathcal{B})$$

is a global weak equivalence inducing an isomorphism on sets of objects.

Proof Choose a well ordering of Z, giving an exhaustion of Z by subsets Z_i indexed by an ordinal $i \in \beta$. Let $\mathcal{B}_i \subset \mathcal{B}$ be the full subobject whose object set is $f(\mathrm{Ob}(A)) \cup Z_i$. By transfinite induction we obtain that the functors

$$T_i : \mathcal{A} \cup^{\coprod_{z\in Z_i}\{\tilde{r}_i\xi(|0)\}} \coprod_{z\in Z_i} \widetilde{\Xi}\left(N_z|N'_z\right) \to \mathbf{Seg}(\mathcal{B}_i)$$

are global weak equivalences, using the previous corollary at each step. At the end of the induction we obtain the required statement. □

We are now ready to show the preservation of global trivial cofibrations under pushouts:

Theorem 18.7.5 *Suppose $\mathcal{A} \to \mathcal{B}$ is a trivial Reedy cofibration. Suppose $\mathcal{A} \to \mathcal{C}$ is any morphism in $\mathcal{PC}(\mathcal{M})$. Then the pushout morphism*

$$\mathcal{C} \to \mathcal{C} \cup^{\mathcal{A}} \mathcal{B}$$

is a global weak equivalence.

Proof We first show this statement assuming that all three objects $\mathcal{A}$, $\mathcal{B}$ and $\mathcal{C}$ satisfy the Segal conditions. Noting that $\mathcal{B} \to \mathbf{Seg}(\mathcal{B})$ is an isomorphism on sets of objects and applying Lemma 14.3.4, it suffices to show that the map

$$\mathcal{C} \to \mathcal{C} \cup^{\mathcal{A}} \mathbf{Seg}(\mathcal{B})$$

is a global weak equivalence. Define

$$\mathcal{F} := \mathcal{A} \cup^{\coprod_{z\in Z}\{\tilde{\iota}_i \xi(|0)\}} \coprod_{z\in Z} \widetilde{\Xi}\left(N_z|N_z'\right),$$

and consider the map $T : \mathcal{F} \to \mathcal{B}$ defined above. By Corollary 18.7.4, T is a global weak equivalence inducing an isomorphism on sets of objects. By Lemma 14.3.4 it follows that the map

$$\mathcal{C} \cup^{\mathcal{A}} \mathcal{F} \to \mathcal{C} \cup^{\mathcal{A}} \mathbf{Seg}(\mathcal{B})$$

is a global weak equivalence, so by 3 for 2 it suffices to show that

$$\mathcal{C} \to \mathcal{C} \cup^{\mathcal{A}} \mathcal{F}$$

is a global weak equivalence. The map $\mathcal{A} \to \mathcal{F}$ is obtained as a transfinite composition of pushouts along things of the form $\{\xi(|0)\} \to \widetilde{\Xi}\left(N_z|N_z'\right)$, and by Corollary 18.7.3 these pushouts are global weak equivalences. Thus, the map $\mathcal{C} \to \mathcal{C} \cup^{\mathcal{A}} \mathcal{F}$ is a global weak equivalence, which finishes this part of the proof.

Starting with $\mathcal{C} \leftarrow \mathcal{A} \to \mathcal{B}$ in general, let $\mathcal{A}' := \mathbf{Seg}(\mathcal{A})$, then

$$\mathcal{B}' := \mathbf{Seg}(\mathcal{A}' \cup^{\mathcal{A}} \mathcal{B}), \quad \mathcal{C}' := \mathbf{Seg}(\mathcal{A}' \cup^{\mathcal{A}} \mathcal{C}).$$

We get a diagram

$$\begin{array}{ccccc}
\mathcal{C} & \longleftarrow & \mathcal{A} & \longrightarrow & \mathcal{B} \\
\downarrow & & \downarrow & & \downarrow \\
\mathcal{C}' & \longleftarrow & \mathcal{A}' & \longrightarrow & \mathcal{B}'
\end{array}$$

such that the bottom row satisfies the hypothesis for the first part of the proof (all objects satisfy the Segal condition and the second map is a global trivial cofibration), and such that the vertical arrows are global weak equivalences inducing isomorphisms on sets of objecs. By Lemma 14.3.4, the bottom map in the diagram

$$\begin{array}{ccc}
\mathcal{C} & \longrightarrow & \mathcal{C}' \\
\downarrow & & \downarrow \\
\mathcal{C} \cup^{\mathcal{A}} \mathcal{B} & \longrightarrow & \mathcal{C}' \cup^{\mathcal{A}'} \mathcal{B}'
\end{array}$$

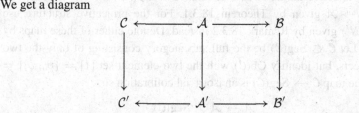

is a global weak equivalence. The top vertical map is a global weak equivalence by construction of $\mathcal{C}'$ and the right vertical map is one too, by the first part of the proof above. By 3 for 2 we conclude that the left vertical map is a global weak equivalence, as required. □

18.8 A versality property

The versality properties for the intervals constructed above yield a similar versality property for any cofibrant replacement of $\mathbf{I}$ if the target $\mathcal{A}$ is fibrant in the diagram structure $\text{FUNC}_c\left(\Delta^i_{\text{Ob}(\mathcal{A})}/\text{Ob}(\mathcal{A}), \mathcal{M}\right)$:

Theorem 18.8.1 *Suppose $\mathcal{A} \in \mathcal{PC}(\mathcal{M})$, and suppose $\mathcal{A}$ is fibrant as an object of $\mathcal{PC}_c(\text{Ob}(\mathcal{A}), \mathcal{M})$, where c indicates either the projective, the Reedy or the injective structures. Let $\mathcal{B} \to \mathbf{I}$ be a cofibrant replacement in $\mathcal{PC}_c([1], \mathcal{M})$, so $\text{Ob}(\mathcal{B})$ is still $[1] = \{\upsilon_0, \upsilon_1\}$. Then if $x, y \in \text{Ob}(\mathcal{A})$ are two objects, they project to equivalent objects in $\tau_{\leq 1}(\mathcal{A})$ if and only if there exists a morphism $\mathcal{B} \to \mathcal{A}$ sending υ_0 to x and υ_1 to y.*

Proof Since $\tau_{\leq 1}\mathcal{B} = \mathbf{I}$ is the category with two isomorphic objects, existence of a map $\mathcal{B} \to \mathcal{A}$ sending υ_0 to x and υ_1 to y implies that x and y are internally equivalent in $\mathcal{A}$.

Suppose x and y are internally equivalent. If $\mathcal{A}$ is a fibrant object for the Reedy or injective model structures relative to $\text{Ob}(\mathcal{A})$, there is a morphism $\Xi(N|N) \to \mathcal{A}$ given by Theorem 18.3.1. For the projective structure use $\Xi^{\text{proj}}(N|N')$ given by Remark 18.3.2 instead. Denote either of these maps by $\mathcal{C} \to \mathcal{A}$. Let $\widetilde{\mathcal{C}} \subset \mathbf{Seg}(\mathcal{C})$ be the full subcategory consisting of only the two main objects, but identify $\text{Ob}(\widetilde{\mathcal{C}})$ with the two-element set $[1] = \{\upsilon_0, \upsilon_1\} = \text{Ob}(\mathbb{I})$. The map $\mathcal{C} \to \mathbf{Seg}(\mathcal{C})$ is an isotrivial cofibration so

$$\mathcal{A} \to \mathcal{A} \cup^{\mathcal{C}} \mathbf{Seg}(\mathcal{C})$$

is an isotrivial cofibration by Theorem 14.3.3; it follows that our map extends to $\mathbf{Seg}(\mathcal{C}) \to \mathcal{A}$. This now restricts to a map

$$\widetilde{\mathcal{C}} \to \mathcal{A}$$

sending υ_0 to x and υ_1 to y. By contractibility, Theorem 18.6.1,

$$\widetilde{\mathcal{C}} \to \mathbf{I}$$

is a weak equivalence inducing an isomorphism on sets of objects. Choose a factorization

$$\widetilde{\mathcal{C}} \xrightarrow{i} \widetilde{\mathcal{C}}' \xrightarrow{p} \mathbf{I},$$

where i is a trivial cofibration and p is a trivial fibration in $\mathcal{PC}([1], \mathcal{M})$. Again our map extends to $\widetilde{\mathcal{C}}' \to \mathcal{A}$, but now since $\mathcal{B}$ is cofibrant and p is a trivial fibration there is a lifting $\mathcal{B} \to \widetilde{\mathcal{C}}'$ inducing the identity on the set of objects. We get the required map $\mathcal{B} \to \mathcal{A}$. □

The importance of this versality property is that it allows us to replace a global weak equivalence by one which is surjective on sets of objects:

Corollary 18.8.2 *Let $\mathcal{B} \to \mathbf{I}$ be a cofibrant replacement in one of the model categories $\mathcal{PC}_c([1], \mathcal{M})$. Suppose $f : \mathcal{A} \to \mathcal{C}$ is a global weak equivalence, and suppose $\mathcal{C}$ is a fibrant object in $\mathcal{PC}_c(\mathrm{Ob}(\mathcal{C}), \mathcal{M})$. Then there exists a pushout $\mathcal{A} \to \mathcal{A}'$ by a collection of copies of $\{\upsilon_0\} \hookrightarrow \mathcal{B}$, and a map $\mathcal{A}' \to \mathcal{C}$ which is a global weak equivalence and a surjection on sets of objects.*

Proof For each object $y \in \mathrm{Ob}(\mathcal{C})$ choose $x \in \mathrm{Ob}(\mathcal{A})$ such that $f(x)$ is internally equivalent to y. For each such pair we get a map $\mathcal{B} \to \mathcal{C}$ sending υ_0 to $f(x)$ and υ_1 to y; attaching a copy of $\mathcal{B}$ to $\mathcal{A}$ by sending υ_0 to x and then doing this for all objects y we obtain the required pushout $\mathcal{A}'$ and extension of the map. □

The global model structure we are constructing will be left proper. Given a cofibration $f : \mathcal{A} \to \mathcal{B}$ and a weak equivalence $g : \mathcal{A} \to \mathcal{C}$ we need to show that the map $h : \mathcal{B} \to \mathcal{B} \cup^{\mathcal{A}} \mathcal{C}$ is a weak equivalence. We may assume that $\mathcal{A}$, $\mathcal{B}$ and $\mathcal{C}$ satisfy the Segal conditions. If $\mathrm{Ob}(g)$ is an isomorphism then so is $\mathrm{Ob}(h)$, and the map g induces a weak equivalence in $\mathcal{M}$ for each sequence of objects. By left properness of $\mathcal{M}$ the same is true for h thus h is a weak equivalence. Using this, the statement becomes invariant under weak equivalences inducing isomorphisms on objects, in the variables $\mathcal{A}$ and $\mathcal{C}$; it is also invariant under weak equivalences in the variable $\mathcal{B}$. For the general case, using the previous corollary we may assume that $\mathrm{Ob}(g)$ is surjective. Define $\mathcal{A}'$ and $\mathcal{B}'$ by throwing out all but one object of $\mathcal{A}$ in the fiber over each object of $\mathcal{C}$, so $\mathcal{A}' \to \mathcal{C}$ is a weak equivalence and an isomorphism on objects. Let $\mathcal{A}''$ be obtained from $\mathcal{A}'$ by pushout over a collection of intervals, so that $\mathcal{A}'' \to \mathcal{A}$ is an isomorphism on objects, and let $\mathcal{B}''$ be the corresponding pushout of $\mathcal{B}'$. Projecting back from $\mathcal{A}''$ to $\mathcal{A}'$ by sending the intervals to points, the map $\mathcal{B}'' \to \mathcal{B}'' \cup^{\mathcal{A}''} \mathcal{A}' = \mathcal{B}'$ is a weak equivalence, and by applying several times the invariance properties described previously, we get to the required statement.

19

The model category of $\mathcal{M}$-enriched precategories

In this chapter, we finish the proof that the category $\mathcal{PC}(\mathcal{M})$ of $\mathcal{M}$-enriched precategories, with variable set of objects, has natural Reedy and projective model structures. Given the product theorem of Chapter 17 and the discussion of intervals in Chapter 18, the proof presents no further obstacles. We also show that the Reedy structure $\mathcal{PC}_{\text{Reedy}}(\mathcal{M})$ is again tractable, left proper and cartesian, allowing us to iterate the operation.

19.1 A standard factorization

Before getting to the construction of the model structure, it will be useful to follow up on Corollary 18.8.2 of the previous chapter, by analyzing further the case of maps which are surjective on the set of objects.

Lemma 19.1.1 *Suppose $f : \mathcal{A} \to \mathcal{B}$ is a morphism in $\mathcal{PC}(\mathcal{M})$ such that $\text{Ob}(f)$ is surjective. Let $\text{Ob}(f)^*(\mathcal{B}) \in \mathcal{PC}(\text{Ob}(\mathcal{A}), \mathcal{M})$ be the precategory obtained by pulling back along $\text{Ob}(f) : \text{Ob}(\mathcal{A}) \to \text{Ob}(\mathcal{B})$. Then f factors as*

$$\mathcal{A} \xrightarrow{f'} \text{Ob}(f)^*(\mathcal{B}) \xrightarrow{f''} \mathcal{B},$$

where the first map f' is an isomorphism on sets of objects, and the second map $f'' : \text{Ob}(f)^*(\mathcal{B}) \to \mathcal{B}$ satisfies the right lifting property with respect to any morphism $g : \mathcal{U} \to \mathcal{V}$ such that $\text{Ob}(g)$ is injective. If $\text{Ob}(g)$ is an isomorphism, the lifting is unique.

Proof In a diagram of the form

$$\begin{array}{ccc} \mathcal{U} & \longrightarrow & \mathrm{Ob}(f)^*(\mathcal{B}) \\ \downarrow & & \downarrow \\ \mathcal{V} & \longrightarrow & \mathcal{B} \end{array}$$

giving a lifting is equivalent to choosing a lifting on the level of objects

$$\mathrm{Ob}(\mathcal{V}) \to \mathrm{Ob}(\mathrm{Ob}(f)^*(\mathcal{B})) = \mathrm{Ob}(\mathcal{A}).$$

This is possible since $\mathrm{Ob}(g)$ is injective and $\mathrm{Ob}(f)$ surjective, and unique if $\mathrm{Ob}(g)$ is an isomorphism. □

Corollary 19.1.2 *In the situation of the lemma, if I is a set of morphisms in $\mathcal{PC}(\mathcal{M})$ which are all injective on the sets of objects, and if the first map $\mathcal{A} \xrightarrow{f'} \mathrm{Ob}(f)^*(\mathcal{B})$ is in $\mathbf{inj}(I)$ then $f \in \mathbf{inj}(I)$. If $I_0 \subset I$ is a subset consisting of morphisms which induce isomorphisms on sets of objects, then $f \in \mathbf{inj}(I_0)$ implies $f' \in \mathbf{inj}(I_0)$.*

Proof Combine the lifting property for $\mathbf{inj}(I)$ with the one of the previous lemma. For the second statement, given a lifting problem for f' with respect to $g \in I_0$, the bottom map can be composed to a map into $\mathcal{B}$, giving a lifting problem for f. So if $f \in \mathbf{inj}(I_0)$, there is a lifting of the composed square, but this constitutes a lifting for f' because of the uniqueness of lifting of g along f'' in the previous lemma. □

Lemma 19.1.3 *Suppose $f : \mathcal{A} \to \mathcal{B}$ is a morphism in $\mathcal{PC}(\mathcal{M})$ such that $\mathrm{Ob}(f)$ is surjective, and $\mathcal{B}$ satisfies the Segal conditions. In the notation of Lemma 19.1.1, the map*

$$\mathcal{A} \xrightarrow{f'} \mathrm{Ob}(f)^*(\mathcal{B})$$

is a local weak equivalence, i.e. a weak equivalence in the model structures of Theorems 12.1.1 and 12.3.2, if and only if f is a global weak equivalence.

Proof Note that $\mathrm{Ob}(f)^*(\mathcal{B})$ satisfies the Segal conditions too. The second map $f'' : \mathrm{Ob}(f)^*(\mathcal{B}) \to \mathcal{B}$, going between Segal $\mathcal{M}$-categories, is essentially surjective from the hypothesis that $\mathrm{Ob}(f)$ is surjective on sets of objects, and fully faithful indeed it induces isomorphisms on the multiple morphism objects. Therefore f'' is a global weak equivalence. By Proposition 12.5.4, it follows that f' is a global weak equivalence if and only if f is. By Lemma 12.4.4, f' is a global weak equivalence if and only if it is a local one. □

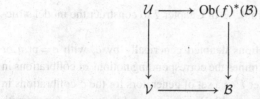

19.2 The model structures

We will be applying Theorem 8.7.3 of Chapter 7 to construct the model structure on $\mathcal{PC}(\mathcal{M})$.

We fix a class of cofibrations denoted generically by c, with $c = \text{proj}$ or $c = \text{Reedy}$. This choice determines the corresponding notions of cofibrations in $\mathcal{PC}_c(\mathcal{M})$ or $\mathcal{PC}_c(X, \mathcal{M})$. Let I be a set of generators for the c-cofibrations in $\mathcal{PC}_c(\mathcal{M})$, as discussed in Chapter 13. We can choose I to consist of maps with c-cofibrant domains, for the Reedy and projective structures, see Chapter 13.

Let K_{loc} denote a set of morphisms in $\mathcal{PC}(\mathcal{M})$ obtained by collecting together the direct localizing systems used in Theorem 12.1.1 or Theorem 12.3.2 to construct $\mathcal{PC}_c([k], \mathcal{M})$, for all $1 \leq k < \infty$. We may assume that the elements of K_{loc} are c-cofibrations with c-cofibrant domains as c is Reedy or projective.

Recall that $\mathbb{I} = \Upsilon(*)$ is the category with a single non-identity morphism $\upsilon_0 \to \upsilon_1$, and $\mathbf{I}$ is the category obtained by inverting this map, that is with a single isomorphism $\upsilon_0 \cong \upsilon_1$. Consider the morphism $\{\upsilon_0\} \hookrightarrow \mathbf{I}$. Choose a factorization into a c-cofibration followed by a local weak equivalence

$$\{\upsilon_0, \upsilon_1\} \to \mathbf{P} \to \mathbf{I}$$

in the model category $\mathcal{PC}_{\text{proj}}([1], \mathcal{M})$, and let

$$\{\upsilon_0\} \stackrel{i_0}{\to} \mathbf{P}$$

denote the inclusion morphism of a single object. This is still a c-cofibration in $\mathcal{PC}_c(\mathcal{M})$ because of Condition (AST) in Definition 7.7.1.

Let $K_{\text{glob}} := K_{\text{loc}} \cup \{i_0\}$. Note that the domain of i_0 is the single object precategory $\{\upsilon_0\}$ which is c-cofibrant for either c.

Theorem 19.2.1 *The class of global weak equivalences is pseudo-generated by K_{glob} in the sense of construction (PG) of Section 8.7. Furthermore, I and K_{glob} satisfy axioms (PGM1) – (PGM6), so they define tractable left proper model structures on $\mathcal{PC}(\mathcal{M})$ by Theorem 8.7.3. For these model structures, the weak equivalences are the global weak equivalences; the cofibrations are the projective (resp. Reedy) cofibrations, and the fibrations are the projective (resp. Reedy) global fibrations.*

Proof The first step is to show that K_{glob} leads to the class of global weak equivalences via prescription (PG). This amounts to showing that a map $f : \mathcal{X} \to \mathcal{Y}$ is a global weak equivalence if and only if there exists a diagram

$$\begin{array}{ccc}
\mathcal{X} & \longrightarrow & \mathcal{A} \\
\downarrow & & \downarrow \\
\mathcal{Y} & \longrightarrow & \mathcal{B}
\end{array}$$

with the horizontal maps in $\mathbf{cell}(K_{\mathrm{glob}})$ and the right vertical map in $\mathbf{inj}(I)$.

The maps in K_{glob} are trivial cofibrations in the projective (resp. Reedy) structure, and the global trivial cofibrations are preserved by pushout (Theorem 18.7.5) and transfinite composition (Lemma 14.3.5), so the maps in $\mathbf{cell}(K_{\mathrm{glob}})$ are global trivial cofibrations.

A map $\mathcal{A} \xrightarrow{p} \mathcal{B}$ in $\mathbf{inj}(I)$ is a global weak equivalence. To see this, note first that $\mathrm{Ob}(p)$ is surjective, by lifting with respect to $\emptyset \to *$. Consider the factorization $p = p'' \circ p'$ of Lemma 19.1.1. The condition $p \in \mathbf{inj}(I)$ implies that p satisfies right lifting with respect to all c-cofibrations, so by Corollary 19.1.2 the first map p' satisfies right lifting with respect to all cofibrations inducing isomorphisms on sets of objects. As $\mathrm{Ob}(p')$ is also an isomorphism, this means that p' is a trivial fibration in the local model category structure of Theorem 12.1.1 or 12.3.2. In particular, p' is a local, hence global, weak equivalence. On the other hand, p'' is a global weak equivalence by Corollary 17.2.8. Thus, p is a global weak equivalence.

By 3 for 2 for global weak equivalences (Proposition 12.5.4) it follows from the previous two paragraphs that if there exists a square diagram as above for $f : \mathcal{X} \to \mathcal{Y}$ then it is a global weak equivalence.

Suppose f is a global weak equivalence, we will construct a square as above. Let $r : \mathcal{Y} \to \mathcal{B}$ be the map given by applying the small object argument to $\mathcal{Y}$ with respect to K_{loc}. Thus $\mathcal{B}$ is K_{loc}-injective. It follows that it satisfies the Segal conditions, and is in fact fibrant in $\mathcal{PC}_c(\mathrm{Ob}(\mathcal{B}), \mathcal{M})$, see Proposition 12.3.3.

By Corollary 18.8.2 of the preceding Chapter 18, there exists a pushout $\mathcal{X} \to \mathcal{X}'$ by a collection of copies of the map $i_0 : \{v_0\} \to \mathbf{P}$ and an extension of rf to a global weak equivalence $g : \mathcal{X}' \to \mathcal{B}$ which is surjective on the set of objects. Note that $\mathcal{X} \to \mathcal{X}'$ is in $\mathbf{cell}(K_{\mathrm{glob}})$. Consider the factorization

$$\mathcal{X}' \to \mathrm{Ob}(g)^*\mathcal{B} \to \mathcal{B}$$

of Lemma 19.1.1 above. The first map is an isomorphism on the set of objects, so it can be considered as a map in $\mathcal{PC}(\mathrm{Ob}(\mathcal{X}'), \mathcal{M})$. Apply the small object argument for the set K_{loc}, to the first map to yield a factorization

$$\mathcal{X}' \to \mathcal{A} \to \mathrm{Ob}(g)^*\mathcal{B},$$

such that $\mathcal{X}' \to \mathcal{A}$ is in $\mathbf{cell}(K_{\mathrm{loc}})$ and $\mathcal{A} \to \mathrm{Ob}(g)^*\mathcal{B}$ is K_{loc}-injective, hence by Proposition 12.3.3 a fibration in $\mathcal{PC}_c(\mathrm{Ob}(\mathcal{X}'), \mathcal{M})$. However, the composed map $\mathcal{X}' \to \mathrm{Ob}(g)^*\mathcal{B}$ is a local weak equivalence by Lemma 19.1.3, so by 3 for 2 in the local model structure, the map $\mathcal{A} \to \mathrm{Ob}(g)^*\mathcal{B}$ is a trivial fibration; hence it is in $\mathbf{inj}(I)$ (see Lemma 13.7.4). Apply now Corollary 19.1.2: note that the factorization of Lemma 19.1.1 for the map $\mathcal{A} \to \mathcal{B}$ is just

$$\mathcal{A} \to \mathrm{Ob}(g)^*\mathcal{B} \to \mathcal{B},$$

where the first map is the same as previously; we know that the first map is in $\mathbf{inj}(I)$ so by Corollary 19.1.2 the full map $\mathcal{A} \to \mathcal{B}$ is in $\mathbf{inj}(I)$. The composition

$$\mathcal{X} \to \mathcal{X}' \to \mathcal{A}$$

of two maps in $\mathbf{cell}(K_{\mathrm{glob}})$ is again in $\mathbf{cell}(K_{\mathrm{glob}})$. Thus we have a square as required, with horizontal maps in $\mathbf{cell}(K_{\mathrm{glob}})$ and the right vertical map in $\mathbf{inj}(I)$. This completes the verification that our global weak equivalence f satisfies the condition (PG), giving the first sentence of the theorem.

We now verify axioms (PGM1) – (PGM6) needed to apply the pseudo-generator recognition theorem 8.7.3:

(PGM1)–by hypothesis $\mathcal{M}$ is locally presentable, and I and K_{glob} are chosen to be small sets of morphisms;

(PGM2)–we have chosen the domains of arrows in I and K_{glob} to be cofibrant, and K_{glob} consists of c-cofibrations, in other words it is contained in $\mathbf{cof}(I)$;

(PGM3)–the class of global weak equivalences is closed under retracts by Proposition 12.5.4 of Chapter 12;

(PGM4)–the class of global weak equivalences satisfies 3 for 2 again by Proposition 12.5.4;

(PGM5)–by Theorem 18.7.5, the class of global trivial c-cofibrations is closed under pushouts;

(PGM6)–the class of global trivial c-cofibrations is closed under transfinite composition, indeed the cofibrations are closed under transfinite composition since they have generating sets, see Chapter 13, and the weak equivalences are too by Lemma 14.3.5.

Theorem 8.7.3 now applies to show that $\mathcal{PC}(\mathcal{M})$ with the given classes of c-cofibrations, global weak equivalences, hence global trivial c-cofibrations whence global c-fibrations, is a tractable left proper (cf 18.8) model category. □

19.3 The cartesian property

Lemma 19.3.1 *Suppose $\mathcal{A} \to \mathcal{B}$ and $\mathcal{C} \to \mathcal{D}$ are global Reedy cofibrations, with the first one being a global weak equivalence. Then the map*

$$\mathcal{A} \times \mathcal{D} \cup^{\mathcal{A}\times\mathcal{C}} \mathcal{B} \times \mathcal{C} \to \mathcal{B} \times \mathcal{D}$$

is a global trivial Reedy cofibration.

Proof It is a Reedy cofibration by Corollary 13.6.13. We just have to show that it is a global weak equivalence. By Corollary 17.2.6, the vertical maps in the diagram

$$\begin{array}{ccc} \mathcal{A} \times \mathcal{C} & \longrightarrow & \mathcal{A} \times \mathcal{D} \\ \downarrow & & \downarrow \\ \mathcal{B} \times \mathcal{C} & \longrightarrow & \mathcal{B} \times \mathcal{D} \end{array}$$

are global weak equivalences. Applying Theorem 18.7.5 to pushout along the left vertical global trivial cofibration, then using 3 for 2, it follows that the map

$$\mathcal{A} \times \mathcal{D} \to \mathcal{A} \times \mathcal{D} \cup^{\mathcal{A}\times\mathcal{C}} \mathcal{B} \times \mathcal{C}$$

is a global weak equivalence. Then by 3 for 2 using the right vertical global weak equivalence, the map in the statement of the lemma is a global weak equivalence. □

Theorem 19.3.2 *Suppose $\mathcal{M}$ is a tractable left proper cartesian model category. Then the model category $\mathcal{PC}_{\mathrm{Reedy}}(\mathcal{M})$ of $\mathcal{M}$-enriched precategories with Reedy cofibrations and global weak equivalences is again a tractable left proper cartesian model category.*

Proof Observe first of all that cartesian product commutes with colimits in $\mathcal{PC}(\mathcal{M})$, as can be seen from the explicit description of products and colimits and using the corresponding condition for $\mathcal{M}$.

Next, note that the map $\emptyset \to *$ is a Reedy cofibration, from the definition.

Proposition 13.6.12 gives cofibrancy of the pushout-product map; and the previous Lemma 19.3.1 gives the trivial cofibration property. This shows that $\mathcal{PC}_{\mathrm{Reedy}}(\mathcal{M})$ is cartesian.

To finish the proof we need to note that it is tractable. This can be seen by inspection of the generating cofibrations for the Reedy structure, given in Proposition 13.6.11. □

Of course, the projective model structure is definitely not cartesian. On the other hand, one can hope to treat the injective model structure. As discussed previously, some further condition is needed in order to insure tractability; for that, it suffices to require that all objects of $\mathcal{M}$ be cofibrant. Similarly, it doesn't seem immediately clear whether $\mathcal{P}\mathbf{C}_{\mathrm{inj}}(\mathcal{M})$ will satisfy condition (PROD) in general, although again this is relatively easy to see in a special case:

Lemma 19.3.3 *If the tractable left proper cartesian model category $\mathcal{M}$ is a presheaf category with monomorphisms as cofibrations, then $\mathcal{P}\mathbf{C}_{\mathrm{inj}}(\mathcal{M})$ is a tractable left proper cartesian model category.*

Proof If $\mathcal{M} = \mathrm{PRESH}(\Phi)$ then one can check over each object of Φ that the map in condition (PROD) is injective. Then proceed as above. □

19.4 Properties of fibrant objects

Bergner [42] made the very interesting observation that one could give an explicit characterization for the fibrant objects in the case of Segal categories $\mathcal{M} = \mathcal{K}$. We get the same kind of property in general:

Proposition 19.4.1 *Let* $c = \mathrm{proj}$ *or* $c = \mathrm{Reedy}$. *In the model category $\mathcal{P}\mathbf{C}_c(\mathcal{M})$ constructed above, an object $\mathcal{A}$ with* $\mathrm{Ob}(\mathcal{A}) = X$ *is fibrant if and only if it is fibrant when considered as an object of the model category $\mathcal{P}\mathbf{C}_c(X, \mathcal{M})$ of Theorem 12.1.1 or 12.3.2. In turn this condition is equivalent to saying that $\mathcal{A}$ satisfies the Segal conditions, and is fibrant as an object of the unital diagram model category* $\mathrm{FUNC}_c\left(\Delta^o_X / X, \mathcal{M}\right)$.

Proof Left to the reader. □

For $c = \mathrm{proj}$ then, an $\mathcal{M}$-precategory $\mathcal{A}$ is fibrant if and only if it satisfies the Segal conditions, and is levelwise fibrant. For $c = \mathrm{Reedy}$ the fibrancy condition is also pretty easy to check: it just means that the standard matching maps are fibrations in $\mathcal{M}$.

19.5 The model category of strict $\mathcal{M}$-enriched categories

Dwyer and Kan proposed, in a series of papers, a model category structure on the category of strict simplicial categories. Their program was finished by Bergner [39]. Lurie [190, Appendix] generalized this to construct a model category of strict $\mathcal{M}$-enriched categories, and then used that to construct the model category of weakly $\mathcal{M}$-enriched precategories as we have done above.

Theorem 19.5.1 *Suppose $\mathcal{M}$ is a tractable left proper cartesian model category such that all objects are cofibrant. Let* $\text{CAT}(\mathcal{M})$ *denote the category of strict $\mathcal{M}$-enriched categories. Define the notion of weak equivalence in the usual way (see Section 12.4). Then* $\text{CAT}(\mathcal{M})$ *has a tractable left proper model structure in which the generating cofibrations are obtained by free additions of generating cofibrations of $\mathcal{M}$ in the morphism space between any two objects. There is a Quillen adjunction*

$$\text{CAT}(\mathcal{M}) \leftrightarrow \mathcal{PC}_{\text{proj}}(\mathcal{M}),$$

and indeed the model structure on $\text{CAT}(\mathcal{M})$ *can be used to generate the model structure on* $\mathcal{PC}_{\text{proj}}(\mathcal{M})$*. However,* $\text{CAT}(\mathcal{M})$ *is not in general cartesian. It follows that any object of* $\mathcal{PC}_{\text{proj}}(\mathcal{M})$ *is equivalent to a strict $\mathcal{M}$-enriched category.*

Proof See Bergner [39] for $\mathcal{M} = \mathcal{K}$ and Lurie [190, 192] for arbitrary $\mathcal{M}$. The strictification result, for the case of Tamsamani n-groupoids, was proven by Paoli [210] using techniques of Cat^n-groups. □

This theorem offers an alternative route to the construction of the model structure on $\mathcal{PC}_{\text{proj}}(\mathcal{M})$, whose proof is somewhat different from ours. The advantage is that it also gives the model structure on $\text{CAT}(\mathcal{M})$ and hence the strictification result; the disadvantage is that it doesn't give the cartesian property. The cartesian question has been treated by Rezk [219] for the case of iterated Rezk categories. We leave it to the reader to explore these different points of view.

PART V

Higher category theory

20

Iterated higher categories

The conclusion of Theorem 19.3.2 matches the hypotheses we imposed that $\mathcal{M}$ be tractable, left proper and cartesian. Therefore, we can iterate the construction to obtain various versions of model categories for n-categories and similar objects. This process is inherent in the definitions of Tamsamani [250] and Pellissier [211]. Rezk [219] considered a modified version of the corresponding iteration of his definition, following Barwick, and Trimble's definition is also iterative (cf. Cheng [74]). Such an iteration is related to Dunn's iteration of the Segal delooping machine [97], and goes back to the well-known iterative presentation of the notion of strict n-category, see Bourn [55] for example.

Iteration leads to what might generically be called *higher category theory*. In the last part of the book, we explore some of the first things which can be said. The next Chapter 21 takes up one route, which is to try to generalize to the higher categorical context the large body of knowledge on usual category theory. The particular question of constructing limits and colimits of higher categories themselves is treated in Chapter 22. Chapter 23 ends the book with a look at the Breen–Baez–Dolan stabilization hypothesis. This provides a first illustration of the interaction between higher morphisms in different dimensions.

The present chapter will set up the basic context and notation, and relate higher groupoids back to homotopy theory. In what follows unless otherwise indicated, the model category denoted $\mathcal{PC}(\mathcal{M})$ will mean by definition $\mathcal{PC}_{\text{Reedy}}(\mathcal{M})$, the global model structure with Reedy cofibrations constructed in Theorem 19.3.2. Weak equivalences in this model structure are the global weak equivalences.

For any $n \geq 0$ define by induction $\mathcal{PC}^0(\mathcal{M}) := \mathcal{M}$ and for $n \geq 1$

$$\mathcal{PC}^n(\mathcal{M}) := \mathcal{PC}(\mathcal{PC}^{n-1}(\mathcal{M})).$$

This is the model category of *$\mathcal{M}$-enriched n-precategories*. Notations for objects therein will be discussed below.

In the iterated situation, we can introduce the following definition:

Definition 20.0.1 An $\mathcal{M}$-enriched n-precategory $\mathcal{A} \in \mathcal{PC}^n(\mathcal{M})$ satisfies the *full Segal condition* if it satisfies the Segal condition as an $\mathcal{PC}^{n-1}(\mathcal{M})$-precategory, and furthermore inductively for any sequence of objects $x_0, \ldots, x_m \in \mathrm{Ob}(\mathcal{A})$ the $\mathcal{M}$-enriched $(n-1)$-precategory

$$\mathcal{A}(x_0, \ldots, x_m) \in \mathcal{PC}^{n-1}(\mathcal{M})$$

satisfies the full Segal condition.

Lemma 20.0.2 *If $\mathcal{A}$ is a fibrant object in the (iterated Reedy) model structure on $\mathcal{PC}^n(\mathcal{M})$, then $\mathcal{A}$ satisfies the full Segal condition.*

Proof The Segal condition for the $\mathcal{PC}^{n-1}(\mathcal{M})$-precategory comes from Proposition 19.4.1 (see Theorem 12.3.2). However, if $\mathcal{A}$ is fibrant then the $\mathcal{A}(x_0, \ldots, x_m)$ are fibrant in $\mathcal{PC}^{n-1}(\mathcal{M})$, so by induction they also satisfy the full Segal condition. □

20.1 Initialization

Here are a few possible choices for $\mathcal{M}$ to start with.

If $\mathcal{M} = \mathrm{SET}$ is the model category of sets, with cofibrations and fibrations being arbitrary morphisms and weak equivalences being isomorphisms, then $\mathcal{PC}^n(\mathrm{SET})$ is the *model category of n-precats* which was considered in my preprint [234] and for which we have now fixed up the proof.

Let $*$ denote the model category with a single object and a single morphism. Then $\mathcal{PC}(*)$ is Quillen equivalent (by a product-preserving map) to the model category $\{\emptyset, *\}$ consisting of the empty set and the one-element set, where weak equivalences are isomorphisms. Iterating again, $\mathcal{PC}^2(*)$ is Quillen-equivalent to $\mathcal{PC}(\{\emptyset, *\})$. These are both model categories of *graphs*, the first allowing multiple edges between nodes and the second allowing only zero or one edge between two nodes. The weak equivalences are defined by requiring isomorphisms on the level of π_0, defined as the set of connected components of a graph. These model categories of graphs are Quillen-equivalent to SET, but have the advantage that the cofibrations are monomorphisms. They are related to the notion of *setoid* in constructive type theory.

In particular, $\mathcal{PC}^{n+2}(*)$ is Quillen-equivalent to $\mathcal{PC}(\mathrm{SET})$ and should perhaps be thought of as the "true" model category of n-categories. It starts with the most canonical initialization possible, and sets appear via the $\mathcal{PC}$ operation.

If we start with $\mathcal{M} = \mathcal{K}$ the Kan–Quillen model category of simplicial sets, then $PC^n(\mathcal{K})$ is the model category of Segal n-precategories introduced by Hirschowitz and myself [145].

One can imagine further constructions starting with $\mathcal{M}$ as a category of diagrams or other such things. Starting with $\mathbb{Z}/2$-equivariant sets should be useful for considering n-categories with duals.

20.2 Notation for n-categories

The two main cases we are interested in for this book are the case of n-categories, obtained iteratively by starting out with $\mathcal{M} = \text{SET}$, and the case of Segal n-categories obtained by starting from the Kan–Quillen model category $\mathcal{K}$ of simplicial sets. In this section, we discuss in some detail the notations which can be established for these cases.

By Lemma 10.7.4, once we start iterating, the hypothesis (DISJ) of 10.7 will be in vigour. Furthermore, most of our examples of starting categories (even $\mathcal{M} = *$) satisfy (DISJ), see Lemma 10.7.5. Whenever such is the case, it is reasonable to introduce an iteration of the notation $A_{n/}$ of Section 10.7.

One can note, on the other hand, that even based on this notation as the general framework, most of what was done in my preprints [234, 235] really used the notation $A(x_0, \ldots, x_n)$ at the crucial places. So, in a certain sense, the notations we introduce here are not really the fundamental objects, nonetheless it is convenient to have them for comparison.

In $\mathcal{PC}^n(\mathcal{M})$ for any $k \leq m$ and any multi-index $m_1, \ldots, m_k$ we can introduce the notation $\mathcal{A}_{m_1,\ldots,m_k/} \in \mathcal{PC}^{n-k}(\mathcal{M})$ defined by induction on k. At the initial $k = 1$ (whenever $n \geq 1$), by noting that $\mathcal{A} \in \mathcal{PC}(\mathcal{PC}^{n-1}(\mathcal{M}))$ we can use the notation

$$\mathcal{A}_{m/} \in \mathcal{PC}^{n-1}(\mathcal{M})$$

considered in Section 10.7. Then for $k \geq 2$ define inductively

$$\mathcal{A}_{m_1,\ldots,m_k/} := (\mathcal{A}_{m_1,\ldots,m_{k-1}/})_{m_k/} \in \mathcal{PC}^{n-k}(\mathcal{M}).$$

For $k < n$, define on the other hand

$$\mathcal{A}_{m_1,\ldots,m_k} := \mathrm{Ob}(\mathcal{A}_{m_1,\ldots,m_k/}) \in \text{SET}.$$

One can remark that the notations $\mathcal{A}_{m_1,\ldots,m_k/} \in \mathcal{PC}^{n-k}(\mathcal{M})$ and $\mathcal{A}_{m_1,\ldots,m_k} \in$ SET make sense for $k < n$ because we have seen that $\mathcal{PC}(\mathcal{M})$ satisfies Condition (DISJ), even if $\mathcal{M}$ itself does not.

In a related direction, notice that if $\mathcal{M}$ is a presheaf category then $\mathcal{PC}(\mathcal{M})$ is also a presheaf category, by the discussion of Section 10.7. Inductively the same is true of $\mathcal{PC}^n(\mathcal{M})$. More precisely, if $\mathcal{M} = \text{PRESH}(\Phi)$ then $\mathcal{PC}^n(\mathcal{M}) = \text{PRESH}(\mathbf{C}^n(\Phi))$ in the notations of Proposition 10.7.6. The basic examples will therefore be presheaf categories, leading to specific notation.

20.2.1 The model category of *n*-prenerves

Start with $\mathcal{M} = \text{SET}$ with the trivial model structure (as Lurie [190] calls it), where the weak equivalences are isomorphisms and the cofibrations and fibrations are arbitrary maps. Iterating the construction of Theorem 19.3.2 we obtain the iterated Reedy model category structure

$$\mathcal{PC}^n(\text{SET}).$$

We call $\mathcal{PC}^n(\text{SET})$ the category of *n-prenerves*, following Tamsamani [250], or *n-precategories*. The terminology "*n*-prenerve" refers more to this specific model, whereas we probably revert to saying "*n*-precategory" when talking about things which should happen in a generic theory.

An *n*-prenerve $\mathcal{A} \in \mathcal{PC}^n$ may be interpreted as a presheaf of sets on the category $\mathbf{C}^n(*)$ considered in Section 10.7:

$$\mathcal{PC}^n(\text{SET}) = \text{PRESH}(\mathbf{C}^n(*)).$$

The underlying category $\mathbf{C}^n(*)$ may be seen as a quotient of $\Delta^n = \Delta\times\cdots\times\Delta$, indeed it is the same as the category which was denoted Θ^n in my preprints on the model structure [234], limits [235] and stabilization [237]. If $\mathcal{A} \in \mathcal{PC}^n(\text{SET})$ then the notation discussed in the previous section applies.

The objects of $\mathbf{C}^n(*)$ are denoted by sequences $(m_1, \ldots, m_k)$ of length $0 \le k \le n$ of integers $m_i \ge 1$ corresponding to objects of Δ. The value of $\mathcal{A}$ on the sequence 0 of length 0 is the set $\mathcal{A}_0 = \text{Ob}(\mathcal{A})$, and the value on $(m_1, \ldots, m_k)$ is denoted $\mathcal{A}_{m_1,\ldots,m_k} \in \text{SET}$. This is the set of objects of the $(n-k)$-prenerve $\mathcal{A}_{m_1,\ldots,m_k/} \in \mathcal{PC}^{n-k}(\text{SET})$. For $k = n$, these coincide: $\mathcal{A}_{m_1,\ldots,m_n} = \mathcal{A}_{m_1,\ldots,m_n/}$ since the model category $\mathcal{M}$ is equal to SET.

That yields a system of notations coinciding with that of *loc. cit.* [234, 235, 237]. A slight difference is that for $\mathcal{A} \in \mathcal{PC}^n(\text{SET})$, and for any sequence of objects $x_0, \ldots, x_m$, what we would be denoting here by

$$\mathcal{A}(x_0, \ldots, x_m) \in \mathcal{PC}^{n-1}(\text{SET})$$

was denoted in those preprints by $\mathcal{A}_{m/}(x_0, \ldots, x_m)$. We have dropped the subscript $(\ )_{m/}$ for brevity.

If $\mathcal{A} \in \mathcal{PC}^n$ is an n-prenerve, in particular for any $m \in \Delta$, we obtain the $(n-1)$-preprenerve $\mathcal{A}_{m/}$, which can be viewed as the presheaf of sets

$$\mathcal{A}_{m/} : (m_2, \ldots, m_k) \mapsto \mathcal{A}_{m,m_2,\ldots,m_k}$$

on $\mathbf{C}^{n-1}(*)$. It comes with a map to the set $(\mathrm{Ob}(\mathcal{A}))^{m+1}$, whose coefficients are the $m+1$ maps $0 \to m$ in Δ, and for $(x_0, \ldots, x_m) \in (\mathrm{Ob}(\mathcal{A}))^{m+1}$ the preimage is a sub-$(n-1)$-preprenerve

$$\mathcal{A}(x_0, \ldots, x_m) \subset \mathcal{A}_{m/}.$$

The functor $(x_0, \ldots, x_m) \mapsto \mathcal{A}(x_0, \ldots, x_m)$ from $\Delta^o_{\mathrm{Ob}(\mathcal{A})}$ to $\mathcal{PC}^{n-1}$ is the standard representation of

$$\mathcal{A} \in \mathcal{PC}^n = \mathcal{PC}(\mathcal{PC}^{n-1})$$

used in most of the book.

As a category, $\mathbf{C}^n(*)$ is the quotient of the cartesian product Δ^n obtained by identifying all of the objects $(M, 0, M')$ for fixed $M = (m_1, \ldots, m_k)$ and variable $M' = (m'_1, \ldots, m'_{n-k-1})$. The object of $\mathbf{C}^n(*)$ corresponding to the class of $(M, 0, M')$ with all $m_i > 0$ is denoted $M = (m_1, \ldots, m_k)$ with $k \leq n$. Here k is called the *length* and is also denoted $k = |M|$. There is a unique object of length zero denoted 0 by convention. Two morphisms from M to M' in Δ^n are identified if they both factor through something of the form $(u_1, \ldots, u_i, 0, u_{i+2}, \ldots, u_n)$ and if their first i components are the same.

Since $\mathcal{PC}^n$ is a category of presheaves of sets, for any object denoted $M = (m_1, \ldots, m_k)$ of the underlying category $\mathbf{C}^n(*)$, we get a representable presheaf $h^n(M)$. In terms of the representable objects in $\mathcal{PC}(\mathcal{PC}^{n-1})$ considered in Section 13.2, we have

$$h^n(m_1, \ldots, m_k) = h([m_1], h^{n-1}(m_2, \ldots, m_k)).$$

20.2.2 The full n-nerve condition

The Segal condition on $\mathcal{PC}(\mathcal{PC}^{n-1})$ just tells us about the first simplicial index. Tamsamani's definition [250] of n-nerves (or n-categories) is inductive on n and includes the condition that the components $\mathcal{A}_{m/}$ themselves be $(n-1)$-nerves:

Definition 20.2.1 (Tamsamani [250]) An *n-nerve* or *n-category* is an n-prenerve $\mathcal{A}$ satisfying the full Segal condition 20.0.1: each $\mathcal{A}_{m/}$ is inductively an $(n-1)$-nerve, and $\mathcal{A}$ satisfies the Segal condition in $\mathcal{PC}(\mathcal{PC}^{n-1})$. A morphism of n-nerves is an *equivalence* if it is a weak equivalence in the model category $\mathcal{PC}^n$.

The condition that $\mathcal{A}_{m/}$ be an $(n-1)$-nerve is equivalent to asking that $\mathcal{A}(x_0, \ldots, x_m)$ be an $(n-1)$-nerve for each sequence of objects $x_0, \ldots, x_m$.

Remark 20.2.2 Suppose $\mathcal{A}$ is a fibrant object of $\mathscr{P}\mathcal{C}^n$, then it is an n-nerve, i.e., an n-category.

Indeed, if $\mathcal{A}$ is fibrant then each $\mathcal{A}(x_0, \ldots, x_m)$ is fibrant in $\mathscr{P}\mathcal{C}^{n-1}$, hence an $(n-1)$-nerve by induction, and $\mathcal{A}$ satisfies the Segal condition at the first level. Note also that, taking the disjoint union over all sequences $x_0, \ldots, x_m$, the $(n-1)$-prenerve $\mathcal{A}_{m/}$ is an $(n-1)$-nerve.

Remark 20.2.3 At $n = 1$, the category of 1-prenerves is the category of simplicial sets, and the 1-nerves are the simplicial sets which are nerves of a 1-category, that is to say the category of 1-nerves is equivalent to CAT. The process $\mathcal{A} \mapsto \mathbf{Seg}(\mathcal{A})$ is the generation of a category by generators and relations discussed in Section 14.8.

If $\mathcal{A}$ is an n-nerve, for two points $x, y \in \mathcal{A}_0 = \mathrm{Ob}(\mathcal{A})$ the $(n-1)$-nerve $\mathcal{A}(x, y)$ is the $(n-1)$-*nerve of morphisms from* x *to* y. This is the essential part of the structure, which corresponds, in the case of categories, to the HOM sets. One could adopt the notation

$$Hom_{\mathcal{A}}(x, y) := \mathcal{A}(x, y).$$

The Segal condition implies that the $\mathcal{A}(x_0, \ldots, x_p)$ are determined up to equivalence by the $\mathcal{A}(x, y)$.

A morphism $\mathcal{A} \xrightarrow{f} \mathcal{B}$ between n-nerves is *fully faithful* if for any $x, y \in \mathrm{Ob}(\mathcal{A})$, $\mathcal{A}(x, y) \to \mathcal{B}(f(x), f(y))$ is an equivalence of $(n-1)$-nerves.

Tamsamani defines a truncation operation T which we denote here by $\tau_{\leq n-1}$ from n-nerves to $(n-1)$-nerves. It is a generalization of the truncation of topological spaces used in the Postnikov tower. Applying it inductively $n-k$ times gives a truncation denoted $\tau_{\leq k}$ from n-nerves to k-nerves. If $k \geq 1$ then $\mathrm{Ob}(\tau_{\leq k}\mathcal{A}) = \mathrm{Ob}(\mathcal{A})$ and for any $x_0, \ldots, x_m \in \mathrm{Ob}(\mathcal{A})$,

$$(\tau_{\leq k}\mathcal{A})(x_0, \ldots, x_m) = \tau_{\leq k-1}(\mathcal{A}(x_0, \ldots, x_m)).$$

The truncation $\tau_{\leq 1}(\mathcal{A})$ is the same 1-category as considered in Section 12.4, and $\tau_{\leq 0}(\mathcal{A})$ is the set of isomorphism classes of this 1-category.

A map $\mathcal{A} \xrightarrow{f} \mathcal{B}$ between n-nerves is *essentially surjective* if the induced map

$$\tau_{\leq 0}(f) : \tau_{\leq 0}\mathcal{A} \to \tau_{\leq 0}\mathcal{B}$$

is a surjection of sets.

Remark 20.2.4 A map f between n-nerves is an equivalence if and only if it is essentially surjective and fully faithful. The notion of equivalence defined using the model structure thus coincides with the inductive definition of Tamsamani [250]. See Proposition 20.3.3.

For an n-nerve $\mathcal{A}$ the simplicial set $p \mapsto \tau_{\leq 0}(\mathcal{A}_{p/})$ satisfies the condition that the Segal maps are isomorphisms, so it is the nerve of the 1-category $\tau_{\leq 1}\mathcal{A}$. One also has

$$\tau_{\leq k}(\mathcal{A})_{p/} := \tau_{\leq k-1}(\mathcal{A}_{p/}).$$

In terms of Tamsamani's notation for truncation [250],

$$\tau_{\leq k}(\mathcal{A}) = T^{n-k}(\mathcal{A}).$$

The truncation operation is associative as implied by this notation:

$$\tau_{\leq k}(\tau_{\leq l}\mathcal{A}) = \tau_{\leq k}(\mathcal{A}).$$

If $\mathcal{A} \xrightarrow{f} \mathcal{B}$ is an equivalence of n-nerves then for any $0 \leq k \leq n$,

$$\tau_{\leq k}(\mathcal{A}) \xrightarrow{\tau_{\leq k}(f)} \tau_{\leq k}(\mathcal{B})$$

is an equivalence of k-nerves.

20.2.3 Cofibrations

The iterated injective and Reedy model structures coincide in the case of $\mathcal{PC}^n$ (SET), by applying Proposition 13.7.2 inductively. A map $\mathcal{A} \to \mathcal{B}$ is a cofibration if and only if, for any multi-index $m_1, \ldots, m_k$ with $k < n$ the map $\mathcal{A}_{m_1,\ldots,m_k} \to \mathcal{B}_{m_1,\ldots,m_k}$ is an injection of sets. The monomorphism condition is not imposed at multi-indices of length $k = n$, indeed at the top level, all maps of SET are cofibrations for its trivial model structure.

Recall that in the trivial model structure on $\mathcal{PC}^0 = $ SET all morphisms are cofibrations. Inductively on $n \geq 1$ the cofibrations of $\mathcal{PC}^n$ are the Reedy cofibrations in $\mathcal{PC}(\mathcal{PC}^{n-1})$, see Section 13.6. The discussion of Proposition 13.7.2 applies, so the Reedy cofibrations are the same as the injective ones. Working out the induction on n, we get the following simple characterization:

Remark 20.2.5 A morphism $\mathcal{A} \to \mathcal{B}$ of n-prenerves is a cofibration if and only if the morphisms $\mathcal{A}_M \to \mathcal{B}_M$ are injective whenever $M \in \mathbf{C}^n(*)$ is an object of nonmaximal length, that is to say when $M = (m_1, \ldots, m_k, 0, \ldots, 0)$ for some $k < n$.

One can use the notation $\mathcal{A} \hookrightarrow \mathcal{B}$ for a cofibration, not meaning to imply injectivity at the top level.

20.2.4 Generating *n*-nerves

Crucial to the model structure is the operation $\mathcal{A} \mapsto \mathbf{Seg}(\mathcal{A})$, which enforces the Segal condition. In view of Definition 20.2.1, it will be convenient to compose this with an operation enforcing the $(n-1)$-nerve condition on the components $\mathcal{A}_{m/}$. The combined operation denoted $\mathcal{A} \mapsto \mathbf{Cat}(\mathcal{A})$ is to be compatible with disjoint unions, preserve discrete objects, and have $\mathrm{Ob}(\mathbf{Cat}(\mathcal{A})) = \mathrm{Ob}(\mathcal{A})$.

Suppose it is known inductively for $(n-1)$-prenerves, then define for any n-prenerve $\mathcal{B}$ a new n-prenerve $\mathbf{Fix}(\mathcal{B})$ with the same set of objects, and

$$\mathbf{Fix}(\mathcal{B})_{m/} := \mathbf{Cat}(\mathcal{B}_{m/}),$$

or

$$\mathbf{Fix}(\mathcal{B})(x_0, \ldots, x_m) := \mathbf{Cat}(\mathcal{B}(x_0, \ldots, x_m)).$$

Using a version of **Seg** which preserves discrete objects, i.e. $\mathbf{Seg}(*) = *$, put

$$\mathbf{Cat}(\mathcal{A}) := \mathbf{Fix}(\mathbf{Seg}(\mathcal{A})).$$

Thus

$$\mathbf{Cat}(\mathcal{A})_{m/} = \mathbf{Cat}(\mathbf{Seg}(\mathcal{A})_{m/})$$

and

$$\mathbf{Cat}(\mathcal{A})(x_0, \ldots, x_m) = \mathbf{Cat}(\mathbf{Seg}(\mathcal{A})(x_0, \ldots, x_m)).$$

An essentially equivalent definition is obtained by letting $\mathbf{Cat}(\mathcal{A})$ be a functorial fibrant replacement for $\mathcal{A}$ in $\mathcal{PC}^n$, by Remark 20.2.2.

Whichever definition is used, there is a natural weak equivalence

$$i_{\mathcal{A}} : \mathcal{A} \to \mathbf{Cat}(\mathcal{A}).$$

A morphism $\mathcal{A} \to \mathcal{B}$ of n-prenerves is a weak equivalence if and only if $\mathbf{Cat}(\mathcal{A}) \to \mathbf{Cat}(\mathcal{B})$ is an equivalence of n-nerves. In particular, $\mathbf{Cat}(i_{\mathcal{A}})$ is an equivalence, see the extensive discussion of this monadic property in Sections 9.2.1 and 9.2.

Remark 20.2.6 Suppose $\mathcal{A}$ is an n-prenerve such that the Segal maps

$$\mathcal{A}_{p/} \to \mathcal{A}_{1/} \times_{\mathcal{A}_0} \cdots \times_{\mathcal{A}_0} \mathcal{A}_{1/}$$

are weak equivalences of $(n-1)$-prenerves. Then $\mathbf{Cat}(\mathcal{A})$ is equivalent to $\mathbf{Fix}(\mathcal{A})$.

Refer to Chapter 14 for a full discussion of **Seg** viewed as a calculus of generators and relations.

The construction **Cat** extends more generally to the case of $\mathcal{M}$-enriched n-precategories. The easiest expression is just to say that $\mathbf{Cat}(\mathcal{A})$ is the fibrant replacement for $\mathcal{A}$ with a trivial cofibration $i_{\mathcal{A}} : \mathcal{A} \to \mathbf{Cat}(\mathcal{A})$. More generally, **Cat** could designate any operation with a weak equivalence $i_{\mathcal{A}}$ such that $\mathbf{Cat}(\mathcal{A})$ satisfies the full Segal condition 20.0.1.

20.2.5 Standard examples

The constructions Υ and $\widetilde{\Upsilon}$ introduced in Sections 14.1 and 16.1 provide important examples. If E is an $(n-1)$-prenerve then we obtain an n-prenerve $\Upsilon(E)$. The main property of this construction is that if $\mathcal{A}$ is any n-prenerve then a morphism of n-prenerves

$$f : \Upsilon(E) \to \mathcal{A}$$

corresponds exactly to a choice of two objects $x = f(0)$ and $y = f(1)$ together with a morphism of $(n-1)$-prenerves $E \to \mathcal{A}(x, y)$.

For example, $\Upsilon(*) = \mathbb{I}$ is the 1-category (which may be considered as an n-nerve for any n) with objects υ_0, υ_1 and with a unique morphism from υ_0 to υ_1. A map $\Upsilon(*) \to \mathcal{A}$ is the same thing as a pair of objects x, y and a 1-morphism from x to y, i.e. an object of $\mathcal{A}(x, y)$. Note here that since $\mathcal{A}(x, y) \in \mathcal{PC}^{n-1}$ the notion of "point" or morphism $* \to \mathcal{A}(x, y)$ is the same as the notion of object of $\mathrm{Ob}(\mathcal{A}(x, y))$.

One can see $\Upsilon(E)$ as the universal n-prenerve $\mathcal{A}$ with two objects x, y and a map $E \to \mathcal{A}(x, y)$.

If E is an $(n-1)$-nerve then $\Upsilon(E)$ is an n-nerve.

The construction Υ can have several arguments, but then the result is no longer an n-nerve. Suppose $E_1, \ldots, E_k$ are $(n-1)$-prenerves. Recall the n-prenerve

$$\widetilde{\Upsilon}_k(E_1, \ldots, E_k)$$

has as object set is the set with $k+1$ elements denoted

$$\widetilde{\Upsilon}_k(E_1, \ldots, E_k)_0 = \{\upsilon_0, \ldots, \upsilon_k\},$$

and morphism object

$$\widetilde{\Upsilon}_k(E_1, \ldots, E_k)(y_0, \ldots, y_p) := E_{y_0+1} \times \cdots \times E_{y_p-1} \times E_{y_p}.$$

When $k = 1$ (and we drop the superscript k in this case) $\widetilde{\Upsilon}_1(E) = \Upsilon(E)$ is the n-prenerve with two objects υ_0, υ_1 and with $(n-1)$-prenerve of morphisms from υ_0 to υ_1 equal to E. Similarly, $\widetilde{\Upsilon}_2(E, F)$ has objects $\upsilon_0, \upsilon_1, \upsilon_2$ and morphisms E from υ_0 to υ_1, F from υ_1 to υ_2 and $E \times F$ from υ_0 to υ_2. We picture $\widetilde{\Upsilon}_k(E_1, \ldots, E_k)$ as a k-simplex (an edge for $k = 1$, a triangle for $k = 2$, a tetrahedron for $k = 3$). The edges are labeled with single E_i, or products $E_i \times \ldots \times E_j$.

There are inclusions of these $\widetilde{\Upsilon}^k$ according to the faces of the k-simplex. The principal faces give inclusions

$$\widetilde{\Upsilon}_{k-1}(E_1, \ldots, E_{k-1}) \hookrightarrow \widetilde{\Upsilon}_k(E_1, \ldots, E_k),$$

$$\widetilde{\Upsilon}_{k-1}(E_2, \ldots, E_k) \hookrightarrow \widetilde{\Upsilon}_k(E_1, \ldots, E_k),$$

and

$$\widetilde{\Upsilon}_{k-1}(E_1, \ldots, E_i \times E_{i+1}, \ldots, E_k) \hookrightarrow \widetilde{\Upsilon}_k(E_1, \ldots, E_k).$$

The inclusions of lower levels are deduced from these by induction. Note that these faces $\widetilde{\Upsilon}_{k-1}$ intersect along appropriate $\widetilde{\Upsilon}_{k-2}$.

See Sections 14.1 and 16.1 for more about Υ and $\widetilde{\Upsilon}$. The properties of $\widetilde{\Upsilon}$ will be developed further in Section 21.4.

20.2.6 Globules

Denote $1^i := (1, \ldots, 1) \in \mathbf{C}^n(*)$. Recall that an *$i$-arrow* in an n-nerve $\mathcal{A}$ means an element of the set $\mathcal{A}_{1^i}$. Inductively, an i-arrow of $\mathcal{A}$ is an $(i-1)$-arrow of $\mathcal{A}_{1/}$.

The collection which associates to $0 \le i \le n$ the set $\mathcal{A}_{1^i}$ is a globular set, in other words a functor $\mathbb{G}^n \to$ SET, see Section 3.2. The "source" and "target" of an i-arrow are $(i-1)$-arrows. They come from the two maps $1^{i-1} \to 1^i$ obtained using the two maps $0 \to 1$ at the i-th coordinate.

Dually, the basic "globules" in $\mathcal{PC}^n$ are

$$\mathbb{O}^i_n := h^n(1^i),$$

which represent an n-prenerve with a single i-morphism together with the various k-morphisms obtained by source and target for $0 \le k < i$.

These can be expressed inductively in terms of the construction Υ:

$$\mathbb{O}^i_n = \Upsilon\left(\mathbb{O}^{i-1}_{n-1}\right),$$

with $\mathbb{O}^0_n = *$. In particular at $i = 1$, $\mathbb{O}^1_n = \Upsilon(*) = \mathbb{I}$ is the 1-category with a single arrow other than the two identities.

One may also note that these globular sets have structures of reflexive globular sets using the degeneracy maps going in the other direction.

The *boundary* of a globule may be defined inductively by

$$\partial \mathbb{O}^i_n = \Upsilon\left(\partial \mathbb{O}^{i-1}_{n-1}\right),$$

starting with $\partial \mathbb{O}^0_n = \emptyset$. The boundary may also be seen as given by a pushout:

$$\partial \mathbb{O}^i_n = \mathbb{O}^{i-1}_n \cup^{\partial \mathbb{O}^{i-1}_n} \mathbb{O}^{i-1}_n.$$

The equivalence with the previous expression in terms of Υ is obtained by induction.

The pushout formula extends to a boundary case which will also be needed: put $\mathbb{O}^{n+1}_n := \mathbb{O}^n_n$, but formally

$$\partial \mathbb{O}^{n+1}_n := \mathbb{O}^n_n \cup^{\partial \mathbb{O}^n_n} \mathbb{O}^n_n.$$

The assignment $i \mapsto \mathbb{O}^i_n$ provides a functor $(\mathbb{G}^n)^o \to \mathcal{P}\mathcal{C}^n$, that is a co-globular object of $\mathcal{P}\mathcal{C}^n$. If $\mathcal{A} \in \mathcal{P}\mathcal{C}^n$ is any n-prenerve, the set of i-arrows of $\mathcal{A}$ is the same as $\mathrm{HOM}_{\mathcal{P}\mathcal{C}^n}\left(\mathbb{O}^i_n, \mathcal{A}\right)$.

20.2.7 The case of Segal *n*-categories

For $\mathcal{M} = \mathcal{K}$, similar notations hold but with an extra simplicial index. Since $\mathcal{K} = \mathrm{PRESH}(\Delta)$, we have

$$\mathcal{P}\mathcal{C}^n(\mathcal{K}) = \mathrm{PRESH}(\mathbf{C}^n(\Delta)).$$

This is the model category of *Segal n-precategories*. Objects satisfying the full Segal condition 20.0.1 will be called *Segal n-categories*.

As before, it is a presheaf category. Denote by

$$\mathbf{c}^n_\Delta : \Delta^{n+1} \to \mathbf{C}^n(\Delta)$$

the composition

$$\Delta^{n+1} \xrightarrow{1 \times \mathbf{c}_\Delta} \Delta^n \times \mathbf{C}(\Delta) \xrightarrow{1 \times \mathbf{c}_{\mathbf{C}(\Delta)}} \cdots \xrightarrow{\mathbf{c}_{\mathbf{C}^{n-1}(\Delta)}} \mathbf{C}^n(\Delta).$$

It expresses $\mathbf{C}^n(\Delta)$ as a quotient category of Δ^{n+1} by identifying all sequences $(m_1, \ldots, m_k, 0, m_{k+2}, \ldots, m_{n+1})$ to a single sequence denoted $(m_1, \ldots, m_k)$.

A Segal n-precategory $\mathcal{A} \in \mathcal{P}\mathcal{C}$ may therefore be considered as a collection of sets $\mathcal{A}_{(m_1,\ldots,m_k)}$ indexed by sequences $(m_1, \ldots, m_k)$ with $k \leq n+1$, such that either $k = 1$ and $m_1 = 0$ (in which case $\mathcal{A}_0$ is just the set of objects $\mathrm{Ob}(\mathcal{A})$) or else all $m_i \geq 1$. These sets are provided with pullback maps corresponding to the images in $\mathbf{C}^n(\Delta)$ of the arrows in Δ^{n+1}.

Putting in a slash means that we look at the functor on the remaining potential indices: for any $k \leq n$ and sequence $(m_1, \ldots, m_k)$ with $m_i \geq 1$, this gives $\mathcal{A}_{m_1,\ldots,m_k/} \in \mathcal{PC}^{n-k}(\Delta)$, which in the notation of the previous paragraph is just the presheaf

$$(m_{k+1}, \ldots, m_l) \mapsto \mathcal{A}_{(m_1,\ldots,m_k,m_{k+1},\ldots,m_l)}$$

on $\mathbf{C}^{n-k}(\Delta)$.

Given that $\mathcal{PC}^n(\mathcal{K})$ is the category of presheaves over $\mathbf{C}^n(\Delta)$, each object $(m_1, \ldots, m_k) \in \mathbf{C}^n(\Delta)$ defines a representable presheaf

$$h(m_1, \ldots, m_k) \in \mathcal{PC}^n(\mathcal{K}).$$

In terms of the representable precategories considered in Lemma 13.2.1, we have

$$h(m_1, \ldots, m_k) = h([m_1], h(m_2, \ldots, m_k)).$$

This provides an alternative inductive expression for the $h(m_1, \ldots, m_k)$. When all the m_i are equal, this will be denoted by $h(m^k)$. In particular, the $h(1^k)$ are the "globules" representing k-morphisms: a map

$$h(1^k) \to \mathcal{A}$$

is a k-morphism of $\mathcal{A}$ (for any $1 \leq k \leq n$). Hence

$$h(1^k) = \Upsilon(\Upsilon(\cdots \Upsilon(*) \cdots)).$$

This notion is not as useful at the last stage; rather one should consider the simplicial set $m \mapsto \mathrm{H}\mathrm{OM}(h(1^n, m), \mathcal{A}) = \mathcal{A}_{(1^n,m)}$ as being the "space of n-morphisms."

Inductively applying Proposition 13.7.2, starting with $\mathcal{K}$ in which the cofibrations are injections of presheaves over Δ, gives:

Remark 20.2.7 In $\mathcal{PC}^n(\mathcal{K})$ the Reedy and injective model structures coincide, and the cofibrations are the monomorphisms of presheaves over $\mathbf{C}^n(\Delta)$.

20.3 Truncation and equivalences

The definition of weak equivalence we have adopted for $\mathcal{PC}(\mathcal{M})$ is designed for enrichment over a general model category $\mathcal{M}$. In Tamsamani's original definition of n-nerves, the notion of equivalence and the truncation operations $\tau_{\leq k}$ were defined inductively along the way. So, in case of $\mathcal{PC}^n(\mathrm{SET})$ there remains the question of equating these two definitions of equivalences. Recall that Tamsamani's definition of equivalence [250] is very similar to the notion of equivalence which was discussed extensively for strict n-groupoids in Chapter 2; here is the definition:

Definition 20.3.1 (Tamsamani) Suppose $f : \mathcal{A} \to \mathcal{B}$ is a morphism of n-nerves, i.e. in $\mathcal{PC}^n(\mathrm{SET})$. Suppose $\mathcal{A}$ and $\mathcal{B}$ both satisfy the full Segal condition. Then we say that f is an *equivalence of n-nerves* if every element of $\mathrm{Ob}(\mathcal{B})$ is equivalent to some $f(x)$ for $x \in \mathrm{Ob}(\mathcal{A})$; and if inductively $\mathcal{A}(x, y) \to \mathcal{B}(f(x), f(y))$ is an equivalence of $(n-1)$-nerves.

There isn't too much difference with the strict case, taking into account the fact that the composition of arrows can be chosen but in a non-unique way, and this definition may be related to other definitions, for example saying that every i-arrow in $\mathcal{B}$ is equivalent to the image of an i-arrow in $\mathcal{A}$, the antecedent being itself unique up to equivalence. This is left as an exercise following the discussion of Chapter 2.

For any tractable left proper cartesian model category $\mathcal{M}$, define the *pretruncation*

$$\tau^p_{\le n} : \mathcal{PC}^n(\mathcal{M}) \to \mathcal{PC}^n(\mathrm{SET})$$

as the functor induced by $\tau_{\le 0} : \mathcal{M} \to \mathrm{SET}$. Applied to $\mathcal{PC}^{n-k}(\mathcal{M})$ for any $0 \le k \le n$, this gives a pretruncation functor

$$\tau^p_{\le k} : \mathcal{PC}^n(\mathcal{M}) \to \mathcal{PC}^k(\mathrm{SET}).$$

If $\mathcal{M} = \mathrm{SET}$ and $n = k$ then it is the identity. Recall that for $\mathcal{A} \in \mathcal{PC}(\mathcal{M})$ the truncation operation was defined by

$$\tau_{\le 1}(\mathcal{A}) := \tau^p_{\le 1}(\mathbf{Seg}(\mathcal{A})).$$

Remark 20.3.2 It doesn't seem to be true in general that the truncation functor from $\mathcal{PC}(\mathcal{M})$ to CAT used starting in Chapter 10 could be expressed in terms of the generators and relations operation from 1-prenerves to 1-nerves as

$$\tau_{\le 1}(\mathcal{A}) \cong \mathbf{Seg}\left(\tau^p_{\le 1}(\mathcal{A})\right).$$

Indeed, the operation $\mathbf{Seg}(\mathcal{A})$ might alter things in a way which isn't seen on the level of 1-truncation.

One should impose the full Segal condition in order to be able to use the pretruncation.

Proposition 20.3.3 *If $\mathcal{A}$ is an $\mathcal{M}$-enriched n-precategory which satisfies the full Segal condition, then for any $k \le n$ the pretruncation $\tau^p_{\le k}(\mathcal{A})$ is a k-nerve, and these truncations may be composed leading at $k = 1$ to the usual truncation $\tau_{\le 1}$. They are compatible with cartesian products.*

Suppose $\mathcal{A} \xrightarrow{f} \mathcal{B}$ is a weak equivalence in $\mathcal{PC}^n(\mathcal{M})$, and $\mathcal{A}$ and $\mathcal{B}$ both satisfy the full Segal condition. Then for any $0 \leq k \leq n$, the truncation $\tau^P_{\leq k}(f)$ is an equivalence of k-nerves in the sense of Definition 20.3.1.

For $\mathcal{M} = \mathrm{SET}$, a morphism $\mathcal{A} \xrightarrow{f} \mathcal{B}$ in $\mathcal{PC}^n(\mathrm{SET})$ between n-nerves, is a weak equivalence if and only if it is an equivalence of n-nerves in the sense of Definition 20.3.1.

Proof For $n=1$, see Lemma 12.4.1. We still should show the compatibility with composing truncation operations. If $\mathcal{P} := \mathcal{PC}(\mathcal{M})$ then we have defined the truncation $\tau_{\leq 0} : \mathcal{P} \to \mathrm{SET}$ using the model structure of $\mathcal{P}$: $\tau_{\leq 0}(\mathcal{A})$ it is the set of morphisms from $*$ to $\mathcal{A}$ in $\mathrm{ho}(\mathcal{P})$. On the other hand, we have defined the truncation denoted also $\tau_{\leq 0} : \mathcal{PC}(\mathcal{M}) \to \mathrm{SET}$ as sending $\mathcal{A}$ to the set of isomorphism classes of $\tau_{\leq 1}(\mathcal{A})$. To show that they are the same, note first that both are invariant under weak equivalences so we may assume that $\mathcal{A}$ is fibrant. Then $\mathrm{HOM}_{\mathrm{ho}(\mathcal{P})}(*, \mathcal{A})$ is the set of morphisms from $*$ to $\mathcal{A}$, up to the relation of homotopy. The set of morphisms is just $\mathrm{Ob}(\mathcal{A})$, and the relation of homotopy says that x is equivalent to y if and only if there exists a map from an interval object to $\mathcal{A}$ sending the endpoints to x and y respectively (see Lemma 18.1.3). If this condition holds then looking at the image of the interval in $\tau_{\leq 1}(\mathcal{A})$ we conclude that the points x and y go to the same isomorphism class. In the other direction, if x and y go to isomorphic objects in $\tau_{\leq 1}(\mathcal{A})$ then by the versality property Theorem 18.3.1 plus the contractibility of the intervals in question, Theorem 18.6.1, the corresponding maps $x, y : * \to \mathcal{A}$ are homotopic. This shows that the two versions of $\tau_{\leq 0}(\mathcal{A})$ coincide.

Assume that $n \geq 2$ and the proposition is known for $\mathcal{PC}^{n-1}(\mathcal{M})$. For $k=1$, the truncation $\tau^P_{\leq 1}(\mathcal{A})$ is the operation of Lemma 12.4.1, which corresponds to the right truncation when applied to $\mathcal{A}$ satisfying the Segal conditions. For $k \geq 2$, we have the truncation functor $\tau^P_{\leq k-1} : \mathcal{PC}^{n-1}(\mathcal{M}) \to \mathcal{PC}^{k-1}(\mathrm{SET})$, taking weak equivalences between objects which satisfy the full Segal condition, to weak equivalences. It follows that if $\mathcal{A} \in \mathcal{PC}^n(\mathcal{M})$ satisfies the full Segal conditions, then applying $\tau^P_{\leq k-1}$ levelwise to $\mathcal{A}$ considered as a diagram from $\Delta^o_{\mathrm{Ob}(\mathcal{A})}$ to $\mathcal{PC}^{n-1}(\mathcal{M})$, it yields a diagram from $\Delta^o_{\mathrm{Ob}(\mathcal{A})}$ to $\mathcal{PC}^{k-1}(\mathrm{SET})$ which again satisifes the Segal conditions, as well as the full Segal conditions levelwise. But $\tau^P_{\leq k-1}$ applied levelwise is by definition $\tau^P_{\leq k}$. This shows that $\tau^P_{\leq k}(\mathcal{A})$ is a k-nerve. The composition of two truncation operations is again a truncation:

$$\tau^P_{\leq r}\left(\tau^P_{\leq k}(\mathcal{A})\right) = \tau^P_{\leq r}(\mathcal{A})$$

whenever $r \leq k$. By induction they are compatible with cartesian products.

Suppose $\mathcal{A} \xrightarrow{f} \mathcal{B}$ is a weak equivalence in $\mathcal{PC}^n(\mathcal{M})$, and $\mathcal{A}$ and $\mathcal{B}$ both satisfy the full Segal condition. Since they satisfy the usual Segal condition, f is essentially surjective, meaning that $\tau_{\leq 0}(\mathcal{A}) \to \tau_{\leq 0}(\mathcal{B})$ is surjective, and induces equivalences $\mathcal{A}(x, y) \to \mathcal{B}(f(x), f(y))$ for all pairs of objects $x, y \in \mathrm{Ob}(\mathcal{A})$. But $\mathcal{A}(x, y)$ and $\mathcal{B}(f(x), f(y))$ also satisfy the full Segal condition, so by the inductive statement known for $\mathcal{PC}^{n-1}(\mathcal{M})$, f induces equivalences of $(k-1)$-nerves

$$\tau_{\leq k-1}\mathcal{A}(x, y) \xrightarrow{\sim} \tau_{\leq k-1}(\mathcal{B}(f(x), f(y))),$$

which are the same as

$$(\tau_{\leq k}\mathcal{A})(x, y) \xrightarrow{\sim} (\tau_{\leq k}\mathcal{B})(f(x), f(y)).$$

This now implies that f induces an equivalence of k-nerves from $\tau_{\leq k}\mathcal{A}$ to $\tau_{\leq k}(\mathcal{B})$.

In the case $\mathcal{M} = \mathrm{SET}$, the above argument works in the other direction to show that if f is an equivalence of n-nerves in the sense of Tamsamani [250], it is essentially surjective and, by applying the inductive statement for $n-1$, it is also fully faithful. □

20.4 Homotopy types and higher groupoids

Come back to the question of relating homotopy types of spaces and n-categories.

Recall that if $\mathcal{M}$ is a tractable left proper model category, we can choose a Reedy cofibrant and levelwise contractible cosimplicial object $\mathbf{R} : \Delta \to \mathcal{M}$, which gives a Quillen adjunction

$$\mathbf{R}_! : \mathcal{K} \leftrightarrow \mathcal{M} : \mathbf{R}^*.$$

Say that an object $A \in \mathcal{M}$ is *spacelike* if $A \cong \mathbf{R}_!(K)$ in $\mathrm{ho}(\mathcal{M})$ for some simplicial set K. This is independent of the choice of resolution $\mathbf{R}$. The initial object $\emptyset \in \mathcal{M}$, the coinitial object $*$ and indeed any contractible object are spacelike, and the spacelike objects are closed under homotopy pushout (i.e. a pushout where one at least of the arrows is a cofibration) and cofibrant transfinite composition.

A weakly $\mathcal{M}$-enriched category $\mathcal{A} \in \mathcal{PC}(\mathcal{M})$ is a *weakly $\mathcal{M}$-enriched groupoid* or just *$\mathcal{M}$-groupoid* if each $\mathcal{A}(x_0, \ldots, x_n)$ is spacelike, and if the category $\tau_{\leq 1}(\mathcal{A})$ is a groupoid.

Lemma 20.4.1 *Suppose $\mathcal{M}$ is a cartesian tractable left proper model category, and consider the model category $\mathcal{PC}(\mathcal{M})$ with Reedy cofibrations.*

Then $\mathcal{A} \in \mathcal{PC}(\mathcal{M})$ *is spacelike if and only if* $\mathbf{Seg}(\mathcal{A})$ *is a weakly* $\mathcal{M}$*-enriched groupoid.*

Proof This is a sketch of proof. Consider the diagram

$$\begin{array}{ccc} \mathcal{K} & \longrightarrow & \mathcal{PC}(\mathcal{K}) \\ & \searrow & \downarrow \\ & & \mathcal{PC}(\mathcal{M}) \end{array}$$

where the top horizontal arrow is an $\mathbf{R}_!$ for $\mathcal{K}$ itself. This is equivalent to the Poincaré–Segal groupoid functor, and induces an equivalence between $\mathrm{ho}(\mathcal{K})$ and the homotopy category of the subcategory of Segal groupoids in $\mathcal{PC}(\mathcal{K})$.

The right vertical arrow is $\mathcal{PC}(\mathbf{R}_!)$, taking Segal groupoids to $\mathcal{M}$-groupoids. This is a left Quillen functor taking contractible objects to contractible objects. Thus the diagonal functor is a left Quillen functor taking contractible objects to contractible objects, so it is equivalent to a $\mathbf{R}_!$ for $\mathcal{PC}(\mathcal{M})$. The spacelike objects of $\mathcal{PC}(\mathcal{M})$ are those which are equivalent to objects in the image of the diagonal functor, equivalently those equivalent to objects coming from the right vertical functor applied to Segal groupoids. These are the $\mathcal{A}$ such that $\mathbf{Seg}(\mathcal{A})$ is an $\mathcal{M}$-groupoid. □

Applying the above discussion to the case of Segal n-categories gives a left Quillen functor from $\mathcal{K}$ to $\mathcal{PC}^n(\mathcal{K})$. This functor depends on the model category structure for $\mathcal{PC}^n(\mathcal{K})$ which has been constructed up until now.

In this case, on the other hand, the Poincaré–Segal groupoid functor

$$\Pi_S : \mathcal{K} \to \mathcal{PC}(\mathcal{K})$$

studied in Section 15.2 may be iterated to give a right Quillen functor

$$\Pi_{n,S} : \mathcal{K} \to \mathcal{PC}^n(\mathcal{K}),$$

which we'll call the *Poincaré–Segal n-groupoid*. For $n = 0$ it is just the identity of $\mathcal{K}$. Supposing that $\Pi_{(n-1),S}$ has already been defined, then for $X \in \mathcal{K}$,

$$\mathrm{Ob}(\Pi_{n,S}(X)) := X_0 = \mathrm{Ob}(\Pi_S(X))$$

is the set of vertices, and

$$\Pi_{n,S}(X)(x_0, \ldots, x_n) := \Pi_{(n-1),S}(\Pi_S(X)(x_0, \ldots, x_n)) \in \mathcal{PC}^{n-1}(\mathcal{K}).$$

The functor $\Pi_{n,S}$ has a left adjoint $|\cdot|$, which is just the multidiagonal realization on $(n+1)$-simplicial sets, similarly to the discussion of Section 15.2.

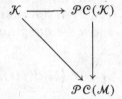

Recall that the Reedy and injective cofibrations are just monomorphisms in $\mathcal{PC}^n(\mathcal{K})$, so these are preserved by $|\cdot|$. Furthermore, the pseudo-generating trivial cofibrations realize to trivial cofibrations of simplicial sets, so

$$|\cdot| : \mathcal{PC}^n(\mathcal{K}) \leftrightarrow \mathcal{K} : \Pi_{n,S}$$

is a Quillen adjunction.

A discussion similar to that of Section 15.3 allows one to give corresponding definitions when X is a topological space, and these reduce to the functors which Tamsamani [249, 250] denoted by $\mathcal{R}^n$ and Ω^n in the course of his argument for n-nerves.

A Segal n-precategory $\mathcal{A} \in \mathcal{PC}^n(\mathcal{K})$ is called a *Segal n-groupoid* if it is a Segal n-category, meaning that it satisfies the full Segal conditions in all dimensions, if each $\mathcal{A}(x_0, \ldots, x_n)$ is a Segal $(n-1)$-groupoid (for which it suffices to check $\mathcal{A}(x_0, x_1)$), and if $\tau_{\leq 1}(\mathcal{A})$ is a 1-groupoid.

Lemma 20.4.2 *If $X \in \mathcal{K}$ is a fibrant simplicial set then $\Pi_{n,S}(X)$ is a Segal n-groupoid.*

Proof It is fibrant in $\mathcal{PC}^n(\mathcal{K})$, so it satisfies the full Segal conditions. Note that $\Pi_S(X)(x_0, \ldots, x_n) \in \mathcal{K}$ are fibrant simplicial sets too, as may be seen from the explicit formula of Section 15.2 using the cartesian property of $\mathcal{K}$. By induction on n, the

$$\Pi_{n,S}(X)(x_0, \ldots, x_n) = \Pi_{(n-1),S}(\Pi_S(X)(x_0, \ldots, x_n))$$

are Segal $(n-1)$-groupoids. Furthermore,

$$\tau_{\leq 0}(\Pi_{n,S}(x_0, \ldots, x_n)) = \pi_0(\Pi_S(X)(x_0, \ldots, x_n)),$$

as will be shown by induction on n, the inductive step for that to be done in a moment. It follows that

$$\tau_{\leq 1}(\Pi_{n,S}(X)) = \tau_{\leq 1}(\Pi_S(X)),$$

and this is a 1-groupoid by Segal's theorem. This completes the verification that $\Pi_{n,S}(X)$ is a Segal n-groupoid. To finish the proof, note that

$$\tau_{\leq 0}(\Pi_{n,S}(X)) = \tau_{\leq 0}(\tau_{\leq 1}(\Pi_S(X))) = \pi_0(X),$$

again applying Segal's theorem. This provides the inductive step for the intermediate statement used above. □

Remark 20.4.3 In a similar way, $\tau_{\leq 1}(\Pi_{n,S}(X))$ is equivalent to the usual Poincaré groupoid $\Pi_1(X)$.

Consider the functor $\mathbf{V} : \Delta \to \mathcal{PC}^n(\mathcal{K})$ defined, in terms of the representable objects of Section 20.2.7, by

$$\mathbf{V}([k]) := h(k^n),$$

with $\mathbf{V}([0]) := h(0) = *$. This functor can play the role of a resolution $\mathbf{R}$ such as was considered at the beginning of this section, although the $\mathbf{V}([k])$ are not contractible. The diagonal realization becomes

$$|\mathcal{A}| = \mathbf{V}^*(\mathcal{A}).$$

This has a right adjoint $\Pi_{n,S} = \mathbf{V}_*$, and a left adjoint $\mathbf{V}_!$. Note, however, that $\mathbf{V}([k])$ is not contractible in $\mathcal{PC}^n(\mathcal{K})$, so $\mathbf{V}_!$ is not a left Quillen functor. The following lemma says that its right adjoint $\mathbf{V}^*$ nevertheless acts like a right Quillen functor, when applied to Segal n-groupoids, which allows us consider the composition $\Pi_{n,S}(|\mathcal{A}|)$ without first taking a fibrant replacement for $|\mathcal{A}|$.

Lemma 20.4.4 *If $\mathcal{A}$ is a fibrant Segal n-groupoid, then the realization $|\mathcal{A}| = \mathbf{V}^*(\mathcal{A})$ is a fibrant simplicial set.*

The basic idea of the proof is that the $\mathbf{V}([k])$ become contractible after group completion when all arrows are inverted, and maps into a fibrant Segal n-groupoid, extend in a homotopically unique way to the group completion. The proof will be discussed in the next part, in Section 21.6.3, since it makes use of the notion of inversion of morphisms.

Tamsamani's iterative version [249, 250] of Segal's theorem [228] is as follows. In order to keep the present discussion self-contained, we use Kan's fibrant replacement functor $\mathbf{Ex}^\infty$ for simplicial sets even though by the preceding lemma it isn't really needed. Once we know Lemma 20.4.4 the occurence of $\mathbf{Ex}^\infty$ can be removed from the statement of the theorem.

Theorem 20.4.5 *If $\mathcal{A} \to \mathcal{B}$ is a morphims of Segal n-groupoids, then it is a weak equivalence if and only if $|\mathcal{A}| \to |\mathcal{B}|$ is a weak equivalence.*

Suppose $X \in \mathcal{K}$ is a fibrant simplicial set. Then

$$|\Pi_{n,S}(X)| \to X$$

is a weak equivalence of simplicial sets. In the other direction, if $\mathcal{A} \in \mathcal{PC}^n(\mathcal{K})$ is a fibrant Segal n-groupoid, then

$$\mathcal{A} \to \Pi_{n,S}(\mathbf{Ex}^\infty |\mathcal{A}|)$$

is a weak equivalence in $\mathcal{PC}^n(\mathcal{K})$.

This pair of functors provides an equivalence between ho($\mathcal{K}$) $\cong$ ho(Top) *and the homotopy category obtained by localizing the subcategory of Segal n-groupoids in $\mathcal{PC}^n(\mathcal{K})$ by weak equivalences.*

Proof For $n = 0$ it is trivial, the functors involved are the identity. For $n = 1$ it is Segal's theorem [228] which we discussed in Section 15.2. Assume therefore that $n \geq 2$ and the theorem is known for $(n-1)$. We'll use Tamsamani's notation [249, 250] $\mathcal{R}$ for the realization and Ω for Π_S. Realizing at the top level induces a map

$$\mathcal{PC}^{n-1}(\mathcal{R}) : \mathcal{PC}^{n-1}(\mathcal{PC}(\mathcal{K})) \to \mathcal{PC}^{n-1}(\mathcal{K}),$$

and applying Π_S at the top level induces

$$\mathcal{PC}^{n-1}(\Omega) : \mathcal{PC}^{n-1}(\mathcal{K}) \to \mathcal{PC}^{n-1}(\mathcal{PC}(\mathcal{K})).$$

These are adjoint.

From the inductive definition of $\Pi_{n,S}$ we get the expression

$$\Pi_{n,S}(X) = \mathcal{PC}^{n-1}(\Omega)(\Pi_{(n-1),S}(X)).$$

Similarly in the other direction, if $\mathcal{A} \in \mathcal{PC}^n(\mathcal{K})$ then

$$|\mathcal{A}| = |\mathcal{PC}^{n-1}(\mathcal{R})(\mathcal{A})|,$$

where the realization on the right is for $\mathcal{PC}^{n-1}(\mathcal{K})$.

We can now prove the first statement of the theorem: if $\mathcal{A} \xrightarrow{f} \mathcal{B}$ is a morphism of Segal n-groupoids, suppose that

$$|\mathcal{PC}^{n-1}(\mathcal{R})(\mathcal{A})| \to |\mathcal{PC}^{n-1}(\mathcal{R})(\mathcal{B})|$$

is a weak equivalence. Note that $\mathcal{PC}^{n-1}(\mathcal{R})(\mathcal{A})$ and $\mathcal{PC}^{n-1}(\mathcal{R})(\mathcal{B})$ are Segal $(n-1)$-groupoids, as may be seen by using induction and noting that the application of realization at the top level doesn't affect the lower truncations. Therefore, the first statement of the theorem for $n-1$ implies that

$$\mathcal{PC}^{n-1}(\mathcal{R})(f) : \mathcal{PC}^{n-1}(\mathcal{R})(\mathcal{A}) \to \mathcal{PC}^{n-1}(\mathcal{R})(\mathcal{B})$$

is a weak equivalence of Segal $(n-1)$-groupoids.

We claim that this implies that f is a weak equivalence; the inductive version of this claim will be needed below. On the first truncation,

$$\tau_{\leq 1}(\mathcal{PC}^{n-1}(\mathcal{R})(\mathcal{A})) = \tau_{\leq 1}(\mathcal{A}),$$

and the same for $\mathcal{B}$, so essential surjectivity of $\mathcal{PC}^{n-1}(\mathcal{R})(f)$ implies essential surjectivity of f. On the other hand, for a sequence of objects $x_0, \ldots, x_n \in \mathrm{Ob}(\mathcal{A})$, we have

$$\mathcal{PC}^{n-1}(\mathcal{R})(\mathcal{A})(x_0, \ldots, x_n) = \mathcal{PC}^{n-2}(\mathcal{R})(\mathcal{A}(x_0, \ldots, x_n)),$$

and similarly for $\mathcal{B}$ at $(f(x_0), \ldots, f(x_n))$. The fact that $\mathcal{PC}^{n-1}(\mathcal{R})(f)$ is fully faithful therefore means that

$$\mathcal{PC}^{n-2}(\mathcal{R})(\mathcal{A}(x_0, \ldots, x_n)) \to \mathcal{PC}^{n-2}(\mathcal{R})(\mathcal{B}(f(x_0), \ldots, f(x_n)))$$

is an equivalence of Segal $(n-2)$-groupoids. By the claim which we are trying to prove, inductively at the case $(n-1)$ of the theorem, this implies that

$$\mathcal{A}(x_0, \ldots, x_n) \to \mathcal{B}(f(x_0), \ldots, f(x_n))$$

is an equivalence of Segal $(n-1)$-groupoids. Therefore, f is fully faithful, completing the proof that it is an equivalence. This finishes the proof of the first statement of the theorem.

Turn next to the proof of the second statement. Segal's theorem, stated for us as Theorem 15.2.1, says that the counit of the adjunction

$$\mathcal{R}\Omega X \to X$$

is an equivalence whenever X is a fibrant simplicial set. Now use this on the top level, after application of the construction $\mathcal{PC}^{n-1}$. Given a Segal $(n-1)$-precategory $\mathcal{B} \in \mathcal{PC}^{n-1}(\mathcal{K})$ with the property that the $\mathcal{B}_{m_1,\ldots,m_{n-1}/} \in \mathcal{K}$ are fibrant, then

$$\mathcal{PC}^{n-1}(\mathcal{R}) \circ \mathcal{PC}^{n-1}(\Omega)(\mathcal{B}) \to \mathcal{B}$$

is a weak equivalence, indeed it is a levelwise weak equivalence considered as a $\mathbf{C}^{n-1}(*)$-diagram in $\mathcal{K}$.

Apply this to $\mathcal{B} = \Pi_{(n-1),S}(X)$, where X is a fibrant simplicial set. Note that

$$\mathcal{PC}^{n-1}(\Omega) \circ \Pi_{(n-1),S} = \Pi_{n,S},$$

and $\Pi_{(n-1),S}(X)$ satisfies the condition of being fibrant at the top level needed for the previous paragraph; therefore

$$\mathcal{PC}^{n-1}(\mathcal{R})\Pi_{n,S}(X) \to \Pi_{(n-1),S}(X)$$

is a weak equivalence in $\mathcal{PC}^{n-1}(\mathcal{K})$. Both sides are Segal $(n-1)$-groupoids.

Applying the realization inductively, and noting that $|\mathcal{PC}^{n-1}(\mathcal{R})(\mathcal{A})| = |\mathcal{A}|$, we conclude that

$$|\Pi_{n,S}(X)| \to |\Pi_{(n-1),S}(X)|$$

is a weak equivalence. Composing with the weak equivalence known by induction for $n-1$, gives the required weak equivalence

$$|\Pi_{n,S}(X)| \to |\Pi_{(n-1),S}(X)| \to X.$$

This proves the second statement.

For the third statement, we would like to show that the map

$$\mathcal{A} \to \Pi_{n,S}(\mathbf{Ex}^{\infty}|\mathcal{A}|)$$

is a weak equivalence. This is a map of Segal n-groupoids, so applying the first statement, it suffices to show that

$$|\mathcal{A}| \to |\Pi_{n,S}(\mathbf{Ex}^{\infty}|\mathcal{A}|)|$$

is a weak equivalence. But this may be composed with the counit. From the second statement of the theorem, already proven previously, the map

$$|\Pi_{n,S}(\mathbf{Ex}^{\infty}|\mathcal{A}|)| \to \mathbf{Ex}^{\infty}|\mathcal{A}|$$

is a weak equivalence of simplicial sets. But the composition fits into a diagram

$$\begin{array}{ccccc}
|\mathcal{A}| & \longrightarrow & |\Pi_{n,S}(|\mathcal{A}|)| & \longrightarrow & |\mathcal{A}| \\
\| & & \downarrow & & \downarrow \\
|\mathcal{A}| & \longrightarrow & |\Pi_{n,S}(\mathbf{Ex}^{\infty}|\mathcal{A}|)| & \longrightarrow & \mathbf{Ex}^{\infty}|\mathcal{A}|
\end{array}$$

where the top composition is the identity and the right square commutes by naturality of the counit of the adjunction. Therefore, the composition along the bottom row is the standard inclusion $|\mathcal{A}| \to \mathbf{Ex}^{\infty}|\mathcal{A}|$, which is a weak equivalence. Now 3 for 2 implies that the first map

$$|\mathcal{A}| \to |\Pi_{n,S}(\mathbf{Ex}^{\infty}|\mathcal{A}|)|$$

is a weak equivalence of simplicial sets. This completes the proof of the third part of the theorem.

The last part follows. □

Tamsamani [249, 250] considered the above theorem for n-simplicial spaces, on his way towards proving the analogous result relating n-nerves and n-truncated homotopy types. When using topological spaces, which are automatically fibrant, Lemma 20.4.4 or $\mathbf{Ex}^{\infty}$ are not needed. The realization functor, which he denoted by $\mathcal{R}^n$, and the Poincaré–Segal n-groupoid functor $\Pi_{n,S}$, which he denoted by Ω^n, have concrete expressions in terms of *globular prisms*, i.e. products of standard simplices quotiented out by relations corresponding to the constancy conditions. These expressions are just the iterative versions of what we discussed in Section 15.3.

20.4.1 Homotopy groups

We can define the *homotopy groups* of an n-groupoid. This is pretty much the same as in the strict case, considered in Section 2.2, and follows the principle described in Section 3.3 in general.

If $\mathcal{A}$ is an n-groupoid, define $\pi_0(\mathcal{A}) := \tau_{\leq 0}(\mathcal{A})$, which is the set of isomorphism classes in the 1-groupoid $\tau_{\leq 1}(\mathcal{A})$. If $x \in \mathrm{Ob}(\mathcal{A})$, define $\pi_1(\mathcal{A}, x)$ to be the group of automorphisms of x in $\tau_{\leq 1}(\mathcal{A})$. Define the remaining homotopy groups for $i \geq 2$ by induction, putting

$$\pi_i(\mathcal{A}, x) := \pi_{i-1}(\mathcal{A}(x, x), \mathbf{1}_x).$$

Here $\mathbf{1}_x \in \mathrm{Ob}(\mathcal{A}(x, x))$ is the image of the unique point of $\mathcal{A}(x) = *$ (that was the "unitality condition" in Definition 10.1.1), by the degeneracy map $\mathcal{A}(x) \to \mathcal{A}(x, x)$. Let $\mathbf{1}^i_x$ be the iterated higher identity, defined inductively by

$$\mathbf{1}^i_x := \mathbf{1}_{\mathbf{1}^{i-1}_x}.$$

It is an i-morphism of $\mathcal{A}$, which is to say a map $\mathbb{O}^i_n \to \mathcal{A}$ in the notation of Section 20.2.6. When $i = 0$, $\mathbf{1}^i_x := x$ by convention.

Lemma 20.4.6 *If $\mathcal{A}$ is an n-groupoid, then $\pi_i(\mathcal{A}, x)$ is equal to the quotient of the set of i-morphisms with source and target $\mathbf{1}^{i-1}_x$, by the relation of inner equivalence. Furthermore, the relation of inner equivalence on i-morphisms in an n-groupoid is equivalent to the relation defined by existence of any $i+1$-morphism with the given source and target. In particular, the homotopy groups vanish for $i \geq n$.*

Proof Left to the reader, following the discussion of Chapter 2. □

The definition of homotopy groups may be extended to the case of Segal n-groupoids, taking into account that we have a space of morphisms in degree n. If $\mathcal{A}$ is a Segal 0-groupoid, which is to say just a simplicial set, then $\pi_0(\mathcal{A})$ is the π_0 of the simplicial set. If x is any vertex then $\pi_i(\mathcal{A}, x)$ are the usual homotopy groups of the simplicial set. Now, if $\mathcal{A}$ is a Segal n-groupoid, and $x \in \mathrm{Ob}(\mathcal{A})$, then we obtain i-morphisms $\mathbf{1}^i_x$ as before for $0 \leq i < n$, and $\mathbf{1}^n_x$ is a vertex of the top-level simplicial set. As before, the homotopy groups are defined by induction on n:

$$\pi_i(\mathcal{A}, x) := \pi_{i-1}(\mathcal{A}(x, x), \mathbf{1}_x),$$

with $\pi_0(\mathcal{A}) := \tau_{\leq 0}(\mathcal{A})$ for an n-groupoid with $n \geq 1$. In the Segal case, they can be nontrivial for all values of i.

Lemma 20.4.7 *The homotopy groups $\pi_i(\mathcal{A}, x)$ are abelian for $i \geq 2$, whenever $\mathcal{A}$ is an n-groupoid or a Segal n-groupoid.*

Proof The truncation $\tau_{\leq 2}(\mathcal{A})$ is a 2-groupoid in the usual sense, equivalent to a strict one, so the Eckmann–Hilton argument (Section 2.1) shows that $\pi_2(\mathcal{A}, x)$ are commutative. For higher i it now follows from the inductive definition. □

20.4.2 Homotopy n-types and n-groupoids

For n-nerves, the right Quillen functor

$$\Pi_n : \mathcal{K} \to \mathcal{PC}^n(\text{SET})$$

is the *Poincaré n-groupoid* whose left adjoint is again denoted $|\cdot|$. These functors were defined explicitly by Tamsamani [250], before having a model structure on $\mathcal{PC}^n(\text{SET})$ or $\mathcal{PC}^n(\mathcal{K})$. Note that the truncation functor $\tau_{\leq 0} : \mathcal{K} \to \text{SET}$ induces

$$\tau_{\leq n} : \mathcal{PC}^n(\mathcal{K}) \to \mathcal{PC}^n(\text{SET})$$

and

$$\Pi_n(X) = \tau_{\leq n}\Pi_{n,S}(X).$$

The truncation $\tau_{\leq 0}$ is left adjoint to the inclusion $\text{SET} \to \mathcal{K}$ given by thinking of sets as discrete simplicial sets, so $\tau_{\leq n}$ is left adjoint to the leftmost first map below:

$$\mathcal{PC}^n(\text{SET}) \to \mathcal{PC}^n(\mathcal{K}) \stackrel{|\cdot|}{\to} \mathcal{K}.$$

The composition of which, with the realization for Segal n-groupoids, gives the realization for n-groupoids (i.e. n-nerves which satisfy the groupoid condition in all dimensions).

The realization and Poincaré groupoid constructions set up an equivalence of homotopy categories between n-truncated homotopy types, and n-groupoids. There are natural isomorphisms between the homotopy groups on both sides. See Tamsamani [250] for more details on this aspect.

20.5 The $(n+1)$-category $nCAT$

The cartesian model category structure on $\mathcal{PC}^n(\mathcal{M})$ allows us to define a structure of $\mathcal{M}$-enriched $(n+1)$-category denoted $n\text{CAT}(\mathcal{M})$. In particular, starting with $\mathcal{M} = \text{SET}$ we obtain the $(n+1)$-nerve $n\text{CAT} = n\text{CAT}(\text{SET})$ discussed in Chapter 3.

In Section 7.9.2, starting with a tractable left proper cartesian model category $\mathcal{P}$ we get the strict $\mathcal{P}$-enriched category $\mathbf{Enr}(\mathcal{P})$. Recall that the objects of $\mathbf{Enr}(\mathcal{P})$ are the cofibrant and fibrant objects of $\mathcal{P}$, and if X, Y are two such objects then the morphism object is given by the internal $\underline{\mathrm{HOM}}$,

$$\mathbf{Enr}(\mathcal{P})(X, Y) = \underline{\mathrm{HOM}}_{\mathcal{P}}(X, Y).$$

The identity and composition operations are the obvious ones.

This general discussion now applies to $\mathcal{P} = \mathcal{PC}^n(\mathcal{M})$. Define

$$n\mathrm{CAT}(\mathcal{M}) := \mathbf{Enr}(\mathcal{PC}^n(\mathcal{M})).$$

Note that $n\mathrm{CAT}(\mathcal{M})$ is a $\mathcal{PC}^n(\mathcal{M})$-enriched category, so

$$n\mathrm{CAT}(\mathcal{M}) \in \mathcal{PC}(\mathcal{PC}^n(\mathcal{M})) = \mathcal{PC}^{n+1}(\mathcal{M}).$$

As $n\mathrm{CAT}(\mathcal{M})$ is a strict $\mathcal{PC}^n(\mathcal{M})$-enriched category, its Segal maps are isomorphisms. Notice that for any fibrant cofibrant objects $\mathcal{A}, \mathcal{B} \in \mathcal{PC}^n(\mathcal{M})$,

$$n\mathrm{CAT}(\mathcal{M})(\mathcal{A}, \mathcal{B}) = \underline{\mathrm{HOM}}_{\mathcal{PC}^n(\mathcal{M})}(\mathcal{A}, \mathcal{B})$$

is also fibrant. By the Segal isomorphisms and the fact that fibrant and cofibrant objects are preserved by cartesian product, it follows that for any sequence $\mathcal{A}_0, \ldots, \mathcal{A}_m$ the object

$$n\mathrm{CAT}(\mathcal{M})(\mathcal{A}_0, \ldots, \mathcal{A}_m) \in \mathcal{PC}^n(\mathcal{M})$$

is fibrant. In particular, $n\mathrm{CAT}(\mathcal{M})$ is a projectively fibrant $\mathcal{PC}^n(\mathcal{M})$-enriched precategory.

One unfortunate consequence of the strictness on the first level is that $n\mathrm{CAT}(\mathcal{M})$ is not Reedy fibrant. Therefore it isn't quite correct to write "$n\mathrm{CAT}(\mathcal{M}) \in \mathrm{Ob}((n+1)\mathrm{CAT}(\mathcal{M}))$" since $n\mathrm{CAT}(\mathcal{M})$ is not a fibrant object of $\mathcal{PC}^{n+1}(\mathcal{M})$. Let $n\mathrm{CAT}(\mathcal{M}) \to n\mathrm{CAT}'(\mathcal{M})$ denote its fibrant replacement, and suppose we are in the usual situation where all objects, including $n\mathrm{CAT}'(\mathcal{M})$, are cofibrant. Then

$$n\mathrm{CAT}'(\mathcal{M}) \in \mathrm{Ob}((n+1)\mathrm{CAT}(\mathcal{M})).$$

The difference between $n\mathrm{CAT}(\mathcal{M})$ and $n\mathrm{CAT}'(\mathcal{M})$ was one of the main obstacles which needed to be overcome in the treatment of limits [235], as will be discussed in Chapter 22.

We have glossed over the question of universes here: various strategies could be envisioned, the easiest of which will be discussed in Section 22.3.

If $\mathcal{A}, \mathcal{B}$ are cofibrant and fibrant $\mathcal{M}$-enriched n-precategories, then a morphism $f : \mathcal{A} \to \mathcal{B}$ corresponds to an object of the $\mathcal{M}$-enriched n-precategory $\underline{\mathrm{HOM}}(\mathcal{A}, \mathcal{B})$, or equivalently to a 1-morphism in $n\mathrm{CAT}(\mathcal{M})$. The morphism f

is a weak equivalence in $\mathscr{P}\mathcal{C}^n(\mathcal{M})$ if and only if it is an internal equivalence in $n\mathrm{CAT}(\mathcal{M})$, i.e. it projects to an isomorphism in $\tau_{\leq 1}(n\mathrm{CAT}(\mathcal{M}))$, and we have an equivalence of categories

$$\tau_{\leq 1}(n\mathrm{CAT}(\mathcal{M})) \cong \mathrm{ho}(\mathscr{P}\mathcal{C}^n(\mathcal{M})).$$

This compatibility was formulated by Tamsamani [250] in asking for nCAT, and may be proven using the same arguments as in the previous section.

The above discussion applies with $\mathcal{M} = \mathrm{SET}$ to give the $(n + 1)$-nerve $n\mathrm{CAT} := n\mathrm{CAT}(\mathrm{SET})$ of n-nerves; and to $\mathcal{M} = \mathcal{K}$ to give the Segal $(n + 1)$-category $n\mathrm{SeCAT} := n\mathrm{CAT}(\mathcal{K})$ of Segal n-categories. These were used in the preprint of Hirschowitz and myself [145] to discuss the notion of higher stacks.

21
Higher categorical techniques

In this chapter we review some of the higher categorical techniques which can be implemented, once we have the model structure. This covers what was done in my preprint on limits [235], but cast in the language of enrichment over a general $\mathcal{M}$.

Fix a tractable left proper cartesian model category $\mathcal{M}$. The terminology *$\mathcal{M}$-precategory* will mean an object of $\mathcal{PC}(\mathcal{M})$ and an *$\mathcal{M}$-category* will mean an $\mathcal{M}$-precategory satisfying the Segal conditions. In other words, an $\mathcal{M}$-category means a weakly $\mathcal{M}$-enriched category. This notation is in effect throughout the remaining chapters.

When speaking of strictly $\mathcal{M}$-enriched categories, that is to say precategories such that the Segal maps are isomorphisms, this will be specified explicitly.

The model structure on $\mathcal{PC}(\mathcal{M})$ is the Reedy model structure constructed in Theorem 19.2.1 unless otherwise specified.

Using the functor $\text{SET} \to \mathcal{M}$, which sends a set A to the colimit of $*$ indexed by the discrete category A, a 1-category may be considered as an $\mathcal{M}$-category.

In some places we will use a notation $\mathcal{P}$ for a cartesian left proper tractable model category. This is usually done in order to cast in general terms some discussions mainly intended for the case $\mathcal{P} = \mathcal{PC}(\mathcal{M})$.

For the case of n-categories, a more iterative notation has been considered in Section 20.2, and this corresponds with the notation which was used in *loc. cit.* [235]. The reader may refer there for a version of the discussion in the old notation, but things seem to be clarified by considering a general enrichment.

The main topics will be:

- the "opposite category";
- inversion of morphisms, giving a construction of "localization";

- the general setup for limits and colimits within an $\mathcal{M}$-category;
- and some invariance properties for limits and colimits.

The construction of limits and colimits of $\mathcal{M}$-categories themselves will be the subject of the next chapter.

21.1 The opposite category

The first construction of category theory is that of the *opposite category*, which reverses the direction of the arrows. We obtain similarly an involution denoted $\mathcal{A} \mapsto \mathcal{A}^o$ of $\mathcal{PC}(\mathcal{M})$, defined by

$$\mathrm{Ob}(\mathcal{A}^o) = \mathrm{Ob}(\mathcal{A}),$$

$$\mathcal{A}^o(x_0, \ldots, x_n) := \mathcal{A}(x_n, \ldots, x_0).$$

Technically speaking this comes from the functor op : $\Delta_X \to \Delta_X$ sending $(x_0, \ldots, x_n)$ to $(x_n, \ldots, x_0)$.

The opposite involution preserves projective, injective and Reedy cofibrations, the levelwise and new weak equivalences of $\mathcal{PC}(X, \mathcal{M})$, and the global weak equivalences of $\mathcal{PC}(\mathcal{M})$, so it preserves our main Reedy global model structure on $\mathcal{PC}(\mathcal{M})$ as well as all of the intermediate ones. In particular, the opposite involution is both left and right Quillen on any of these model structures.

In the iterative situation we obtain an action of $(\mathbb{Z}/2)^n$ on $\mathcal{PC}^n(\mathcal{M})$ with opposite involutions at each stage; conjecturally this $(\mathbb{Z}/2)^n$ should be the full ∞-categorical group of automorphisms of the theory of n-categories, see Toën [256].

21.2 Equivalent objects

Recall from Section 18.2 that $\mathbb{I} := \Upsilon(*)$ denotes the $\mathcal{M}$-precategory with two objects υ_0, υ_1 and a single morphism between them. It is already an $\mathcal{M}$-category, and is the image of the usual category $[0 \to 1]$ under the map $\mathcal{PC}(\mathrm{SET}) \to \mathcal{PC}(\mathcal{M})$.

Then, also, $\mathbf{I}$ denotes the image of the category with two isomorphic objects υ_0, υ_1, with unique morphisms going in either direction between them, whose compositions are the identity. Thus $\mathbf{I}(x_0, \ldots, x_p) = *$ for any sequence of objects. It contains $\mathbb{I}$ as subcategory.

Lemma 21.2.1 *The object* **I** *is Reedy cofibrant in* $\mathcal{PC}(\mathcal{M})$, *and indeed the map*

$$\{\upsilon_0, \upsilon_1\} \to \mathbf{I}$$

is a Reedy cofibration in $\mathcal{PC}_{\text{Reedy}}([1], \mathcal{M})$. *Thus* **I** *is an interval object for* $\mathcal{PC}(\mathcal{M})$. *Furthermore, the inclusion* $\mathbb{I} \to \mathbf{I}$ *is a Reedy cofibration.*

Proof Fix a sequence $y. = (y_0, \ldots, y_k)$. To see whether a map $\mathcal{A} \to \mathcal{B}$ (of precategories with the same set X of objects) is a Reedy cofibration, we should look at the diagram

$$\begin{array}{ccc} \operatorname{colim}_{y.\to z.} \mathcal{A}(z.) & \to & \operatorname{colim}_{y.\to z.} \mathcal{B}(z.) \\ \downarrow & & \downarrow \\ \mathcal{A}(y.) & \longrightarrow & \mathcal{B}(y.) \end{array}$$

where the colimits are taken over surjections $y. \to z.$ which are not the identity in Δ_X. This diagram induces the latching map from the pushout of the upper left angle, to the bottom right. The Reedy cofibration condition says that those maps should be cofibrations for all $y.$.

The colimits which occur are taken over indexing categories which, if nonempty, are connected. Indeed, for any sequence $y.$, let $w.$ be the same sequence of objects but without repetitions, then the tautological surjection $y. \to w.$ is a coinitial object in the category of all surjections (going to an initial object in the indexing category for the colimit since our precategories are contravariant functors on Δ_X). Note that if $y.$ doesn't contain any repetitions then there is no surjection which isn't the identity, so then the indexing category is empty.

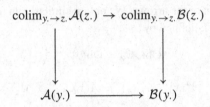

In the present case, $\mathcal{A} = \{\upsilon_0, \upsilon_1\} = \mathbf{disc}(X)$ and $\mathcal{B} = \mathbf{I} = \mathbf{codsc}(X)$ for $X = [1] = \{\upsilon_0, \upsilon_1\}$. In particular, $\mathcal{B}(z.) = *$ for any sequence $z.$. If the indexing category is nonempty hence connected, this implies that $\operatorname{colim}_{y.\to z.} \mathcal{B}(z.) = *$ (see Lemma 8.1.9). If the indexing category is empty then the colimit is $\emptyset$. For $\mathcal{A}$, $\mathcal{A}(z.) = \emptyset$ if $\bar{z}.$ is a nonconstant sequence, or $\mathcal{A}(z.) = *$ if $z.$ is a constant sequence. These possibilities are the same for all $z.$ with a surjection from a given $y.$, so $\operatorname{colim}_{y.\to z.} \mathcal{A}(z.)$ is either $\emptyset$ or $*$. We conclude that the four objects in the above diagram are either $\emptyset$ or $*$, and furthermore the configuration which would yield a latching map $* \cup^{\emptyset} * \to *$ does not occur. Indeed, if the indexing category of surjections is nonempty then both vertical maps are isomorphisms and the latching map is an isomorphism. If the indexing category is empty then the latching map is just $\mathcal{A}(y.) \to \mathcal{B}(y.)$ which is either an isomorphism, or else $\emptyset \to *$. In the last case, it is a cofibration by Condition (AST) in Definition 7.7.1. This completes the proof that our map $\mathcal{A} \to \mathcal{B}$ is a Reedy cofibration.

To show that $\mathbb{I} \to \mathbf{I}$ is a Reedy cofibration, use the same argument but this time with $\mathcal{A} := \mathbb{I}$. In this case if $z.$ is an anywhere decreasing sequence then $\mathcal{A}(z.) = \emptyset$, whereas if $z.$ is a nondecreasing sequence then $\mathcal{A}(z.) = *$. Hence again, $\mathrm{colim}_{y. \to z.} \mathcal{A}(z.)$ is either $\emptyset$ or $*$. And if the indexing category for the colimit is nonempty then this answer is the same as for $\mathcal{A}(y.)$ because, if there is a surjection $y. \to z.$, then the anywhere decreasing/nondecreasing dichotomy is the same for $y.$ and $z.$. The rest of the proof goes through the same way. □

An object $x \in \mathrm{Ob}(\mathcal{C})$ in an $\mathcal{M}$-precategory, may also be viewed as a point $* \xrightarrow{x} \mathcal{C}$, hence projects to an element of $\tau_{\leq 0}\mathcal{C} = \mathrm{ho}(\mathcal{PC}(\mathcal{M}))(*, \mathcal{C})$. Two objects x, y are *inner-equivalent* if they project to the same thing in $\tau_{\leq 0}\mathcal{C}$. The following lemma extends and simplifies the criterion of Lemma 18.1.2 which was for any model category:

Lemma 21.2.2 *If $\mathcal{C}$ is a fibrant $\mathcal{M}$-category, then two objects $x, y \in \mathrm{Ob}(\mathcal{C})$ are inner-equivalent, if and only if there exists a morphism $\mathbf{I} \to \mathcal{C}$ sending υ_0 to x and υ_1 to y.*

Proof One direction is obvious: if there exists such a morphism then by functoriality of $\tau_{\leq 0}$ x and y are equivalent (because $\tau_{\leq 0}(\mathbf{I}) = *$). For the other direction, suppose x and y are inner-equivalent. Use Theorem 18.8.1 which is based on the construction of versal intervals in Chapter 18. The theorem starts with a cofibrant replacement $\mathcal{B} \to \mathbf{I}$ in $\mathcal{PC}_{\mathrm{Reedy}}([1], \mathcal{M})$, but by the previous lemma we can use $\mathcal{B} = \mathbf{I}$ which is already Reedy cofibrant. Note also that our $\mathcal{C}$ is a fibrant object in $\mathcal{PC}_{\mathrm{Reedy}}(\mathrm{Ob}(\mathcal{C}), \mathcal{M})$ (see Proposition 19.4.1), so we can apply Theorem 18.8.1 to our inner-equivalent objects x and y to conclude that there is a morphism $\mathbf{I} \to \mathcal{C}$ sending υ_0 to x and υ_1 to y. □

This leads to a lifting property for inner equivalences along a fibration. Bergner [40] shows that this kind of lifting property can actually be used to give a characterization of fibrations between Segal categories.

Corollary 21.2.3 *Suppose $f : \mathcal{A} \to \mathcal{B}$ is a fibration between fibrant $\mathcal{M}$-categories. Suppose that $a \in \mathrm{Ob}(\mathcal{A})$ and $b \in \mathrm{Ob}(\mathcal{B})$ are objects such that $f(a)$ is inner-equivalent to b. Then there is a different object $a' \in \mathrm{Ob}(\mathcal{A})$ inner-equivalent to a such that $f(a') = b$ in $\mathrm{Ob}(\mathcal{B})$.*

If $f : \mathcal{A} \to \mathcal{B}$ is a fibration between fibrant $\mathcal{M}$-categories and if f is an equivalence then f is surjective on objects.

Proof The equivalence between $f(a)$ and b comes, using the previous lemma, from a morphism $\mathbf{I} \to \mathcal{B}$ sending υ_0 to $f(a)$ and υ_1 to b. We have a lifting a over $\{\upsilon_0\}$. The inclusion

$$\{\upsilon_0\} \subset \mathbf{I}$$

is a trivial cofibration, so the fibration property of f means that there is a lifting to a morphism $\mathbf{I} \to \mathcal{A}$. The image of υ_1 by this map is an object a' equivalent to a and projecting to b.

In the situation of the second paragraph, note that essential surjectivity of f means that every object b is equivalent to some $f(a)$, then apply the previous statement. □

21.3 Homotopies and the homotopy 2-category

In this section we consider a left proper tractable cartesian model category denoted by $\mathscr{P}$, much of what we say being destined towards the case $\mathscr{P} = \mathscr{P}\mathcal{C}(\mathcal{M})$. A "point" of $\mathcal{A} \in \mathscr{P}$ means a morphism $* \to \mathcal{A}$. In the case $\mathscr{P} = \mathscr{P}\mathcal{C}(\mathcal{M})$, this is the same thing as an element of $\mathrm{Ob}(\mathcal{A})$.

Let $\mathcal{I} \in \mathscr{P}$ be an *interval object*, that is to say an object fitting into a diagram

$$* \sqcup * \xrightarrow{i_0 \sqcup i_1} \mathcal{I} \to *,$$

where the first map is a cofibration and the second (tautological) map is a weak equivalence. The second condition just means that $\mathcal{I}$ is contractible. In case $\mathscr{P} = \mathscr{P}\mathcal{C}(\mathcal{M})$, the typical example will be $\mathcal{I} := \mathbf{I}$ the 1-category with two isomorphic objects. In view of this example we will sometimes use the notations υ_0, υ_1 for the points $i_0, i_1 : * \to \mathcal{I}$.

The category of fibrant and cofibrant objects yields a $\mathscr{P}$-category $\mathbf{Enr}(\mathscr{P})$, which may be considered as an object of $\mathscr{P}\mathcal{C}(\mathscr{P})$ satisfying strictly the Segal conditions. At the end of this section we will study the 2-truncation $\tau_{\leq 2}(\mathbf{Enr}(\mathscr{P}))$ and relate it to the 2-truncation of the Dwyer–Kan localization of $\mathscr{P}$. Looking at such a homotopy 2-category dates back to the end of Gabriel and Zisman's book [117].

21.3.1 Homotopic morphisms

We can relate several different versions of the notion of two morphisms being homotopic. Suppose $\mathcal{A}$ is a cofibrant object and $\mathcal{B}$ is a fibrant object in $\mathscr{P}$. Recall from Quillen's definition [215] that two maps $f, g : \mathcal{A} \to \mathcal{B}$ are *homotopic* if there is a "cylinder object" or diagram

$$\mathcal{A} \rightrightarrows \mathcal{C} \to \mathcal{A},$$

such that all morphisms are weak equivalences, the first two morphisms are cofibrations, and such that the compositions are the identity of $\mathcal{A}$, plus a morphism $\mathcal{C} \xrightarrow{h} \mathcal{B}$ inducing f and g on the two copies of $\mathcal{A}$.

In our situation, if $\mathcal{B}$ is fibrant then $\underline{\mathrm{HOM}}(\mathcal{A}, \mathcal{B})$ is a fibrant object of $\mathcal{P}$, whose points are the morphisms $\mathcal{A} \to \mathcal{B}$.

Proposition 21.3.1 *Suppose $\mathcal{B}$ is a fibrant and $\mathcal{A}$ is cofibrant in $\mathcal{P}$. For two morphisms $f, g : \mathcal{A} \to \mathcal{B}$ corresponding to points also denoted $f, g : * \to \underline{\mathrm{HOM}}(\mathcal{A}, \mathcal{B})$, the following conditions are equivalent:*

(a) f and g project to the same morphism from $\mathcal{A}$ to $\mathcal{B}$ in $\mathrm{ho}(\mathcal{P})$;
(b) there exists a homotopy in Quillen's sense between f and g;
(c) (in case $\mathcal{P} = \mathcal{PC}(\mathcal{M})$) f and g are are inner-equivalent objects in the fibrant $\mathcal{M}$-category $\underline{\mathrm{HOM}}(\mathcal{A}, \mathcal{B})$;
(d) there exists a map $\mathcal{A} \times \mathcal{I} \to \mathcal{B}$ inducing f and g on $\mathcal{A} \times \{\upsilon_i\}$ for $i = 0, 1$ respectively; and
(e) f and g project to the same morphism from $*$ to $\underline{\mathrm{HOM}}(\mathcal{A}, \mathcal{B})$ in $\mathrm{ho}(\mathcal{P})$.

Proof The equivalence between (a) and (b) is from Quillen [215].

Given (b) with

$$\mathcal{A} \rightrightarrows \mathcal{C} \to \mathcal{A}$$

and a map $h : \mathcal{C} \to \mathcal{B}$ providing a homotopy from f to g, we get a diagram

$$\begin{array}{ccc} \mathcal{A} \times \{\upsilon_0, \upsilon_1\} & \longrightarrow & \mathcal{C} \\ \downarrow & & \downarrow \\ \mathcal{A} \times \mathcal{I} & \longrightarrow & \mathcal{A} \end{array}$$

where the upper and left downward maps are cofibrations, the upper one by the Quillen homotopy, and the left one using Lemma 21.2.1 and the cartesian condition. Choose a factorization

$$\mathcal{A} \times \mathcal{I} \cup^{\mathcal{A} \times \{\upsilon_0, \upsilon_1\}} \mathcal{C} \xrightarrow{i} \mathcal{F} \xrightarrow{p} \mathcal{A},$$

with i a cofibration and p a trivial fibration. Then the map $\mathcal{C} \to \mathcal{F}$ is a trivial cofibration, so the fibrant property for $\mathcal{B}$ allows us to extend h to a map $\mathcal{F} \to \mathcal{B}$; restricting gives the map $\mathcal{A} \times \mathcal{I} \to \mathcal{B}$ required for (d). On the other hand, a maps as in (d) gives a Quillen homotopy between f and g of a particularly simple form, implying (d). Thus (b) and (d) are equivalent.

Applying Lemma 21.2.2 to $\underline{\mathrm{HOM}}(\mathcal{A}, \mathcal{B})$ gives the equivalence between (c) and (d).

Statement (d) for the maps $\mathcal{A} \to \mathcal{B}$ is equivalent to statement (d) for the maps $* \to \underline{\mathrm{HOM}}(\mathcal{A}, \mathcal{B})$ since a map $* \times \mathcal{I} \to \underline{\mathrm{HOM}}(\mathcal{A}, \mathcal{B})$ is the same thing as a map $\mathcal{A} \times \mathcal{I} \to \mathcal{B}$. Therefore, the equivalence between (d) and (e) follows from the already proven equivalence between (a) and (d) for the case of two maps $* \to \underline{\mathrm{HOM}}(\mathcal{A}, \mathcal{B})$. □

Given two maps $f, g : \mathcal{A} \to \mathcal{B}$, an $\mathcal{I}$-*homotopy* between f and g is a map $\mathcal{A} \times \mathcal{I} \stackrel{h}{\to} \mathcal{B}$ restricting to f and g over the endpoints, as in (d). If $\mathcal{A}' \stackrel{i}{\to} \mathcal{A}$ is a map such that $fi = gi$ then a homotopy *relative to* $\mathcal{A}'$ is an h which restricts to the constant map fip_1 on $\mathcal{A}' \times \mathcal{I}$ where p_1 denotes the first projection.

21.3.2 Homotopy lifting properties

It can be useful to have several different characterizations of weak equivalences between fibrant objects by homotopy lifting properties. We state some equivalent conditions, but don't claim any originality for these clasical variants which are part of the model-category folklore. One of the versions is an extension to model categories of a well-known characterization of weak equivalences between spaces pointed out just recently by R. Vogt [263].

Extending the preliminary discussion of Section 7.5, we write everything here in terms of a given interval object $\mathcal{I}$ inside a cartesian model category $\mathcal{P}$, keeping in mind the main application using the interval object $\mathbf{I}$ in $\mathcal{P} = \mathcal{PC}(\mathcal{M})$. The various lifting properties could really be written down in terms of abstract Quillen homotopies in any model category, a translation which is left to the reader.

Proposition 21.3.2 *Suppose $\mathcal{P}$ is a tractable cartesian model category and $* \sqcup * \xrightarrow{i_0 \sqcup i_1} \mathcal{I}$ is an interval object. For a map $p : \mathcal{X} \to \mathcal{Y}$ between fibrant objects, the following conditions are equivalent:*

(a) The map p is a weak equivalence.

(b) For any generating cofibration (resp. cofibration between cofibrant objects) $\mathcal{A} \stackrel{a}{\to} \mathcal{B}$ and any commutative diagram

$$\begin{array}{ccc} \mathcal{A} & \xrightarrow{f} & \mathcal{X} \\ {\scriptstyle a}\downarrow & & \downarrow{\scriptstyle p} \\ \mathcal{B} & \xrightarrow{g} & \mathcal{Y} \end{array}$$

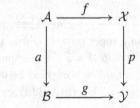

there is an $\mathcal{I}$-homotopy from g to a new morphism $g_1 : \mathcal{B} \to \mathcal{Y}$, a lifting $u_1 : \mathcal{B} \to \mathcal{A}$ of g_1, and an $\mathcal{I}$-homotopy from f to $f_1 := u_1 \circ a$ lifting the previous homotopy from $g \circ a$ to $g_1 \circ a$.

(c) *For any generating cofibration (resp. cofibration between cofibrant objects) $\mathcal{A} \overset{a}{\to} \mathcal{B}$ and any commutative diagram*

$$\begin{array}{ccc} \mathcal{A} & \xrightarrow{f} & \mathcal{X} \\ {\scriptstyle a}\downarrow & & \downarrow{\scriptstyle p} \\ \mathcal{B} & \xrightarrow{g} & \mathcal{Y} \end{array}$$

there exists a map $u : \mathcal{B} \to \mathcal{X}$ such that $u \circ a = f$, and an $\mathcal{I}$-homotopy between pu and g which restricts along a to the identity homotopy of maps $\mathcal{A} \to \mathcal{Y}$.

(d) *(Vogt [263]) for any generating cofibration (resp. cofibration between cofibrant objects) $\mathcal{A} \overset{a}{\to} \mathcal{B}$ and any diagram*

$$\begin{array}{ccc} \mathcal{A} & \xrightarrow{f} & \mathcal{X} \\ {\scriptstyle a}\downarrow & & \downarrow{\scriptstyle p} \\ \mathcal{B} & \xrightarrow{g} & \mathcal{Y} \end{array}$$

not necessarily commutative *but with a homotopy $h : \mathcal{A} \times \mathcal{I} \to \mathcal{Y}$ such that $h \circ (1 \times i_0) = ga$ and $h \circ (1 \times i_1) = pf$, there exists a map $r : \mathcal{B} \to \mathcal{X}$ and a homotopy $h' : \mathcal{B} \times \mathcal{I} \to \mathcal{Y}$ such that $h' \circ (1 \times i_0) = g$, $h' \circ (1 \times i_1) = f$, and $h' \circ (a \times 1) = h$.*

Proof For (d), see Vogt [263] using $\underline{\text{HOM}}(\mathcal{I}, \mathcal{Y}) = \mathcal{Y}^{\mathcal{I}}$ for the path-object. The other proofs are similar and are left to the reader. □

21.3.3 The homotopy 2-category

Continue with the notations of a cartesian model category $\mathscr{P}$, together with a choice of interval object $\mathcal{I}$. Define $\Pi_1 : \mathscr{P}_f \to \text{GPD}$ by sending $\mathcal{A} \in \mathscr{P}_f$ to the following groupoid. Its objects are the points $x : * \to \mathcal{A}$. Given $x, y : * \to \mathcal{A}$, the morphisms are the homotopy classes of homotopies $u : \mathcal{I} \to \mathcal{A}$ with $u \circ i_0 = x$ and $u \circ i_1 = y$. A homotopy from u to v is a map $k : \mathcal{I} \times \mathcal{I} \to \mathcal{A}$ such that $k \circ (i_0 \times 1) = x \circ q$ and $k \circ (i_1 \times 1) = y \circ q$ (denoting here $q : \mathcal{I} \to *$

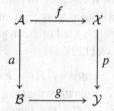

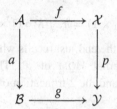

the projection) as well as $k \circ (1 \times i_0) = u$ and $k \circ (1 \times i_1) = v$. Another way of saying this is that the morphisms from x to y are the homotopy classes of maps from $*$ to $\underline{\mathrm{HOM}}^{x,y}(\mathcal{I}, \mathcal{A})$. It is left as an exercise using the fibrant property of $\mathcal{A}$ to construct the composition and inverse for this groupoid, by juxtaposing homotopies.

Now $\Pi_1 \circ \mathbf{Enr}(\mathscr{P})$ is a GPD-category, in particular a 2-category. Its objects are the cofibrant and fibrant objects $\mathcal{A} \in \mathscr{P}_{cf}$, and the groupoid of maps from $\mathcal{A}$ to $\mathcal{B}$ is $\Pi_1(\underline{\mathrm{HOM}}(\mathcal{A}, \mathcal{B}))$. This kind of 2-category was mentioned at the end of Gabriel and Zisman's book [117].

Proposition 21.3.3 *The 2-category $\Pi_1 \circ \mathbf{Enr}(\mathscr{P})$ constructed above is naturally equivalent to the 2-truncation $\tau_{\leq 2} L_{DK}(\mathscr{P})$ of the Dwyer–Kan localization of $\mathscr{P}$. Given $\mathcal{A}, \mathcal{B} \in \mathscr{P}_{cf}$ the 1-morphisms from $\mathcal{A}$ to $\mathcal{B}$ in this 2-category are the same as those of $\mathscr{P}$; given $f, g : \mathcal{A} \to \mathcal{B}$ the 2-morphisms between f and g are the homotopies $h : \mathcal{A} \times \mathcal{I} \to \mathcal{B}$ with $h_0 = f$ and $h_1 = g$, up to the relation of homotopy between homotopies: h and h' correspond to the same 2-morphism if there exists $K : \mathcal{A} \times \mathcal{I} \times \mathcal{I} \to \mathcal{B}$ such that*

$$K \circ (1 \times 1 \times i_0) = f \circ \mathbf{pr}_1, \quad K \circ (1 \times 1 \times i_1) = g \circ \mathbf{pr}_1,$$

$$K \circ (1 \times i_0 \times 1) = h, \quad K \circ (1 \times i_1 \times 1) = h'.$$

Proof The discussion at the end just recalls what was said above, in view of the existence of the internal $\underline{\mathrm{HOM}}$ of $\mathscr{P}$. The equivalence between the 2-category $\Pi_1 \circ \mathbf{Enr}(\mathscr{P})$ and the 2-truncation of the Dwyer–Kan localization is left as an exercise. □

Corollary 21.3.4 *Suppose $\mathscr{P}$ is a tractable left proper cartesian model category. Suppose given two fibrant and cofibrant objects $\mathcal{A}, \mathcal{B} \in \mathscr{P}_{\mathrm{cf}} = \mathrm{Ob}(\mathbf{Enr}(\mathscr{P}))$, and a map $f : \mathcal{A} \to \mathcal{B}$ which may also be seen as a point $\mathbf{f} : * \to \mathbf{Enr}(\mathscr{P})(\mathcal{A}, \mathcal{B})$ or a morphism $\mathbb{I} \to \mathbf{Enr}(\mathscr{P})$. The point $\mathbf{f}$ is an inner equivalence if and only if f is a weak equivalence in $\mathscr{P}$. If $\mathbf{Enr}(\mathscr{P}) \to \mathbf{Enr}(\mathscr{P})'$ is a fibrant replacement, then f is an inner or weak equivalence, if and only if the corresponding morphism $\mathbb{I} \to \mathbf{Enr}(\mathscr{P})'$ extends to a map $\mathbf{I} \to \mathbf{Enr}(\mathscr{P})'$.*

Proof The 1-truncation $\tau_{\leq 1}(\mathbf{Enr}(\mathscr{P}))$ obtained by applying $\tau_{\leq 0}$ to the morphism objects, is equivalent to $\mathrm{ho}(\mathscr{P})$. This may be seen directly using Proposition 21.3.1. Thus, f is an inner equivalence if and only if it projects to an isomorphism in $\mathrm{ho}(\mathscr{P})$ which in turn says that it is a weak equivalence. For the last statement, note that $\tau_{\leq 1}(\mathbf{Enr}(\mathscr{P})) \cong \tau_{\leq 1}(\mathbf{Enr}(\mathscr{P})')$, so if there exists an extension from $\mathbb{I}$ to $\mathbf{I}$ then f projects to an isomorphism in $\mathrm{ho}(\mathscr{P})$. In the other direction, proceed as in the proof of Theorem 18.8.1 with $\mathcal{B} = \mathbf{I}$. □

21.4 Constructions with $\widetilde{\Upsilon}$

Return to the notation $\mathcal{M}$ for a tractable left proper cartesian model category. Recall from Section 14.1 that for any $E \in \mathcal{M}$ we have an $\mathcal{M}$-precategory $\Upsilon(E)$. It has two objects denoted υ_0 and υ_1 and the given $E \in \mathcal{M}$ of morphisms from υ_0 to υ_1, but no morphisms in the other direction and only identity endomorphisms of υ_0 and υ_1. The main property of this construction is that if $\mathcal{A}$ is any $\mathcal{M}$-precategory then a morphism in $\mathcal{PC}(\mathcal{M})$

$$f : \Upsilon(E) \to \mathcal{A}$$

corresponds exactly to a choice of two objects $x = f(\upsilon_0)$ and $y = f(\upsilon_1)$ in $\mathrm{Ob}(\mathcal{A})$ together with a morphism $E \to \mathcal{A}(x, y)$ in $\mathcal{M}$. In other words $\Upsilon(E)$ is the universal $\mathcal{M}$-precategory $\mathcal{A}$ with two objects x, y and a map $E \to \mathcal{A}(x, y)$.

More generally, the construction of Section 14.1 gives precategories $\Upsilon(E_1, \ldots, E_k)$, and in Chapter 16 we have defined their associated strict categories $\widetilde{\Upsilon}_k(E_1, \ldots, E_k)$. For $k = 1$ there is no change: $\widetilde{\Upsilon}_1(E) = \Upsilon(E)$ is already a strict $\mathcal{M}$-enriched category.

Next for example $\widetilde{\Upsilon}_2(E, F)$ has objects $\upsilon_0, \upsilon_1, \upsilon_2$ and morphisms E from υ_0 to υ_1, F from υ_1 to υ_2 and $E \times F$ from υ_0 to υ_2. We picture $\widetilde{\Upsilon}_k(E_1, \ldots, E_k)$ as a k-simplex (an edge for $k = 1$, a triangle for $k = 2$, a tetrahedron for $k = 3$). The edges are labeled with single E_i, or products $E_i \times \cdots \times E_j$.

There are inclusions of these $\widetilde{\Upsilon}_k$ according to the faces of the k-simplex. The principal faces give inclusions

$$\widetilde{\Upsilon}_{k-1}(E_1, \ldots, E_{k-1}) \hookrightarrow \widetilde{\Upsilon}_k(E_1, \ldots, E_k),$$

$$\widetilde{\Upsilon}_{k-1}(E_2, \ldots, E_k) \hookrightarrow \widetilde{\Upsilon}_k(E_1, \ldots, E_k),$$

and

$$\widetilde{\Upsilon}_{k-1}(E_1, \ldots, E_i \times E_{i+1}, \ldots, E_k) \hookrightarrow \widetilde{\Upsilon}_k(E_1, \ldots, E_k).$$

The inclusions of lower levels are deduced from these by induction. Note that these faces $\widetilde{\Upsilon}_{k-1}$ intersect along appropriate $\widetilde{\Upsilon}_{k-2}$.

Remark 21.4.1 $\Upsilon(*) = \mathbb{I}$ is the category with objects υ_0, υ_1 and with a unique morphism from υ_0 to υ_1. A map $\Upsilon(*) \to \mathcal{A}$ is the same as a pair of objects x, y and a 1-morphism from x to y, i.e. a point of $\mathcal{A}(x, y)$.

21.4.1 Cellular structure of $\widetilde{\Upsilon}$

Another way of constructing the $\widetilde{\Upsilon}_k(E_1, \ldots, E_k)$ is by the explicit addition of representable cells, as follows. These considerations might be related to Johnson's work [155].

For $E \in \mathcal{M}$, recall from Section 13.2 that $h([p], E)$ denotes the universal $\mathcal{M}$-precategory with objects $\upsilon_0, \ldots, \upsilon_p$ and with a morphism

$$E \to h([p], E)(\upsilon_0, \ldots, \upsilon_p).$$

In Lemma 13.2.1 this can be described explicitly by saying that for a sequence of objects $\upsilon_{i_1}, \ldots, \upsilon_{i_k}$ the object $h([p], E)(\upsilon_{i_1}, \ldots, \upsilon_{i_k})$ is empty if some $i_j < i_{j+1}$, is equal to $*$ if all i_j are equal, and is equal to E if some $i_j < i_{j+1}$. It has the universal property that a morphism $h([p], E) \to \mathcal{A}$ corresponds to a collection of objects $x_0, \ldots, x_p \in \mathrm{Ob}(\mathcal{A})$ together with a map $E \to \mathcal{A}(x_0, \ldots, x_p)$ in $\mathcal{M}$.

One has $h([1], E) = \Upsilon(E)$. The construction of the higher $\widetilde{\Upsilon}_k$ may be described inductively as follows: we will construct $\widetilde{\Upsilon}_k(E_1, \ldots, E_k)$ together with a morphism

$$\mathbf{t} : h([k], E_1 \times \cdots \times E_k) \to \widetilde{\Upsilon}_k(E_1, \ldots, E_k).$$

Suppose we have constructed these maps up to $k - 1$. Note that the first and last face morphisms coupled with the projections onto the first and last $k - 1$ factors give a map

$$h([k-1], E_1 \times \cdots \times E_k) \cup^{h([k-2], E_1 \times \cdots \times E_k)} h([k-1], E_1 \times \cdots \times E_k)$$
$$\xrightarrow{\alpha} h([k], E_1 \times \cdots \times E_k),$$

but on the other hand the projections onto subsets of factors of the product $E_1 \times \cdots \times E_k$ together with the maps $\mathbf{t}$ in our inductive construction for $k - 1$ and $k - 2$ give a map

$$h([k-1], E_1 \times \cdots \times E_k) \cup^{h([k-2], E_1 \times \cdots \times E_k)} h([k-1], E_1 \times \cdots \times E_k)$$
$$\xrightarrow{\beta} \widetilde{\Upsilon}_{k-1}(E_1, \ldots, E_{k-1}) \cup^{\widetilde{\Upsilon}_{k-2}(E_2, \ldots, E_{k-1})} \widetilde{\Upsilon}_{k-1}(E_2, \ldots, E_k).$$

Lemma 21.4.2 *The maps α and β fit into a cocartesian diagram expressing $\widetilde{\Upsilon}_k(E_1, \ldots, E_k)$ as a pushout:*

$$\begin{array}{ccc} \mathcal{P} & \longrightarrow & h([k], E_1 \times \cdots \times E_k) \\ \downarrow & & \downarrow \\ \mathcal{Q} & \longrightarrow & \widetilde{\Upsilon}_k(E_1, \ldots, E_k) \end{array}$$

where

$$\mathcal{P} := h([k-1], E_1 \times \cdots \times E_k) \cup^{h([k-2], E_1 \times \cdots \times E_k)} h([k-1], E_1 \times \cdots \times E_k)$$

and

$$\mathcal{Q} := \widetilde{\Upsilon}_{k-1}(E_1, \ldots, E_{k-1}) \cup^{\widetilde{\Upsilon}_{k-2}(E_2, \ldots, E_{k-1})} \widetilde{\Upsilon}_{k-1}(E_2, \ldots, E_k).$$

If the E_i are cofibrant, the bottom map is a trivial Reedy cofibration.

Proof The fact that the diagram is a pushout is verified levelwise on sequences of objects. For sequences contained either in $\{\upsilon_0, \ldots, \upsilon_{k-1}\}$ or $\{\upsilon_1, \ldots, \upsilon_k\}$ the horizontal maps are isomorphisms, by an argument distinguishing the case of sequences contained in $\{\upsilon_1, \ldots, \upsilon_{k-1}\}$. Whereas for sequences starting with υ_0 and ending with υ_k the vertical maps are isomorphisms, between $\emptyset$ and itself on the left, between $E_1 \times \ldots \times E_k$ and itself on the right.

Suppose the E_i are cofibrant. An inductive argument shows that

$$\mathcal{P} \to h(\partial[k], E_1 \times \cdots \times E_k)$$

is a Reedy cofibration hence that the top map of the previous diagram is a Reedy cofibration. It follows that the bottom map is one too. It is a weak equivalence because both sides are sequentially free and take the same values E_i on adjacent objects. □

We can think of this as saying that $\widetilde{\Upsilon}_k(E_1, \ldots, E_k)$ is obtained by adding on the cell $h([k], E_1 \times \cdots \times E_k)$ to the amalgamated sum

$$\widetilde{\Upsilon}_{k-1}(E_1, \ldots, E_{k-1}) \cup^{\widetilde{\Upsilon}_{k-2}(E_2, \ldots, E_{k-1})} \widetilde{\Upsilon}_{k-1}(E_2, \ldots, E_k)$$

of the earlier things we have inductively constructed.

The case $k = 2$ is simpler to write down and is worth mentioning separately. Recall that for $k = 1$ we just had $\Upsilon(E) = h([1], E)$. The next step is

$$\widetilde{\Upsilon}_2(E, F) = h([2], E \times F) \cup^{h([1], E \times F) \cup^* h([1], E \times F)} (h([1], E) \cup^* h([1], F)).$$

A somewhat more complicated situation is obtained when we throw out all but one of the inner faces. For this we introduce more generally the following notation: if $S \subset [k] = \{\upsilon_0, \ldots, \upsilon_k\}$ is a subset of objects, let $\widetilde{\Upsilon}_{k;S}(E_1, \ldots, E_k)$ denote the full sub-$\mathcal{M}$-category of $\widetilde{\Upsilon}_k(E_1, \ldots, E_k)$ containing exactly the objects of S. Let $\mathrm{P}([k])$ denote the lattice of subsets of $[k]$ with the inclusion relation, and suppose $\Lambda \subset \mathrm{P}([k])$ is a saturated sublattice, thus it is a collection of subsets of $[k]$ closed under taking subsets. Put

$$\widetilde{\Upsilon}_k(\Lambda; E_1, \ldots, E_k) := \mathrm{colim}_{S \in \Lambda} \widetilde{\Upsilon}_{k;S}(E_1, \ldots, E_k).$$

This is particulary useful when Λ is an *inner horn*, that is to say for some $0 < j < k$, $\Lambda = \Lambda^j_k$ is P([k]) minus [k] itself and minus the j-th face $\{\upsilon_0, \ldots, \widehat{\upsilon_j}, \ldots, \upsilon_k\}$.

Lemma 21.4.3 *Suppose E_i are cofibrant and $\Lambda \subset \Lambda'$ is an inclusion of saturated sublattices in* P([k]). *Then the map*

$$\widetilde{\Upsilon}_k(\Lambda; E_1, \ldots, E_k) \to \widetilde{\Upsilon}_k(\Lambda'; E_1, \ldots, E_k)$$

is a Reedy cofibration. If Λ is an inner horn then

$$\widetilde{\Upsilon}_k(\Lambda; E_1, \ldots, E_k) \to \widetilde{\Upsilon}_k(E_1, \ldots, E_k)$$

is a trivial Reedy cofibration.

Proof Denote by $\partial[k]$ the sublattice which is the "boundary" of [k], that is to say P([k]) minus [k] itself. It suffices to show that

$$\widetilde{\Upsilon}_k(\partial[k]; E_1, \ldots, E_k) \to \widetilde{\Upsilon}_k([k]; E_1, \ldots, E_k)$$

is a Reedy cofibration. Given a sequence of objects $\upsilon_{i_0}, \ldots, \upsilon_{i_m}$ with $i_0 \leq \cdots \leq i_m$, if there are any repetitions then the relative latching maps are isomorphisms, whereas if there are no repetitions then the relative latching maps are either identities or $\emptyset \to E_{i_0+1} \times \cdots \times E_{i_m}$, the latter being a cofibration using the cartesian property of $\mathcal{M}$ because E_i are cofibrant.

From the boundary case, the other cases can be built up by induction, noting that if $\Lambda' = \Lambda \cup \{S\}$ then $\widetilde{\Upsilon}_k(\Lambda'; E_1, \ldots, E_k)$ can be expressed as a pushout of Λ along the boundary lattice ∂S of all strict subsets of S.

If Λ is an inner horn, let $[k-1] \cup^{[k-2]} [k-1]$ denote the lattice which is the union of the first and last faces along their intersection. Then

$$[k-1] \cup^{[k-2]} [k-1] \hookrightarrow \Lambda$$

can be built up by a sequence of inner horn extensions for smaller values $k' < k$. Using Lemma 21.4.2 for the union of first and last faces (for which the corresponding $\widetilde{\Upsilon}$ is sequentially free), and the present statement inductively for k' for the smaller inner horns, gives that

$$\Upsilon_k(E_1, \ldots, E_k) \to \widetilde{\Upsilon}_k(\Lambda; E_1, \ldots, E_k)$$

is a weak equivalence. By 3 for 2 it follows that the map stated in the lemma is a weak equivalence. Note here that $\widetilde{\Upsilon}_k(\Lambda; E_1, \ldots, E_k)$ is not itself sequentially free in general, a phenomenon occuring starting with $k = 3$ (see below), thus the reference to Lemma 21.4.2. □

21.4.2 Some trivial inclusions

The above lemmas tell when an inclusion from a union of faces, into the whole $\widetilde{\Upsilon}_k$, is a trivial cofibration. In practice this will mostly be used for $k = 2, 3$.

For $k = 2$ the only inclusion which is a trivial cofibration is

$$\Upsilon(E_1) \cup^{\{v_1\}} \Upsilon(E_2) \hookrightarrow \widetilde{\Upsilon}_2(E_1, E_2).$$

For $k = 3$ we denote our inclusions in shorthand notation where 0, 1, 2, 3 refer to the vertices. To fix notations, the above inclusion for $k = 2$ would be noted

$$(01) + (12) \subset (012).$$

Now for $k = 3$ the main inclusion which is a trivial cofibration is

$$(01) + (12) + (23) \subset (0123),$$

meaning

$$\Upsilon(E_1) \cup^{\{v_1\}} \Upsilon(E_2) \cup^{\{v_2\}} \Upsilon(E_3) \hookrightarrow \widetilde{\Upsilon}_3(E_1, E_2, E_3).$$

This is the standard one, corresponding basically to the Segal maps. For $k = 3$ there are then some others which we obtain from this standard one by adding in triangles on the right, keeping equivalence with $(01) + (12) + (23)$ according to the result for $k = 2$. The following will be Reedy cofibrations, as can be seen by an inductive application of Lemma 21.4.2. For example,

$$(01) + (123) \subset (0123),$$

meaning

$$\Upsilon(E_1) \cup^{\{v_1\}} \widetilde{\Upsilon}_2(E_2, E_3) \hookrightarrow \widetilde{\Upsilon}_3(E_1, E_2, E_3),$$

and similarly

$$(012) + (23) \subset (0123).$$

Then also

$$(012) + (123) \subset (0123),$$

meaning

$$\widetilde{\Upsilon}_2(E_1, E_2) \cup^{\Upsilon_1(E_2)} \widetilde{\Upsilon}_2(E_2, E_3) \hookrightarrow \widetilde{\Upsilon}_3(E_1, E_2, E_3),$$

which was given by Lemma 21.4.2.

Corollary 21.4.4 *The inclusion denoted*

$$(012) + (023) \subset (0123),$$

meaning

$$\widetilde{\Upsilon}_2(E_1, E_2) \cup^{\Upsilon_1(E_1 \times E_2)} \widetilde{\Upsilon}_2(E_1 \times E_2, E_3) \hookrightarrow \widetilde{\Upsilon}_3(E_1, E_2, E_3)$$

is a trivial Reedy cofibration. Similarly for that denoted

$$(013) + (123) \subset (0123),$$

as well as

$$(012) + (013) + (123) \subset (0123)$$

and

$$(012) + (023) + (123) \subset (0123).$$

Proof This follows from Lemma 21.4.3. □

In the case

$$\Lambda = (012) + (023) \subset (0123),$$

if we look at the full sequence of objects the value is empty:

$$\widetilde{\Upsilon}(\Lambda; E_1, E_2, E_3)(\upsilon_0, \upsilon_1, \upsilon_2, \upsilon_3) = \emptyset,$$

whereas

$$\widetilde{\Upsilon}(\Lambda; E_1, E_2, E_3)(\upsilon_0, \upsilon_3) = E_1 \times E_2 \times E_3,$$

because υ_0, υ_3 is a sequence for the face (023); thus $\widetilde{\Upsilon}(\Lambda; E_1, E_2, E_3)$ is not sequentially free, an example of what was pointed out at the end of the proof of Lemma 21.4.3.

Our main examples of inclusions which are *not* trivial cofibrations are when we leave out the first or the last faces. At $k = 2$,

$$(01) + (02) \subset (012),$$

i.e. $\Upsilon(E_1) \cup^{\{\upsilon_0\}} \Upsilon(E_1 \times E_2) \to \widetilde{\Upsilon}_2(E_1, E_2)$, is not a weak equivalence in general. Similarly

$$(02) + (12) \subset (012) \quad \text{not a w.e..}$$

At $k = 3$ these are

$$(012) + (023) + (013) \subset (0123) \quad \text{not a w.e.;}$$

$$(013) + (023) + (123) \subset (0123) \quad \text{not a w.e..}$$

We call these the *left and right shells*. We shall meet both of them and denote the left shell as

$$(012) + (023) + (013) = \Lambda^L \widetilde{\Upsilon}_3(E_1, E_2, E_3),$$

and the right shell as

$$(013) + (023) + (123) = \Lambda^R \widetilde{\Upsilon}_3(E_1, E_2, E_3).$$

The main parts of our arguments for limits will consist of saying that under certain circumstances we have an extension property for morphisms with respect to these cofibrations which are not trivial.

Starting with the above examples of trivial inclusions and applying 3 for 2 and pushout along trivial cofibrations, we obtain many of the trivial cofibrations which will be used in what follows.

We will often be considering morphisms of the form

$$f : \widetilde{\Upsilon}_k(E_1, \ldots, E_k) \to \mathcal{C}.$$

When we would like to restrict this to a face (or higher order face such as an edge) then, denoting the face by $i_1, \ldots, i_j$ we denote the restriction of f to the face by

$$r_{i_1 \ldots i_j}(f).$$

For example, when $k = 2$ the restriction of

$$f : \widetilde{\Upsilon}_2(E, F) \to \mathcal{C}$$

to the edge (02) (which is a $\Upsilon(E \times F)$) would be denoted $r_{02}(f)$. The object $f(1)$ could also be denoted $r_1(f)$.

We make the same convention for restricting maps of the form

$$\mathcal{A} \times \widetilde{\Upsilon}_k(E_1, \ldots, E_k) \to \mathcal{C},$$

to maps on $\mathcal{A}$ times some face of the $\widetilde{\Upsilon}_k(E_1, \ldots, E_k)$.

21.5 Acyclicity of inversion

In this section we consider the question of inverting morphisms in an $\mathcal{M}$-category by pushout with a standard interval. This corresponds to Theorem 2.5.1 of my preprint on limits [235]. This kind of theory perhaps first appeared in the work of Dwyer and Kan [101, 102, 103]. A modern treatment for simplicial categories, Segal categories and the like, is Bergner [41].

It is classical that if a morphism in a 1-category has an inverse, then the inverse is unique. In other words, specifying the inverse doesn't add any additional data. In our present situation we can formulate this property as follows:

Theorem 21.5.1 *For any fibrant $\mathcal{C} \in \mathcal{PC}(\mathcal{M})$ the morphism of restriction from* $\mathbf{I}$ *to* $\mathbb{I}$*:*

$$r : \underline{\mathrm{HOM}}(\mathbf{I}, \mathcal{C}) \to \underline{\mathrm{HOM}}(\mathbb{I}, \mathcal{C})$$

is fully faithful, so $\underline{\mathrm{HOM}}(\mathbf{I}, \mathcal{C})$ *is equivalent to the full sub-$\mathcal{M}$-category of invertible elements of* $\underline{\mathrm{HOM}}(\mathbb{I}, \mathcal{C})$.

Proof We first construct some trivial cofibrations.

Suppose B is cofibrant in $\mathcal{M}$. The morphism

$$\Upsilon(B) \cup^{\{v_1\}} \mathbb{I} \to \widetilde{\Upsilon}_2(B, *)$$

is a trivial cofibration (Section 21.4.2), so the pushout along

$$\Upsilon(B) \cup^{\{v_1\}} \mathbb{I} \to \Upsilon(B) \cup^{\{v_1\}} \mathbf{I}$$

gives a trivial cofibration

$$\Upsilon(B) \cup^{\{v_1\}} \mathbf{I} \to \widetilde{\Upsilon}_2(B, *) \cup^{\mathbb{I}} \mathbf{I}.$$

The morphism

$$\Upsilon(B) \to \Upsilon(B) \cup^{\{v_1\}} \mathbf{I}$$

is a weak equivalence (again because it is pushout of the trivial cofibration $* \to \mathbf{I}$). Therefore, the composed morphism

$$i_{01} : \Upsilon(B) \to \widetilde{\Upsilon}_2(B, *) \cup^{\mathbb{I}} \mathbf{I}$$

corresponding to the edge (01) is an equivalence. Thus the projection

$$\widetilde{\Upsilon}_2(B, *) \cup^{\mathbb{I}} \mathbf{I} \to \Upsilon(B)$$

is an equivalence (by condition CM2). This in turn implies that the morphism corresponding to the edge (02)

$$i_{02} : \Upsilon(B) \to \widetilde{\Upsilon}_2(B, *) \cup^{\mathbb{I}} \mathbf{I} \tag{21.5.1}$$

is a trivial cofibration.

A similar argument shows that

$$i_{02} : \Upsilon(B) \to \mathbf{I} \cup^{\mathbb{I}} \widetilde{\Upsilon}_2(*, B) \tag{21.5.2}$$

is a trivial cofibration.

Next, note that

$$\Upsilon(B) \times \mathbb{I} = \widetilde{\Upsilon}_2(B, *) \cup^{\Upsilon(B)} \widetilde{\Upsilon}_2(*, B)$$

(the square decomposes as a union of two triangles). The morphisms in the amalgamated sum are both i_{02}. Thus if we attach $\mathbf{I}$ to each of the intervals $\mathbb{I}$ on the two opposite sides of this square, the result

$$\Upsilon(B) \times \mathbb{I} \cup^{\{\upsilon_0, \upsilon_1\} \times \mathbb{I}} (\{\upsilon_0, \upsilon_1\} \times \mathbf{I})$$

can be seen as an amalgamated sum of the two maps considered in (21.5.1) and (21.5.2). Combining with those trivial cofibrations, the morphism from the diagonal

$$\Upsilon(B) \to \Upsilon(B) \times \mathbb{I} \cup^{\{\upsilon_0, \upsilon_1\} \times \mathbb{I}} (\{\upsilon_0, \upsilon_1\} \times \mathbf{I})$$

is an equivalence, which in turn implies that the projection

$$\Upsilon(B) \times \mathbb{I} \cup^{\{\upsilon_0, \upsilon_1\} \times \mathbb{I}} (\{\upsilon_0, \upsilon_1\} \times \mathbf{I}) \to \Upsilon(B)$$

is an equivalence or, equally well, that the inclusion

$$(\Upsilon(B) \times \mathbb{I}) \cup^{\{\upsilon_0, \upsilon_1\} \times \mathbb{I}} (\{\upsilon_0, \upsilon_1\} \times \mathbf{I}) \hookrightarrow \Upsilon(B) \times \mathbf{I} \tag{21.5.3}$$

is a trivial cofibration. Cofibrancy of this map comes from the cartesian property for Reedy cofibrations in $\mathcal{PC}(\mathcal{M})$, using the last statement of Lemma 21.2.1.

Suppose now that $B' \to B$ is a cofibration. Let $\mathcal{G}$ denote the pushout fitting into the cocartesian diagram

$$\begin{array}{ccc} (\Upsilon(B') \times \mathbb{I}) \cup^{\{\upsilon_0, \upsilon_1\} \times \mathbb{I}} (\{\upsilon_0, \upsilon_1\} \times \mathbf{I}) & \to & \Upsilon(B') \times \mathbf{I} \\ \downarrow & & \downarrow \\ (\Upsilon(B) \times \mathbb{I}) \cup^{\{\upsilon_0, \upsilon_1\} \times \mathbb{I}} (\{\upsilon_0, \upsilon_1\} \times \mathbf{I}) & \longrightarrow & \mathcal{G}. \end{array}$$

The trivial cofibration (21.5.3) and the usual properties of model categories imply that the morphism

$$\mathcal{G} \hookrightarrow \Upsilon(B) \times \mathbf{I} \tag{21.5.4}$$

is a trivial cofibration. Note, however, the simpler expression

$$\mathcal{G} = (\Upsilon(B) \times \mathbb{I}) \cup^{\Upsilon(B') \times \mathbb{I}} (\Upsilon(B') \times \mathbf{I}).$$

We are now ready to prove the theorem. Fix u, v objects of $\underline{\text{Hom}}(\mathbf{I}, \mathcal{C})$. Suppose $B' \hookrightarrow B$ is any cofibration, and suppose given a morphism

$$B \to \underline{\text{Hom}}(\mathbb{I}, \mathcal{C})(r(u), r(v))$$

provided with lifting

$$B' \to \underline{\textsc{Hom}}(\mathbf{I}, \mathcal{C})(u, v).$$

These correspond exactly to a morphism

$$\mathcal{G} \to \mathcal{C},$$

which since $\mathcal{C}$ is fibrant extends along the trivial cofibration of (21.5.4) to a morphism

$$\Upsilon(B) \times \mathbf{I} \to \mathcal{C}.$$

This is exactly the lifting to a map

$$B \to \underline{\textsc{Hom}}(\mathbf{I}, \mathcal{C})(u, v)$$

needed to establish the statement that the morphism induced by r

$$\underline{\textsc{Hom}}(\mathbf{I}, \mathcal{C})(u, v) \to \underline{\textsc{Hom}}(\mathbb{I}, \mathcal{C})(r(u), r(v))$$

is a trivial fibration, in particular an equivalence. This proves the theorem. □

Corollary 21.5.2 *Suppose $f : U \to V$ is an arrow in a fibrant $\mathcal{M}$-category $\mathcal{C}$. It may be viewed as a map $* \to \mathcal{C}(U, V)$ in $\mathcal{M}$ or equivalently $\mathbb{I} \to \mathcal{C}$ in $\mathcal{PC}(\mathcal{M})$. Then the $\mathcal{M}$-category of morphisms $\mathbf{I} \to \mathcal{C}$ restricting on $\mathbb{I} \subset \mathbf{I}$ to f is contractible.*

Proof The $\mathcal{M}$-category in question is just the fiber of the morphism in the theorem, over the object $f : * \to \underline{\textsc{Hom}}(\mathbb{I}, \mathcal{C})$. □

The next corollary says that equivalences may be inverted with dependence on parameters. Suppose $\mathcal{C}$ is a fibrant $\mathcal{M}$-precategory. Suppose $\psi, \psi' : \mathcal{A} \to \mathcal{C}$ are two morphisms and suppose f is a morphism from ψ to ψ'. This means

$$f : \mathbb{I} \to \underline{\textsc{Hom}}(\mathcal{A}, \mathcal{C}),$$

sending υ_0 to ψ and υ_1 to ψ'.

Corollary 21.5.3 *In the above situation, suppose that for every object $a \in \mathrm{Ob}(\mathcal{A})$ the induced morphism $f_a : \psi(a) \to \psi'(a)$ is an equivalence in $\mathcal{C}$. Then f is an equivalence considered as a 1-morphism in $\underline{\textsc{Hom}}(\mathcal{A}, \mathcal{C})$.*

Proof The morphism f is a map

$$f : \mathcal{A} \times \mathbb{I} \to \mathcal{C},$$

which we can also think of as a map

$$f_1 : \mathcal{A} \to \underline{\textsc{Hom}}(\mathbb{I}, \mathcal{C}).$$

From Theorem 21.5.1 the morphism

$$\underline{\mathrm{HOM}}(\mathbf{I}, \mathcal{C}) \to \underline{\mathrm{HOM}}(\mathbb{I}, \mathcal{C})$$

is a fibrant equivalence onto the full subcategory of invertible objects. The hypothesis of the corollary says exactly that the morphism f_1 lands in this full subcategory. Therefore, it lifts to a morphism

$$g : \mathcal{A} \to \underline{\mathrm{HOM}}(\mathbf{I}, \mathcal{C}),$$

in other words to

$$\mathcal{A} \times \mathbf{I} \to \mathcal{C},$$

or equally well

$$\mathbf{I} \to \underline{\mathrm{HOM}}(\mathcal{A}, \mathcal{C}).$$

This shows that f was an equivalence. □

21.6 Localization and interior

21.6.1 Localization of $\mathcal{M}$-categories

Using the previous results, we can define the *localization* of an $\mathcal{M}$-category $\mathcal{A} \in \mathcal{PC}(\mathcal{M})$ along a collection of morphisms $f_i \in \mathcal{A}(x_i, y_i)$, by which we mean that f_i is a point of $\mathcal{A}(x_i, y_i)$, or equivalently $f_i : * \to \mathcal{A}(x_i, y_i)$.

The idea is to view f_i as a morphism $\mathbb{I} \to \mathcal{A}$, and to simply take the pushout with $\mathbb{I} \to \mathbf{I}$. Although this makes use of all our previous complicated machinery, at this point it is conceptually much simpler than the Dwyer–Kan simplicial localization.

Given such a collection, we obtain by the universal property of $\mathbb{I} = \Upsilon(*)$, maps $\tilde{f}_i : \mathbb{I} \to \mathcal{A}$ sending v_0 to x_i and v_1 to y_i. Put

$$\mathrm{Loc}^{\mathrm{pre}}(\mathcal{A}; \{f_i\}) := \mathcal{A} \cup^{\coprod_i \mathbb{I}} \coprod_i \mathbf{I},$$

$$\mathrm{Loc}(\mathcal{A}; \{f_i\}) := \mathbf{Seg}(\mathrm{Loc}^{\mathrm{pre}}(\mathcal{A}; \{f_i\})).$$

Now $\mathrm{Loc}(\mathcal{A}; \{f_i\})$ is an $\mathcal{M}$-category with a morphism $\mathcal{A} \to \mathrm{Loc}(\mathcal{A}; \{f_i\})$. The images of f_i are clearly inner equivalences, since they factor through objects of the form $\mathbf{I}$ by construction. The following theorem states the universal property of this localization.

Theorem 21.6.1 *With the above constructions, if $\mathcal{A} \to \mathcal{A}'$ is a weak equivalence, letting f_i' denote the images of the f_i, it induces a weak equivalence*

$$\mathrm{Loc}(\mathcal{A}; \{f_i\}) \to \mathrm{Loc}\left(\mathcal{A}'; \{f_i'\}\right).$$

If $\mathcal{A}$ is cofibrant then so is $\mathrm{Loc}(\mathcal{A}; \{f_i\})$. *Suppose $\mathcal{A}$ is cofibrant and suppose $\mathcal{B}$ is a fibrant object of $\mathcal{PC}(\mathcal{M})$. Then the map*

$$\underline{\mathrm{HOM}}(\mathrm{Loc}(\mathcal{A}; \{f_i\}), \mathcal{B}) \to \underline{\mathrm{HOM}}(\mathcal{A}, \mathcal{B})$$

is a fully faithful morphism whose essential image is the full sub-$\mathcal{M}$-category of $\underline{\mathrm{HOM}}(\mathcal{A}, \mathcal{B})$ consisting of morphisms $\mathcal{A} \to \mathcal{B}$ which send all of the f_i to inner equivalences.

Proof Apply Theorem 21.5.1, noting that since

$$\mathrm{Loc}^{\mathrm{pre}}(\mathcal{A}; \{f_i\}) \to \mathrm{Loc}(\mathcal{A}; \{f_i\})$$

is a trivial cofibration, for any fibrant $\mathcal{B}$ it induces a weak equivalence

$$\underline{\mathrm{HOM}}(\mathrm{Loc}(\mathcal{A}; \{f_i\}), \mathcal{B}) \to \underline{\mathrm{HOM}}(\mathrm{Loc}^{\mathrm{pre}}(\mathcal{A}; \{f_i\}), \mathcal{B}).$$

□

In the case $\mathcal{M} = \mathcal{K}$ of Segal categories which correspond to simplicial categories as in Bergner [40], the above localization is the same as the Dwyer–Kan simplicial localization.

Theorem 21.6.2 *Suppose $\mathcal{A} \in \mathcal{PC}(\mathcal{K})$ comes from a simplicial category. Let $\mathrm{Loc}'(\mathcal{A}; \{f_i\})$ be a fibrant replacement in $\mathcal{PC}(\mathcal{K})$; and let $L_{DK}(\mathcal{A}, \{f_i\})$ denote the Dwyer–Kan localization of the simplicial category corresponding to $\mathcal{A}$. Then there is a morphism completing the bottom row of the commutative square*

$$\begin{array}{ccc} \mathcal{A} & \longrightarrow & \mathrm{Loc}(\mathcal{A}; \{f_i\}) \\ \downarrow & & \downarrow \\ L_{DK}(\mathcal{A}, \{f_i\}) & \to & \mathrm{Loc}'(\mathcal{A}; \{f_i\}) \end{array}$$

and the bottom and right vertical arrows are equivalences.

Proof See Bergner [41]. □

21.6.2 Localization of n-categories

In an n-nerve or Segal n-category, we can speak of localizing i-morphisms for $1 \le i \le n$. Recall that

$$h(1^i) := \Upsilon(\Upsilon(\cdots\Upsilon(*)\cdots)) = \Upsilon^{i-1}(\mathbb{I})$$

in the notation of Section 20.2. Here the superscript on Υ refers to iterating this operation $i-1$ times. An i-morphism in $\mathcal{A}$ is a morphism $h(1^i) \to \mathcal{A}$. Let $\overline{h(1^i)} := \Upsilon^{i-1}(\mathbf{I})$. If $u : h(1^i) \to \mathcal{A}$ is an i-morphism, then the pushout

$$\mathcal{A}' := \mathcal{A} \cup^{h(1^i)} \overline{h(1^i)}$$

is the *localization of $\mathcal{A}$ inverting u*. Given a collection $U := \{u_j : h(1^{i(j)}) \to \mathcal{A}\}_{j\in F}$ of i-morphisms for various $i(j)$ depending on the index $j \in F$, define $\mathrm{Loc}^{\mathrm{pre}}(\mathcal{A}; U)$ to be the composition of the above pushouts for each u_j, and $\mathrm{Loc}(\mathcal{A}; U) := \mathbf{Cat}(\mathrm{Loc}^{\mathrm{pre}}(\mathcal{A}; U))$.

Corollary 21.6.3 *The operation $\mathcal{A} \mapsto \mathrm{Loc}(\mathcal{A}; U)$ preserves weak equivalences in the variable $\mathcal{A}$. Suppose $\mathcal{A}$ is a Segal n-precategory, and $\mathcal{B}$ is a fibrant Segal n-category. Then*

$$\underline{\mathrm{HOM}}(\mathrm{Loc}(\mathcal{A}; U), \mathcal{B}) \to \underline{\mathrm{HOM}}(\mathcal{A}, \mathcal{B})$$

is an equivalence onto the full sub-Segal n-category of $\underline{\mathrm{HOM}}(\mathcal{A}, \mathcal{B})$ consisting of morphisms f such that $f(u_j)$ is an invertible $i(j)$-morphism in $\mathcal{B}$ for each $u_j \in U$.

Proof Apply Theorem 21.6.1 iteratively. □

Corollary 21.6.4 *Let U^{tot} be the union of the sets of i-morphisms of $\mathcal{A}$ for all $1 \le i \le n$. Then $\mathrm{Loc}(\mathcal{A}; U^{\mathrm{tot}})$ is a Segal n-groupoid. The map from $\mathcal{A}$ induces a weak equivalence of realizations*

$$|\mathcal{A}| \xrightarrow{\sim} |\mathrm{Loc}(\mathcal{A}; U^{\mathrm{tot}})|,$$

from which it follows that $\mathrm{Loc}(\mathcal{A}; U^{\mathrm{tot}}) \sim \Pi_{n,S}(|\mathcal{A}|)$. If $\mathcal{B}$ is a fibrant Segal n-groupoid, then the map

$$\underline{\mathrm{HOM}}(\mathrm{Loc}(\mathcal{A}; U^{\mathrm{tot}}), \mathcal{B}) \to \underline{\mathrm{HOM}}(\mathcal{A}, \mathcal{B})$$

is an equivalence.

Proof By construction all arrows in $\mathrm{Loc}(\mathcal{A}; U^{\mathrm{tot}})$ are invertible so it is a Segal n-groupoid. The prelocalization process and the operation **Cat** involve pushouts along morphisms which induce trivial cofibrations of realizations, so the map $\mathcal{A} \to \mathrm{Loc}(\mathcal{A}; U^{\mathrm{tot}})$ induces a weak equivalence of realizations.

A Segal n-groupoid is weak equivalent to the $\Pi_{n,S}$ of its realization, which gives the next statement. The last statement follows from the previous corollary, since all arrows in $\mathcal{B}$ are invertible. □

21.6.3 Proof of Lemma 20.4.4

We can now give the proof of Lemma 20.4.4, with notation from there. Suppose given a fibrant Segal n-groupoid $\mathcal{A}$. Its realization $|\mathcal{A}|$ is the simplicial set $[k] \mapsto \mathrm{HOM}(\mathbf{V}([k]), \mathcal{A})$ where $\mathbf{V}([k]) = h(k^n)$. One can see geometrically that the passage from the product Δ^{n+1} to the quotient category $\mathbf{C}^n(\Delta)$ induces a globularization of the representable objects $h(k^n)$, basically contracting some subcomplexes to points. But these are disjoint and contractible subcomplexes, so the realization $|\mathbf{V}([k])| = |h(k^n)|$ remains contractible. Suppose $X \to Y$ is a generating trivial cofibration of simplicial sets, i.e. a horn inclusion. Then Y is a standard simplex so $|\mathbf{V}_!(Y)| = |h(k^n)|$ is contractible, and inductively using the expression for X as a pushout of standard simplices along smaller horn inclusions it follows also that $|\mathbf{V}_!(X)|$ is contractible. Therefore

$$|\mathbf{V}_!(X)| \to |\mathbf{V}_!(Y)|$$

is a trivial cofibration of simplicial sets. Hence, by the previous corollary,

$$\mathrm{Loc}(\mathbf{Cat}(\mathbf{V}_!(X)), U^{\mathrm{tot}}) \to \mathrm{Loc}(\mathbf{Cat}(\mathbf{V}_!(Y)), U^{\mathrm{tot}})$$

is a weak equivalence of Segal n-groupoids. It is also an injective, hence Reedy, cofibration. Here U^{tot} means the full collection of morphisms on either side.

We now show that $|\mathcal{A}| = \mathbf{V}^*(\mathcal{A})$ satisfies the right lifting property with respect to $X \to Y$. A map $X \to \mathbf{V}^*(\mathcal{A})$ corresponds to a map $\mathbf{V}_!(X) \to \mathcal{A}$, which extends to a map

$$\mathbf{Cat}(\mathbf{V}_!(X)) \to \mathcal{A},$$

since $\mathcal{A}$ is fibrant. Since $\mathcal{A}$ is a Segal n-groupoid, this extends to a map from the localization:

$$\mathrm{Loc}(\mathbf{Cat}(\mathbf{V}_!(X)), U^{\mathrm{tot}}) \to \mathcal{A}.$$

That then extends along the trivial cofibration considered above, to give

$$\mathrm{Loc}(\mathbf{Cat}(\mathbf{V}_!(Y)), U^{\mathrm{tot}}) \to \mathcal{A}.$$

which in restricts back to $\mathbf{V}_!(Y) \to \mathcal{A}$, the required extension. This completes the proof of Lemma 20.4.4.

It is worth pointing out that Berger [36] constructs a model category where the fibrant objects are n-groupoids. Although he works in a context more closely related to Batanin's definition, that basic idea should be applicable in any theory. In particular, in our case there should be a similar model structure, which would be useful for dealing with realization, total localization, and things like Lemma 20.4.4. We haven't done that because of limitations of time and space.

21.6.4 Interiors

Going from $\mathcal{A}$ to a localization corresponds to adding inverses to morphisms in $\mathcal{A}$, gaining invertibility by making $\mathcal{A}$ bigger. Going in the other direction, one can gain invertibility by making $\mathcal{A}$ smaller, i.e. by throwing out the non-invertible morphisms. Consider the case of Segal n-categories, and define an operation $\mathbf{Int}^k$ which throws out all non-invertible i-morphisms for $i > k$. Recall that a Segal n-category is said to be $^{>}k$*-groupic* if the i-arrows are invertible up to equivalence, for $i > k$.

Begin by taking the point of view that the *interior* will be a subobject $\mathbf{Int}^k(\mathcal{A}) \subset \mathcal{A}$ in the same category $\mathcal{PC}^n(\mathcal{K})$. If $k \geq n$ then $\mathbf{Int}^k(\mathcal{A}) := \mathcal{A}$, indeed we think of a Segal n-category as being an ∞-category in which the i-morphisms are already invertible for $i > n$. So, assume $k < n$, and start first with the case $k > 0$. For $\mathcal{A} \in \mathcal{PC}^n(\mathcal{K})$ satisfying the full Segal conditions (Definition 20.0.1), define the Segal n-category $\mathbf{Int}^k(\mathcal{A})$ by induction as follows: keep the same set of objects

$$\mathrm{Ob}(\mathbf{Int}^k(\mathcal{A})) := \mathrm{Ob}(\mathcal{A}),$$

but define

$$(\mathbf{Int}^k(\mathcal{A}))(x_0, \ldots, x_m) := \mathbf{Int}^{k-1}(\mathcal{A}(x_0, \ldots, x_m)).$$

By induction, $\mathbf{Int}^{k-1}(\mathcal{A}(x_0, \ldots, x_m)) \subset \mathcal{A}(x_0, \ldots, x_m)$ so this defines a subobject $\mathbf{Int}^k(\mathcal{A}) \subset \mathcal{A}$.

It remains to give the definition in the case $k = 0 < n$. The idea is to restrict to the subset of morphisms which are inner equivalences, i.e. project to invertible morphisms in $\tau_{\leq 1}(\mathcal{A})$. For any sequence $(x_0, \ldots, x_m)$, the object $\mathcal{A}(x_0, \ldots, x_m) \in \mathcal{PC}^{n-1}(\mathcal{K})$ is a Segal $(n-1)$-category. By induction on n, we may assume known its interior $\mathbf{Int}^0(\mathcal{A}(x_0, \ldots, x_m))$, which is a Segal $(n-1)$-groupoid. The truncation $\tau_{\leq 0}(\mathcal{A}(x_0, \ldots, x_m))$ is the π_0 of this Segal $(n-1)$-groupoid (isomorphic to the π_0 of its topological realization). As a functor of $(x_0, \ldots, x_m) \in \Delta_{\mathrm{Ob}(\mathcal{A})}$, these sets fit together to form the nerve of the category $\tau_{\leq 1}(\mathcal{A})$. Consider the interior groupoid $\mathbf{Int}^0(\tau_{\leq 1}(\mathcal{A})) \subset \tau_{\leq 1}(\mathcal{A})$, defined as the sub-1-category consisting only of invertible morphisms. Now there is a morphism

$$\mathbf{Int}^0(\mathcal{A}(x_0,\ldots,x_m)) \to \mathbf{disc}(\tau_{\leq 1}(\mathcal{A})(x_0,\ldots,x_m))$$

going towards the discrete $(n-1)$-precategory corresponding to the set $\tau_{\leq 1}(\mathcal{A})(x_0,\ldots,x_m)$, so we can pull back the subset corresponding to invertible morphisms to give a subobject fitting into a pullback square

$$\begin{array}{ccc}(\mathbf{Int}^0(\mathcal{A}))(x_0,\ldots,x_m) & \to & \mathbf{disc}(\mathbf{Int}^0(\tau_{\leq 1}(\mathcal{A}))(x_0,\ldots,x_m)) \\ \downarrow & & \downarrow \\ \mathbf{Int}^0(\mathcal{A}(x_0,\ldots,x_m)) & \longrightarrow & \mathbf{disc}(\tau_{\leq 1}(\mathcal{A})(x_0,\ldots,x_m))\end{array}$$

where the vertical arrows are inclusions of subobjects. This diagram defines the upper left corner as a subobject

$$(\mathbf{Int}^0(\mathcal{A}))(x_0,\ldots,x_m) \hookrightarrow \mathbf{Int}^0(\mathcal{A}(x_0,\ldots,x_m)) \hookrightarrow \mathcal{A}(x_0,\ldots,x_m),$$

and these fit together functorially in $(x_0,\ldots,x_m)$ to define the subobject

$$\mathbf{Int}^0(\mathcal{A}) \subset \mathcal{A}.$$

An alternative definition of $\mathbf{Int}^k(\mathcal{A}) \subset \mathcal{A}$ is as follows, using the notation $\mathcal{A}_{m_1,\ldots,m_l}$ introduced in Section 20.2. An element of $\mathcal{A}_{m_1,\ldots,m_l}$ projects to various i-morphisms of $\mathcal{A}$ by the Segal maps and source and target maps. This element is defined to be in $\mathbf{Int}^k(\mathcal{A})_{m_1,\ldots,m_l}$ if and only if all of its i-morphism projections are inner equivalences, for $i > k$. One can unwind the previous inductive definition and see that it is the same.

Theorem 21.6.5 *If $\mathcal{A}$ is a Segal n-category, the subobject $\mathbf{Int}^k(\mathcal{A}) \subset \mathcal{A}$ constructed above is a $^{>}k$-groupic Segal n-category. If $\mathcal{B}$ is any other $^{>}k$-groupic Segal n-category, then any morphism $\mathcal{B} \to \mathcal{A}$ factors through $\mathbf{Int}^k(\mathcal{A})$. If $\mathcal{A}$ is fibrant, then so is $\mathbf{Int}^k(\mathcal{A})$. If $\mathcal{B}$ is a $^{>}k$-groupic Segal n-category and $\mathcal{A}$ is fibrant,*

$$\underline{\mathrm{HOM}}(\mathcal{B}, \mathbf{Int}^k(\mathcal{A})) \to \mathbf{Int}^k(\underline{\mathrm{HOM}}(\mathcal{B}, \mathcal{A}))$$

is an equivalence.

Proof By construction the i-arrows of $\mathbf{Int}^k(\mathcal{A})$ are inner equivalences, for $i > k$. Furthermore, if $f : \mathcal{B} \to \mathcal{A}$ is a morphism with $\mathcal{B}$ being $^{>}k$-groupic, then any i-arrow u of $\mathcal{B}$ is an inner equivalence for $i > k$. The image of an inner equivalence by a morphism is again an inner equivalence, so $f(u)$ is in $\mathbf{Int}^k(\mathcal{A})$. It follows from the Segal conditions that all of $\mathcal{B}$ goes into $\mathbf{Int}^k(\mathcal{A})$. If $\mathcal{A}$ is fibrant, if $\mathcal{B} \to \mathcal{B}'$ is a trivial cofibration and we are given a map $\mathcal{B} \to \mathbf{Int}^k(\mathcal{A})$, then it extends to $\mathcal{B}' \to \mathcal{A}$. We can also choose an extension to $\mathbf{Cat}(\mathcal{B}') \to \mathcal{A}$, but all the arrows of $\mathbf{Cat}(\mathcal{B}')$ go to arrows which are equivalent to ones coming from $\mathbf{Cat}(\mathcal{B})$, and these go into $\mathbf{Int}^k(\mathcal{A})$, so $\mathcal{B}'$ maps into $\mathbf{Int}^k(\mathcal{A})$. This shows that the latter is fibrant.

One can remark that if $\mathcal{A}$ is fibrant, then $\mathbf{Int}^k(\mathcal{A})$ can be characterized more easily as the subobject consisting of elements of $\mathcal{A}_{m_1,\ldots,m_l}$, all of whose Segal projections are *invertible* in the sense that there exist left and right inverses with choices of composition equal to the identities on the nose. If now $\mathcal{B}$ is a ${}^{>}k$-groupic Segal n-category, any i-arrow of $\underline{\mathrm{HOM}}(\mathcal{B}, \mathbf{Int}^k(\mathcal{A}))$ is invertible in the same sense for $i > k$ (this takes an argument using the fibrant property of $\mathcal{A}$, similar to what was done in Section 21.6.3). Therefore it maps into $\mathbf{Int}^k(\underline{\mathrm{HOM}}(\mathcal{B}, \mathcal{A}))$. The map is in fact an isomorphism. □

We get a higher categorical adjunction between $\mathbf{Int}^k$ and the process of localizing all the i-morphisms for $i > k$:

Corollary 21.6.6 *Suppose $\mathcal{A}$ is a fibrant Segal n-category, and $\mathcal{B}$ is a Segal n-precategory. Let $\mathrm{ARR}^{>k}(\mathcal{B})$ be the set consisting of all the i-morphisms of $\mathcal{B}$ for $i > k$, giving the localization $\mathcal{B} \to \mathrm{Loc}(\mathcal{B}, \mathrm{ARR}^{>k}(\mathcal{B}))$ inverting all of these. Then, in the diagram*

$$\begin{array}{ccc}
\underline{\mathrm{HOM}}(\mathrm{Loc}(\mathcal{B}, \mathrm{ARR}^{>k}(\mathcal{B})), \mathbf{Int}^k(\mathcal{A})) & \longrightarrow & \underline{\mathrm{HOM}}(\mathcal{B}, \mathbf{Int}^k(\mathcal{A})) \\
\downarrow & & \downarrow \\
\mathbf{Int}^k((\underline{\mathrm{HOM}}(\mathrm{Loc}(\mathcal{B}, \mathrm{ARR}^{>k}(\mathcal{B})), \mathcal{A})) & \longrightarrow & \mathbf{Int}^k(\underline{\mathrm{HOM}}(\mathcal{B}, \mathcal{A}))
\end{array}$$

the top and left arrows are equivalences; and the right and bottom arrows are equivalences onto the full sub-Segal n-category of $\mathbf{Int}^k(\underline{\mathrm{HOM}}(\mathcal{B}, \mathcal{A}))$ consisting of morphisms $\mathcal{B} \to \mathcal{A}$ which send all i-morphisms in $\mathcal{B}$ to invertible i-morphisms in $\mathcal{A}$.

Proof Combine Theorem 21.6.5 and Corollary 21.6.3. □

For $k \le n$, the ${}^{>}k$-groupic Segal n-categories are to be viewed as essentially equivalent to Segal k-categories. In view of this, we define a different operation $\mathcal{A} \mapsto \mathbf{Int}^k_S(\mathcal{A})$ which takes a Segal n-category to a Segal k-category. This is defined inductively by setting, when $k \ge 1$,

$$\mathrm{Ob}\left(\mathbf{Int}^k_S(\mathcal{A})\right) := \mathrm{Ob}(\mathcal{A}),$$

and, for any sequence of objects $(x_0, \ldots, x_m)$,

$$\left(\mathbf{Int}^k_S(\mathcal{A})\right)(x_0, \ldots, x_m) := \mathbf{Int}^{k-1}_S(\mathcal{A}(x_0, \ldots, x_m)).$$

Note that $\mathbf{Int}^1_S(\mathcal{A})$ is a Segal category, whose interior in the previous sense is a Segal groupoid $\mathbf{Int}^0\left(\mathbf{Int}^1_S(\mathcal{A})\right)$; define $\mathbf{Int}^0_S(\mathcal{A})$ to be the diagonal realization

$$\mathbf{Int}^0_S(\mathcal{A}) := \left|\mathbf{Int}^0\left(\mathbf{Int}^1_S(\mathcal{A})\right)\right|,$$

which is a simplicial set.

We leave it to the reader to state the higher categorical universal properties of $\mathbf{Int}^k_S$, corresponding to those of $\mathbf{Int}^k$ which were given above.

The above definitions have been given for the case of Segal n-categories, but the same discussion works equally well for n-nerves.

21.7 Limits

Among the most useful tools in homotopy theory are the notions of homotopy limit and colimit. Our purpose is to define the corresponding notions in an $\mathcal{M}$-category (i.e. weakly $\mathcal{M}$-enriched category) $\mathcal{C}$. This takes up what was done in the first part of my preprint [235]. The notion of limit in an n-category has been considered by many authors, particularly in the case $n = 2$ or in the $(\infty, 1)$-category of spaces. The latter case goes classically under the terminology "homotopy limit." A main starting point is Bousfield and Kan [58], with also Boardman and Vogt [50], Vogt [262], Illusie [150] and Thomason [252]. Abstract formulations were given by Bourn and Cordier [56], Dror Farjoun and Zabrodsky [94], Heller [140, 141] and others, culminating in Lurie's theory of ∞-topoi [190]. For 2-limits, see Street [242, 243], Bourn [53, 54], Gray [126, 128], Grandis and Paré [125], Gambino [118], Fiore [113] and many others. The notion of 2-limit was popularized by its lack of explanation where it was used by Deligne and Mumford in their paper on moduli stacks of curves [91]. The link between the homotopy limit and 2-limit directions is studied by Gray [127] for example. This rich subject has been the subject of many works by many authors and the above list is necessarily incomplete; it is left to the reader to fill in further references starting from here.

We always suppose that the target $\mathcal{C}$ is fibrant in $\mathcal{PC}(\mathcal{M})$. When this is not the case we first have to take a fibrant replacement. The model structure used throughout is the cartesian one constructed in Theorem 19.2.1, whose cofibrations are the Reedy cofibrations.

The terminologies "inverse limit" and "limit" could be used interchangeably, as could "direct limit" and "colimit." The more modern "limit/colimit" will be chosen here.

Suppose $\mathcal{C}$ is a fibrant $\mathcal{M}$-category (which is to say, a fibrant object of $\mathcal{PC}(\mathcal{M})$ which is hence automatically an $\mathcal{M}$-category), and suppose $\mathcal{A}$ is a cofibrant $\mathcal{M}$-precategory. Suppose $\varphi : \mathcal{A} \to \mathcal{C}$ is a morphism. For any object $U \in \mathrm{Ob}(\mathcal{C})$ let $U_{\mathcal{A}}$ denote the constant morphism from $\mathcal{A}$ to $\mathcal{C}$ with value U. Define the *object of cones*

$$\underline{\mathrm{Cone}}(U, \varphi) := \underline{\mathrm{HOM}}(\mathcal{A}, \mathcal{C})(U_{\mathcal{A}}, \varphi) \in \mathcal{M},$$

and the *set of cones*

$$\mathrm{Cone}(U, \varphi) := \mathrm{HOM}_{\mathcal{M}}(*, \underline{\mathrm{Cone}}(U, \varphi)).$$

An element $\eta \in \mathrm{Cone}(U, \varphi)$ is therefore the same thing as a map

$$\eta : \Upsilon(*) \times \mathcal{A} \to \mathcal{C},$$

such that $\eta|_{\upsilon_0 \times \mathcal{A}} = U_{\mathcal{A}}$ and $\eta|_{\upsilon_1 \times \mathcal{A}} = \varphi$.

If V is another object of $\mathcal{C}$ then we have a morphism

$$\mathcal{C}(V, U) \to \underline{\mathrm{HOM}}(\mathcal{A}, \mathcal{C})(V_{\mathcal{A}}, U_{\mathcal{A}}),$$

and we use this to define

$$\underline{\mathrm{Cone}}(V, U, \varphi) := \underline{\mathrm{HOM}}(\mathcal{A}, \mathcal{C})(V_{\mathcal{A}}, U_{\mathcal{A}}, \varphi) \times_{\underline{\mathrm{HOM}}(\mathcal{A},\mathcal{C})(V_{\mathcal{A}},U_{\mathcal{A}})} \mathcal{C}(V, U),$$

or, more generally, if $V^0, \ldots, V^p \in \mathcal{C}_0$ we define

$$\underline{\mathrm{Cone}}(V^0, \ldots, V^p, \varphi)$$
$$:= \underline{\mathrm{HOM}}(\mathcal{A}, \mathcal{C}) \left(V^0_{\mathcal{A}}, \ldots, V^p_{\mathcal{A}}, \varphi\right) \times_{\underline{\mathrm{HOM}}(\mathcal{A},\mathcal{C})(V^0_{\mathcal{A}},\ldots,V^p_{\mathcal{A}})} \mathcal{C}(V^0, \ldots, V^p).$$

Notice now that, since $\mathcal{A}$ is cofibrant and $\mathcal{C}$ fibrant, $\underline{\mathrm{HOM}}(\mathcal{A}, \mathcal{C})$ is fibrant and in particular it is an $\mathcal{M}$-category. The Segal condition together with fibrancy of the restriction maps which make up the Segal map at the triple $(V_{\mathcal{A}}, U_{\mathcal{A}}, \varphi)$, imply that the morphism

$$\underline{\mathrm{Cone}}(V, U, \varphi) \to \mathcal{C}(V, U) \times \underline{\mathrm{Cone}}(U, \varphi)$$

is an equivalence and indeed a trivial fibration. On the other hand, we have a projection

$$\underline{\mathrm{Cone}}(V, U, \varphi) \to \underline{\mathrm{Cone}}(V, \varphi).$$

It is in this sense that we have a "weak morphism" from $\mathcal{C}(V, U) \times \underline{\mathrm{Cone}}(U, \varphi)$ to $\underline{\mathrm{Cone}}(V, \varphi)$.

Definition 21.7.1 We say that an object $U \in \mathcal{C}_0$ together with element $f \in \mathrm{Cone}(U, \varphi)$ is a *limit of* φ if for any $V \in \mathrm{Ob}(\mathcal{C})$ the resulting weak morphism from $\mathcal{C}(V, U)$ to $\underline{\mathrm{Cone}}(V, \varphi)$ is an equivalence. More precisely, this means that the morphism

$$\underline{\mathrm{Cone}}(V, U, \varphi) \times_{\underline{\mathrm{Cone}}(U,\varphi)} \{f\} \to \underline{\mathrm{Cone}}(V, \varphi)$$

should be an equivalence, where the right term in the fiber product denotes $f : * \to \underline{\mathrm{Cone}}(U, \varphi)$.

If such a limit exists we say that φ *admits a limit* (we will discuss uniqueness below). If any morphism $\varphi : \mathcal{A} \to \mathcal{C}$ from an $\mathcal{M}$-precategory $\mathcal{A}$ to $\mathcal{C}$ admits a limit then we say that $\mathcal{C}$ *admits limits*.

Lemma 21.7.2 *Suppose $f \in \mathrm{Cone}(U, \varphi)$ and $g \in \mathrm{Cone}(V, \varphi)$ are two different limits of φ. Then there is an essentially canonical map $U \to V$ corresponding to a contractible object of $\mathcal{M}$ mapping to $\mathcal{C}(U, V)$, and which is an equivalence between U and V in $\mathcal{C}$.*

Proof Then the inverse image of $\{g\}$ for the morphism

$$\underline{\mathrm{Cone}}(V, U, \varphi) \times_{\underline{\mathrm{Cone}}(U,\varphi)} \{f\} \to \underline{\mathrm{Cone}}(V, \varphi)$$

is contractible. This gives a contractible object of $\mathcal{M}$ mapping to $\mathcal{C}(V, U)$. We also have a contractible object of $\mathcal{M}$ mapping to $\mathcal{C}(U, V)$. A similar argument with $p = 3$ gives a contractible object of $\mathcal{M}$ mapping to $\mathcal{C}(V, U, V)$ which maps into the contractible things for V, U, for U, V and for V, V. The image at the end includes the identity. This shows that the composition of the morphisms in the two directions is the identity. The same works in the other direction. This shows that the essentially well defined morphisms $U \to V$ and $V \to U$ are equivalences. □

The condition of being a limit may also be interpreted in terms of the construction Υ described in the previous section. Start by noting the following universal property of $\underline{\mathrm{Cone}}$:

Lemma 21.7.3 *For $E \in \mathcal{M}$ a morphism*

$$E \to \underline{\mathrm{Cone}}(U, \varphi)$$

is the same thing as a morphism

$$u : \mathcal{A} \times \Upsilon(E) \to \mathcal{C},$$

such that $r_0(u) = U_{\mathcal{A}}$ and $r_1(u) = \varphi$.

Proof A map

$$E \to \underline{\mathrm{Hom}}(\mathcal{A}, \mathcal{C})(U_{\mathcal{A}}, \varphi)$$

corresponds by the universal property of Υ to a map

$$\Upsilon(E) \to \underline{\mathrm{Hom}}(\mathcal{A}, \mathcal{C}),$$

sending υ_0 to $U_{\mathcal{A}}$ and υ_1 to φ. In turn this corresponds to a map $\mathcal{A} \times \Upsilon(E) \to \mathcal{C}$ as required. □

Using the above descriptions we can describe explicitly the lifting property of the previous paragraph and thus obtain the following characterization which is the one we shall use in our proofs later:

Lemma 21.7.4 *A morphism $f \in \mathrm{Cone}(U, \varphi)$ is a limit if and only if for every cofibration $E' \to E$ between cofibrant objects (resp. every cofibration in the generating set), for every morphism*

$$v : \mathcal{A} \times \Upsilon(E) \to \mathcal{C},$$

with $r_0(v) = V_{\mathcal{A}}$ for $V \in \mathcal{C}_0$ and $r_1(v) = \varphi$, and for every extension over $\mathcal{A} \times \Upsilon(E')$ to a morphism

$$w' : \mathcal{A} \times \widetilde{\Upsilon}_2(E', *) \to \mathcal{C},$$

with $r_{12}(w') = f$ and $r_{01}(w')$ coming from a morphism $z' : \Upsilon(E') \to \mathcal{C}$ with $r_0(z') = V$ and $r_1(z') = U$, there exists a common extension of these two: a morphism

$$w : \mathcal{A} \times \widetilde{\Upsilon}_2(E, *) \to \mathcal{C},$$

with $r_{12}(w) = f$ and $r_{01}(w)$ coming from a morphism $z : \Upsilon(E) \to \mathcal{C}$ with $r_0(z) = V$ and $r_1(z) = U$; such that the restriction of w to $\mathcal{A} \times \widetilde{\Upsilon}_2(E', *)$ is equal to w'; and such that $r_{02}(w) = v$.

Proof In view of the discussion of Section 21.4.1, a morphism

$$E \to \underline{\mathrm{Cone}}(V, U, \varphi)$$

is the same thing as a morphism

$$g : \mathcal{A} \times h([2], E) \to \mathcal{C},$$

with $r_0(g) = V_{\mathcal{A}}$, $r_1(g) = U_{\mathcal{A}}$ and $r_2(g) = \varphi$ and such that $r_{01}(g)$ comes from a morphism $\Upsilon(E) \to \mathcal{C}$. To see this, use the definition of $h([2], E)$ by universal property.

Similarly, a morphism

$$E \to \underline{\mathrm{Cone}}(V, U, \varphi) \times_{\underline{\mathrm{Cone}}(U, \varphi)} \{f\}$$

is the same thing as a morphism

$$g : \mathcal{A} \times \widetilde{\Upsilon}_2(E, *) \to \mathcal{C},$$

such that $r_2(g) = \varphi$ and $r_{01}(g)$ comes from a morphism $g_{01} : \Upsilon(E) \to \mathcal{C}$ with $r_0(g_{01}) = V$ and $r_1(g_{01}) = U$.

Noting that the morphism

$$\underline{\mathrm{Cone}}(V, U, \varphi) \times_{\underline{\mathrm{Cone}}(U, \varphi)} \{f\} \to \underline{\mathrm{Cone}}(V, \varphi)$$

is a fibration, it is an equivalence if and only if it satisfies the lifting property for all cofibrations $E' \to E$ in the generating set. Since $\mathcal{M}$ is tractable these are cofibrations between cofibrant objects. In view of the previous translations this gives the required statement. □

21.8 Colimits

We obtain the notion of direct limit or colimit, by "reversing the arrows" in the above discussion. Formally this can be obtained by just applying the *opposite* construction of Section 21.1, but it will be convenient to have a specific notation.

Suppose $\mathcal{C}$ is a fibrant $\mathcal{M}$-category, and suppose $\mathcal{A}$ is an $\mathcal{M}$-precategory. Suppose $\varphi : \mathcal{A} \to \mathcal{C}$ is a morphism. If $U \in \mathcal{C}_0$ is an object then we define the *cocone object*

$$\underline{\mathrm{Cocone}}(\varphi, U) := \underline{\mathrm{HOM}}(\mathcal{A}, \mathcal{C})(\varphi, U_{\mathcal{A}}),$$

where again $U_{\mathcal{A}}$ denotes the constant morphism with value U. Define the *set of cocones* by

$$\mathrm{Cocone}(\varphi, U) := \mathrm{HOM}_{\mathcal{M}}(*, \underline{\mathrm{Cocone}}(\varphi, U)),$$

and, as before, an element $f \in \underline{\mathrm{Cocone}}(\varphi, U)$ is the same thing as a map $\Upsilon(*) \times \mathcal{A} \to \mathcal{C}$ sending $\upsilon_0 \times \mathcal{A}$ to φ and $\upsilon_1 \times \mathcal{A}$ to the constant $U_{\mathcal{A}}$.

If V is another object of $\mathcal{C}$ then we have a morphism

$$\mathcal{C}(U, V) \to \underline{\mathrm{HOM}}(\mathcal{A}, \mathcal{C})(U_{\mathcal{A}}, V_{\mathcal{A}}),$$

and we use this to define

$$\underline{\mathrm{Cocone}}(\varphi, U, V) := \underline{\mathrm{HOM}}(\mathcal{A}, \mathcal{C})(\varphi, U_{\mathcal{A}}, V_{\mathcal{A}}, \varphi) \times_{\underline{\mathrm{HOM}}(\mathcal{A},\mathcal{C})(U_{\mathcal{A}}, V_{\mathcal{A}})} \mathcal{C}(U, V),$$

or, more generally, if $V^0, \ldots, V^p \in \mathcal{C}_0$ we define

$$\underline{\mathrm{Cocone}}(\varphi, V^0, \ldots, V^p)$$
$$:= \underline{\mathrm{HOM}}(\mathcal{A}, \mathcal{C}) \left(\varphi, V^0{}_{\mathcal{A}}, \ldots, V^p{}_{\mathcal{A}}\right) \times_{\underline{\mathrm{HOM}}(\mathcal{A},\mathcal{C})(V^0{}_{\mathcal{A}}, \ldots, V^p{}_{\mathcal{A}})} \mathcal{C}(V^0, \ldots, V^p).$$

As before, since $\mathcal{A}$ is cofibrant and $\mathcal{C}$ is fibrant, $\underline{\mathrm{HOM}}(\mathcal{A}, \mathcal{C})$ is a fibrant $\mathcal{M}$-category, and the morphism

$$\underline{\mathrm{Cocone}}(\varphi, U, V) \to \mathcal{C}(U, V) \times \underline{\mathrm{Cocone}}(\varphi, U)$$

is a trivial fibration. Using the projection

$$\underline{\mathrm{Cocone}}(\varphi, U, V) \to \underline{\mathrm{Cocone}}(V, \varphi)$$

gives a weak morphism from $\mathcal{C}(U, V) \times \underline{\mathrm{Cocone}}(\varphi, U)$ to $\underline{\mathrm{Cocone}}(\varphi, V)$.

Definition 21.8.1 We say that an element $f \in \mathrm{Cocone}(U, \varphi)_0$ is a *colimit* (or *direct limit*) of φ if for any $V \in \mathrm{Ob}(\mathcal{C})$ the resulting weak morphism from $\mathcal{C}(U, V)$ to $\underline{\mathrm{Cocone}}(\varphi, V)$ is an equivalence. More precisely, this means that the morphism

$$\underline{\mathrm{Cocone}}(\varphi, U, V) \times_{\underline{\mathrm{Cocone}}(\varphi, U)} \{f\} \to \underline{\mathrm{HOM}}(\varphi, V)$$

should be an equivalence. If such a colimit exists we say that φ *admits a colimit*. Exactly the same discussion of uniqueness as above (21.7.2) holds here too. If any morphism $\varphi : \mathcal{A} \to \mathcal{C}$ from any $\mathcal{M}$-precategory $\mathcal{A}$ to $\mathcal{C}$ admits a colimit then we say that $\mathcal{C}$ *admits colimits*.

We have the following characterization analogue to Lemma 21.7.4. Again, this is the characterization which we shall use in the proofs:

Lemma 21.8.2 *A morphism $f \in \mathrm{Cocone}(\varphi, U)$ is a colimit if and only if for every cofibration $E' \to E$ between cofibrant objects (resp. every cofibration in the generating set), for every morphism*

$$v : \mathcal{A} \times \Upsilon(E) \to \mathcal{C},$$

with $r_0(v) = \varphi$ and $r_1(v) = V_{\mathcal{A}}$ with $V \in \mathcal{C}_0$, and for every extension over $\mathcal{A} \times \Upsilon(E')$ to a morphism

$$w' : \mathcal{A} \times \widetilde{\Upsilon}_2(*, E') \to \mathcal{C},$$

with $r_{01}(w') = f$ and $r_{12}(w')$ coming from a morphism $z' : \Upsilon(E') \to \mathcal{C}$ with $r_1(z') = U$ and $r_2(z') = V$, there exists a common extension of these two: a morphism

$$w : \mathcal{A} \times \widetilde{\Upsilon}_2(*, E) \to \mathcal{C},$$

with $r_{01}(w) = f$ and $r_{12}(w)$ coming from a morphism $z : \Upsilon(E) \to \mathcal{C}$ with $r_1(z) = U$ and $r_2(z) = V$; such that the restriction of w to $\mathcal{A} \times \widetilde{\Upsilon}_2(*, E')$ is equal to w'; and such that $r_{02}(w) = v$.

Proof Dualize the proof of Lemma 21.7.4. □

21.9 Invariance properties

The following statements give some invariance properties for limits and colimits, analogs of the standard properties in usual 1-categories.

Lemma 21.9.1 *Suppose $f : \mathcal{A}' \hookrightarrow \mathcal{A}$ is a trivial cofibration of cofibrant $\mathcal{M}$-categories. Suppose that $\varphi : \mathcal{A} \to \mathcal{C}$ is a morphism and that $u : \varphi \to U$*

is a morphism from φ *to* $U \in \mathrm{Ob}(\mathcal{C})$ *which is a colimit (resp.* $u : U \to \varphi$ *is a morphism from* U *to* φ *which is a limit). This corresponds to a diagram*

$$\epsilon : \mathcal{A} \times \Upsilon(*) \to \mathcal{C},$$

and pullback by f *gives a diagram*

$$\epsilon' : \mathcal{A}' \times \Upsilon(*) \to \mathcal{C}.$$

Then ϵ' *is a morphism from* $\varphi \circ f$ *to* U *which is a colimit (resp. is a morphism from* U *to* $\varphi \circ f$ *which is a limit).*

Proof The proof is written for the colimit case. Suppose we are given

$$u : \mathcal{A}' \times \Upsilon(E) \to \mathcal{C},$$

and an extension over $E' \subset E$ to a diagram

$$v_1 : \mathcal{A}' \times \widetilde{\Upsilon}_2(*, E') \to \mathcal{C},$$

whose restriction to the edge (01) is ϵ' and whose restriction to the edge (02) is u. Then we can first extend v_1 to a diagram

$$\mathcal{A} \times \Upsilon(*, E') \to \mathcal{C},$$

because

$$\mathcal{A}' \times \widetilde{\Upsilon}_2(*, E') \hookrightarrow \mathcal{A} \times \widetilde{\Upsilon}_2(*, E')$$

is a trivial cofibration (and note also that we can assume that the extension satisfies the relevant properties as in the definition of limit); then we can also extend our above morphism u to a diagram

$$\mathcal{A} \times \Upsilon(E) \to \mathcal{C},$$

which is compatible with the extension of v_1, because

$$\mathcal{A} \times \Upsilon(E') \cup^{\mathcal{A}' \times \Upsilon(E')} \mathcal{A}' \times \Upsilon(E) \to \mathcal{A} \times \Upsilon(E)$$

is a trivial cofibration. Apply the colimit property of ϵ to conclude that there is an extension to a diagram

$$v : \mathcal{A} \times \widetilde{\Upsilon}_2(*, E) \to \mathcal{C}.$$

This restricts over $\mathcal{A}'$ to a diagram of the form we would like, showing that ϵ' is a colimit. □

Lemma 21.9.2 *Suppose that $f : \mathcal{A}' \hookrightarrow \mathcal{A}$ is a trivial cofibration between cofibrant objects in $\mathcal{PC}(\mathcal{M})$ and suppose $\varphi : \mathcal{A} \to \mathcal{C}$ is a morphism to a fibrant $\mathcal{M}$-category $\mathcal{C}$, and suppose now that we know that $\varphi \circ f$ has a colimit*

$$\epsilon' : \varphi \circ f \to U,$$

for an object $U \in \mathcal{C}$. Then φ has a colimit. Similarly for limits.

Proof Again write the proof in the colimit case. The morphism ϵ' may be considered as a diagram

$$\epsilon' : \mathcal{A}' \times \Upsilon(*) \to \mathcal{C}.$$

This extends along $\mathcal{A}' \times \{\upsilon_0\}$ to

$$\varphi : \mathcal{A} \times \{\upsilon_0\} \to \mathcal{C},$$

and it extends along $\mathcal{A}' \times \{\upsilon_1\}$ to

$$U_{\mathcal{A}} : \mathcal{A} \times \{\upsilon_1\} \to \mathcal{C}.$$

Putting these all together we obtain a morphism

$$\mathcal{A} \times \{\upsilon_0\} \cup^{\mathcal{A}' \times \{\upsilon_0\}} \mathcal{A}' \times \Upsilon(*) \cup^{\mathcal{A}' \times \{\upsilon_1\}} \mathcal{A} \times \{\upsilon_1\} \to \mathcal{C}.$$

Since $\mathcal{A}' \subset \mathcal{A}$ is a trivial cofibration, the morphism

$$\mathcal{A} \times \{\upsilon_0\} \cup^{\mathcal{A}' \times \{\upsilon_0\}} \mathcal{A}' \times \Upsilon(*) \cup^{\mathcal{A}' \times \{\upsilon_1\}} \mathcal{A} \times \{\upsilon_1\} \to \mathcal{A} \times \Upsilon(*)$$

is a trivial cofibration (applying the first part of CM4 two times), so by the fibrant property of $\mathcal{C}$ our morphism extends to a morphism

$$\epsilon : \mathcal{A} \times \Upsilon(*) \to \mathcal{C}$$

with the required properties of being constant along $\mathcal{A} \times \{\upsilon_1\}$ and restricting to φ along $\mathcal{A} \times \{\upsilon_0\}$. Thus we may write $\epsilon : \varphi \to U$.

We claim that this map is a colimit of φ. Given a diagram

$$u : \mathcal{A} \times \Upsilon(E) \to \mathcal{C}$$

going from φ to a constant object B, the restriction u' to $\mathcal{A}' \times \Upsilon(E)$ admits (by the hypothesis that ϵ' is a colimit) an extension to

$$v' : \mathcal{A}' \times \Upsilon(*, E) \to \mathcal{C},$$

restricting along the edge (01) to ϵ' and restricting along the edge (12) to the pullback of a diagram $\Upsilon(E) \to \mathcal{C}$. Then, using as usual the fibrant property of $\mathcal{C}$ and the fact that $\mathcal{A}' \to \mathcal{A}$ is a trivial cofibration, we can extend v' to a morphism

$$v : \mathcal{A} \times \Upsilon(*, E) \to \mathcal{C},$$

again restricting along the edge (01) to ϵ, restricting along the edge (02) to our given diagram u and restricting along the edge (12) to the pullback of a diagram $\Upsilon(E) \to \mathcal{C}$.

If $E' \subset E$ and we are already given an extension $v_{E'}$ over $\mathcal{A} \times \Upsilon(*, E')$ then (as before, using the fibrant property of $\mathcal{C}$ applied to an appropriate cofibration) we can assume that our extension v above restricts to $v_{E'}$. This completes the proof that ϵ is a colimit. □

Corollary 21.9.3 *Suppose $\varphi, \varphi' : \mathcal{A} \to \mathcal{C}$ are two morphisms from a cofibrant to a fibrant $\mathcal{M}$-category. Suppose that they are homotopic, i.e. correspond to inner equivalent points of $\underline{\mathrm{HOM}}(\mathcal{A}, \mathcal{C})$. Then φ admits a limit (resp. colimit) if and only if φ' does.*

Proof The homotopy may be represented as a map $\psi : \mathcal{A} \times \mathbb{I} \to \mathcal{C}$. By the previous two lemmas, φ admits a limit (resp. colimit) if and only if ψ does, and the same holds for φ'. □

Proposition 21.9.4 *Suppose $f : \mathcal{A}' \to \mathcal{A}$ is an equivalence between cofibrant $\mathcal{M}$-categories and suppose $\mathcal{C}$ is a fibrant $\mathcal{M}$-category. Suppose $\varphi : \mathcal{A} \to \mathcal{C}$ is a morphism. Then the limit (resp. colimit) of φ exists if and only if the limit (resp. colimit) of $\varphi \circ f$ exists.*

Proof Combine the two previous lemmas to give the claimed invariance statement when f is a trivial cofibration.

For arbitrary f, factorize the map $(f, 1_{\mathcal{A}})$ as

$$\mathcal{A}' \sqcup \mathcal{A} \xrightarrow{(j', j)} \mathcal{A}'' \xrightarrow{p} \mathcal{A},$$

where p is a trivial fibration and (j', j) a cofibration. Since $\mathcal{A}$ and $\mathcal{A}'$ are cofibrant, each component j and j' is a cofibration, but the equations $pj' = f$ and $pj = 1_{\mathcal{A}}$ imply that j and j' are trivial cofibrations. The invariance statement for the case of cofibrations, says on the one hand that φp admits a limit (resp. colimit) if and only if $\varphi p j = \varphi$ does; but on the other hand that φp admits a limit (resp. colimit) if and only if $\varphi p j' = \varphi f$ does. This completes the proof. □

Proposition 21.9.5 *Suppose $g : \mathcal{C} \to \mathcal{C}'$ is an equivalence between fibrant $\mathcal{M}$-categories. Then the limit (resp. colimit) of $\varphi : \mathcal{A} \to \mathcal{C}$ exists if and only if the limit (resp. colimit) of $g \circ \varphi$ exists. In particular $\mathcal{C}$ admits limits (resp. colimits) if and only if $\mathcal{C}'$ does.*

Proof In general if $\mathcal{F}' \subset \mathcal{F}$ is any cofibration and if $\mathcal{F}' \to \mathcal{C}$ is a morphism, then there exists an extension to $\mathcal{F} \to \mathcal{C}$ if and only if the composed morphism

$\mathcal{F}' \to \mathcal{C}'$ extends over $\mathcal{F}$. This can be seen as a corollary of the discussion in Section 21.3.1, but alternatively look at the commutative diagram

$$\begin{array}{ccc} \underline{\mathrm{HOM}}(\mathcal{F}, \mathcal{C}) & \longrightarrow & \underline{\mathrm{HOM}}(\mathcal{F}', \mathcal{C}) \\ \downarrow & & \downarrow \\ \underline{\mathrm{HOM}}(\mathcal{F}, \mathcal{C}') & \to & \underline{\mathrm{HOM}}(\mathcal{F}', \mathcal{C}'). \end{array}$$

The horizontal arrows are fibrations and the vertical arrows are equivalences. If an element $a \in \underline{\mathrm{HOM}}(\mathcal{F}', \mathcal{C})$ maps to something b which is hit from $c \in \underline{\mathrm{HOM}}(\mathcal{F}, \mathcal{C}')$ then there is $d \in \underline{\mathrm{HOM}}(\mathcal{F}, \mathcal{C})$ mapping to something equivalent to c; thus the image e of d in $\underline{\mathrm{HOM}}(\mathcal{F}', \mathcal{C})$ maps to something equivalent to b. This implies (since the right vertical arrow is an equivalence) that e is equivalent to a. Since the top morphism is a fibration, there is another element $d' \in \underline{\mathrm{HOM}}(\mathcal{F}, \mathcal{C})$ which maps directly to a.

Consider still our equivalence $g : \mathcal{C} \to \mathcal{C}'$ between fibrant $\mathcal{M}$-categories. Using the general lifting principle of the previous paragraph, and the fact that the property of being a limit is expressed in terms of extending morphisms across certain cofibrations $\mathcal{F}' \subset \mathcal{F}$, we conclude that a functor $\mathcal{A} \to \mathcal{C}$ has a limit or colimit only if the composition $\mathcal{A} \to \mathcal{C}'$ does. This proves the first sentence of the proposition.

If $g : \mathcal{C} \to \mathcal{C}'$ is an equivalence between fibrant $\mathcal{M}$-categories and if $\mathcal{C}'$ admits colimits (resp. limits) then any functor $\mathcal{A} \to \mathcal{C}$ admits a colimit (resp. limit) by the result proven in the previous paragraph.

Suppose, on the other hand, that we know that $\mathcal{C}$ admits colimits (resp. limits). Suppose that $\varphi : \mathcal{A} \to \mathcal{C}'$ is a functor. Since g induces an equivalence from $\underline{\mathrm{HOM}}(\mathcal{A}, \mathcal{C})$ to $\underline{\mathrm{HOM}}(\mathcal{A}, \mathcal{C}')$ there is a morphism $\psi : \mathcal{A} \to \mathcal{C}$ such that $g \circ \psi$ and φ are homotopic, i.e. inner equivalent objects of $\underline{\mathrm{HOM}}(\mathcal{A}, \mathcal{C}')$. By Corollary 21.9.3, φ admits colimits (resp. limits) if and only if $g \circ \psi$ does. Then, as discussed above, $g \circ \psi$ admits colimits (resp. limits) if and only if ψ does. Now by hypothesis ψ has a colimit (resp. limit), so φ does too. This shows that $\mathcal{C}'$ admits colimits (resp. limits). □

In view of the above invariance properties, it makes sense to speak of when a morphism $\mathcal{A} \to \mathcal{C}$ in $\mathrm{ho}(\mathcal{PC}(\mathcal{M}))$ admits limits or colimits, by first taking a cofibrant replacement of $\mathcal{A}$ and a fibrant replacement of $\mathcal{C}$. The question is independent of the choice of replacements.

There are variance properties for other situations too. Suppose $h : \mathcal{C} \to \mathcal{C}'$ is a morphism between fibrant $\mathcal{M}$-categories, and suppose $\varphi : \mathcal{A} \to \mathcal{C}$ is a morphism. If

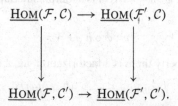

$$u : \varphi \to U$$

is a colimit then $h(u) : \varphi \circ h \to h(U)$ is a morphism. Suppose that $\varphi \circ h$ admits a colimit

$$v : \varphi \circ h \to V.$$

Then by the limit property there is a factorization, i.e. a diagram

$$[v, w] : \varphi \circ h \to V \to h(U),$$

whose (02)-edge is $h(u)$. We say that *the morphism h commutes with the colimit of φ* if the colimit of $\varphi \circ h$ exists and if the factorization morphism $w : V \to h(U)$ is an equivalence.

Suppose that $\mathcal{C}$ and $\mathcal{C}'$ admit colimits. We say that *the morphism h commutes with colimits* if h commutes with the colimit of any $\varphi : \mathcal{A} \to \mathcal{C}$ in the previous sense.

We have similar definitions for limits, which we repeat for the record. Suppose again that $h : \mathcal{C} \to \mathcal{C}'$ is a morphism between fibrant $\mathcal{M}$-categories, and suppose $\varphi : \mathcal{A} \to \mathcal{C}$ is a morphism. If

$$u : U \to \varphi$$

is a limit then $h(u) : h(U) \to \varphi \circ h$ is a morphism. Suppose that $\varphi \circ h$ admits a limit

$$v : V \to \varphi \circ h.$$

Then by the limit property there is a factorization, i.e. a diagram

$$[w, v] : h(U) \to V \to \varphi \circ h,$$

whose (02)-edge is $h(u)$. We say that *the morphism h commutes with the limit of φ* if the limit of $\varphi \circ h$ exists and if the factorization morphism $w : h(U) \to V$ is an equivalence.

Suppose that $\mathcal{C}$ and $\mathcal{C}'$ admit limits. We say that *the morphism h commutes with limits* if h commutes with the limit of any $\varphi : \mathcal{A} \to \mathcal{C}$ in the above sense.

21.10 Limits of diagrams

A natural theorem to look for is the generalization of the classical result in category theory, saying that if a category $\mathcal{C}$ admits limits, then so does the diagram category $\underline{\text{HOM}}(\mathcal{B}, \mathcal{C})$.

In preparation, it will be useful to study certain situations of what happens when we take fiber products or other limits in the category $\mathcal{PC}(\mathcal{M})$, of the target $\mathcal{M}$-categories $\mathcal{C}_i$. We study what happens to limits in the $\mathcal{C}_i$. We could also say the same things about colimits in $\mathcal{C}$ but the limit case is the one we need, so we state that case here and leave it to the reader to make the corresponding statements for colimits.

Lemma 21.10.1 *Suppose $\{\mathcal{C}_i\}_{i\in S}$ is a collection of fibrant $\mathcal{M}$-categories indexed by a set S. Let $\mathcal{C} := \prod_{i\in S}\mathcal{C}_i$ and suppose $\varphi = \{\varphi_i\}$ is a morphism from $\mathcal{A}$ to $\mathcal{C}$. Suppose that the φ_i admit limits*

$$u_i : U_i \to \varphi_i$$

in $\mathcal{C}_i$. Then $U = \{U_i\}$ is an object of $\mathcal{C}$ and we have a morphism

$$u : U \to \varphi$$

composed of the factors u_i. This morphism is a limit of φ in $\mathcal{C}$.

Proof The property that u be a limit consists of a collection of extension properties that have to be satisfied. The morphisms u_i admit the corresponding extensions and putting these together we get the required extensions for u. □

Lemma 21.10.2 *Suppose $f : \mathcal{C} \to \mathcal{D}$ and $g : \mathcal{B} \to \mathcal{D}$ are morphisms of fibrant $\mathcal{M}$-categories with f a fibration. Suppose that $\varphi : \mathcal{A} \to \mathcal{C}\times_{\mathcal{D}}\mathcal{B}$ is a morphism such that the component morphisms $\varphi_{\mathcal{C}} : \mathcal{A} \to \mathcal{C}$, $\varphi_{\mathcal{D}} : \mathcal{A} \to \mathcal{D}$ and $\varphi_{\mathcal{B}} : \mathcal{A} \to \mathcal{B}$ have limits $\lambda_{\mathcal{C}}$, $\lambda_{\mathcal{D}}$ and $\lambda_{\mathcal{B}}$ respectively. Suppose furthermore that f and g preserve these limits, which means that the projections of $\lambda_{\mathcal{C}}$ and $\lambda_{\mathcal{B}}$ into $\mathcal{D}$ are equivalent (as objects with morphisms to $\varphi_{\mathcal{D}}$) to $\lambda_{\mathcal{D}}$. Then we may (by changing the $\lambda_{\mathcal{C}}, \lambda_{\mathcal{D}}, \lambda_{\mathcal{B}}$ by equivalences) assume that $\lambda_{\mathcal{C}}$ and $\lambda_{\mathcal{B}}$ project to $\lambda_{\mathcal{D}}$; and the resulting object $\lambda \in \mathcal{C}\times_{\mathcal{D}}\mathcal{B}$ is a limit of φ.*

Proof Set $\lambda'_{\mathcal{B}} := \lambda_{\mathcal{B}}$ and let $\lambda'_{\mathcal{D}} := g(\lambda_{\mathcal{B}})$ be the projection to $\mathcal{D}$. Note that by hypothesis $\lambda'_{\mathcal{D}}$ is a limit of $\varphi_{\mathcal{D}}$. Now $\lambda_{\mathcal{C}}$ (considered as an object with morphism to $\varphi_{\mathcal{C}}$) projects in $\mathcal{D}$ to something equivalent to $\lambda_{\mathcal{D}}$ and hence equivalent to $\lambda'_{\mathcal{D}}$ (equivalence of the diagrams including the morphism to $\varphi_{\mathcal{D}}$). Since f is a fibration, we can modify $\lambda_{\mathcal{C}}$ by an equivalence, to obtain $\lambda'_{\mathcal{C}}$ projecting directly to $\lambda'_{\mathcal{D}}$. Note that the equivalent $\lambda'_{\mathcal{C}}$ is again a limit of $\varphi_{\mathcal{C}}$. Together these give an element $\lambda \in \mathcal{C}\times_{\mathcal{D}}\mathcal{B}$ with a map

$$u : \lambda \to \varphi,$$

and we claim that u is a limit. Use the criterion of Lemma 21.7.4. Suppose $F' \to F$ is a cofibration between cofibrant objects in $\mathcal{M}$ and suppose

$$v : V \xrightarrow{F} \varphi$$

is any F-morphism, by which we mean a diagram

$$\mathcal{A} \times \Upsilon(F) \xrightarrow{v} \mathcal{C} \times_{\mathcal{D}} \mathcal{B},$$

restricting on $\mathcal{A} \times \{\upsilon_0\}$ to the constant $V_{\mathcal{A}}$ and restricting on $\mathcal{A} \times \{\upsilon_1\}$ to φ. Suppose v is provided with an extension over F' to a diagram

$$w' : \mathcal{A} \times \widetilde{\Upsilon}_2(F, *) \to \mathcal{C} \times_{\mathcal{D}} \mathcal{B},$$

restricting on (02) to v' (the restriction of v to F') and on (12) to u. We look for an extension of w' to a diagram

$$w : \mathcal{A} \times \widetilde{\Upsilon}_2(F, *) \to \mathcal{C} \times_{\mathcal{D}} \mathcal{B},$$

restricting on (02) to v and on (12) to u. Denoting with subscripts the components in $\mathcal{C}$, $\mathcal{D}$ and $\mathcal{B}$, we have that the pairs $(v_{\mathcal{C}}, w'_{\mathcal{C}})$ and $(v_{\mathcal{B}}, w'_{\mathcal{B}})$ admit extensions $w_{\mathcal{C}}$ and $w_{\mathcal{B}}$ respectively. The projections of these extensions in $\mathcal{D}$ give diagrams which we denote

$$w_{\mathcal{C}/\mathcal{D}}, w_{\mathcal{B}/\mathcal{D}} : \mathcal{A} \times \widetilde{\Upsilon}_2(F, *) \to \mathcal{D},$$

both restricting on (02) to $v_{\mathcal{D}}$ and on (12) to $u_{\mathcal{D}}$, and extending $w'_{\mathcal{D}}$. Applying again the limit property for $u_{\mathcal{D}}$ to the cofibration

$$F \times \{\upsilon_0\} \cup^{F' \times \{\upsilon_0\}} F' \times \mathbf{I} \cup^{F' \times \{\upsilon_1\}} F \times \{\upsilon_1\} \hookrightarrow F \times \mathbf{I},$$

we find that there is a diagram

$$z_{\mathcal{D}} : \mathcal{A} \times \Upsilon^2(F \times \mathbf{I}, *) \to \mathcal{D},$$

giving a homotopy in Quillen's sense [215] between $w_{\mathcal{C}/\mathcal{D}}$ and $w_{\mathcal{B}/\mathcal{D}}$. It is reasonable to call this a homotopy, because the diagram

$$\widetilde{\Upsilon}_2(F, *) \rightrightarrows \widetilde{\Upsilon}_2(F \times \mathbf{I}, *) \to \widetilde{\Upsilon}_2(F, *)$$

is a cylinder object (except that the second map isn't a fibration but that doesn't matter here). Such a homotopy can be changed into one of the more classical form

$$\mathcal{A} \times \widetilde{\Upsilon}_2(F, *) \times \mathbf{I} \to \mathcal{D}$$

(see Proposition 21.3.1).

Now apply the lifting property for the morphism $\mathcal{C} \to \mathcal{D}$, for the above map $z_{\mathcal{D}}$, with respect to the trivial cofibration

$$\mathcal{A} \times \widetilde{\Upsilon}_2(F \times \{\upsilon_0\} \cup^{F' \times \{\upsilon_0\}} F' \times \mathbf{I}, *) \to \mathcal{A} \times \widetilde{\Upsilon}_2(F \times \mathbf{I}, *).$$

We get a morphism

$$z_{\mathcal{C}} : \mathcal{A} \times \Upsilon^2(F \times \mathbf{I}, *) \to \mathcal{C},$$

providing a homotopy between $w_{\mathcal{C}}$ and a new morphism $w^n_{\mathcal{C}}$ projecting to $w_{\mathcal{B}/\mathcal{D}}$ in $\mathcal{D}$. The new $w^n_{\mathcal{C}}$ may be chosen as a solution of the required extension problem. The fact that it projects to $w_{\mathcal{B}/\mathcal{D}}$ means that the pair $w = \left(w^n_{\mathcal{C}}, w_{\mathcal{B}}\right)$ is a solution of the required extension problem to show that u is a limit. This completes the proof. □

Lemma 21.10.3 *Suppose $\mathcal{C}_i$ is a collection of fibrant objects of $\mathcal{PC}(\mathcal{M})$ for $i = 0, 1, 2, \ldots$ and suppose that $f_i : \mathcal{C}_i \to \mathcal{C}_{i-1}$ are fibrations. Let $\mathcal{C}$ be the limit of this system. Then $\mathcal{C}$ is also fibrant in $\mathcal{PC}(\mathcal{M})$. Suppose that we have $\varphi : \mathcal{A} \to \mathcal{C}$ projecting to the $\varphi_i : \mathcal{A} \to \mathcal{C}_i$ and suppose that the φ_i admit limits $u_i : U_i \to \varphi_i$. Suppose finally that the f_i commute with the limits of the φ_i. Then φ admits a limit and the projections $\mathcal{C} \to \mathcal{C}_i$ commute with the limit of φ.*

Proof The fact that $\mathcal{C}$ is fibrant may be directly checked by producing liftings of trivial cofibrations.

First, we construct a morphism $u : U \to \varphi$ projecting in each $\mathcal{C}_i$ to a limit of φ_i. To do this, note by 21.9.3 that it suffices to have u project to a morphism equivalent to u_i. On the other hand, by the hypothesis that the f_i commute with the limits u_i, we have that $f_i(u_i)$ is a limit of φ_{i-1}. In particular, $f_i(u_i)$ is equivalent to u_{i-1} as a diagram from $\mathcal{A} \times \Upsilon(*)$ to $\mathcal{C}_{i-1}$. The morphism from such diagrams in $\mathcal{C}_i$, to such diagrams in $\mathcal{C}_{i-1}$, is a fibration (since it comes from the fibration f_i). Therefore, we can change u_i to an equivalent diagram with $f_i(u_i) = u_{i-1}$. Do this successively for $i = 1, 2, \ldots$, yielding a system of morphisms u_i with $f_i(u_i) = u_{i-1}$. These now form a morphism

$$u : U \to \varphi.$$

We claim that u is a limit of φ. Suppose $E' \subset E$ is a cofibration in $\mathcal{M}$ and suppose

$$w : W \xrightarrow{E} \varphi$$

is an E-morphism (i.e. a diagram of the form

$$\mathcal{A} \times \Upsilon(E) \to \mathcal{C},$$

being constant equal to W on $\mathcal{A} \times \{\upsilon_0\}$), provided over E' with an extension to a diagram

$$[v', u] : W \xrightarrow{E'} U \to \varphi,$$

that is a morphism

$$\mathcal{A} \times \widetilde{\Upsilon}_2(E, *) \to \mathcal{C},$$

restricting to u on the (12)-edge and restricting to $w|_{E'}$ on the (02)-edge. We would like to extend this to a diagram $[v, u]$ giving w on the (02)-edge. Let w_i

(resp. v'_i) be the projections of these diagrams in $\mathcal{C}_i$. These admit extensions v_i. The projection of v_i to $\mathcal{C}_{i-1}$ is an extension of the desired sort for w_{i-1} and v'_{i-1}. The extensions v_i are unique up to equivalence–which means a diagram

$$\mathcal{A} \times \widetilde{\Upsilon}_2(E, *) \times \mathbf{I} \to \mathcal{C}_i,$$

satisfying appropriate boundary conditions–and from this and the usual argument constructing a trivial cofibration and using the fibration property of f_i, we conclude that v_i may be modified by an equivalence so that it projects to v_{i-1}. As before, do this successively for $i = 1, 2, \ldots$ to obtain a system of extensions v_i with $f_i(v_i) = v_{i-1}$. This system corresponds to the desired extension v of w and v'. This shows that the u is a limit.

Note from our construction the projection of the limit u to $\mathcal{C}_i$ is a limit u_i for φ_i so the projections $\mathcal{C} \to \mathcal{C}_i$ commute with the limit of φ. □

The above results are designed to assist in the proof of the following theorem, which says that existence of limits passes to diagram $\mathcal{M}$-categories. In order to avoid a discussion of universes, we formulate "existence of limits" as meaning "existence of limits indexed by $\mathcal{A}$ in a given collection."

Theorem 21.10.4 *Suppose $\mathcal{B}, \mathcal{C} \in \mathcal{PC}(\mathcal{M})$ with $\mathcal{B}$ cofibrant and $\mathcal{C}$ fibrant. Then if $\mathcal{C}$ admits limits (resp. colimits) indexed by cofibrant $\mathcal{M}$-categories $\mathcal{A}$ in a certain collection, so does $\underline{\mathrm{HOM}}(\mathcal{B}, \mathcal{C})$.*

More precisely, suppose $\varphi : \mathcal{A} \to \underline{\mathrm{HOM}}(\mathcal{B}, \mathcal{C})$ is a morphism such that for each $b \in \mathrm{Ob}(\mathcal{B})$, the restriction $\varphi(b) : \mathcal{A} \to \mathcal{C}$ admits a limit (resp. colimit). Then φ admits a limit (resp. colimit).

The morphisms of functoriality for $\mathcal{B}' \to \mathcal{B}$ commute with such limits (resp. colimits).

Proof Suppose given $\varphi : \mathcal{A} \to \underline{\mathrm{HOM}}(\mathcal{B}, \mathcal{C})$. We will construct a limit $\lambda \in \underline{\mathrm{HOM}}(\mathcal{B}, \mathcal{C})$ such that for any $b \in \mathrm{Ob}(\mathcal{B})$ the restriction $\lambda(b)$ is equivalent (via the natural morphism) to the limit of $\varphi(b) : \mathcal{A} \to \mathcal{C}$. This condition implies that the restriction morphism for any $\mathcal{B}' \to \mathcal{B}$ commutes with the limit. In effect, there is a morphism from the limit over $\mathcal{B}$ (pulled back to $\mathcal{B}'$) to the limit over $\mathcal{B}'$, and this morphism is an equivalence over every object in $\mathcal{B}'$ by the condition, which implies that it is an equivalence by Lemma 21.5.3.

The first remark is that if $\mathcal{B} \to \mathcal{B}'$ and $\mathcal{B} \to \mathcal{B}''$ are Reedy cofibrations betweeen cofibrant objects in $\mathcal{PC}(\mathcal{M})$ such that $\underline{\mathrm{HOM}}(\mathcal{B}, \mathcal{C})$, $\underline{\mathrm{HOM}}(\mathcal{B}', \mathcal{C})$, and $\underline{\mathrm{HOM}}(\mathcal{B}'', \mathcal{C})$ admit limits complying with the above condition, then $\underline{\mathrm{HOM}}$ $(\mathcal{B}' \cup^{\mathcal{B}} \mathcal{B}'', \mathcal{C})$ admits limits again complying with the above condition. To see this we apply Lemma 21.10.2 to the cartesian diagram

$$\begin{array}{ccc} \underline{\mathrm{Hom}}(\mathcal{B}' \cup^{\mathcal{B}} \mathcal{B}'', \mathcal{C}) & \rightarrow & \underline{\mathrm{Hom}}(\mathcal{B}', \mathcal{C}) \\ \downarrow & & \downarrow \\ \underline{\mathrm{Hom}}(\mathcal{B}'', \mathcal{C}) & \longrightarrow & \underline{\mathrm{Hom}}(\mathcal{B}, \mathcal{C}). \end{array}$$

The only thing that we need to know is that the images of the limits by the right and bottom arrows in the diagram, are again equivalent to the limit in $\underline{\mathrm{Hom}}(\mathcal{B}, \mathcal{C})$. These arrows are restrictions along the given cofibrations, and the restriction of a limit is again a limit, by the discussion of the first paragraph of the proof.

One can remark that weak equivalences of cofibrant $\mathcal{M}$-precategories $\mathcal{B}$ are turned into equivalences of the $\underline{\mathrm{Hom}}(\mathcal{B}, \mathcal{C})$, so the statement of the theorem is invariant under trivial cofibrations between cofibrant objects.

As pointed out at the start of the proof, morphisms of restriction between any $\mathcal{B}$'s for which we know that the limits exist (and satisfying the condition of the first paragraph), commute with the limits. In particular, when we apply 21.10.2 and 21.10.3 the hypotheses about commutation with the limits will hold.

As seen in Section 13.1 and Proposition 13.5.1, any Reedy cofibrant object $\mathcal{B} \in \mathcal{PC}(\mathcal{M})$ may be expressed as a countable direct union of its skeleta

$$\mathcal{B} = \lim_{\rightarrow} \mathbf{sk}_m(\mathcal{B}),$$

and, furthermore, $\mathbf{sk}_m(\mathcal{B})$ is obtained from $\mathbf{sk}_{m-1}(\mathcal{B})$ by pushouts along the standard Reedy cofibrations

$$h([m], \partial[m];\, f) \xrightarrow{R([m], f)} h([m], F),$$

for cofibrations between cofibrant objects $F' \xrightarrow{f} F$. Recall that $h([m], \partial[m];\, f)$ is a sort of relative boundary object of $h([m], F)$ coming from the cofibration f, defined at the end of Section 13.2.

Using Lemma 21.10.3 for the colimit over m which is transformed to a limit by $\underline{\mathrm{Hom}}(-, \mathcal{C})$, and using Lemmas 21.10.1 and 21.10.2 for the families of pushouts at each stage, we conclude that it suffices to know the theorem for $\mathcal{B} = h([m], \partial[m];\, f)$ and $\mathcal{B}' = h([m], E)$.

However, the boundary $h([m], \partial[m];\, f)$ may be expressed in turn as a composition of pushouts of the same type for smaller values of m. By induction on m using Lemmas 21.10.1 and 21.10.2, it suffices to know the theorem for just the standard representables $\mathcal{B} = h([m], E)$ as E runs over the cofibrant objects of $\mathcal{M}$.

The map $\Upsilon_m(E,\ldots,E) \to h([m],F)$ is a trivial cofibration as pointed out in Lemma 17.2.1, so

$$\underline{\mathrm{HOM}}(h([m],F),\mathcal{C}) \to \underline{\mathrm{HOM}}(\Upsilon_m(E,\ldots,E),\mathcal{C})$$

is a weak equivalence compatible with the restrictions to individual objects. By the invariance statement of Proposition 21.9.5, it suffices to prove the theorem for $\mathcal{B} = \Upsilon_m(E,\ldots,E)$, then again expressing this as a pushout of m copies of $\Upsilon(E)$, we have reduced to the basic case: it suffices to prove the theorem for $\mathcal{B} = \Upsilon(E)$ for E a cofibrant object of $\mathcal{M}$.

Suppose we have a morphism

$$\mathcal{A} \to \underline{\mathrm{HOM}}(\Upsilon(E),\mathcal{C}),$$

which may be viewed as $\varphi : \mathcal{A} \times \Upsilon(E) \to \mathcal{C}$. Let $\varphi(0)$ (resp. $\varphi(1)$) denote the restriction of φ to $\mathcal{A} \times \{\upsilon_0\}$ (resp. $\mathcal{A} \times \{\upsilon_1\}$). Let $(\lambda(0),\epsilon(0))$ and $(\lambda(1),\epsilon(1))$ denote limits of $\varphi(0)$ and $\varphi(1)$ in $\mathcal{C}$. Thus $\lambda(j) \in \mathrm{Ob}(\mathcal{C})$ and $\epsilon(j)$ are points of $\underline{\mathrm{HOM}}(\mathcal{A},\mathcal{C})(\lambda(j)_{\mathcal{A}},\varphi(j))$.

The morphism φ may be viewed in the other direction as a map $\Upsilon(E) \to \underline{\mathrm{HOM}}(\mathcal{A},\mathcal{C})$ so it corresponds to a morphism

$$E \stackrel{\varphi^t}{\to} \underline{\mathrm{HOM}}(\mathcal{A},\mathcal{C})(\varphi(0),\varphi(1)).$$

We can extend the pair $(\epsilon(0),\varphi^t)$ to a morphism

$$\widetilde{\Upsilon}_2(*,E) \to \underline{\mathrm{HOM}}(\mathcal{A},\mathcal{C}),$$

whose main cell $h([2],E)$ corresponds to

$$E \to \underline{\mathrm{HOM}}(\mathcal{A},\mathcal{C})(\lambda(0)_{\mathcal{A}},\varphi(0),\varphi(1)).$$

The resulting (02) edge $E \to \underline{\mathrm{HOM}}(\mathcal{A},\mathcal{C})(\lambda(0)_{\mathcal{A}},\varphi(1))$ extends to

$$\widetilde{\Upsilon}_2(E,*) \to \underline{\mathrm{HOM}}(\mathcal{A},\mathcal{C}),$$

whose top cell is

$$E \to \underline{\mathrm{HOM}}(\mathcal{A},\mathcal{C})(\lambda(0)_{\mathcal{A}},\lambda(1)_{\mathcal{A}},\varphi(1)),$$

projecting to $\epsilon(1)$ on the second edge, by the limit property for $\epsilon(1)$. By definition of the limit property, the first edge of this comes from a morphism $\lambda : \Upsilon(E) \to \mathcal{C}$.

Noting that the product $\Upsilon(E) \times \mathbb{I}$ is a pushout of two triangles $\widetilde{\Upsilon}_2(E,*)$ and $\widetilde{\Upsilon}_2(*,E)$, the above morphisms glue together to give a morphism

$$\Upsilon(E) \times \mathbb{I} \stackrel{\epsilon^t}{\to} \underline{\mathrm{HOM}}(\mathcal{A},\mathcal{C}),$$

with the following effect on the corners (to help orient ourselves):

$$\begin{matrix} (v_0, v_0) & (v_0, v_1) \\ (v_1, v_0) & (v_1, v_1) \end{matrix} \mapsto \begin{matrix} \lambda(0)_{\mathcal{A}} & \varphi(0) \\ \lambda(1)_{\mathcal{A}} & \varphi(1) \end{matrix}.$$

The left edge $\Upsilon(E) \times \{v_0\}$ maps to $\underline{\mathrm{HOM}}(\mathcal{A}, \mathcal{C})$ by a map factoring through $\lambda : \Upsilon(E) \to \mathcal{C}$. The right edge is φ^t.

Looking at ϵ^t in a different way, we may consider it as a map

$$\mathbb{I} \xrightarrow{\epsilon} \underline{\mathrm{HOM}}(\mathcal{A}, \underline{\mathrm{HOM}}(\Upsilon(E), \mathcal{C})),$$

or, equivalently, as a point

$$\epsilon : * \to \underline{\mathrm{HOM}}(\mathcal{A}, \underline{\mathrm{HOM}}(\Upsilon(E), \mathcal{C}))(\lambda_{\mathcal{A}}, \varphi).$$

In other words, ϵ is a morphism from $\lambda_{\mathcal{A}}$ to φ considered as families over $\mathcal{A}$ with values in $\underline{\mathrm{HOM}}(\Upsilon(E), \mathcal{C})$. Thus $\epsilon \in \mathrm{Cone}(\lambda, \varphi)$.

To finish the proof we show that ϵ is a limit of φ. Suppose we have $\mu \in \underline{\mathrm{HOM}}(\Upsilon(E), \mathcal{C})$ and suppose given a morphism $f : \Upsilon(F) \times \mathcal{A} \times \Upsilon(E) \to \mathcal{C}$ restricting over 0^F to $\mu_{\mathcal{A}}$ and over 1^F to φ (here 0^F and 1^F denote the endpoints of $\Upsilon(F)$ and we will use similar notation for E). This extends to morphisms

$$f'(0^E) : \widetilde{\Upsilon}_2(F, *) \times \mathcal{A} \to \mathcal{C},$$

$$f'(1^E) : \widetilde{\Upsilon}_2(F, *) \times \mathcal{A} \to \mathcal{C},$$

by the limit properties of $\lambda(0)$ and $\lambda(1)$ (the above morphisms restricting to $\epsilon(0)$ and $\epsilon(1)$ on the second edges).

Now we try to extend to a morphism on all of $\widetilde{\Upsilon}_2(F, *) \times \mathcal{A} \times \Upsilon(E) \to \mathcal{C}$. For this we use notation of the form (i, j) for the objects of $\widetilde{\Upsilon}_2(F, *) \times \Upsilon(E)$, where $i = 0, 1, 2$ (objects in $\widetilde{\Upsilon}_2(F, *)$ and $j = 0, 1$ (objects in $\Upsilon(E)$). The product $\widetilde{\Upsilon}_2(F, *) \times \Upsilon(E)$ should be pictured as a prism:

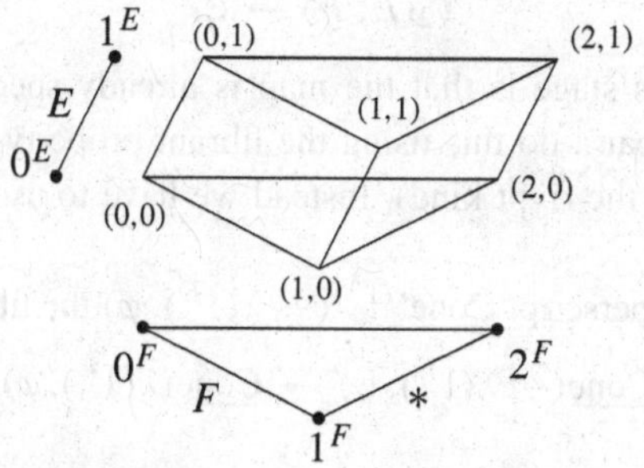

The idea is to divide the prism up into three tetrahedra.

We are already given maps defined over the triangles $(012, 0)$ and $(012, 1)$ (these are $f'(0^E)$ and $f'(1^E)$), as well as over the squares $(02, 01)$ (our given map f) and $(12, 01)$ (the map ϵ). First extend using the fibrant property of $\mathcal{C}$ to a map on the tetrahedron $(0, 0)(1, 0)(2, 0)(2, 1)$:

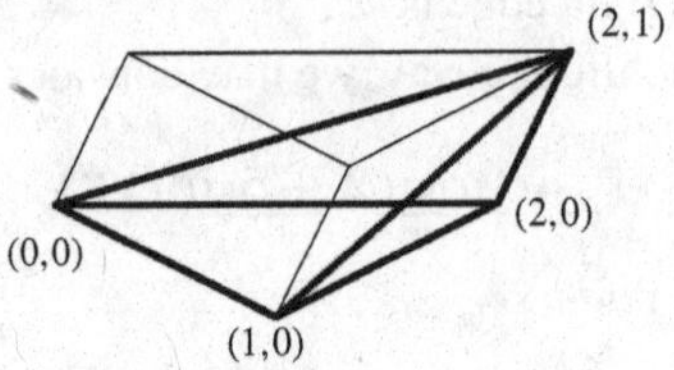

Then extend again using the fibrant property of $\mathcal{C}$ to a map on the tetrahedron $(0, 0)(1, 0)(1, 1)(2, 1)$:

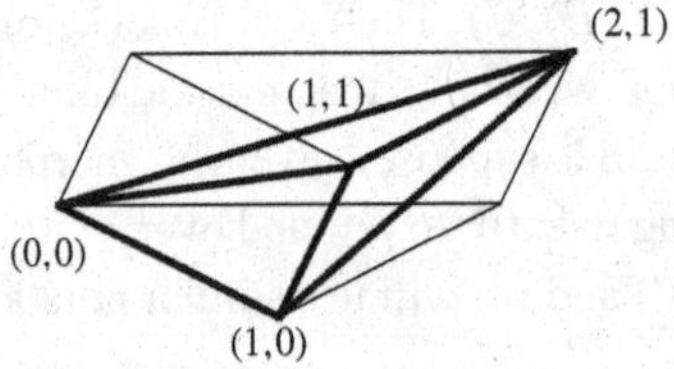

Here note that on the face $(0, 0)(1, 0)(1, 1)$ the map is chosen first as coming from a map $\widetilde{\Upsilon}_2(F, E) \to \mathcal{C}$. Finally, we have to find an extension over the tetrahedron $(0, 0)(0, 1)(1, 1)(2, 1)$:

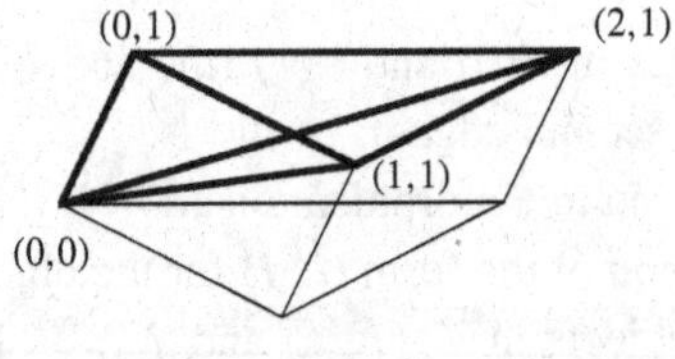

Again we require that the map on the first face $(0, 0)(0, 1)(1, 1)$ come from a map

$$\widetilde{\Upsilon}_2(E, F) \to \mathcal{C}.$$

Our problem at this stage is that the map is already specified on all of the other faces, so we can't do this using the fibrant property of $\mathcal{C}$ (the face that is missing is not of the right kind). Instead we have to use the limit property of $\epsilon(1^E)$.

Denote with a superscript $\underline{\mathrm{Cone}}^{\epsilon(1^E)}(-, \lambda(1^E), \varphi)$ the fiber of the map

$$\underline{\mathrm{Cone}}(-, \lambda(1^E), \varphi) \to \underline{\mathrm{Cone}}(\lambda(1^E), \varphi)$$

over $\epsilon(1^E)$.

The limit condition on $\lambda(1^E)$ means that the morphism

$$\underline{\mathrm{Cone}}^{\epsilon(1^E)}(\mu(0^E), \mu(1^E), \lambda(1^E), \varphi) \to \underline{\mathrm{Cone}}(\mu(0^E), \mu(1^E), \varphi)$$

is a trivial fibration. The morphisms

$$\underline{\mathrm{Cone}}^{\epsilon(1^E)}(\mu(0^E), \lambda(1^E), \varphi) \to \underline{\mathrm{Cone}}(\mu(0^E), \varphi)$$

and

$$\underline{\mathrm{Cone}}^{\epsilon(1^E)}(\mu(1^E), \lambda(1^E), \varphi) \to \underline{\mathrm{Cone}}(\mu(1^E), \varphi)$$

are trivial fibrations too. This implies that the morphism (which is a fibration)

$$\begin{aligned}
&\underline{\mathrm{Cone}}^{\epsilon(1^E)}(\mu(0^E), \mu(1^E), \lambda(1^E), \varphi)\\
&\to \underline{\mathrm{Cone}}^{\epsilon(1^E)}(\mu(0^E), \lambda(1^E), \varphi) \times_{\underline{\mathrm{Cone}}(\mu(0^E),\varphi)}\\
&\underline{\mathrm{Cone}}(\mu(0^E), \mu(1^E), \varphi) \times_{\underline{\mathrm{Cone}}(\mu(1^E),\varphi)}\\
&\underline{\mathrm{Cone}}^{\epsilon(1^E)}(\mu(1^E), \lambda(1^E), \varphi)
\end{aligned}$$

is an equivalence. This exactly implies that the restriction to the shell that we are interested in is an equivalence. The fact that this equivalence is a fibration implies that it is surjective on objects, giving finally the extension that we need.

To prove the limit property we really had to consider a cofibration $F' \to F$ between cofibrant objects, and suppose already given an extension over F'. Following through the same argument as above, one can choose the extension over F in a compatible way. The relevant morphisms involved are trivial cofibrations, requiring a bit more room to write down. □

Corollary 21.10.5 *Suppose $\varphi : \mathcal{A} \to \mathcal{C}$ is a morphism in $\mathcal{PC}(\mathcal{M})$ to a fibrant object $\mathcal{C}$, and suppose that $\mathcal{B} \in \mathcal{PC}(\mathcal{M})$ is cofibrant. Let $\varphi_{\mathcal{B}} : \mathcal{A} \to \mathrm{H{\scriptstyle OM}}(\mathcal{B}, \mathcal{C})$ denote the morphism constant along $\mathcal{B}$. Suppose that φ admits a limit $u : U \to \varphi$. Then $u_{\mathcal{B}} : U_{\mathcal{B}} \to \varphi_{\mathcal{B}}$ (the pullback of u along $\mathcal{B} \to *$) is a limit of $\varphi_{\mathcal{B}}$.*

Proof This is just commutativity for pullbacks for $\mathcal{B} \to *$. □

Corollary 21.10.6 *Suppose $\mathcal{A}$ and $\mathcal{B}$ are cofibrant, and $\mathcal{C}$ fibrant $\mathcal{P}$-categories. Suppose $F : \mathcal{A} \times \mathbb{I} \times \mathcal{B} \to \mathcal{C}$ is a morphism such that the restriction to $\mathcal{A} \times \{\upsilon_0\} \times \mathcal{B}$ is the pullback of a morphism $\varphi : \mathcal{A} \to \mathcal{C}$, and the restriction to $\mathcal{A} \times \{\upsilon_1\} \times \mathcal{B}$ is the pullback of a morphism $\psi : \mathcal{B} \to \mathcal{C}$. Suppose that ψ admits a limit $\epsilon \in \mathrm{Cone}(\psi, \lambda)$. Then there is a natural element $g \in \mathrm{Cocone}(\varphi, \lambda)$ fitting into a diagram*

$$\mathcal{A} \times \widetilde{\Upsilon}_2(*, *) \times \mathcal{B} \to \mathcal{C},$$

restricting on (01) to g, on (12) to ϵ, and on (02) to F.

Proof Think of F as being a map $\mathbb{I} \times \mathcal{B} \to \underline{\mathrm{HOM}}(\mathcal{A}, \mathcal{C})$ which restricts over υ_1 to the constant $\psi_{\mathcal{A}}$, and which restricts over υ_0 to $\varphi_{\mathcal{B}}$. Hence it is an element of $\mathrm{Cone}(\varphi, \psi_{\mathcal{A}})$ which we denote $\tilde{F}$. The constant $\lambda_{\mathcal{A}} \in \mathrm{HOM}(\mathcal{A}, \mathcal{C})$ is a limit of $\psi_{\mathcal{A}}$ by the previous corollary. Therefore there is an element of $\mathrm{Cone}(\varphi, \lambda_{\mathcal{A}}, \psi_{\mathcal{A}})$ restricting over (01) to a map from φ to $\lambda_{\mathcal{A}}$ in $\underline{\mathrm{HOM}}(\mathcal{A}, \mathcal{C})$ (which gives an element of $\mathrm{Cocone}(\varphi, \lambda)$), over (12) to the limit cone ϵ, and over (02) to $\tilde{F}$. This gives the required diagram. □

We can use the result of Theorem 21.10.4 to obtain the variation of the limit depending on the family. Suppose $\mathcal{C}$ is a fibrant $\mathcal{M}$-category in which limits exist, and suppose $\mathcal{A} \in \mathcal{PC}(\mathcal{M})$. Let $\mathcal{B} = \underline{\mathrm{HOM}}(\mathcal{A}, \mathcal{C})$. We have a tautological morphism

$$\zeta : \mathcal{A} \to \underline{\mathrm{HOM}}(\mathcal{B}, \mathcal{C}).$$

By the previous theorem, limits exist in $\underline{\mathrm{HOM}}(\mathcal{B}, \mathcal{C})$. Thus we obtain the limit of ζ which is an element of $\underline{\mathrm{HOM}}(\mathcal{B}, \mathcal{C})$: it is a morphism λ from $\mathcal{B} = \underline{\mathrm{HOM}}(\mathcal{A}, \mathcal{C})$ to $\mathcal{C}$, which is the morphism which to $\varphi \in \underline{\mathrm{HOM}}(\mathcal{A}, \mathcal{C})$ associates $\lambda(\varphi)$ which is the limit of φ. The same remark holds for colimits.

The following statement says that limits over a product can be computed one factor at a time. It would be interesting to extend this to some kind of higher fibered categories.

Theorem 21.10.7 *Suppose $\mathcal{A}$, $\mathcal{B}$ and $\mathcal{C}$ are $\mathcal{M}$-precategories, with $\mathcal{C}$ fibrant. Suppose $F : \mathcal{A} \times \mathcal{B} \to \mathcal{C}$ is a functor. Then letting $\psi : \mathcal{A} \to \underline{\mathrm{HOM}}(\mathcal{B}, \mathcal{C})$ denote the corresponding functor, suppose that ψ admits a limit $\lambda \in \underline{\mathrm{HOM}}(\mathcal{B}, \mathcal{C})$. Suppose now that λ (considered as a morphism $\mathcal{B} \to \mathcal{C}$) admits a limit $\mu \in \mathcal{C}$. Then μ is a limit of $F : \mathcal{A} \times \mathcal{B} \to \mathcal{C}$. In particular if the intermediate limits exist going in the other direction then the composed limits are canonically equivalent. Thus if $\mathcal{C}$ admits limits then limits commute with each other.*

Proof The general proof is left to the reader. In the case $\mathcal{C} = \mathbf{Enr}(\mathcal{P})'$ this will be easy to see from our explicit construction of the limits in the next chapter. □

22
Limits of weak enriched categories

In this chapter, our goal is to prove that the $(n+1)$-category nCAT admits limits and colimits. This corresponds to the second part of my preprint [235]. We can consider the more general situation of categories enriched over a model category.

The theory of homotopy limits and colimits in a model category is of course an old subject. Without being at all exhaustive, some recent references include Dwyer *et al.* [100] and Lurie [190].

Suppose $\mathcal{P}$ is a tractable left proper cartesian model category. Then we obtain a $\mathcal{P}$-category $\mathbf{Enr}(\mathcal{P}) \in \mathcal{PC}(\mathcal{P})$. This is a strictly associative $\mathcal{P}_f$-enriched category. In particular, it is levelwise fibrant considered as a $\Delta_{\mathrm{Ob}(\mathcal{P}_{cf})}$-diagram, so it is fibrant for the projective model structure on $\mathcal{PC}(\mathcal{P})$. However, it is not fibrant for the Reedy structure which is the one we are using. Let

$$\mathbf{Enr}(\mathcal{P}) \stackrel{\nu}{\to} \mathbf{Enr}(\mathcal{P})'$$

denote a fibrant replacement. We show that $\mathbf{Enr}(\mathcal{P})'$ admits limits (and later, colimits) indexed by any other $\mathcal{P}$-category.

If $\mathcal{M}$ is a tractable left proper cartesian model category, we can set $\mathcal{P} := \mathcal{PC}^n(\mathcal{M})$ with the Reedy structure, which is again tractable left proper and cartesian, and define

$$n\mathrm{CAT}(\mathcal{M}) := \mathbf{Enr}(\mathcal{P})$$

as in Section 20.5. It is a strictly $\mathcal{PC}^n(\mathcal{M})$-enriched category with fibrant replacement denoted $n\mathrm{CAT}(\mathcal{M})'$. See Section 22.3 for a discussion of the universe issues involved. The $(n+1)$-category nCAT is $n\mathrm{CAT}(\mathrm{S{\small ET}})$, that is to say it is $\mathbf{Enr}(\mathcal{P})$ for the case when $\mathcal{P} = \mathcal{PC}^n(\mathrm{S{\small ET}})$ is the model category of n-precategories.

The notations CAT and **Enr** are equivalent in that $\mathcal{PC}^0(\mathcal{P}) = \mathcal{P}$ so $\mathbf{Enr}(\mathcal{P}) = 0\mathrm{CAT}(\mathcal{P})$. Everything will be written in terms of the notation $\mathbf{Enr}(\mathcal{P})$.

Before getting to the construction of limits, we consider the more general problem of how to look at functors $\mathcal{A} \to \mathbf{Enr}(\mathcal{P})'$. One approach involves the notion of *cartesian family*; the advantage of this point of view is that it is the most natural one for defining the Yoneda functors, however that requires some further work and our discussion in Sections 22.1 and 22.2 will be incomplete and partly conjectural. In Section 22.3 we discuss the problem of universes inherent in consideration of $\mathbf{Enr}(\mathcal{P})'$. Then, for the discussion of limits, we develop a notion of *quasifibrant precategory* in Section 22.4 and use this property of $\mathbf{Enr}(\mathcal{P})$ as a bridge to $\mathbf{Enr}(\mathcal{P})'$. The construction of limits takes place in Section 22.6. Subsequent sections are devoted to deducing the existence of colimits, from the existence of limits together with some considerations of cardinality and a direct telescope construction for splittings of idempotents.

22.1 Cartesian families

The need to use a fibrant replacement $\mathbf{Enr}(\mathcal{P})'$ indicates that there should be other more natural ways of looking at functors $\mathcal{A} \to \mathbf{Enr}(\mathcal{P})'$. One of these is the notion of *cartesian family*. Suppose $\mathcal{A} = (X, \mathcal{A})$ is a $\mathcal{P}$-precategory, with set of objects $X = \mathrm{Ob}(\mathcal{A})$. The remaining structure consists of a functor

$$\mathcal{A} : \Delta_X^o \to \mathcal{P},$$

subject to the unitality condition $\mathcal{A}(x_0) = *$ as well as the Segal conditions. A *$\mathcal{P}$-family over $\mathcal{A}$* is a morphism of Δ_X^o-diagrams in $\mathcal{P}$,

$$\mathcal{B} \to \mathcal{A} \quad \text{in } \mathrm{FUNC}\left(\Delta_X^o, \mathcal{P}\right),$$

which we will denote by $\mathcal{B}/\mathcal{A}$. A morphism between families is a morphism of diagrams compatible with the maps to $\mathcal{A}$.

A family $\mathcal{B}/\mathcal{A}$ is *cartesian* if, for any sequence of objects $(x_0, \ldots, x_m) \in \Delta_X$, the map

$$\mathcal{B}(x_0, \ldots, x_m) \to \mathcal{B}(x_0) \times \mathcal{A}(x_0, \ldots, x_m)$$

is a weak equivalence in $\mathcal{P}$. This condition, which is a sort of relative Segal condition for $\mathcal{B}/\mathcal{A}$, will usually be imposed only when $\mathcal{A}$ itself satisfies the Segal conditions. Notice most importantly that $\mathcal{B}$ is not supposed to be unital, indeed the $\mathcal{B}(x_0)$ are the main starting points for the structure.

A family $\mathcal{B}/\mathcal{A}$ is said to be *fibrant* if the morphism $\mathcal{B} \to \mathcal{A}$ is a Reedy fibration in $\text{FUNC}(\Delta_X, \mathcal{P})$.

If $E \in \mathcal{P}$ and $\mathcal{B}/\mathcal{A}$ is a $\mathcal{P}$-family, then we obtain a new $\mathcal{P}$-family denoted $E \otimes \mathcal{B}/\mathcal{A}$ defined by

$$(E \otimes \mathcal{B})(x_0, \ldots, x_m) := E \times (\mathcal{B}(x_0, \ldots, x_m)).$$

If $\mathcal{B}/\mathcal{A}$ is cartesian then so is $E \otimes \mathcal{B}/\mathcal{A}$; and if E is fibrant and $\mathcal{B}/\mathcal{A}$ is fibrant then $E \otimes \mathcal{B}/\mathcal{A}$ is fibrant too. Clearly $* \otimes \mathcal{B}/\mathcal{A} = \mathcal{B}/\mathcal{A}$.

A $\mathcal{P}$-family over $\mathcal{A}$ is *cocartesian* if, for any sequence of objects $(x_0, \ldots, x_m) \in \Delta_X$, the map

$$\mathcal{B}(x_0, \ldots, x_m) \to \mathcal{A}(x_0, \ldots, x_m) \times \mathcal{B}(x_m)$$

is a weak equivalence. If $\mathcal{B}/\mathcal{A}$ is a cartesian (resp. cocartesian) family then putting $\mathcal{B}^o(x_0, \ldots, x_m) := \mathcal{B}(x_m, \ldots, x_0)$ results in a cocartesian (resp. cartesian) family over $\mathcal{A}^o$.

22.1.1 The category of cartesian families

We construct a strictly $\mathcal{P}$-enriched category denoted $\mathbf{Fam}(\mathcal{A})$ whose objects are the fibrant cartesian families $\mathcal{B}/\mathcal{A}$. Given two fibrant cartesian families $\mathcal{B}/\mathcal{A}$ and $\mathcal{B}'/\mathcal{A}$, the internal mapping object $\underline{\text{HOM}}(\mathcal{B}/\mathcal{A}, \mathcal{B}'/\mathcal{A})$ is constructed as follows.

Lemma 22.1.1 *Given two $\mathcal{P}$-families $\mathcal{B}/\mathcal{A}$ and $\mathcal{B}'/\mathcal{A}$ there exists an internal mapping object $\underline{\text{HOM}}(\mathcal{B}/\mathcal{A}, \mathcal{B}'/\mathcal{A}) \in \mathcal{P}$ satisfying the universal property that for any $E \in \mathcal{P}$, a map*

$$E \to \underline{\text{HOM}}(\mathcal{B}/\mathcal{A}, \mathcal{B}'/\mathcal{A})$$

is the same thing as a map $E \otimes \mathcal{B}/\mathcal{A} \to \mathcal{B}'/\mathcal{A}$. If $\mathcal{P}'$ is a fibrant cartesian family then $\underline{\text{HOM}}(\mathcal{B}/\mathcal{A}, \mathcal{B}'/\mathcal{A})$ is fibrant in $\mathcal{P}$.

Proof To prove that this universal property defines $\underline{\text{HOM}}(\mathcal{B}/\mathcal{A}, \mathcal{B}'/\mathcal{A})$, note that the functor $\mathcal{P}^o \to \text{SET}$ which to E associates the set of morphisms $E \otimes \mathcal{B}/\mathcal{A} \to \mathcal{B}'/\mathcal{A}$ sends colimits of E to limits of sets.

The functor satisfies the lifting property along trivial cofibrations $E \to E'$ so the representing object also satisfies this lifting property, in other words it is fibrant. □

Corollary 22.1.2 *If $\mathcal{B}/\mathcal{A}$, $\mathcal{B}'/\mathcal{A}$ and $\mathcal{B}''/\mathcal{A}$ are three fibrant cartesian families, then we have a natural composition*

$$\underline{\text{HOM}}(\mathcal{B}'/\mathcal{A}, \mathcal{B}''/\mathcal{A}) \times \underline{\text{HOM}}(\mathcal{B}/\mathcal{A}, \mathcal{B}'/\mathcal{A}) \to \underline{\text{HOM}}(\mathcal{B}/\mathcal{A}, \mathcal{B}''/\mathcal{A}),$$

and this is strictly associative. Using this composition defines a strictly $\mathcal{P}_{\mathrm{f}}$-enriched category **Fam**$(\mathcal{A})$ *whose objects are the fibrant cartesian families $\mathcal{B}/\mathcal{A}$, with*

$$\mathbf{Fam}(\mathcal{A})(\mathcal{B}/\mathcal{A}, \mathcal{B}'/\mathcal{A}) := \underline{\mathrm{HOM}}(\mathcal{B}/\mathcal{A}, \mathcal{B}'/\mathcal{A}).$$

Proof Dual to the fact that $E_1 \otimes (E_2 \otimes \mathcal{B}/\mathcal{A}) = (E_1 \times E_2) \otimes \mathcal{B}/\mathcal{A}$. □

Putting $\mathbf{Fam}^o(\mathcal{A}) := \mathbf{Fam}(\mathcal{A}^o)$ gives the category of *cocartesian families* whose elements will be written as cocartesian families $\mathcal{B}/\mathcal{A}$ corresponding to cartesian families $\mathcal{B}^o/\mathcal{A}^o$.

22.1.2 Homotopy invariance properties

The reader may elaborate the homotopy invariance properties of the notion of cartesian family: if $\mathcal{A}$ is equivalent to $\mathcal{A}'$ then $\mathbf{Fam}(\mathcal{A})$ is naturally equivalent to $\mathbf{Fam}(\mathcal{A}')$, and similarly there are appropriate extension properties.

Similarly, given a cartesian family, it can be replaced by an equivalent fibrant cartesian family.

If $f : \mathcal{A}' \to \mathcal{A}$ and $\mathcal{B}/\mathcal{A}$ is a fibrant cartesian family, then $f^*(\mathcal{B})/\mathcal{A}'$ is a fibrant cartesian family defined by

$$f^*(\mathcal{B})\left(x_0', \ldots, x_p'\right) := \mathcal{B}\left(f\left(x_0'\right), \ldots, f\left(x_p'\right)\right) \times_{\mathcal{A}(f(x_0'),\ldots)} \mathcal{A}'\left(x_0', \ldots, x_p'\right).$$

22.1.3 Relation with morphisms to $\mathbf{Enr}(\mathcal{P})'$

Some work is needed in order to relate $\mathbf{Fam}(\mathcal{A})$ to $\underline{\mathrm{HOM}}(\mathcal{A}, \mathbf{Enr}(\mathcal{P})')$. There is a tautological cartesian family $\mathcal{U}/\mathbf{Enr}(\mathcal{P})$ defined by

$$\mathcal{U}(\mathcal{A}_0, \ldots, \mathcal{A}_p) := \mathcal{A}_0 \times \underline{\mathrm{HOM}}(\mathcal{A}_0, \mathcal{A}_1) \times \cdots \times \underline{\mathrm{HOM}}(\mathcal{A}_{p-1}, \mathcal{A}_p),$$

with transition morphisms defined using the composition operations of $\underline{\mathrm{HOM}}$. By the homotopy extension property there is a cartesian family $\mathcal{U}'/\mathbf{Enr}(\mathcal{P})'$ together with an equivalence

$$\mathcal{U}/\mathbf{Enr}(\mathcal{P}) \xrightarrow{\sim} \nu^*(\mathcal{U}')/\mathbf{Enr}(\mathcal{P})$$

of cartesian families over $\mathbf{Enr}(\mathcal{P})$.

There is a tautological cartesian family

$$\mathcal{T}/\mathbf{Fam}(\mathcal{A}) \times \mathcal{A},$$

defined by

$$\begin{aligned}&\mathcal{T}((\mathcal{B}_0/\mathcal{A}, x_0), \ldots, (\mathcal{B}_p/\mathcal{A}, x_p))\\ &:= \mathcal{B}(x_0, \ldots, x_p) \times \underline{\mathrm{HOM}}(\mathcal{B}_0/\mathcal{A}, \mathcal{B}_1/\mathcal{A}) \times \cdots \times \underline{\mathrm{HOM}}(\mathcal{B}_{p-1}/\mathcal{A}, \mathcal{B}_p/\mathcal{A}).\end{aligned}$$

The transition morphisms are obtained by using the composition operations of the internal $\underline{\mathrm{HOM}}$ factors.

The first step is to give a weak statement saying that a cartesian family $\mathcal{B}/\mathcal{A}$ corresponds to a map $\mathcal{A} \to \mathbf{Enr}(\mathcal{P})'$. The proof is left to the reader.

Lemma 22.1.3 *If $\mathcal{B}/\mathcal{A}$ is a cartesian family, there is a morphism $f : \mathcal{A} \to \mathbf{Enr}(\mathcal{P})'$ together with an equivalence of cartesian families*

$$\mathcal{B}/\mathcal{A} \xrightarrow{\sim} f^*(\mathcal{U}')/\mathcal{A}.$$

Applying this lemma to the tautological cartesian family $\mathcal{T}/\mathbf{Fam}(\mathcal{A}) \times \mathcal{A}$ gives a map $t : \mathbf{Fam}(\mathcal{A}) \times \mathcal{A} \to \mathbf{Enr}(\mathcal{P})'$ together with an equivalence $\mathcal{T} \xrightarrow{\sim} t^*(\mathcal{U}')$.

The map t now corresponds to a map

$$\mathbf{t} : \mathbf{Fam}(\mathcal{A}) \to \underline{\mathrm{HOM}}(\mathcal{A}, \mathbf{Enr}(\mathcal{P})').$$

The following statement is the strong version of the relationship between cartesian families and maps to $\mathbf{Enr}(\mathcal{P})'$:

Theorem 22.1.4 *For any $\mathcal{P}$-category $\mathcal{A}$, the above map $\mathbf{t}$ is a weak equivalence between $\mathbf{Fam}(\mathcal{A})$ and $\underline{\mathrm{HOM}}(\mathcal{A}, \mathbf{Enr}(\mathcal{P})')$, in the model category $\mathcal{PC}(\mathcal{P})$ of $\mathcal{P}$-precategories.*

22.2 The Yoneda embeddings

In this section we consider the Yoneda maps which may be defined using the notion of cartesian family. Yoneda maps for higher and enriched categories have been considered by many authors (for a recent example, Elgueta [109] evokes the Yoneda principle in the 2-categorical context). In our situation, one route would be to apply the strictification theory of Bergner and Lurie to first strictify a $\mathcal{P}$-category $\mathcal{A}$ (see Theorem 19.5.1), then use the strict Yoneda maps. In the present section we prefer to indicate, however without too many proofs, how one can treat the question in the weakly enriched context. For complete Segal spaces, this has been treated by Barwick [24]. For quasicategories, it is one of the main subjects of Nichols-Barrer's thesis [209].

Suppose $\mathcal{A} \in \mathcal{PC}(\mathcal{P})$. For $x \in \mathcal{A}$ the goal is to try to view $y \mapsto \mathcal{A}(x, y)$ as a "functor" from $\mathcal{A}$ to $\mathbf{Enr}(\mathcal{P})'$, the $\mathcal{P}$-enriched analogue of the usual Yoneda functors. Similarly in the variable x and in fact it is convenient to consider both variables at once. The notion of cartesian family is designed to help in giving a canonical formulation.

22.2.1 The arrow family

If $\mathcal{A}$ is a $\mathcal{P}$-category, define the *arrow family*

$$\mathbf{Arr}(\mathcal{A}) \in \mathbf{Fam}(\mathcal{A}^o \times \mathcal{A})$$

by

$$\mathbf{Arr}(\mathcal{A})((x_0, y_0), \ldots, (x_n, y_n)) := \mathcal{A}(x_n, x_{n-1}, \ldots, x_0, y_0, \ldots, y_n).$$

This maps to

$$(\mathcal{A}^o \times \mathcal{A})((x_0, y_0), \ldots, (x_n, y_n)) = \mathcal{A}(x_n, x_{n-1}, \ldots, x_0) \times \mathcal{A}(y_0, \ldots, y_n)$$

by the structural maps for $\mathcal{A}$, and the Segal conditions for $\mathcal{A}$ imply that the map

$$\mathcal{A}(x_n, x_{n-1}, \ldots, x_0, y_0, \ldots, y_n) \\ \rightarrow \mathcal{A}(x_0, y_0) \times \mathcal{A}(x_n, x_{n-1}, \ldots, x_0) \times \mathcal{A}(y_0, \ldots, y_n)$$

is a weak equivalence, giving the cartesian property for $\mathbf{Arr}(\mathcal{A})$.

By the discussion of Section 22.1.3, $\mathbf{Arr}(\mathcal{A})$ corresponds to a map

$$\mathcal{A}^o \times \mathcal{A} \rightarrow \mathbf{Enr}(\mathcal{P})'.$$

In the context of weak enrichment this map doesn't have a canonical incarnation, as can be seen already by the choice of fibrant replacement $\mathbf{Enr}(\mathcal{P})'$; this is the reason for introducing the notion of cartesian family.

Restricting to one or the other variables, gives the *Yoneda functors*

$$\mathbf{Yon}^{\mathcal{A}} : \mathcal{A} \rightarrow \underline{\mathrm{HOM}}(\mathcal{A}^0, \mathbf{Enr}(\mathcal{P})'),$$

which can be written heuristically $\mathbf{Yon}^{\mathcal{A}}(x) : y \mapsto \mathcal{A}(y, x)$, and

$$\mathbf{Yon}_{\mathcal{A}} : \mathcal{A}^o \rightarrow \underline{\mathrm{HOM}}(\mathcal{A}, \mathbf{Enr}(\mathcal{P})'),$$

written heuristically as $\mathbf{Yon}_{\mathcal{A}}(x) : y \mapsto \mathcal{A}(x, y)$.

The heuristic expressions don't actually define the morphisms; however, they are correct up to inner equivalence in $\mathbf{Enr}(\mathcal{P})'$, which is the same as homotopy equivalence in $\mathcal{P}$:

$$\mathbf{Yon}^{\mathcal{A}}(x)(y) \sim \mathcal{A}(y, x),$$

$$\mathbf{Yon}_{\mathcal{A}}(x)(y) \sim \mathcal{A}(x, y).$$

By Theorem 22.1.4, $\underline{\mathrm{HOM}}(\mathcal{A}^0, \mathbf{Enr}(\mathcal{P})')$ is naturally equivalent to $\mathbf{Fam}(\mathcal{A}^0)$. Let $\mathbf{Fam}(\mathcal{A}^0) \to \mathbf{Fam}(\mathcal{A}^0)'$ denote a weak equivalence towards a fibrant replacement. Then if $\mathcal{A}$ is Reedy cofibrant, the Yoneda map may be replaced by an equivalent version

$$\mathbf{Yon}^{\mathcal{A}} : \mathcal{A} \to \mathbf{Fam}(\mathcal{A}^o)' = \mathbf{Fam}^o(\mathcal{A})',$$

and if $\mathcal{A}$ is projectively cofibrant we can even assume that this goes to $\mathbf{Fam}^o(\mathcal{A})$. Similarly for the other one:

$$\mathbf{Yon}_{\mathcal{A}} : \mathcal{A}^o \to \mathbf{Fam}(\mathcal{A})',$$

and again if $\mathcal{A}$ is projectively cofibrant it can go to $\mathbf{Fam}(\mathcal{A})$.

Conjecture 22.2.1 *The Yoneda maps are fully faithful.*

Another possible route towards understanding the Yoneda embeddings, would be to strictify and replace $\mathcal{A}$ by a strict $\mathcal{P}$-enriched category, using the Bergner–Lurie theorem 19.5.1. Then the arrow family can be defined more easily.

22.2.2 Representable cocartesian families

The essential images of the Yoneda maps can be described explicitly by a notion of *representable (co)cartesian family*. Suppose $\mathcal{A}$ is a $\mathcal{P}$-category. Given a triple $(\mathcal{B}/\mathcal{A}, x, \eta)$, where $\mathcal{B}/\mathcal{A}$ is a cocartesian family, $x \in \mathrm{Ob}(\mathcal{A})$ and $\eta : P \to \mathcal{B}(x)$ is a fibration in $\mathcal{P}$ from a contractible object P, we can look at the maps

$$\mathcal{B}(z, x) \times_{\mathcal{B}(x)} P \to \mathcal{B}(z),$$

for any $z \in \mathrm{Ob}(\mathcal{A})$. Say that $(\mathcal{B}/\mathcal{A}, x, \eta)$ is a *representable cocartesian family* if the above maps are equivalences for any $z \in \mathrm{Ob}(\mathcal{A})$. Notice that the cocartesian condition provides an equivalence

$$\mathcal{B}(z, x) \stackrel{\sim}{\to} \mathcal{A}(z, x) \times \mathcal{B}(x),$$

so

$$\mathcal{B}(z, x) \times_{\mathcal{B}(x)} P \stackrel{\sim}{\to} \mathcal{A}(z, x) \times P \sim \mathcal{A}(z, x)$$

is an equivalence. In other words, for a representable cocartesian family we have equivalences

$$\mathcal{A}(z, x) \stackrel{\sim}{\leftarrow} \mathcal{B}(z, x) \times_{\mathcal{B}(x)} P \stackrel{\sim}{\to} \mathcal{B}(z),$$

making more precise the heuristic statement given previously for the Yoneda functor.

Two different representable cocartesian families at the same object $x \in \mathrm{Ob}(\mathcal{A})$ are equivalent, and the families in the image of the Yoneda functor are representable.

Lemma 22.2.2 *The essential image of the Yoneda functor*

$$\mathbf{Yon}^{\mathcal{A}} : \mathcal{A} \to \mathbf{Fam}^{o}(\mathcal{A})',$$

consists of the cocartesian families which can be completed to a representable cocartesian triple by appropriate choice of (x, η).

We won't go into the proof of that here, the reader may consider it to be an exercise or else a part of Conjecture 22.2.1. A similar conjecture is made by Nichols-Barrer [209].

Suppose $(\mathcal{B}/\mathcal{A}, x, \eta)$ and $(\mathcal{B}'/\mathcal{A}, x', \eta')$ are two representable cocartesian families. The sequence of morphisms

$$\begin{array}{ccc} P \times \underline{\mathrm{HOM}}(\mathcal{B}/\mathcal{A}, \mathcal{B}'/\mathcal{A}) & & \mathcal{A}(x, x') \\ \downarrow & & \uparrow \\ \mathcal{B}(x) \times \underline{\mathrm{HOM}}(\mathcal{B}/\mathcal{A}, \mathcal{B}'/\mathcal{A}) & \longrightarrow \mathcal{B}'(x) \xleftarrow{\sim} & \mathcal{B}'(x, x') \times_{\mathcal{B}'(x')} P' \end{array}$$

provides a morphism

$$\underline{\mathrm{HOM}}(\mathcal{B}/\mathcal{A}, \mathcal{B}'/\mathcal{A}) \to \mathcal{A}(x, x') \tag{22.2.1}$$

in $\mathrm{ho}(\mathcal{P})$. Once we know Lemma 22.2.2, Conjecture 22.2.1 is equivalent to:

Conjecture 22.2.3 *If* $(\mathcal{B}/\mathcal{A}, x, \eta)$ *and* $(\mathcal{B}'/\mathcal{A}, x', \eta')$ *are two representable cocartesian families, the map* (22.2.1) *is a weak equivalence, i.e. an isomorphism in the homotopy category.*

We leave it to the reader to formulate the corresponding statements for representable cartesian families and the Yoneda functor $\mathbf{Yon}_{\mathcal{A}}$.

22.2.3 The interior of *nCAT*

Consider first the general case of a tractable left proper cartesian model category $\mathcal{P}$. Let $\mathbf{R}^* : \mathcal{P} \to \mathcal{K}$ be the functor obtained by choosing a cofibrant cosimplicial resolution of $*$ (see Section 18.5). This induces a functor $\mathcal{PC}(\mathbf{R}^*) : \mathcal{PC}(\mathcal{P}) \to \mathcal{PC}(\mathcal{K})$.

Lemma 22.2.4 *The Segal precategory $\mathcal{PC}(\mathbf{R}^*)(\mathbf{Enr}(\mathcal{P})) \in \mathcal{PC}(\mathcal{K})$ is a Segal category, weak equivalent to the Dwyer–Kan simplicial localization $L_{DK}(\mathcal{P})$ inverting weak equivalences of $\mathcal{P}$.*

Proof The morphism objects of $\mathbf{Enr}(\mathcal{P})$ are fibrant, so applying the right Quillen functor $\mathbf{R}^*$ preserves homotopy types and it is compatible with products. Thus $\mathcal{PC}(\mathbf{R}^*)(\mathbf{Enr}(\mathcal{P}))$ is a simplicial category, in particular it is a Segal category. The set of objects is the (large) set of fibrant and cofibrant objects of $\mathcal{P}$. These provide representatives for any weak equivalence class in $\mathcal{P}$, hence for any equivalence class in $L_{DK}(\mathcal{P})$.

Between any two objects $X, Y \in \mathcal{P}$, the morphism space is given by $\mathbf{R}^*(\underline{\text{HOM}}(X, Y))$. This is the simplicial set which to $[k] \in \Delta$ associates $\text{HOM}(X \times \mathbf{R}([k]), Y)$. However, by the cartesian property of $\mathcal{P}$ the cosimplicial object

$$k \mapsto X \times \mathbf{R}([k])$$

is a cofibrant resolution of X in the sense of Hirschhorn [144], so it can serve to compute the simplicial function complex from X to Y. Thus, the simplicial set $\mathbf{R}^*(\underline{\text{HOM}}(X, Y))$ is one possible choice of function complex from X to Y, as constructed by Hirschhorn [144], and this is weak equivalent to the simplicial set of maps from X to Y in $L_{DK}(\mathcal{P})$. We refer to *loc. cit.* [144] for the compatibility of this construction with composition, which shows that it provides the desired equivalence. □

Lemma 22.2.5 *In the case $\mathcal{P} = \mathcal{PC}^n(\text{SET})$, if $\mathcal{A}$ is a fibrant n-nerve then $\mathbf{R}^*(\mathcal{A})$ is equivalent to the realization of the ${}^{>}0$-groupic interior $|\mathbf{Int}^0(\mathcal{A})|$. Hence, if $\mathcal{B} \in \mathcal{PC}(\mathcal{P})$ is fibrant at least in the projective model structure, then $\mathcal{PC}(\mathbf{R}^*)(\mathcal{B})$ is a Segal 1-category equivalent to the ${}^{>}1$-groupic interior of $\mathcal{B}$.*

Proof For any $[k] \in \Delta$, $\mathbf{R}([k])$ is contractible, so $\mathbf{Cat}(\mathbf{R}([k]))$ is an n-groupoid, which is to say ${}^{>}0$-groupic. Therefore, maps from $\mathbf{R}([k])$ to a fibrant $\mathcal{A}$ must go into the ${}^{>}0$-groupic interior of $\mathcal{A}$. When applied to fibrant n-groupoids, the functor $\mathbf{V}^*$ defined above Lemma 20.4.4 gives a simplicial set equivalent to the result of $\mathbf{R}^*$, as may be seen from the theory of localization using the fact that $|\mathbf{V}([k])|$ is contractible. Hence

$$\mathbf{R}^*(\mathcal{A}) = \mathbf{R}^*(\mathbf{Int}^0(\mathcal{A})) \sim \mathbf{V}^*(\mathbf{Int}^0(\mathcal{A})) = |\mathbf{Int}^0(\mathcal{A})|.$$

Applying $\mathcal{PC}$ to this statement, we get that $\mathcal{PC}(\mathbf{R}^*)(\mathcal{B}) \sim \mathbf{Int}^1(\mathcal{B})$. This last equivalence should be interpreted by considering $\mathbf{Int}^1(\mathcal{B})$, which is a ${}^{>}1$-groupic n-category, as a Segal 1-category with $(n-1)$-truncated morphism spaces. □

Corollary 22.2.6 *There is a natural equivalence between the Dwyer–Kan localization and the interior of* nCAT

$$L_{DK}(\mathcal{PC}^n(\text{SET})) \sim \mathbf{Int}^1(n\text{CAT}).$$

Proof Put together the two previous lemmas, applied to $\mathcal{P} = \mathcal{PC}^n(\text{SET})$. □

22.3 Universe considerations

We've been mostly ignoring universe questions up until now. Fix a nested pair of universes $\mathbb{U} \in \mathbb{V}$ and suppose that $\mathcal{P}$ is a tractable left proper cartesian model category which is $(\mathbb{U}, \mathbb{V})$-big. This means that the morphism sets are $\mathbb{U}$-small, but the set of objects $\text{Ob}(\mathcal{P})$ is only $\mathbb{V}$-small. And $\mathcal{P}$ is required to be closed under limits and colimits indexed by $\mathbb{U}$-small categories.

The model category $\mathcal{PC}(\mathcal{P})$ constructed in Chapter 19 is, to be more precise, the category of $\mathcal{P}$-precategories $(X, \mathcal{A})$ where $X \in \mathbb{U}$ is a small set, and $\mathcal{A} : \Delta_X^o \to \mathcal{P}$ is a diagram. The resulting category $\mathcal{PC}(\mathcal{P})$ is again $(\mathbb{U}, \mathbb{V})$-big, and is closed under $\mathbb{U}$-small limits and colimits.

In the previous sections of the present chapter we have been looking at $\mathbf{Enr}(\mathcal{P})$ and its fibrant replacement. However, these $\mathcal{P}$-precategories have sets of objects which are $\mathbb{V}$-small but not $\mathbb{U}$-small, in particular $\mathbf{Enr}(\mathcal{P})$ is not actually an object of $\mathcal{PC}(\mathcal{P})$. As such, it doesn't make sense to take its fibrant replacement without extending the theory to cover big $\mathcal{P}$-precategories. Getting such an extension in general looks to be a nontrivial question, which is not treated here. This is not a problem, since the discussion up until now has been essentially heuristic as regards $\mathbf{Enr}(\mathcal{P})'$.

On the other hand, the next sections are directed towards a more precise construction of limits and, later, colimits in $\mathbf{Enr}(\mathcal{P})'$. It will therefore be convenient to pass to a full $\mathcal{P}$-subcategory which has a $\mathbb{U}$-small set of objects. If κ is a regular cardinal, let $\mathbf{Enr}(\mathcal{P})_\kappa$ denote the subcategory of cofibrant and fibrant κ-presentable objects of $\mathcal{P}$ which are equal to chosen representatives of their isomorphism classes. Now $\text{Ob}(\mathbf{Enr}(\mathcal{P})_\kappa)$ is $\mathbb{U}$-small so it makes sense to write

$$\mathbf{Enr}(\mathcal{P})_\kappa \in \mathcal{PC}(\mathcal{P}),$$

and to let

$$\mathbf{Enr}(\mathcal{P})_\kappa \to \mathbf{Enr}(\mathcal{P})'_\kappa$$

be a fibrant replacement in the Reedy model structure of $\mathcal{PC}(\mathcal{P})$.

The family of $\mathbf{Enr}(\mathcal{P})'_\kappa$ is an ind-object of $\mathcal{PC}(\mathcal{P})$. Let $\mathbf{Reg}(\mathbb{U})$ denote the well-ordered $\mathbb{V}$-small set of regular cardinals in $\mathbb{U}$. If $\kappa \leq \eta$ in $\mathbf{Reg}(\mathbb{U})$ then

$\mathbf{Enr}(\mathcal{P})_\kappa$ is a full strictly $\mathcal{P}$-enriched subcategory of $\mathbf{Enr}(\mathcal{P})_\eta$. The functoriality of fibrant replacement gives a transition morphism

$$\mathbf{i}_{\kappa,\eta} : \mathbf{Enr}(\mathcal{P})'_\kappa \to \mathbf{Enr}(\mathcal{P})'_\eta,$$

and these are compatible in case of three $\kappa \leq \eta \leq \mu$. We get a functor

$$\mathbf{Enr}(\mathcal{P})'_{\bullet} : \mathbf{Reg}(\mathbb{U}) \to \mathcal{PC}(\mathcal{P}),$$

which can be viewed as an ind-object having some good properties. For example, if $\kappa \leq \eta$ then the transition map $\mathbf{i}_{\kappa,\eta}$ is an inclusion on sets of objects:

$$\mathrm{Ob}(\mathbf{i}_{\kappa,\eta}) : \mathrm{Ob}\left(\mathbf{Enr}(\mathcal{P})'_\kappa\right) \hookrightarrow \mathrm{Ob}\left(\mathbf{Enr}(\mathcal{P})'_\eta\right).$$

Furthermore, the map

$$\mathbf{Enr}(\mathcal{P})'_\kappa \to \mathrm{Ob}(\mathbf{i}_{\kappa,\eta})^*\left(\mathbf{Enr}(\mathcal{P})'_\eta\right) \tag{22.3.1}$$

is a weak equivalence (hence a levelwise weak equivalence). By a careful choice of fibrant replacement, this map (22.3.1) may be assumed to be an isomorphism as shown in the following proposition. The proof uses the characterization of Reedy fibrant object of Proposition 19.4.1, due to Bergner [42] in the case of Segal categories.

Proposition 22.3.1 *The fibrant replacements* $\mathbf{Enr}(\mathcal{P})'_\kappa$ *may be chosen such that for any* $\kappa \leq \eta$*,* (22.3.1) *are isomorphisms; in other words*

$$\mathbf{Enr}(\mathcal{P})'_\kappa \hookrightarrow \mathbf{Enr}(\mathcal{P})'_\eta$$

is the full sub-$\mathcal{P}$*-category with* $\mathrm{Ob}\left(\mathbf{Enr}(\mathcal{P})'_\kappa\right) \subset \mathrm{Ob}\left(\mathbf{Enr}(\mathcal{P})'_\eta\right)$ *as subset of objects.*

Proof Recall that a model category, in Quillen's original definition, was only supposed to be closed under finite limits and colimits. In this sense, $\mathcal{P}$ may be considered as a $\mathbb{V}$-small model category. Now $\mathcal{E} := \mathbf{Enr}(\mathcal{P})$ is a strictly $\mathcal{P}$-enriched category whose set of objects $Z := \mathrm{Ob}(\mathcal{P}_{\mathrm{cf}})$ is $\mathbb{V}$-small. Thus $\mathcal{E}$ may be viewed as an object of $\mathcal{PC}(Z, \mathcal{P})$ or equivalently a diagram $\Delta^o_Z \to \mathcal{P}$. It satisfies the Segal condition, even strictly.

For any $\kappa \in \mathbf{Reg}(\mathbb{U})$ let Z_κ denote the set of representatives of isomorphism classes of κ-presentable objects of $\mathcal{P}$, and let $\mathcal{E}_\kappa$ denote the restricted Z_κ-precategory. By Proposition 19.4.1, in order to obtain a fibrant replacement for $\mathcal{E}_\kappa$ it suffices to find a morphism $\mathcal{E}_\kappa \to \mathcal{E}'_\kappa$ which is a levelwise weak equivalence, such that $\mathcal{E}'_\kappa$ is a Reedy fibrant diagram over $\Delta^o_{Z_\kappa}$.

The Reedy fibrant replacements can be constructed all at once. Indeed, the procedure for constructing a Reedy fibrant replacement doesn't involve the small object argument, and can be done for diagrams in any model category in Quillen's original sense. It involves choosing successively fibrant replacements for the matching maps. Hence, we may work in the universe $\mathbb{V}$ and choose a Reedy fibrant replacement

$$\mathcal{E} \to \mathcal{E}' \text{ in } \mathrm{FUNC}_{\mathrm{Reedy}}\left(\Delta^o_Z/Z, \mathscr{P}\right).$$

The Reedy condition is defined by looking at sequences of objects, so the restriction of a Reedy fibrant replacement in $\mathscr{P}\mathcal{C}(Z, \mathscr{P})$ to any $\mathscr{P}\mathcal{C}(Z_\kappa, \mathscr{P})$ is again a Reedy fibrant replacement. So we can let $\mathcal{E}'_\kappa$ be the restriction of $\mathcal{E}'$ to the set Z_κ. This gives our collection of fibrant replacements, such that by construction the restriction maps (22.3.1) are isomorphisms. □

In the following discussion, we start by using the above notation $\mathbf{Enr}(\mathscr{P})'_\kappa$ with a regular cardinal κ fixed. When we get to the constructions of limits and colimits, the choice of κ will depend on the size of the indexing category, so the philosophy of the ind-object $\mathbf{Enr}(\mathscr{P})'_\bullet$ guides the statements of the theorems on existence of limits and colimits. The strong property of the previous proposition will be used in Lemma 22.6.2 to facilitate the passage between different regular cardinals.

Alternatively, one could use Proposition 22.3.1 to define $\mathbf{Enr}(\mathscr{P})'$ as the union of its pieces $\mathbf{Enr}(\mathscr{P})'_\kappa$, as a diagram $\Delta^o_Z \to \mathscr{P}$ in the universe $\mathbb{V}$. As pointed out above, it isn't easy to see this specific construction as fitting into a general theory.

22.4 Diagrams in quasifibrant precategories

For our construction of limits of $\mathscr{P}$-categories, we now describe a way of understanding diagrams in $\mathbf{Enr}(\mathscr{P})'_\kappa$, different from the method of cartesian families. Recall that, before taking the fibrant replacement, $\mathbf{Enr}(\mathscr{P})_\kappa$ was already projectively fibrant. We look for an interpretation of the statement that $\mathbf{Enr}(\mathscr{P})'_\kappa$ is a fibrant replacement for $\mathbf{Enr}(\mathscr{P})_\kappa$ in the injective or Reedy structure, without refering to the theory of cartesian families. The following property of $\mathbf{Enr}(\mathscr{P})_\kappa$ shows that it is close to being fibrant in a useful way:

Definition 22.4.1 We say that a $\mathscr{P}$-category $\mathcal{C}$ (that is, a $\mathscr{P}$-precategory satisfying the Segal conditions) is *quasifibrant* if for any sequence of objects $x_0, \ldots, x_p$ the morphism

$$\mathcal{C}(x_0, \ldots, x_p) \to \mathcal{C}(x_0, \ldots, x_{p-1}) \times_{\mathcal{C}(x_1,\ldots,x_{p-1})} \mathcal{C}(x_1, \ldots, x_p)$$

is a fibration of objects of $\mathscr{P}$.

Note inductively that the morphisms involved in the fiber product here are themselves fibrations, and we get that the projections

$$\mathcal{C}(x_0, \ldots, x_p) \to \mathcal{C}(x_0, \ldots, x_{p-1})$$

and

$$\mathcal{C}(x_0, \ldots, x_p) \to \mathcal{C}(x_1, \ldots, x_p)$$

are fibrations.

Remark 22.4.2 The Segal conditions for $\mathcal{C}$ imply that the morphism in the definition of quasifibrant is an equivalence whenever $p \geq 2$. Thus, if $\mathcal{C}$ is quasifibrant, the morphism in question is actually a trivial fibration.

Remark 22.4.3 If $\mathcal{C}'$ is a fibrant $\mathcal{P}$-precategory then it is quasifibrant.

This is because the morphisms

$$h([p-1], E) \cup^{h([p-2],E)} h([p-1], E) \to h([p], E)$$

are trivial cofibrations.

Lemma 22.4.4 *The $\mathcal{P}$-category* $\mathbf{Enr}(\mathcal{P})_\kappa$ *is quasifibrant. If $E_1, \ldots, E_k \in \mathcal{P}$, then a map*

$$\widetilde{\Upsilon}_k(E_1, \ldots, E_k) \xrightarrow{f} \mathbf{Enr}(\mathcal{P})_\kappa$$

is the same thing as a collection of objects $U_i = f(v_i) \in \mathcal{P}_{\mathrm{cf},\kappa}$ for $i = 0, \ldots, k$, together with maps $U_{i-1} \times E_i \to U_i$.

Proof The morphisms in question are actually isomorphisms for $p \geq 2$ and for $p = 1$ they are just projections from the $\underline{\mathrm{HOM}}(\mathcal{A}_0, \mathcal{A}_1)$–which are fibrant–to $*$.

For maps from $\widetilde{\Upsilon}_k$ to $\mathbf{Enr}(\mathcal{P})_\kappa$, note that both are strictly $\mathcal{P}$-enriched categories and $\widetilde{\Upsilon}_k(E_1, \ldots, E_k)$ is freely generated by the E_i. □

Consider the following internal notion of *sequential diagram*:

Definition 22.4.5 Suppose $\mathcal{C} \in \mathcal{PC}(\mathcal{P})$. Then for $E_1, \ldots, E_k \in \mathcal{P}$ we define

$$Diag(E_1, \ldots, \underline{E_i}, \ldots, E_k; \mathcal{C})$$

to be the object of $\mathcal{P}$ representing the functor

$$F \mapsto \mathrm{HOM}_{\mathcal{PC}(\mathcal{P})}(\widetilde{\Upsilon}_k(E_1, \ldots, E_i \times F, \ldots, E_k), \mathcal{C}).$$

In other words, a map

$$F \to Diag(E_1, \ldots, \underline{E_i}, \ldots, E_k; \mathcal{C})$$

is the same thing as a map

$$\widetilde{\Upsilon}_k(E_1, \ldots, E_i \times F, \ldots, E_k) \to \mathcal{C}.$$

Here are some basic properties:

Remark 22.4.6 The $\mathcal{P}$-precategory $Diag(E_1, \ldots, \underline{E_i}, \ldots, E_k; \mathcal{C})$ decomposes as a disjoint union of

$$Diag^{a,b}(E_1, \ldots, \underline{E_i}, \ldots, E_k; \mathcal{C})$$

over all pairs (a, b), where

$$a : \widetilde{\Upsilon}_{i-1}(E_1, \ldots, E_{i-1}) \to \mathcal{C}$$

and

$$b : \widetilde{\Upsilon}_{k-1-i}(E_{i+1}, \ldots, E_k) \to \mathcal{C}$$

are the restrictions of the diagram to the first and last faces separated by the i-th edge.

In the notation $Diag^{a,b}$, if we don't wish to specify b for example, then denote this by the superscript $Diag^{a,\cdot}$.

In particular, note that we can decompose into a disjoint union over the $(k+1)$-tuples of objects which are the images of the vertices $0, \ldots, k$ (these objects are all specified either as a part of a or as a part of b).

Remark 22.4.7 In case $\mathcal{C} = \mathbf{Enr}(\mathcal{P})_\kappa$ we have

$$Diag^{a,b}(E_1, \ldots, \underline{E_i}, \ldots, E_k; \mathbf{Enr}(\mathcal{P})_\kappa) = \underline{\mathrm{HOM}}(U_{i-1} \times E_i, U_i),$$

where U_j are the objects of $\mathcal{P}_{\mathrm{cf},\kappa}$ which are the images of the vertices

$$v_j \in \mathrm{Ob}(\widetilde{\Upsilon}_k(E_1, \ldots, E_k))$$

by the maps a (if $j \leq i-1$) or b (if $j \geq i$).

This comes from the description of maps in Lemma 22.4.4.

Remark 22.4.8 For $k = 1$, a and b are just objects of $\mathcal{C}$ and

$$Diag^{a,b}(\underline{E}, \mathcal{C}) = \underline{\mathrm{HOM}}_{\mathcal{P}}(E, \mathcal{C}(a, b)).$$

Thus, for the case $\mathcal{C} = \mathbf{Enr}(\mathcal{P})_\kappa$ with objects $U, V \in \mathcal{P}_{\mathrm{cf},\kappa}$ we have

$$Diag^{U,V}(\underline{E}, \mathbf{Enr}(\mathcal{P})_\kappa) = \underline{\mathrm{HOM}}_{\mathcal{P}}(U \times E, V).$$

When we are only interested in the set of points, it doesn't matter which E_i is underlined and we denote this set of points by

$$Diag(E_1, \ldots, E_k; \mathcal{C}) = \text{HOM}(\widetilde{\Upsilon}_k(E_1, \ldots, E_k), \mathcal{C}).$$

We can put a superscript $Diag^{a,b}$ here if we want (with the obvious meaning as above). The edge i dividing between a and b should be understood from the data of a and b.

Look now at how to pass between something quasifibrant such as $\mathcal{C} = \mathbf{Enr}(\mathcal{P})_\kappa$ and its fibrant completion. A morphism

$$u : \widetilde{\Upsilon}_k(E_1, \ldots, E_k) \to \mathcal{C}$$

may be described inductively as the triple

$$u = (\tilde{u}, u^-, u^+), \tag{22.4.1}$$

where

$$u^- : \widetilde{\Upsilon}_{k-1}(E_1, \ldots, E_{k-1}) \to \mathcal{C}$$

and

$$u^+ : \widetilde{\Upsilon}_{k-1}(E_2, \ldots, E_k) \to \mathcal{C}$$

are morphisms which agree on $\widetilde{\Upsilon}_{k-1}(E_1, \ldots, E_{k-1})$, and where

$$\tilde{u} : E_1 \times \cdots \times E_k \to \mathcal{C}(x_0, \ldots, x_k)$$

is a lifting of the morphism

$$(\tilde{u}^-, \tilde{u}^+) : E_1 \times \cdots \times E_k \to \mathcal{C}(x_0, \ldots, x_{k-1}) \times_{\mathcal{C}(x_1, \ldots, x_k)} \mathcal{C}(x_1, \ldots, x_{k-1})$$

along

$$\mathcal{C}(x_0, \ldots, x_k) \to \mathcal{C}(x_0, \ldots, x_{k-1}) \times_{\mathcal{C}(x_1, \ldots, x_k)} \mathcal{C}(x_1, \ldots, x_{k-1}).$$

The first component $\tilde{u}^-$ comes by composing the projection

$$E_1 \times \cdots \times E_k \to E_1 \times \cdots \times E_{k-1}$$

with the map

$$\widetilde{\Upsilon}_{k-1}(E_1, \ldots, E_{k-1})(v_0, \ldots, v_{k-1}) \to \mathcal{C}(x_0, \ldots, x_{k-1}).$$

The second component $\tilde{u}^+$ is the same at the $x_1, \ldots, x_k$.

Suppose $\mathcal{C}$ is quasifibrant. Then the morphisms involved in the previous description are fibrations.

Corollary 22.4.9 *If $\mathcal{C}$ is quasifibrant and if $E'_i \subset E_i$ are trivial cofibrations then any diagram*

$$\widetilde{\Upsilon}_k\left(E'_1, \ldots, E'_k\right) \to \mathcal{C}$$

extends to a diagram

$$\widetilde{\Upsilon}_k(E_1, \ldots, E_k) \to \mathcal{C}.$$

Proof Proving this by induction on k reduces to showing the lifting property for the trivial cofibration

$$E'_1 \times \cdots \times E'_k \hookrightarrow E_1 \times \cdots \times E_k$$

along the morphism

$$\mathcal{C}(x_0, \ldots, x_k) \to \mathcal{C}(x_0, \ldots, x_{k-1}) \times_{\mathcal{C}(x_1, \ldots, x_k)} \mathcal{C}(x_1, \ldots, x_{k-1}).$$

This morphism being fibrant by hypothesis, the lifting property holds. □

Corollary 22.4.10 *Suppose $\mathcal{C}$ is quasifibrant. Then for a, b fixed as in Remark 22.4.6 the morphisms*

$$Diag^{a,b}(E_1, \ldots, \underline{E_i}, \ldots, E_k; \mathcal{C}) \to Diag^{a(i-1),b(i)}(\underline{E_i}; \mathcal{C})$$

are trivial fibrations, where $a(i-1)$ and $b(i)$ are the images by a and b of the $(i-1)$-th and i-th vertices.

Proof To prove this we use the description (22.4.1), inductively reducing k. Note first that one can see that the map is a fibration, using the trivial inclusions discussed in Section 21.4.2, in an argument similar to some more essential considerations in the next sections and which needn't be expanded upon here. To show that it is an equivalence, Remark 22.4.2 says that for any $k \geq 2$ the choice of lifting $\tilde{u}$ doesn't change the equivalence type of the $Diag$ object of $\mathcal{P}$. This reduces down to the case $k = 1$, which gives exactly that the restriction to the i-th edge is an equivalence (the restrictions to the other edges are fixed and don't contribute anything because we fix a, b). □

Proposition 22.4.11 *Suppose $\mathcal{C}$ is quasifibrant. Then the diagram object $Diag(E_1, \ldots, \underline{E_i}, \ldots, E_k; \mathcal{C})$ is fibrant in $\mathcal{P}$. Furthermore in this case given cofibrations $E'_j \hookrightarrow E_j$, the morphism of restriction*

$$Diag\left(E'_1, \ldots, \underline{E'_i}, \ldots, E'_k; \mathcal{C}\right) \to Diag(E_1, \ldots, \underline{E_i}, \ldots, E_k; \mathcal{C})$$

is fibrant.

If $\mathcal{C}$ quasifibrant and if $\mathcal{C} \to \mathcal{C}'$ is an equivalence to a fibrant $\mathcal{C}'$ then the morphism

$$Diag^{a,b}(E_1, \ldots, \underline{E_i}, \ldots, E_k; \mathcal{C}) \to Diag^{a',b'}(E_1, \ldots, \underline{E_i}, \ldots, E_k; \mathcal{C}')$$

is an equivalence of fibrant objects of $\mathcal{P}$. Here a, b are fixed as in Remark 22.4.6, and a', b' denote the images in $\mathcal{C}'$.

Proof The first statements are direct consequences of the lifting property of Corollary 22.4.9.

To prove the second paragraph, in view of Corollary 22.4.10 it suffices to consider the case $k = 1$. Now

$$Diag^{U,V}(\underline{E}; \mathcal{C}) = \underline{\text{HOM}}(E, \mathcal{C}(U, V)).$$

If $\mathcal{C} \to \mathcal{C}'$ is an equivalence between quasifibrant $\mathcal{P}$-categories, in particular it is "fully faithful," i.e. it induces equivalences of fibrant objects of $\mathcal{P}$

$$\mathcal{C}(U, V) \to \mathcal{C}'(U, V).$$

Therefore,

$$Diag^{U,V}(\underline{E}; \mathcal{C}) \to Diag^{U,V}(\underline{E}; \mathcal{C}')$$

are equivalences by Remark 22.4.8. This completes the proof for $k = 1$ and hence by Corollary 22.4.10 for any k. □

22.5 Extension properties

In this section we consider the essential construction which links up the objects of $\mathcal{P}$ occuring in the enrichment with the objects of $\mathcal{P}$ occuring in $\mathbf{Enr}(\mathcal{P})_\kappa$.

22.5.1 A tautological diagram

Suppose $B \in \mathcal{P}$ is any object. By the universal property of $\Upsilon(B)$, a morphism

$$\Upsilon(B) \xrightarrow{f} \mathbf{Enr}(\mathcal{P})_\kappa$$

is the same thing as a pair of objects $U := f(\upsilon_0)$ and $V := f(\upsilon_1)$ in $\mathcal{P}_{\mathrm{cf},\kappa}$, together with a map

$$B \to \mathbf{Enr}(\mathcal{P})_\kappa(U, V).$$

By the construction of $\mathbf{Enr}(\mathcal{P})_\kappa$, this is a map $B \to \underline{\text{HOM}}_{\mathcal{P}}(U, V)$, which in turn by the adjunction property of $\underline{\text{HOM}}_{\mathcal{P}}$, is the same thing as a map $U \times B \to V$.

This applies particularly to the identity morphism $* \times B \to B$, if B is already a fibrant and cofibrant κ-presentable object. Denote the resulting natural morphism composed with $\mathbf{Enr}(\mathcal{P})_\kappa \to \mathbf{Enr}(\mathcal{P})'_\kappa$, by

$$1_B \in \mathrm{HOM}^{*,B}\left(\Upsilon(B), \mathbf{Enr}(\mathcal{P})'_\kappa\right).$$

This will enter into many of the formulae below.

Corollary 22.5.1 *Fix a, b as in Remark 22.4.6, and suppose that the restriction of a to $\Upsilon(B)$ is equal to 1_B. Then the morphism*

$$\begin{aligned} &Diag^{a,b}(B, E_2, \ldots, \underline{E_i}, \ldots, E_k; \mathbf{Enr}(\mathcal{P})_\kappa) \\ &\to Diag^{a,b}\left(B, E_2, \ldots, \underline{E_i}, \ldots, E_k; \mathbf{Enr}(\mathcal{P})'_\kappa\right) \end{aligned}$$

is an equivalence between fibrant objects of $\mathcal{P}$.

Proof Apply Proposition 22.4.11 to the map $\mathbf{Enr}(\mathcal{P})_\kappa \to \mathbf{Enr}(\mathcal{P})'_\kappa$, which is a fibrant replacement of a quasifibrant $\mathcal{P}$-category by Lemma 22.4.4. □

22.5.2 Horn extension properties

For any $E_1, E_2, \ldots, E_k$ use the terminology of Section 21.4.1 to define the *left shell*

$$\Lambda^L\widetilde{\Upsilon}_k(E_1, \ldots, E_k) := \widetilde{\Upsilon}_k\left(\Lambda^0_k; E_1, \ldots, E_k\right).$$

It can be thought of as a union

$$\begin{aligned} &\Lambda^L\widetilde{\Upsilon}_k(E_1, \ldots, E_k) \\ &= \widetilde{\Upsilon}_{k-1}(E_1, \ldots, E_{k-1}) \cup \bigcup_{i=1}^{k-1} \widetilde{\Upsilon}_{k-1}(\ldots, E_i \times E_{i+1}, \ldots), \end{aligned}$$

consisting of all of the "faces" except the first one $\Upsilon^{k-1}(E_2, \ldots, E_k)$. Assuming that the E_i are cofibrant, the map

$$\Lambda^L\widetilde{\Upsilon}_k(E_1, \ldots, E_k) \to \widetilde{\Upsilon}_k(E_1, \ldots, E_k)$$

is a cofibration.

Now set $E_1 = B$ and let $\underline{\mathrm{HOM}}^{1_B}\left(\widetilde{\Upsilon}_k(B, E_2, \ldots, E_k), \mathbf{Enr}(\mathcal{P})'_\kappa\right)$ denote the fiber of

$$\underline{\mathrm{HOM}}\left(\widetilde{\Upsilon}_k(B, E_2, \ldots, E_k), \mathbf{Enr}(\mathcal{P})'_\kappa\right) \to \underline{\mathrm{HOM}}\left(\Upsilon(B), \mathbf{Enr}(\mathcal{P})'_\kappa\right)$$

over 1_B. Let $\underline{\mathrm{HOM}}^{1_B}\left(\Lambda^L\widetilde{\Upsilon}_k(B, E_2, \ldots, E_k), \mathbf{Enr}(\mathcal{P})'_\kappa\right)$ denote the fiber of

$$\underline{\mathrm{HOM}}\left(\Lambda^L\widetilde{\Upsilon}_k(1_B, E_2, \ldots, E_k), \mathbf{Enr}(\mathcal{P})'_\kappa\right) \to \underline{\mathrm{HOM}}\left(\Upsilon B, \mathbf{Enr}(\mathcal{P})'_\kappa\right)$$

over 1_B.

The first lemma is where we use the quasifibrant property to pass from $\mathbf{Enr}(\mathcal{P})_\kappa$ to $\mathbf{Enr}(\mathcal{P})'_\kappa$:

Lemma 22.5.2 *Suppose $B \in \mathcal{P}_{\mathrm{cf},\kappa}$, $E \in \mathcal{P}_c$, and U an object of $\mathbf{Enr}(\mathcal{P})'_\kappa$ (it is also an object of $\mathbf{Enr}(\mathcal{P})_\kappa$). The morphism*

$$Diag^{1_B,U}\left(B, \underline{E}; \mathbf{Enr}(\mathcal{P})'_\kappa\right) \to Diag^{*,U}\left(\underline{B \times E}, \mathbf{Enr}(\mathcal{P})'_\kappa\right)$$

is a trivial fibration in $\mathcal{P}$.

Proof To show that the map is a fibration, one checks directly the lifting property for a trivial cofibration $F' \hookrightarrow F$, using the fibrant property of $\mathbf{Enr}(\mathcal{P})'_\kappa$.

To show it is a weak equivalence, in view of Proposition 22.4.11 it suffices to prove the same thing for diagrams in $\mathbf{Enr}(\mathcal{P})_\kappa$. In this case, use the calculation of Remark 22.4.7: both sides become equal to $\underline{\mathrm{HOM}}(B \times E, U)$. □

We get shell extension properties for $k = 2$:

Corollary 22.5.3 *The morphism*

$$\mathrm{HOM}^{1_B}\left(\widetilde{\Upsilon}_2(B, E), \mathbf{Enr}(\mathcal{P})'_\kappa\right) \to \mathrm{HOM}^*\left(\Upsilon(B \times E), \mathbf{Enr}(\mathcal{P})'_\kappa\right)$$

is surjective.

Proof We can fix an object U for the image of the last vertex. The morphism

$$Diag^{1_B,U}\left(B, \underline{E}; \mathbf{Enr}(\mathcal{P})'_\kappa\right) \to Diag^{*,U}\left(\underline{B \times E}, \mathbf{Enr}(\mathcal{P})'_\kappa\right)$$

is a trivial fibration by Lemma 22.5.2. This implies that it is surjective on objects (21.2.3). □

Corollary 22.5.4 *Suppose $E' \to E$ is a cofibration between cofibrant objects of $\mathcal{P}$. Suppose we are given an element of*

$$\mathrm{HOM}^{1_B}\left(\widetilde{\Upsilon}_2(B, E'), \mathbf{Enr}(\mathcal{P})'_\kappa\right),$$

and an extension over the shell to an element of

$$\mathrm{HOM}^{1_B}\left(\Lambda^L\widetilde{\Upsilon}_2(B, E), \mathbf{Enr}(\mathcal{P})'_\kappa\right).$$

Then these two have a common extension to an element of

$$\mathrm{HOM}^{1_B}\left(\widetilde{\Upsilon}_2(B, E), \mathbf{Enr}(\mathcal{P})'_\kappa\right).$$

Proof Again we can fix U. By Lemma 22.5.2, the morphism

$$Diag^{1_B,U}\left(B,\underline{*};\mathbf{Enr}(\mathcal{P})'_\kappa\right) \to Diag^{*,U}\left(\underline{B},\mathbf{Enr}(\mathcal{P})'_\kappa\right)$$

is a trivial fibration. Therefore, it has the lifting property with respect to any cofibration $E' \to E$ between cofibrant objects. This lifting property gives exactly what we want to show–this is because a morphism

$$E \to Diag^{1_B,U}\left(B,\underline{*};\mathbf{Enr}(\mathcal{P})'_\kappa\right)$$

is the same thing as an object of

$$Diag^{1_B,U}\left(B,E;\mathbf{Enr}(\mathcal{P})'_\kappa\right),$$

or, equivalently, an element of $\text{H}\textsc{om}^{1_B}\left(\widetilde{\Upsilon}_2(B,E),\mathbf{Enr}(\mathcal{P})'_\kappa\right)$. □

Now we treat a similar type of extension problem for shells with $k = 3$. Suppose $b \in Diag\left(F;\mathbf{Enr}(\mathcal{P})'_\kappa\right)$, and let

$$Diag^{1_B,b}_{\Lambda^L}\left(B,\underline{E},F;\mathbf{Enr}(\mathcal{P})'_\kappa\right)$$

be the object of $\mathcal{P}$ representing the functor

$$G \mapsto \text{H}\textsc{om}^{1_B,b}\left(\Lambda^L\widetilde{\Upsilon}_3(B,E\times G,F),\mathbf{Enr}(\mathcal{P})'_\kappa\right),$$

where the superscript on the $\text{H}\textsc{om}$ means that we look only at morphisms restricting to 1_B on the edge 01 and to b on the edge 23.

The shell $\Lambda^L\widetilde{\Upsilon}_3(B,E\times G,F)$ has three faces. We call the faces (012) and (023) the *front faces*:

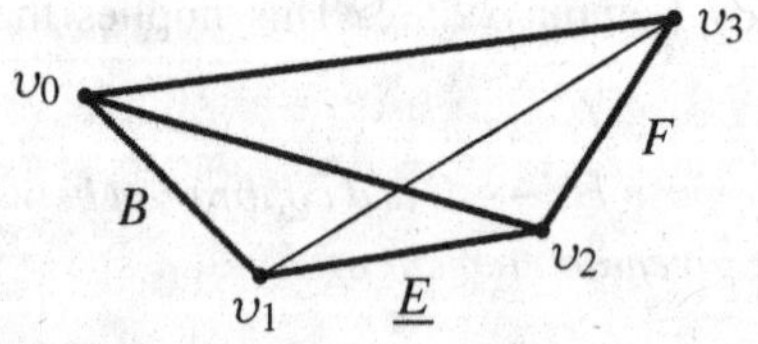

and the face (013) the *back face*:

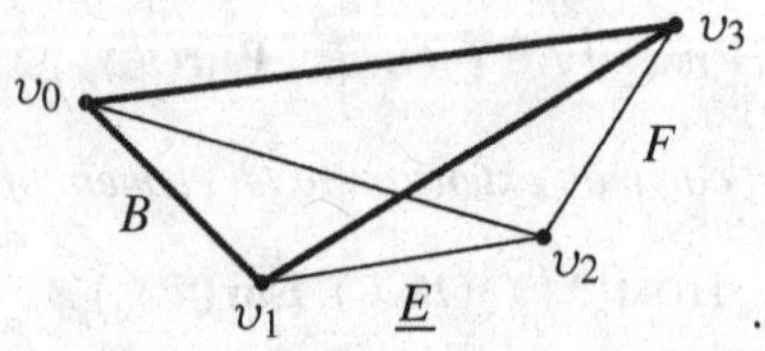

.

Restriction to the front faces (which meet along the edge (02)) gives a diagram

$$\begin{array}{ccc} Diag^{1_B,b}_{\Lambda L}\left(B,\underline{E},F;\mathbf{Enr}(\mathcal{P})'_\kappa\right) & \longrightarrow & Diag^{1_B,b_2}\left(B,\underline{E};\mathbf{Enr}(\mathcal{P})'_\kappa\right) \\ \downarrow & & \downarrow \\ Diag^{*,b}\left(\underline{B\times E},F;\mathbf{Enr}(\mathcal{P})'_\kappa\right) & \rightarrow & Diag^{*,b_2}\left(\underline{B\times E};\mathbf{Enr}(\mathcal{P})'_\kappa\right) \end{array}$$

where b_2 denotes the object image of υ_2 under the map b. A similar definition will hold for b_3 below, and recall that the image of υ_0 under the map 1_B is $*$.

Lemma 22.5.5 *In the above diagram, the map from the upper left corner to the fiber product D of the complementary angle is a trivial fibration.*

Proof Restriction to the edge (03) is a map

$$D \rightarrow Diag^{*,b_3}\left(\underline{B\times E\times F};\mathbf{Enr}(\mathcal{P})'_\kappa\right).$$

Everything restricts by definition to 1_B on the edge (01), so adding in also the restriction to the back face gives an isomorphism

$$Diag^{1_B,b}_{\Lambda L}\left(B,\underline{E},F;\mathbf{Enr}(\mathcal{P})'_\kappa\right)$$
$$\stackrel{\cong}{\rightarrow} D\times_{Diag^{*,b_3}\left(\underline{B\times E\times F};\mathbf{Enr}(\mathcal{P})'_\kappa\right)} Diag^{1_B,b_3}\left(B,\underline{E\times F};\mathbf{Enr}(\mathcal{P})'_\kappa\right).$$

However, the second morphism in this fiber product is

$$Diag^{1_B,b_3}\left(B,\underline{E\times F};\mathbf{Enr}(\mathcal{P})'_\kappa\right) \rightarrow Diag^{*,b_3}\left(\underline{B\times E\times F};\mathbf{Enr}(\mathcal{P})'_\kappa\right),$$

which is a trivial fibration by Lemma 22.5.2. It follows that the morphism

$$Diag^{1_B,b}_{\Lambda L}\left(B,\underline{E},F;\mathbf{Enr}(\mathcal{P})'_\kappa\right) \rightarrow D$$

is a trivial fibration. □

Corollary 22.5.6 *The morphism*

$$Diag^{1_B,b}\left(B,\underline{E},F;\mathbf{Enr}(\mathcal{P})'_\kappa\right) \rightarrow Diag^{1_B,b}_{\Lambda L}\left(B,\underline{E},F;\mathbf{Enr}(\mathcal{P})'_\kappa\right)$$

is a trivial fibration.

Proof It is a fibration because $\mathbf{Enr}(\mathcal{P})'_\kappa$ is fibrant. In view of the previous Lemma 22.5.5 it suffices to note that the map

$$Diag^{1_B,b}\left(B,\underline{E},F;\mathbf{Enr}(\mathcal{P})'_\kappa\right) \rightarrow D$$

is an equivalence. This holds by the fibrant property of $\mathbf{Enr}(\mathcal{P})'_\kappa$ since the union of the front faces (012) and (023) is one of the admissible ones in our list of 21.4.2, see Corollary 21.4.4. □

Corollary 22.5.7 *Suppose $E' \subset E$ is a cofibration between cofibrant objects in $\mathcal{P}$. Then for any morphism*

$$\widetilde{\Upsilon}_3(B, E', F) \to \mathbf{Enr}(\mathcal{P})'_{\kappa}$$

sending the edge (01) *to* 1_B, *and any extension of this over the left shell to a morphism*

$$\Lambda^L \widetilde{\Upsilon}_3(B, E, F) \to \mathbf{Enr}(\mathcal{P})'_{\kappa},$$

again restricting to 1_B *on the edge* (01), *there exists a common extension to a morphism*

$$\widetilde{\Upsilon}_3(B, E, F) \to \mathbf{Enr}(\mathcal{P})'_{\kappa}.$$

Proof By the previous corollary the morphism

$$Diag^{1_B,b}\left(B, \underline{*}, F; \mathbf{Enr}(\mathcal{P})'_{\kappa}\right) \to Diag^{1_B,b}_{\Lambda^L}\left(B, \underline{*}, F; \mathbf{Enr}(\mathcal{P})'_{\kappa}\right)$$

is a trivial fibration. Therefore, it satisfies the lifting property for any cofibration $E' \hookrightarrow E$, and as before (22.5.4) a map

$$E \to Diag^{1_B,b}\left(B, \underline{*}, F; \mathbf{Enr}(\mathcal{P})'_{\kappa}\right)$$

is the same thing as an element of $Diag^{1_B,b}\left(B, E, F; \mathbf{Enr}(\mathcal{P})'_{\kappa}\right)$. Similarly for E' and $Diag^{1_B,b}_{\Lambda^L}$. This gives the required statement. □

We only need these properties for the cases $k = 2, 3$. The formulations and proofs of the corresponding horn extension properties for general $k > 3$ are left to the reader.

22.5.3 Extension properties for internal HOM

Now we take the above extension properties and recast them in terms of internal $\underline{\text{HOM}}$. This is because we will need them for products of our precategories Υ with an arbitrary $\mathcal{A}$. Note that there is a difference between the internal $\underline{\text{HOM}}$ refered to in this section (which are $\mathcal{P}$-categories) and the $Diag$ objects of $\mathcal{P}$ above.

We state the following lemma for any value of k, but give the proof only for $k = 2$ and $k = 3$. It is left to the reader to fill in the combinatorial details for arbitrary k, on the model of the proof for $k = 3$.

Lemma 22.5.8 *For any cofibrant objects $E_2, \ldots, E_k$ of $\mathcal{P}$, the morphism*

$$\underline{\mathrm{HOM}}^{1_B}\left(\widetilde{\Upsilon}_k(B, E_2, \ldots, E_k), \mathbf{Enr}(\mathcal{P})'_\kappa\right)$$
$$\downarrow$$
$$\underline{\mathrm{HOM}}^{1_B}\left(\Lambda^L\widetilde{\Upsilon}_k(B, E_2, \ldots, E_k), \mathbf{Enr}(\mathcal{P})'_\kappa\right)$$

is a trivial fibration of $\mathcal{P}$-categories.

Proof The morphism in question is a fibration, as is dual to the statement of Lemma 21.4.3.

The proof is divided into several paragraphs. The first few give the proof for $k = 2$. Then at the end, we give the proof for $k = 3$.

We begin the proof for $k = 2$. Corollary 22.5.3 implies that the morphism in question

$$\underline{\mathrm{HOM}}^{1_B}\left(\widetilde{\Upsilon}_2(B, E_2), \mathbf{Enr}(\mathcal{P})'_\kappa\right) \to \underline{\mathrm{HOM}}^{*}\left(\Upsilon(B \times E_2), \mathbf{Enr}(\mathcal{P})'_\kappa\right)$$

is surjective on objects.

The source and target are already $\mathcal{P}$-categories (indeed they are fibrant), so to prove that the map is an equivalence it suffices to prove that it induces equivalences between the morphism objects in $\mathcal{P}$. Suppose

$$f, g : \widetilde{\Upsilon}_2(B, E_2) \to \mathbf{Enr}(\mathcal{P})'_\kappa$$

are two maps restricting to 1_B on (01). Then the internal mapping object $\underline{\mathrm{HOM}}^{1_B}\left(\widetilde{\Upsilon}_2(B, E_2), \mathbf{Enr}(\mathcal{P})'_\kappa\right)(f, g) \in \mathcal{P}$ of morphisms between them represents the functor

$$F \mapsto \mathrm{HOM}^{f,g;1_B}\left(\Upsilon(F) \times \widetilde{\Upsilon}_2(B, E_2), \mathbf{Enr}(\mathcal{P})'_\kappa\right),$$

where the superscript means morphisms restricting to f and g over $\upsilon_0, \upsilon_1 \in \Upsilon(F)$ and restricting to 1_B over $\Upsilon(F) \times \Upsilon B$. This maps (by restricting to the edge 02) to the functor

$$F \mapsto \mathrm{HOM}^{f,g;*}\left(\Upsilon(F) \times \Upsilon(B \times E_2), \mathbf{Enr}(\mathcal{P})'_\kappa\right).$$

We would like to prove that this restriction map of functors is an equivalence. In order to prove this it suffices to prove that it has the lifting property for any generating cofibrations between cofibrant objects $F' \to F$. Thus, suppose that we have a morphism

$$\eta : \Upsilon(F) \times \Upsilon(B \times E_2) \to \mathbf{Enr}(\mathcal{P})'_\kappa,$$

restricting appropriately to f and g and to $*$, as well as a morphism

$$\zeta' : \Upsilon(F') \times \widetilde{\Upsilon}_2(B, E_2) \to \mathbf{Enr}(\mathcal{P})'_{\kappa},$$

compatible with η and restricting appropriately to f, g and 1_B. We would like to extend this latter to a map defined on F and compatible with the previous one. This extension will complete the proof for $k = 2$.

To prove the extension statement, proceed in the same way as at the end of the proof of Theorem 21.10.4: consider the diagram as a prism, the product of an interval (01) and a triangle (012), and denote the points by (i, j) for $i = 0, 1$ and $j = 0, 1, 2$. More generally, for example, (ab, cd) denotes the square which is the edge (ab) crossed with the edge (cd). We are provided with maps on the end triangles f on $(0, 012)$ and g on $(1, 012)$ as well as η on the top square $(01, 02)$. We fix the map on the square $(01, 01)$ (which is $\Upsilon(F) \times \Upsilon(B)$), pullback of 1_B, calling the pullback again 1_B. We are also provided with a map ζ' defined on the whole diagram with respect to F' and we would like to extend this all to ζ defined on the whole diagram.

Note that we can write $\Upsilon(F) \times \widetilde{\Upsilon}_2(B, E)$ as the amalgamated sum of three tetrahedra, which we denote

$$(0,0)\ (1,0)\ (1,1)\ (1,2), \text{ i.e. } \widetilde{\Upsilon}_3(F, B, E),$$

$$(0,0)\ (0,1)\ (0,2)\ (1,2), \text{ i.e. } \widetilde{\Upsilon}_3(B, E, F),$$

$$(0,0)\ (0,1)\ (1,1)\ (1,2), \text{ i.e. } \widetilde{\Upsilon}_3(B, F, E).$$

The first step is to use the fibrant property of $\mathbf{Enr}(\mathcal{P})'_{\kappa}$ and the case $(012) + (013) + (123) \subset (0123)$ of Corollary 21.4.4 to extend our given morphisms g, the restriction of 1_B to the triangle $(0, 0)$, $(1, 0)$, $(1, 1)$, and the restriction of η to the triangle $(0, 0)$, $(1, 0)$, $(1, 2)$, to a map on the tetrahedron

$$(0,0)\ (1,0)\ (1,1)\ (1,2).$$

We can do this in a way which extends the map ζ'.

Next we again use the fibrant property of $\mathbf{Enr}(\mathcal{P})'_{\kappa}$ together with the case $(012) + (023) \subset (0123)$ of Corollary 21.4.4, to extend across the tetrahedron

$$(0,0)\ (0,1)\ (0,2)\ (1,2).$$

Note that we are provided with the map f on the triangle $(0, 012)$, and the restriction of the η on the triangle $(0, 0)$, $(0, 2)$, $(1, 2)$. We can find our extension again in a way extending the given map ζ'.

Finally, we come to the tetrahedron

$$(0,0)\ (0,1)\ (1,1)\ (1,2),$$

which is of the form $\widetilde{\Upsilon}_3(B, F, E)$. Here we are given maps on all of the faces except the last one, i.e. on the left shell of this tetrahedron, and we would like to extend it. The given maps are the pullback of 1_B on the first face, and the maps coming from the two previous paragraphs on the other two faces. Furthermore, we already have a map ζ' over the tetrahedron $\widetilde{\Upsilon}_3(B, F', E)$. The given map on the left shell restricts on the first edge to 1_B, so this is an extension problem of the type which we have already treated in Corollary 22.5.7 above. (Note, however, that the notations E and F are interchanged between 22.5.7 and the present situation.) Thus Corollary 22.5.7 provides the extension we are looking for, and we have finished making our extension across the three tetrahedra. This completes the proof of "fully faithfulness" so the morphism in the lemma is an equivalence in the case $k = 2$.

Here is the proof for $k = 3$. First of all, the morphism from the upper left to the fiber product in the diagram

$$\begin{array}{ccc}
\underline{\mathrm{HOM}}^{1_B}\left(\widetilde{\Upsilon}_3(B, E_2, E_3), \mathbf{Enr}(\mathcal{P})'_\kappa\right) & \rightarrow & \underline{\mathrm{HOM}}^{1_B}\left(\widetilde{\Upsilon}_2(B, E_2), \mathbf{Enr}(\mathcal{P})'_\kappa\right) \\
\downarrow & & \downarrow \\
\underline{\mathrm{HOM}}\left(\widetilde{\Upsilon}_2(B \times E_2, E_3), \mathbf{Enr}(\mathcal{P})'_\kappa\right) & \longrightarrow & \underline{\mathrm{HOM}}\left(\Upsilon(B \times E_2), \mathbf{Enr}(\mathcal{P})'_\kappa\right)
\end{array}$$

is an equivalence, by the fibrant property for $\mathbf{Enr}(\mathcal{P})'_\kappa$ and Corollary 21.4.4 for the inclusion of faces $(012) + (023) \subset (0123)$.

By the case $k = 2$ applied to the face (013), in the same way as in the proof of Lemma 22.5.5, the morphism from the upper left to the fiber product in the diagram

$$\begin{array}{ccc}
\underline{\mathrm{HOM}}^{1_B}\left(\Lambda^L\widetilde{\Upsilon}_3(B, E_2, E_3), \mathbf{Enr}(\mathcal{P})'_\kappa\right) & \rightarrow & \underline{\mathrm{HOM}}^{1_B}\left(\widetilde{\Upsilon}_2(B, E_2), \mathbf{Enr}(\mathcal{P})'_\kappa\right) \\
\downarrow & & \downarrow \\
\underline{\mathrm{HOM}}\left(\widetilde{\Upsilon}_2(B \times E_2, E_3), \mathbf{Enr}(\mathcal{P})'_\kappa\right) & \longrightarrow & \underline{\mathrm{HOM}}\left(\Upsilon(B \times E_2), \mathbf{Enr}(\mathcal{P})'_\kappa\right)
\end{array}$$

is an equivalence. This implies that the morphism

$$\begin{array}{c}
\underline{\mathrm{HOM}}^{1_B}\left(\widetilde{\Upsilon}_3(B, E_2, E_3), \mathbf{Enr}(\mathcal{P})'_\kappa\right) \\
\downarrow \\
\underline{\mathrm{HOM}}^{1_B}\left(\Lambda^L\widetilde{\Upsilon}_3(B, E_2, E_3), \mathbf{Enr}(\mathcal{P})'_\kappa\right) *
\end{array}$$

is an equivalence. This finishes the proof of the lemma for the case $k = 3$. For $k \geq 4$ the proof is the same using higher dimensional diagrams. □

The main part of the argument for Theorem 22.5.8 below will only use the cases $k = 2$ and $k = 3$, although, technically speaking, the next few cases are useful for checking the compatibility of the homotopy idempotent.

Corollary 22.5.9 *Suppose $\mathcal{A}$ is a cofibrant $\mathcal{P}$-precategory and suppose $E_i \in \mathcal{P}$ are cofibrant objects for $i = 2, \ldots, k$. Suppose we are given a morphism*

$$\mathcal{A} \times \Lambda^L \widetilde{\Upsilon}_k(B, E_2, \ldots, E_k) \to \mathbf{Enr}(\mathcal{P})'_\kappa$$

restricting to $1_{B,\mathcal{A}}$ on $\mathcal{A} \times \Upsilon B$. Then there is an extension to a morphism

$$\mathcal{A} \times \widetilde{\Upsilon}_k(B, E_2, \ldots, E_k) \to \mathbf{Enr}(\mathcal{P})'_\kappa.$$

Proof The restriction morphism on the $\underline{\text{HOM}}$ is a fibration and an equivalence by the previous lemma, therefore it is surjective on objects. □

What we really need to know is a relative version of this for cofibrations $E'_i \to E_i$.

Corollary 22.5.10 *Suppose $\mathcal{A}$ is a cofibrant $\mathcal{P}$-precategory, and suppose $E'_i \to E_i$ are cofibrations between cofibrant objects of $\mathcal{P}$ for $i = 2, \ldots, k$. Suppose we are given a morphism*

$$\mathcal{A} \times \Lambda^L \widetilde{\Upsilon}_k(B, E_2, \ldots, E_k) \to \mathbf{Enr}(\mathcal{P})'_\kappa,$$

restricting to $1_{B,\mathcal{A}}$ on $\mathcal{A} \times \Upsilon B$, together with a filling-in

$$\mathcal{A} \times \widetilde{\Upsilon}_k\left(B, E'_2, \ldots, E'_k\right) \to \mathbf{Enr}(\mathcal{P})'_\kappa,$$

then there is an extension of all of this to a morphism

$$\mathcal{A} \times \widetilde{\Upsilon}_k(B, E_2, \ldots, E_k) \to \mathbf{Enr}(\mathcal{P})'_\kappa.$$

Proof Let H_E denote the $\underline{\text{HOM}}$ for the full Υ and let H_E^{Sh} denote the $\underline{\text{HOM}}$ for $\Lambda^L \Upsilon$. The morphism

$$H_E \to H_E^{Sh} \times_{H_{E'}^{Sh}} H_{E'}$$

is an equivalence (as is seen by applying the lemma for both E and E') and it is a fibration (since it comes from $\underline{\text{HOM}}$ applied to a cofibration). Therefore it is surjective on objects, which exactly means that we have the above extension property. □

22.6 Limits of weak enriched categories

The goal of this section is to prove the following theorem:

Theorem 22.6.1 *The* ind*-object of fibrant $\mathcal{P}$-categories (cf. Section 22.3) $\mathbf{Enr}(\mathcal{P})'_{\bullet}$ admits limits. More precisely, if $\mathcal{A}$ is any cofibrant $\mathcal{P}$-precategory with $\mathbb{U}$-small set of objects, and*

$$\varphi_{\mu} : \mathcal{A} \to \mathbf{Enr}(\mathcal{P})'_{\mu}$$

is a morphism for some $\mu \in \mathbf{Reg}(\mathbb{U})$, then there is $\kappa_0 \geq \mu$ such that for any $\kappa \geq \kappa_0$, $\varphi := \mathbf{i}_{\mu,\kappa} \circ \varphi_{\mu} : \mathcal{A} \to \mathbf{Enr}(\mathcal{P})'_{\kappa}$ admits a limit. Furthermore, the transition morphisms preserve the limits constructed here.

The rest of this section is devoted to the proof. As a preliminary remark notice that by Proposition 21.9.5 the statement doesn't depend on which choices of $\mathbf{Enr}(\mathcal{P})'_{\mu}$ and $\mathbf{Enr}(\mathcal{P})'_{\kappa}$ were made. Accordingly, the strong restriction condition of Proposition 22.3.1 will be assumed.

Bergner [46, 47] has considered the homotopy limits of $(\infty, 1)$-categories, and even provides a lifting to a construction of homotopy limits of model categories.

22.6.1 Construction of the limiting object

Suppose $\mathcal{A}$ is a cofibrant $\mathcal{P}$-category. If B is a cofibrant and fibrant κ-presentable object of $\mathcal{P}$ we have denoted by $B_{\mathcal{A}}$ the constant morphism $\mathcal{A} \to \mathbf{Enr}(\mathcal{P})_{\kappa}$ with value B, which may also be considered as a morphism $\mathcal{A} \to \mathbf{Enr}(\mathcal{P})'_{\kappa}$ in $\mathcal{PC}(\mathcal{P})$.

We now give an explicit construction of the object which is supposed to be the limit of our given functor. Suppose $\mu \in \mathbf{Reg}(\mathbb{U})$ and $\varphi_{\mu} : \mathcal{A} \to \mathbf{Enr}(\mathcal{P})'_{\mu}$ is a morphism. Consider

$$\lambda' := \underline{\text{HOM}}\left(\mathcal{A}, \mathbf{Enr}(\mathcal{P})'_{\mu}\right)(*_{\mathcal{A}}, \varphi_{\mu}).$$

It has the universal property that, for any B in $\mathcal{P}$,

$$\text{HOM}(B, \lambda') = \text{HOM}^{*_{\mathcal{A}}, \varphi_{\mu}}\left(\mathcal{A} \times \Upsilon(B), \mathbf{Enr}(\mathcal{P})'_{\mu}\right). \qquad (22.6.1)$$

The notation on the right means the fiber of the map

$$\text{HOM}\left(\mathcal{A} \times \Upsilon B, \mathbf{Enr}(\mathcal{P})'_{\mu}\right) \xrightarrow{(r_0, r_1)} \text{HOM}\left(\mathcal{A}, \mathbf{Enr}(\mathcal{P})'_{\mu}\right)^2$$

over $(*_{\mathcal{A}}, \varphi_{\mu})$.

Notice that λ' is a fibrant object of $\mathcal{P}$. Let $\lambda \xrightarrow{r} \lambda'$ be a cofibrant replacement, in other words a trivial fibration from a cofibrant object. Let $\kappa \in \mathbf{Reg}(\mathbb{U})$

be any regular cardinal such that $\kappa \geq \mu$ and λ is κ-presentable. Then $\lambda \in \mathcal{P}_{\mathrm{cf},\kappa} = \mathrm{Ob}\left(\mathbf{Enr}(\mathcal{P})'_{\kappa}\right)$.

Put $\varphi := \mathbf{i}_{\mu,\kappa} \circ \varphi_{\mu}$. The problem below will be to prove that λ is a limit of φ in $\mathbf{Enr}(\mathcal{P})'_{\kappa}$.

The universal property of λ' applied to the map $\lambda \xrightarrow{r} \lambda'$, then followed by $\mathbf{i}_{\mu,\kappa}$, gives a morphism

$$\eta : \mathcal{A} \times \Upsilon(\lambda) \to \mathbf{Enr}(\mathcal{P})'_{\kappa}$$

sending $\mathcal{A} \times \{\upsilon_0\}$ to $*_{\mathcal{A}}$ and sending $\mathcal{A} \times \{\upsilon_1\}$ to φ.

The universal property of λ' lifts along r:

Lemma 22.6.2 *Suppose B is a cofibrant object of $\mathcal{P}$, and*

$$g : \mathcal{A} \times \Upsilon(B) \to \mathbf{Enr}(\mathcal{P})'_{\kappa}$$

is a morphism restricting to $*_{\mathcal{A}}$ on $\mathcal{A} \times \{\upsilon_0\}$ and to φ on $\mathcal{A} \times \{\upsilon_1\}$. Then there is a map $g^+ : B \to \lambda$ such that

$$\eta \circ (1 \times \Upsilon(g^+)) = g.$$

If $B' \to B$ is a cofibration and we are already given g^+ on B', then it can be extended to g^+ on B in a compatible way.

Proof The conditions on the restrictions of g to $\mathcal{A} \times \{\upsilon_0\}$ and $\mathcal{A} \times \{\upsilon_1\}$ imply that it goes into the full sub-$\mathcal{P}$-category of $\mathbf{Enr}(\mathcal{P})'_{\kappa}$ consisting of objects of $\mathcal{P}_{\mathrm{cf},\mu}$. By the condition of Proposition 22.3.1 which is in force here, this full subcategory is exactly $\mathbf{Enr}(\mathcal{P})'_{\mu}$, in other words g may be viewed as a map

$$g : \mathcal{A} \times \Upsilon(B) \to \mathbf{Enr}(\mathcal{P})'_{\mu}.$$

The universal property (22.6.1) of λ' gives a map $g' : B \to \lambda'$. Since r is a trivial fibration, it lifts to a map $g^+ : B \to \lambda$. Note that η includes composition with $\Upsilon(r)$ so composition of $\Upsilon(g^+)$ with η gets us back to g.

In the case of a cofibration $B' \to B$ if $g^+|_{B'}$ is already given, we can lift along the trivial fibration r and get an extension to g^+ on B. □

22.6.2 Construction of the limit cone

The next step is to find the structural morphism $\epsilon : \lambda_{\mathcal{A}} \to \varphi$. By Corollary 22.5.9 (for $k = 2$), there is a morphism

$$\epsilon^{(2)} : \mathcal{A} \times \widetilde{\Upsilon}_2(\lambda, *) \to \mathbf{Enr}(\mathcal{P})'_{\kappa},$$

such that

$$r_{02}(\epsilon^{(2)}) = \eta$$

and

$$r_{01}(\epsilon^{(2)}) = 1_{\lambda,\mathcal{A}}.$$

Note that $r_{12}(\epsilon^{(2)})$ is a morphism from $\mathcal{A} \times \Upsilon(*) = \mathcal{A} \times \mathbb{I}$ into $\mathbf{Enr}(\mathcal{P})'_\kappa$ restricting to $\lambda_{\mathcal{A}}$ and φ, which by definition means a morphism $\lambda_{\mathcal{A}} \to \varphi$, i.e. an element of $\mathrm{Cone}(\lambda, \varphi)$. Call this morphism ϵ. Here is a picture of $\epsilon^{(2)}$:

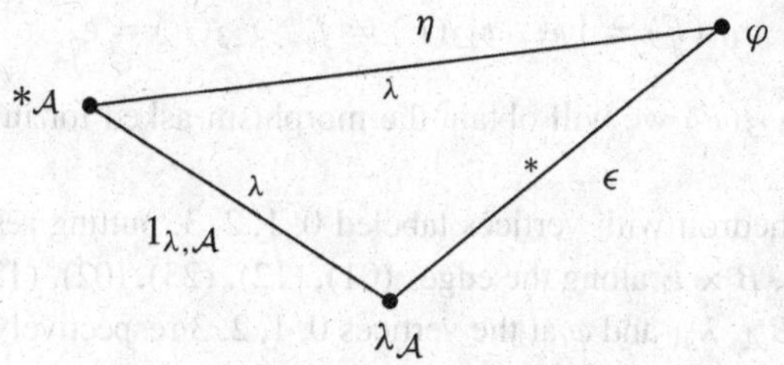

22.6.3 The universal property

To say that $\epsilon : \lambda_{\mathcal{A}} \to \varphi$ is a limit, basically means that, for any other object B and map $f : B_{\mathcal{A}} \to \varphi$, there is a map $g : B \to \lambda$ such that $\epsilon \circ g = f$. The composition relation is expressed more precisely as the triangular diagram

$$\mathcal{A} \times \widetilde{\Upsilon}_2(*, *) \to \mathbf{Enr}(\mathcal{P})'_\kappa,$$

restricting on (01) to $g_{\mathcal{A}}$ (which means g parametrized in a constant way by $\mathcal{A}$), restricting on (12) to ϵ and restricting on (02) to f. The statement of the following lemma says the same thing but for a family of maps parametrized by E. The full universal property saying that ϵ is a limiting cone, corresponding in the 1-categorical case to uniqueness of the factorization, really involves the situation of a cofibration $E' \to E$. However, it will be easier to understand the argument by starting with a single E.

Lemma 22.6.3 *Given any cofibrant E, and any cofibrant, fibrant and κ-presentable object $B \in \mathcal{P}_{\mathrm{cf},\kappa}$, suppose $f : E \to \underline{\mathrm{Cone}}(B, \varphi)$, in other words*

$$f : \mathcal{A} \times \Upsilon(E) \to \mathbf{Enr}(\mathcal{P})'_\kappa$$

is a morphism with

$$r_0(f) = B_{\mathcal{A}}, \quad r_1(f) = \varphi.$$

Then there is a morphism $E \to \underline{\mathrm{HOM}}(B, \lambda)$, i.e. $B \times E \xrightarrow{g^+} \lambda$, corresponding to $\Upsilon(E) \xrightarrow{n} \mathbf{Enr}(\mathcal{P})'_\kappa$ with $n(\upsilon_0) = B$ and $n(\upsilon_1) = \lambda$, and a morphism

$$f' : \mathcal{A} \times \widetilde{\Upsilon}_2(E, *) \to \mathbf{Enr}(\mathcal{P})'_\kappa,$$

with $r_{01}(f') = n_{\mathcal{A}}$, $r_{02}(f') = f$, *and* $r_{12}(f') = \epsilon$.

Proof The basic idea is to use what we know up until now to construct a morphism

$$F : \mathcal{A} \times \widetilde{\Upsilon}_3(B, E, *) \to \mathbf{Enr}(\mathcal{P})'_\kappa$$

with

$$r_{01}(F) = 1_B, \quad r_{13}(F) = f, \quad r_{23}(F) = \epsilon.$$

Setting $f' = r_{123}(F)$ we will obtain the morphism asked for in the previous paragraph.

Draw a tetrahedron with vertices labeled 0, 1, 2, 3, putting respectively B, E, $*$, $B \times E$, E, $B \times E$ along the edges (01), (12), (23), (02), (13), (03); then putting in $*_{\mathcal{A}}$, $B_{\mathcal{A}}$, $\lambda_{\mathcal{A}}$ and φ at the vertices 0, 1, 2, 3 respectively. And finally putting in $1_{B,\mathcal{A}}$ along edge (01), f along edge (13) and ϵ along edge (23). The elements g along (03), $g^+_{\mathcal{A}}$ along (02), then $n_{\mathcal{A}}$ along (12), are to be filled in by the subsequent discussion.

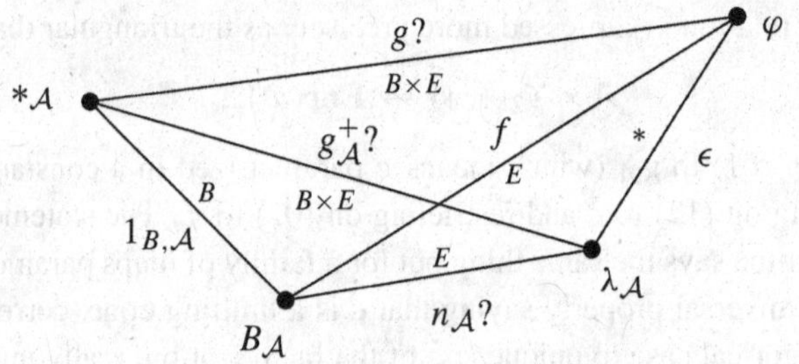

Our strategy is to fill in all of the faces except (123), then call upon Corollary 22.5.9 to fill in the tetrahedron thus getting face (123).

The first step is the face (01)3. This we fill in using simply the fact that $\mathbf{Enr}(\mathcal{P})'_\kappa$ is a fibrant $\mathcal{P}$-category. The edges (01) and (13) are specified so we can fill in to a morphism $\mathcal{A} \times \widetilde{\Upsilon}_2(B, E) \to \mathbf{Enr}(\mathcal{P})'_\kappa$ restricting to $1_{B,\mathcal{A}}$ and f on the edges (01) and (13) respectively. Now the restriction of this morphism to edge (03) provides a morphism $g : \mathcal{A} \times \Upsilon(B \times E) \to \mathbf{Enr}(\mathcal{P})'_\kappa$ restricting to $*_{\mathcal{A}}$ and φ.

The next step is to notice that by the universal property of λ (Lemma 22.6.2) there is a morphism $g^+ : B \times E \to \lambda$ such that g is deduced from η by pullback via $\Upsilon(E \times B) \to \Upsilon(\lambda)$. This same morphism yields

$$\widetilde{\Upsilon}_2(E \times B, *) \to \widetilde{\Upsilon}_2(\lambda, *),$$

and we can use this to pull back the above morphism $\epsilon^{(2)}$. This gives a morphism

$$h : \mathcal{A} \times \widetilde{\Upsilon}_2(E \times B, *) \to \mathbf{Enr}(\mathcal{P})'_{\kappa},$$

where (adopting exceptionally for obvious reasons here the notations 0, 2 and 3 for the vertices of this $\widetilde{\Upsilon}_2$)

$$r_{03}(h) = g, \quad r_{02}(h) = g^+_{\mathcal{A}}, \quad r_{23}(h) = \epsilon.$$

This treats the face (023).

Finally, for the face (012) we have a morphism

$$g^+ : \Upsilon(E \times B) \to \mathbf{Enr}(\mathcal{P})'_{\kappa},$$

restricting to $*$ and λ on endpoints. By Corollary 22.5.9 applied with $k = 2$ for the map

$$\Lambda^L \widetilde{\Upsilon}_2(B, E) \to \mathbf{Enr}(\mathcal{P})'_{\kappa},$$

given by 1_B and g^+ we get a morphism

$$m : \widetilde{\Upsilon}_2(B, E) \to \mathbf{Enr}(\mathcal{P})'_{\kappa},$$

with $r_{01}(m) = 1_B$ and $r_{02}(m) = g^+$. Put $n := r_{12}(m)$. Insert this triangle into the previous diagram by making it constant in the $\mathcal{A}$-direction.

Putting all of these together we obtain a morphism

$$F' : \mathcal{A} \times \Lambda^L \widetilde{\Upsilon}_3(B, E, *) \to \mathbf{Enr}(\mathcal{P})'_{\kappa},$$

restricting to $1_{B,\mathcal{A}}$ on edge (01) and restricting to f on edge (13) and ϵ on edge (23). Corollary 22.5.9 applied with $k = 3$ gives an extension over the tetrahedron to a morphism

$$F : \mathcal{A} \times \widetilde{\Upsilon}_3(B, E, *) \to \mathbf{Enr}(\mathcal{P})'_{\kappa},$$

again restricting to 1_B on edge (01) and restricting to f on edge (13) and ϵ on edge (23). The restriction to the last face r_{123} yields the filling-in desired. □

Lemma 22.6.3 almost gives the required property to show that $\lambda \stackrel{\epsilon}{\to} \varphi$ is a limit, by the criterion of Lemma 21.7.4. Technically speaking, we should also show the corresponding statement in the relative situation where $E' \to E$ is a cofibration between cofibrant objects, and where we already have a filling-in of the face (123) for E'. We would like to obtain a filling-in of this face for E. Basically the only difficulty is that we don't yet know that the filling-in of face (123) for E' comes from a filling-in of the whole tetrahedron compatible with the above process. In particular this causes a problem at the step where we fill in face (023).

To get started, use the fibrant property of $\mathbf{Enr}(\mathcal{P})'_\kappa$ to obtain a morphism

$$\mathcal{A} \times \widetilde{\Upsilon}_3(B, E', *) \to \mathbf{Enr}(\mathcal{P})'_\kappa, \tag{22.6.2}$$

restricting to our given morphism on the face (123), and restricting to $1_{B,\mathcal{A}}$ on the edge (01). Furthermore, the restriction to the face (012) should be constant in the $\mathcal{A}$-direction, that is it should come from a morphism

$$\widetilde{\Upsilon}_2(B, E') \to \mathbf{Enr}(\mathcal{P})'_\kappa,$$

by pulling back along the projection $\mathcal{A} \to *$. In order to do this notice that the restriction of the given map to the edge (12) comes from $\Upsilon(E') \to \mathbf{Enr}(\mathcal{P})'_\kappa$. Thus we can first extend this map combined with 1_B to a morphism $\widetilde{\Upsilon}_2(B, E') \to \mathbf{Enr}(\mathcal{P})'_\kappa$. Now the morphism

$$(\mathcal{A} \times \widetilde{\Upsilon}_2(B, E')) \cup^{\mathcal{A}\times\Upsilon(E')} \mathcal{A} \times \widetilde{\Upsilon}_2(E', *) \to \widetilde{\Upsilon}_3(B, E', *)$$

is a trivial cofibration, so we can extend from here to obtain

$$\mathcal{A} \times \widetilde{\Upsilon}_3(B, E', *) \to \mathbf{Enr}(\mathcal{P})'_\kappa,$$

with restriction to the face (012), being constant in the $\mathcal{A}$-direction, coming from the already chosen $\widetilde{\Upsilon}_2(B, E') \to \mathbf{Enr}(\mathcal{P})'_\kappa$. This is the point of departure for the rest of the argument.

The first step following the previous outline is to fill in the face (013). We note that the morphism $\Upsilon(B) \cup^{\{v_1\}} \Upsilon(E) \to \Upsilon^2(B, E)$ is a trivial cofibration. Thus also the morphism

$$(\Upsilon(B) \cup^{\{v_1\}} \Upsilon(E)) \cup^{\Upsilon(B)\cup^{\{v_1\}}\Upsilon(E')} \widetilde{\Upsilon}_2(B, E') \to \widetilde{\Upsilon}_2(B, E) \tag{22.6.3}$$

is a trivial cofibration, so given the edges (01) and (13) (for E) with filling-in over the face (013) with respect to E', we can fill in (013) with respect to E.

22.6.4 The face (023)

Now we treat the face (023). Let

$$g : \mathcal{A} \times \Upsilon(E \times B) \to \mathbf{Enr}(\mathcal{P})'_\kappa$$

be the restriction of the map (22.6.3) to the edge (03). Let g' denote its restriction to $\mathcal{A} \times \Upsilon(E' \times B)$. The map (22.6.2) constructed above restricts on (023) to a morphism

$$h' : \mathcal{A} \times \widetilde{\Upsilon}_2(E' \times B, *) \to \mathbf{Enr}(\mathcal{P})'_\kappa,$$

where (using as above the notations 0, 2 and 3 for the vertices of this $\widetilde{\Upsilon}_2$)

$$r_{03}(h') = g', \quad r_{23}(h') = \epsilon.$$

Let $a' = r_{02}(h')$. It is a morphism

$$a'_{\mathcal{A}} : \mathcal{A} \times \Upsilon(E' \times B) \to \mathbf{Enr}(\mathcal{P})'_{\kappa},$$

with $r_0\left(a'_{\mathcal{A}}\right) = *_{\mathcal{A}}$ and $r_2\left(a'_{\mathcal{A}}\right) = \lambda_{\mathcal{A}}$. By hypothesis on our map over the full tetrahedron for E', $a'_{\mathcal{A}}$ comes from a map

$$a' : \Upsilon(E' \times B) \to \mathbf{Enr}(\mathcal{P})'_{\kappa},$$

again with values $*$ and λ on the endpoints. This map corresponds to one also denoted

$$a' : E' \times B \to \mathbf{Enr}(\mathcal{P})'_{\kappa}(*, \lambda).$$

In the process used for the proof of Lemma 22.6.3, the map g^+ which shows up in the same place goes from $E \times B$ to $\lambda = \mathbf{Enr}(\mathcal{P})_{\kappa}(*, \lambda)$. We would like g^+ to be compatible with the given map a' on $E' \times B$, but if we try to use a' "as is," the difference between $\mathbf{Enr}(\mathcal{P})'_{\kappa}$ and $\mathbf{Enr}(\mathcal{P})_{\kappa}$ will be an obstruction. The solution to this problem is to first modify our initial datum so that a' comes from a map $E' \times B \to \lambda$ before continuing to proceed along the same lines as in the proof of Lemma 22.6.3. The argument which follows is a standard exercise in making this kind of modification within a model category, using the interval object $\mathbf{I}$; the cartesian property is used without mention to say that various pushout-product maps are cofibrations.

The morphism $\mathbf{Enr}(\mathcal{P})_{\kappa} \to \mathbf{Enr}(\mathcal{P})'_{\kappa}$ is an equivalence so a' is equivalent to a different morphism $b' : E' \times B \to \mathbf{Enr}(\mathcal{P})_{\kappa}(*, \lambda)$. The two resulting morphisms $\Upsilon(E' \times B) \to \mathbf{Enr}(\mathcal{P})'_{\kappa}$ are equivalent so by Proposition 21.3.1 there is a morphism

$$\mathbf{I} \times \Upsilon(E' \times B) \to \mathbf{Enr}(\mathcal{P})'_{\kappa}$$

sending the endpoints $\upsilon_0, \upsilon_1 \in \mathbf{I}$ to a' and b', and constant on $\mathbf{I} \times \{\upsilon_i\}$.

Using this different morphism b' (which is now the same thing as a map $E' \times B \to \lambda$) we pull back our standard

$$\eta \in \mathrm{HOM}\left(\mathcal{A} \times \widetilde{\Upsilon}_2(\lambda, *), \mathbf{Enr}(\mathcal{P})'_{\kappa}\right),$$

to get a morphism

$$\mathcal{A} \times \widetilde{\Upsilon}_2(E' \times B, *) \to \mathbf{Enr}(\mathcal{P})'_{\kappa},$$

restricting on the edges to b' and ϵ respectively.

Now we have a map from

$$(\mathcal{A} \times \{\upsilon_0, \upsilon_1\} \times \widetilde{\Upsilon}_2(E' \times B, *)) \cup \left(\mathcal{A} \times \mathbf{I} \times \left[\Upsilon(E' \times B) \cup \Upsilon(*)\right]\right)$$

to $\mathbf{Enr}(\mathcal{P})'_\kappa$, where we have omitted in the notation the objects along which the glueing takes place.

The morphism from the above domain to

$$\mathcal{A} \times \mathbf{I} \times \widetilde{\Upsilon}_2(E' \times B, *)$$

is a trivial cofibration, so since $\mathbf{Enr}(\mathcal{P})'_\kappa$ is fibrant there exists an extension of the above to a morphism

$$\mathcal{A} \times \mathbf{I} \times \widetilde{\Upsilon}_2(E' \times B, *) \to \mathbf{Enr}(\mathcal{P})'_\kappa.$$

This morphism is a standard one coming from $b' : E' \times B \to \lambda$ on the end $\upsilon_1 \in \mathbf{I}$, and it is our given h' on the end $\upsilon_0 \in \mathbf{I}$.

We now go to the edge (03) of the triangle (023). We are also given an extension of g' to $g : \mathcal{A} \times \Upsilon(E \times B) \to \mathbf{Enr}(\mathcal{P})'_\kappa$ along the edge (03) of the triangle and υ_0 of the interval $\mathbf{I}$. Thus, using the face (03) $\times$ $\mathbf{I}$, we have a morphism

$$(\mathcal{A} \times \Upsilon(E \times B)) \cup^{\mathcal{A} \times \Upsilon(E' \times B)} (\mathcal{A} \times \mathbf{I} \times \Upsilon(E' \times B)) \to \mathbf{Enr}(\mathcal{P})'_\kappa.$$

Fill this in along the trivial cofibration

$$(\mathcal{A} \times \Upsilon(E \times B)) \cup^{\mathcal{A} \times \Upsilon(E' \times B)} (\mathcal{A} \times \mathbf{I} \times \Upsilon(E' \times B)) \hookrightarrow \mathcal{A} \times \mathbf{I} \times \Upsilon(E \times B),$$

to give on the whole a morphism

$$\mathcal{A} \times \mathbf{I} \times \Upsilon(E \times B) \cup^{\mathcal{A} \times \mathbf{I} \times \Upsilon(E' \times B)} (\mathcal{A} \times \mathbf{I} \times \widetilde{\Upsilon}_2(E' \times B, *) \to \mathbf{Enr}(\mathcal{P})'_\kappa,$$

where the morphism $\Upsilon(E' \times B) \to \widetilde{\Upsilon}_2(E' \times B, *)$ in question is the one coming from the edge (03).

Next we extend down along the triangle (023) times the end $\upsilon_1 \in \mathbf{I}$. To do this, notice that our extension from the previous paragraph over $\upsilon_1 \in \mathbf{I}$ gives an extension of the morphism $b' : \Upsilon(E' \times B) \to \mathbf{Enr}(\mathcal{P})'_\kappa$ to a morphism $b : \Upsilon(E \times B) \to \mathbf{Enr}(\mathcal{P})'_\kappa$ (these being along the top edge (03)). By the universal property of λ, Lemma 22.6.2, this lifts to an extension $E \times B \to \lambda$. Now the morphism that we already have on the end $\upsilon_1 \in \mathbf{I}$ comes by pulling back the standard $\eta : \mathcal{A} \times \Upsilon^2(\lambda, *) \to \mathbf{Enr}(\mathcal{P})'_\kappa$ via the map $E' \times B \to \lambda$ so our extension allows us to pull back η to get a map $b : \mathcal{A} \times \widetilde{\Upsilon}_2(E \times B, *) \to \mathbf{Enr}(\mathcal{P})'_\kappa$ extending the previous b'.

Now we have our map

$$\mathcal{A} \times \mathbf{I} \times \widetilde{\Upsilon}_2(E' \times B, *) \to \mathbf{Enr}(\mathcal{P})'_\kappa,$$

which is provided with an extension from E' to E, over the faces (03) $\times$ $\mathbf{I}$ and (023) $\times$ $\{\upsilon_1\}$ of the product of the triangle with the interval. Another small step is to notice that along the face (02) $\times$ $\mathbf{I}$ the morphism is pulled back along

$\mathcal{A} \to *$ from a morphism $\mathbf{I} \times \Upsilon(E' \times B) \to \mathbf{Enr}(\mathcal{P})'_{\kappa}$. On the other hand at the edge $(02) \times \{\upsilon_1\}$ the extension from E' to E again comes from a morphism $\Upsilon(E \times B) \to \mathbf{Enr}(\mathcal{P})'_{\kappa}$. We get a morphism

$$\mathbf{I} \times \Upsilon(E' \times B) \times^{\{\upsilon_1\}\times\Upsilon(E'\times B)} \Upsilon(E \times B) \to \mathbf{Enr}(\mathcal{P})'_{\kappa},$$

which can be extended along the trivial cofibration

$$\mathbf{I} \times \Upsilon(E' \times B) \times^{\{\upsilon_1\}\times\Upsilon(E'\times B)} \Upsilon(E \times B) \hookrightarrow \mathbf{I} \times \Upsilon(E \times B),$$

to give a map

$$\mathbf{I} \times \Upsilon(E \times B) \to \mathbf{Enr}(\mathcal{P})'_{\kappa}.$$

Similarly, we note that the map on the face $(23) \times \mathbf{I}$ is pulled back from our map $\epsilon : \Upsilon(*) \to \mathbf{Enr}(\mathcal{P})'_{\kappa}$.

All together on the triangular icosahedron $(023) \times \mathbf{I}$ we have a morphism defined for E' plus, along the faces

$$(03) \times \mathbf{I}, \quad (02) \times \mathbf{I}, \quad (23) \times \mathbf{I}, \quad (023) \times \{\upsilon_1\},$$

extensions from E' to E (all compatible on intersections of the faces and having the required properties along (02) and (23)). The inclusion of this $\mathcal{P}$-precategory (which we will call $\mathcal{G}$ instead of writing it out) into

$$\mathcal{A} \times \mathbf{I} \times \widetilde{\Upsilon}_2(E \times B, *)$$

is a trivial cofibration. Indeed $\mathcal{G}$ comes by attaching to the end

$$\mathcal{A} \times \{\upsilon_1\} \times \widetilde{\Upsilon}_2(E \times B, *),$$

something of the form

$$\mathcal{A} \times \partial\widetilde{\Upsilon}_2(E \times B, *) \cup^{\mathcal{A}\times\partial\widetilde{\Upsilon}_2(E'\times B,*)} \mathcal{A} \times \widetilde{\Upsilon}_2(E' \times B, *),$$

where $\partial\widetilde{\Upsilon}_2(E \times B, *)$ denotes the amalgamated sum of the three "edges" $\Upsilon(E \times B)$ (two times) and $\Upsilon(*)$. The inclusion of the end $\mathcal{A} \times \{\upsilon_1\} \times \widetilde{\Upsilon}_2$ $(E \times B, *)$ into $\mathcal{G}$ is an equivalence, as is the inclusion of this end into the full product

$$\mathcal{A} \times \mathbf{I} \times \widetilde{\Upsilon}_2(E \times B, *),$$

which proves that the map in question (from $\mathcal{G}$ to the above full product) is a trivial cofibration.

Now we again make use of the fibrant property to extend our map from $\mathcal{G}$ to a morphism

$$\mathcal{A} \times \mathbf{I} \times \widetilde{\Upsilon}_2(E \times B, *) \to \mathbf{Enr}(\mathcal{P})'_{\kappa}.$$

When restricted to $\mathcal{A} \times \{v_0\} \times \widetilde{\Upsilon}_2(E \times B, *)$ this gives the extension h desired in order to complete our treatment of the face (023).

22.6.5 End of the proof of Theorem 22.6.1

For the face (012) the argument is the same as in the previous case but applying Corollary 22.5.10 rather than 22.5.9 in view of our relative situation $E' \to E$.

This completes the construction of a morphism

$$F' : \mathcal{A} \times \Lambda^L \widetilde{\Upsilon}_3(B, E, *) \to \mathbf{Enr}(\mathcal{P})'_\kappa,$$

restricting to $1_{B,\mathcal{A}}$ on edge (01) and restricting to f on edge (13) and ϵ on edge 23. Furthermore, by construction it restricts to our already-given morphism over E'. Corollary 22.5.10 applied with $k=3$ gives an extension over the tetrahedron to a morphism

$$F : \mathcal{A} \times \widetilde{\Upsilon}_3(B, E, *) \to \mathbf{Enr}(\mathcal{P})'_\kappa,$$

again restricting to $1_{B,\mathcal{A}}$ on edge (01) and restricting to f on edge (13) and ϵ on edge 23, and restricting to the already-given morphism over E'. The restriction to the last face $r_{123}(F)$ yields the filling-in desired.

Recall the argument of Lemma 21.7.4. A map $E \to \underline{\mathrm{Cone}}(B, \varphi)$ is the same thing as a map $\mathcal{A} \times \Upsilon(E, *) \to \mathbf{Enr}(\mathcal{P})'_\kappa$ with appropriate restrictions to various vertices and edges, by Lemma 21.7.3. Similarly, a map $E \to \underline{\mathrm{Cone}}(B, \lambda, \varphi)$ is the same thing as a map $\mathcal{A} \times \widetilde{\Upsilon}_2(E, *) \to \mathbf{Enr}(\mathcal{P})'_\kappa$ again with appropriate restrictions. The above argument therefore shows that, given a diagram

$$\begin{array}{ccc} E' & \longrightarrow & \underline{\mathrm{Cone}}(B, \lambda, \varphi) \\ \downarrow & & \downarrow \\ E & \longrightarrow & \underline{\mathrm{Cone}}(B, \varphi) \end{array}$$

such that the top map projects to $\epsilon \in \mathrm{Cone}(\lambda, \varphi)$, there exists a lifting $E \to \underline{\mathrm{Cone}}(B, \lambda, \varphi)$ also projecting to ϵ. This proves that the right vertical arrow is a trivial fibration, which says that $\lambda \xrightarrow{\epsilon} \varphi$ is a limit of φ.

Given a second regular cardinal $\kappa \leq \kappa'$, the limit ϵ of φ in $\mathbf{Enr}(\mathcal{P})'_\kappa$ maps by $\mathbf{i}_{\kappa,\kappa'}$ to the corresponding object for κ', indeed the image of $\epsilon^{(2)}$ and hence of ϵ will be possible choices for these in $\mathbf{Enr}(\mathcal{P})'_\kappa$. So the image $\mathbf{i}_{\kappa,\kappa'}(\epsilon)$ is a limit at the κ'-stage, by the same proof. This shows that the transition maps preserve the limits which have been constructed, finishing the proof of Theorem 22.6.1. □

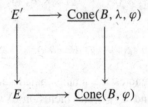

Corollary 22.6.4 *If $F : \mathcal{A} \times \mathcal{B} \to \mathbf{Enr}(\mathcal{P})'_{\cdot}$ is a functor from the product of two $\mathcal{P}$-categories, then taking the limits first in one direction and then in the other, is independent of which direction is chosen first.*

Proof This is a consequence of Theorem 21.10.7 but can also be seen directly from the construction of the limit λ in Section 22.6.1. □

The general construction applies to give limits of n-categories and (∞, n)-categories.

Theorem 22.6.5 *The $(n+1)$-category nCAT of all small n-categories, admits limits indexed by small $(n+1)$-categories. The Segal $(n+1)$-category of all small Segal n-categories nSeCAT admits limits indexed by small Segal $(n+1)$-categories.*

Proof These are just the special cases of Theorem 22.6.1 where $\mathcal{P} = \mathcal{PC}^n(\mathrm{SET})$ or $\mathcal{P} = \mathcal{PC}^n(\mathcal{K})$. □

22.7 Cardinality

In Section 22.9 we would like to use the existence of limits in order to construct colimits in $\mathbf{Enr}(\mathcal{P})'_{\cdot}$. This is a standard technique in category theory, but needs a theory of cardinality. For motivation the reader may wish to consult Section 22.9.1 where the technique is explained as it applies to construct colimits in a 1-category.

22.7.1 Cardinality in $\mathcal{P}$

If $\mathcal{P}$ is a tractable model category, in particular it is locally presentable, and there are several possible definitions of cardinality of an object $\mathcal{A} \in \mathcal{P}$. The easiest is to define the *regular cardinality* $\#^{\mathrm{reg}}(\mathcal{A})$ to be the smallest regular cardinal λ such that $\mathcal{A}$ is λ-presentable and $\mathcal{P}$ itself is locally λ-presentable.

Here are some other more precise ways. Let κ be a regular cardinal such that $\mathcal{P}$ is locally κ-presentable. Then any object $\mathcal{A} \in \mathcal{P}$ may be canonically written as a κ-filtered colimit indexed by the category $\mathcal{P}_\kappa/\mathcal{A}$, see Theorem 8.1.2. Define $\#_\kappa(\mathcal{A})$ to be the essential cardinality of the indexing category $\mathcal{P}_\kappa/\mathcal{A}$, that is to say the sum of the cardinalities of the morphism sets over all pairs of isomorphism classes of objects. Let $\#'_\kappa(\mathcal{A}) \leq \#_\kappa(\mathcal{A})$ be the smallest cardinality of a κ-filtered category ϕ such that $\mathcal{A}$ is the colimit of a family of κ-presentable objects indexed by ϕ. Let $\#''_\kappa(\mathcal{A})$ be the sum over all isomorphism classes of κ-presentable objects $Y \in \mathcal{P}$, of the cardinality of $\mathcal{P}(Y, \mathcal{A})$.

We leave it to the reader to investigate the relationships between these notions, which are undoubtedly considered already in the literature. For our purposes the much cruder regular cardinality will be sufficient.

Lemma 22.7.1 *For any λ the set of isomorphism classes of objects $\mathcal{A}$ with $\#^{\mathrm{reg}}(\mathcal{A}) \leq \lambda$ is small. Given any small set of objects $\{\mathcal{A}_i\}_{i\in I} \subset \mathcal{P}$, there is an upper bound λ such that $\#^{\mathrm{reg}}(\mathcal{A}_i) \leq \lambda$ for all i. Given such a family with the bound λ, if $\# I \leq \lambda$ too, then*

$$\#^{\mathrm{reg}}\left(\coprod_{i\in I} \mathcal{A}_i\right) \leq \lambda.$$

Suppose $\mathcal{B} \in \mathcal{P}$ is another object, and suppose $f_i : \mathcal{A}_i \to \mathcal{B}$ is a family of morphisms. Then there is a family of factorizations

$$\mathcal{A}_i \stackrel{f_i'}{\to} \mathcal{B}' \stackrel{g}{\to} \mathcal{B},$$

with the same morphism g working for all i, such that $\#^{\mathrm{reg}}(\mathcal{B}') \leq \lambda$.

Proof The first statement follows from Definition 8.1.1 of the local presentability property of $\mathcal{P}$, see notably the last statement of Theorem 8.1.2. Any small family of cardinals has an upper bound. For the direct sum, use the definition of λ-presentability: given a map from the direct sum into a λ-filtered colimit $\mathrm{colim}_j \mathcal{B}_j$, since each $\mathcal{A}$ is λ-presentable, for each i there is $j(i)$ with $\mathcal{A}_i$ factoring through $\mathcal{B}_j$. But then since $\# I \leq \lambda$, there is an upper bound for these $j(i)$ in the λ-filtred indexing category, which shows that λ-presentable property for the direct sum. For the last statement, write $\mathcal{B}$ as a λ-filtered colimit of λ-presentable objects, and apply the previous statement on direct sums. □

If $\mathcal{A} \in \mathcal{PC}(\mathcal{P})$ define $\#^{\mathrm{PC,reg}}(\mathcal{A})$ to be the sum over all objects $x. \in \Delta_{\mathrm{Ob}(\mathcal{A})}$ of $\#^{\mathrm{reg}}(\mathcal{A}(x.))$.

22.7.2 Factorization of cocones

The factorization statement needed for the construction of colimits says that cocones from functors $\psi : \mathcal{A} \to \mathbf{Enr}(\mathcal{P})'_{.}$ into arbitrary objects, factor through objects of regular cardinality depending on ψ:

Lemma 22.7.2 *Suppose $\mathcal{A}$ is a fixed cofibrant $\mathcal{P}$-precategory, and $\psi : \mathcal{A} \to \mathbf{Enr}(\mathcal{P})'_{.}$ is a functor. Then there is a regular cardinal κ such that, for any $U \in \mathrm{Ob}(\mathcal{P})$ and any element $u \in \mathrm{Cocone}(\psi, U)$, there exists a factorization of u through an object $W \in \mathrm{Ob}(\mathcal{P}_\kappa)$, which is to say an element of $\mathrm{Cocone}(\psi, W, U)$ restricting to u on the (02) edge.*

Proof We may assume that ψ factors through a functor denoted by the same letter

$$\psi : \mathcal{A} \to \mathbf{Enr}(\mathcal{P})_{\bullet},$$

because homotopy equivalences of functors ψ lead to equivalences between the cocones. Alternatively here one could note that the problem of constructing colimits where this factorization will be applied, is invariant under equivalences of ψ.

Put

$$\mathbf{V}(\mathcal{A}) := \mathcal{A} \times \mathbb{I} \cup^{\mathcal{A}\times\{v_1\}} *.$$

As for any element of Cocone(ψ, U), notice that u may be viewed as a morphism

$$u : \mathbf{V}(\mathcal{A}) \to \mathbf{Enr}(\mathcal{P})'_{\bullet},$$

restricting to ψ over $\mathcal{A} \times \{v_0\}$ and to U over $*$. The map $\mathcal{A} \sqcup * \xrightarrow{i} \mathbf{V}(\mathcal{A})$ is a cofibration in the usual Reedy structure. Choose a projective cofibrant replacement fitting into a diagram

$$\mathcal{A} \sqcup * \xrightarrow{j} \widetilde{\mathbf{V}}(\mathcal{A}) \xrightarrow{p} \mathbf{V}(\mathcal{A}),$$

with $p \circ j = i$, such that j is a projective cofibration and p a projective fibration. The composition gives a map

$$\widetilde{\mathbf{V}}(\mathcal{A}) \xrightarrow{u \circ p} \mathbf{Enr}(\mathcal{P})'_{\bullet}.$$

Because of the projective cofibrant property of j and the fact that even before its fibrant replacement $\mathbf{Enr}(\mathcal{P})_{\bullet}$ was projectively (i.e. levelwise) fibrant, $u \circ p$ is homotopically equivalent relative to $\mathcal{A} \sqcup *$, to a map

$$\tilde{u} : \widetilde{\mathbf{V}}(\mathcal{A}) \to \mathbf{Enr}(\mathcal{P})_{\bullet}.$$

Note that, according to the hypothesis made at the start of the proof, u restricted to $\mathcal{A} \sqcup *$ factors through $\mathbf{Enr}(\mathcal{P})_{\bullet}$.

Define now $\widehat{\mathbf{V}}(\mathcal{A}) \hookrightarrow \widetilde{\mathbf{V}}(\mathcal{A})$ to be the sub-precategory such that

$$\widehat{\mathbf{V}}(\mathcal{A})(*, \ldots, *) = *,$$

but it is the same as $\widetilde{\mathbf{V}}(\mathcal{A})$ on other sequences of objects. The maps

$$\widehat{\mathbf{V}}(\mathcal{A}) \to \widetilde{\mathbf{V}}(\mathcal{A}) \to \mathbf{V}(\mathcal{A})$$

are weak equivalences. The structural data of a map

$$\widehat{\mathbf{V}}(\mathcal{A}) \to \mathbf{Enr}(\mathcal{P})_{\bullet},$$

sending $*$ to U and equa to ψ on $\mathcal{A}$, consists of a collection of maps from various objects, depending only on $\mathcal{A}$, ψ and the choice of $\widehat{\mathbf{V}}(\mathcal{A})$, into U. These are subject to some relations again depending on the same things. As in the previous lemma, there is a regular cardinal κ such that the objects involved are κ-presentable, and the total number of maps to look at is $< \kappa$. By local presentability, U is a κ-filtered colimit of κ-presentable objects, so the collection of data factors through some $W \to U$ with $W \in \mathrm{Ob}(\mathscr{P}_\kappa)$. Let

$$\widehat{\mathbf{V}}_2(\mathcal{A}) \to \widetilde{\mathbf{V}}_2(\mathcal{A}) \to \mathbf{V}_2(\mathcal{A})$$

be the weak equivalences obtained by analogous constructions starting with $\mathcal{A} \times \widetilde{\Upsilon}_2(*, *)$. Our factorization gives a map $\widehat{\mathbf{V}}_2(\mathcal{A}) \xrightarrow{\hat{f}} \mathbf{Enr}(\mathscr{P})$, restricting to $\tilde{u}$ on the (02) edge and restricting to the map $W \to U$ on the (12) edge. Using the fibrant property of $\mathbf{Enr}(\mathscr{P})'$ in the usual Reedy structure, $\hat{f}$ is homotopy equivalent relative to $\mathcal{A} \sqcup * \sqcup *$ to a map $\mathbf{V}_2(\mathcal{A}) \xrightarrow{f} \mathbf{Enr}(\mathscr{P})'$, which is the desired element of $\mathrm{Cocone}(\psi, W, U)$. □

22.7.3 Cardinality of *n*-precategories

In the special case of n-categories, one can give some more sensitive and precise notions of cardinality. While not necessary for the construction of colimits, this discussion may be interesting for other reasons.

Suppose $\mathcal{A}$ is an n-precategory, which is to say an object of $\mathscr{PC}^n(\mathrm{SET})$. We define the *cardinal* of $\mathcal{A}$, denoted $\#\mathcal{A}$ in the following way. Choose for every $y \in \pi_0(\mathcal{A}) = \tau_{\leq 0}(\mathcal{A})$ (the set of equivalence classes of objects) a lifting to an object $\tilde{y} \in \mathrm{Ob}(\mathcal{A})$. Then

$$\#\mathcal{A} := \sum_{y,z \in \pi_0(\mathcal{A})} \#\mathcal{A}(\tilde{y}, \tilde{z}).$$

The sum of cardinals is of course the cardinal of the disjoint union of representing sets. This definition is recursive, as the formula refers inductively to the cardinals of the $(n-1)$-categories $\mathcal{A}(\tilde{y}, \tilde{z})$. At the start we define the cardinal of a 0-category (i.e. a set) in the usual way.

Lemma 22.7.3 *The above definition of $\#\mathcal{A}$ doesn't depend on the choice of representatives. If $\mathcal{A} \to \mathcal{B}$ is an equivalence of n-categories then $\#\mathcal{A} = \#\mathcal{B}$.*

Proof Left to the reader. □

An easier and more obvious notion is the *precardinality of* $\mathcal{A}$. If $\mathcal{A} \in \mathcal{PC}^n$ (SET) we define

$$\#^{\mathrm{pre}}\mathcal{A} := \sum_{M \in \mathbf{C}^n(*)} \#(\mathcal{A}_M).$$

For infinite cardinalities the precardinal of $\mathcal{A}$ is also the maximum of the cardinalities of the sets $\mathcal{A}_M$. In any case note that the precardinality is infinite unless $\mathcal{A}$ is empty.

Remark 22.7.4 Let $\mathcal{A} \mapsto \mathbf{Seg}(\mathcal{A})$ denote the operation of replacing an n-precategory by the associated fibrant n-category. Then the precardinal of $\mathbf{Seg}(\mathcal{A})$ is bounded by the maximum of ω and the precardinal of $\mathcal{A}$. Similarly by the argument of (section 6, proof of CM5(1) in my preprint [234]), for any n-precategory $\mathcal{A}$ there is a replacement by a fibrant n-category $\mathcal{A} \hookrightarrow \mathcal{A}'$ with

$$\#^{\mathrm{pre}}\mathcal{A}' = \#^{\mathrm{pre}}\mathcal{A}.$$

Note trivially that

$$\#\mathcal{A} \leq \#^{\mathrm{pre}}\mathcal{A}.$$

The following lemma gives a converse up to equivalence:

Lemma 22.7.5 *Suppose* $\mathcal{A}$ *is an* n*-category with* $\#\mathcal{A} \leq \alpha$ *for an infinite cardinal* α. *Then* $\mathcal{A}$ *is equivalent to an* n*-category* $\mathcal{A}'$ *of precardinality* $\leq \alpha$.

Proof Left to the reader. □

In the case of n-precategories, one can see the factorization statement of Lemma 22.7.2 by just counting all of the elements of data going into a cocone u.

22.8 Splitting idempotents

Another main element of the argument to be used in Section 22.9 to construct colimits is the splitting of idempotents. Again this is reviewed in Section 22.9.1 for the case of 1-categories. Notice that the basic construction for splitting an idempotent p is to take the homotopy colimit of an infinite sequence of compositions of p, by a mapping telescope construction. It isn't surprising that we need to treat some special case of colimits first.

Suppose C is a usual 1-category. An *idempotent* or *projector* on an object $x \in \mathrm{Ob}(C)$ is an element $p \in C(x,x)$ such that $p^2 = p$. If $\mathcal{A} \in \mathcal{PC}(\mathcal{P})$ is a $\mathcal{P}$-category, then a *homotopy idempotent for* $x \in \mathrm{Ob}(\mathcal{A})$ is a point

$p : * \to \mathcal{A}(x, x)$ such that $p \circ p \sim p$, where $p \circ p$ is any choice of composition of p with itself and $\sim$ is the relation of homotopy. If $p \sim p'$ and p is a homotopy idempotent then so is p', and up to the relation of homotopy, a homotopy idempotent is the same thing as an idempotent in the truncated 1-category $\tau_{\leq 1}(\mathcal{A})$. If $\mathcal{A}'$ is equivalent to $\mathcal{A}$ then $\tau_{\leq 1}(\mathcal{A})$ is equivalent to $\tau_{\leq 1}(\mathcal{A}')$ and, assuming both are fibrant, say, the notions of homotopy idempotents are the same.

Similarly in a model category $\mathcal{P}$, a homotopy idempotent for an object X is a morphism $p : X \to X$ such that $p \circ p \sim p$; if X is fibrant and cofibrant then the set of homotopy idempotents up to homotopy is the same as the set of idempotents on X in $\text{ho}(\mathcal{P})$. Furthermore, $\text{ho}(\mathcal{P})$ is equivalent to $\tau_{\leq 1}\mathbf{Enr}(\mathcal{P})_K$ or $\tau_{\leq 1}\mathbf{Enr}(\mathcal{P})'_K$, so up to homotopy everywhere, homotopy idempotents for the model category are the same as homotopy idempotents for the enriched category.

An idempotent (X, p) in a 1-category C is *split* if there exists a diagram

$$X \xrightarrow{a} T \xrightarrow{b} X,$$

such that $b \circ a = p$ and $a \circ b = 1_T$. It is an exercise to show that the splitting is unique up to unique isomorphism compatible with all of the maps.

This notion extends to $\mathcal{P}$-categories or model categories by asking for homotopies rather than equalities, and again splitting a homotopy idempotent is the same thing as splitting the corresponding idempotent in the homotopy category or the truncated 1-category.

Unfortunately, the notion of homotopy idempotent doesn't contain enough information to produce a splitting in general. This has been pointed out, and the phenomenon studied, by Dydak [106, 107], Freyd and Heller [114], Hastings and Heller [138, 139] and others. The basic and in some sense universal example is an endomorphism p of $K(F, 1)$ where F is *Thompson's group*. While p is a pointed endomorphism, the homotopy between p^2 and p doesn't preserve the basepoint and the cited references show that p cannot be split. Kaledin pointed me to Toën's paper [258] which in turn pointed out the above references. A splitting is constructed for idempotents in the model category of dg-algebras in *loc. cit.* [258]. Satoshi Mochizuki [206] proves that the higher derived categories of abelian categories are idempotent complete.

In spite of these problems, we don't need to include too much extra data. Indeed, the natural candidate for a splitting of a homotopy idempotent p is the "mapping telescope" made from p and which calculates the homotopy colimit of the diagram

$$X \xrightarrow{p} X \xrightarrow{p} X \xrightarrow{p} X \xrightarrow{p} \ldots .$$

Homotopies on the mapping telescope may be defined stage-by-stage, which limits the number of coherence conditions which are needed, to a single one as follows: if H denotes the homotopy from p^2 to p, then we obtain two different homotopies pH and Hp from p^3 to p^2; these should be the same. An idempotent together with a homotopy satisfying this condition, will be called a *compatible homotopy idempotent*. In the other direction, a split homotopy idempotent can always be transformed, using some cofibrant and fibrant replacements and a lifting, into a strict idempotent p' in $\mathcal{P}$, in particular it has a compatible homotopy which is just the constant homotopy between $(p')^2 = p'$ and itself.

Given a compatible homotopy idempotent, we will show that the underlying homotopy idempotent splits. Notice, however, that we don't claim that the compatible homotopy coming from the splitting is the same as the given one; such a statement would undoubtedly require a further coherence hypothesis. The compatibility condition for the homotopy was missing from the "sketch of proof" given in my preprint [235] and the present discussion corrects that problem.

The reader may notice that the compatibility condition on the homotopy H is much less subtle than the necessary and sufficient obstruction discussed by Freyd and Heller [114] and Hastings and Heller [138]. They show that a homotopy idempotent (p, H) on a space X induces a map from Thompson's group $F \xrightarrow{\Phi} \pi_1(X)$ and p splits if and only if this map is not injective. If H satisfies the compatibility condition that pH and Hp are the same, then the map Φ sends already the second generator of F (in the notations of Hastings and Heller [138]) to the trivial element, so it is *a fortiori* noninjective. The compatibility condition allows us to build up the required homotopies in an elementary way.

The splitting object T will be obtained by a mapping telescope construction. Introduce first some notations in preparation. Suppose

$$* \sqcup * \xrightarrow{i_0, i_1} \mathcal{I}$$

is a cofibration giving an interval object in $\mathcal{P}$, in other words $\mathcal{I} \to *$ is a trivial fibration. It follows from the cartesian property for $\mathcal{P}$ that if X is a cofibrant object, then

$$X \sqcup X = X \times (* \sqcup *) \to X \times \mathcal{I}$$

is a cofibration. If $a : X \to X$ is a morphism in $\mathcal{P}$, define the *telescope of length m* $\mathbf{Tel}_m(X, a)$ by induction on the positive integer m. Set

$$\mathbf{Tel}_0(X, a) := X,$$

and let $\mathbf{Tel}_1(X, a)$ be the pushout in the diagram

$$\begin{array}{ccc} X & \xrightarrow{a} & X \\ {\scriptstyle 1\times i_1}\downarrow & & \downarrow{\scriptstyle \nu_1} \\ X\times\mathcal{I} & \xrightarrow{\mu_1} & \mathbf{Tel}_1(X,a). \end{array}$$

We have labeled the tautological morphisms μ_1 and ν_1 respectively. By definition $\mu_1 \circ (1 \times i_1) = \nu_1 \circ a$. Let $\nu_0 := \mu_1 \circ (1 \times i_0) : X \to \mathbf{Tel}_1(X, a)$ and use this to view $\mathbf{Tel}_0(X, a) = X$ as a subobject.

Suppose now that $\mathbf{Tel}_m(X, a)$ is already defined, together with maps

$$\nu_i : X \to \mathbf{Tel}_m(X, a) \text{ for } 0 \le i \le m,$$

$$\mu_i : X \times \mathcal{I} \to \mathbf{Tel}_m(X, a) \text{ for } 1 \le i \le m,$$

satisfying

$$\mu_i \circ (1 \times i_1) = \nu_i \circ a, \quad \mu_i \circ (1 \times i_0) = \nu_{i-1}.$$

Define $\mathbf{Tel}_{m+1}(X, a)$ to be the pushout in the diagram

$$\begin{array}{ccc} X & \xrightarrow{\nu_0} & \mathbf{Tel}_1(X,a) \\ {\scriptstyle \nu_m}\downarrow & & \downarrow \\ \mathbf{Tel}_m(X,a) & \to & \mathbf{Tel}_{m+1}(X,a). \end{array}$$

Use the bottom map to define the tautological maps

$$\nu_i : X \to \mathbf{Tel}_{m+1}(X, a), \quad \mu_i : X \times \mathcal{I} \to \mathbf{Tel}_{m+1}(X, a),$$

for $0 \le i \le m$ and $1 \le i \le m$ respectively; let ν_{m+1} (resp. μ_{m+1}) be the composition of the right vertical map with ν_1 (resp. μ_1). This defines $\mathbf{Tel}_.(X, a)$ and $\nu_.$, $\mu_.$ by induction. Note furthermore that the maps ν_m and μ_m are cofibrations, as are the transitions $\mathbf{Tel}_m(X, a) \to \mathbf{Tel}_n(X, a)$ for $m \le n$.

Put $\mathbf{Tel}(X, a) := \lim_{\to, m} \mathbf{Tel}_m(X, a)$. It is the homotopy colimit of the diagram

$$X \xrightarrow{a} X \xrightarrow{a} X \xrightarrow{a} X \xrightarrow{a} \ldots,$$

indeed $\nu_m : X \to \mathbf{Tel}_m(X, a)$ is a trivial cofibration as may be seen by induction, and the transition maps are homotopic to a. Let $\nu_i : X \to \mathbf{Tel}(X, a)$ and $\mu_i : X \times \mathcal{I} \to \mathbf{Tel}(X, a)$ be the maps induced by the same ones at the finite stages, which are compatible with the transition maps.

Lemma 22.8.1 *Suppose $Z \in \mathcal{P}$. A map $\varphi : \mathbf{Tel}(X, a) \to Z$ is the same thing as the data of a collection of maps $\varphi_i : X \times \mathcal{I} \to Z$ such that*

$$\varphi_{i+1} \circ (1 \times i_0) = \varphi_i \circ (1 \times i_1) \circ a, \tag{22.8.1}$$

for all $i \geq 1$. Also the telescope construction is compatible with products: $\mathbf{Tel}(X, a) \times Y = \mathbf{Tel}(X \times Y, a \times 1)$.

Proof Given φ, put $\varphi_i := \varphi \circ \mu_i$. This satisfies the required conditions. On the other hand, given a collection of φ_i, construct $\varphi|_{\mathbf{Tel}_m(X,a)}$ by induction and take the limit over m. □

Using this lemma, we can define various maps between telescopes:

Lemma 22.8.2 *Suppose $a : X \to X$, $b : Y \to Y$, $n \geq 0$, and suppose given a map $h : X \times \mathcal{I} \to Y$ such that, denoting $h_0 := h \circ (1 \times i_0)$ and $h_1 := h \circ (1 \times i_1)$, we require*

$$b \circ h_1 = h_0 \circ a. \tag{22.8.2}$$

Then there exists a map $\varphi_n(h) : \mathbf{Tel}(X, a) \to \mathbf{Tel}(Y, b)$ such that the diagrams

$$\begin{array}{ccc} X \times \mathcal{I} & \xrightarrow{\mu_i} & \mathbf{Tel}(X, a) \\ \downarrow{\scriptstyle (h, \mathbf{pr}_2)} & & \downarrow{\scriptstyle \varphi_n(h)} \\ Y \times \mathcal{I} & \xrightarrow{\mu_{i+n}} & \mathbf{Tel}(Y, b) \end{array}$$

and

$$\begin{array}{ccccc} X & \xrightarrow{a} & X & \xrightarrow{\nu_i} & \mathbf{Tel}(X, a) \\ \downarrow{\scriptstyle h_1} & & \downarrow{\scriptstyle h_0} & & \downarrow{\scriptstyle \varphi_n(h)} \\ Y & \xrightarrow{b} & Y & \xrightarrow{\nu_{i+n}} & \mathbf{Tel}(Y, b) \end{array}$$

commute.

Proof Use $(h, \mathbf{pr}_2)$ as a map from $X \times \mathcal{I}$ to $Y \times \mathcal{I}$, and by Lemma 22.8.1 we can construct $\varphi_n(h) : \mathbf{Tel}(X, a) \to \mathbf{Tel}(Y, b)$ by setting

$$\varphi_i := \mu_{i+n} \circ (h, \mathbf{pr}_2).$$

It sends the i-th piece of $\mathbf{Tel}(X, a)$ to the $(i + n)$-th piece of $\mathbf{Tel}(Y, b)$, and the required diagrams follow from the construction. □

There are two basic constructions of homotopies between these maps. To simplify the writing we will pretend that $\mathcal{I}$ is a real interval with coordinate t and endpoints $t = 0, 1$, and use x for the coordinate on X. In these terms, the condition (22.8.2) on a homotopy h for the construction of $\varphi_n(h)$ in Lemma 22.8.2 above may be written as

$$b(h(x, 1)) = h(a(x), 0).$$

Suppose now given a map

$$K : X \times \mathcal{I} \times \mathcal{I} \to Y,$$

such that "for all $t \in \mathcal{I}$,"

$$b(K(x, t, 1)) = K(a(x), t, 0).$$

This may be written more precisely,

$$b \circ K \circ (1 \times 1 \times i_1) = K \circ (1 \times 1 \times i_0) \circ (a \times 1),$$

as maps from $X \times \mathcal{I}$ to Y, so if we consider K as a map from $(X \times \mathcal{I}) \times \mathcal{I}$ to Y, it gives a map between telescopes by the construction of Lemma 22.8.2:

$$\varphi_n(K) : \mathbf{Tel}(X \times \mathcal{I}, a \times 1) = \mathbf{Tel}(X, a) \times \mathcal{I} \to \mathbf{Tel}(Y, b).$$

If $K \circ (1 \times i_0 \times 1) = h$ and $K \circ (1 \times i_1 \times 1) = h'$ then $\varphi_n(K)$ becomes a homotopy between $\varphi_n(h)$ and $\varphi_n(h')$. This gives the following:

Corollary 22.8.3 *Suppose h and h' are maps from $X \times \mathcal{I}$ to Y satisfying the condition of Lemma 22.8.2, such that there exists a homotopy K between these homotopies satisfying the condition (22.8.2) all along. Then $\varphi_n(h)$ and $\varphi_n(h')$ are homotopic maps from* **Tel**(X, a) *to* **Tel**(Y, b).

□

The other construction of a homotopy is the "slide"; to do this we need to use a fibrant replacement for the target. Recall that $\mathbf{pr}_1 : X \times \mathcal{I} \to X$ is the constant homotopy; in particular $\varphi_0(\mathbf{pr}_1)$ is the identity.

Lemma 22.8.4 *Suppose $(X, a)$is as above. Suppose f : $\mathbf{Tel}(X, a) \to T'$ is a map to a fibrant object (typically a fibrant replacement). Then for any n, $f \circ \varphi_n(\mathbf{pr}_1)$ is homotopic to $f \circ \varphi_{n+1}(a \circ \mathbf{pr}_1)$.*

Proof Suppose $\mathcal{I} \cup^* \mathcal{I} \to \mathbf{Tel}_2(*, 1)'$ is a fibrant replacement, denoting by $u, v : \mathcal{I} \to \mathbf{Tel}_2(*, 1)'$ the two corresponding maps. Then $\mathcal{I}'$ is contractible. It follows that we can choose a map

$$\sigma : \mathcal{I} \times \mathcal{I} \to \mathcal{I}',$$

such that $\sigma \circ (1 \times i_0) = u, \sigma \circ (1 \times i_1) = v, \sigma \circ (i_0 \times 1) = u, \sigma \circ (i_1 \times 1) = v$. In other words, σ is the first cell of the "slide homotopy" we are looking for in the case $X = *$,

$$\mathbf{Tel}_1(*, 1) \times \mathcal{I} \to \mathbf{Tel}_2(*, 1)'.$$

Taking the product with X, this gives a map

$$\mathbf{Tel}_1(X, 1_X) \times \mathcal{I} \to \mathbf{Tel}_2(*, 1)' \times X.$$

On the other hand, using a on the second cell gives a map

$$\mathbf{Tel}_2(*, 1) \times X = \mathbf{Tel}_2(X, 1_X) \to \mathbf{Tel}_2(X, a),$$

so if $\mathbf{Tel}_2(X, a) \to \mathbf{Tel}_2(X, a)'$ is a fibrant replacement, we can extend to $\mathbf{Tel}_2(*, 1)' \times X \to \mathbf{Tel}_2(X, a)'$. We get a map $X \times \mathcal{I} \times \mathcal{I} \to \mathbf{Tel}_2(X, a)'$, which extends along the additional copy of X to give the first cell of the slide homotopy

$$\sigma_a : \mathbf{Tel}_1(X, a) \times \mathcal{I} \to \mathbf{Tel}_2(X, a)'.$$

The restrictions to the two endpoints are the two inclusions of $\mathbf{Tel}_1(X, a)$ into $\mathbf{Tel}_2(X, a)$, whereas on the two copies of $X \times \mathcal{I}$ it restricts to the two maps $X \times \mathcal{I} \to \mathbf{Tel}_2(X, a)$. In particular, note that, on the "boundary," σ_a maps into $\mathbf{Tel}_2(X, a)$. At each cell indexed by a positive integer, we have a copy of $\mathbf{Tel}_2(X, a)$; choose an extension of each of these to $\mathbf{Tel}_2(X, a)' \to T'$. The σ_a composed with these extensions, which along the boundary go into the various copies of $\mathbf{Tel}_2(X, a)$ and along the two copies of X coincide up to a shift, then glue together on the cells of $\mathbf{Tel}(X, a)$ (which are all the same as $\mathbf{Tel}_1(X, a)$). This gives a homotopy

$$\mathbf{Tel}(X, a) \times \mathcal{I} \to T',$$

going from $f \circ \varphi_0(\mathbf{pr}_1) = f$ to $f \circ \varphi_1(a \circ \mathbf{pr}_1)$. The same construction shifted by n produces the required homotopy. □

Suppose now that we are given an idempotent up to homotopy $p : X \to X$, and a compatible homotopy $H : X \times \mathcal{I} \to X$ from p^2 to p. Compatibility means that pH and Hp are homotopic as homotopies from p^3 to p^2. Note that H satisfies the condition of Lemma 22.8.2 required to obtain $\varphi_0(H) : \mathbf{Tel}(X, 1_X) \to \mathbf{Tel}(X, p)$. Let $\overline{H}$ denote the inverse homotopy from p to p^2, which exists in the truncated 2-category discussed in Section 21.3.3. Then it satisfies the condition of Lemma 22.8.2 required to obtain $\varphi_0(\overline{H}) : \mathbf{Tel}(X, p) \to \mathbf{Tel}(X, 1_X)$. Noting that the first inclusion $X \to \mathbf{Tel}(X, 1_X)$ is a weak equivalence, we can try to use $\varphi_0(H)$ and $\varphi_0(\overline{H})$ to split the idempotent p in $\mathrm{ho}(\mathscr{P})$.

Lemma 22.8.5 *Fix $r \geq 3$. With the above notations, $\varphi_n(p^r \circ \mathbf{pr}_1)$ and $\varphi_n(p^2 \circ \mathbf{pr}_1)$ are homotopic, as maps from $\mathbf{Tel}(X, p)$ to $\mathbf{Tel}(X, p)$.*

Proof Consider first the case $r = 3$. Use Corollary 22.8.3, with $Y = X$ and $a = b = p$. The map K is the homotopy between pH and Hp, but viewed in the other direction as a homotopy between $p^3 \circ \mathbf{pr}_1$ and $p^2 \circ \mathbf{pr}_1$. In the informal language used above 22.8.3,

$$K(x,t,0) = p(H(x,t)), \quad K(x,t,1) = H(p(x),t),$$

$$K(x,0,t) = p^3(x), \quad K(x,1,t) = p^2(x).$$

Notice that $b(K(x,t,1)) = p(H(p(x),t)) = K(a(x),t,0)$ so the condition (22.8.2) is satisfied all along. The construction of Corollary 22.8.3 thus gives a homotopy between $\varphi_n(p^3 \circ \mathbf{pr}_1)$ and $\varphi^n(p^2 \circ \mathbf{pr}_1)$. For $r \geq 4$ notice that by composing several copies of Hp^j we obtain a homotopy H' between p^r and p^2, such that there exists K' a homotopy between pH' and $H'p$, then apply the same argument. □

We can now construct the splitting. As pointed out at the beginning of the section, this result in the case of topological spaces is much weaker than the known criterion; for the case of a model category, it is most undoubtedly already known too and we can't claim any originality.

Theorem 22.8.6 *With the above assumptions and notations, the composition*

$$\mathbf{Tel}(X, 1_X) \xrightarrow{\varphi_0(H)} \mathbf{Tel}(X, p) \xrightarrow{\varphi_0(\overline{H})} \mathbf{Tel}(X, 1_X)$$

is homotopic to $\varphi_0(p \circ \mathbf{pr}_1)$, whereas the composition

$$\mathbf{Tel}(X, p) \xrightarrow{\varphi_0(\overline{H})} \mathbf{Tel}(X, 1_X) \xrightarrow{\varphi_0(H)} \mathbf{Tel}(X, p)$$

is homotopic to the identity. These diagrams give a splitting of p in $\mathrm{ho}(\mathcal{P})$ as a result of the hypothesis of compatibility of the homotopy H. For a regular cardinal, $\kappa > \omega$ if $X \in \mathcal{P}_\kappa$ then $\mathbf{Tel}(X, p) \in \mathcal{P}_\kappa$.

Proof We can compose homotopies in the following way, corresponding to horizontal composition in the 2-category discussed in Section 21.3.3:

$$X \times \mathcal{I} \xrightarrow{(H, \mathbf{pr}_2)} X \times \mathcal{I} \xrightarrow{\overline{H}} X,$$

giving a homotopy denoted $\overline{H} \cdot H$ from p^3 to itself. Similarly in the other direction we get $H \cdot \overline{H}$ again from p^3 to itself. With these notations, the first composition of the statement is

$$\varphi_0(\overline{H}) \circ \varphi_0(H) = \varphi_0(\overline{H} \cdot H),$$

going from $\mathbf{Tel}(X, 1_X)$ to itself; and the second one is

$$\varphi_0(H) \circ \varphi_0(\overline{H}) = \varphi_0(H \cdot \overline{H}),$$

going from $\mathbf{Tel}(X, p)$ to itself.

On the other hand, the interchange rule in a 2-category says that $\overline{H} \cdot H$ is the same as the vertical composition of $\overline{H}p$ with pH, which is a composition of homotopies going first from p^3 to p^2, then back to p^3. By the compatibility hypothesis pH is the same as Hp, and the compatibility between vertical and horizontal composition in a 2-category says that the vertical composition of $\overline{H}p$ with Hp is the same as $(\overline{H} * H)p$; by construction $\overline{H}$ is inverse to H so this composition is just the identity of p^3. We conclude from this argument that $\overline{H} \cdot H$ is homotopic to the identity homotopy of p^3. In particular, $\varphi_0(\overline{H} \cdot H)$ is homotopic to $\varphi_0(p^3 \circ \mathbf{pr}_1)$.

The same argument in the other direction says that $\varphi_0(H \cdot \overline{H})$ is homotopic to $\varphi_0(p^3 \circ \mathbf{pr}_1)$ but this time considered as a map from $\mathbf{Tel}(X, p)$ to itself.

We treat the first composition first, by looking at the diagram

$$\begin{array}{ccc} X & \xrightarrow{\nu_0} & \mathbf{Tel}(X, 1_X) \\ {\scriptstyle p^3}\downarrow & & \downarrow{\scriptstyle \varphi_0(p^3 \circ \mathbf{pr}_1)} \\ X & \xrightarrow{\nu_0} & \mathbf{Tel}(X, 1_X). \end{array}$$

Since the horizontal arrows are weak equivalences, we conclude that the first composition in the statement of the theorem is homotopic to p^3, which in turn is homotopic to p. This via the weak equivalence between X and $\mathbf{Tel}(X, 1_X)$.

For the second composition, we claim that $\varphi_0(p^3 \circ \mathbf{pr}_1)$ is homotopic to the identity on $\mathbf{Tel}(X, p)$. For this use the slide homotopies. These say that $1_{\mathbf{Tel}(X,p)} = \varphi_0(\mathbf{pr}_1)$ is homotopic to $\varphi_2(p^2 \circ \mathbf{pr}_1)$, whereas $\varphi_0(p^3 \circ \mathbf{pr}_1)$ is homotopic to $\varphi_2(p^5 \circ \mathbf{pr}_1)$. But Lemma 22.8.5 for $r = 5$ and $n = 2$ says that $\varphi_2(p^5 \circ \mathbf{pr}_1)$ is homotopic to $\varphi_2(p^2 \circ \mathbf{pr}_1)$. Putting these together we conclude that the second map in the statement of the theorem, $\varphi_0(p^3 \circ \mathbf{pr}_1)$, is homotopic to the identity.

This completes the proof that our diagrams constitute a splitting of the idempotent p in $\mathrm{ho}(\mathscr{P})$. □

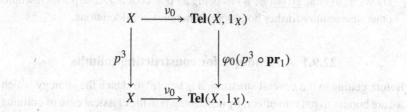

22.9 Colimits of weak enriched categories

In this section we give a somewhat revised version of the construction of colimits from my preprint on limits [235]. This corrects some fuzzy aspects of the argument from there, for instance we have included the compatibility condition for the homotopy in our construction of splitting idempotents above. Nonetheless the present discussion remains sketchy, for example on why the compatibility condition holds when we apply Theorem 22.8.6. It should be viewed as a somewhat heuristic description of one possible but not very concrete way of getting colimits. Perhaps the main interest of the present discussion is that it allows us to gauge the possibility of transposing a traditional category-theoretic argument into the higher categorical setting. Obviously it would be good to have a more explicit construction too.

We will adopt the simplifying notations and assumptions:

- Use the viewpoint that $\mathbf{Enr}(\mathcal{P})'_{\cdot}$ is an ind-object of $\mathcal{PC}(\mathcal{P})$ having the strong property of Proposition 22.3.1. This is reasonable since we only consider maps into $\mathbf{Enr}(\mathcal{P})'_{\cdot}$, and the regular cardinalities of objects will constitute a separate element of our discussion.
- Assume that all objects are cofibrant. This holds for n-categories $\mathcal{P} = \mathcal{PC}^n(\text{SET})$ or Segal n-categories $\mathcal{P} = \mathcal{PC}^n(\mathcal{K})$ for example. It allows us to view $\underline{\text{HOM}}(E, B)$ as being an object of $\mathcal{P}_{\text{cf}}$, a step which would otherwise require further homotopy coherence considerations.

22.9.1 A technique for constructing colimits

Before getting to the general situation, it is helpful to sketch the strategy which we are hoping to put into effect, by describing it in the classical case of colimits in 1-categories.

Consider the following heuristic argument, which shows that if $\mathcal{C}$ is a 1-category in which all limits exist and in which projectors are effective, then $\mathcal{C}$ admits colimits. For a functor $\psi : \mathcal{A} \to \mathcal{C}$ let $(\psi/\mathcal{C})$ be the category whose objects are pairs (X, x), where X is an object of $\mathcal{C}$ and $x : \psi \to c_{\mathcal{A}}$ is a morphism, i.e. a natural transformation from ψ to the constant functor with values X. There is a forgetful functor $f : (\psi/\mathcal{C}) \to \mathcal{C}$. Let $U \in \mathcal{C}$ be the limit of f. This comes with its limit cone, which is a natural transformation

$$r(X, x) : U \to X = f(X, x),$$

which is natural in $(X, x) \in (\psi/\mathcal{C})$.

For any $a \in \text{Ob}(\mathcal{A})$ there is a uniquely defined morphism $\psi(a) \to f$. By the limit property this factors through r via a morphism $\psi(a) \to U$, and uniqueness implies that it is functorial in a. Thus we get a morphism $u:\psi \to U_{\mathcal{A}}$ and (U, u) is in $(\psi/\mathcal{C})$ (see below, however). There is a morphism from U considered as the limit of f, to U considered as an object of $(\psi/\mathcal{C})$

$$p := r(U, u) : U \to U.$$

This is itself a morphism in $(\psi/\mathcal{C})$, so we get $p \circ p = p$ from naturality of r with respect to p. Thus p is a projector. Let T be the direct factor of U given by p. Composition $\psi \to U \to T$ gives a map $\psi \to T$ and we get a factorization $\psi \to t \to U$. Now T is seen to be an initial object of $(\psi/\mathcal{C})$, hence $\psi \to T$ is a colimit.

The problem with this heuristic argument is a set-theoretic one. When one speaks of "limits" it is presupposed that the indexing category $\mathcal{A}$ is small with respect to the given universe. However, our category $\mathcal{C}$ is likely to be a class. Thus, in the above argument, $(\psi/\mathcal{C})$ is not small and we are not allowed to take the limit over $(\psi/\mathcal{C})$.

Let's see how to fix this up in the case where $\mathcal{C} = \text{SET}$ is the category of sets. Suppose we have a functor $\psi : \mathcal{A} \to \text{SET}$ from a small category $\mathcal{A}$. Let α be a cardinal number bigger than $|\mathcal{A}|$ and bigger than the cardinal of any set in the image of ψ. Let $(\psi/\mathcal{C})_{\alpha}$ be the category of pairs (X, x) as above where X is contained in a fixed set of cardinality α. Note that $(\psi/\mathcal{C})_{\alpha}$ has cardinality $\leq 2^{\alpha}$. Let (U, u) be as above. The cardinality of U seems *a priori* only to be bounded by $2^{2^{\alpha}}$. Let $W \subset U$ be the smallest subset through which the map $u : \psi \to U$ factors, denoting by $w : \psi \to W$ the factorization. Note that the cardinality of W cannot be bigger than the sum of the cardinals of the $\psi(a)$ over $a \in \text{Ob}(\mathcal{A})$, in particular W has cardinal $\leq \alpha$. We obtain the map $r(W, w) : U \to W$ and composition

$$W \to U \to W$$

is seen to be a projector p on W. The direct factor T defined by p is the colimit of ψ. The arguments used in Lemma 22.9.12 below could profitably be written down as an exercise in this simpler case first.

More generally, if we can define the subcategories $(\psi/\mathcal{C})_{\alpha}$ and if we know for some reason that every object $B \in (\psi/\mathcal{C})$ admits a map $B' \to (\psi/\mathcal{C})$ from an object $B' \in (\psi/\mathcal{C})_{\alpha}$ then we can fix up the argument. This well-known kind of condition is known as the "solution set condition."

We would like to do the same thing for limits in $\mathbf{Enr}(\mathcal{P})'_{\kappa}$, namely show that colimits exist just using a general argument working from the existence of limits. This calls upon some kind of notion of cardinality for objects of $\mathcal{P}$, discussed in Section 22.7 above, and the splitting of compatible homotopy idempotents discussed in Section 22.8.

22.9.2 A criterion for colimits in $\mathbf{Enr}(\mathcal{P})'$.

Before getting to the application of the theory of cardinality we give a criterion which simplifies the problem of finding colimits in $\mathbf{Enr}(\mathcal{P})'_{\cdot}$.

For this we need another type of universal morphism. Suppose $E, B \in \mathcal{P}$, with B fibrant. Recall that now E is cofibrant by hypothesis. Then $\underline{\mathrm{HOM}}(E, B)$ is fibrant and we have a canonical morphism

$$\underline{\mathrm{HOM}}(E, B) \times E \to B.$$

This may be interpreted as an object

$$\nu_{E,B} \in Diag^{\underline{\mathrm{HOM}}(E,B),B}(E; \mathbf{Enr}(\mathcal{P})_{\cdot}),$$

which yields by composition with $\mathbf{Enr}(\mathcal{P})_{\cdot} \to \mathbf{Enr}(\mathcal{P})'_{\cdot}$ the element which we denote by the same symbol

$$\nu_{E,B} \in Diag^{\underline{\mathrm{HOM}}(E,B),B}\left(E; \mathbf{Enr}(\mathcal{P})'_{\cdot}\right).$$

Lemma 22.9.1 *The element $\nu_{E,B}$ has the following universal property: for any $F \in \mathcal{P}$ the morphism*

$$Diag^{U,\nu_{E,B}}\left(\underline{F}, E; \mathbf{Enr}(\mathcal{P})'_{\cdot}\right) \to Diag^{U,B}\left(\underline{F \times E}; \mathbf{Enr}(\mathcal{P})'_{\cdot}\right)$$

is a trivial fibration of fibrant objects of $\mathcal{P}$.

Proof To prove this note that the fibrant property comes from the fact that $\mathbf{Enr}(\mathcal{P})'_{\cdot}$ is fibrant. Note that both sides are fibrant by Proposition 22.4.11. The fact that it is an equivalence may be checked using diagrams in $\mathbf{Enr}(\mathcal{P})_{\cdot}$ rather than diagrams in $\mathbf{Enr}(\mathcal{P})'_{\cdot}$, according to the second part of Proposition 22.4.11. Using Remark 22.4.7 for diagrams in $\mathbf{Enr}(\mathcal{P})_{\cdot}$, both sides become equal to

$$\underline{\mathrm{HOM}}(U \times F \times E, B),$$

where U is the image of the first object $0 \in \widetilde{\Upsilon}_2(F, E)$, hence the morphism is an equivalence. □

Corollary 22.9.2 *Given a morphism*

$$f : \Upsilon(F \times E) \to \mathbf{Enr}(\mathcal{P})'_{\cdot},$$

with the image of the second vertex equal to B, there is an extension to a morphism

$$g : \widetilde{\Upsilon}_2(F, E) \to \mathbf{Enr}(\mathcal{P})'_{\cdot},$$

such that $r_{02}(g) = f$ *and* $r_{12}(f) = \nu_{E,B}$. *Similarly, if* $E' \to E$ *is a cofibration between cofibrant objects and we are already given the extension* g' *for* E' *then we can assume that* g *is compatible with* g'.

We also have a version of this universal property for shell-extension in higher degree. This concerns the *right shell* $\Lambda^R\widetilde{\Upsilon}_k$ corresponding to the horn Λ^k_k obtained by discarding the face $(01\cdots(k-1))$.

Corollary 22.9.3 *Suppose we are given a morphism*

$$f : \Lambda^R\widetilde{\Upsilon}_k(F_1, \ldots, F_{k-1}, E) \to \mathbf{Enr}(\mathcal{P})'_{.}$$

such that f *restricts on the last edge to* $\nu_{E,B}$. *Then there is a filling-in to a morphism*

$$g : \widetilde{\Upsilon}_k(F_1, \ldots, F_{k-1}, E) \to \mathbf{Enr}(\mathcal{P})'_{.}.$$

If g' *is already given over* $F'_1, \ldots, F'_{k-1}, E'$ *then we can assume that* g *is compatible with* g'.

Proof This is the analogue of the arguments given in Section 22.5.2. □

The above property also works in a family. Given a morphism

$$f : \mathcal{A} \times \Upsilon(F \times E) \to \mathbf{Enr}(\mathcal{P})'_{.},$$

sending the last vertex to the constant object $B_{\mathcal{A}}$, there is an extension to a morphism

$$g : \mathcal{A} \times \widetilde{\Upsilon}_2(F, E) \to \mathbf{Enr}(\mathcal{P})'_{.},$$

such that $r_{02}(g) = f$ and $r_{12}(f) = \nu_{E,B,\mathcal{A}}$ is the morphism pulled back from $\nu_{E,B}$. Again if an extension g' is already given on $E' \hookrightarrow E$ then g may be chosen compatibly with g'.

Similarly there is a shell-extension property as in Corollary 22.9.3 in a family.

For the proof one has to go through a procedure analogous to the passage from diagrams to internal <u>HOM</u> as in the proof of Lemma 22.5.8. This discussion of the universal morphism $\nu_{E,B}$ is parallel to the discussion of the discussion of the universal 1_B, but with "arrows reversed."

We now come to our simplified criterion for limits in $\mathbf{Enr}(\mathcal{P})'_{.}$. One should be careful that the following lemma only applies to colimits taken in $\mathbf{Enr}(\mathcal{P})'_{.}$ and not to an arbitrary $\mathcal{P}$-category $\mathcal{C}$. The proof uses in an essential way the fact that the morphism spaces for the "category" $\mathbf{Enr}(\mathcal{P})'_{.}$ are objects of $\mathcal{P}$ which are also basically the same thing as the objects of $\mathbf{Enr}(\mathcal{P})'_{.}$. Of course it is possible that the same techniques of proof might work in a limited range of other closely related circumstances.

Proposition 22.9.4 *Suppose given a cofibrant $\mathcal{P}$-precategory $\mathcal{A}$ and a functor $\psi : \mathcal{A} \to \mathbf{Enr}(\mathcal{P})'_{\bullet}$ is a morphism. Suppose that $U \in \mathbf{Enr}(\mathcal{P})'_{\bullet}$ and $\epsilon \in \mathrm{Cocone}(\psi, U)$ has the following weak limit-like property: for any other morphism $f \in \mathrm{Cocone}(\psi, V)$ there exists a morphism $g : U \to V$ such that the composition $g\epsilon$ (well defined up to homotopy) is homotopic to f; and furthermore that such a factorization is unique up to a (not necessarily unique) homotopy of the factorization. Then $\psi \xrightarrow{\epsilon} U$ is a colimit.*

Proof First, we explain more precisely what the existence and uniqueness of the factorization mean. Given an element $f \in \mathrm{Cocone}(\psi, V)$ there exists an element $g' \in \mathrm{HOM}^{\epsilon}(\psi, U, V)$ projecting via r_{02} to a morphism equivalent to f. This equivalence may be measured in $\underline{\mathrm{Cocone}}(\psi, V) \in \mathcal{P}$. Note that since $\mathbf{Enr}(\mathcal{P})'_{\bullet}$ is fibrant, the projection

$$\underline{\mathrm{Cocone}^{\epsilon}}(\psi, U, V) \to \underline{\mathrm{Cocone}}(\psi, V)$$

is a fibration, so if an object equivalent to f is in the image then f is in the image. Thus we can restate the criterion as saying simply that there exists an element g' projecting via r_{02} to f.

Suppose given two such factorizations g'_1 and g'_2. By "homotopy of the factorization" we mean a homotopy between $r_{12}(g'_1)$ and $r_{12}(g'_2)$ such that the resulting homotopy between f and itself (this homotopy being well defined up to 2-homotopy) is 2-homotopic to the identity 1_f. Again using the fibrant condition of $\mathbf{Enr}(\mathcal{P})'_{\bullet}$ we obtain that this condition implies the simpler statement that there exists a morphism

$$\mathcal{A} \times \widetilde{\Upsilon}_2(*, *) \times \mathbf{I} \to \mathbf{Enr}(\mathcal{P})'_{\bullet},$$

restricting to g'_1 and g'_2 on the two endpoints $0, 1 \in \mathbf{I}$; restricting to the pullback ϵ on the edge (01) of the $\widetilde{\Upsilon}_2$, this edge being $\mathcal{A} \times \Upsilon(*) \times \mathbf{I}$; and restricting to the pullback of f on the edge (02) which is $\mathcal{A} \times \Upsilon(*) \times \mathbf{I}$.

The proof of the proposition is based on the idea of using the universal maps $\nu_{E,B}$ to transform maps $E \to \underline{\mathrm{Cocone}}(\psi, B)$ into elements of Cocone $(\psi, \underline{\mathrm{HOM}}(E, B))$.

Simple factorization

We start by showing the simple version of the factorization property necessary to show that ϵ is a colimit; we will treat the relative case for $E' \hookrightarrow E$ below. For now, suppose that we are given a morphism

$$u : \mathcal{A} \times \Upsilon(E) \to \mathbf{Enr}(\mathcal{P})'_{\bullet},$$

restricting to ψ on $\mathcal{A} \times \{\upsilon_0\}$ and restricting to a constant object $B \in \mathbf{Enr}(\mathcal{P})'_{.}$ (i.e. to the pullback $B_{\mathcal{A}}$) on $\mathcal{A} \times \{\upsilon_1\}$. This corresponds to a map

$$E \to \underline{\text{Cocone}}(\psi, B).$$

We would like to extend it to a morphism

$$v : \mathcal{A} \times \widetilde{\Upsilon}_2(*, E) \to \mathbf{Enr}(\mathcal{P})'_{.},$$

restricting to our given morphism u on the edge (02), restricting to ϵ on the edge (01), and restricting on (12) to the pullback of a map $\Upsilon(E) \to \mathbf{Enr}(\mathcal{P})'_{.}$. Our given morphism u corresponds, by the remarks following Corollary 22.9.3, to a morphism $w : \mathcal{A} \times \Upsilon(*) \to \mathbf{Enr}(\mathcal{P})'_{.}$ restricting to ψ on $\mathcal{A} \times \{\upsilon_0\}$ and restricting to the constant object $\underline{\text{HOM}}(E, B)$ (pulled back to $\mathcal{A}$) on $\mathcal{A} \times \{\upsilon_1\}$. More precisely there is a morphism

$$w' : \mathcal{A} \times \widetilde{\Upsilon}_2(*, E) \to \mathbf{Enr}(\mathcal{P})'_{.},$$

restricting to u over the edge (02), and restricting to the universal morphism $\nu_{E,B}$ in Lemma 22.9.1, over the edge (12). The restriction to the edge (01) is the morphism w.

Now w is an element of $\underline{\text{Cocone}}(\psi, B)$, so by hypothesis there is a diagram

$$g : \mathcal{A} \times \widetilde{\Upsilon}_2(*, *) \to \mathbf{Enr}(\mathcal{P})'_{.},$$

sending the edge (01) to ϵ and sending the edge (02) to w. Putting this together with the diagram w' and using the fibrant property of $\mathbf{Enr}(\mathcal{P})'_{.}$ (i.e. composing these together) we obtain existence of a diagram

$$\mathcal{A} \times \widetilde{\Upsilon}_3(*, *, E) \to \mathbf{Enr}(\mathcal{P})'_{.},$$

restricting to g on the face (012) and restricting to w' on the face (023). The face (013) yields a diagram

$$\mathcal{A} \times \widetilde{\Upsilon}_2(*, E) \to \mathbf{Enr}(\mathcal{P})'_{.},$$

restricting to ϵ on the first edge and restricting to our original morphism u on the edge (03): this is the morphism v we are looking for.

Uniqueness of these factorizations

The homotopy uniqueness property for factorization of morphisms implies a similar property for the factorizations of E-morphisms obtained in the previous paragraph. Suppose that we are given a morphism

$$u : \mathcal{A} \times \Upsilon(E) \to \mathbf{Enr}(\mathcal{P})'_{.}$$

corresponding to $E \to \underline{\text{Cocone}}(\psi, B)$ as above, and suppose that we are given two extensions

$$v_1, v_2 : \mathcal{A} \times \widetilde{\Upsilon}_2(*, E) \to \mathbf{Enr}(\mathcal{P})'_{\cdot},$$

restricting to our given morphism on the edge (02), and restricting to ϵ on the edge (01). Complete the v_i to diagrams

$$z_i : \mathcal{A} \times \widetilde{\Upsilon}_3(*, *, E) \to \mathbf{Enr}(\mathcal{P})'_{\cdot},$$

restricting to v_i on the faces (013) and restricting to the universal morphism $\nu_{E,B}$ of Lemma 22.9.1 on the edge (23). To do this, use the universal property of $\nu_{E,B}$ discussed at Corollary 22.9.3 and after, to fill in the faces (023) and (123); then we have a map defined on the shell and by the universal property of $\nu_{E,B}$ which gives shell extension (Corollary 22.9.3 and the following remarks) we can extend to the whole tetrahedron.

Note furthermore that we can assume that the restrictions to the faces (023) are the same for z_1 and z_2 (because we have chosen these faces using only the map u and not refering to the v_i). Call these $r_{023}(z)$. In particular the restrictions to (02) give the same map $w : \psi \to \underline{\text{HOM}}(E, B)$. Now the restrictions of the above diagrams z_i to the faces (012) give two different factorizations of this map w so by hypothesis there is a homotopy between these factorizations: it is a morphism

$$\mathcal{A} \times \widetilde{\Upsilon}_2(*, *) \times \mathbf{I} \to \mathbf{Enr}(\mathcal{P})'_{\cdot},$$

restricting to $r_{012}(z_i)$ on the endpoints $i = 0, 1$ of $\mathbf{I}$, restricting to the pullback of ϵ along $(01) \times \mathbf{I}$ and restricting to the pullback of our morphism w along $(02) \times \mathbf{I}$. Attach this homotopy to the constant homotopy which is the pullback of $r_{023}(z)$ from $\mathcal{A} \times \widetilde{\Upsilon}_2(*, E)$ to $\mathcal{A} \times \widetilde{\Upsilon}_2(*, E) \times \mathbf{I}$. We obtain a homotopy defined on the union of the faces (012) and (023) and going between z_0 and z_1. Using the fact that the inclusion of this union of faces into the tetrahedron is a trivial cofibration (see the list 21.4.2 above) we get that the inclusion, written in an obvious shorthand notation where (0123) stands for $\mathcal{A} \times \Upsilon^3(*, *, E)$ and $(012 + 023)$ for the union of the two faces,

$$\begin{array}{c} (0123) \times \{v_0, v_1\} \cup^{(012+023)\times\{v_0,v_1\}} (012 + 023) \times \mathbf{I} \\ \downarrow \\ (0123) \times \mathbf{I} \end{array}$$

is a trivial cofibration. We have a map from the domain here into $\mathbf{Enr}(\mathcal{P})'_{\cdot}$ so it extends to a map

$$\mathcal{A} \times \widetilde{\Upsilon}_3(*, *, E) \times \mathbf{I} \to \mathbf{Enr}(\mathcal{P})'_{\cdot}.$$

The restriction of this map to the face (013) is a homotopy

$$\mathcal{A} \times \widetilde{\Upsilon}_2(*, E) \times \mathbf{I} \to \mathbf{Enr}(\mathcal{P})'_{\cdot}$$

between our factorizations v_1 and v_2.

The relative case

To actually prove the proposition, we need to obtain a factorization property as above in the relative situation $E' \hookrightarrow E$ where we already have the factorization over E' and we would like to extend to E. This is where we use the homotopy uniqueness of factorization which was in the hypothesis of the lemma. This property is used in the form given in the previous segment of the proof. Suppose we are given

$$v' : \mathcal{A} \times \widetilde{\Upsilon}_2(*, E') \to \mathbf{Enr}(\mathcal{P})'_{\cdot}$$

in Cocone(ψ, U, B), restricting to ϵ on the first edge, and suppose we are given

$$u : \mathcal{A} \times \Upsilon(E) \to \mathbf{Enr}(\mathcal{P})'_{\cdot}$$

in Cocone(ψ, B), hence restricting to ψ on $\mathcal{A} \times \{\upsilon_0\}$ and restricting to a constant object $B \in \mathbf{Enr}(\mathcal{P})'_{\cdot}$ (i.e. to the pullback $B_{\mathcal{A}}$) on $\mathcal{A} \times \{\upsilon_1\}$. Suppose that the restriction of u to $\mathcal{A} \times \Upsilon(E')$ is equal to the restriction of v' to the edge (02). In the paragraph on simple factorization, we constructed an extension

$$v_0 : \mathcal{A} \times \widetilde{\Upsilon}_2(*, E) \to \mathbf{Enr}(\mathcal{P})'_{\cdot},$$

which restricts to ϵ on the first edge and to u on the edge (02). Let v_0' denote the restriction of v_0 to $\mathcal{A} \times \widetilde{\Upsilon}_2(*, E')$. By the uniqueness property of the factorization as proven in the previous segment for E', there exists a homotopy

$$\mathcal{A} \times \widetilde{\Upsilon}_2(*, E') \times \mathbf{I} \to \mathbf{Enr}(\mathcal{P})'_{\cdot}$$

between v_0' and v', constant along edges (01) and (02). Let $\mathcal{D}$ be the amalgamated sum of $\widetilde{\Upsilon}_2(*, E')$ and $\Upsilon(E)$ with the latter attached along the edge (02) (i.e. the amalgamation is taken over the copy of $\Upsilon(E') \hookrightarrow \widetilde{\Upsilon}_2(*, E')$ corresponding to the edge (02)). Our homotopy glues with the constant map u to give a morphism

$$\mathcal{A} \times \mathcal{D} \times \mathbf{I} \to \mathbf{Enr}(\mathcal{P})'_{\cdot},$$

and this glues with u to obtain

$$\mathcal{A} \times \widetilde{\Upsilon}_2(*, E) \times \{\upsilon_0\} \cup^{\mathcal{A}\times\mathcal{D}\times\{\upsilon_0\}} \mathcal{A} \times \mathcal{D} \times \mathbb{I} \to \mathbf{Enr}(\mathcal{P})'_{\cdot}.$$

The inclusion

$$\mathcal{A} \times \mathcal{D} \times \{\upsilon_0\} \hookrightarrow \mathcal{A} \times \mathcal{D} \times \mathbb{I}$$

is a trivial cofibration, so the inclusion

$$\mathcal{A} \times \widetilde{\Upsilon}_2(*, E) \times \{\upsilon_0\} \cup^{\mathcal{A}\times\mathcal{D}\times\{\upsilon_0\}} \mathcal{A} \times \mathcal{D} \times \mathbb{I} \hookrightarrow \mathcal{A} \times \widetilde{\Upsilon}_2(*, E) \times \mathbf{I}$$

is a trivial cofibration, and by the fibrant property of $\mathbf{Enr}(\mathcal{P})'_{\cdot}$ there exists an extension of the above morphism to a morphism

$$\mathcal{A} \times \widetilde{\Upsilon}_2(*, E) \times \mathbf{I} \to \mathbf{Enr}(\mathcal{P})'_{\cdot}.$$

The value of this over $1 \in \mathbf{I}$ is the extension

$$\upsilon : \mathcal{A} \times \widetilde{\Upsilon}_2(*, E) \to \mathbf{Enr}(\mathcal{P})'_{\cdot}$$

we are looking for: it restricts to ϵ on the edge (01), it restricts to u on the edge (02), and it restricts to υ' over $\mathcal{A} \times \widetilde{\Upsilon}_2(*, E')$. This completes the proof of Proposition 22.9.4. □

Remark 22.9.5 One also obtains a criterion similar to 22.9.4 for limits in $\mathbf{Enr}(\mathcal{P})'_{\cdot}$. The proof is the same as above but using the universal diagram

$$B \xrightarrow{E} B \times E$$

in the place of $\underline{\mathrm{HOM}}(E, B) \xrightarrow{E} B$. We didn't need this in the proof of 22.6.1; it didn't seem to make any substantial savings and would probably have complicated the notation in many places.

22.9.3 Construction of the colimit

Theorem 22.9.6 *Assuming that all objects of $\mathcal{P}$ are cofibrant, the* ind-$\mathcal{P}$-*category* $\mathbf{Enr}(\mathcal{P})'_{\cdot}$ *admits colimits.*

The proof takes up the rest of this section. Suppose $\psi : \mathcal{A} \to \mathbf{Enr}(\mathcal{P})'_{\cdot}$ is a functor.

Fix a regular cardinal κ as in Lemma 22.7.2 above.

Let $\mathcal{C}_\kappa := \mathbf{Enr}(\mathcal{P})'_{\kappa}$, in particular ψ may be considered as a functor

$$\psi : \mathcal{A} \to \mathcal{C}_\kappa.$$

In order to replicate the proof that was given in Section 22.9.1 for the category of sets, we need to know what the category of "objects under ψ" is. Put

$$\mathbf{V}(\mathcal{A}) := (\mathcal{A} \times \mathbb{I}) \cup^{\mathcal{A}\times\{\upsilon_1\}} *,$$

as was introduced in the proof of Lemma 22.7.2. Define the $\mathcal{P}$-category $\psi/\mathcal{C}_\kappa$ of *objects under* ψ to be the $\mathcal{P}$-category of morphisms

$$\mathbf{V}(\mathcal{A}) \to \mathcal{C}_\kappa,$$

restricting to ψ on $\mathcal{A} \times \{\upsilon_0\}$. In other words, it is the fiber of the morphism

$$\underline{\text{HOM}}(\mathbf{V}(\mathcal{A}), \mathcal{C}_\kappa) \to \underline{\text{HOM}}(\mathcal{A}, \mathcal{C}_\kappa)$$

over ψ.

There is a forgetful morphism ev_1 given by evaluation on $\{\upsilon_1\}$, from $\psi/\mathcal{C}_\kappa$ to $\mathcal{C}_\kappa$.

An object of $\psi/\mathcal{C}_\kappa$ is a pair (V, v) where $V \in \text{Ob}(\mathcal{C}_\kappa)$ and $v : \mathcal{A}\times\mathbb{I} \to \mathcal{C}_\kappa$ is a morphism restricting to $V_\mathcal{A}$ on $\mathcal{A} \times \{\upsilon_1\}$ and ψ on $\mathcal{A} \times \{\upsilon_0\}$. In other words, v is an element of $\text{Cocone}(\psi, V)$.

In what follows we often designate $v \in \text{Cocone}(\psi, V)$ by the notation $\psi \xrightarrow{v} V$.

A point of the $\mathcal{P}$-object of morphisms between (V_1, v_1) and (V_2, v_2) in $\psi/\mathcal{C}_\kappa$ is a diagram $\mathcal{A}\times\mathbb{I}\times\mathbb{I} \to \mathcal{C}_\kappa$ restricting to the given ones over $\mathcal{A}\times\mathbb{I}\times\{\upsilon_i\}$, restricting to the pullback of ψ on $\mathcal{A} \times \{\upsilon_0\} \times \mathbb{I}$, and restricting to the pullback of $\mathbb{I} \to \mathcal{C}_\kappa$, a morphism from V_1 to V_2, on the edge $\mathcal{A} \times \{\upsilon_1\} \times \mathbb{I}$.

Let $\tau_{\leq 1}(\psi/\mathcal{C}_\kappa)$ denote the 1-truncation. It has the same objects, but a morphism is a point, as discussed in the previous paragraph, up to homotopy. Two points corresponding to maps as above are homotopic if there is a morphism

$$\mathcal{A} \times \mathbb{I} \times \mathbb{I} \times \mathbb{I} \to \mathcal{C}_\kappa,$$

restricting to appropriate things on the boundary.

Varying the regular cardinal, gives an ind-1-category $\tau_{\leq 1}(\psi/\mathcal{C}_.)$. The criterion of 22.9.4 can be restated in these terms:

Corollary 22.9.7 *If $\psi \xrightarrow{t} T$ projects to an initial object in $\tau_{\leq 1}(\psi/\mathcal{C}_.)$, then it is a colimit of ψ.*

Proof The condition that t be an initial object means that for any V and $v \in \text{Cocone}(\psi, V)$ there is a morphism from (T, t) to (V, v) which is unique up to homotopy. Such a morphism is a map from $\mathcal{A} \times \mathbb{I} \times \mathbb{I}$ to $\mathcal{C}_\kappa$ (for κ big enough), but restricting to $\widetilde{\Upsilon}_2(*, *) \hookrightarrow \mathbb{I} \times \mathbb{I}$ we get an element of $\text{Cocone}(\psi, T, V)$ of the required form for the criterion of Proposition 22.9.4. By the standard arguments of extension along trivial cofibrations, the uniqueness up to homotopy according to the present definition, implies the uniqueness up to homotopy of the kind, albeit slightly different, entering into Proposition 22.9.4. □

The initial object T will be constructed as the mapping telescope of a compatible homotopy idempotent, following the strategy outlined in Section 22.9. To start, Theorem 22.6.1 gives a limit U of the forgetful functor

$$f : (\psi/\mathcal{C})_\kappa \to \mathbf{Enr}(\mathcal{P})'_{.}.$$

Let $r \in \mathrm{Cone}(U, f)$ be the limit cone.

By Corollary 21.10.5, the pullback of U to a constant family $U_{\mathcal{A}}$ over $\mathcal{A}$ is again a limit of the functor

$$f_{\mathcal{A}} : \mathcal{A} \times (\psi/\mathcal{C})_\kappa \to \mathbf{Enr}(\mathcal{P})'_{.},$$

so, as in Corollary 21.10.6, the morphism of families over $\mathcal{A} \times (\psi/\mathcal{C})_\kappa$, from ψ to $f_{\mathcal{A}}$, factorizes into

$$\psi \xrightarrow{a} U_{\mathcal{A}} \xrightarrow{r_{\mathcal{A}}} f_{\mathcal{A}}, \tag{22.9.1}$$

which is notation for a map

$$\mathcal{A} \times \widetilde{\Upsilon}_2(*, *) \times (\psi/\mathcal{C})_\kappa \to \mathbf{Enr}(\mathcal{P})'_{.}.$$

The morphism $\psi \to U_{\mathcal{A}}$, which is to say an element $u \in \mathrm{Cocone}(\psi, U)$, is automatically provided with the following data:

Lemma 22.9.8 *With the above notations, our object $U \in \mathbf{Enr}(\mathcal{P})'_{.}$ and its morphism (i.e. cocone) $u : \psi \to U$ are provided with:*

(A) for every morphism $\psi \to V$ where $\#^{\mathrm{reg}} V \leq \kappa$, a factorization which we call the official factorization

$$\psi \xrightarrow{a} U \to V,$$

in other words a diagram

$$\mathcal{A} \times \widetilde{\Upsilon}_2(*, *) \to \mathbf{Enr}(\mathcal{P})'_{.},$$

restricting to u on the edge (01) *and restricting to our given morphism on the edge* (02)*; and*

(B) for every diagram

$$\psi \to V \to V',$$

a completion of this and the official factorization diagrams

$$\psi \to U \to V, \quad \psi \to U \to V'$$

to a diagram, the official commutativity diagram,

$$\psi \to U \to V \to V',$$

which again means a morphism

$$\mathcal{A} \times \widetilde{\Upsilon}_3(*, *, *) \to \mathbf{Enr}(\mathcal{P})'_{\bullet},$$

restricting to our given diagrams on the faces (023), (012) *and* (013).

Proof These come from the limit cone r and the factorization (22.9.1). □

Let

$$\psi \stackrel{w}{\to} W \stackrel{a}{\to} U$$

be a factorization of a as in Lemma 22.7.2 with $\#^{\text{reg}} W \leq \kappa$. This means a diagram whose (02)-edge is equal to u.

In what follows, the corresponding element of $\tau_{\leq 1}(\psi/\mathcal{C}_{\bullet})$ will be denoted with a tilde as $\tilde{a}$. It should be noted that the functor

$$\tau_{\leq 1}(\psi/\mathcal{C}_{\bullet}) \to \tau_{\leq 1}(\mathcal{C}_{\bullet})$$

is not necessarily faithful; this means that the lift $\tilde{a}$ corresponding to the above diagram, is not necessarily uniquely determined by just the morphism $W \stackrel{a}{\to} U$ in $\mathcal{C}_{\bullet}$. This difficulty is specific to the higher categorical situation, coming from the "homotopy of commutativity" $aw \sim u$.

As was explained in Section 22.9, at this point we have no control over the choice of W, so the initial object T which we would like to choose may be a direct factor of this W. Notice that by Lemma 22.9.8 (A) there is a morphism

$$q : U \to W,$$

giving a factorization

$$[a, q] : \psi \stackrel{u}{\to} U \stackrel{q}{\to} W.$$

The edge (02) of this diagram is b, so we can write $w \sim qq$. Again, this corresponds to a map denoted

$$\tilde{q} : (U, u) \to (W, w)$$

in $\tau_{\leq 1}(\psi/\mathcal{C}_{\bullet})$.

Using the fibrant property of $\mathbf{Enr}(\mathcal{P})'_{\bullet}$ we can glue the diagrams $[b, i]$ and $[a, q]$ together to give a diagram

$$\psi \stackrel{w}{\to} W \stackrel{a}{\to} U \stackrel{q}{\to} W,$$

in other words a morphism

$$\mathcal{A} \times \widetilde{\Upsilon}_3(*, *, *) \to \mathbf{Enr}(\mathcal{P})'_{\bullet},$$

restricting to $\tilde{a}$ on the face (012), restricting to $\tilde{q}$ on the face (023), and satisfying the usual condition that the restriction to the face (123) be constant in the $\mathcal{A}$ direction. Denote by p the restriction to the edge (13), and denote by $\tilde{p}$ the restriction to the face (013). Thus

$$\tilde{p} : \psi \xrightarrow{w} W \xrightarrow{p} W$$

is a diagram whose restrictions to the edges (01) and (02) are both equal to the morphism w. The above may be rewritten in terms of $\tau_{\leq 1}(\psi/\mathcal{C}_{\cdot})$ as

$$\tilde{q}\tilde{a} = \tilde{p}.$$

The official commutativity diagram for $\tilde{p}$ is a diagram of the form

$$\psi \xrightarrow{u} U \xrightarrow{q} W \xrightarrow{p} W.$$

The restriction of this diagram to the face (023) is the diagram $\tilde{p}$. The restrictions to (012) and (013) are both equal, by Lemma 22.9.8 (B), to the official factorization diagram $\tilde{q}$. In particular, the face (123) gives a diagram

$$(U, u) \xrightarrow{q} (W, w) \xrightarrow{\tilde{p}} (W, w),$$

whose (13)-edge is again the morphism $\tilde{q}$. Homotopically we get an equation

$$\tilde{p} \circ \tilde{q} \sim \tilde{q}.$$

In view of the fact that $\tilde{p} \sim \tilde{q}\tilde{a}$ we get $\tilde{p}^2 = \tilde{p}$ in $\tau_{\leq 1}(\psi/\mathcal{C}_{\cdot})$, which after forgetting ψ can be written with a homotopy

$$p \circ p \overset{H}{\sim} p.$$

This equation says that, up to homotopy, p is a projector or idempotent. It is the projector onto the answer that we are looking for.

Lemma 22.9.9 *The homotopy H satisfies the compatibility condition $pH \sim Hp$.*

Proof It is naturally defined using the universal properties. Consideration of naturality as in Lemma 22.9.8 allows one to show that $pH * H \sim Hp * H$, where $*$ is the vertical composition in the 2-category of Proposition 21.3.3. This requires some work, including the extension of Lemma 22.9.8 to diagrams of 2 and 3 maps, and the utilization of shell extension properties as in Lemma 22.5.8 for $k = 4, 5$, and we don't do the details here. □

In Lemma 22.8.6, the "image" of the homotopy idempotent p was therefore constructed as a mapping telescope which we now denote by $T := \mathbf{Tel}(W, p)$.

As W is κ-presentable, the telescope T is also κ-presentable. Splitting the idempotent p means that T comes equipped with morphisms

$$W \xrightarrow{j} T \xrightarrow{r} W \xrightarrow{j} T,$$

such that $rj \sim p$ and $jr \sim 1_T$, and furthermore $jp \sim j$ and $pr \sim r$. However, these are not yet morphisms under ψ.

Compose the diagrams $\psi \xrightarrow{w} W$ and $W \xrightarrow{j} T$ to get a diagram

$$\psi \xrightarrow{w} W \xrightarrow{j} T,$$

whose (02)-edge is a cocone $\psi \xrightarrow{t} T$. This gives an object (T, t) of $\tau_{\leq 1}(\psi/\mathcal{C}_.)$, together with a morphism

$$\tilde{j}' : (W, w) \to (T, t'),$$

tautologically defined by j since t was obtained as a composition. Set

$$\tilde{j} := \tilde{j}' \circ \tilde{p} : (W, w) \to (T, t).$$

This projects to the same underlying morphism j, due to the fact that $jp \sim j$.

Let $\tilde{r}$ be obtained as the composed diagram

$$\psi \xrightarrow{t} T \xrightarrow{r} W,$$

where the (02) edge is w, obtained by choosing a homotopy between rt and w. Note that $t \sim jw$ by definition so $rt \sim rjw \sim w$. The homotopy may be chosen so that $\tilde{p}\tilde{r} = \tilde{r}$ (if necessary using the same trick of composing with an extra $\tilde{p}$ as for $\tilde{j}$).

Lemma 22.9.10 *With the above definitions, the equalities*

$$\tilde{j}\tilde{p} = \tilde{j}, \quad \tilde{j}\tilde{r} = 1_{(T,t)},$$

and

$$\tilde{r}\tilde{j} = \tilde{p} = \tilde{q}\tilde{a}$$

hold in $\tau_{\leq 1}(\psi/\mathcal{C}_.)$.

Proof For the first, $\tilde{j}\tilde{p} = \tilde{j}'\tilde{p}^2 = \tilde{j}'\tilde{p} = \tilde{j}$.

For the second, $\tilde{j}\tilde{r} = \tilde{j}'\tilde{p}\tilde{r} = \tilde{j}'\tilde{r}$. This lifts 1_T so it is invertible; but using the third equation of the lemma which will be proven next,

$$\tilde{j}\tilde{r}\tilde{j}\tilde{r} = \tilde{j}\tilde{p}\tilde{r} = \tilde{j}'\tilde{p}^2\tilde{r} = \tilde{j}\tilde{r}.$$

It follows that $\tilde{j}\tilde{r}$ is the identity.

Now for the third, There are two maps $\tilde{r}\tilde{j}'$ and $\tilde{p}$ from (W, w) to itself in $\psi/\mathcal{C}_\kappa$. As U is the limit of the forgetful functor over $\psi/\mathcal{C}_\kappa$, and the map u from ψ to U is obtained by the limit property, the compositions of these two maps with q are the same. More precisely, there are two different corresponding homotopies $pq \sim q$ and $jrq \sim q$. The compositions of these homotopies with $\tilde{q} : (U, u) \to (W, w)$ are diagrams

$$\psi \to U \to W \to W,$$

meaning as usual morphisms $\mathcal{A} \times \widetilde{\Upsilon}_3(*, *, *) \to \mathcal{C}_\cdot$. Both diagrams share the edges (01) equal to u, (02) and (03) equal to w, (12) and (13) equal to q, and the triangle (013) equal to $\tilde{q}$. These diagrams give, on the one hand, a homotopy between $\tilde{r}\tilde{j}'\tilde{q}$ and $\tilde{q}$, and, on the other hand, a homotopy between $\tilde{p}\tilde{q}$ and $\tilde{q}$. Composing furthermore with $\tilde{a}$ this gives the relation

$$\tilde{r}\tilde{j} = \tilde{r}\tilde{j}'\tilde{q}\tilde{a} = \tilde{p}\tilde{q}\tilde{a} = \tilde{p}$$

in $\tau_{\leq 1}(\psi/\mathcal{C}_\cdot)$. □

Notice that $\#^{\mathrm{reg}}(T) \leq \kappa$ so Lemma 22.9.8(A) applies for $(T, t) \in \psi/\mathcal{C}_\kappa$ to give an official factorization diagram

$$\psi \xrightarrow{u} U \xrightarrow{s} T,$$

which is equivalently written $\tilde{s} : (U, u) \to (T, t)$.

Lemma 22.9.11 *The morphism $\psi \xrightarrow{t} T$ has the unique homotopy factorization property of 22.9.4 with respect to any morphism $\psi \to B$ such that $\#^{\mathrm{reg}}(B) \leq \kappa$.*

Proof Any such $\psi \xrightarrow{b} B$ is an object of $\psi/\mathcal{C}_\kappa$. Applying Lemma 22.9.8, we get a morphism of objects under ψ from $\tilde{z} : (U, u) \to (B, b)$. Composing with $(T, t) \xrightarrow{\tilde{a}\tilde{r}} (U, u)$ gives existence of a factorization

$$\tilde{z}\tilde{a}\tilde{r} : (T, t) \to (B, b).$$

Suppose now given a different map $\tilde{e} : (T, t) \to (B, b)$ of objects under ψ. As in part (B) of Lemma 22.9.8, the composition $\tilde{e} \circ \tilde{s}$ is equal to $\tilde{z}$. Thus

$$\tilde{e}\tilde{s}\tilde{a}\tilde{r} = \tilde{z}\tilde{a}\tilde{r}.$$

However, $\tilde{j}\tilde{q}\tilde{a} : (W, w) \to (T, t)$ is a map in $\psi/\mathcal{C}_\kappa$, and as in the proof of the previous lemma,

$$\tilde{j}\tilde{q}\tilde{a}\tilde{q} = \tilde{s}.$$

In particular

$$\tilde{s}\tilde{a}\tilde{r} = \tilde{j}\tilde{q}\tilde{a}\tilde{q}\tilde{a}\tilde{r} = \tilde{j}\tilde{p}\tilde{r} = \tilde{j}\tilde{r} = 1_{(T,t)}.$$

It follows that

$$\tilde{e} = \tilde{z}\tilde{a}\tilde{r}.$$

This proves unicity of the factorization. □

Lemma 22.9.12 *The morphism $\psi \xrightarrow{t} T$ has the unique homotopy factorization property of 22.9.4 with respect to any morphism $\psi \to B$ without requiring a bound on the cardinality of B, in other words (T,t) is an initial object of $\tau_{\leq 1}(\psi/\mathcal{C}_{\cdot})$.*

Proof Suppose $\psi \xrightarrow{b} B$ is an object of $\psi/\mathcal{C}_{\cdot}$. Then there is a factorization through $\psi \to B' \to B$ with $\#^{\mathrm{reg}} B' \leq \kappa$, by Lemma 22.7.2. An argument similar to the one of Lemma 22.7.2 also shows that any two such factorizations may in turn be factored through a third one.

Applying the previous lemma to $\psi \to B'$ we obtain the existence of a factorization

$$\psi \to T \to B' \to B.$$

Suppose $\tilde{e}_1, \tilde{e}_2 : (T,t) \to (B,b)$ are two factorizations. Then they factor through a common $B' \to B$ with $\#^{\mathrm{reg}} B' \leq \kappa$, and the uniqueness in the previous lemma implies that $\tilde{e}_1 = \tilde{e}_2$ in $\tau_{\leq 1}(\psi/\mathcal{C}_{\cdot})$. □

Corollary 22.9.13 *In the situation of Lemma 22.9.12 the map $t : \psi \to T$ is a colimit.*

Proof By Lemma 22.9.12 this map satisfies the condition of Proposition 22.9.4 or equivalently Corollary 22.9.7, so it is a colimit. □

This corollary completes the proof of Theorem 22.9.6.
We get colimits of n-categories and (∞, n)-categories.

Theorem 22.9.14 *The $(n+1)$-category nCAT of small n-categories admits colimits indexed by small $(n+1)$-categories. The Segal $(n+1)$-category nSeCAT of all small Segal n-categories admits colimits indexed by small Segal $(n+1)$-categories.*

Proof These are just the special cases of Theorem 22.9.6 where $\mathcal{P} = \mathcal{PC}^n(\mathrm{SET})$ or $\mathcal{P} = \mathcal{PC}^n(\mathcal{K})$. The additional hypothesis saying that all objects should be cofibrant, holds in these cases. □

22.10 Fiber products and amalgamated sums

Taking $\mathcal{A}$ to be the category with three objects a, b and c and morphisms $a \to b$ and $c \to b$, a functor $\mathcal{A} \to \mathbf{Enr}(\mathcal{P})_{.}$ is just a triple X, Y, Z of objects of $\mathcal{P}$ with maps $u : X \to Y$ and $v : Z \to Y$. The limit of the projection into $\mathbf{Enr}(\mathcal{P})'_{.}$ is the *homotopy fiber product* denoted $X \times^{\mathrm{ho}}_{Y} Z$.

Lemma 22.10.1 *Suppose $\mathcal{A}$ is as above and $\varphi : \mathcal{A} \to \mathbf{Enr}(\mathcal{P})_{.}$ is a morphism corresponding to a pair of maps $u : X \to Y$ and $v : Z \to Y$ of objects of $\mathcal{P}$ such that u is a fibration. Then the usual fiber product $X \times_Y Z$ is a limit of φ so we can write*

$$X \times^{\mathrm{ho}}_{Y} Z = X \times_Y Z.$$

Proof One way to prove this is to use our explicit construction of the limit (Section 22.6.1). The second way is to show that $U := X \times_Y Z$ satisfies the required universal property as follows. First of all, note that the commutative square

$$\begin{array}{ccc} U & \longrightarrow & X \\ \downarrow & & \downarrow \\ Z & \longrightarrow & Y \end{array}$$

corresponds to a map $\mathbb{I} \times \mathbb{I} \to \mathbf{Enr}(\mathcal{P})_{.}$, which we can project into $\mathbb{I} \times \mathbb{I} \to \mathbf{Enr}(\mathcal{P})'_{.}$. Then combine this with the projection

$$\mathcal{A} \times \mathbb{I} \to \mathbb{I} \times \mathbb{I},$$

which sends $\mathcal{A} \times \{v_0\}$ to $(0, 0)$ and sends $\mathcal{A} \times \{v_1\}$ to the copy of $\mathcal{A} \subset \mathbb{I} \times \mathbb{I}$ corresponding to the sides $(1, 01)$ and $(01, 1)$ of the square. We get a map $\mathcal{A} \times \mathbb{I} \to \mathbf{Enr}(\mathcal{P})'_{.}$ having the required constancy property to give an element $\epsilon \in \mathrm{HOM}(U, \varphi)$. This is the map which we claim is a limit.

In passing note that since X, Y, Z are elements of $\mathbf{Enr}(\mathcal{P})_{.}$ they are by definition fibrant, and since by hypothesis the map $X \to Y$ is a fibration, the map $U \to Z$ is a fibration too, and so U is fibrant.

We now fix a fibrant object of $\mathcal{P}$ V and study the functor which to an n-precat F associates the set of morphisms

$$g : \mathcal{A} \times \Upsilon(F) \to \mathbf{Enr}(\mathcal{P})'_{.},$$

with $r_0(g) = V_{\mathcal{A}}$ and $r_1(g) = \varphi$. This is of course just the functor represented by $\mathrm{HOM}(V, \varphi)$. Recalling that $\underline{\mathrm{HOM}}(V, U)$ is the morphism set in $\mathbf{Enr}(\mathcal{P})_{.}$, we obtain by composition with ϵ a morphism

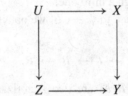

$$C_\epsilon : \underline{\mathrm{HOM}}(V, U) \to \underline{\mathrm{HOM}}(V, \varphi).$$

In this case, since ϵ comes from $\mathbf{Enr}(\mathcal{P})_{.}$ in which the composition at the first stage is strict, the morphism C_ϵ is strictly well defined rather than being a weak morphism as usual in the notion of limit. We would like to show that C_ϵ is an equivalence (which would prove the lemma).

A morphism $g : \mathcal{A} \times \Upsilon(F) \to \mathbf{Enr}(\mathcal{P})'_{.}$ decomposes as a pair of morphisms (g_1, g_2) with

$$g_i : \mathbb{I} \times \Upsilon(F) \to \mathbf{Enr}(\mathcal{P})'_{.};$$

in turn these decompose as pairs g_i^+ and g_i^- where

$$g_i^+ : \Upsilon(*, F) \to \mathbf{Enr}(\mathcal{P})'_{.},$$

$$g_i^- : \Upsilon(F, *) \to \mathbf{Enr}(\mathcal{P})'_{.}.$$

(Decompose the square $\mathbb{I} \times \Upsilon(F)$ into two triangles, drawing the edge $\mathbb{I}$ vertically with vertex 0 on top.) The conditions on everything to correspond to a morphism g are that

$$r_{12}\left(g_i^+\right) = r_{01}\left(g_i^-\right),$$

and

$$r_{02}\left(g_1^-\right) = r_{02}\left(g_2^-\right).$$

The endpoint conditions on g correspond to the conditions

$$r_{12}\left(g_1^-\right) = u, \quad r_{12}\left(g_2^-\right) = v,$$

and

$$r_{01}\left(g_i^+\right) = 1_V.$$

Putting these all together we see that our functor of F is of the form a fiber product of four diagram objects of $\mathcal{P}$ as defined in 22.4.5. More precisely, define M_u to be

$$Diag^{1_V, X}\left(*, \underline{*}; \mathbf{Enr}(\mathcal{P})'_{.}\right) \times_{Diag^{V,X}(\underline{*}; \mathbf{Enr}(\mathcal{P})'_{.})} Diag^{V,u}\left(\underline{*}, *; \mathbf{Enr}(\mathcal{P})'_{.}\right),$$

where the morphisms in the fiber product are r_{12} then r_{01}; and define M_v similarly. Then

$$\underline{\mathrm{HOM}}(V, \varphi) = M_u \times_{Diag^{V,Y}(\underline{*}; \mathbf{Enr}(\mathcal{P})'_{.})} M_v,$$

where here the morphisms in the fiber product are the restrictions r_{02} on the second factors of the M.

Refer now to the calculation of Remark 22.4.7 in view of the comparison result of Proposition 22.4.11 applied to $\mathbf{Enr}(\mathcal{P})_{.} \to \mathbf{Enr}(\mathcal{P})'_{.}$.

By this calculation the restriction morphism

$$r_{12} : Diag^{1_V,X}\left(\underline{*}, *; \mathbf{Enr}(\mathcal{P})'_{\cdot}\right) \to Diag^{V,Z}\left(\underline{*}; \mathbf{Enr}(\mathcal{P})'_{\cdot}\right)$$

is a trivial fibration. Therefore the second projections are equivalences

$$M_u \to Diag^{V,u}\left(*, \underline{*}; \mathbf{Enr}(\mathcal{P})'_{\cdot}\right),$$

and similarly for v. Using these second projections in each of the factors M we get an equivalence

$$\underline{\mathrm{HOM}}(V, \varphi)$$
$$\to Diag^{V,u}\left(\underline{*}, *; \mathbf{Enr}(\mathcal{P})'_{\cdot}\right) \times_{Diag^{V,Y}(\underline{*};\mathbf{Enr}(\mathcal{P})'_{\cdot})} Diag^{V,u}\left(\underline{*}, *; \mathbf{Enr}(\mathcal{P})'_{\cdot}\right),$$

where the morphisms in the fiber product are r_{02}. There is a morphism from the same fiber product taken with respect to $\mathbf{Enr}(\mathcal{P})_{\cdot}$, into here. In the case of the fiber product taken with respect to $\mathbf{Enr}(\mathcal{P})_{\cdot}$ the calculation of Remark 22.4.7 gives directly that it is equal to

$$\underline{\mathrm{HOM}}(V, X) \times_{\underline{\mathrm{HOM}}(V,Y)} \underline{\mathrm{HOM}}(V, Z),$$

which is just $\underline{\mathrm{HOM}}(V, U)$. The morphism

$$\underline{\mathrm{HOM}}(V, X) = Diag^{V,u}(\underline{*}, *; \mathbf{Enr}(\mathcal{P})_{\cdot}) \to Diag^{V,u}\left(\underline{*}, *; \mathbf{Enr}(\mathcal{P})'_{\cdot}\right)$$

is an equivalence by Proposition 22.4.11, and similarly for the other factors in the fiber product.

Now we are in the general situation that we have equivalences of fibrant n-precategories $P \to P'$, $Q \to Q'$ and $R \to R'$ compatible with diagrams

$$P \to Q \leftarrow R, \quad P' \to Q' \leftarrow R'.$$

If we know that the morphisms $P \to Q$ and $P' \to Q'$ are fibrations then we can conclude that these induce an equivalence

$$P \times_Q R \to P' \times_{Q'} R'.$$

Prove this in several steps using pushout along trivial cofibrations:

$$P' \times_{Q'} R' \xrightarrow{\sim} P' \times_{Q'} R = (P' \times_{Q'} Q) \times_Q R,$$

and

$$P' \times_{Q'} Q \xrightarrow{\sim} P',$$

so

$$P \xrightarrow{\sim} P' \times_{Q'} Q,$$

giving finally

$$P \times_Q R \xrightarrow{\sim} (P' \times_{Q'} Q) \times_Q R;$$

then apply CM2.

Applying this general fact to the previous situation gives that the morphism

$$\underline{\mathrm{HOM}}(V, X) \times_{\underline{\mathrm{HOM}}(V,Y)} \underline{\mathrm{HOM}}(V, Z)$$
$$\to Diag^{V,u}_{\cdot}\left(\underline{*}, *; \mathbf{Enr}(\mathcal{P})'_{\cdot}\right) \times_{Diag^{V,Y}_{\cdot}(\underline{*};\mathbf{Enr}(\mathcal{P})'_{\cdot})} Diag^{V,u}_{\cdot}\left(\underline{*}, *; \mathbf{Enr}(\mathcal{P})'_{\cdot}\right)$$

is an equivalence. By CM2 this implies that

$$C_\epsilon : \underline{\mathrm{HOM}}(U, V) \to \underline{\mathrm{HOM}}(V, \varphi)$$

is an equivalence. □

Lemma 22.10.1 basically says that for calculating homotopy fiber products we can forget about the whole limit machinery and go back to our usual way of assuming that one of the morphisms is a fibration.

Taking $\mathcal{A}$ to be the opposite of the category in the previous paragraph, a functor $\mathcal{A} \to \mathbf{Enr}(\mathcal{P})_{\cdot}$ is a triple U, V, W with morphisms $f : V \to U$ and $g : V \to W$. The colimit is the *homotopy pushout* of U and W over V, denoted $U \cup^V_{\mathrm{ho}} W$.

Lemma 22.10.2 *Suppose that f is a cofibration. Let P denote a fibrant replacement*

$$U \cup^V W \hookrightarrow P.$$

Then there is a natural morphism $u : \varphi \to P$ which is a colimit. Thus we can say that the morphism

$$U \cup^V W \to U \cup^V_{\mathrm{ho}} W$$

is a weak equivalence, or equivalently that the morphism of objects of $\mathcal{P}$

$$\mathbf{Seg}(U \cup^V W) \to U \cup^V_{\mathrm{ho}} W$$

is an equivalence.

The proof is similar to the proof of 22.10.1 and is left as an exercise. It provides justification *a posteriori* for thinking of $\mathbf{Seg}(U \cup^V W)$ as the "categorical pushout of U and W over V." It also shows that this pushout, which occurs in the generalized Seifert–Van Kampen theorem [234], is the same as the homotopy pushout, an observation going back to Whitehead [266].

23

Stabilization

With the very recent work of Hopkins and Lurie, the *Baez–Dolan conjectures* have come to the forefront of research on n-categories. Baez and Dolan [8] proposed a whole series of definitions and properties to be expected of a good theory of n-categories. The main elements were their conjectures on the universal property of certain n-categories defined by looking at cobordisms. These generalize the known ideas of topological and conformal field theories associated to knot invariants. It now appears that Hopkins and Lurie have proven a good portion of these conjectures, see Lurie's expository account [193].

In the present chapter, we will consider a small and beginning piece of the picture, which Baez and Dolan called the *stabilization hypothesis*. It is the analogue for n-categories of the well-known stabilization theorems in homotopy theory. This chapter constitutes a revised version of my preprint [237], which had been occasioned by Baez and Dolan's paper "Categorification" [12].

We work in the context of weak n-categories obtained by iterating our weak enrichment construction starting from the model category SET. This is the first place where we really consider the higher iterates all together, rather than just using one step of the iteration procedure. It will therefore be convenient to stick to the case $\mathcal{PC}^n(\text{SET})$ although parts of the discussion could undoubtedly be done for iterations of the form $\mathcal{PC}^n(\mathcal{M})$ starting from a different model category. See Section 20.2 for notations; however, we shall revert here to the terminologies "n-precategory" and "n-category" rather than "n-prenerve" and "n-nerve," since the general outlines of the proof should in principle work for a wide array of possible theories.

To explain the statement, recall that Baez and Dolan introduce the notion of *k-uply monoidal n-category*, which is an $(n+k)$-category having only one i-morphism for all $i < k$. This includes the notions previously defined and examined by many authors, of monoidal (resp. braided monoidal, symmetric monoidal) category (resp. 2-category) and so forth, as is explained by Baez and

Dolan [8, 12]. See the bibliographies of those preprints as well as that of the the recent preprint of Breen [60] for many references concerning these types of objects. In the case where the n-category in question is an n-groupoid, this notion is–except for truncation at n–the same thing as the notion of k-fold iterated loop space, or "E_k-space," which appears in Dunn [97] (see also some anterior references from there). The fully stabilized notion of k-uply monoidal n-categories for $k \gg n$ is what Grothendieck [132] calls *Picard n-categories.*

The *stabilization hypothesis* of Baez and Dolan [8] states that for $n + 2 \leq k \leq k'$, the k-uply monoidal n-categories are the same thing as the k'-uply monoidal n-categories. What we will show is that a k-uply monoidal n-category can be "delooped" to a $(k + 1)$-uply monoidal n-category, when $k \geq n + 2$.

This statement first appeared in a preliminary way in Breen [59]; also there is some related correspondence between Breen and Grothendieck in "Pursuing stacks" [132]. For this reason, it seems good to call it the *Breen–Baez–Dolan stabilization hypothesis.*

Before giving the precise statement, we make a change of indexing. A *k-connected n-category* is an n-category which has up to equivalence only one i-morphism for each $i \leq n$. More precisely, this means that the truncation $\tau_{\leq k}(\mathcal{A})$ is trivial, equivalent to $*$. Note that a $(k-1)$-connected $(n+k)$-category is equivalent to a k-uply monoidal n-category (see Proposition 23.1.6 below).

We prove the following theorem (Corollary 23.2.4):

Theorem 23.0.1 *If $\mathcal{A}$ is a k-connected weak n-category and if $2k \geq n$ then there is a $(k + 1)$-connected weak $(n + 1)$-category $\mathcal{Y}$, with an object denoted by $y \in \mathcal{Y}_0$, together with an equivalence $\mathcal{A} \cong \mathcal{Y}_1(y, y)$.*

Translated back into Baez and Dolan's notation [8] this says that if $\mathcal{A}$ is a k-uply monoidal n-category and if $k \geq n + 2$ then there exists a "delooping" $\mathcal{Y}$ of $\mathcal{A}$ which is a $(k + 1)$-uply monoidal n-category.

The statement of this theorem is the main content of the "stabilization hypothesis," but it still leaves much further work to be done. For example, one would like to show that the stabilization construction $\mathcal{A} \mapsto \mathcal{Y}$ induces an equivalence of the higher categories which parametrize these objects. One of the problems here is to pin down the right definition of these parametrizing categories.

The technique used for proving Theorem 23.0.1 is to remark that one can reason with "dimensions of cells" for n-categories, in exactly the same way as for topological spaces. The eventual noninvertibility of the i-morphisms up to equivalence doesn't interfere. With this line of reasoning available, the same arguments as in the classical topological case work. A similar proof should be available for any theory of n-categories, but the idea relies heavily on the existence of some kind of formalism for calculating with pushouts along cell additions, such as the model-category formalism which has been developed in this book.

23.1 Minimal dimension

Recall that the model category of n-precategories is $\mathcal{PC}^n(\text{SET})$ (or $\mathcal{PC}^n$ for short) obtained by iterating our basic construction of a Reedy model category $\mathcal{PC}$ starting with the trivial model structure on SET. In this case the Reedy and injective model structures are the same. See Section 20.2 for notations, but "(pre)nerve" is replaced here by "(pre)category."

Lemma 23.1.1 *If $f : \mathcal{A} \to \mathcal{B}$ is a morphism of n-precategories, there exists a diagram*

$$\begin{array}{ccc} \mathcal{A} & \longrightarrow & \mathcal{A}' \\ {\scriptstyle f}\big\downarrow & & \big\downarrow \\ \mathcal{B} & \longrightarrow & \mathcal{B}' \end{array}$$

such that the horizontal arrows are weak equivalences, and the right vertical arrow is a successive transfinite pushout by trivial cofibrations and cofibrations of the form $\partial\mathbb{O}^i_n \hookrightarrow \mathbb{O}^i_n$ for $0 \le i \le n+1$, cf. Section 20.2.6.

Proof The property of admitting a diagram of the desired form, is closed under weak equivalences and transfinite composition. Any cofibration of n-precategories is a successive pushout of cofibrations of the form $\partial h^n(M) \to h^n(M)$ for $M \in \mathbf{C}^n(*)$, this is basically the expression for the skeleta obtained by attaching maps of Section 13.1. If $M = (m, M')$ then

$$\partial h^n(m, M') \hookrightarrow h^n(m, M')$$

is weakly equivalent to an amalgamated sum of inclusions of the form

$$\partial h^n(1, M') \hookrightarrow h^n(1, M'),$$

and we have

$$h^n(1, M') = \Upsilon(h^{n-1}(M')), \quad \partial h^n(1, M') = \Upsilon(\partial h^{n-1}(M')).$$

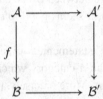

Noting that Υ commutes with pushouts and preserves trivial cofibrations, and that by induction on n we have the desired statement for $h^{n-1}(M')$, we get that $\partial h^n(1, M') \hookrightarrow h^n(1, M')$ admits a diagram as required. Therefore $\partial h^{n-1}(M) \to h^{n-1}(M)$ admits an diagram of the required form, so any cofibration of n-precategories admits such an expression. By the model structure any morphism is equivalent to a cofibration, so any morphism admits a diagram as claimed. □

Definition 23.1.2 A morphism of n-precategories $f : \mathcal{A} \to \mathcal{B}$ has *minimal dimension* m (usually denoted $m(f)$) if m is the largest integer such that there exists a diagram as in Lemma 23.1.1 such that the right vertical arrow is a successive (eventually transfinite) pushout by either trivial cofibrations or cofibrations of the form $\partial \mathbb{O}^i_n \hookrightarrow \mathbb{O}^i_n$ for $i \geq m$.

Make the convention that the minimal dimension of a weak equivalence f is $m(f) := \infty$.

The minimal dimension exists, by Lemma 23.1.1. It takes values in the set $\{0, 1, \ldots, n, n+1, \infty\}$. In particular, if $m(f) \geq n+2$ then f is an equivalence. To say that f has minimal dimension $\geq m$ mean that homotopically, $\mathcal{B}$ is obtained from $\mathcal{A}$ by adding on cells of dimension $\geq m$.

For example, if $n = 0$, an n-precategory is just a set. The minimal dimension of a morphism f of sets is $m(f) = \infty$ if f is an isomorphism, $m(f) = 1$ if f is surjective but not an isomorphism, and $m(f) = 0$ otherwise.

Lemma 23.1.3 *If $f : \mathcal{A} \to \mathcal{B}$ is a morphism of minimal dimension $m(f)$ and if $g : \mathcal{A} \to \mathcal{C}$ is any morphism, and if either f or g is a cofibration, then the minimal dimension of the morphism*

$$\mathcal{C} \to \mathcal{B} \cup^{\mathcal{A}} \mathcal{C}$$

is at least $m(f)$.

Proof Use left properness of $\mathcal{PC}^n$. □

23.1.1 The Whitehead construction

We say that an n-category $\mathcal{A}$ is *k-connected* (for $0 \leq k \leq n$) if the truncation $\tau_{\leq k}(\mathcal{A})$ is a contractible k-category (i.e. the morphism $\tau_{\leq k}(\mathcal{A}) \to *$ is an equivalence of k-categories).

Say that an n-precategory $\mathcal{A}$ is *k-connected* if the n-category **Cat**$(\mathcal{A})$ is k-connected. This notion of k-connectedness is preserved by equivalence of n-precategories.

If $\mathcal{A}$ is k-connected for some $k \geq 0$ then $\tau_{\leq 0}\mathcal{A}$ is the one-point set, i.e. there is a unique equivalence class of objects. Thus the choice of an object $a \in \mathrm{Ob}(\mathcal{A})$ will be well defined up to equivalence.

We now show how to replace a k-connected n-category by an equivalent n-category which is trivial (i.e. *equal* to $*$) in degrees $\leq k$.

More generally, suppose that $\mathcal{A}$ is any n-category. Choose an object $a \in \mathrm{Ob}(\mathcal{A})$. Let $\mathcal{A}' \subset \mathcal{A}$ be the n-precategory defined by setting (for $M \in \mathbf{C}^n(*)$)

$$\mathcal{A}'_M \subset \mathcal{A}_M$$

equal to the subset of elements α such that for any morphism $u : U \to M$ in $\mathbf{C}^n(*)$ with $|U| \leq k$, the image $u^*(\alpha)$ is equal to the degeneracy $d^*(a)$ where $d : U \to 0$ is the unique map in $\mathbf{C}^n(*)$. By construction we have $\mathcal{A}'_M = *$ whenever $|M| \leq k$.

Introduce the following notation for this construction:

$$\mathbf{Wh}_{>k}(\mathcal{A}, a) := \mathcal{A}'.$$

The letters Wh refer to the fact that in the context of spaces, this is the *Whitehead tower*, and we call it the "Whitehead construction."

It has a single object $\mathrm{Ob}(\mathbf{Wh}_{>k}(\mathcal{A}, a)) = \{a\}$. For $k = 0$, $\mathbf{Wh}_{>0}(\mathcal{A}, a)$ is just the pullback of $\mathcal{A}$ by the functor

$$\Delta^o_{\{a\}} \to \Delta^o_{\mathrm{Ob}(\mathcal{A})}.$$

For all k an alternate definition is obtained by induction, setting

$$\mathbf{Wh}_{>0}(\mathcal{A}, a)(a, \ldots, a) := \mathbf{Wh}_{>0}(\mathcal{A}(a, \ldots, a), d_p(a)),$$

where $d_p(a)$ is the degeneracy of a, an object of $\mathcal{A}_{p/}(a, \ldots, a)$.

Lemma 23.1.4 *If $\mathcal{A}$ is an n-category, $a \in \mathrm{Ob}(\mathcal{A})$ is an object, and $0 \leq k \leq n$, then $\mathbf{Wh}_{>k}(\mathcal{A}, a)$ is again an n-category. If $f : (\mathcal{A}, a) \to (\mathcal{B}, b)$ is an equivalence sending a to b then the induced morphism $\mathbf{Wh}_{>k}(f)$ is an equivalence. The construction* **Wh** *is compatible with cartesian products.*

Proof For the proof by induction on n, note first that for $k = 0$ we are just taking the full sub-n-category of $\mathcal{A}$ containing only the object a, which is again an n-category. The statement about equivalences is equally clear in this case. If $k \geq 1$, note as above that

$$\mathbf{Wh}_{>k}(\mathcal{A}, a)_{p/} = \mathbf{Wh}_{>(k-1)}(\mathcal{A}_{p/}(a, \ldots, a), d_p(a)).$$

By the first claim for $n - 1$ the components $\mathbf{Wh}_{>k}(\mathcal{A}, a)_{p/}$ are $(n-1)$-categories. The fact that $\mathbf{Wh}_{>k}$ is compatible with cartesian products – obtained directly from the definition – together with the second of our claims for $n - 1$, imply that the Segal maps for the $\mathbf{Wh}_{>k}(\mathcal{A}, a)_{p/}$ are equivalences. This proves that $\mathbf{Wh}_{>k}(\mathcal{A}, a)$ is an n-category. The second claim about equivalences follows from the fact that $\mathbf{Wh}_{>k}(f)$ is evidently essentially surjective, and full faithfulness comes (via the above formula for $\mathbf{Wh}_{>k}(\mathcal{A}, a)_{1/}(a, a)$) from the inductive statement for $n - 1$. This completes the proof of the two claims. □

Lemma 23.1.5 *If furthermore $\mathcal{A}$ is k-connected then the morphism*

$$i : \mathbf{Wh}_{>k}(\mathcal{A}, a) \to \mathcal{A}$$

is an equivalence of n-categories.

Proof For any $k \geq 0$, k-connectedness implies that $\tau_{\leq 0}(\mathcal{A})$ has only one object, so the morphism i is essentially surjective. In the case $k = 0$, $\mathbf{Wh}_{>0}(\mathcal{A}, a)_{1/}(a, a) = \mathcal{A}_{1/}(a, a)$ so i is fully faithful; this treats the case $k = 0$. For $k \geq 1$ we proceed by induction on n and use the same formula as in Lemma 23.1.4 to get that i is fully faithful. □

The Whitehead operation provides a useful way of constructing k-uply monoidal n-categories. Indeed, one can start with *any* $(n + k)$-category $\mathcal{A}$, choose an object $a \in \mathrm{Ob}(\mathcal{A})$, and taking $\mathbf{Wh}_{>k-1}(\mathcal{A}, a)$ yields a k-uply monoidal n-category. Heuristically, it is the $(n + k)$-subcategory of $\mathcal{A}$ containing as i-morphisms only the higher identities 1^i_a for all $i \leq k-1$ but containing the full n-category of endomorphisms of 1^{k-1}_a from $\mathcal{A}$. This construction intervenes in Baez and Dolan [8, 12], for example in the construction of "generalized center." It is also related to Dunn's Segal-type E_k-machine [97], see the discussion with Hirschowitz [145].

If $\mathcal{A}$ is an $(n + k)$-groupoid, corresponding to an $(n + k)$-truncated homotopy type, then $\mathbf{Wh}_{>k-1}(\mathcal{A}, a)$ is the $(k - 1)$-connected $(n + k)$-groupoid corresponding to the k-th stage in the *Whitehead tower* (see Whitehead [265] or the exposition in Bott and Tu [52]), in other words it is the homotopy-fiber of the morphism

$$\mathcal{A} \to \tau_{\leq k}(\mathcal{A}).$$

In particular, the *homotopy groups* (which can be defined directly in terms of the structure of $(n + k)$-groupoid, see Section 20.4.1 and Tamsamani [250]) of $\mathbf{Wh}_{>k-1}(\mathcal{A}, a)$ are trivial in degrees $i \leq k$, and are the same as the homotopy groups of $\mathcal{A}$ (based at a) in degrees $i > k$. This explains the notation $\mathbf{Wh}_{>k}$.

The Whitehead operation allows us to establish the relationship between k-connectedness and minimal dimension:

Proposition 23.1.6 *Suppose $\mathcal{A}$ is an n-category, and choose an object $a \in \mathrm{Ob}(\mathcal{A})$. Then $\mathcal{A}$ is k-connected if and only if the minimal dimension of the morphism $\{a\} \to \mathcal{A}$ is $\geq k + 1$.*

Proof Suppose $\mathcal{A}$ is a k-connected n-precategory, and let

$$\mathcal{B} := \mathcal{A} \cup^{\partial \mathbb{O}^i_n} \mathbb{O}^i_n,$$

with $i \geq k+1$. We will show that $\mathcal{B}$ is also k-connected. Note that

$$\mathbf{Cat}(\mathcal{B}) \cong \mathbf{Cat}\left(\mathbf{Cat}(\mathcal{A}) \cup^{\partial \mathbb{O}^i_n} \mathbb{O}^i_n\right)$$

so we may assume that $\mathcal{A}$ is an n-category. Fix objects $x_0, \ldots, x_p \in \mathrm{Ob}(\mathcal{A})$. Note that $\mathcal{A}(x_0, \ldots, x_p)$ is $(k-1)$-connected, in fact

$$\tau_{\leq k}(\mathcal{A})(x_0, \ldots, x_p) = \tau_{\leq k-1}(\mathcal{A}(x_0, \ldots, x_p)),$$

as follows from Tamsamani's definition of the truncation operations "starting from the top and going down" [250].

Now the morphism of $(n-1)$-precategories

$$\left(\partial \mathbb{O}^i_n\right)(y_0, \ldots, y_p) \to \mathbb{O}^i_n(y_0, \ldots, y_p)$$

has minimal dimension $\geq i-1$ (this can be seen by the expressions $\mathbb{O}^i_n = \Upsilon\left(\mathbb{O}^{i-1}_n\right)$ and $\partial \mathbb{O}^i_n = \Upsilon\left(\partial \mathbb{O}^{i-1}_n\right)$ together with the definition of Υ and Theorem 23.1.14 for $(n-1)$-categories). Therefore by the inductive version of the statement of the present lemma for $(n-1)$-precategories, we get that

$$\left(\mathcal{A} \cup^{\partial \mathbb{O}^i_n} \mathbb{O}^i_n\right)(x_0, \ldots, x_p)$$

is $(k-1)$-connected. Finally, recall that **Cat** is, up to equivalence, a successive composition of operations denoted **Fix** and **Gen**$[m]$. A pushout of $(k-1)$-connected $(n-1)$-precategories remains $(k-1)$-connected, so the operation $\mathcal{A} \mapsto \mathbf{Gen}[m](\mathcal{A})$ preserves the property of $(k-1)$-connectedness of the components $\mathcal{A}(x_0, \ldots, x_p)$; the operation **Fix** also clearly does. Therefore

$$\mathbf{Cat}\left(\mathcal{A} \cup^{\partial \mathbb{O}^i_n} \mathbb{O}^i_n\right)(x_0, \ldots, x_p)$$

is $(k-1)$-connected, which implies that $\mathbf{Cat}\left(\mathcal{A} \cup^{\partial \mathbb{O}^i_n} \mathbb{O}^i_n\right)$ is k-connected.

For the purposes of the above argument when $k=0$ we use "-1-connected" to mean nonempty.

We have now obtained one half of the proposition: if the minimal dimension of the morphism $\{a\} \to \mathcal{A}$ is $\geq k+1$ then $\mathcal{A}$ is k-connected.

For the other half, suppose $\mathcal{A}$ is k-connected. Set $\mathcal{A}' = \mathbf{Wh}_{>k}(\mathcal{A}, a)$. The morphism $\{a\} \to \mathcal{A}'$ is obtained by a sequence of cofibrations of the form $\partial h^n(M) \to h^n(M)$ for $|M| \geq k+1$. By the same argument, as in Lemma 23.1.1, this morphism is obtained up to weak equivalence by adding on cofibrations of the form $\partial \mathbb{O}^i_n \hookrightarrow \mathbb{O}^i_n$ for $i \geq k+1$. Thus the minimal dimension of $\{a\} \to \mathcal{A}'$ is at least $k+1$, hence by Lemma 23.1.5 the same is true of $\{a\} \to \mathcal{A}$. This completes the proof of the proposition. □

In the above discussion we have used in an essential way that the source of the map in question consists of only one point. For a general map, being of minimal dimension $\geq k$ is related to the behavior on truncations but not exactly the same thing:

Lemma 23.1.7 *Suppose $f : \mathcal{A} \to \mathcal{B}$ is a morphism of n-categories. If $m(f) \geq k+1$ then $\tau_{\leq k-1}(f)$ is an equivalence; and in the other direction, if $\tau_{\leq k}(f)$ is an equivalence then $m(f) \geq k+1$.*

Proof Left as an exercise. □

One cannot in general make a more precise statement than that. This may be understood by reference to classical homotopy theory (which is the case where the n-categories in question are n-groupoids). In this case, $f : \mathcal{A} \to \mathcal{B}$ has minimal dimension $\geq k+1$ if and only if f induces an isomorphism on π_i for $i \leq k-1$, and a surjection on π_k–the minimal dimension here corresponds to the dimensions of the cells which f adds to $\mathcal{A}$ to obtain $\mathcal{B}$. Of course, if $\mathcal{A}$ consists of only one point, then f being surjective on π_k is the same thing as $\pi_k(f)$ being an isomorphism and we recover the characterization of Proposition 23.1.6.

23.1.2 The main estimate

The main estimate will be a compatibility of the notion of minimal dimension with pushout-products. For expositional reasons it is convenient to formulate this property as a definition, indeed it will be supposed inductively in the preliminary lemmas.

Definition 23.1.8 Suppose $f : \mathcal{A} \to \mathcal{B}$ and $g : \mathcal{C} \to \mathcal{D}$ are cofibrations of n-precategories with minimal dimensions $m(f)$ and $m(g)$ respectively. We say that *f and g satisfy the pushout-product estimate*, if the minimal dimension of the morphism

$$f \wedge g : \mathcal{A} \times \mathcal{D} \cup^{\mathcal{A} \times \mathcal{C}} \mathcal{B} \times \mathcal{C} \to \mathcal{B} \times \mathcal{D}$$

is greater than or equal to $m(f) + m(g)$. We say that *the pushout-product estimate holds for n-precategories*, if any pair of cofibrations f and g between n-precategories satisfies the pushout-product estimate.

The goal of this section will be to prove the pushout-product estimate for n-precategories. By induction we may assume that the estimate holds for $(n-1)$-precategories, starting with the following lemma which initiates the induction at $n = 0$, i.e. for sets.

Lemma 23.1.9 *The pushout-product estimate holds for 0-precategories, i.e. for morphisms of sets.*

Proof It suffices to consider the morphisms

$$a : \partial h(1^0) = \emptyset \to * = \mathbb{O}^0_0,$$

and

$$b : \partial \mathbb{O}^0_0 = 2* \to * = \mathbb{O}^0_0,$$

where $2*$ is the set with two elements. Note that $m(a) = 0$ and $m(b) = 1$. Obviously $m(a \wedge a) = 0$. We have

$$a \wedge b = \left(\emptyset \cup^{\emptyset} 2* \to *\right),$$

so $m(a \wedge b) = 1$, and similarly $m(b \wedge a) = 1$. Finally

$$b \wedge b = \left(2* \cup^{2* \times 2*} 2* \to *\right),$$

where the two maps in the amalgamated sum are the two projections; thus $b \wedge b$ is an isomorphism so $m(b \wedge b) = \infty \geq 1 + 1$. This completes the verification of Theorem 23.1.14 for $n = 0$. □

Lemma 23.1.10 *If f and g are cofibrations in $\mathcal{PC}^n$, and if one of them is a weak equivalence, then they satisfy the pushout-product estimate.*

Proof A cartesian product with a weak equivalence is again a weak equivalence, by the cartesian property of $\mathcal{PC}^n$. In this case one of $m(f)$ or $m(g)$ is by convention ∞, and the pushout-product morphism is a weak equivalence so its minimal dimension is ∞ too. □

Lemma 23.1.11 *For any morphism $h : \mathcal{A} \to \mathcal{B}$ of $(n-1)$-precategories of minimal dimension $m(h)$, the minimal dimension of*

$$\Upsilon(h) : \Upsilon(\mathcal{A}) \to \Upsilon(\mathcal{B})$$

is at least equal to $m(h) + 1$.

Proof Apply Υ, which commutes with pushouts and takes weak equivalences to weak equivalences, to the sequence of cell extensions calculating the minimal dimension of h; this gives a sequence showing that the minimal dimension of $\Upsilon(h)$ is at least $m(h) + 1$. □

Lemma 23.1.12 *Suppose that the pushout-product estimate is known for $(n-1)$-precategories. If $i : \mathcal{I} \hookrightarrow \mathcal{F}$ is a cofibration of $(n-1)$-precategories of minimal dimension $m(i)$ then the minimal dimension of the morphism*

$$\mathcal{F} \cup^{\mathcal{I}} \mathcal{F} \to \mathcal{F}$$

is at least $m(i) + 1$.

Proof The statement of the claim is easily verified by hand for $n-1=0$, i.e. when $\mathcal{I}$ and $\mathcal{F}$ are sets. Therefore we may assume here that $n \geq 2$, i.e. $\mathcal{I}$ and $\mathcal{F}$ are $(n-1)$-precategories with $n-1 \geq 1$.

Let $\mathbf{I}$ be the 1-category with two objects υ_0, υ_1 and a single isomorphism between them; we shall consider it as an $(n-1)$-precategory. The minimal dimension of the cofibration of $(n-1)$-precategories

$$j : \{\upsilon_0, \upsilon_1\} \to \mathbf{I}$$

is 1. Therefore, applying the inductive hypothesis that Theorem 23.1.14 holds for $(n-1)$-precategories, we get that the minimal dimension of the map

$$i \wedge j : \mathcal{I} \times \mathbf{I} \cup^{\mathcal{I}\times\{\upsilon_0,\upsilon_1\}} \mathcal{F} \times \{\upsilon_0, \upsilon_1\} \to \mathcal{F} \times \mathbf{I}$$

is at least $m(i)+1$. Finally, note that

$$\begin{aligned} &\mathcal{I} \times \mathbf{I} \cup^{\mathcal{I}\times\{\upsilon_0,\upsilon_1\}} \mathcal{F} \times \{\upsilon_0, \upsilon_1\} \\ &\quad = \left(\mathcal{F} \times \{\upsilon_0\} \cup^{\mathcal{I}\times\{\upsilon_0\}} \mathcal{I} \times \mathbf{I}\right) \cup^{\mathcal{I}\times\mathbf{I}} \left(\mathcal{F} \times \{\upsilon_1\} \cup^{\mathcal{I}\times\{\upsilon_1\}} \mathcal{I} \times \mathbf{I}\right), \end{aligned}$$

and this latter expression maps to the amalgamated sum $\mathcal{F} \cup^{\mathcal{I}} \mathcal{F}$ by a map which comes from the three equivalences

$$\left(\mathcal{F} \times \{\upsilon_0\} \cup^{\mathcal{I}\times\{\upsilon_0\}} \mathcal{I} \times \mathbf{I}\right) \to \mathcal{F},$$

$$\mathcal{I} \times \mathbf{I} \to \mathcal{I},$$

and

$$\left(\mathcal{F} \times \{\upsilon_1\} \cup^{\mathcal{I}\times\{\upsilon_1\}} \mathcal{I} \times \mathbf{I}\right) \to \mathcal{F}.$$

To see that these maps are equivalences, use the preservation of weak equivalences by cartesian product, and preservation of trivial cofibrations by pushout. Because of left properness of $\mathscr{P}\mathscr{C}^n$, Corollary 7.3.2 says that a map of pushout diagrams composed of three weak equivalences, induces a weak equivalence on pushouts. Therefore we have a diagram

$$\begin{array}{ccc} \mathcal{I} \times \mathbf{I} \cup^{\mathcal{I}\times\{\upsilon_0,\upsilon_1\}} \mathcal{F} \times \{\upsilon_0, \upsilon_1\} & \longrightarrow & \mathcal{F} \times \mathbf{I} \\ \downarrow & & \downarrow \\ \mathcal{F} \cup^{\mathcal{I}} \mathcal{F} & \longrightarrow & \mathcal{F} \end{array}$$

in which the vertical arrows are weak equivalences, and in which the top map has (as we have seen previously) minimal dimension at least $m(i)+1$. Therefore the bottom map has minimal dimension at least $m(i)+1$, as claimed. □

Using 21.10.1 and the fact that $\mathbb{O}^i_n = \Upsilon\left(\mathbb{O}^{i-1}_{n-1}\right)$ and $\partial\mathbb{O}^i_n = \Upsilon\left(\partial\mathbb{O}^{i-1}_{n-1}\right)$ it will be easy to deduce the pushout-product estimate in general from the following special case:

Lemma 23.1.13 *Suppose the pushout-product estimate is known for $(n-1)$-precategories. Suppose $f : \mathcal{A} \to \mathcal{B}$ and $g : \mathcal{C} \to \mathcal{D}$ are cofibrations of $(n-1)$-precategories with minimal dimensions $m(f)$ and $m(g)$ respectively. Then the minimal dimension of*

$$\Upsilon(\mathcal{A}) \times \Upsilon(\mathcal{D}) \cup^{\Upsilon(\mathcal{A})\times\Upsilon(\mathcal{C})} \Upsilon(\mathcal{B}) \times \Upsilon(\mathcal{C}) \to \Upsilon(\mathcal{B}) \times \Upsilon(\mathcal{D})$$

is at least $m(f) + m(g) + 2$.

Proof Draw a square divided into two triangles, labeling the horizontal edges with $\mathcal{B}$, the vertical edges with $\mathcal{D}$, and the hypotenuse with $\mathcal{B} \times \mathcal{D}$). From this picture can be read the expression

$$\Upsilon(\mathcal{B}) \times \Upsilon(\mathcal{D}) = \widetilde{\Upsilon}_2(\mathcal{B}, \mathcal{D}) \cup^{\Upsilon(\mathcal{B}\times\mathcal{D})} \widetilde{\Upsilon}_2(\mathcal{D}, \mathcal{B}). \tag{23.1.1}$$

The maps in the amalgamated sum are cofibrations. The equation can be checked directly from the definition of Υ.

We have similar equations for the other products occuring in the statement of the lemma.

Next, note that

$$\left(\Upsilon(\mathcal{A}) \cup^* \Upsilon(\mathcal{D})\right) \cup^{(\Upsilon(\mathcal{A})\cup^*\Upsilon(\mathcal{C}))} \left(\Upsilon(\mathcal{B}) \cup^* \Upsilon(\mathcal{C})\right)$$

$$= \Upsilon(\mathcal{B}) \cup^* \Upsilon(\mathcal{D}). \tag{23.1.2}$$

In this formula the amalgamated sums are taken over the appropriate 1-object sets $*$, for example in $\Upsilon(\mathcal{A}) \cup^* \Upsilon(\mathcal{D})$, $1 \in \Upsilon(\mathcal{A})$ is joined with $v_0 \in \Upsilon(\mathcal{D})$.

Introduce the notation

$$Q(\mathcal{A}, \mathcal{B}, \mathcal{C}, \mathcal{D}) := \widetilde{\Upsilon}_2(\mathcal{A}, \mathcal{D}) \cup^{\widetilde{\Upsilon}_2(\mathcal{A},\mathcal{C})} \widetilde{\Upsilon}_2(\mathcal{B}, \mathcal{C}).$$

The morphisms

$$i(\mathcal{A}, \mathcal{D}) : \Upsilon(\mathcal{A}) \cup^* \Upsilon(\mathcal{D}) \to \widetilde{\Upsilon}_2(\mathcal{A}, \mathcal{D}),$$

and similarly $i(\mathcal{A}, \mathcal{C})$ and $i(\mathcal{B}, \mathcal{C})$ are all equivalences (21.4.2). Using the equation (23.1.2) we obtain a morphism

$$\Upsilon(\mathcal{B}) \cup^* \Upsilon(\mathcal{D}) \to Q(\mathcal{A}, \mathcal{B}, \mathcal{C}, \mathcal{D}),$$

and by "Reedy's lemma," Corollary 7.3.2, this morphism is an equivalence. Then using the fact that $i(\mathcal{B},\mathcal{D})$ is an equivalence and 3 for 2 we get that the morphism

$$u : Q(\mathcal{A},\mathcal{B},\mathcal{C},\mathcal{D}) \to \widetilde{\Upsilon}_2(\mathcal{B},\mathcal{D})$$

is an equivalence. Similarly the map

$$v : Q(\mathcal{C},\mathcal{D},\mathcal{A},\mathcal{B}) = \widetilde{\Upsilon}_2(\mathcal{D},\mathcal{A}) \cup^{\widetilde{\Upsilon}_2(\mathcal{C},\mathcal{A})} \widetilde{\Upsilon}_2(\mathcal{C},\mathcal{B}) \to \widetilde{\Upsilon}_2(\mathcal{D},\mathcal{B})$$

is an equivalence.

Using the equation (23.1.1) and its other similar versions, and a manipulation of pushout formulae, we obtain

$$\Upsilon(\mathcal{A}) \times \Upsilon(\mathcal{D}) \cup^{\Upsilon(\mathcal{A})\times\Upsilon(\mathcal{C})} \Upsilon(\mathcal{B}) \times \Upsilon(\mathcal{C})$$
$$= Q(\mathcal{A},\mathcal{B},\mathcal{C},\mathcal{D}) \cup^{\mathcal{Y}} Q(\mathcal{C},\mathcal{D},\mathcal{A},\mathcal{B}),$$

where

$$\mathcal{Y} := \Upsilon(\mathcal{A}\times\mathcal{D} \cup^{\mathcal{A}\times\mathcal{C}} \mathcal{B}\times\mathcal{C}).$$

In particular, the left-hand side of the morphism in the lemma is

$$Q(\mathcal{A},\mathcal{B},\mathcal{C},\mathcal{D}) \cup^{\mathcal{Y}} Q(\mathcal{C},\mathcal{D},\mathcal{A},\mathcal{B}).$$

Now the equivalences u and v above give an equivalence

$$Q(\mathcal{A},\mathcal{B},\mathcal{C},\mathcal{D}) \cup^{\mathcal{Y}} Q(\mathcal{C},\mathcal{D},\mathcal{A},\mathcal{B}) \to \widetilde{\Upsilon}_2(\mathcal{B},\mathcal{D}) \cup^{\mathcal{Y}} \widetilde{\Upsilon}_2(\mathcal{D},\mathcal{B}).$$

Rewrite the right-hand side of this morphism as

$$\mathcal{V} := \widetilde{\Upsilon}_2(\mathcal{B},\mathcal{D}) \cup^{\mathcal{Y}} \widetilde{\Upsilon}_2(\mathcal{D},\mathcal{B}) = \widetilde{\Upsilon}_2(\mathcal{B},\mathcal{D}) \cup^{\Upsilon(\mathcal{B}\times\mathcal{D})} \mathcal{Z} \cup^{\Upsilon(\mathcal{D}\times\mathcal{B})} \widetilde{\Upsilon}_2(\mathcal{D},\mathcal{B}),$$

where

$$\mathcal{Z} := \Upsilon(\mathcal{B}\times\mathcal{D}) \cup^{\mathcal{Y}} \Upsilon(\mathcal{B}\times\mathcal{D}).$$

(The second map $\Upsilon(\mathcal{D}\times\mathcal{B}) \to \mathcal{Z}$ in the expression for $\mathcal{V}$ is obtained using the standard isomorphism $\mathcal{D}\times\mathcal{B} \cong \mathcal{B}\times\mathcal{D}$.)

For the lemma, it suffices to show that the morphism

$$w : \widetilde{\Upsilon}_2(\mathcal{B},\mathcal{D}) \cup^{\mathcal{Y}} \widetilde{\Upsilon}_2(\mathcal{D},\mathcal{B}) \to \Upsilon(\mathcal{B}) \times \Upsilon(\mathcal{D}),$$

has minimal dimension at least $m(f) + m(g) + 2$.

We have a morphism

$$\alpha : \mathcal{Z} \to \Upsilon(\mathcal{B}\times\mathcal{D})$$

(the identity on both components). Taking the pushout of the expression $\mathcal{V}$ from above along the morphism α gives

$$\mathcal{V} \cup^{\mathcal{Z}} \Upsilon(\mathcal{B}\times\mathcal{D}) = \widetilde{\Upsilon}_2(\mathcal{B},\mathcal{D}) \cup^{\Upsilon(\mathcal{B}\times\mathcal{D})} \widetilde{\Upsilon}_2(\mathcal{D},\mathcal{B}) = \Upsilon(\mathcal{B}) \times \Upsilon(\mathcal{D}).$$

In other words, the morphism w is obtained by a pushout of a cofibration $\mathcal{Z} \hookrightarrow \mathcal{V}$ along α. In particular the minimal dimension of w, and hence of the arrow in the lemma, is at least as big as the minimal dimension $m(\alpha)$, which we shall now bound.

By the inductive hypothesis, the pushout-product estimate is known for $\mathcal{A}, \mathcal{B}, \mathcal{C}, \mathcal{D}$ which are $(n-1)$-categories. Therefore, setting

$$\mathcal{W} := \mathcal{A} \times \mathcal{D} \cup^{\mathcal{A}\times\mathcal{C}} \mathcal{B} \times \mathcal{C},$$

the minimal dimension of the cofibration

$$\mathcal{W} \to \mathcal{B} \times \mathcal{D} \tag{23.1.3}$$

is at least $m(f) + m(g)$.

With this notation $\mathcal{Y} = \Upsilon(\mathcal{W})$, and $\mathcal{Z} = \Upsilon(\mathcal{B} \times \mathcal{D} \cup^{\mathcal{W}} \mathcal{B} \times \mathcal{D})$. The morphism α is obtained by applying Υ to the map

$$\beta : \mathcal{B} \times \mathcal{D} \cup^{\mathcal{W}} \mathcal{B} \times \mathcal{D} \to \mathcal{B} \times \mathcal{D}.$$

To complete the proof, apply Lemma 23.1.12 to the cofibration (23.1.3) which has minimal dimension at least $m(f) + m(g)$. By Lemma 23.1.12 we find that the minimal dimension of the map β above is at least $m(f) + m(g) + 1$. Since $\alpha = \Upsilon(\beta)$, applying Lemma 23.1.11 we get that $m(\alpha) \geq m(f) + m(g) + 2$. This proves the lemma. □

Theorem 23.1.14 *For any $n \geq 0$, the pushout-product estimate holds for n-precategories. In other words, if $f : \mathcal{A} \to \mathcal{B}$ and $g : \mathcal{C} \to \mathcal{D}$ are cofibrations of n-precategories with minimal dimensions $m(f)$ and $m(g)$ respectively, the minimal dimension of the morphism*

$$f \wedge g : \mathcal{A} \times \mathcal{D} \cup^{\mathcal{A}\times\mathcal{C}} \mathcal{B} \times \mathcal{C} \to \mathcal{B} \times \mathcal{D}$$

is greater than or equal to $m(f) + m(g)$.

Proof In view of Lemma 23.1.10, it suffices to consider the case when f and g are cofibrations of the form $\partial \mathbb{O}^i_n \to \mathbb{O}^i_n$ and $\partial \mathbb{O}^j_n \to \mathbb{O}^j_n$ respectively. But the basic cells are constructed inductively using Υ, in the sense that $\mathbb{O}^i_n = \Upsilon(\mathbb{O}^{i-1}_{n-1})$ and $\partial \mathbb{O}^i_n = \Upsilon(\partial \mathbb{O}^{i-1}_{n-1})$. So, this case was treated in Lemma 23.1.13. □

23.2 The stabilization hypothesis

We now show how to prove Theorem 23.0.1. By a *pointed n-precategory* we mean an n-precategory $\mathcal{A}$ with chosen object a or equivalently with a morphism $a : * \to \mathcal{A}$. The *minimal dimension* of a pointed n-precategory, denoted $m(\mathcal{A}, a)$, is defined as the minimal dimension of the map a.

We have defined the *Whitehead operation* $(\mathcal{A}, a) \mapsto \mathbf{Wh}_{>k}(\mathcal{A}, a)$ which gives a subobject of $\mathcal{A}$ containing as i-morphisms only the 1^i_a for $i \leq k$. Recall also from Lemma 23.1.5 that if $\mathcal{A}$ is k-connected then the morphism

$$\mathcal{A} \to \mathbf{Wh}_{>k}(\mathcal{A}, a)$$

is an equivalence.

Remark 23.2.1 If $(\mathcal{A}, a)$ is a pointed $(n+k)$-category of minimal dimension k then $\mathcal{A}$ is $(k-1)$-connected by Proposition 23.1.6 and equivalent to its $(n+k)$-subcategory $\mathbf{Wh}_{>k-1}(\mathcal{A}, a)$. This latter, in Baez–Dolan's terminology, is a k-uply monoidal n-category. Conversely it is clear that a k-uply monoidal n-category, considered as a pointed $(n+k)$-category, has minimal dimension k.

In what follows we trade $n+k$ for n and shall look at pointed n-categories $(\mathcal{A}, a)$ which are $(k-1)$-connected or, equivalently, of minimal dimension k.

Corollary 23.2.2 *Suppose $(\mathcal{A}, a)$, $(\mathcal{B}, b)$ are pointed n-precategories with minimal dimensions $m(\mathcal{A}, a)$ and $m(\mathcal{B}, b)$ respectively. Then the minimal dimension of the morphism*

$$\mathcal{A} \cup^* \mathcal{B} \to \mathcal{A} \times \mathcal{B}$$

is at least $m(\mathcal{A}, a) + m(\mathcal{B}, b)$. In particular, if $m(\mathcal{A}, a) + m(\mathcal{B}, b) \geq n+2$ then the above morphism is a weak equivalence. (The amalgamated sum in the display identifies $* = \{a\} \subset \mathcal{A}$ with $* = \{b\} \subset \mathcal{B}$.)

Proof Apply Theorem 23.1.14 to the inclusions $\{a\} \hookrightarrow \mathcal{A}$ and $\{b\} \hookrightarrow \mathcal{B}$. The map of Theorem 23.1.14 for these two cofibrations is exactly the map $\mathcal{A} \cup^* \mathcal{B} \to \mathcal{A} \times \mathcal{B}$. For the second statement, recall that a morphism of n-categories with minimal dimension $\geq n+2$, in fact has minimal dimension ∞ and is a weak equivalence. □

Now suppose $(\mathcal{A}, a)$ is a pointed n-precategory. We define a simplicial n-precategory $\mathcal{R}$ by the following:

$$\mathcal{R}_{p/} := \mathcal{A} \cup^* \mathcal{A} \cup^* \cdots \cup^* \mathcal{A} \quad (k \text{ times}).$$

In particular $\mathcal{R}_{0/}$ (which we denote $\mathcal{R}_0$) is the set $*$. The maps in the simplicial structure are obtained by using the identity map or else the projection $p : \mathcal{A} \to *$ on each component in an appropriately organized way. For example,

$$\mathcal{R}_{2/} = \mathcal{A} \cup^* \mathcal{A},$$

the first face map is the identity on the first component and the projection p on the second component; the second face map is the identity on the second component and the projection on the first component; and finally the (02) face-map is the identity on both components.

The simplicial object $\mathcal{R} : \Delta^o \to \mathcal{PC}^n$ can be considered as an $(n+1)$-precategory. Note that $\mathcal{R}_0 = *$ so it has a single object. Let $\mathbf{Cat}(\mathcal{R})$ denote the replacement of $\mathcal{R}$ (considered as an $(n+1)$-precategory) by an $(n+1)$-category.

Corollary 23.2.3 *Suppose $(\mathcal{A}, a)$ is a pointed n-category which has minimal dimension k. Suppose that $2k \geq n+2$. Then the simplicial n-precategory $\mathcal{R}$ defined above satisfies the Segal condition that the maps*

$$\mathcal{R}_{p/} \to \mathcal{R}_{1/} \times_{\mathcal{R}_0} \cdots \times_{\mathcal{R}_0} \mathcal{R}_{1/}$$

are weak equivalences. The morphisms $\mathcal{R}_{p/} \to \mathbf{Cat}(\mathcal{R})_{p/}$ are weak equivalences. In particular we have the equivalence of n-categories

$$\mathcal{A} \overset{\cong}{\to} \mathbf{Cat}(\mathcal{R})_{1/}(x, x),$$

where x is the unique object of $\mathcal{R}$, and $(\mathbf{Cat}(\mathcal{R}), x)$ has minimal dimension $k+1$.

Proof For the first part it follows from the previous corollary, noting that $\mathcal{R}_0 = *$. The second part follows from the fact that the operation **Cat** applied to an $(n+1)$-precategory such that the Segal maps are weak equivalences of n-precategories, is the same thing as $\mathbf{Fix}(\mathcal{R})$ consisting of applying **Cat** to each of the $\mathcal{R}_{p/}$. □

The following corollary was stated as Theorem 23.0.1 at the start of the chapter:

Corollary 23.2.4 *Suppose $\mathcal{A}$ is a k-connected n-category with $2k \geq n$. Then there exists a $(k+1)$-connected $(n+1)$-category $\mathcal{Y}$ with $\mathcal{Y}_{1/}(y, y) \cong \mathcal{A}$ (for the essentially unique object y of $\mathcal{Y}$).*

Proof Set $\mathcal{Y} := \mathbf{Cat}(\mathcal{R})$ in the previous corollary. Use Proposition 23.1.6 to compare k-connectedness with minimal dimension. □

This corollary says that $\mathcal{Y}$ is a "delooping" of $\mathcal{A}$. That basically gives the Baez–Dolan stabilization hypothesis [8, 12], although there are a lot of further details which need to be considered: one would like to give an equivalence between the basepointed k-connected n-categories and the basepointed $(k+1)$-connected $(n+1)$-categories.

Numerology: In Baez–Dolan's notation a "k-uply monoidal n-category" is a $(k-1)$-connected $(n+k)$-category cf. Remark 23.2.1. Thus, the corollary says that if $\mathcal{A}$ is a k-uply monoidal n-category, and if $2(k-1) \geq n+k$ (i.e. $k \geq n+2$), then there exists a delooping $\mathcal{Y}$ of $\mathcal{A}$.

23.3 Suspension and free monoidal categories

The stabilization construction of Theorem 23.0.1 may be interpreted as the "suspension" of a pointed n-category, a construction refered to by Baez and Dolan in their paper on categorification [12].

Suppose $\mathcal{A}$ is an n-precategory and $a \in \mathrm{Ob}(\mathcal{A})$ is an object. Recall that $\Upsilon(\{a\}) = \mathbb{I}$ is the 1-category with objects υ_0, υ_1 and a morphism $0 \to 1$. We can use pushout along the projection $\mathbb{I} \to *$ to define the *suspension*:

$$\Sigma(\mathcal{A}, a) := \Upsilon(\mathcal{A}) \cup^{\Upsilon(\{a\})} *.$$

Note that there is an inclusion $\mathbb{I} \subset \mathbf{I}$ (where $\mathbf{I}$ is the 1-category with an isomorphism between 0 and 1) which is homotopic to the projection $\mathbb{I} \to *$, so by left properness and Corollary 7.3.2 we get an equivalence

$$\Sigma(\mathcal{A}, a) \stackrel{\cong}{\leftarrow} \Upsilon(\mathcal{A}) \cup^{\Upsilon(\{a\})} \mathbf{I}.$$

The suspension is an $(n+1)$-precategory. To get an $(n+1)$-category apply the operation **Cat**. The construction $\Sigma(\mathcal{A}, a)$ is invariant under weak equivalences in the variable $\mathcal{A}$ (again by 7.3.2).

Let 0 denote the unique object of $\Sigma(\mathcal{A}, a)$. Note of course that we have a morphism

$$\mathcal{A} \to [\mathbf{Cat}(\Sigma(\mathcal{A}, a))]_{1/}(0, 0).$$

If $\mathcal{A}$ is an n-groupoid, then the suspension $\mathbf{Cat}(\Sigma(\mathcal{A}, a))$ is an $(n+1, 1)$-category, i.e. the i-morphisms are invertible up to equivalence for $i \geq 1$. However, the suspension may or may not be an $(n+1)$-groupoid in this case. We have the following criterion: if $\mathcal{A}$ is an n-groupoid then the suspension $\mathbf{Cat}(\Sigma(\mathcal{A}, a))$ is an $(n+1)$-groupoid if and only if $\mathcal{A}$ is 0-connected (i.e. the set of equivalence classes of objects $\tau_{\leq 0}\mathcal{A}$ has only one element). Let Gr denote the group-completion to an $(n+1)$-groupoid if it isn't already one. Then $Gr(\mathbf{Cat}(\Sigma(\mathcal{A}, a)))$ is the $(n+1)$-groupoid corresponding to the topological suspension of the space which corresponds to the n-groupoid $\mathcal{A}$.

We return now to the consideration of an arbitrary $\mathcal{A}$.

Lemma 23.3.1 *If $(\mathcal{A}, a)$ is a pointed n-category of minimal dimension k then its suspension $(\mathbf{Cat}(\Sigma(\mathcal{A}, a)), 0)$ is a pointed $(n+1)$-category of minimal dimension $k+1$.*

Proof The minimal dimension of the map

$$\Upsilon(\{a\}) \to \Upsilon(\mathcal{A})$$

is at least $m(\mathcal{A}, a)+1$, cf. Lemma 23.1.11. Preservation of minimal dimension under amalgamated sums implies that the minimal dimension of

$$* \to \Upsilon(\mathcal{A}) \cup^{\Upsilon(\{a\})} * = \Sigma(\mathcal{A}, a)$$

is the same as that of the previous map, thus it is $\geq m(\mathcal{A}, a) + 1$. □

The "delooping" operation of Theorem 23.0.1 may be seen as a suspension:

Lemma 23.3.2 *Suppose $\mathcal{A}$ is a pointed n-category with minimal dimension k such that $2k \geq n+2$. Then the $(n+1)$-precategory $\mathcal{R}$ constructed in 23.2.3 is equal to $\Sigma(\mathcal{A}, a)$, therefore the "delooping" $\mathcal{Y}$ of 23.2.4 is equal to* **Cat**$(\Sigma(\mathcal{A}, a))$.

Proof Analyzing closely the construction Υ we see that $\Upsilon(\mathcal{A})_{p/}$ is a disjoint union of one copy of $\mathcal{A}$ for each $i = 1, \ldots, p$ (these are indexed by the sequences of objects $\epsilon_0, \ldots, \epsilon_p$, where $\epsilon_i = 0, 1$ and $\epsilon_i \leq \epsilon_{i+1}$). Similarly, $\Upsilon(\{a\})$ is a disjoint union of one copy of $\{a\}$ for each $i = 1, \ldots, p$. Taking the pushout by the map

$$\Upsilon(\{a\}) \to *$$

amounts exactly to forming the amalgamated sum used in the construction of $\mathcal{R}$. □

We have the following variant of 23.2.3, 23.2.4 and Theorem 23.0.1, concerning the case where the minimal dimension k is not necessarily big. Basically it says that the part of $\mathcal{A}$ which is in the stable range is preserved by suspension:

Proposition 23.3.3 *Suppose $(\mathcal{A}, a)$ is a pointed n-precategory of minimal dimension $m(\mathcal{A}, a)$. Then the morphism*

$$\mathcal{A} \to [\mathbf{Cat}(\Sigma(\mathcal{A}, a))]_{1/}(0, 0)$$

has minimal dimension at least $2m(\mathcal{A}, a)$. In particular, by Lemma 23.1.7 this morphism induces an equivalence on truncations

$$\tau_{\leq 2m(\mathcal{A},a)-2}(\mathcal{A}) \cong \tau_{\leq 2m(\mathcal{A},a)-2}[\mathbf{Cat}(\Sigma(\mathcal{A}, a))]_{1/}(0, 0).$$

If $2m(\mathcal{A}, a) \geq n+2$ we recover the statement of 23.2.3.

Proof For any $(n+1)$-precategory $\mathcal{B}$, view it as a simplicial object in the category of n-precategories. Using minimal dimension, we can say that $\mathcal{B}$ is (m,k)-*arranged* if the Segal map

$$\mathcal{B}_{m/} \to \mathcal{B}_{1/} \times_{\mathcal{B}_0} \cdots \times_{\mathcal{B}_0} \mathcal{B}_{1/}$$

has minimal dimension at least $k+1$. This is the same as the notion which was considered in Chapter 15. Now, the procedure introduced for the proof of Theorem 15.4.5 may be used as a refined version of the operation **Cat**. From this we obtain that if $\mathcal{B}$ is (m,k)-arranged for all $m+k \le q$ then the morphism $\mathcal{B}_{1/} \to \mathbf{Cat}(\mathcal{B})_{1/}$ has minimal dimension at least $q-1$. As a corollary, if $\mathcal{B}$ is (m,k)-arranged for all $k \le p$ then the above applies with $q = p+2$, so the morphism $\mathcal{B}_{1/} \to \mathbf{Cat}(\mathcal{B})_{1/}$ has minimal dimension at least $p+1$.

Apply the above to $\mathcal{R} = \Sigma(\mathcal{A}, a)$, which is the same $(n+1)$-precategory as used in 23.2.3. Our main estimate 23.1.14 (cf. Corollary 23.2.2) implies that the Segal maps of $\mathcal{R}$ have minimal dimension at least $2m(\mathcal{A}, a)$. Thus $\mathcal{R}$ is (m,k)-arranged for all $k \le 2m(\mathcal{A}, a) - 1$ so the morphism $\mathcal{A} = \mathcal{R}_{1/} \to \mathbf{Cat}(\mathcal{R})_{1/}$ has minimal dimension at least $2m(\mathcal{A}, a)$. □

The next step into the domain of Baez and Dolan's conjectures [12] is to look at the free k-uply monoidal $(n-k)$-category on a single generator. Using the basic globules $\mathbb{O}^i_n = h(1^i)$ of Section 20.2.6, consider the pushout of $\mathbb{O}^k_n$ along the unique morphism $\partial\mathbb{O}^k_n \to *$, to get an n-precategory

$$\sigma^k_n := \mathbb{O}^k_n \cup^{\partial\mathbb{O}^k_n} *.$$

Call $\mathbf{Cat}\left(\sigma^k_n\right)$ the *free k-uply monoidal $(n-k)$-category on one generator.*

It is not hard to see that it satisfies the requisite universal property: a morphism $\sigma^k_n \to \mathcal{A}$ is the same thing as specification of an object $a \in \mathcal{A}$ and a k-endomorphism of the $(k-1)$-fold identity map of a. Let s be the base object of $\sigma^k - n$.

We note that

$$\sigma^{k+1}_{n+1} = \Sigma\left(\sigma^k_n, s\right),$$

so that σ^k_n is an iteration of the suspension operation starting with $\sigma^0_m = * \sqcup * \in \mathcal{PC}^m$. To see this, note that

$$\begin{aligned}
\Sigma\left(\sigma^k_n, s\right) &= \Upsilon\left(\mathbb{O}^k_n \cup^{\partial\mathbb{O}^k_n} *\right) \cup^{\Upsilon(*)} * \\
&= \Upsilon\left(\mathbb{O}^k_n\right) \cup^{\Upsilon(\partial\mathbb{O}^k_n)} \Upsilon(*) \cup^{\Upsilon(*)} * \\
&= \Upsilon\left(\mathbb{O}^k_n\right) \cup^{\Upsilon(\partial\mathbb{O}^k_n)} *
\end{aligned}$$

$$= \mathbb{O}^{k+1}_{n+1} \cup^{\partial \mathbb{O}^{k+1}_{n+1}} *$$
$$= \sigma^{k+1}_{n+1}.$$

Baez and Dolan make the following conjecture [12]:

Conjecture 23.3.4 (Baez and Dolan [12], section 4) *The $(n-k)$-category of endomorphisms of the $(k-1)$-fold identity of s in σ^k_n, is the Poincaré $(n-k)$-category $\Pi_{n-k}(X_k)$ of the space*

$$X_k = \coprod_{\ell=0}^{\infty} \mathcal{C}(k)_\ell / S_\ell,$$

where $\mathcal{C}(k)_\ell / S_\ell$ is the configuration space of ℓ. distinct unordered points in $\mathbb{R}^k$.

They already give a sketch of an argument for this conjecture in their paper [12], pointing out that for an operad $O = \{O_\ell\}$, May [203] constructs the free O-algebra on one point as

$$\coprod_{\ell=0}^{\infty} O_\ell / S_\ell.$$

Applied to the "little k-cubes" operad $\mathcal{C}(k)$, this gives the space X_k defined above. Baez and Dolan [12] argue that since $\mathcal{C}(k)$-algebras are E_k-spaces, i.e. spaces with k-fold delooping, the free $\mathcal{C}(k)$-algebra on one point should be the same as the free k-uply monoidal ∞-groupoid.

This correspondence may be made precise using the results of Dunn [97], which takes us closer to a rigorous proof of Conjecture 23.3.4. In effect, Dunn compares different k-fold delooping machines; and his model for the k-fold version of Segal's machine is the same as a Segal k-category with only one object in degrees $< k$. Applying the Poincaré groupoid construction Π_n in the top simplicial degree we obtain a correspondence between $(n+k)$-categories with one object in degrees $< k$ and which are $^{>}k$-groupic (i.e. the i-arrows are invertible up to equivalence for $i > k$), and n-truncated E_k-spaces for Dunn's Segal-machine. In particular the n-truncation of the free E_k-space on one point for Dunn's Segal-machine, is the $(n+k)$-category σ^k_{n+k} defined above. Now the only thing we need to know is that in Dunn's comparison [97] between different machines for E_k-spaces, the free E_k-spaces on one point are the same. This should follow directly from Dunn [97] using the universal properties of the free objects. Assuming this, we would get that the n-truncations of the free

Segal E_k-space and the free $\mathcal{C}(k)$-algebra are the same, thus that σ^k is the k-fold delooping of $\Pi_n(X_k)$.

The paper of Balteanu *et al.* [15] proves a result similar to Conjecture 23.3.4, starting from a different definition.

This conjecture is an intermediate step which could provide a bridge towards understanding the full Baez–Dolan conjectures, which concern the free k-uply monoidal n-category *with duals* on one generator, compared to a category defined by cobordisms in Lurie's work [193].

Epilogue

We have given a thorough treatment of some of the main first elements of the theory of higher categories, adopting Segal's method of weak enrichment in an iterative way. The fundamental step is the construction of the cartesian model category $\mathcal{PC}(\mathcal{M})$ in a way which can be iterated to give $\mathcal{PC}^n(\mathcal{M})$. The case of n-categories is obtained by starting with $\mathcal{M} = \text{SET}$. The case of Segal n-categories, which are (∞, n)-categories in Lurie's terminology, is obtained by starting with the standard model category $\mathcal{K}$ of simplicial sets.

The internal $\underline{\text{HOM}}$ within one of these iteratively constructed cartesian model categories $\mathcal{P}$ provides the morphism spaces which go together to create a $\mathcal{P}$-enriched category $\mathbf{Enr}(\mathcal{P})$. For $\mathcal{P} = \mathcal{PC}^n(\text{SET})$ this gives nCAT, the $(n+1)$-category of weakly associative n-categories. For $\mathcal{P} = \mathcal{PC}^n(\mathcal{K})$ this gives nSeCAT, the $(\infty, n+1)$-category of (∞, n)-categories.

In the last part of the book, we have indicated how to start towards the development of various aspects of the theory of higher categories using this model structure. This is only a start, touching the questions of inverting morphisms and localization, and limits and colimits. It leaves open a vast expanse of questions, many of which are the subject of recent and ongoing research by many people. Some of the main directions which need to be understood are: the Yoneda embeddings; adjoints of morphisms – a key ingredient of the Baez–Dolan notion of higher category with duals; an explicit construction of colimits; applications of limits and colimits to various kinds of universal constructions; and specially the limiting case of ∞-categories with noninvertible morphisms at all levels.

Applications of the higher categorical philosophy are now abundant. In some cases these follow the motivations originally introduced by Grothendieck – such is the case for the theory of higher stacks, which is enjoying a widening influence on geometry. In other cases these stem from the interaction with physics, particularly the Baez–Dolan conjectures as proven by Hopkins and Lurie. And there are some fascinating new directions which hadn't been foreseen!

The applications and technical developments currently in progress use various different approaches to higher category theory, including the other approaches we have been able to mention only briefly in the first part of the book. One of the main open questions is how to unify and compare these different approaches; progress is currently being made in that direction too.

The approach we have presented here has the advantage of being closely related to the very early historical approaches to category theory and homotopy theory, but also being well adapted to the requirements of a modern development using model categories. The basic facets of a higher category: its set of objects, and the morphism spaces, are transparently present in our definition. We have given a comprehensive treatment of the process of generating a higher category by generators and relations, a process which represents the essence of the higher algebra involved. I hope this can illuminate the inner workings of the theory, and encourage the development of this historically inspirational subject.

References

[1] J. Adámek, H. Herrlich, G. Strecker. *Abstract and Concrete Categories–The Joy of Cats*. John Wiley and Sons (1990), online edition available at http://katmat.math.uni-bremen.de/acc/.

[2] J. Adámek, J. Rosický. *Locally Presentable and Accessible Categories*. London Math. Soc. Lecture Note Series **189**, Cambridge University Press (1994).

[3] J. Adams. *Infinite Loop Spaces*. Annals of Math. Studies **90**, Princeton University Press (1978).

[4] M. Artin. Versal deformations and algebraic stacks, *Invent. Math.* **27** (1974), 165–189.

[5] D. Ara. *Sur les ∞-groupoïdes de Grothendieck et une variante ∞-catégorique*. Doctoral thesis, Université de Paris 7 (2010).

[6] H. Bacard. Segal enriched categories I. Arxiv preprint arXiv:1009.3673 (2010).

[7] B. Badzioch. Algebraic theories in homotopy theory. *Ann. Math.* **155** (2002), 895–913.

[8] J. Baez, J. Dolan. Higher-dimensional algebra and topological quantum field theory. *J. Math. Phys.* **36** (1995), 6073–6105.

[9] J. Baez, J. Dolan. n-Categories, sketch of a definition. Letter to R. Street, 29 Nov. and 3 Dec. 1995. Available online at http://math.ucr.edu/home/baez/ncat.def.html.

[10] J. Baez. An introduction to n-categories. In *Category Theory and Computer Science (Santa Margherita Ligure 1997)*. Lect. Notes in Computer Science **1290**, Springer-Verlag (1997), 1–33.

[11] J. Baez, J. Dolan. Higher-dimensional algebra III: n-categories and the algebra of opetopes. *Adv. Math.* **135** (1998), 145–206.

[12] J. Baez, J. Dolan. Categorification. In *Higher Category Theory (Evanston, 1997)*. Contemp. Math. **230**, A.M.S. (1998), 1–36.

[13] J. Baez, P. May. *Towards Higher Categories*. IMA Volumes Math. Appl. **152**, Springer-Verlag (2009).

[14] I. Baković, B. Jurčo. The classifying topos of a topological bicategory. *Homology, Homotopy Appl.* **12** (2010), 279–300.

[15] C. Balteanu, Z. Fiedorowicz, R. Schwänzl, R. Vogt. Iterated monoidal categories. *Adv. Math.* **176** (2003), 277–349.

[16] C. Barwick. $(\infty, n) - $ *Cat as a closed model category*. Doctoral dissertation, University of Pennsylvania (2005).

[17] C. Barwick. ∞-groupoids, stacks, and Segal categories. Seminars 2004–2005 of the Mathematical Institute, University of Göttingen (Y. Tschinkel, ed.). Universitätsverlag Göttingen (2005), 155–195.

[18] C. Barwick. On (enriched) left Bousfield localization of model categories. Arxiv preprint arXiv:0708.2067 (2007), now in [20].

[19] C. Barwick. On Reedy model categories. Preprint arXiv: 0708.2832 (2007), now in [20].

[20] C. Barwick. On left and right model categories and left and right Bousfield localizations. *Homology, Homotopy Appl.* **1** (2010), 1–76.

[21] C. Barwick, D. Kan. Relative categories: Another model for the homotopy theory of homotopy theories. Preprint arXiv:1011.1691 (2010).

[22] C. Barwick, D. Kan. A Thomason-like Quillen equivalence between quasi-categories and relative categories. Preprint arXiv:1101.0772 (2011).

[23] C. Barwick, D. Kan. n-relative categories: a model for the homotopy theory of n-fold homotopy theories. Preprint arXiv:1102.0186 (2011).

[24] C. Barwick. On the Yoneda lemma and the strictification theorem for homotopy theories. Preprint (2008).

[25] M. Batanin. On the definition of weak ω-category. Macquarie mathematics report number 96/207, Macquarie University, Australia.

[26] M. Batanin. Monoidal globular categories as a natural environment for the theory of weak n-categories. *Adv. Math.* **136** (1998), 39–103.

[27] M. Batanin. Homotopy coherent category theory and A_∞ structures in monoidal categories. *J. Pure Appl. Alg.* **123** (1998), 67–103.

[28] M. Batanin. On the Penon method of weakening algebraic structures. *J. Pure Appl. Alg.* **172** (2002), 1–23.

[29] M. Batanin. The Eckmann–Hilton argument and higher operads. *Adv. Math.* **217** (2008), 334–385.

[30] M. Batanin, D. Cisinski, M. Weber. Algebras of higher operads as enriched categories II. Preprint arXiv:0909.4715v1 (2009).

[31] F. Bauer, T. Datuashvili. Simplicial model category structures on the category of chain functors. *Homology, Homotopy Appl.* **9** (2007), 107–138.

[32] H. Baues. *Combinatorial Homotopy and 4-Dimensional Complexes.* de Gruyter, Berlin (1991).

[33] T. Beke. Sheafifiable homotopy model categories. *Math. Proc. Cambridge Phil. Soc.* **129** (2000), 447–475.

[34] J. Bénabou. *Introduction to Bicategories.* Lect. Notes in Math. **47**, Springer-Verlag (1967).

[35] C. Berger. Double loop spaces, braided monoidal categories and algebraic 3-type of space. *Contemp. Math.* **227** (1999), 49–65.

[36] C. Berger. A cellular nerve for higher categories. *Adv. Math.* **169** (2002), 118–175.

[37] C. Berger. Iterated wreath product of the simplex category and iterated loop spaces. *Adv. Math.* **213** (2007), 230–270.

[38] C. Berger, I. Moerdijk. On an extension of the notion of Reedy category. *Math. Z.* DOI 10.1007/s00209-010-0770-x (2010), 1–28.

[39] J. Bergner. A model category structure on the category of simplicial categories. *Trans. Amer. Math. Soc.* **359** (2007), 2043–2058.

[40] J. Bergner. Three models for the homotopy theory of homotopy theories. *Topology* **46** (2007), 397–436.

[41] J. Bergner. Adding inverses to diagrams encoding algebraic structures. *Homology, Homotopy Appl.* **10** (2008), 149–174.

[42] J. Bergner. A characterization of fibrant Segal categories. *Proc. Amer. Math. Soc.* **135** (2007), 4031–4037.

[43] J. Bergner. Rigidification of algebras over multi-sorted theories. *Alg. Geom. Topol.* **6** (2006), 1925–1955.

[44] J. Bergner. A survey of $(\infty, 1)$-categories. In *Towards Higher Categories* (J. Baez, P. May, eds.). IMA VOLUMES MATH APPL. **152**, Springer-Verlag (2009).

[45] J. Bergner. Simplicial monoids and Segal categories. *Contemp. Math.* **431** (2007), 59–83.

[46] J. Bergner. Homotopy fiber products of homotopy theories. Arxiv preprint arXiv:0811.3175 (2008).

[47] J. Bergner. Homotopy limits of model categories and more general homotopy theories. Arxiv preprint arXiv:1010.0717 (2010).

[48] J. Bergner. Models for (∞, n)-categories and the cobordism hypothesis. Arxiv preprint arXiv:1011.0110 (2010).

[49] B. Blander. Local projective model structures on simplicial presheaves. *K-theory* **24** (2001), 283–301.

[50] J. Boardman, R. Vogt. *Homotopy Invariant Algebraic Structures on Topological Spaces.* LECTURE NOTES IN MATH. **347**, Springer-Verlag (1973).

[51] A. Bondal, M. Kapranov. Enhanced triangulated categories. *Math. U.S.S.R. Sb.* **70** (1991), 93–107.

[52] R. Bott, L. Tu. *Differential Forms in Algebraic Topology*. GRADUATE TEXTS IN MATH. **82**, Springer-Verlag (1982).

[53] D. Bourn. Anadèses et catadèses naturelles. *C.R. Acad. Sci. Paris Sér. A–B* **276** (1973), A1401–A1404.

[54] D. Bourn. Sur les ditopos. *C. R. Acad. Sci. Paris* **279** (1974), 911–913.

[55] D. Bourn. La tour de fibrations exactes des n-catégories. *Cah. Top. Géom. Differ. Catég.* (1984).

[56] D. Bourn, J.-M. Cordier. A general formulation of homotopy limits. *J. Pure Appl. Alg.* **29** (1983), 129–141.

[57] A. Bousfield. Cosimplicial resolutions and homotopy spectral sequences in model categories. *Geometry & Topology* **7** (2003) 1001–1053.

[58] A. Bousfield, D. Kan. *Homotopy Limits, Completions and Localizations.* LECTURE NOTES IN MATH. **304**, Springer-Verlag (1972).

[59] L. Breen. *On the Classification of* 2*-Gerbs and* 2*-Stacks.* ASTÉRISQUE **225**, S.M.F. (1994).

[60] L. Breen. Monoidal categories and multiextensions. *Compositio Math.* **117** (1999), 295–335.

[61] E. Brown, Jr. Finite computability of Postnikov complexes. *Ann. Math.* **65** (1957), 1–20.

[62] K. Brown. Abstract homotopy theory and generalized sheaf cohomology. *Trans. Amer. Math. Soc.* **186** (1973), 419–458.

[63] K. Brown, S. Gersten. *Algebraic K-theory as Generalized Sheaf Cohomology*. LECTURE NOTES IN MATH. **341**, Springer-Verlag (1973), 266–292.

[64] R. Brown. Groupoids and crossed objects in algebraic topology. *Homology Homotopy Appl.* **1** (1999), 1–78.

[65] R. Brown. Computing homotopy types using crossed n-cubes of groups. *Adams Memorial Symposium on Algebraic Topology*, Vol. 1 (N. Ray, G Walker, eds.). Cambridge University Press (1992) 187–210.

[66] R. Brown, N.D. Gilbert. Algebraic models of 3-types and automorphism structures for crossed modules. *Proc. London Math. Soc.* (3) **59** (1989), 51–73.

[67] R. Brown, P. Higgins. The equivalence of ∞-groupoids and crossed complexes. *Cah. Top. Géom. Différ. Catég.* **22** (1981), 371–386.

[68] R. Brown, P. Higgins. The classifying space of a crossed complex. *Math. Proc. Cambridge Phil. Soc.* **110** (1991), 95–120.

[69] R. Brown, J.-L. Loday. Van Kampen theorems for diagrams of spaces. *Topology* **26** (1987), 311–335.

[70] R. Brown, J.-L. Loday. Homotopical excision, and Hurewicz theorems, for n-cubes of spaces. *Proc. London Math. Soc.* **54** (1987), 176–192.

[71] J. Cabello, A. Garzon. Closed model structures for algebraic models of n-types. *J. Pure Appl. Alg.* **103** (1995), 287–302.

[72] E. Cheng. The category of opetopes and the category of opetopic sets. *Th. Appl. Cat.* **11** (2003), 353–374.

[73] E. Cheng. An omega-category with all duals is an omega groupoid. *Appl. Cat. Struct.* **15** (2007), 439–453.

[74] E. Cheng. Comparing operadic theories of n-category. Preprint arXiv:0809.2070 (2008).

[75] E. Cheng, A. Lauda. *Higher-Dimensional Categories: an Illustrated Guidebook* (2004). http://cheng.staff.shef.ac.uk/guidebook/guidebook-new.pdf

[76] E. Cheng, M. Makkai. A note on the Penon definition of n-category. Preprint arXiv:0907.3961 (2009).

[77] D. Cisinski. *Les Préfaisceaux Comme Modèles des Types d'Homotopie.* ASTÉRISQUE **308**, S.M.F. (2006).

[78] D. Cisinski. Batanin higher groupoids and homotopy types. In *Categories in Algebra, Geometry and Mathematical Physics, Proceedings of Streetfest* (M. Batanin *et al.*, eds.), *Contemporary Math.* **431** (2007), 171–186.

[79] D. Cisinski. Propriétés universelles et extensions de Kan dérivées. *Th. Appl. Cat.* **20** (2008), 605–649.

[80] F. Cohen, T. Lada, J. P. May. *The homology of iterated loop spaces.* LECTURE NOTES IN MATH. **533**, Springer-Verlag (1976).

[81] J. Cordier. Comparaison de deux catégories d'homotopie de morphismes cohérents. *Cah. Top. Géom. Différ. Catég.* **30** (1989), 257–275.

[82] J. Cordier, T. Porter. Vogt's theorem on categories of homotopy coherent diagrams. *Math. Proc. Cambridge Phil. Soc.* **100** (1986), 65–90.

[83] J. Cordier, T. Porter. Homotopy coherent category theory. *Trans. Amer. Math. Soc.* **349** (1997), 1–54.

[84] J. Cranch. *Algebraic theories and* $(\infty, 1)$*-categories*. PhD thesis, University of Sheffield, arXiv:1011.3243 (2010).

[85] S. Crans. Quillen closed model structures for sheaves. *J. Pure Appl. Alg.* **101** (1995), 35–57.

[86] S. Crans. A tensor product for Gray-categories. *Th. Appl. Cat.* **5** (1999), 12–69.

[87] S. Crans. On braidings, syllapses and symmetries. *Cah. Top. Géom. Différ. Catég.* **41** (2000), 2–74.

[88] E. Curtis. Lower central series of semisimplicial complexes. *Topology* **2** (1963), 159–171.

[89] E. Curtis. Some relations between homotopy and homology. *Ann. Math.* **82** (1965), 386–413.

[90] P. Deligne. Théorie de Hodge, III. *Publ. Math. I.H.E.S.* **44** (1974), 5–77.

[91] P. Deligne, D. Mumford. On the irreducibility of the space of curves of a given genus. *Publ. Math. I.H.E.S.* **36** (1969), 75–109.

[92] P. Deligne, A. Ogus, J. Milne, K. Shih. *Hodge Cycles, Motives, and Shimura Varieties*. LECTURE NOTES IN MATH. **900**, Springer-Verlag (1982).

[93] V. Drinfeld. DG quotients of DG categories. *J. Alg.* **272** (2004), 643–691.

[94] E. Dror Farjoun, A. Zabrodsky. The homotopy spectral sequence for equivariant function complexes. In *Algebraic Topology, Barcelona, 1986*. LECTURE NOTES IN MATH. **1298**, Springer-Verlag (1987), 54–81.

[95] D. Dugger. Combinatorial model categories have presentations. *Adv. Math.* **164** (2001), 177–201.

[96] D. Dugger, D. Spivak. Mapping spaces in quasi-categories. Preprint arXiv:0911.0469 (2009).

[97] G. Dunn. Uniqueness of n-fold delooping machines. *J. Pure Appl. Alg.* **113** (1996), 159–193.

[98] J. Duskin. Simplicial matrices and the nerves of weak n-categories I: nerves of bicategories. *Th. Appl. Cat.* **9** (2002), 198–308.

[99] W. Dwyer, P. Hirschhorn, D. Kan. Model categories and more general abstract homotopy theory, a work in what we like to think of as progress. (This historically important manuscript was later integrated into the next reference.)

[100] W. Dwyer, P. Hirschhorn, D. Kan, J. Smith. *Homotopy Limit Functors on Model Categories and Homotopical Categories*. MATH. SURVEYS AND MONOGRAPHS **113**, A.M.S. (2004).

[101] W. Dwyer, D. Kan. Simplicial localizations of categories. *J. Pure Appl. Alg.* **17** (1980), 267–284.

[102] W. Dwyer, D. Kan. Calculating simplicial localizations. *J. Pure Appl. Alg.* **18** (1980), 17–35.

[103] W. Dwyer, D. Kan. Function complexes in homotopical algebra. *Topology* **19** (1980), 427–440.

[104] W. Dwyer, D. Kan, J. Smith. Homotopy commutative diagrams and their realizations. *J. Pure Appl. Alg.* **57** (1989), 5–24.

[105] W. Dwyer, J. Spalinski. Homotopy theories and model categories. In *Handbook of Algebraic Topology* (I. M. James, ed.), Elsevier (1995).

[106] J. Dydak. A simple proof that pointed, connected FANR spaces are regular fundamental retracts of ANRs. *Bull. Acad. Polon. Sci. Ser. Sci. Math. Phys.* **25** (1977), 55–62.

[107] J. Dydak. 1-movable continua need not be pointed 1-movable. *Bull. Acad. Polon. Sci. Ser. Sci. Math. Phys.* **25** (1977), 485–488.

[108] E. Dyer, R. Lashoff. Homology of iterated loop spaces. *Amer. J. Math.* **84** (1962), 35–88.

[109] J. Elgueta. On the regular representation of an (essentially) finite 2-group. Preprint arXiv:0907.0978 (2009).

[110] G. Ellis. Spaces with finitely many nontrivial homotopy groups all of which are finite. *Topology* **36** (1997), 501–504.

[111] Z. Fiedorowicz. Classifying spaces of topological monoids and categories. *Amer. J. Math.* **106** (1984), 301–350.

[112] Z. Fiedorowicz, R. Vogt. Simplicial n-fold monoidal categories model all n-fold loop spaces. *Cah. Top. Géom. Différ. Catég.* **44** (2003), 105–148.

[113] T. Fiore. Pseudo limits, biadjoints, and pseudo algebras: categorical foundations of conformal field theory. *Mem. Amer. Math. Soc.* **182** (2006).

[114] P. Freyd, A. Heller. Splitting homotopy idempotents, II. *J. Pure Appl. Alg.* **89** (1993), 93–106.

[115] K. Fukaya. Morse homotopy, A_∞-category and Floer homologies. In *Proceedings of GARC Workshop on Geometry and Topology* (H. J. Kim, ed.), Seoul National University, (1993).

[116] C. Futia. Weak omega categories I. Preprint arXiv:math/0404216 (2004).

[117] P. Gabriel, M. Zisman. *Calculus of Fractions and Homotopy Theory.* Springer-Verlag (1967).

[118] N. Gambino. Homotopy limits for 2-categories. *Math. Proc. Cambridge Phil. Soc.* **145** (2008), 43–63.

[119] R. Garner, N. Gurski. The low-dimensional structures that tricategories form. Preprint arXiv:0711.1761 (2007).

[120] P. Gaucher. Homotopy invariants of higher dimensional categories and concurrency in computer science. *Math. Struct. Comp. Sci.* **10** (2000), 481–524.

[121] J. Giraud. *Cohomologie Nonabélienne.* GRUNDLEHREN DER WISSENSCHAFTEN IN EINZELDARSTELLUNG **179**, Springer-Verlag (1971).

[122] P. Goerss, R. Jardine. *Simplicial Homotopy Theory.* PROGRESS IN MATH. **174**, Birkhäuser (1999).

[123] R. Gordon, A.J. Power, R. Street. Coherence for tricategories. *Memoirs A.M.S.* **117** (1995), 558 ff.

[124] M. Grandis. Directed homotopy theory, I. The fundamental category. *Cah. Top. Géom. Différ. Catég.* **44** (2003), 281–316.

[125] M. Grandis, R. Paré. Limits in double categories. *Cah. Top. Géom. Différ. Catég.* **40** (1999), 162–220.

[126] J. Gray. *Formal Category Theory: Adjointness for 2-Categories.* LECTURE NOTES IN MATH. **391**, Springer-Verlag (1974).

[127] J. Gray. Closed categories, lax limits and homotopy limits. *J. Pure Appl. Alg.* **19** (1980), 127–158.

[128] J. Gray. The existence and construction of lax limits. *Cah. Top. Géom. Différ. Catég.* **21** (1980), 277–304.

[129] A. Grothendieck. Sur quelques points d'algèbre homologique, I. *Tohoku Math. J.* **9** (1957), 119–221.

[130] A. Grothendieck. Techniques de construction et théorèmes d'existence en géométrie algébrique. III. Préschemas quotients. *Séminaire Bourbaki, 13e année, 1960/61* **212** (1961).

[131] A. Grothendieck. *Revetements Etales et Groupe Fondamental (SGA I)*, LECTURE NOTES IN MATH. **224**, Springer-Verlag (1971).

[132] A. Grothendieck. *Pursuing Stacks*.

[133] A. Grothendieck. *Les Dérivateurs* (G. Maltsiniotis, ed.) Available online at http://people.math.jussieu.fr/ maltsin/textes.html.

[134] N. Gurski. An algebraic theory of tricategories. Ph.D. thesis, University of Chicago (2006).

[135] R. Hain. Completions of mapping class groups and the cycle $C - C^-$. In *Mapping Class Groups and Moduli Spaces of Riemann Surfaces: Proceedings of Workshops held in Göttingen and Seattle*. CONTEMPORARY MATH. **150**, A.M.S. (1993), 75–106.

[136] R. Hain. The de rham homotopy theory of complex algebraic varieties I. *K-theory* **1** (1987), 271–324.

[137] R. Hain. The Hodge de Rham theory of relative Malcev completion. *Ann. Sci. de l'E.N.S.* **31** (1998), 47–92.

[138] H. Hastings, A. Heller. Splitting homotopy idempotents. In *Shape Theory and Geometric Topology (Dubrovnik, 1981)*. LECTURE NOTES IN MATH. **870**, Springer-Verlag (1981), 23–36.

[139] H. Hastings, A. Heller. Homotopy idempotents on finite-dimensional complexes split. *Proc. Amer. Math. Soc.* **85** (1982), 619–622.

[140] A. Heller. Homotopy in functor categories. *Trans. Amer. Math. Soc.* **272** (1982), 185–202.

[141] A. Heller. Homotopy theories. *Mem. Amer. Math. Soc.* **71** (388) (1988).

[142] C. Hermida, M. Makkai, A. Power. On weak higher-dimensional categories I. *J. Pure Appl. Alg.* Part 1: **154** (2000), 221–246; Part 2: **157** (2001), 247–277; Part 3: **166** (2002), 83–104.

[143] V. Hinich. Homological algebra of homotopy algebras. *Comm. Alg.* **25** (1997), 3291–3323.

[144] P. Hirschhorn. *Model Categories and their Localizations*. MATH. SURVEYS AND MONOGRAPHS **99**, A.M.S. (2003).

[145] A. Hirschowitz, C. Simpson. Descente pour les n-champs. Preprint math/9807049 (1998).

[146] J. Hirsh, J. Millès. Curved Koszul duality theory. Preprint, University of Nice (2010).

[147] S. Hollander. A homotopy theory for stacks. *Israel J. Math.* **163** (2008) 93–124.

[148] M. Hovey. Monoidal model categories. Arxiv preprint math/9803002 (1998).

[149] M. Hovey. *Model Categories*. MATH. SURVEYS AND MONOGRAPHS **63**, A.M.S. (1999).

[150] L. Illusie. *Complexe Cotangent et Déformations, II*. LECTURE NOTES IN MATH. **283**, Springer-Verlag (1972).

[151] I. James. Reduced product spaces. *Ann. Math.* **62** (1955), 170–197.

[152] G. Janelidze. *Precategories and Galois theory*. Springer-Verlag (1990).

[153] J.F. Jardine. Simplicial presheaves, *J. Pure Appl. Alg.* **47** (1987), 35–87.

[154] M. Johnson. The combinatorics of n-categorical pasting. *J. Pure Appl. Alg.* **62** (1989), 211–225.

[155] M. Johnson. On modified Reedy and modified projective model structures. Preprint arXiv:1004.3922v1 (2010).

[156] A. Joyal. Letter to A. Grothendieck (referred to in Jardine's paper).

[157] A. Joyal. Quasi-categories and Kan complexes. *J. Pure Appl. Alg.* **175** (2002), 207–222.

[158] A. Joyal. Disks, duality and θ-categories. Preprint (1997).

[159] A. Joyal, J. Kock. Weak units and homotopy 3-types. In *Categories in Algebra, Geometry and Mathematical Physics: Conference and Workshop in Honor of Ross Street's 60th Birthday.* CONTEMPORARY MATH. **431**, A.M.S (2007), 257–276.

[160] A. Joyal, J. Kock. Coherence for weak units. Preprint arXiv:0907.4553 (2009).

[161] A. Joyal, M. Tierney. Algebraic homotopy types. Occurs as an entry in the bibliography of [8].

[162] A. Joyal, M. Tierney. Quasi-categories vs Segal spaces. In *Categories in Algebra, Geometry and Mathematical Physics: Conference and Workshop in Honor of Ross Street's 60th Birthday*. CONTEMPORARY MATH. **431**, A.M.S. (2007), 277–326.

[163] D. Kan. On c.s.s. complexes. *Amer. J. Math.* **79** (1957), 449–476.

[164] D. Kan. A combinatorial definition of homotopy groups. *Ann. Math.* **67** (1958), 282–312.

[165] D. Kan. On homotopy theory and c.s.s. groups. *Ann. Math.* **68** (1958), 38–53.

[166] D. Kan. On c.s.s. categories. *Bol. Soc. Math. Mexicana* (1957), 82–94.

[167] M. Kapranov. On the derived categories of coherent sheaves on some homogeneous spaces. *Invent. Math.* **92** (1988), 479–508.

[168] M. Kapranov, V. Voevodsky. ∞-groupoids and homotopy types. *Cah. Top. Géom. Differ. Catég.* **32** (1991), 29–46.

[169] L. Katzarkov, T. Pantev, B. Toën. Algebraic and topological aspects of the schematization functor. *Compositio Math.* **145** (2009), 633–686.

[170] B. Keller. Deriving DG categories. *Ann. Sci. E.N.S.* **27** (1994), 63–102.

[171] G. Kelly, *Basic concepts of enriched category theory*. LONDON MATH. SOC. LECTURE NOTES **64**, Cambridge University Press (1982).

[172] J. Kock. Weak identity arrows in higher categories. *Int. Math. Res. Papers* (2006).

[173] J. Kock. Elementary remarks on units in monoidal categories. *Math. Proc. Cambridge Phil. Soc.* **144** (2008), 53–76.

[174] J. Kock, A. Joyal, M. Batanin, J. Mascari. Polynomial functors and opetopes. *Adv. Math.* **224** (2010), 2690–2737.

[175] G. Kondratiev. Concrete duality for strict infinity categories. Preprint arXiv:0807.4256 (2008) (see also arXiv:math/0608436).

[176] M. Kontsevich. Homological algebra of mirror symmetry. In *Proceedings of I.C.M.-94, Zurich.* Birkhäuser (1995), 120–139.

[177] J.-L. Krivine. *Théorie Axiomatique des Ensembles*, Presses Universitaires de France (1969); English translation by D. Miller, *Introduction to Axiomatic Set Theory*, D. Reidel Publishing Co. (1971).

[178] S. Lack. A Quillen model structure for Gray-categories. Preprint arXiv:1001.2366 (2010).

[179] Y. Lafont, F. Métayer, K. Worytkiewicz. A folk model structure on omega-cat. *Adv. Math.* **224** (2010), 1183–1231.

[180] G. Laumon, L. Moret-Bailly. *Champs Algébriques*. Springer-Verlag (2000).

[181] F. W. Lawvere. Functorial semantics of algebraic theories. *Proc. Nat. Acad. Sci.* **50** (1963), 869–872.

[182] F. W. Lawvere. *Functorial semantics of algebraic theories*, Dissertation, Columbia University 1963; reprint in *Th. Appl. Cat.* **5** (2004), 23–107.

[183] T. Leinster. A survey of definitions of n-category. *Th. Appl. Cat.* **10** (2002), 1–70.

[184] T. Leinster. *Higher Operads, Higher Categories*. LONDON MATH. SOC. LECTURE NOTES **298**, Cambridge University Press (2004).

[185] T. Leinster. Up-to-homotopy monoids. Arxiv preprint math/9912084 (1999).

[186] O. Leroy. Sur une notion de 3-catégorie adaptée à l'homotopie. Preprint Univ. de Montpellier 2 (1994).

[187] L. G. Lewis. Is there a convenient category of spectra? *J. Pure Appl. Alg.* **73** (1991), 233–246.

[188] J.-L. Loday. Spaces with finitely many non-trivial homotopy groups. *J. Pure Appl. Alg.* **24** (1982), 179–202.

[189] J. Lurie. On infinity topoi. Preprint arXiv:math/0306109 (2003).

[190] J. Lurie. Higher topos theory. *Ann. Math. Studies* **170** (2009).

[191] J. Lurie. Derived Algebraic Geometry II–VI. Arxiv preprints (2007–2009).

[192] J. Lurie. (Infinity,2)-Categories and the Goodwillie Calculus I. Preprint arXiv:0905.0462v2 (2009).

[193] J. Lurie. On the classification of topological field theories. *Current Developments in Mathematics* Vol. 2008 (2009), 129–280.

[194] D. McDuff. On the classifying spaces of discrete monoids. *Topology* **18** (1979), 313–320.

[195] M. Mackaay. Spherical 2-categories and 4-manifold invariants. *Adv. Math.* **143** (1999), 288–348.

[196] S. Mac Lane. *Categories for the Working Mathematician.* GRADUATE TEXTS IN MATH. **5**, Springer-Verlag (1971).

[197] M. Makkai, R. Paré. *Accessible Categories: The Foundations of Categorical Model Theory*. CONTEMPORARY MATH. **104**, A.M.S. (1989).

[198] G. Maltsiniotis. *La Théorie de l'Homotopie de Grothendieck.* ASTÉRISQUE **301**, S.M.F. (2005).

[199] G. Maltsiniotis. Infini groupoïdes non stricts, d'après Grothendieck. Preprint (2007).

[200] G. Maltsiniotis. Infini catégories non strictes, une nouvelle définition. Preprint (2007).

[201] W. Massey. *Algebraic Topology: An Introduction*. GRADUATE TEXTS IN MATH. **56**, Springer-Verlag (1977).

[202] J. P. May. *Simplicial Objects in Algebraic Topology.* Van Nostrand (1967).

[203] J.P. May. *The Geometry of Iterated Loop Spaces*. LECTURE NOTES IN MATH. **271**, Springer-Verlag (1972).

[204] J. P. May. Classifying spaces and fibrations. *Mem. Amer. Math. Soc.* **155** (1975).

[205] J. P. May, R. Thomason. The uniqueness of infinite loop space machines. *Topology* **17** (1978), 205–224.

[206] S. Mochizuki. Idempotent completeness of higher derived categories of abelian categories. K-theory preprint archives n. 970 (2010).

[207] F. Morel, V. Voevodsky. $\mathbb{A}^1$-homotopy theory of schemes. *Publ. Math. I.H.E.S.* **90** (1999), 45–143.

[208] S. Moriya. Rational homotopy theory and differential graded category. *J. Pure Appl. Alg.* **214** (2010), 422–439.

[209] J. Nichols-Barrer. On quasi-categories as a foundation for higher algebraic stacks. Ph.D. thesis, M.I.T. (2007).

[210] S. Paoli. Weakly globular cat^n-groups and Tamsamani's model. *Adv. Math.* **222** (2009), 621–727.

[211] R. Pellissier. Catégories enrichies faibles. Thesis, Université de Nice (2002), available online at http://tel.archives-ouvertes.fr/tel-00003273/fr/.

[212] J. Penon. Approche polygraphique des ∞-catégories non strictes. *Cah. Top. Géom. Différ. Catég.* **40** (1999), 31–80.

[213] A. Power. Why tricategories? *Information and Computation* **120** (1995), 251–262.

[214] J. Pridham. Pro-algebraic homotopy types. *Proc. London Math. Soc.* **97** (2008), 273–338.

[215] D. Quillen. *Homotopical Algebra*. LECTURE NOTES IN MATH. **43**, Springer-Verlag (1967).

[216] D. Quillen. Rational homotopy theory. *Ann. Math.* **90** (1969), 205–295.

[217] C. Reedy. Homotopy theory of model categories. Preprint (1973) available from P. Hirschhorn.

[218] C. Rezk. A model for the homotopy theory of homotopy theory. *Trans. Amer. Math. Soc.* **353** (2001), 973–1007.

[219] C. Rezk. A cartesian presentation of weak n-categories. *Geometry & Topology* **14** (2010), 521–571.

[220] E. Riehl. On the structure of simplicial categories associated to quasi-categories. Preprint arXiv:0912.4809 (2009).

[221] J. Rosický and W. Tholen, Left-determined model categories and universal homotopy theories. *Trans. Amer. Math. Soc.* **355** (2003), 3611–3623.

[222] J. Rosický. On homotopy varieties. *Adv. Math.* **214** (2007), 525–550.

[223] J. Rosický. On combinatorial model categories. *Appl. Cat. Structures* **17** (2009) 303–316.

[224] R. Schwänzl, R. Vogt. Homotopy homomorphisms and the hammock localization. *Papers in Honor of José Adem, Bol. Soc. Mat. Mexicana* **37** (1992), 431–448.

[225] G. Segal. Homotopy everything H-spaces. Preprint.

[226] G. Segal. Classifying spaces and spectral sequences. *Publ. Math. IHES* **34** (1968), 105–112.

[227] G. Segal. Configuration spaces and iterated loop spaces. *Invent. Math.* **21** (1973), 213–221.

[228] G. Segal. Categories and cohomology theories. *Topology* **13** (1974), 293–312.

[229] B. Shipley, S. Schwede. Equivalences of monoidal model categories. *Algebr. Geom. Topol.* **3** (2003), 287–334.

[230] C. Simpson. Homotopy over the complex numbers and generalized de Rham cohomology. In *Moduli of Vector Bundles* (M. Maruyama, ed.). LECTURE NOTES IN PURE AND APPLIED MATH. **179**, Marcel Dekker (1996), 229–263.

[231] C. Simpson. Flexible sheaves. Preprint q-alg/9608025 (1996).

[232] C. Simpson. The topological realization of a simplicial presheaf. Preprint q-alg/9609004 (1996).

[233] C. Simpson. Algebraic (geometric) n-stacks. Preprint alg-geom/9609014 (1996).

[234] C. Simpson. A closed model structure for n-categories, internal Hom, n-stacks and generalized Seifert-Van Kampen. Preprint alg-geom/9704006 (1997).

[235] C. Simpson. Limits in n-categories. Preprint alg-geom 9708010 (1997).

[236] C. Simpson. Effective generalized Seifert–Van Kampen: how to calculate ΩX. Preprint q-alg/9710011 (1997).

[237] C. Simpson. On the Breen–Baez–Dolan stabilization hypothesis. Preprint math.CT/9810058 (1998).

[238] C. Simpson. Homotopy types of strict 3-groupoids. Preprint, math.CT/9810059 (1998).

[239] J. Smith. Combinatorial model categories. Unpublished manuscript referred to in [95].

[240] A. Stanculescu. A homotopy theory for enrichment in simplicial modules. Preprint arXiv:0712.1319 (2007).

[241] J. Stasheff. Homotopy associativity of H-spaces, I, II. *Trans. Amer. Math. Soc.* **108** (1963), 275–292, 293–312.

[242] R. Street. Elementary cosmoi, I. In *Category Seminar (Sydney, 1972/1973)*. LECTURE NOTES IN MATH. **420**, Springer-Verlag (1974), 134–180.

[243] R. Street. Limits indexed by category-valued 2-functors. *J. Pure Appl. Alg.* **8** (1976), 149–181.

[244] R. Street. The algebra of oriented simplexes. *J. Pure Appl. Alg.* **49** (1987), 283–335.

[245] R. Street. Weak ω-categories. *Diagrammatic Morphisms and Applications (San Francisco, 2000) Contemporary Mathematics* **318**, A.M.S. (2003), 207–213.

[246] G. Tabuada. Differential graded versus Simplicial categories. *Top. Appl.* **157** (2010), 563–593.

[247] G. Tabuada. Homotopy theory of spectral categories. *Adv. Math.* **221** (2009), 1122–1143.

[248] Z. Tamsamani. Sur des notions de n-catégorie et n-groupoïde non-stricte via des ensembles multi-simpliciaux. Thesis, Université Paul Sabatier, Toulouse (1996), first part available as alg-geom/9512006.

[249] Z. Tamsamani. Equivalence de la théorie homotopique des n-groupoïdes et celle des espaces topologiques n-tronqués. Preprint alg-geom alg-geom/9607010 (1996).

[250] Z. Tamsamani. Sur des notions de n-catégorie et n-groupoïde non-stricte via des ensembles multi-simpliciaux. *K-theory* **16** (1999), 51–99.

[251] D. Tanre. *Homotopie Rationnelle: Modèles de Chen, Quillen, Sullivan.* LECTURE NOTES IN MATH. **1025**, Springer-Verlag (1983).

[252] R. Thomason. Homotopy colimits in the category of small categories. *Math. Proc. Cambridge Phil. Soc.* **85** (1979), 91–109.

[253] R. Thomason. Uniqueness of delooping machines. *Duke Math. J.* **46** (1979), 217–252.

[254] R. Thomason. Algebraic K-theory and étale cohomology. *Ann. Sci. E.N.S.* **18** (1985), 437–552.

[255] B. Toën. Champs affines. *Selecta Math.* **12** (2006), 39–134.

[256] B. Toën. Vers une axiomatisation de la théorie des catégories supérieures. *K-theory* **34** (2005), 233–263.

[257] B. Toën. The homotopy theory of dg-categories and derived Morita theory. *Invent. Math.* **167** (2007), 615–667.

[258] B. Toën. Anneaux de définition des dg-algèbres propres et lisses. *Bull. Lond. Math. Soc.* **40** (2008), 642–650.

[259] T. Trimble. Notes on tetracategories. Available online at http://math.ucr.edu/home/baez/trimble/tetracategories.html.

[260] D. Verity. Weak complicial sets I. Basic homotopy theory. *Adv. Math.* **219** (2008), 1081–1149.

[261] V. Voevodsky. The Milnor conjecture. Preprint (1996).

[262] R. Vogt. Homotopy limits and colimits. *Math. Z.* **134** (1973), 11–52.

[263] R. Vogt. The HELP-Lemma and its converse in Quillen model categories. Preprint arXiv:1004.5249v1 (2010).

[264] M. Weber. Yoneda Structures from 2-toposes. *Appl. Cat. Struct.* **15** (2007), 259–323.

[265] G. Whitehead. *Elements of Homotopy Theory*. Springer-Verlag (1978).

[266] J. H. C. Whitehead. On the asphericity of regions in a 3-sphere. *Fund. Math.* **32** (1939), 149–166.

[267] M. Zawadowski. Lax Monoidal Fibrations. Preprint arXiv:0912.4464 (2009).

Index

Printed in the United States
By Bookmasters